Methods in Neurobiology

VOLUME 2

Methods in Neurobiology

VOLUME 2

Edited by
Robert Lahue
University of Waterloo
Waterloo, Ontario, Canada

PLENUM PRESS • NEW YORK AND LONDON

Library of Congress Cataloging in Publication Data

Main entry under title:

Methods in neurobiology.

Includes index.
1. Neurophysiology—Technique. 2. Neurobiology—Technique. I. Lahue, Robert.
QP357.M47 591.1'88'028 80-15623
ISBN 0-306-40518-0 (v. 2)

A Division of Plenum Publishing Corporation
233 Spring Street, New York, N. Y. 10013

Printed in the United States of America

Preface

Rapid advances in knowledge have led to an increasing interest in neurobiology over the last several years. These advances have been made possible, at least in part, by the use of increasingly sophisticated methodology. Furthermore, research in the most rapidly advancing areas is essentially multidisciplinary and is characterized by contributions from many investigators employing a variety of techniques. While a grasp of fundamental neurobiological concepts is an obvious prerequisite for those who wish to follow or participate in this field, critical awareness and evaluation of neurobiological research also requires an understanding of sophisticated methodologies.

The objective of *Methods in Neurobiology* is the development of such critical abilities. The reader is exposed to the basic concepts, principles, and instrumentation of key methodologies, and the application of each methodology is placed in the special context of neurobiological research. The reader wil! gain familiarity with the terminology and procedures of each method and the ability to evaluate results in light of the particular features of neurobiological preparations and applications.

Robert Lahue

Waterloo

Contributors

Hymie Anisman, Department of Psychology, Carleton University, Ottawa, Ontario, Canada

Edward L. Bennett, Laboratory of Chemical Biodynamics, Lawrence Berkeley Laboratory, University of California, Berkeley, California 94720

Anders Björklund, Departments of Histology and Neurology, University of Lund, Lund, Sweden

A. Boesten, Department of Neuroanatomy, University of Leiden, Leiden, The Netherlands

D. De Groot, Department of Electron Microscopy, Mental Hospital "Endegeest," Oegstgeest, The Netherlands

Allan M. Erickson, Division of Morphological Science, Faculty of Medicine, The University of Calgary, Calgary, Alberta, Canada

Bengt Falck, Departments of Histology and Neurology, University of Lund, Lund, Sweden

H.K.P. Feirabend, Laboratory of Anatomy and Embryology, University of Leiden, Leiden, The Netherlands

Harold Gainer, Section on Functional Neurochemistry, Laboratory of Developmental Neurobiology, National Institute of Child Health and Human Development, National Institutes of Health, Bethesda, Maryland

N.M. Gerrits, Laboratory of Anatomy and Embryology, Department of Neuroanatomy, University of Leiden, Wassenaarseweg 62, Leiden, The Netherlands

H.J. Groenewegen, Department of Anatomy and Embryology, Free University, v.d. Boechorststraat 7, Amsterdam, The Netherlands

Martin J. Hollenberg, Division of Morphological Science, Faculty of Medicine, The University of Calgary, Calgary, Alberta, Canada

P. Kontro, Department of Biomedical Sciences, University of Tampere, Tampere, Finland

Robert Lahue, Department of Psychology, Renison College, University of Waterloo, Waterloo, Ontario, Canada

Olle Lindvall, Departments of Histology and Neurology, University of Lund, Lund, Sweden

Y. Peng Loh, Section on Functional Neurochemistry, Laboratory of Developmental Neurobiology, National Institute of Child Health and Human Development, National Institutes of Health, Bethesda, Maryland

S.S. Oja, Department of Biomedical Sciences, University of Tampere, Tampere, Finland

Mark R. Rosenzweig, Department of Psychology, University of California, Berkeley, California 94720

Reinhard Rüchel, Biophysics Branch, Armed Forces Institute of Pathology, Washington, D.C. 20306. Present address: Department of Microbiology, Institute of Hygiene, Kreuzbergring 57, D-3400 Göttingen, West Germany

Angelo Santi, Department of Psychology, Wilfrid Laurier University, Waterloo, Ontario, Canada

J. Voogd, Laboratory of Anatomy and Embryology, University of Leiden, Leiden, The Netherlands

G. Vrensen, Department of Electron Microscopy, Mental Hospital "Endegeest," Oegstgeest, The Netherlands. Present address: The Netherlands Ophthalmic Research Institute, P.O. Box 6411, 1005 EK Amsterdam, The Netherlands

Contents

Chapter 1

Behavioral Techniques in Pharmacological and Neuropharmacological Analysis

Hymie Anisman and Angelo Santi

1. Introduction 1
2. Basic Principles 2
   2.1. Routes of Administration 3
   2.2. Central vs. Peripheral Drug Effects 4
   2.3. Dose–Response Relationships 6
   2.4. Drug Receptors 6
   2.5. Drug Effects on Neurotransmitter Activity 7
   2.6. Transmitter Storage: Drug–Behavior Relationships 8
   2.7. Transmitter Turnover 11
   2.8. Drug Interactions—Effects of Chronic Depletion 12
   2.9. Specificity of Drug Effects—Neurochemical Interactions . . 12
   2.10. Drug Synergism 13
   2.11. Species and Strain Factors 14
3. Locomotor Excitation and Inhibition 15
   3.1. Vertical–Horizontal Activity 16
   3.2. Response Excitation and Response Disinhibition 17
   3.3. Activity–Reactivity 17
   3.4. Stereotypy 19
   3.5. Circling–Rotational Behavior 20
   3.6. Perseveration 21
4. Habituation 22
   4.1. Habituation and Sensitization 23
   4.2. Temporal Evaluation 24
   4.3. Spontaneous Alternation 25
   4.4. Carry-Over Designs 29
   4.5. Exploration and Startle 30
   4.6. Summary 32

5. Appetitively Motivated Operant Behaviors 33
5.1. Operant-Appetitive Paradigms 33
5.2. Schedules of Reinforcement 34
5.3. Adjunctive Behavior 46
5.4. Observing Behavior 48
5.5. Drug Effects and the Experimental Analysis of Behavioral Processes 48
5.6. Summary 59
6. Aversively Motivated Behaviors 59
6.1. Components of Avoidance Behavior 60
6.2. Avoidance Techniques: Task Manipulations and Multiple-Testing Procedures 63
6.3. Discriminated Avoidance 72
6.4. Transfer Designs 77
6.5. Summary 82
7. Overview 83
References 84

Chapter 2

Behavioral and Biochemical Methods to Study Brain Responses to Environment and Experience

Edward L. Bennett and Mark R. Rosenzweig

1. Introduction 101
2. Environmental and Training Techniques 102
2.1. Differential Environments 102
3. Behavioral Techniques to Test Effects of Prior Experience . . . 112
3.1. Hebb–Williams Maze 113
4. Biochemical Techniques 117
4.1. General Considerations 117
4.2. Weights of Brain Regions 118
4.3. RNA and DNA Content 123
4.4. AChE and ChE Activities 128
References 138

Chapter 3

Cell Fractionation

Robert Lahue

1. Introduction 143
1.1. Brief History of Centrifugal Fractionation 143
2. Basic Information on Centrifugation 145
2.1. Centrifugal Force 145

2.2. Centrifugal Force and Sedimentation 147
3. Centrifuges 148
3.1. Determination of Relative Centrifugal Force 149
3.2. Analytical vs. Preparative Centrifugation 150
4. Homogenization 152
4.1. Coaxial Homogenizers 152
4.2. Blenders 157
4.3. Cellular Heterogeneity and Tissue Preparation 158
4.4. Homogenization Media 161
5. Differential Centrifugation 167
5.1. Normal-Rate Separation 167
5.2. Factors That Influence Sedimentation 170
5.3. Density Gradients 173
6. Analytical vs. Preparative Strategies 208
6.1. Analytical Approach 210
6.2. Validation of Fractionation Schemes 214
6.3. Balance Sheets 216
References 227
Selected Bibliography of Fractionation Literature 236
Whole Cells: Neurons and Glia 236
Nuclei 237
Mitochondria 238
Lysosomes 238
Synaptosomes 239
Plasma Membranes: Synaptic and Glial Membranes 240
Synaptic Vesicles 241
Endoplasmic Reticulum; Ribosomes, Polysomes 241
Filaments, Tubules, and Other Neuronal Inclusions 242
Myelin 243
Soluble Phase 243

Chapter 4

Polyacrylamide Gel Electrophoresis: Principles, Techniques, and Micromethods

Reinhard Rüchel, Y. Peng Loh, and Harold Gainer

1. Introduction 245
2. Basic Concepts 246
3. Outline of Electrophoretic Techniques 249
3.1. General Terminology 249
3.2. Moving-Boundary Technique 250
3.3. Zone Electrophoresis 250
3.4. Disk Electrophoresis 251
3.5. Isotachophoresis 257

3.6. Isoelectric Focusing 258
3.7. Two-Dimensional Techniques 260
3.8. Buffer Systems 260
3.9. Electrophoresis in Sodium Dodecyl Sulfate 261
4. Practical Aspects of Electrophoresis and Micromethods 262
4.1. The Case for Miniaturization 262
4.2. Zone Microelectrophoresis 263
4.3. Disk Microelectrophoresis 263
4.4. Gradient Microelectrophoresis 265
4.5. Microisoelectric Focusing 268
4.6. Two-Dimensional Microelectrophoresis 269
4.7. Microelectrophoresis in Sodium Dodecyl Sulfate 272
4.8. Microelectrophoresis of Nucleic Acids 277
4.9. Equipment for Microelectrophoresis 280
4.10. Recent Advances 282
5. Related Techniques and Analysis of Gel Patterns 283
5.1. Qualitative and Quantitative Analysis of Gel Patterns . . . 283
5.2. Determination of Molecular Weight 284
5.3. Sources of Artifacts 286
6. Interfacing Electrophoresis with Other Techniques 290
6.1. Elution and *in Situ* Histochemistry 290
6.2. Immunological and Autoradiographic Techniques 291
6.3. Peptide Mapping and Amino Acid Analysis of Proteins Following Electrophoresis 292
7. Concluding Remarks 292
References 293

Chapter 5

Classic Methods in Neuroanatomy

J. Voogd and H.K.P. Feirabend

1. Introduction 301
2. Dissection and Gross Anatomy of the Brain 303
3. Reconstruction Methods 307
4. Fixation and General Histological Procedure 310
4.1. Fixation 310
4.2. Frozen Sections 313
4.3. Paraffin Embedding 314
4.4. Celloidin and Plastic Embedding 315
5. Nissl Stains 315
5.1. Nissl's Original Method and Its Modifications 315
5.2. Cytoarchitectonics 317
5.3. Axonal Reaction of the Perikaryon 318
6. Myelin Stains 322

6.1. Staining of the Lipids in the Myelin Sheath 322
6.2. Mordanting of Myelin: Weigert and Häggqvist Methods . . 323
6.3. Klüver–Barrera Method 328
6.4. Myeloarchitecture: Counting and Measuring of Nerve Fibers 328
6.5. Myelogenesis, Myelin Degeneration, and Demyelinization . 331
7. Reduced-Silver Methods 333
7.1. Neurofibrillar Stains 333
7.2. Silver Impregnation of Degenerating Axons and Axon Terminals 334
8. Golgi Method 342
9. Histological Methods 345
9.1. Gelatin Embedding 345
9.2. Paraffin Embedding 346
9.3. Celloidin Embedding 347
9.4. Nissl Stains 349
9.5. Myelin Stains 350
9.6. Reduced-Silver Methods 354
9.7. Golgi Methods 357
References 358
Bibliography 364

Chapter 6

Fluorescence Microscopy of Biogenic Monoamines

Olle Lindvall, Anders Björklund, and Bengt Falck

1. Introduction 365
2. Fluorescence Microscopy and Microspectrofluorometry 366
2.1. The Fluorescence Microscope 367
2.2. The Microspectrofluorometer 368
3. Visualization of Biogenic Monoamines 372
3.1. Chemical Background 372
3.2. Practical Performance of the Formaldehyde and Glyoxylic Acid Methods 384
4. Fluorescence-Histochemical Techniques for Identification and Differentiation of Biogenic Monoamines 411
4.1. Tests for Specificity 411
4.2. Identification and Differentiation of Biogenic Amines and Related Compounds after Formaldehyde Treatment . . . 412
4.3. Identification and Differentiation of Biogenic Amines and Related Compounds after Glyoxylic Acid Treatment . . . 418
5. General Conclusion 422
References 422

Chapter 7

Electron Microscopy in Neurobiology

G. Vrensen, D. De Groot, and A. Boesten

1. Introduction 433
2. The Transmission Electron Microscope: Basic Principles and Design 435
   2.1. Introduction 435
   2.2. Resolving Power 435
   2.3. Image Formation and Contrast 438
   2.4. Depth of Field and Depth of Focus 440
   2.5. Magnification 441
   2.6. Design of the Transmission Electron Microscope 441
   2.7. Special Applications 448
3. Electron-Microscopic Histology 450
   3.1. Introduction 450
   3.2. Fixation 450
   3.3. Dehydration 453
   3.4. Embedding 453
   3.5. Ultramicrotomy 455
   3.6. Staining Procedures 461
   3.7. Other Techniques 464
   3.8. Some Critical Remarks on Electron-Microscopic Histology . 464
4. Applications of Electron Microscopy in Neurobiology 466
   4.1. Ultrastructural Studies in Neuroanatomy and Neurocytology 466
   4.2. Electron Microscopy and Synaptic Transmission 477
   4.3. Electron Microscopy in Cell and Tissue Culture 484
   4.4. Electron Microscopy and Cell Fractionation 485
   4.5. Electron Microscopy: Relevance and Limitations in Neurobiological Research 486
5. Quantitative Stereology in Electron Microscopy 488
   5.1. Introduction 488
   5.2. Basic Principles and Terminology 489
   5.3. Some Practical Aspects of Quantitative Stereology 493
   References 496
   Bibliography 499

Chapter 8

Scanning Electron Microscopy: Applications to Neurobiology

Martin J. Hollenberg and Allan M. Erickson

1. Fundamentals of Scanning Electron Microscopy 501
   1.1. Introduction 501
   1.2. History 502

1.3. The Instrument 502
1.4. The Sample 504
1.5. Advances in Techniques of Specimen Examination and Characterization 517
2. Specific Applications of Scanning Electron Microscopy to Neurobiology 521
2.1. Introduction 521
2.2. Central Nervous System 521
2.3. Peripheral Nervous System and Individual Neurons . . . 525
2.4. Receptors 528
2.5. Conclusion 533
References 534

Chapter 9

Autoradiography in the Nervous System

H.J. Groenewegen, N.M. Gerrits, and G. Vrensen

1. Introduction 543
2. Application of Autoradiography in Different Fields of Neurobiological Research 545
2.1. Introduction 545
2.2. Cell Biology 545
2.3. Embryology 546
2.4. Neuronal Connectivity 546
3. Basic Principles of Autoradiography 554
3.1. Introduction 554
3.2. Radioisotopes 555
3.3. Photographic Emulsions 557
3.4. Resolution of Autoradiographs 558
3.5. Efficiency of Autoradiographs 561
4. Methods 563
4.1. Administration of Labeled Compounds 563
4.2. Preparation of Autoradiographs 567
4.3. Analysis of Autoradiographs 571
5. Critical Evaluation of Anterograde Tracing Techniques 577
5.1. Introduction 577
5.2. The Problem of Passing Fibers 577
5.3. Injection Site 581
5.4. Termination Area 582
5.5. Differential Uptake of Precursors 586
5.6. Sensitivity of the Autoradiographic Tracing Method . . . 588
5.7. Transneuronal Transport 588
5.8. Some Technical Advantages and Limitations of the Autoradiographic Tracing Method 589

5.9. Combination of the Autoradiographic Tracing Method with Other Techniques 591
References 592

Chapter 10

Isotope Methods

S.S. Oja and P. Kontro

1. Introduction 599
2. Physical Background 600
   2.1. Properties of the Nucleus 600
   2.2. Types of Radioactive Decay 601
   2.3. Rate of Radioactive Decay 604
   2.4. Units and Definitions 605
3. Isotopes of Neurobiological Interest 607
4. Labeled Compounds 609
   4.1. Availability 609
   4.2. Position of Label 612
   4.3. Manufacturing of Labeled Compounds 613
   4.4. Purity of Labeled Compounds 614
   4.5. Stability and Storage of Labeled Compounds 615
5. Measurement of Isotopes 617
   5.1. Stable Isotopes 617
   5.2. Radioactive Isotopes 620
6. Statistics of Radioactivity Detection 634
7. Data Analysis in Neurobiological Applications 636
   7.1. Assay of Brain Free Amino Acids with Labeled Dansyl Chloride 636
   7.2. Radioimmunoassay for Myelin Basic Protein 638
   7.3. Compartmental Analysis of Tracer Kinetics 638
   7.4. Determination of the Blood–Brain Exchange of Taurine 640
   7.5. Determination of the Rate Constants for Amino Acid Efflux from Brain Slices 643
   References 646

Index 649

1

Behavioral Techniques in Pharmacological and Neuropharmacological Analysis

Hymie Anisman and Angelo Santi

1. Introduction

Performance in complex tasks likely reflects the additive or interactive effects of a relatively large set of simple behaviors with a variety of organismic and environmental variables. By the same token, the neurochemical basis for such behaviors cannot be ascribed to any single neurochemical; rather, a number of neurochemical and hormonal systems, acting in series or in parallel, or both, subserve ultimate levels of performance. Owing to the complexity of interactions among these systems, analysis of pharmacological treatments may be exceedingly difficult. Indeed, the situation is reminiscent of a complex chemical reaction in which elimination of any one step (or several steps) in the reaction precludes the attainment of the end product. That is, elimination or reduction of any one of the components of a complex behavior may prevent the outcome that might otherwise be observed. A treatment may, for example, affect performance in a learning task by disrupting associative processes, including memory consolidation and retrieval. On the other hand, a comparable performance change can be induced via nonassociative effects of the drug, such as variation in response inhibition–disinhibition,* motor depression, or motivation. In effect, a nonassocia-

* The terms inhibition and disinhibition have often been misleading, since they appear to mean different things in discussions of neuropharmacology, on the one hand, and of conditioning processes, on the other. Indeed, even within the behavioral discipline, the terminology is inconsistent and used loosely to describe various phenomena. In the context of this chapter, "inhibition" will refer to the reduction of active behaviors following stimulation (e.g., after habituation or shock), while "disinhibition" will refer to the attenuation of these inhibitory effects.

Hymie Anisman • Department of Psychology, Carleton University, Ottawa, Ontario, Canada. *Angelo Santi* • Department of Psychology, Wilfrid Laurier University, Waterloo, Ontario, Canada.

tive type of change can mimic associative types of variations (see review in Bignami, 1977), and similarly, two different nonassociative biases may lead to comparable behavioral consequences (Kokkinidis and Anisman, 1977).

While it is realized that emergent properties in complex behavioral situations are not necessarily predictable from the knowledge of isolated elements comprising the behavior, it is equally true that evaluation of treatments in terms of gross behavioral change also is not necessarily representative of the mechanisms subserving the behavior. In effect, it is necessary that the nature of the behavioral change be examined in terms of the composite of elements subserving the behavior, as well as in a series of situations in which subtle manipulations of the parameters of the task are carried out. Such an approach may be effective in delineating the necessary and sufficient conditions for a particular behavior to become apparent. In addition, the precise source of a pharmacological or neurophysiological treatment may be elucidated.

Advances in pharmacology and neurophysiology have made it possible to deduce the neurochemical substrates for some drug–behavior relationships. Similarly, on the behavioral side, numerous techniques have been developed to assess the concomitants for drug effects on complex behaviors. It is therefore somewhat surprising that there still remains a paucity of data attempting to evaluate behavioral changes with the aim of dissociating the effects of drugs on the molecular components subserving these behaviors. The purpose of this chapter is to illustrate methods that can be used to this end. Clearly, this review cannot be all-encompassing, and only a broad sampling of methods are described. Out of necessity, caveats that need to be considered will be elucidated as well.

This chapter is divided into five primary sections. First, some general pharmacological principles will be introduced. This will be followed by the behavioral techniques in order of the *apparent* complexity of the behavior, i.e.: (1) locomotor activity and related behaviors; (2) habituation; (3) appetitively motivated behaviors; and (4) aversively motivated behaviors. Since this chapter is intended for both the behaviorist, who lacks sophistication on the pharmacological side, and the pharmacologist, with little experience on the side of behavioral analyses, large portions of this chapter will no doubt be redundant to one group or the other.

2. *Basic Principles*

Prior to embarking on a discussion of the behavioral analysis of drug effects, it is necessary to elucidate several characteristics of pharmacological studies that need to be considered. This will prove particularly redundant to the pharmacologically sophisticated, and is meant for the behaviorist who is interested in the pharmacological analysis of behavior. The descriptions given here will be relatively brief, since this chapter is not intended as a review of mechanisms of drug action. Moreover, several excellent sources are available (Cooper *et al.*, 1974; Fingl and Woodbury, 1970; S.D. Iversen and L.L. Iversen, 1975; Rech and Moore, 1971) that cover these facts in detail. The latter three references are of particular relevance to the novice, and the first is somewhat more in depth and should be dealt with afterward.

2.1. Routes of Administration

The effectiveness of a drug in modifying a particular behavior involves a number of factors, including the quantity of drug administered and its absorption rate, distribution within the body, binding with tissue, and rate of degradation. Since the route of drug administration influences some of these factors, it needs to be considered further. The particular route of administration selected is, of course, dependent on a variety of factors. Table 1 indicates various methods of drug administration, together with some characteristic benefits and drawbacks.

Following administration, a drug reaches its site of action by entering the bloodstream and then proceeding to peripheral sites, where equilibration with extracellular fluid occurs rapidly. Passage into the brain or spinal cord occurs through capillaries or through cerebrospinal fluid (CSF). Unlike peripheral effects, which occur fairly rapidly, entry into brain proceeds relatively slowly owing to three factors, collectively described as the *blood–brain barrier:* (1) brain capillaries lack pores, thus necessitating passage of the drug through endothelial cells; (2) permeability of the drug is restricted by glial tissue that surrounds brain capillaries; and (3) extracellular space is limited owing to the close proximity of neurons and glial cells.

As seen in Table 1, the quantity and latency of drug reaching the brain are dependent on the nature of the injection. Of the systemic routes, injection directly into the bloodstream [intravenous (i.v.)] produces the most rapid effect and requires a relatively small amount of drug, since the drug need not be absorbed into the bloodstream. Drug administration by mouth [*per os* (p.o.)] or injection into the peritoneal cavity [intraperitoneal (i.p.)], into large muscles [intramuscular (i.m.)], and beneath the skin [subcutaneous (s.c.)] all produce slower drug effects, but as seen in Table 1, have specific advantages that cannot be obtained via the i.v. route. One can of course use i.v. drip for sustained effects, but this is cumbersome, particularly in the free-moving animal.

Drugs can be directly injected into CSF by infusion through cannulae inserted into the ventricles [intraventricular (i.vent.)]. Similarly, using the cannulation method, or through intracranial injection, drugs can be applied to specific brain areas. To reduce diffusion of drugs in the brain, a more precise technique, iontophoresis, may be used (McLennan, 1970). In this technique, application of electric current through fine glass electrodes moves charged drug molecules. As indicated by S.D. Iversen and L.L. Iversen (1975), this is a useful technique to evaluate responses of single neurons, but is of questionable value in behavioral studies, since insufficient numbers of neurons are stimulated.

Table 1. Routes of Drug Administration

Route of administration	Latency to behavioral change	Characteristic uses
Intravenous (i.v.)	Rapid	Used only with soluble compounds. More predictable and faster action than with other systemic injections. Unlike slower methods, it cannot be reversed. In rodents, injection may prove difficult. May prove dangerous if drug is injected quickly or if drug precipitates out while in the bloodstream.
Per os (oral) (p.o.)	Slow	Used when drug is not soluble or when injection produces local irritation. Imprecise dose–effect relationship.
Intraperitoneal (i.p.)	Intermediate	Peak times variable. Not recommended for repeated injection or among young animals. Although drugs can be administered in a suspension, equivocal results have been observed. Irritants may cause behavioral side effects.
Subcutaneous (s.c.)	Intermediate	Absorption slow, giving relatively slow, constant effect. With non-readily soluble drugs, or when administered with a vasoconstrictor, drug effects are further protracted. Should be avoided with irritants. Recommended for repeated injection and with young animals.
Intramuscular (i.m.)	Intermediate	Gives slow, protracted effect, particularly when suspended in oil. Irritants and insoluble compounds should be avoided.
Intraventricular (i.vent.)	Exceptionally rapid	Access to many brain areas by diffusion through ventricular system. Leaves brain quickly. When used with substance that does not pass blood–brain barrier, central effects can be determined. Liquid rather than granular substance preferred.
Intracranial	Exceptionally rapid	Drugs may be applied to specific brain regions to determine type of neurons involved in behavioral effects. Considerable diffusion with increasing volume of drug.
Iontophoresis	Exceptionally rapid	Affects only a small number of neurons. Of limited use in behavioral evaluation.

2.2. Central vs. Peripheral Drug Effects

The fact that drug entry into brain tissue may be restricted might, on the one hand, prove to be problematic in pharmacological work; however, on the other hand, it may prove to be a useful methodological tool. For example, agents such as scopolamine readily pass into the brain. In evaluating the behavioral effects of this anticholinergic, though, it must be remembered that the drug also exerts peripheral effects. One may employ the quaternary congener of this drug, methylscopolamine, which does not readily pass into brain, to evaluate the peripheral effects of scopolamine. In particular, if scopolamine alters behavior and methylscopolamine does not, it is likely that the cholinergic effects are restricted to central actions of the drug or to the interaction between central and peripheral effects. It must be remembered, though, that methylscopolamine may have indirect central effects by affecting peripheral organs, which in turn influence central processes. To dissociate central and peripheral action further, methylscopolamine and scopolamine may be directly injected into the brain. If the drugs influence behavior in the same way, then the effects would be due to their central action (e.g., see J.M. Williams *et al.*, 1974; Abeelen *et al.*, 1970).

In some cases, drugs may exert considerable peripheral effects and affect the central nervous system only after high doses, but not because of blood–brain factors. Rather, enzymatic breakdown occurs in the peripheral system, thus resulting in minimal passage into brain. For example, 3,4-dihydroxyphenylalanine (L-dopa) when injected systematically is degraded in the peripheral nervous system by dopa-decarboxylase. Consequently, high dosages of the drug are necessary in order that central effects be obtained. To reduce the peripheral effects, carbidopa (MK-486), an extracerebral decarboxylase inhibitor, may be administered to assure that greater amounts of L-dopa will reach the brain (Bartholini *et al.*, 1969; Butcher *et al.*, 1970; Stromberg, 1970). Similarly, in the case of 5-hydroxytryptophan (5-HTP), the drug is preceded by injections of an extracerebral decarboxylase inhibitor, e.g., MK-486. In the absence of MK-486, treatment with 5-HTP reduces locomotor stimulation, whereas the opposide is true if 5-HTP treatment is given after MK-486 (Modigh, 1974*a,b*). Parenthetically, 5-hydroxytryptamine [(5-HT) serotonin] is not administered systemically because it does not pass into the brain, whereas 5-HTP does, after which it is transformed to 5-HT.

In considering the route of injection, solubility of the drug needs to be considered, since drugs that are not readily soluble may have adverse effects. In some cases, these may, in fact, be quite subtle. For example, α-methyl-*p*-tyrosine (α-MpT) and *p*-chlorophenylalanine (PCPA) are not readily soluble and are often injected as a suspension. The behavioral effects of such treatments may be very different from those of the counterparts of these agents, α-MpT methylester and PCPA methylester, which are freely soluble in water. Indeed, some of the differential results reported in the literature with these agents may well be a consequence of subtle differences evoked by the treatments, such as local irritation induced by the nonsoluble compounds (for discussion of the effects of the dopamine-β-hydroxylase inhibitor FLA-63, see Thornburg and Moore, 1973).

One final caveat needs to be interjected here. Namely, in dealing with young organisms, drug effects on behavior may be particularly large, since, among other things, the blood–brain barrier is not fully developed (see Weiner, 1974). As a result, drugs that do not pass into the brain in adult animals may do so in young animals. Moreover, because of the differential permeability, the appropriateness of dose comparisons across developmental levels is questionable. Incidentally, the problems encountered with young organisms are not restricted to this one point. Other variables, such as temperature regulation, peripheral degradation, and motoric abilities, need to be considered.

2.3. Dose–Response Relationships

The effectiveness of a drug in modifying a given behavior is dependent on the dosage of the drug employed. Some drugs exert linear changes with increasing dosage; other dose–response effects are sigmoid, U-shaped, or even bimodal. Fingl and Woodbury (1970) have characterized four primary parameters regarding the dose–response relationship. These are potency, slope, maximum efficacy, and variability. Typically, when dosage (in terms of log) is plotted against a biological response, a sigmoid curve is obtained. The major portion of the curve is the linear portion, thereby permitting drug–dose comparisons for specified behaviors. The *potency* of a drug usually is expressed in terms of the dosage necessary to produce one half its maximum effect (ED50). The term LD50 represents the dose lethal to one half of individual subjects. The greater the *slope* of the curve, the smaller the margin of safety (i.e., the smaller the range between the effective and lethal doses). *Efficacy* refers to the maximal effectiveness of a drug in producing a behavioral change. Finally, in dealing with drug effects, it needs to be remembered that among different individuals, a given drug dosage may produce very different behavioral effects. The smaller the between-subject *variability*, the greater the confidence in predicting a specified behavioral change.

2.4. Drug Receptors

In most instances, drugs interact with sites referred to as drug receptors. Those drugs that stimulate neurons by directly interacting with receptors are *agonists.* Those agents that combine with receptors but do not initiate action are considered *antagonists.* Antagonists may be competitive or noncompetitive; in the former case, the effects of the antagonist can be reversed by increasing agonist dosages, whereas this is not the case for the latter type.

Drugs may influence behavior by any number of ways. These are reviewed nicely in Cooper *et al.* (1974) and S.D. Iversen and L.L. Iversen (1975) and need not be reiterated here. Suffice it to say that receptor stimulation can be augmented by (1) direct agonists, (2) increasing release of a neurotransmitter, (3) increasing synthesis of a neurotransmitter, (4) inhibiting enzymatic destruction of a neurotransmitter, (5) blocking reuptake of a transmitter, or (6) administering precursors of a particular neurochemical. On the other side, receptor stimulation can be reduced by (1) administration of antagonists, (2) increasing reuptake, (3) increasing enzymatic degradation, (4) blocking axonal flow, (5) depleting storage pools (vesicles) of a transmitter, (6) destroying terminals by chemical methods, or (7) injecting drugs that mimic neurotransmitters, but have weaker potentiating properties (pseudotransmitters).

2.5 Drug Effects on Neurotransmitter Activity

Neuronal activity may be modified by any number of different drug treatments. With the increasing clinical and laboratory use of pharmacological agents, greater numbers of drugs are being developed. This refers not only to drugs with broad-spectrum effects that typically are meant for therapeutic purposes, but also to relatively specific agents intended primarily for research purposes. With respect to the latter types of agents, there are numerous techniques available to achieve relatively comparable effects on levels or activity of transmitters. Tables 2–5 present a limited listing of drug effects on ACh, norepinephrine (NE), dopamine (DA), and 5-HT activity. These agents are described in a fairly abbreviated form, and the reader should be cautioned that the effects of these agents are not necessarily as selective as

Table 2. Drug Effects on Cholinergic Activity

Drug	Effect
Scopolamine Atropine Benztropine	Block cholinergic (muscarinic) receptors centrally and peripherally.
Methylscopolamine Methylatropine	Block cholinergic receptors peripherally.
β-Methylcholine Arecoline Oxotremorine Pilocarpine	Receptor stimulants; the first two agents do not readily penetrate the CNS.
Physostigmine Diisopropylfluorophosphate Sarin	Inhibit acetylcholinesterase (AChE), thereby increasing ACh levels.
Neostigmine	Inhibits AChE, but does not readily penetrate the CNS.
Nicotine Carbamylcholine	Stimulate cholinergic (nicotinic) receptors.
d-Tubocurarine Gallamine Decamethonium	Block nicotinic receptors at the neuromuscular junction.
Hexamethonium Mecamylamine Pempidine Chlorisondamine	Ganglion blocker.
Hemicholinium-3 Triethylcholine	Block choline uptake.
Cytochalasin B Botulinum toxin Collagenase	Block ACh release at neuromuscular junction.
β-Bungarotoxin	Promotes ACh release at neuromuscular junction.

Table 3. Drug Effects on Dopaminergic Activity

Drug	Effect
α-Methyl-*p*-tyrosine (α-MpT)	Reduces dopamine (DA) by inhibiting tyrosine hydroxylase, the rate-limiting enzyme in the synthesis of DA and norepinephrine (NE).
3,4-Dihydroxyphenylalanine (L-dopa)	Precursor that results in increased DA levels.
Catechol Pyrogallol Isopropyltropolone	Increase DA by inhibiting catechol-*O*-methyl transferase (COMT).
Iproniazid Phenelzine Pheniprazine Pargyline Tranylcypromine	Increase DA by inhibiting monoamine oxidase (MAO).
Amphetamine Cocaine	Block reuptake of DA, and also increase release.
Reserpine Tetrabenazine	Deplete the vesicular storage system.
6-Hydroxydopamine (6-OHDA)	Produces permanent degeneration of terminals. Effects on NE vs. DA are manipulable.
Apomorphine Piribedil	Selective agonists.
Chlorpromazine Haloperidol Pimozide Clozapine	DA antagonists.

described in the tables. For further information, the reader is encouraged to consult recent journal articles or at least see the texts mentioned in Section 2.

2.6. Transmitter Storage: Drug–Behavior Relationships

Nominally, treatments that inhibit synthesis of transmitters or result in their rapid utilization should produce profound behavioral changes in their own right, and also should antagonize the effects of indirectly acting stimulants. This, however, is not always the case. There are data available that indicate that NE and 5-HT exist in several storage forms. Specifically, work with labeled neurotransmitters (e.g., [^{3}H]-NE) has shown that when the compound is injected intraventricularly, its disappearance occurs in a multiphasic fashion (Glowinski *et al.*, 1965; L.L. Iversen and Glowinski, 1966). That is, at first a phase of rapid disappearance of [^{3}H]-NE occurs, followed by a stabilization period, and then a further period of [^{3}H]-NE decrease. On the basis of these data, it has been proposed that the amine is stored in two separate pools. NE newly synthesized or newly taken up is stored in a functional compartment, while previously synthesized amines are stored in a main compartment. On neuronal stimulation, amine from the former pool is released preferentially.

It seems that some treatments elicit NE release from the functional storage pools, whereas other treatments cause release from both pools.

Table 4. Drug Effects on Noradrenergic Activity

Drug	Effect
FLA-63 U-14,624	Reduce NE by inhibiting dopamine-β-hydroxylase, the enzyme that converts DA to NE.
3,4-Dihydroxyphenylserine (DOPS)	Is converted to NE by decarboxylation.
α-Methyldopa α-Methy-*m*-tyrosine	Forms false transmitters released from adrenergic neurons.
Catechol Pyrogallol Isopropyltropolone	Increase NE by COMT inhibition.
Iproniazid Phenelzine Pheniprazine Pargyline Tranylcypromine	Increase NE by MAO inhibition.
Amphetamine	Blocks NE reuptake and increases release.
Desmethylimipramine Amitriptyline	Inhibit reuptake of NE.
Reserpine Tetrabenazine	Deplete the vesicular storage system.
Phenethylamine Octopamine Tyramine	Displace NE and mimic its postsynaptic action.
6-OHDA	Produces permanent degeneration of terminals.
Guanethedine Bretylium	Block release of NE.
Phenoxybenzamine Dibenamine Phentolamine Tolazoline Yohimbine	α-Receptor blockers.
Dichloroisoprenaline Propranolol	β-Receptor blockers.
Clonidine	Receptor agonist.

Table 5. Drug Effects on Serotonergic Activity

L-Tryptophan	5-HT precursor.
Tryptamine α-Methyltryptamine α-Methyl-5-HT Chlorimipramine Imipramine	Inhibit 5-HT reuptake.
Iproniazid	Increases 5-HT by inhibiting MAO.
PCPA	Reduces 5-HT via tryptophan hydroxylase inhibition.
p-Chloroamphetamine	Inhibits tryptophan hydroxylase and blocks 5-HT uptake.
Lysergic acid diethylamide Methysergide Cyproheptadine	Receptor antagonists.
Reserpine	Interferes with the uptake–storage mechanism of the amine granules.
5,6-Dihydroxytryptamine 5,7-Dihydroxytryptamine	Produce permanent degeneration of terminals.

Moderate levels of stress, for instance, result in increased utilization of [^{3}H]-NE administered 10 min earlier, but do not affect utilization of this amine administered 180 min earlier. Severe stress, on the other hand, produces a decline in [^{3}H]-NE administered either 10 or 180 min earlier (Thierry, 1973; Thierry *et al.*, 1968, 1971). Apparently, moderate stress preferentially releases newly synthesized NE from functional pools, whereas severe stress affects NE stored in the main compartment as well as in the functional pools.

There are indications that (1) different behaviors elicited by a single drug and (2) comparable behavioral changes induced by different drugs may be subserved by newly synthesized amines on the one hand and previously stored amines on the other. Typically, it is observed that locomotor stimulation and stereotypy induced by D-amphetamine are antagonized after inhibition of tyrosine hydroxylase by α-MpT (Carlsson, 1970; Randrup and Munkvad, 1970). This presumably is a result of DA and NE synthesis being inhibited. In contrast to these behaviors, however, amphetamine-induced perseveration (discussed in detail in Section 3.6) is not affected by α-MpT, suggesting that perseveration is not solely dependent on amine synthesis, but may be produced by amines present in the main storage pools. Predictably, if the pools are depleted by reserpine, subsequent administration of α-MpT antagonizes the perseveration (Anisman and Kokkinidis, 1975). In the same vein,

Breese *et al.* (1975) recently observed that although amphetamine-induced locomotor stimulation is antagonized by α-MpT, this is not true of methylphenidate-induced motor effects. For α-MpT to effectively counter the effects of the latter agent, preinjection of reserpine is necessary. Thus, it is possible that the two agents are affecting amines from different pools.

Although there is some controversy concerning the existence of multiple storage pools (Doteuchi *et al.*, 1974), it is clear that the effects of drug treatments cannot always be explained on the basis of a single-pool model. Regardless of whether the assumptions of Glowinski *et al.* (1965) and L.L. Iversen and Glowinski (1966) are correct, the fact is that drug effects are subject to complex treatment interactions.

2.7. *Transmitter Turnover*

It is believed that on the release of a transmitter substance from storage pools, the synthesis rate of the transmitter is increased. Any one of several possible mechanisms might be responsible for the increased synthesis, including presynaptic mechanisms or effects related to storage pools of the transmitter. Regardless of the mechanism, it seems that transmitters can be maintained at fairly stable levels. In this case, the rate of synthesis should be equal to its utilization. Under steady-state conditions (i.e., when synthesis equals utilization), turnover of the transmitter is equal to the rate of synthesis or utilization.

Various treatments, such as moderate levels of stress (Anisman, 1975; Stone, 1975), will increase turnover rates, but without dramatic changes in amine content. As a result, evaluation of treatment effects in terms of levels of transmitter may not be an accurate index of transmitter activity. Turnover rate, on the other hand, may be a more sensitive index of neuronal activity. It is noteworthy, though, that when the actual levels of a transmitter change, it is not sufficient to evaluate either synthesis or utilization alone, but rather, both measures are necessary.

A number of techniques are available to assess turnover rate:

1. The rate of loss of labeled transmitter (e.g., [^{3}H]-NE or [^{14}C]-NE) administered exogenously (L.L. Iversen and Glowinski, 1966) can be used as an estimate of turnover under steady-state conditions.

2. The rate of formation of labeled transmitter after administration of labeled precursors (tyrosine and dopa in the case of DA and NE, or 5-HTP in the case of 5-HT) gives an indication of the rate of synthesis (Roth *et al.*, 1966).

3. The rate of transmitter decline after synthesis inhibition, e.g., NE levels after tyrosine hydroxylase or dopamine-β-hydroxylase inhibition (Brodie *et al.*, 1966; Corrodi *et al.*, 1970), is another technique for measurement of utilization, and thus under steady-state conditions indicates turnover rate.

4. By measuring the metabolite levels of degraded transmitters, utili-

zation rate can be estimated. Inasmuch as several routes of degradation may exist, e.g., by catechol-*O*-methyltransferase (COMT) and monoamine oxidase (MAO) in the case of amines, it is necessary to evaluate several metabolites. In some instances (e.g., NE degradation by COMT), the product, normetanephrine, is rapidly degraded to 3-methoxy-4-hydroxphenylglycol. Thus, both these metabolites need to be evaluated (Stone, 1975).

5. Increase in levels of transmitter after inhibition of the degrading enzymes (e.g., MAO in the case of NE or 5-HT) yields an estimate of utilization (Neff and Costa, 1966).

2.8. *Drug Interactions—Effects of Chronic Depletion*

Although one might expect that the effectiveness of drugs that excite neurons would be antagonized by treatments that reduce synthesis or deplete the pools of neurotransmitters, this is not consistently the case. To the contrary, prior depletion of a transmitter may result in the augmentation of some drug effects.

If transmitter activity is severely depleted for prolonged periods, subsequent drug injections may elicit paradoxical results. For example, after 6-hydroxydopamine (6-OHDA) injection into the substantia nigra, which destroys DA nerve terminals, the locomotor-stimulating effects of *d*-amphetamine are antagonized, but locomotor excitation induced by apomorphine is exaggerated (see Creese and Iversen, 1973, 1975*a,b*; Fibiger and Grewaal, 1974; D.C.S. Roberts *et al.*, 1975). It seems that after chemical lesion of this sort, receptor supersensitivity occurs, such that the effectiveness of direct agonists (e.g., apomorphine) is increased. Similar supersensitivity effects have also been observed after prolonged haloperidol (a DA antagonist) treatment (e.g., Sayers *et al.*, 1975; see also Reid, 1975; Moore and Thornburg, 1975). It is worthy of note that unlike chemical lesions, supersensitivity is not observed after electrolytic lesions, which are not specific to storage sites (see discussions in Ungerstedt *et al.*, 1975; Moore and Thornburg, 1975).

Finally, one further interjection is necessary, a caveat on a caveat as it were. With chemical-lesion techniques, as well as with other drug treatments suspected of altering neurochemical levels, the validity of conclusions is enhanced considerably if the behavioral work is accompanied by assay procedures. Often, substantial variations in transmitter levels are necessary before certain behavioral changes are observed. In the absence of direct verification of neurochemical levels, negative results may lead to confusing interpretations with respect to previous work (for discussion of this issue, see D.C.S. Roberts *et al.*, 1975).

2.9. *Specificity of Drug Effects—Neurochemical Interactions*

As indicated earlier, it is not likely that complex behaviors are subserved by a single neurochemical, but rather involve the interplay of several neurotransmitters (cf. Anisman, 1975*a*; Hamburg *et al.*, 1975; Richardson,

1974). In his assessment of the cholinergic system, Karczmar (1975) stated that he did not know of a single behavior that was not affected by treatments that influence ACh. While this is a fairly strong statement, to which some may take exception, it probably is not far from the truth. Likewise, one could propose an equally strong statement regarding the involvement of other transmitters in behavioral regulation. In effect, we have here a situation in which consideration should be given, not to which transmitter mediates a given behavior, but rather to the proportion of the variance accountable by a particular neurochemical.

A second consideration, and this basically is the other side of the same coin, is that variation in one neurochemical may promote change in a second system. This may be the case with relatively "specific" agents as well as "nonspecific" agents. In the case of *d*-amphetamine, for example, behavioral changes observed are thought to be a consequence of increased release and blocked reuptake of DA or NE or both (D.C.S. Roberts *et al.*, 1975; Creese and Iversen, 1975*a*). However, it has also been observed that cortical ACh is increased after *d*-amphetamine (Pepeu and Bartolini, 1968; Nistri *et al.*, 1972). While the source for the cholinergic change is not fully understood, the limbic system apparently is involved, since septal lesioning eliminates the change in cortical ACh (Nistri *et al.*, 1972). By way of a second example, there now exist numerous reports indicative of a DA–ACh balance in the nigrostriatal system. Basically, DA neurons tonically inhibit ACh neurons, whereas ACh neurons excite DA neurons. As a result, changes in DA activity via drug treatments effectively modify ACh activity (cf. Bartholini *et al.*, 1973; Stadler *et al.*, 1973; Javoy *et al.*, 1974) (but see also Groves and Rebec, 1976).

2.10. Drug Synergism

The effectiveness of a particular drug treatment can be modified not only by drugs that affect the same neurochemical, as in the case of 6-OHDA antagonism of amphetamine, but also by drugs that affect other transmitters. For instance, *d*-amphetamine-induced locomotor activity and stereotypy, which are apparently mediated by dopamine, also are enhanced by treatment with an anticholinergic (Carlton, 1961*a*,*b*). Indeed, when both agents are administered in doses that by themselves do not affect behavior, a substantial change in behavior may be observed. Comparable effects have been seen in other behavioral situations including pseudoconditioning (Izquierdo, 1974) and perseveration (Kokkinidis and Anisman, 1977).

Drug synergism may come about because of any one of several factors. Some of these are listed below, though the list is not exhaustive:

1. A drug changes the potency of an agonist, in which the dose–response curve shifts to the left (potentiation).
2. A given behavior is subserved by excitation of two neurochemical systems. The behavior in question does not occur with either agonist alone, regardless of dosage.

3. A given drug elicits two distinct behavioral changes that are antagonistic with respect to one another. By employing a second drug, one of the antagonistic behaviors is suppressed, permitting the other behavior to be manifest.
4. The behavioral effect of a drug is inhibited by the presence of situational or organismic variables. The second drug antagonizes these competing tendencies, thus permitting the expression of the drug effect.
5. Peripheral factors do not permit central effects of a drug to be expressed. The second drug antagonizes the peripheral effects.

2.11. *Species and Strain Factors*

The effectiveness of drug treatments in modifying behavior have been found to vary across species. The species × drug interaction may be due to a variety of factors including levels of neurochemicals, turnover rate, speed of degradation, and even route of degradation. For example, in the case of *d*-amphetamine, degradation occurs via oxidative deamination or aromatic hydroxylation. In some species, the major route involves oxidation, while in other species, hydroxylation is the primary route (Axelrod, 1970). This, in and of itself, may account for species differences. However, when it is realized that metabolites of *d*-amphetamine may also affect behavior, the importance of species differences becomes further exaggerated. One of the metabolites of aromatic hydroxylation is *p*-hydroxynorephedrine. This substance may act as a false transmitter at NE receptors, but with weaker potentiating effects than NE (Brodie *et al.*, 1970). Thus, the effectiveness of D-amphetamine may be a product of modification of DA and NE, together with the value contributed by the false transmitter.

Other agents may differentially affect behavior across species owing to the rate of enzymatic degradation. Goldstein *et al.* (1969) reported an inverse relationship between liver-enzyme activity and sleeping time in mice, rabbits, rats, and dogs, after administration of the sedative Hexobarbital.

That species differences should be considered important, in terms of both methodological factors and generalizability of data, would be apparent to even the least pharmacologically sophisticated researchers. Thus, it is enigmatic that *strain* differences often are not given the same consideration. Increasingly, it has become apparent that strains differ in levels and turnover rates of several transmitters (Mandel and Ebel, 1974; Mandel *et al.*, 1974; Wimer *et al.*, 1973; Pryor *et al.*, 1966) and respond differentially to drug treatments (Abeelen and Strijbosch, 1969; Oliverio, 1974; Oliverio *et al.*, 1973). Moreover, the effects of drugs interact not only with strain, but also with other treatment variables (Anisman, 1975*b*,*c*, 1976; Anisman and Cygan, 1975; Anisman *et al.*, 1975).

It is unfortunate that pharmacogenetics has only recently achieved some degree of popularity (see reviews in Broadhurst, 1976; Oliverio, 1974;

Abeelen, 1974), since the use of strain factors may serve as an ideal way of drawing fracture lines between drug–behavior relationships and also occasions where a drug influences more than a single behavior (i.e., are the behaviors independently affected by the drug, or is one behavior an artifact of the other?). To be more explicit, strains of mice can be obtained that are differentially sensitive to drug effects as concerns one particular behavior. These strains can then be examined with respect to drug-induced changes on a second behavioral index. Comparable performance across strains in the second behavior would suggest that the two behaviors and the drug-induced chemical change are not in strict relation to one another. A second type of simple genetic analysis involves the use of hybrid mice. For example, the finding that the F_1 generation resembles one parent strain so far as one phenotype is concerned (e.g., locomotor activity), but the other parent strain on a second phenotype (e.g., locomotion after drug treatment), would serve as *prima facie* evidence indicating that the two behaviors are independent of one another and are likely subserved by different mechanisms (e.g., see Anisman, 1976). There exist a variety of useful genetic techniques that cannot be dealt with at this point. The reader is, however, encouraged to see Broadhurst (1976) for a review of recent literature.

3. Locomotor Excitation and Inhibition

A variety of pharmacological treatments may elicit changes in locomotor activity. Such behavioral variations have been employed for purposes of drug classification, as well as for evaluation of drug efficacy. Indeed, evaluations of locomotor activity and related behaviors to be discussed in this section are used extensively in drug-screening programs and as an evaluative tool in pharmacological analyses.*

It needs to be recognized, though, that several drug effects may lead to comparable behavioral outcomes. To be more specific, a particular treatment may retard locomotor activity by (1) affecting the organism's motoric abilities; (2) increasing the rate of habituation (see later); (3) decreasing the propensity to initiate locomotor responding, but without affecting motor ability *per se*;

* A tangential issue must be briefly addressed at this point. There have been numerous attempts to ascribe neurochemical–behavioral relationships on the basis of behavioral change induced by relatively nonspecific drugs (e.g., the ratio obtained with the optical isomers of amphetamine). While this type of tautological reasoning has been challenged from the biochemical side (see the excellent report by Bunney *et al.*, 1975), it is worthwhile noting that this type of approach also fails from the purely behavioral vantage. Specifically, it is well known that the magnitude of drug-induced changes in locomotor activity is dependent on, among other things, the dimensions of the apparatus employed (cf. Stewart, 1975*a,b*). In the case of the potency ratio of *d*- and *l*-amphetamine, work in this laboratory (Anisman and Kokkinidis, unpublished report) has revealed that the ratio varies from 10 : 1 to 3 : 1, depending on apparatus dimensions. Thus, we are faced with the question addressed previously by Bunney *et al.* (1975) of when a particular potency ratio is a meaningful one.

(4) increasing response inhibition, or inhibition of responding to certain stimuli; and (5) inducing variations in emotionality or, for that matter, illness. On the other side, a treatment may augment locomotor activity by (1) reducing inhibition elicited by certain stimuli (i.e., response disinhibition); (2) retarding the course of habituation; and (3) eliciting *genuine* locomotor stimulation, beyond that of disinhibition.

Although some complex behaviors may be influenced by changes in locomotor activity (Anisman, 1975*a*; Anisman and Waller, 1973; Bignami, 1977), as will be seen later in this chapter, modification of locomotor activity does not assure a concomitant change in the complex behavior. To the contrary, a major determinant in this respect involves the *types* of variation induced by the drug treatment. In the following sections, methods to assess the nature of drug-induced changes in locomotor activity are described. This is followed by a discussion of the potential effects of these changes on more complex behaviors.

3.1. Vertical–Horizontal Activity

While a treatment may produce locomotor stimulation or increased exploration, in some instances a measure such as open-field activity may not be an adequate assessment of the drug effects. For example, in an animal that is relatively active in an open field (defined here as square-crossing or photocell counts), the effect of drug treatment may decrease this type of locomotion, but increase vertical movement, such as rearing (e.g., see the work of Stewart, 1975*a*,*b*; Stewart and Blain, 1975). Thus, a drug may increase the frequency of rearing at the expense of changes in horizontal locomotion (cf. Abeelen, 1974). Alternatively, a drug may influence both rearing and locomotion in a comparable manner (cf. Abeelen and Strijbosch, 1969; Oliverio *et al.*, 1973).

Some drug treatments, such as *d*-amphetamine or apomorphine, will elicit locomotor excitation as well as stereotypic behavior. The relationship between these behaviors is competitive, such that locomotor stimulation is absent during the stereotypy phase (D.S. Segal, 1975). In rats, a dose of 5.0 mg/kg of *d*-amphetamine will initially result in locomotor stimulation, followed later by reduced locomotion but increased stereotypy (continuous sniffing, grooming). Later yet, increased locomotor activity is observed (Lyon and Randrup, 1972). Thus, the effect of amphetamine on locomotor stimulation varies temporally. Needless to say, the drug effects reported may be biased one way or another depending on *when* observations are made. Clearly, it is necessary to evaluate behavior over a prolonged time period, and in addition, stereotypic behaviors and locomotor activity need to be examined concurrently. D.S. Segal (1975) and D.S. Segal and Mandell (1974) clearly indicated the importance of concurrent measurement of these behaviors in the evaluation of amphetamine tolerance. Similarly, Lyon and Randrup (1972) have indicated the value of such measurement in describing drug-induced changes in avoidance behavior. It is not inappropriate to suggest that a similar procedure would be useful in the evaluation of other poorly understood behaviors.

3.2. *Response Excitation and Response Disinhibition*

As indicated earlier, drugs may increase locomotor activity by reducing response inhibition engendered by the experimental situation (e.g., novel stimuli, stress, and even the injection procedure itself). Alternatively, a drug may elicit *genuine* locomotor stimulation quite apart from effects on behavioral inhibition. At first blush, this appears to be a relatively minor distinction; however, in other behavioral situations (viz., tasks involving inhibition), these subtle differences may affect performance in vastly divergent ways (e.g., Anisman *et al.*, 1976; Kokkinidis and Anisman, 1976*a,b*). In particular, if the task is one in which response inhibition is strong, as after exposure to a stressor, then both agents should promote an increase in activity. In contrast, when little response inhibition is present, the effects of a disinhibitory type of agent should elicit less marked effects than an excitatory agent.

In the case of agents that retard locomotor activity, it is necessary to determine whether or not the behavioral change is a result of response suppression without changes in locomotor ability (e.g., activity after application of a stressor). Parenthetically, in this category of response change, one can distinguish between lethargy in terms of motor ability *per se* and a reduction in the animals' ability to *initiate* responses. Fibiger *et al.* (1974, 1975), for instance, have shown that dopaminergic receptor blockade or depletion will affect performance in learning tasks by disrupting response initiation. It is noteworthy, though, that dopaminergic manipulations may also induce catalepsy–dyskinesia, which probably involves locomotor dysfunction as opposed to changes in response initiation.

3.3. *Activity–Reactivity*

It may be of importance to distinguish between agents that ordinarily induce locomotor excitation and agents that produce such an effect only after external stimulation (cases in which drug effects are enhanced by stimulus factors). For example, as seen in Fig. 1, scopolamine and *d*-amphetamine may increase locomotor activity in mice, depending on the strains employed. In the case of scopolamine, activity is increased in A and DBA/2, but not in C57BL/6 mice. Although it has been suggested that the lack of scopolamine effect in C57BL/6 involves a single major gene mediating cholinergic mechanisms affected by scopolamine (Oliverio *et al.*, 1973), it is clear that after shock, activity is affected in C57BL/6, just as it is in other strains of mice (see also Anisman, 1976; Anisman and Cygan, 1975; Anisman *et al.*, 1975). Although, the source for the drug × shock interaction is not apparent (but for provisional speculations see Anisman, 1976), the important

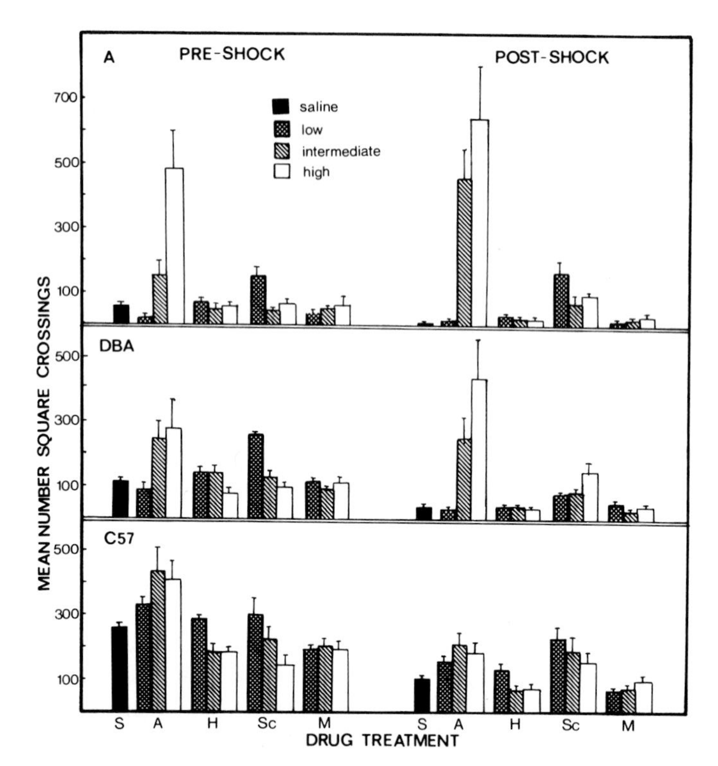

Fig. 1. Mean square-crossings (±S,E.M.) as a function of strain and drug treatment prior to and following exposure to electric shock. (S) Saline; (A) (+)-amphetamine (1.0, 5.0, or 10.0 mg/kg); (H) *p*-hydroxyamphetamine (0.63, 3.15, or 6.30 mg/kg); (Sc) scopolamine (1.0, 5.0, or 10.0 mg/kg); (M) methylscopolamine (0.9, 4.5, or 9.0 mg/kg). From Anisman and Cygan (1975); reprinted with permission; © Pergamon Press.

point for the present purpose is that drug effects that might otherwise not be apparent can be elicited by strong stimuli.

With respect to the effects of *d*-amphetamine, it is clear from Fig. 1 that the locomotor excitation elicited by the drug can be increased substantially in A and DBA/2 mice after the inception of shock. Again, the source of the drug × shock interaction has not been definitively deduced. However, inasmuch as dopamine-β-hydroxylase inhibition eliminates the synergism, but does not affect the locomotor stimulation induced by amphetamine alone, it is likely that the behavioral interaction involves the combination of dopaminergic plus noradrenergic stimulation (for details, see Anisman and Cygan, 1975). This notion, incidentally, is supported by data indicating that shock enhances NE turnover elicited by *d*-amphetamine (Javoy *et al.*, 1968). In any event, it seems that *d*-amphetamine may well elicit *genuine* locomotor stimulation, and these effects are stimulus-dependent, at least to some degree.

Finally, there are indications that the behavioral effects of serotonergic manipulations may also be modifiable by stimulus factors. That is, although

PCPA and raphe lesions will modify locomotor activity, it seems that these effects are more prominent in the presence of strong stimulation (Davis and Sheard, 1974; J.M. Williams *et al.*, 1975).

Taking these findings together, it appears that certain pharmacological treatments will affect general activity. However, in some instances, these effects are prominent only under conditions of strong stimulation. From the vantage of complex behavioral analyses, these stimulus × drug interactions are of utmost importance. It is not sufficient simply to determine whether a drug treatment modifies locomotor activity; rather, the necessary conditions for these effects need to be outlined. As will be seen later, these subtle drug differences (i.e., changes in activity vs. reactivity) play a major role in affecting performance in complex tasks. One final point that may be of value should be introduced here. Namely, in certain disorders, such as hyperactivity (hyperkinesis) in children, there are indications that the children are hyperreactive as opposed to hyperactive. Thus, it may be of value to examine the effects of those drugs that affect reactivity as well as those that affect activity in evaluation of this syndrome.

3.4. Stereotypy

As briefly mentioned earlier, a wide variety of drugs that directly or indirectly stimulate DA activity (e.g., *d*- and *l*-amphetamine, apomorphine, L-dopa, methylphenidate) will produce stereotypic behavior in rodents (i.e., continuous sniffing, grooming, or gnawing). It seems likely that DA rather than NE stimulation is responsible for these effects, since the stereotypy is antagonized by tyrosine hydroxylase inhibition, but not by inhibition of dopamine-β-hydroxylase (Creese and Iversen, 1975*b*; Pedersen and Christensen, 1972; Randrup and Munkvad, 1966). Similarly, chemical lesioning of DA pathways by 6-OHDA eliminates stereotypy, whereas lesions of NE pathways have negligible effects (Creese and Iversen, 1975*a,b*). Other transmitters likely are involved in an amplifying capacity in eliciting stereotypy, in that anticholinergics and serotonin depletors, which do not by themselves produce stereotypy, enhance the effects of *d*-amphetamine (see Costall and Olley, 1971; Janowsky *et al.*, 1972; Scheel-Kruger, 1970).

The situation, however, has been complicated by several recent findings. Specifically, the type of stereotypy observed is dependent on the specific drug employed. Whereas amphetamine stereotypy is characterized by licking and rapid movements of the head and forelegs, methylphenidate and L-dopa produce intense stereotyped gnawing (Pedersen, 1968; Pedersen and Christensen, 1972; Molander and Randrup, 1974), while apomorphine produces stereotypy with little evidence of gnawing (Pedersen, 1968; Scheel-Kruger, 1970; Molander and Randrup, 1974). Further to the point, although amphetamine stereotypy is not antagonized by NE inhibition (Randrup *et al.*, 1963; Randrup and Scheel-Kruger, 1966; Scheel-Kruger and Randrup, 1967; Creese and Iversen, 1975*a*), dopamine-β-hydroxylase inhibition antagonizes

L-dopa-induced gnawing (Molander and Randrup, 1974). In contrast, in the case of gnawing produced by methylphenidate, FLA-63 is without effect (Pedersen and Christensen, 1972).

Thus, we have the possibility that in fact NE stimulation may affect stereotypy, at least in the case of the gnawing produced by L-dopa. In effect, the possibility must be entertained that different types of stereotypy exist, some of which are modifiable by NE, and possibly still others by the interaction of catecholamines with either ACh or 5-HT. Indeed, Costall and Naylor (1972, 1974) found that lesions of the nucleus amygdaloideus laterales, which is primarily noradrenergic, resulted in abolition of gnawing, biting, and licking induced by apomorphine, but did not affect other forms of stereotypy, including sniffing and front-limb movement. These behaviors were in fact enhanced by the lesion. While this probably represents enhanced sniffing at the expense of gnawing—and thus does not really represent antagonism of stereotypy *per se*—the alternative notion that different mechanisms subserve different stereotypies needs to be considered.

3.5. Circling–Rotational Behavior

After unilateral electrolytic or chemical lesion of the substantia nigra, systemic injection of indirectly acting catecholaminergic stimulants, such as *d*-amphetamine, will produce rotational behavior ipsilateral to the site of the lesion (Ungerstedt, 1971*a*; Ungerstedt *et al.*, 1969). If the lesion is of a chemical nature (i.e., by 6-OHDA), such that receptor sites are not damaged, systemic injection of apomorphine or L-dopa will induce contralateral circling (Ungerstedt *et al.*, 1969; Van Voigtlander and Moore, 1973). It is likely that these differential drug effects reflect supersensitivity of dopaminergic receptors. Given the apparent balances and interactions among transmitter systems, it is not altogether surprising to learn that agents that affect ACh and 5-HT successfully elicit circling in lesioned animals (see review in deFeudis, 1974).

Inasmuch as circling can be dissociated pharmacologically (i.e., direct vs. indirect receptor stimulants) and neuroanatomically, these techniques may serve as a good preparation in the evaluation of drug–behavior relationships. It is also possible that it may be useful in evaluation of response biases (e.g., side preferences) (Kokkinidis and Anisman, 1977) (see later).

Recently, Glick and his associates (Glick, 1973; Glick *et al.*, 1974; Jerussi and Glick, 1975) indicated that systemic injection of *d*-amphetamine or apomorphine may induce circling or side preference in the absence of unilateral lesions. These investigators indicated that there exists an asymmetrical imbalance of DA that is exaggerated by drug treatment. Accordingly, animals turn contralateral to the side containing the greater DA levels. Kokkinidis and Anisman (1977) recently argued that circling seen after systemic injection likely is an artifact of drug-induced perseverative tendencies, rather than genuine circling seen in lesioned animals (see the next section). Regardless of the source for the side preference, it is clear that drug-induced response biases may exist, and consequently may affect performance in various behavioral tasks, particularly those involving response discrimination.*

3.6. Perseveration

Some pharmacological treatments may induce perseverative behavior, i.e., repetitive sequences of responses to specific stimuli. For example, in a simultaneous discrimination task involving two response alternatives, animals treated with *d*-amphetamine tend to respond repeatedly to a single alternative (see reviews in Bignami, 1977; Carlton, 1963). Similarly, several different types of lesions may induce such perseverative effects, though these effects probably reflect other factors such as stimulus–response disinhibition (Frontali *et al.*, 1977). Depending on the nature of the task, these biases may mimic associative changes and augment performance, or alternatively elicit chance-level performance (e.g., in simultaneous discrimination tasks). This being the case, it is necessary to determine whether a perseverative bias exists, and in addition, whether the bias is of the response type (e.g., side preference based on such things as rotational behavior) or involves stimulus factors or the interface between stimulus and response factors.

To evaluate the nature of the bias induced by *d*-amphetamine, a free-running Y-maze-exploratory task was employed in our investigations (Anisman and Kokkinidis, 1975; Kokkinidis and Anisman, 1976*a,b*). When animals are placed in a Y-maze and permitted to explore freely, more often than not they tend to enter the least-recently visited arm. Thus, a series of arm entries would be 1,2,3,1,2,3. After *d*-amphetamine (5–10 mg/kg), animals tend primarily to visit successively only two arms. Once a novel arm is visited, this arm is employed in the perseverative sequence. Thus, a sequence of arm entries would appear as 1,2,1,2,1,3,1,3,1,3. The perseveration induced by *d*-amphetamine is distinct from stereotyped behavior in several respects. First, perseveration is apparently dependent on stimulus factors, whereas stereotypy is not (Kokkinidis and Anisman, 1976*a,b*). Second, apomorphine, which induces stereotypy, will reduce the propensity to visit novel arms of the Y-maze, but will not produce perseveration (Anisman, unpublished report). Third, α-MpT, which antagonizes stereotypy, does not affect perseveration

* It is curious that some investigators have computed circling in terms of number of turns to the contralateral side minus turns to the ipsilateral side. This has the effect of biasing the data in favor of more active animals, in which circling occurs to a greater extent than among inactive animals. An alternative measure of circling is number of turns to the contralateral side/total turns (contralateral plus ipsilateral) for each individual subject. This effectively reduces the bias produced by active animals. Moreover, by including an additional measure, namely, total contralateral turns/total number of turns for each group, an analysis can be computed as to whether the turns in a particular direction exceed chance (50%), i.e., by χ^2 analyses. While the latter measure is biased by extreme scores (i.e., animals that make many turns in a single direction), this will be detected in the proportion score of individual animals. Finally, it may be wise to pretest animals prior to drug treatments to determine whether turn biases exist. Subsequently, drug data should be corrected for bias. Elaboration of this discussion is available in Swonger and Rech (1972).

unless animals are pretreated with reserpine (Anisman and Kokkinidis, 1975). Finally, after pretreatment with reserpine, dopamine-β-hydroxylase inhibition antagonizes the amphetamine-induced perseveration, suggesting involvement of NE, rather than DA, which presumably subserves stereotypy (Anisman and Kokkinidis, 1975).

Perseveration in the Y-maze probably involves stimulus factors, rather than response feedback (Kokkinidis and Anisman, 1977). Specifically, after unilateral i.vent. injection of *d*-amphetamine, mice tend to exhibit contralateral turning. In the Y-maze, the consistent turning occurs at the choice point and at the ends of the alley, resulting in consistent arm alternation. Intraperitoneal injection, although producing rotational behavior in a circular runway, elicits perseveration in the Y-maze (see Fig. 2).

Had the perseveration been due to response factors, then performance in the Y-maze should have been comparable after i.p. and i.vent. injection. Parenthetically, these data speak directly to the data on circling induced by systemic injection of *d*-amphetamine (cf. Glick *et al.*, 1974). That is, if i.p. injection induced *genuine* circling, as does i.vent. injection, then the circling should have been observed in the Y-maze (as in i.vent. injected mice). The fact that this was not the case suggests that amphetamine-induced perseveration is manifested as rotational behavior in a circular tub. Indeed, since genuine circling and perseverative tendencies mimic one another in a circular runway, it may be more appropriate to evaluate circling in a Y-maze type of task (see discussion in Kokkinidis and Anisman, 1977).

4. Habituation

With the repeated presentation of non-biologically-significant stimuli, the response to these stimuli habituates. That is, the intensity of the response elicited by these stimuli is typically observed to wane. Habituation has received

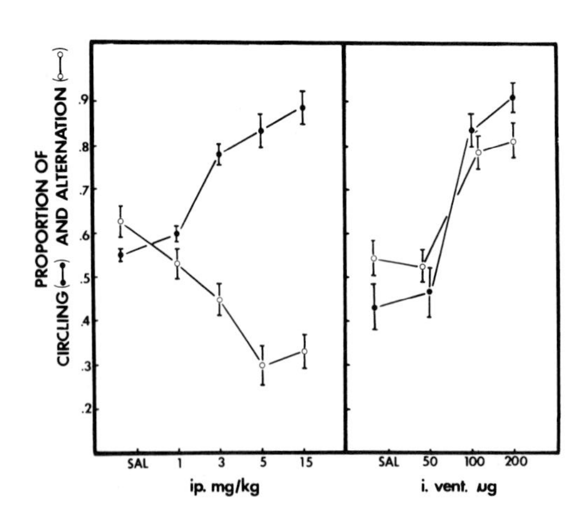

Fig. 2. Mean proportion of rotational and alternation responses (±S.E.M.) after intraperitoneal (i.p.) or intraventricular (i.vent.) injection of *d*-amphetamine. From Kokkinidis and Anisman (1977); reprinted with permission; © Springer-Verlag.

considerable attention over the past few years because of several factors: (1) its phylogenetic ubiquity, occurring, for example, in humans (Callaway, 1973; Hill, 1973) and invertebrates (R.B. Clark, 1965; Eisenstein, 1967; Eisenstein and Peretz, 1975), as well as surgically simplified organisms (R.F. Thompson and Spencer, 1966); and (2) its heuristic value in terms of the potential use in the investigation of other phenomena, e.g., simple learning processes and pharmacological effects (Bignami, 1976; Bruner and Tauc, 1966; Carlton and Markiewicz, 1971).

Recently, the criteria for habituation have been extended to include factors such as spontaneous recovery, potentiation of habituation, intensity and frequency of stimulation, subzero responding, dishabituation, habituation of dishabituation, and stimulus generalization (Groves and Thompson, 1970; R.F. Thompson and Spencer, 1966; see also Groves and Lynch, 1972; Groves *et al.*, 1974). This group of investigators also has indicated that given a relatively strong stimulus, habituation may be preceded by a period of sensitization, i.e., where responsivity to the stimulus is increased (Groves and Thompson, 1970, 1973; R.F. Thompson *et al.*, 1973). Moreover, it seems that the processes of habituation and sensitization are independent of one another.

Among rodents, evidence for habituation has been established for numerous behavioral measures, including exploratory behavior (Carlton, 1968, 1969), startle response (Carlton and Advokat, 1973; Groves *et al.*, 1974), head-raising of pups (File and Plotkin, 1974), and flexor withdrawal reflex (Pearson, 1973). While the characteristics typical of the habituation process are apparent in each of these behaviors (cf. Groves and Thompson, 1973; R.F. Thompson and Spencer, 1966), there seem to be distinct variations in the course of habituation in conditions involving different behavioral manipulations. Indeed, it appears likely that some pharmacological treatments will affect habituation in one task but not another, and vice versa. The habituation process may be mediated by several neurochemical changes, and the effectiveness of any particular drug treatment is situation-specific.

Given the complex nature of habituation from the model set out by R.F. Thompson and Spencer (1966) and Groves and Thompson (1970, 1973), to say nothing of the fact that habituation may involve a classically conditioned component (Stein, 1966) or effects on information-processing (Whitlow, 1975), the phenomenon cannot be considered in a unimodal fashion, as it often is.

4.1. Habituation and Sensitization

Inasmuch as the processes of habituation and sensitization are subserved by different mechanisms, it may be inappropriate to evaluate the effects of a drug treatment on habituation *per se*. In the previous sections, it was shown repeatedly that several treatments may produce comparable gross behavioral changes, but for entirely different reasons. Just as activity may be increased because of either disinhibition or genuine excitation, habituation may be disrupted because of dishabituation or sensitization. Indeed, it would not be altogether surprising to learn that the neurochemical substrates subserving excitation and sensitization are common ones, and similarly that disinhibition and dishabituation involve at least some overlapping mechanisms. This notion is supported by the fact that excitation is augmented by strong stimuli (cf. Anisman *et al.*, 1975; Anisman and Cygan, 1975), whereas this is not the case for disinhibitory drugs. Moreover, drugs that produce disinhibition also produce dishabituation, whereas this is not the case for drugs that produce response excitation (Kokkinidis and Anisman, 1976*a*,*b*) (see later).*

Since the course of habituation likely reflects the additive effects of *bona fide* habituation together with sensitization (see Fig. 1 in Groves and Thompson, 1970), it may prove difficult to ascertain (1) the relative contributions of these two processes within a given habituation function and (2) which process is affected by a given treatment.† In the following sections, techniques are described that may be effective in discerning whether a treatment affects habituation *per se*, or whether response change reflects variations in locomotor excitation or other types of nonassociative changes.

4.2. Temporal Evaluation

Inasmuch as the processes of habituation and sensitization are subserved by different mechanisms, it is necessary to evaluate the effects of drug treatments in terms of temporal changes. To be more explicit, if a treatment disrupts the course of habituation owing to increased sensitization, then the habituation function should be altered by a constant amount, regardless of which portion of the curve is examined. If, on the other hand, the drug truly disrupts habituation, these effects should be more apparent as training progresses. After all, prior to habituation occurring (i.e., during the first few trials), the drug should not affect behavior. After some inhibition has accrued, the drug effect should become more apparent. Carlton and Advokat (1973), for instance, demonstrated that PCPA does not affect habituation in a startle task until late in training, suggesting a specific effect on habituation.

* It is recognized that dishabituation according to Groves and Thompson (1970) and R.F. Thompson *et al.* (1973) is simply a case of sensitization. While the data presented here by no means provide strong contradictory evidence, they do suggest that different drug treatments may alter the course of habituation—one because of apparent dishabituation, the other without effects on habituation and its subsequent behavioral effects (e.g., see Kokkinidis and Anisman, 1976*b*).

† In the evaluation of locomotor activity, the use of temporal (horizontal) evaluation is typically encouraged to ascertain time of peak drug effects. While this is a reasonable desire, one should be aware that the magnitude of a particular drug effect may be confounded by the course of habituation. Thus, 15 min after injection and placement in an open field, habituation of exploration may be greater than it is 5 min after treatment. Thus, the level of activity will be greater in the latter than in the former instance, although the change in the drug effect *per se* is minimal.

Although simply evaluating individual curves may lead to valid conclusions, it needs to be remembered that even early in training, say after only a few stimulus presentations, inhibition may be present. Thus, adequate evaluation of drug effects may be limited to the few trials during which habituation has not occurred.

4.3. Spontaneous Alternation

Spontaneous alternation appears to involve the ability to selectively inhibit responses to nonreinforced stimuli (Douglas and Isaacson, 1965; Drew *et al.*, 1973; Swonger and Rech, 1972). That is, if animals have been exposed to a particular set of stimuli, then in a subsequent choice situation they tend to select novel stimuli in preference to those that they already have been exposed to. On this basis, it has been suggested that when animals are exposed to particular stimuli, the rewarding properties of these cues wane (i.e., habituation), and consequently alternative stimuli that may be present in a given situation are sought out.

Typically, two types of alternation tasks have been employed. The more commonly used T-maze task involves the placement of the animal in a particular compartment; the gate separating this compartment from the rest of the maze is then removed, thus allowing the animal to respond by visiting one of the two empty compartments. When the choice is made, the animal is removed, and some time later reintroduced to the first compartment. Since habituation to one arm of the maze occurs during the first trial, the animal typically enters the compartment that had not been visited previously (Douglas, 1966; Douglas and Isaacson, 1965; Squire, 1969). The second form of alternation utilizes a free-running Y-maze task in which animals are placed in the maze for a set period of time, and the order of arm entries is recorded. Theoretically, in such a situation, it would be expected that the organism would enter the least recently visited arm. Thus, if an animal visited arms 1 and 2, respectively, then it would be expected that the animal would next enter arm 3 (i.e., the least recently visited arm). The succeeding response would similarly involve entering the least recently visited arm, in this case, arm 1. If the animal returns instead to arm 2, this response would be considered a nonalternation (Anisman and Kokkinidis, 1975).

When the behaviors in the T- and Y-maze tasks are considered together, it appears likely that animals alternate on the basis of position cues as opposed to response feedback, i.e., alternation of left–right turns. More specifically, in a T-maze task, position responding and right–left responding are confounded, in that a right turn always takes the animal to one compartment, while a left turn takes the animal to the other compartment. The behavior observed in the Y-maze clarifies the role of stimulus and response feedback in the mediation of alternation responding. If, for example, alternation were dependent on internal feedback (left–right turns), then alternation between compartments would not be observed in a Y-maze, since under these conditions, animals would alternate successively between two compartments only. The fact that alternation occurs in both tasks suggests that animals alternate on the basis of position of stimulus factors (see also Glanzer, 1953). Further evidence indicating that alternation is not dependent on internal feedback has also been derived from studies in which it was observed that contextual or olfactory cues will affect alternation performance (Douglas, 1966).

The free-running Y-maze is preferred to the T-maze, since this task not only yields a score for alternation, but also concomitantly permits the measurement of locomotor activity simply by recording the frequency of arm entries. These behaviors lend themselves to scrutiny in order to assess whether a particular treatment affects both, or only one, of these behaviors. Thus, the independence of response biases (i.e., motorigenic effects) elicited by a drug and the effects of the drug on alternation behavior can be established. This factor is particularly important, since some of the mechanisms that may subserve alternation behavior apparently also may be involved in the mediation of locomotor activity (cf. Anisman and Kokkinidis, 1975). A second major advantage of the Y-maze task is that it does not involve a between-trials handling procedure as in the case of the T-maze. Accordingly, alternation performance is not confounded by handling effects (see Griffiths and Wahlsten, 1974).

As in other forms of habituation, spontaneous alternation is also modifiable by cholinergic manipulations. It has been observed that various anticholinergics such as scopolamine and atropine reduce alternation to chance levels, i.e., the pattern of visiting the novel arm is eliminated (Anisman and Kokkinidis, 1975; Kokkinidis and Anisman, 1976*a*; Meyers and Domino, 1964; Squire, 1969), whereas administration of an anticholinesterase, such as eserine, produces elevated alternation scores (Kokkinidis and Anisman, 1976*a*; Squire, 1969). In agreement with this notion, it has been reported that lesions to the hippocampus, which is rich in ACh (Dudar, 1975; Mellgren and Srebro, 1973; Srebro and Mellgren, 1974), attenuated spontaneous alternation performance (Douglas and Isaacson, 1965; Stevens and Cowey, 1973). Finally, developmental studies have also provided evidence for cholinergic involvement in spontaneous alternation. Animals younger than 20 days of age, in which the forebrain cholinergic system is not yet developed (Campbell *et al.*, 1969; Campbell and Mabry, 1973; Moorcroft, 1971; Thornburg and Moore, 1973), did not exhibit alternation in a Y-maze. Instead, a high level of perseveration was observed wherein animals successively tended to enter only two arms of the maze. However, if tested after 20 days, when cholinergic maturation had occurred, the level of alternation approached that of adult animals (Egger *et al.*, 1973). Taken together, the available data implicate the involvement of cholinergic activity, possibly mediated by the forebrain limbic system, in the modulation of alternation behavior (Carlton, 1969; Carlton and Markiewicz, 1971). Apparently, when an organism lacks the benefits of ongoing cholinergic activity, it is incapable of inhibiting a nonreinforced response, and consequently spontaneous alternation is disrupted.

Due to the rather complex relationship between neurotransmission and behavioral change, it is not surprising that manipulations of neurotransmitter systems other than ACh also modify alternation. For example, it has been suggested that brain serotonin is involved in the ability to inhibit nonreinforced responses. Depletion of serotonin by PCPA results in the inability of the organism to habituate in some situations (Carlton and Advokat, 1973), and also produces increased motility (Brody, 1970; Swonger *et al.*, 1970; Tenen, 1967). Conversely, administration of 5-HTP results in improved habituation (Fechter, 1974), presumably because of elevated levels of serotonin. In accordance with the pharmacological experiments, reduction in levels of serotonin via lesions to the raphe nuclei, an area particularly rich in serotonin-containing cell bodies (Cooper *et al.*, 1974; Modigh, 1974*a,b*), results in disruption of habituation (Davis and Sheard, 1974; Groves *et al.*, 1974).

In considering the role of serotonin in the modulation of spontaneous alternation behavior, the available data are somewhat more complex than for other tasks. Specifically, manipulation of endogenous levels of serotonin by various pharmacological agents, such as PCPA, methysergide, or LSD, does not directly alter spontaneous alternation scores (Swonger and Rech, 1972). In combination with other pharmacological agents (e.g., scopolamine), alternation scores are influenced in a predictable fashion; i.e., the effects of scopolamine are magnified (Swonger and Rech, 1972), suggesting that an interaction between two or more neurotransmitter systems may exist (Kokkinidis and Anisman, 1976*a*; Carlton and Advokat, 1973; Drew *et al.*, 1973). It seems likely that although ACh and serotonin may both play an inhibitory role in behavior, the effects of these transmitters are different qualitatively or quantitatively, or both, since manipulation of the former neurochemical exerts effects on behavior, while manipulation of the latter transmitter does not.

4.3.1. Dissociation of Habituation and Perseveration

Since a wide variety of drug treatments may affect alternation performance, it is of importance to determine whether this is a result of effects on habituation or, alternatively, of changes in other behavior (e.g., perseveration). To be more explicit, alternation levels may be reduced by disrupting habituation (which results in random selection). Alternatively, a drug may introduce behavioral change in which animals successively visit only two arms of the maze for a prolonged number of trials. Clearly, this is quite distinct from that of random responding.

Work in this laboratory (Kokkinidis and Anisman, 1976*a*) has revealed one potentially useful tool in differentiating perseveration and effects on habituation. In particular, when saline-treated animals are placed in the Y-maze, the frequency of alternation declines monotonically. Presumably, habituation occurs to each of the arms, thereby reducing the alternation tendency. It follows that if a drug effect is restricted to modification of the habituation tendency, then preexposing the animal to the Y-maze will not

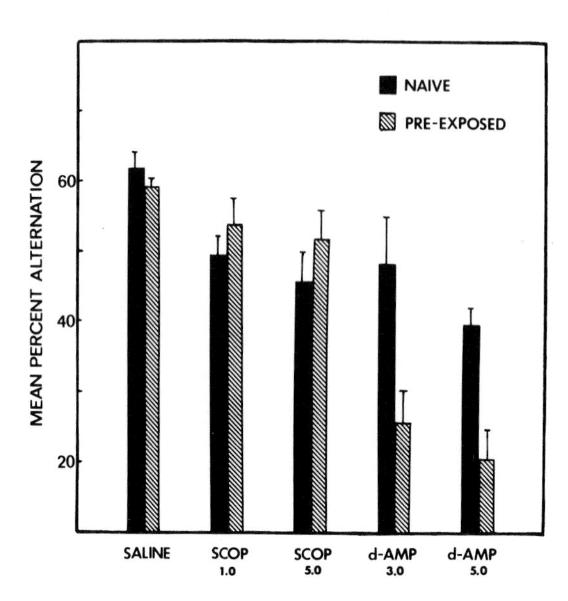

Fig. 3. Mean proportion of alternation (±S.E.M.) after scopolamine (1.0, 3.0, or 5.0 mg/kg) or (+)-amphetamine (1.0, 3.0, or 5.0 mg/kg) as a function of preexposure (5 min) to the Y-maze. Note that alternation below chance levels indicates perseveration. From Kokkinidis and Anisman (1976*a*); reprinted with permission; © Springer-verlag.

further augment the drug effects, or at least will not produce perseveration. As seen in Fig. 3, this is precisely what happens in the case of scopolamine. In contrast, in the case of *d*-amphetamine, the perseverative effects of the drug are greatly magnified. In fact, 3.0 mg/kg, which otherwise does not produce perseveration, does so after Y-maze preexposure. Thus, it seems that *d*-amphetamine affects perseveration tendencies without modifying the habituation process. In effect, there exist two competing tendencies—the natural tendency to alternate and the drug-induced perseverative tendency. When the alternation tendency is reduced by preexposure, the perseverative effects of *d*-amphetamine are permitted to surface. Parenthetically, it is not surprising to learn that *d*-amphetamine-induced perseveration is enhanced by scopolamine. In fact, scopolamine will produce perseveration after low doses of *d*-amphetamine that ordinarily produce only a small reduction in alternation (Kokkinidis and Anisman, 1976*a*).

As indicated earlier, the free-running Y-maze task also permits assessment of locomotor activity (as gauged by number of arm entries). Locomotor stimulation and alternation level are apparently independent of one another, in that manipulation of one behavior will not necessarily lead to change in the other (Anisman and Kokkinidis, 1975). Indeed, it is likely that these behaviors are subserved by different mechanisms (Anisman and Kokkinidis, 1975; Kokkinidis and Anisman, 1976*a*,*b*). Since both these behaviors can be measured concurrently, the amount of information derived is far in excess of that obtained in the typically employed locomotor-activity task. At the very least, it may lead to some indication as to whether qualitative differences among drug effects exist.

4.4. *Carry-Over Designs*

In a habituation situation, responsivity declines not only within a session but also between sessions. Although spontaneous recovery occurs between sessions, the level of reactivity on, say, the second day of testing is initially lower than the initial level seen on Day 1. If an agent disrupts the course of habituation when administered on Day 1 of training, then on retesting in the nondrug state, performance should be comparable to that of animals not previously exposed to the habituation stimuli (i.e., comparable to Day 1 performance in nondrugged subjects).

4.4.1. *Spontaneous Alternation*

In accordance with the criteria for habituation proposed by Groves and Thompson (1970), alternation declines within a session, and some degree of recovery of alternation occurs between days (Kokkinidis and Anisman, 1976*b*; Swonger and Rech, 1972). If scopolamine is administered on Day 1, then performance in the nondrugged state 24 hr afterward is comparable to that seen on Day 1 in saline-treated animals, and considerably higher than seen in these same animals on Day 2. Administration of *d*-amphetamine, which produces locomotor excitation, and apparently does not affect habituation (see earlier), does not have a carry-over effect on Day 2. Rather, these animals behave much like those initially treated with saline. Incidentally, the effects reported here are not attributable to drug effects on consolidation, drug-dissociated learning, or proactive drug effects (cf. Kokkinidis and Anisman, 1976*b*). Thus, it seems that although drug effects may initially elicit comparable changes in alternation, which may lead one to suspect comparable underlying mechanisms, the effects of these agents may be quite different, and can be readily dissociated employing a carry-over design.

4.4.2. *Water-Lick*

If animals are exposed to a particular chamber, then water-deprived and reintroduced to this chamber with a water bottle present, these animals approach the water more readily than animals not preexposed to the chamber. Presumably, preexposure produces habituation to the stimulus complex, thereby reducing the exploratory drive that competes with the approach response. As in the alternation task, carry-over effects of drugs are observed employing this technique. Carlton (1968, 1969) observed that scopolamine administered during preexposure eliminated the effects of the initial habituation session. Administration of *d*-amphetamine had no such effects. Two cautionary notes are needed here. First, when relatively high doses of *d*-amphetamine are used, carry-over effects of the drug may be observed (Anisman *et al.*, 1976). These carry-over effects apparently are not due to effects on habituation, but rather involve carry-over of excitation.

Second, the water-lick carry-over paradigm involves a deprivation procedure, and thus there are hazards inherent in this paradigm. Although the scopolamine and amphetamine effects do not appear to be affected by water deprivation *per se* (Anisman *et al.*, 1976), other pharmacological treatments may well be exaggerated or diminished due to deprivation procedures.

4.4.3. *Other Carry-Over Designs*

The carry-over technique has been used fairly frequently with tasks other than spontaneous alternation and water-lick. These data have been consistent with those described in previous sections. A series of experiments by Leaton (1968*a,b*) provide additional support for the position that scopolamine produces dishabituation. It was observed that animals will frequent an arm of a T-maze that contains novel objects more often than the opposite goal box that remains empty. Administration of scopolamine increased the number of entries into the arm containing the novel objects. Apparently, habituation to the novel objects was retarded, consequently maintaining the high level of responding to these stimuli. Utilizing yet a different approach, Carlton (1966) demonstrated that animals will bar-press to receive the rewarding value offered by the switching on of a dim light. Over sessions, a decline of bar-pressing is observed owing to the habituation of the rewarding properties of the stimulus. With the administration of scopolamine, a reliable dose-dependent increase in responding is observed, suggesting that this anticholinergic results in behavioral dishabituation. Finally, Oliverio (1968) reported that whereas prior habituation to apparatus cues may disrupt subsequent avoidance behavior, administration of scopolamine during the initial habituation session eliminates these disruptive effects. Again, it appears that scopolamine attenuates the habituation process, thereby reducing the disruptive effects of prior stimulus exposure on other behaviors. It is noteworthy that the aforementioned effects of anticholinergics probably do not involve peripheral factors, since the quaternary analogue of scopolamine (i.e., methylscopolamine) is without effect on habituation (cf. Carlton, 1969). Similarly, the drug does not affect memory processes, since posttrial injection has negligible effects (Anisman *et al.*, 1976; see reviews by Bignami, 1976; Carlton, 1969). Moreover, the effects of the drug treatment cannot be attributed to drug-dissociated responding (Carlton, 1969) or to the potential motorigenic effects of anticholinergics (Anisman, 1975*a*).

4.5. Exploration and Startle

As indicated previously, habituation occurs to exploratory behavior and startle response. Interestingly, though, drug treatments that modify exploration do not necessarily affect startle habituation. J.M. Williams *et al.* (1974) (see Fig. 4) observed that scopolamine attenuated habituation of head-poke exploration, but did not modify habituation in a startle task. As can be seen in Fig. 4, when animals were tested subsequently 1 week later, saline animals

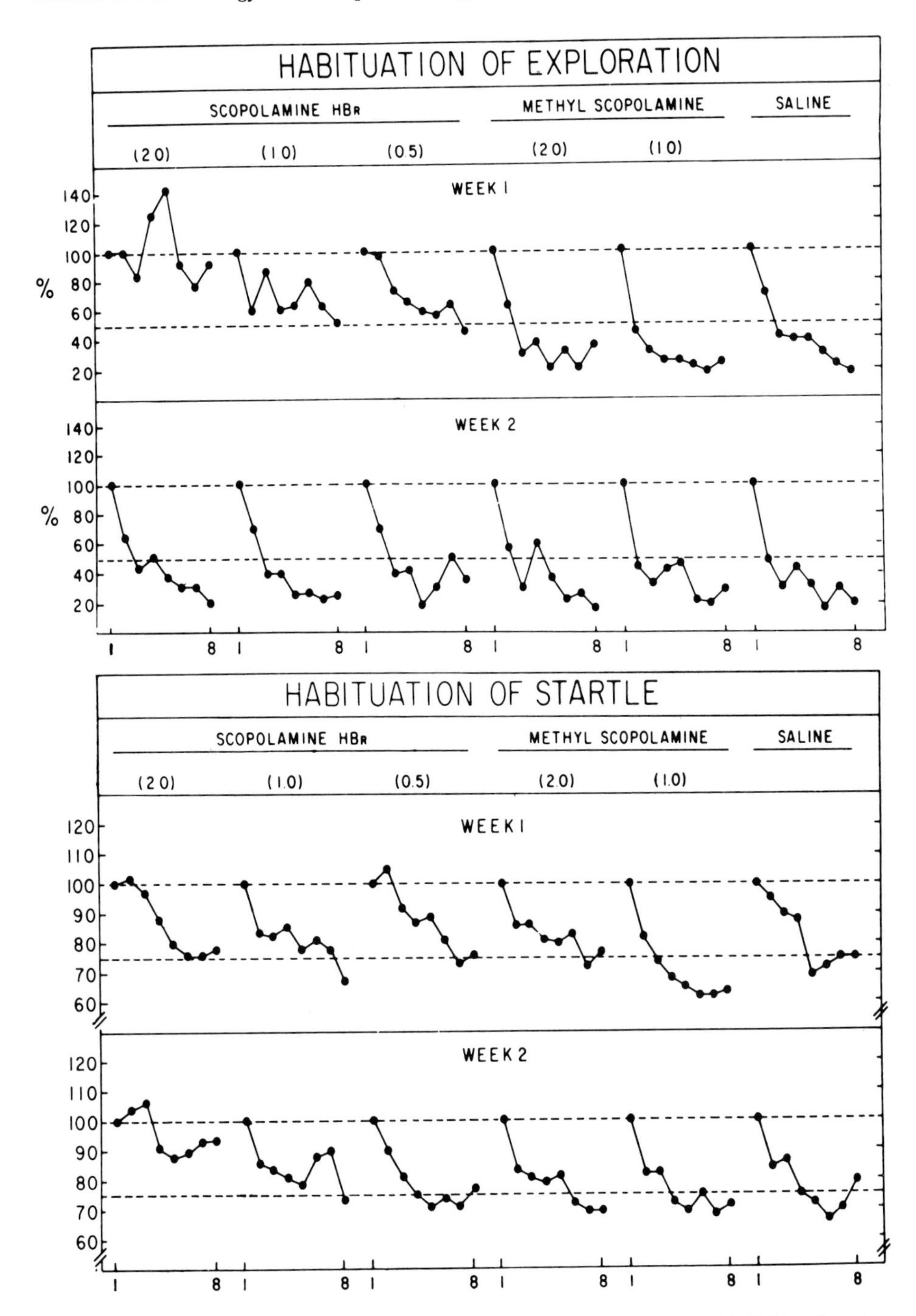

Fig. 4. Habituation of exploration and of startle plotted as a percentage of initial level to show differential decline as a function of the different treatments. Drug injections were administered during Week 1. Trials refer to consecutive 7- to 5-min intervals. From J. M. Williams *et al.* (1974); reprinted with permission; © American Psychological Association.

exhibited complete recovery, and performance was comparable to that seen in animals that previously received scopolamine. Unlike scopolamine, medial septal lesions attenuated habituation in both the exploratory and startle tasks. Moreover, Carlton and Advokat (1973) previously observed that PCPA effectively modified habituation in the startle task. Since habituation can be dissociated at a pharmacological level, it probably is not appropriate to consider habituation in terms of unitary physiological processes.

It is of interest to note that numerous investigators have considered the septo-hippocampal axis primarily in terms of cholinergic mechanisms. However, septal lesion may also induce 5-HT depletion (cf. Ungerstedt, 1971*b*; Lints and Harvey, 1969). Accordingly, lesions of the medial septum may provoke effects comparable to those of PCPA as well as scopolamine.

J.M. Williams *et al.* (1974, 1975) have differentiated between habituation of emitted and elicited responses. Exploration, an emitted response, is subserved by mechanisms different from those that govern startle, which is an elicited response. In the former case, the exploratory response has some reinforcing value, and inhibition is dependent on response-contingent feedback. This is not the case for startle. Considering these facts, it seems apparent that there exist several forms of habituation that can be dissociated both pharmacologically and neuroanatomically. Moreover, treatments may affect performance in habituation tasks, but because of factors that mimic treatments that directly influence the course of habituation (e.g., locomotor excitation, perseveration). In the evaluation of habituation, it is necessary to divorce such factors from genuine modification of habituation.*

4.6. Summary

It is clear that evaluation of habituation may be a more promising approach to pharmacological analysis than is measurement of locomotor activity. Aside from the fact that habituation tests may provide information regarding locomotion, these tasks indicate several more subtle characteristics of behavior. This being so, it becomes possible to (1) determine several types of biases, e.g., perseveration and circling, induced by the drug; and (2) make an evaluation as to whether changes in locomotor activity are a reflection of disinhibition, on the one hand, or genuine response excitation, on the other. Further to the same point, habituation tasks provide the opportunity of assessing possible drug effects on stimulus–response interfaces as opposed to effects that are purely on the response side.

In closing, one final consideration needs to be mentioned. In a series of

* File (1976) and File and Wardell (1975) have indicated that the effects of scopolamine on habituation of hole-board exploration are dependent on a variety of stimulus factors (e.g., presence of novel stimuli, extraapparatus cues). This being the case, the explanation offered by J.M. Williams *et al.* (1975) needs to be either revised or extended. This notwithstanding, the double dissociation between treatments and test variables provides fairly strong evidence, and deserves continued attention.

review papers, both Bignami and Carlton (Bignami, 1976, 1977; Bignami *et al.*, 1975; Carlton, 1963, 1969; Carlton and Markiewicz, 1971) have advocated the notion that habituation and "inhibition" processes may indirectly affect learning. Indeed, Bignami (1976) has indicated clearly that inhibitory processes may augment or disrupt performance, depending on a variety of variables, including stimulus factors, prior experience, and organismic variables. Given this state of affairs, analyses of complex behaviors involving learning, memory, and performance need to take into account the notions of inhibition and habituation.

5. *Appetitively Motivated Operant Behaviors*

The division of this chapter into aversively motivated and appetitively motivated behaviors was for convenience only. However, it is noteworthy that there are some points of disagreement among researchers concerning the significance of motivational concepts in analyzing drug effects. For example, while Miller and Barry (1960) argue that an important determinant of selective drug effects involves differences in behaviors of varying motivations, Dews (1970) has suggested that differences in drug effects may be due to the differences in the behaviors developed under different motivators rather than to direct motivational specificity. According to this position, dependence of drug effects on motivational processes is small compared to the critical dependence of drug effects on the pattern of responding determined by schedules of reinforcement. The research supporting this position is derived from two sources: (1) similarly motivated behaviors are not similarly affected by drugs if the control rate of responding varies (Dews, 1955, 1958; Kelleher and Morse, 1968) and (2) differently motivated behaviors will be similarly affected by drugs if the control rate of responding is equivalent (Kelleher and Morse, 1964; Cook and Catania, 1964). The concept of rate dependency is pervasive and will be expanded on in this section.

For purposes of this chapter, though, it is hoped that the division into appetitive–aversive procedures has no inherent theoretical overtones.* The purpose of this section of the chapter is to outline operant techniques and procedures that may be of use to researchers in behavioral pharmacology regardless of their theoretical persuasion.

5.1. *Operant-Appetitive Paradigms*

In operant-appetitive paradigms, the behavior emitted by an organism (e.g., lever-pressing in rats or key-pecking in pigeons) is increased in

* Nevertheless, it needs to be realized that treatments in the appetitive and aversive situations may produce different types of neurochemical consequences. For example, stress employed in the aversively motivated paradigms has repeatedly been found to elicit increased turnover of NE and 5-HT together with a decline in endogenous levels of these amines, as well as an increase in ACh (see reviews in Anisman, 1976; Stone, 1975).

frequency or maintained at a stable rate of occurrence by the presentation of a reinforcing stimulus following the emission of the behavior. Reinforcing stimuli are empirically established. That is, if a stimulus is presented contingent on some response and the response increases in frequency, then the stimulus is called a positive reinforcing stimulus. If removal of a stimulus is made contingent on some response and the response increases in frequency, then the stimulus is called a negative reinforcing stimulus. Reinforcing stimuli can be identified only in terms of their ability to increase the frequency of responses on which they are made contingent.

Reinforcement is a term that refers to the operation of presenting a reinforcing stimulus following the occurrence of a response. For example, a pellet is dropped into a food cup whenever a rat presses a lever or a pigeon is given access to a hopper of grain after a response key has been pecked. The rules that specify the basis on which reinforcing stimuli are presented in an experimental situation are referred to as schedules of reinforcement.

5.2. Schedules of Reinforcement

5.2.1. Simple Schedules

Reinforcing stimuli can be programmed on the basis of responding alone, on the basis of time alone, or on the basis of time and responding. In addition, the requirement can be fixed or constant from one reinforcement to the next, or it can be variable. A classification of the most basic schedules is given in Table 6. For example, if a reinforcer is contingent on the occurrence of a fixed number of responses, then this defines a fixed-ratio (FR) schedule. If the number of responses required varies from one reinforcement to the next, then a variable-ratio (VR) schedule is defined. When the reinforcer is contingent on both time and responding, then an interval schedule is operating. In a fixed-interval (FI) schedule, the reinforcer is delivered following the first response that occurs after a fixed period of time has elapsed. If the required time period varies from one reinforcement to the next, then a variable-interval (VI) schedule is operating.

These four basic schedules—FR, VR, FI, and VI—have characteristic performances that they generate and maintain. Representative samples of these performances are presented in Fig. 5. It can be seen in this figure that both FR and VR schedules maintain very high rates of responding. However, FR performances have a characteristic pause in responding following delivery

Table 6. A Classification of Simple Schedules of Reinforcement

	Constancy of the requirement	
Reinforcer contingency	Fixed	Variable
Responses alone	Fixed-ratio (FR)	Variable-ratio (VR)
Time and responses	Fixed-interval (FI)	Variable-interval (VI)
Time alone	Fixed-time (FT)	Variable-time (VT)

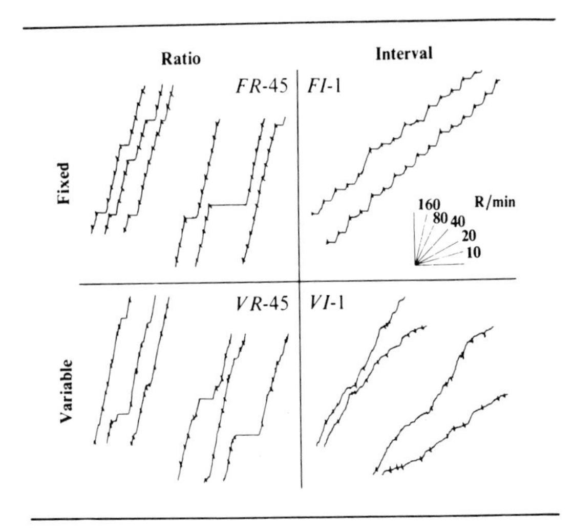

Fig. 5. Cumulative records of lever-pressing by rats on various schedules of reinforcement. Water was presented after every 45th response (FR-45), after every 45th response on the average (VR-45), for the first response occurring 1 min after the previous reinforcement (FI-1), or at variable times averaging 1 min after the previous reinforcement (VI-1). Water presentations are indicated by diagonal marks. The left-hand records in each panel are taken from the 47th session of training, the right-hand records from the 48th session. From Nevin (1973); reprinted with permission; © Scott, Foresman.

of the reinforcer. In VR schedules, the pauses are less frequent and not correlated with reinforcements. The rates of responding on interval schedules are generally lower than on ratio schedules. Note also that in FI performance, there is a marked pause after each reinforcement. From time to time, a gradual acceleration in response rate occurs within an interval. This is referred to as the FI scallop, and it may be quite pronounced under some experimental conditions. In contrast to the FI and other performances, responding on a VI schedule is maintained throughout the intervals between reinforcements. Although the VI record illustrated in Fig. 5 shows wide fluctuations in rate from moment to moment, this is not characteristic of VI performance. Under the appropriate experimental conditions, organisms will respond at a fairly constant rate on VI schedules.

Increasing attention has been given to schedules, e.g., fixed-time (FT), in which the reinforcer is contingent on time alone because of the ability of these schedules to maintain characteristic patterns of responding, even though the reinforcer is presented in a response-independent fashion. For example, an FI schedule may produce more responding than an FT schedule, but yield similar patterns of responding. Thus, the shift from a response-dependent FI schedule to a response-independent FT schedule may lower the overall response rate but leave the response scallop intact.

It should be noted that response rate and response-patterning are two distinctly different dependent variables. Response rate refers to the number of responses occurring in a given unit of time. Response-patterning refers to the way in which these responses are distributed in time. As Zeiler (1977) has recently shown, these two aspects of responding may be controlled by different variables. Response-patterning can be explained by one variable, namely, the placement of the reinforcer in time. However, response rate appears to reflect the operation of several other variables. These are hypothesized by Zeiler (1977) to be: (1) response dependency, which refers

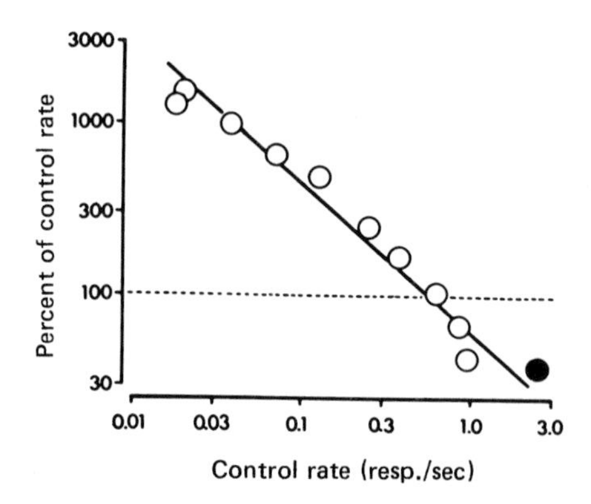

Fig. 6. Effects of *d*-amphetamine on responding during individual 1-min segments of the FI-FR schedule of food presentation. *Abscissa:* Control rate in 1-min segments. *Ordinate:* Rate in 1-min segments after *d*-amphetamine, expressed as a percentage of the control rate. Coordinates are logarithmic. Each point represents a single observation. Note that there is an inverse linear relationship between control rate and amount of increase after *d*-amphetamine. Note that the point for FR responding (●) falls along the same regression line as the points for FI responding (○). From McKearney (1972); reprinted with permission; © University of Toronto Press.

to the close temporal contiguity between the reinforced response and the reinforcing stimulus; (2) the number of responses per reinforcer; (3) the time between reinforcer presentations (interreinforcer time); and (4) the availability of the reinforcer when responding weakens.

5.2.2. *Schedules of Reinforcement and Behavioral Pharmacology*

There is some controversy as to the role that reinforcement schedules play in the experimental analysis of behavior (Jenkins, 1970; Mackintosh, 1974; Platt, 1974; Weisman, 1975). Nevertheless, it is generally agreed that they are convenient tools for the determination of a number of behavioral, perceptual, motivational, and pharmacological processes. One of the main advantages of schedules of reinforcement in the analysis of drug effects is that they can be used to generate a wide variety of different rates and patterns of responding in many species. These rates and patterns of responding provide sensitive and stable baselines from which the effects of pharmacological treatments can be determined (Boren, 1966; T. Thompson and Schuster, 1968; Harvey, 1971).

The use of schedule-controlled responding to assess drug effects was pioneered by Dews (1955), who demonstrated that while pentobarbital reduced responding on an FI schedule, it increased responding on an FR schedule. This finding suggested to Dews (1958) that an important determinant of drug effects was the predrug rate of responding. A great deal of empirical research has established that the effects of amphetamine and barbiturates are rate-dependent. Drug-induced changes in responding are inversely related to the rate at which the response occurs in the drug's absence; i.e., amphetamine increases low rates of responding and decreases high rates of responding (Dews, 1964; Kelleher and Morse, 1968; McKearney, 1970; Barrett, 1974). A similar effect has been obtained with barbiturates.

In the empirical literature, rate dependencies are usually depicted as in Fig. 6. This figure reveals the effects of *d*-amphetamine on the responding of a pigeon during individual 1-min segments of a 10-min FI schedule of food reinforcement. At rates of 0.3 response per second and lower, there is a graded increase in rate, while at 1 response per second and higher, there is a reduction in response rate relative to the control rate. It has been shown that the effects of other drugs such as scopolamine (Bignami and Gatti, 1969), meprobamate and chlordiazepoxide (Cook and Catania, 1964; Kelleher and Morse, 1968), and imipramine (Dews, 1962; C.B. Smith, 1964; Vaillant, 1964) may also be rate-dependent.

Obviously, schedules do more than merely provide a baseline; they play a significant role in determining how a particular drug will affect behavior. Recent investigations of rate-dependent effects have examined a number of methodological weaknesses in early studies that predominantly used FI and FR schedules. For example, Branch and Gollub (1974) have indicated that

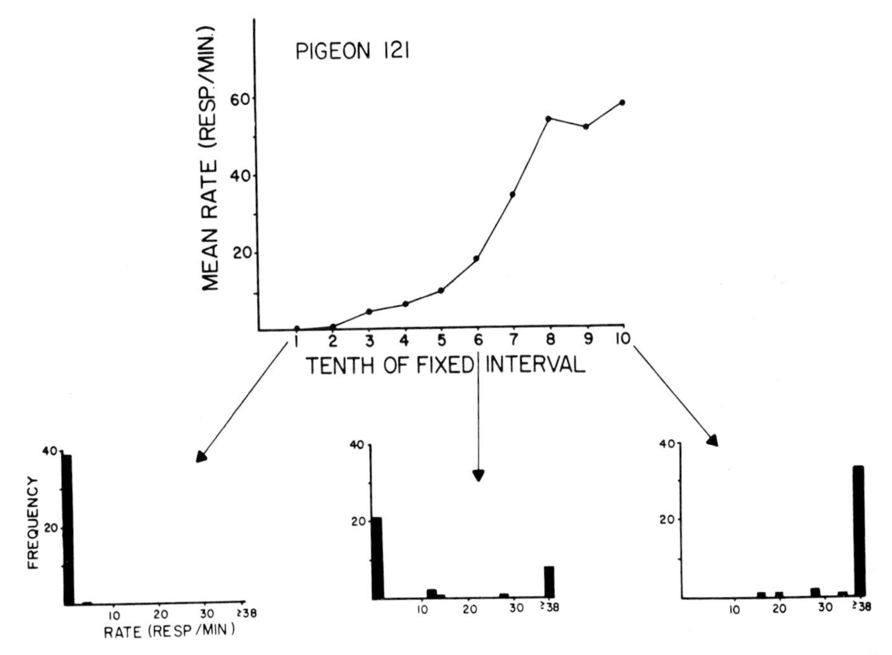

Fig. 7. *Top:* Mean response rate in each tenth of the FI 300-sec schedule in one session for pigeon 121 under control conditions. *Bottom:* Three frequency histograms showing the frequencies of occurrence of specific rates in the first, sixth, and tenth tenths of the interval. The frequencies were obtained from the 40 repetitions of the interval in a session. See the text for further explanation. From Branch and Gollub (1974); reprinted with permission; © Society for the Experimental Analysis of Behavior.

a detailed analysis of FI responding yields information about drug effects that are not found when average rates are employed. Their analysis arose from the finding of Schneider (1969) that FI performance consists of a period of nonzero responding followed by an abrupt transition to a high, constant rate until reinforcement occurs. Average rates within segments of fixed intervals could be affected by the number of occurrences of a zero rate in a particular segment. This can be seen in Fig. 7. Note that in the upper figure, the mean rate from the sixth tenth is about 20 responses per minute. However, the frequency histogram in the lower figure indicates that this mean was produced by the averaging of zero rates and rates greater than 38 responses per minute. A response rate of 20 pecks per minute did not occur in the sixth tenth of any interval presented in the histogram.

The effects of different doses of *d*-amphetamine are particularly interesting when examined in segments of a fixed interval. In early parts of the interval, *d*-amphetamine decreases the frequency of pausing. Moreover, intermediate rates of responding are much more frequent under *d*-amphetamine than under control conditions. On the basis of their data, Branch and Gollub (1974) questioned the use of average local rates in FI schedules to examine "rate-dependent" effects of drugs. There are two reasons for this. First, average rate does not adequately represent performance in tenths of fixed intervals. Second, it is possible that periods of pausing and periods of responding are controlled by different independent variables. According to Branch and Gollub, it is necessary to subject the "rate-dependency" hypothesis to an examination in which the procedures employed generate a variety of response rates that are unambiguously rates. Procedures that could be used for this purpose will be described in the next few sections.*

Schedules of reinforcement can also be employed in behavioral pharmacology to develop drug profiles. If a sufficiently large dosage range is employed, different drugs can reveal quite distinct profiles in terms of their effects on behavior maintained by different schedules of reinforcement (Bignami and Gatti, 1969). For example, in both FR and FI schedules, two drugs might yield similar profiles in terms of their ability to reduce overall response rates. However, these effects might be dissociated if it could be shown that with the FI schedule one drug disrupts the temporal pattern of responding while the second drug leaves it intact. Once distinct drug profiles are established, they may be of use in classifying new pharmacological agents in drug-screening programs (Ando, 1975).

* The data of Branch and Gollub (1974) have been selected as a representative sample. It should be apparent, though, that the position expressed by these investigators is applicable to drugs other than amphetamine. For example, in a recent analysis of ethanol, hypnotic-sedatives, and minor tranquilizers, Bignami (1976) indicated that much of the inconsistent data involving conditioned emotional response, as well as lack of parallelism between these data and behavior in conflict tasks, might be due to averaging of data. In fact, he indicates that drug effects that *appear* to be dependent on the CS–US interval may in fact reflect the effects of the inappropriate use of averaging.

5.2.3. Differential Reinforcement Schedules

Along with simple schedules, there are many refined techniques for generating different rates of responding. Differences between interval and ratio schedules in terms of the average response rates that they produce have been hypothesized to result from the differential reinforcement of interresponse times (IRTs), i.e., the time between two successive responses (Reynolds, 1968; Nevin, 1973). A reinforcer is presumed to increase not only the probability of a reinforced response itself, but also the temporal interval between the reinforced response and the preceding response. Schedules can be viewed as ways of differentially reinforcing responses on the basis of their distribution or patterning in time. Certain schedules that differentially reinforce spaced responding make this relationship between responses and reinforcers explicit.

For example, in a differential-reinforcement-of-low-rate (DRL) schedule, a reinforcer is delivered only if a response follows a specified time period of no responding. DRL schedules have been employed in several pharmacological investigations (Sidman, 1955; E.F. Segal, 1962; Laties *et al.*, 1965; Ando, 1975). Laties *et al.* (1965) indicated that mediating behavior (i.e., behavior emitted in the interval between occurrences of the reinforced response, which makes reinforcement more probable in DRL schedules) may play a crucial role in allowing organisms to efficiently space their responses in time. For example, in the rat, tail-nibbling could be said to mediate spaced responding on a lever if it results in the next lever-press occurring late enough after the last lever press to be reinforced. The disruption of DRL responding by amphetamine does not appear to be produced by interference with an internal timing mechanism. Rather, the disruption appears to be a secondary effect of increasing the rate of emission of all overt behavior (E.F. Segal, 1962). This hypothesis has been supported by data that indicate that on a DRL schedule, mediating behavior decreases in duration after the administration of amphetamine (Laties *et al.*, 1965). Much of the research demonstrating a shift in the IRT distribution toward shorter durations with amphetamine has employed rats as experimental subjects. However, drug effects on DRL schedules may be dependent on the species studied, since, in pigeons, amphetamine appears to increase the frequency of long IRTs (Hearst and Vane, 1967; McMillan and Campbell, 1970). The interaction between species-specific behaviors and the opportunity for mediating behavior may be important determinants of the differential drug effects (Schwartz and Williams, 1971).

A recent study by Ando (1975) employed a DRL schedule to provide profiles on a number of psychotropic drugs, including *d*-amphetamine, methamphetamine, pipradrol, nicotine, diazepam, chlorpromazine, chlordiazepoxide, pentobarbital, imipramaine, and others. In addition to providing drug profiles, it would appear that differential reinforcement schedules could be used to examine the "rate-dependency" hypothesis more extensively.

The precision of control over response rates can be increased by reinforcing responses in a narrow band of IRTs. For example, one could require that a response terminate an IRT greater than 1.5 sec but less than 3 sec in order to be reinforced. This type of requirement is termed response-pacing, and it allows precise control of both average response rate and form of the IRT distribution (Mallot and Cumming, 1964; Shimp, 1967; Blackman, 1968). Response-pacing techniques would be ideal mechanisms for varying reinforcement frequency and response rate, independent of one another, to determine the dependence of drug effects on both response rate and reinforcement frequency. Such an application will be considered in more detail in the following section.

5.2.4. *Multiple Schedules of Reinforcement*

A multiple schedule of reinforcement is one in which different reinforcement schedules, each associated with a different stimulus, can be programmed in succession. Such a schedule can be used to develop two different performances by a single subject, within a single experiment. This technique is of particular value in studies examining the "rate-dependency" hypothesis, because it allows one to evaluate the effects of a drug on different rates of responding with a single subject in the same experimental session. An example of the use of a multiple schedule in pharmacological research is presented in Fig. 8.

Many experiments showing rate dependencies have failed to keep factors such as reinforcement density and schedule of reinforcement constant while manipulating response rate. To examine drug effects on different rates of responding but equivalent reinforcement frequencies, in the same subject, a multiple schedule of reinforcement could be employed. D.M. Thompson and Corr (1974), for example, employed a multiple schedule in which a VI schedule operated in one component, while in the other component either a VI-plus-signal or a variable-time (VT) schedule was operational. (A VI-plus-signal schedule is the same as the simple VI schedule discussed earlier, except that a signal is presented whenever the reinforcer is available. A VT schedule is one in which the reinforcer is presented at varying time intervals, independent of responding.) Employing multiple schedules of this nature allowed Thompson and Corr to generate a fairly high response rate in the VI component and a low response rate in either the VI-plus-signal or the VT component. At the same time, equivalent rates of reinforcement were maintained in both components of the multiple schedule by choosing equal temporal values for VI and VT schedules. Thompson and Corr found that in the VI, small doses of *d*-amphetamine increased response rate, while larger doses decreased responding. In the components generating a low rate of responding, no rate-increasing effect of *d*-amphetamine was found at any dose. This result was not due to baseline insensitivity, since pentobarbital increased response rate in both the VI-plus-signal and the VT components at a dose that decreased responding in the VI component.

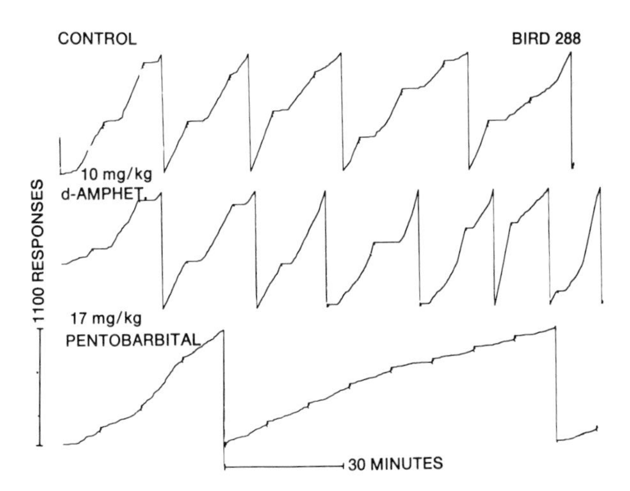

Fig. 8. Cumulative records of responding under a multiple 10-min FI 31-response FR schedule of food presentation in the pigeon. *Abscissa:* Time. *Ordinate:* Cumulative responses. FI and FR components alternated throughout the session. If no response was made within 60 sec of the end of the FI, or if 31 responses were not emitted within 60 sec during the FR, the alternate schedule component was presented (60-sec limited-hold). Note that different patterns of responding were maintained under the two schedules: under the FI, there was a pause followed by an increasing rate until food presentation, whereas under the FR, there was a high and steady rate until food was presented. *Top:* Control. *Middle:* Performance at about 3½ hr after administration of 10.0 mg/kg *d*-amphetamine. *Bottom:* Performance at about 2 hr after administration of 17.0 mg/kg pentobarbital. Note that after *d*-amphetamine, FI responding was increased, but FR responding was completely abolished (flat portions after FI food presentation); after pentobarbital, FI responding was markedly decreased, but FR responding continued to occur normally. From McKearney (1972); reprinted with permission; © University of Toronto Press.

Combining response-pacing techniques with interval schedules in a multiple-schedule paradigm would also be an ideal mechanism for varying reinforcement frequency and response rate. For example, in one group of rats, the value of the VI schedules could be equivalent in the two components of a multiple schedule, while the response-pacing requirement added to the VI schedule could differ for the two components. This would result in response rates differing between components while reinforcement frequencies would be comparable. In a second group of rats, by varying the value of the VI between components and maintaining equivalent pacing requirements, one can generate equivalent response rates in both components while reinforcement frequency varies. An example of the effectiveness of such scheduling procedures can be found in Blackman (1968).

5.2.5. *Discriminative Stimuli and Drug Effects*

Since multiple schedules bring different response rates under the control of discrete environmental stimuli, it can be said that a successive discrimination has been established. However, to specify the nature of the discriminative stimuli controlling behavior, certain methodological problems must be avoided. For example, if the components of a multiple schedule alternate regularly in time, the subject may learn to respond differentially on the basis of either the passage of time, the change in stimulation, or the discriminative aspects of reinforcement–nonreinforcement. To establish that the properties of the discriminative stimuli themselves control the tendency to respond differentially, these methodological problems must be avoided (see Jenkins, 1965). The importance of this issue for behavioral pharmacologists was brought to bear by McKearney (1970), who demonstrated that the properties of discriminative stimuli, such as their physical intensity and mode of presentation, attenuate the magnitude of drug effects. This suggests that in some cases, discrepancies among studies in terms of the drug effects obtained may be due to failure of the researcher to clearly isolate the nature of the discriminative stimuli controlling a behavior.

When responding is under the control of powerful discriminative stimuli, the control rate of responding does not appear to be the sole determiner of drug effects (B. Weiss and Laties, 1964; Laties and Weiss, 1966; Laties, 1975). Specifically, Laties and Weiss (1966) provided external stimuli during performance on an FI schedule and compared this to FI performance in the absence of external stimuli, when, presumably, behavior is being controlled by internal stimuli. The external stimuli consisted of five symbols that changed once per minute during a 5-min FI. With this external "clock," virtually no responding occurred until the fifth stimulus appeared. Amphetamine and scopolamine had minimal effects on responding in the FI with added clocks, whereas chlorpromazine and promazine affected responding even in the FI with added clock.

Interpreting the role of discriminative stimuli in modulating drug effects is complicated because of the fact that discriminative stimuli can change the rate and pattern of responding. Thus, the modulation of a drug effect cannot be attributed solely to the presence of external stimuli in these situations. To circumvent this, Laties (1972) has attempted to compare performances with and without external discriminative stimuli in such a way that the baseline rate of responding remains equivalent. He found differential sensitivity to drug effects similar to that reported earlier. Several other studies have also exhibited an attenuation of drug effects as a result of the presence of external discriminative stimuli (McKearney, 1970; Carey and Kritkausky, 1972; D.M. Thompson and Corr, 1974).

It is evident, then, that the effects of pharmacological treatments on behavior under the control of discriminative stimuli do not simply reflect rate dependencies. In fact, Dews (1958) has suggested that attenuation of stimulus control is the main mode of drug action. That is, pharmacological treatments may affect behavior by modifying the extent to which changes in the physical characteristics of stimuli control different rates of responding or different response probabilities. Most of the research concerned with stimulus control has centered on discriminations that are based on variations along a simple stimulus dimension. For example, with a multiple schedule, responding may be reinforced in the presence of a green light, but never reinforced in the presence of a red light. This can be contrasted with discriminations based on more complex stimulus relationships. These are referred to as conditional discriminations, since reinforcement is not contingent simply on the presence or absence of a particular stimulus, but rather on the relationship of that stimulus to another stimulus. Conditional discriminations may be more susceptible to drug effects than simple discriminations (Dews, 1955).

An evaluation of the effects of certain drugs on stimulus control has led Dews (1971) to the following tentative conclusions: scopolamine does not appear to affect stimulus control; the effect of amphetamine on stimulus control appears to be entirely accounted for by the control rate of responding; and pentobarbital has effects on stimulus control that cannot be entirely accounted for in terms of the control rate of responding. This suggests that while behavior under strong stimulus control may be less sensitive to drug effects, it may also be possible that different drugs have different effects on stimulus control.* The evidence on this point is still somewhat equivocal. Heise and Lilie (1970) indicated that both scopolamine and amphetamine had large effects on behavior under weak stimulus control, but only scopolamine affected behavior under strong stimulus control. Ksir (1975) recently demonstrated that both scopolamine and *d*-amphetamine show no drug effect under strong stimulus control, but they do show a dose-related increase in error rate in the weak-stimulus-control condition. These discrepancies may be the result of any number of differences between these studies in terms of species (rats vs. pigeons), task employed (simultaneous vs. successive discrimination), and stimulus dimension used (shape and color vs. brightness). Clearly, a series of systematic studies varying these parameters is necessary to demonstrate that different pharmacological treatments interact differentially with the strength of stimulus control.

Thus far, it has been indicated that drug effects on behavior under stimulus control can be interpreted as reflecting either a rate dependency or a selective effect on the process of discriminative control. Dews (1971, p. 40) has referred to the latter as the effect of a drug on "the reasonably discrete function of 'being controlled by discriminative stimuli.'" A third interpretation is that drugs interfere with sensory or perceptual processes. The methodology for assessing drug effects on such processes is considered in a later section.

5.2.6. Conjunctive Schedules of Reinforcement

A conjunctive schedule is one in which a reinforcer is presented when each of two (or more) schedule requirments is satisfied. For example, in a conjunctive FR20 FI 1-min schedule, a reinforcer is presented following the

* In this context, it should be noted that what is being considered is the effects of drugs on behaviors already under stimulus control, rather than on the development or acquisition of stimulus control.

first response after 1 min if at least 19 responses have already occurred. Many studies of rate-dependent drug effects have employed patterns of responding in which response rates immediately after food are considerably lower than those prior to food reinforcement. To test the generality of rate-dependent drug effects, Barrett (1974) employed a conjunctive FR-FI schedule to generate rates of responding that were highest early in the interval and then shifted to a lower rate until the reinforcer was delivered. With this schedule, the familiar rate-dependent effects of pentobarbital and amphetamine were obtained. However, the drugs differentially affected the pattern of responding. Barrett's study highlights two points that behavioral pharmacologists may find useful. First, complex schedules or combinations of schedules can be used to generate virtually any pattern of responding, and this may be helpful in establishing the generality of rate-dependent drug effects. Second, response rate and pattern of responding may be affected by drugs differentially. This may reflect, in part, the fact that response rate and response patterning are controlled by different variables (Zeiler, 1977).

5.2.7. *Concurrent Schedules of Reinforcement*

As noted earlier, in a multiple schedule, two different schedules are programmed successively and are associated with different stimuli. In a concurrent schedule, two response alternatives are available, and responses are reinforced according to two VI schedules that run simultaneously and are mutually independent. Concurrent schedules are especially useful to behavioral pharmacologists who wish to assess the reinforcing properties of a drug.

T. Thompson and Pickens (1972) and Goldberg and Kelleher (1976) have reviewed a number of studies in which drugs are presented contingent on an operant response to determine the reinforcing efficacy of various pharmacological agents. For a number of stimulant drugs, an inverse relationship between response rate and magnitude of drug reinforcement (dose per infusion) has been found (Balster and Schuster, 1973). This appears to be due to the immediate effects of the drugs that preclude responding for a period of time. That is, the inverse relationship found between response rate and dose size of cocaine, for example, is not due to a decline in the reinforcing properties of different doses of cocaine. Rather, it is apparently due to the dose-related disruption of behavior that occurs immediately after cocaine injection (Pickens and Thompson, 1968; Dougherty and Pickens, 1973). It has been demonstrated that these response-disrupting drug effects can be minimized by employing a 15-min time-out following reinforcement (Balster and Schuster, 1973), i.e., a period of nonreinforcement, in this case 15 min in duration, in which stimulus conditions are altered and responses have no programmed consequences.

The response-disrupting effects of certain drugs have led investigators to suggest that the use of nonrate measures of reinforcer efficacy, such as choice, may be valuable (Dougherty and Pickens, 1973; Balster and Schuster,

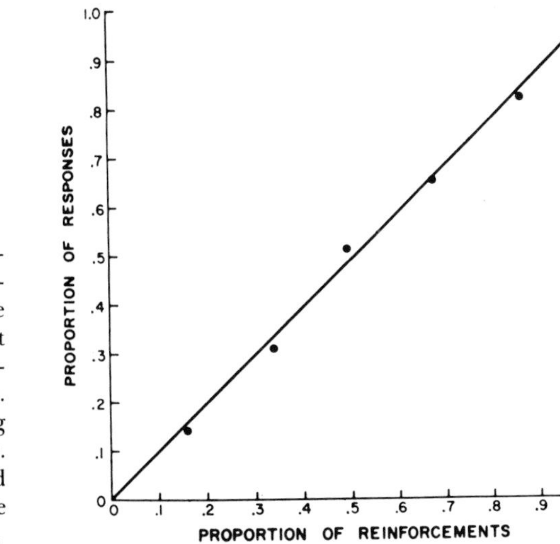

Fig. 9. Relative frequency of responding to one alternative in a two-choice procedure as a function of the relative frequency of reinforcement thereon. VI schedules governed reinforcements for both alternatives. The diagonal line shows matching between the relative frequencies. From Herrnstein (1970); reprinted with permission; © Society for the Experimental Analysis of Behavior.

1973; Iglauer and Woods, 1974). It is well known that absolute rates of responding in single-response situations are extremely insensitive to variations in the parameters of reinforcement such as frequency, magnitude, and delay (Herrnstein, 1970). However, concurrent-scheduling procedures have shown themselves to be extremely sensitive to reinforcement parameters. Preference, in concurrent schedules, is operationally defined as the relative proportion of responses or the relative proportion of time spent responding on either response alternative. In general, a matching relationship, as shown in Fig. 9, is found between the relative frequency of responding to one alternative and the relative frequency of reinforcement of that alternative. That is, $P_1/(P_1 + P_2) = r_1/(r_1 + r_2)$, where P_1 and P_2 are the absolute rates of responding at alternatives 1 and 2, and r_1 and r_2 are rates of reinforcement (Herrnstein, 1970). A similar matching relationship occurs with regard to the relative magnitude of reinforcement (Catania, 1963; Neuringer, 1967).

A concurrent chain schedule, illustrated in Fig. 10, was employed by Iglauer and Woods (1974) to assess the reinforcing efficacy of different doses of cocaine. Note that the initial link conditions are identical VI schedules that are presented simultaneously on different response levers. Responding in the initial links leads to one of two terminal links on a VI 1-min schedule. The terminal links contain the independent variable, which is differing doses of cocaine. In the concurrent chain procedure, the dependent variable is the relative rate of responding in the initial link. Iglauer and Woods (1974) found that preference was always for the larger dose of cocaine and relative response rates roughly matched relative drug intake, as can be seen in Fig. 11. However, the empirical significance of the matching observed was questionable, since the VI response rates were unusually low and there were exclusive preferences shown in which responding occurred and reinforcers

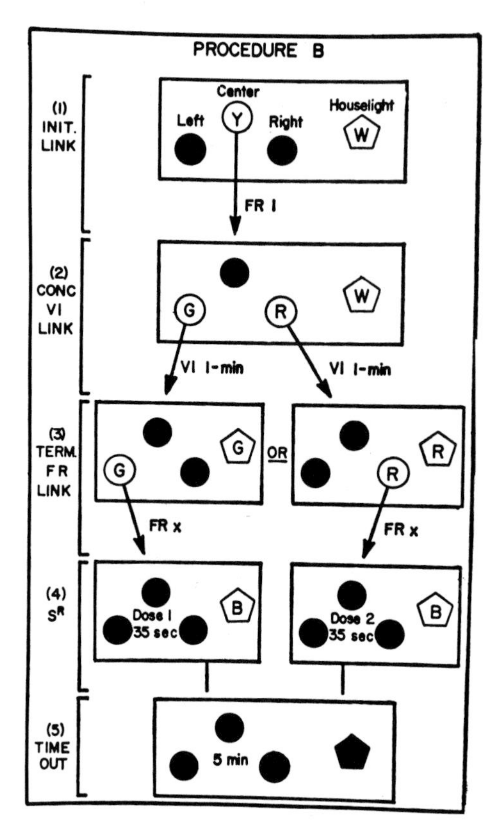

Fig. 10. Diagram of one cycle of a concurrent chain schedule. Each box represents one possible state. Numbers on the left refer to successive experimental conditions. Responding on either side lever during the concurrent VI component leads, on a VI 1-min schedule, to one of two terminal-link states (Condition 3). During a terminal link on one side lever, the houselight color is changed to match that lever light, and the other side-lever light is darkened. The completion of an FR requirement on the lighted side lever is then reinforced with one of two drug doses (Condition 4). Ratio values are the same on both levers, while drug doses usually differ. From Iglauer and Woods (1974); reprinted with permission; © Society for the Experimental Analysis of Behavior.

were obtained almost exclusively on the lever producing a higher dose of cocaine. To circumvent this problem, a similar study was conducted in which the independence of the concurrent schedules was modified to ensure delivery of an equal number of cocaine injections on both schedules (Llewellyn *et al.*, 1976). This resulted in preferences being less extreme than previously reported and yet still corresponding to the matching relation. Both these studies indicate that the reinforcing efficacy of cocaine is a function of dose per injection.

One other potential application of concurrent schedules in behavioral pharmacology should be mentioned. That is, it may be possible to scale drugs against each other by assuming that the subject must be exhibiting matching in this type of situation and by adjusting the measure of drug reinforcement accordingly. The analysis of choice between two qualitatively different reinforcers is exemplified in a study by Hollard and Davison (1971), who examined choice between food and electrical brain stimulation.

5.3. Adjunctive Behavior

Certain types of simple schedules of reinforcement generate behavioral "by-products" that are themselves worthy of study. Food-deprived rats responding under a VI schedule of pellet reinforcement will exhibit, under appropriate experimental conditions, a concurrent pattern of excessive fluid intake. Polydipsia, or schedule-induced drinking, has been observed with VI (Falk, 1961; F.C. Clark, 1962; Stricker and Adair, 1966), with FI (Stein, 1964; Fal, 1966), and with DRL schedules (E.F. Segal and Holloway, 1963). It appears to be just one of a number of behaviors that might occur under similar schedule conditions. Depending on the options available in the experimental chamber, behavior such as wheel-running, hair-licking, bar-biting, attack, or escape can be observed to occur at specific times with respect to the presentation of reinforcers that are not dependent on the occurrences of these behaviors. This so-called adjunctive behavior is defined by Falk (1972, p. 172) as "behavior maintained at high probability by stimuli whose reinforcing properties in the situation are derived primarily as a function of schedule parameters governing the availability of another class of reinforcers." Extended discussion of the nature and determinants of adjunctive

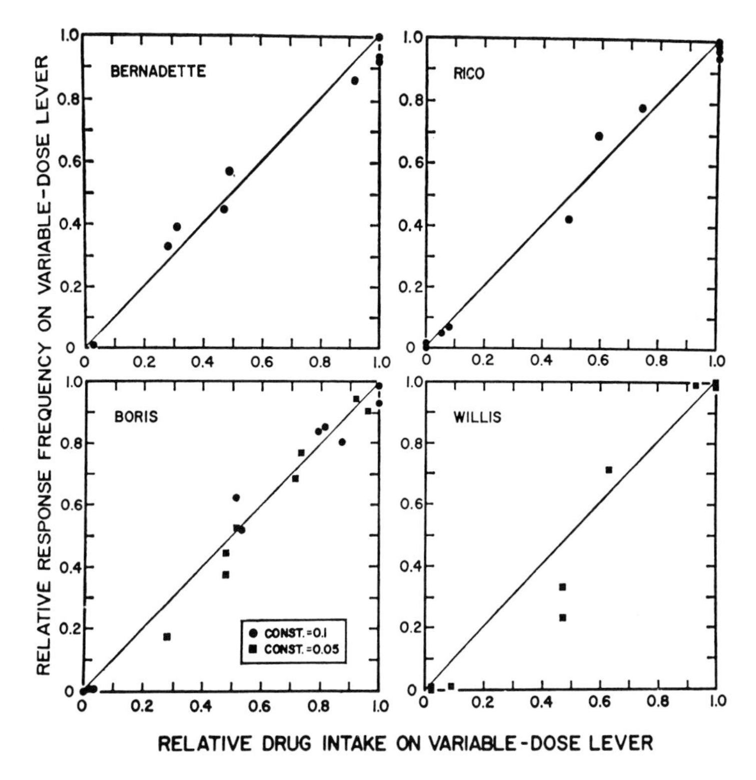

Fig. 11. Relative response frequencies on the variable-dose lever plotted against relative drug intake on the variable-dose lever. Drug intake on a lever is the number of reinforcements obtained on that lever multiplied by the dose available on it. Relative drug intake on the variable-dose lever is the drug intake on this lever divided by the sum of the intakes on both levers. The diagonal line represents the locus of perfect matching. Data are drawn from the criterion sessions (five or one) at each determination. From Iglauer and Woods (1974); reprinted with permission; © Society for the Experimental Analysis of Behavior.

behavior can be found in Falk (1972), Hawkins *et al.* (1972), and Keehn (1972).

The behavioral effect of drugs on adjunctive behaviors has not been extensively examined (Wuttke and Innis, 1972; J.B. Smith and Clark, 1975), though there is some evidence indicating that they may be of value in pharmacological research. For example, dose–effect relationships for amphetamine have been shown to be different for schedule-controlled drinking and adjunctive drinking (McKearney, 1973; Wuttke, 1970), even though a common mechanism for these two kinds of drinking has been suggested by Keehn *et al.* (1976). They have shown, for example, that haloperidol, which reduced physiologically induced drinking by sated animals, also reduced schedule-induced drinking by rats in direct proportion to drug dose.

Finally, there is some indication that comparative drug effects on schedule-controlled lever-pressing and adjunctive drinking are different for amphetamine and chlorpromazine (Byrd, 1973, 1974), suggesting that adjunctive behavior may be of value in providing additional data for drug profiles.

5.4. Observing Behavior

Observing behavior is behavior the only direct consequence of which is the production of stimuli related to reinforcement contingencies. For example, in a procedure reported by Kendall (1972), pigeons were exposed to FI schedules with added "clock" stimuli. A second response alternative was available that briefly produced the appropriate "clock" stimuli. If the second alternative was not responded to, the simple FI schedule operated with only one stimulus presented throughout.

Using this type of procedure, Branch (1975) assessed the effects of chlorpromazine and *d*-amphetamine on observing responses during an FI schedule. While drug effects on food-producing responses could be accounted for by a "rate-dependency" interpretation, this was not the case for observing responses. Branch (1975) suggested that the observing responses in his study were adjunctive in nature and that the effects of amphetamine and chlorpromazine on the rate of adjunctive behavior may not be predictable from control rates. However, there was one aspect in which both food-producing and observing responses were similarly affected by drugs. That is, the way in which the temporal distribution of these responses was disrupted was the same regardless of how overall rates were affected. Thus, it is not unlikely that a drug's interaction with variables governing overall response rate may be independent of interactions with variables responsible for the temporal distribution of responses. This point has been raised in earlier sections (Barrett, 1974; Zeiler, 1977).

5.5. Drug Effects and the Experimental Analysis of Behavioral Processes

The pharmacological effects obtained in many behavioral situations are often attributed to specific influences on one or more of a number of diverse processes. For example, certain drugs are said to have specific effects on motivational, associative, memorial, attentional, inhibitory, or sensory processes, and so on. Unfortunately, these processes have usually been inferred in a *post hoc* fashion, rather than precisely defined through appropriate experimental and conceptual analysis. Since many of these concepts may be of value in providing avenues of attack for behavioral pharmacologists, it may be worthwhile to review recent developments in the experimental analysis of these processes. As Sidman (1956) has noted, the more systematic is the behavioral investigation that occurs prior to the investigation of drug effects, the greater is the likelihood that there will be some order to the drug–behavior interactions that are found.

5.5.1. *Motivational Processes*

In Section 5, it was indicated that differences in drug effects may be due to variations in the rate and pattern of responding developed under different motivators, rather than to direct motivational specificity (Dews, 1970; T. Thompson and Schuster, 1968; McKearney, 1972). The support for this position was derived from two sources. First, similarly motivated behaviors are *not* similarly affected by drugs if the control rate of responding varies (Dews, 1955, 1958; Kelleher and Morse, 1968). Second, differently motivated behaviors are similarly affected by drugs if the control rate of responding is equivalent. The most frequently cited support for this statement is a study by Kelleher and Morse (1964). Figure 12 displays the results of their study. Using a multiple-schedule procedure, they were successful in producing similar rates and patterns of responding regardless of whether the reinforcer was food presentation or electric-shock termination. From the figure, it is clear that both FI and FR responding were affected in a similar fashion by doses of *d*-amphetamine, regardless of the nature of the reinforcer.

Because of the responsiveness of thermoregulatory behavior to homeostatic variables, B. Weiss and Laties (1963) investigated the effects of amphetamine, chlorpromazine, and pentobarbital on behavior maintained by heat reinforcement. The effects of these drugs on this behavior were similar to their effects on behavior maintained by food or water reinforcement and therefore independent of the motivational state maintaining the behavior. Consistent with these findings, Waller and Waller (1962) and Cook and Catania (1964) have shown that the effect of chlorpromazine is more dependent on the pattern of responding than on the nature of the reinforcer (i.e., food or electric shock). There are, of course, studies that do show differential drug effects on behavior maintained by different reinforcers. However, as T. Thompson and Schuster (1968) have indicated, these contradictory findings may have resulted from differences in species, drug dose, type of reinforcement schedule employed, and so on.

The attempt to establish the motivational specificity of drug effects does not appear to be a very promising approach at present. Those who wish to attempt it must bear in mind that they must clearly show that motivational

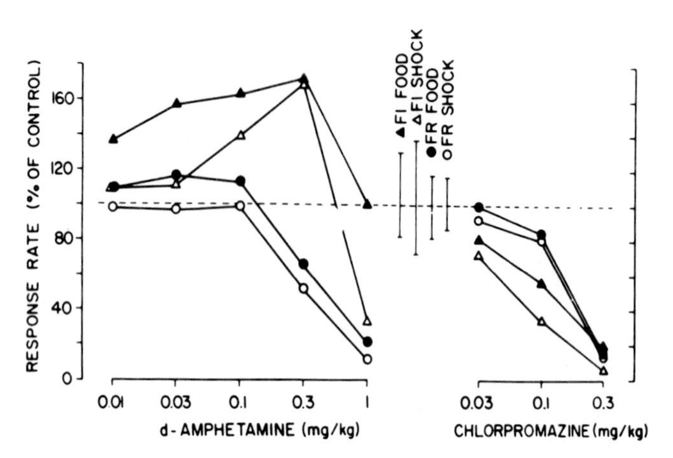

Fig. 12. Effects of *d*-amphetamine and chlorpromazine on responding under multiple FI-FR schedules of food presentation and stimulus–shock termination. Each point represents two or more observations in three monkeys under each multiple schedule. Sessions were 2½ hr long. Vertical lines signify the range of control (nondrug) observations. Note the similarity of the pairs of dose–effect curves for FI and for FR components, regardless of the type of reinforcing event. From Kelleher and Morse (1964); reprinted with permission; © Federation of American Societies for Experimental Biology.

effects are not secondary to differences in behavior generated by different motivators (Dews, 1970). Also, if the behavioral effect of a drug is attributed to a selective motivational effect, it must be demonstrated that manipulation of the appropriate motivational factor itself will mimic the drug effect. If the pharmacologically induced changes in behavior do not agree with those resulting from manipulation of the motivational factor, then the involvement of different mechanisms is suggested (Franklin and Quartermain, 1970; Kirkstone and Levitt, 1970).

An example of this approach is provided in a recent study by Sanger and Blackman (1975). They examined the effects of chlordiazepoxide, phenobarbitone, and chlorpromazine on DRL responding in albino rats. Chlordiazepoxide and phenobarbitone both disrupted DRL responding by increasing overall response rate. This was mainly due to an increase in the percentage of IRTs shorter than 1.5 sec. It has been suggested that the rate-increasing effects of chlordiazepoxide may be related to the appetite-stimulating action of this drug (Bainbridge, 1968). To evaluate this motivational interpretation, Sanger and Blackman (1975) reduced the body weight of their rats from 85 to 75% of their preexperimental values. The reduction of body weight had little effect on either overall response rate or the percentage of interresponse times shorter than 1.5 sec. Hence, they concluded that the effects of chlordiazepoxide on DRL responding was not related to the action of the drug on motivational processes.

5.5.2. Associative Processes

The attempt to uncover drug effects that are specifically related to the selective process of learning in appetitive situations is particularly problematical. Part of the problem is the traditional one of being able to disentangle all those processes such as attention, learning, motivation, memory, and so on that are presumed to underlie observed performance (Catania, 1973, p. 31). In behavioral pharmacology, the situation is even more complex (B. Weiss and Heller, 1969). One must be able to discriminate between associative and nonassociative effects of pharmacological manipulations (Anisman, 1975*a*), and also be aware of the possible nonspecific effects on behavior produced by the toxic action of certain agents such as inhibitors of protein synthesis.

Dews (1970) has contended that there is no adequate evidence indicating that drugs selectively affect the process of learning. In addition, he maintains that learning may not be the most sensitive function of the CNS for assessing drug effects. In this regard, he cites a study by Glassman (1969) indicating that learning has survived rather drastic treatments, such as an injection of puromycin sufficient to drastically reduce protein synthesis in the brain. Jarvik (1972), in a review of the pharmacological literature on learning and memory, has also indicated that these functions are not very sensitive to drugs.

Despite these difficulties, the attempt to link drug effects to alterations in associative processes will probably continue. The traditional experimental design in this situation is the "independent-groups" design. That is, some behavioral task is selected, and then a control group given saline and various experimental groups given differing doses of a drug are compared in terms of either response speed, trials to criterion, percentage correct, or another parameter. Cogent criticisms of this approach have been made (Skinner, 1950; Sidman, 1960, p. 117), which have resulted in the development of alternative methods of examining acquisition with "individual-subject" designs. Boren and Devine (1968) have employed the acquisition of behavioral chains to study transitional phenomena. They required rhesus monkeys to acquire a different four-response chain of lever-pressing in each session. After continued training, it was found that the pattern of learning and the number or errors per session reached a steady state. This steady state of repeated acquisition was then used as a baseline to study the effects of various independent variables. The advantages of this approach, as outlined by Boren and Devine (1968), are that intergroup variability is eliminated, direct behavioral measures of individual performance are employed rather than statistical derivations, and the results are more applicable to the behavior of the individual.

The use of this technique in behavioral pharmacology can be seen in the work of D.M. Thompson (1973, 1974, 1975), in which the effects of acute (D.M. Thompson, 1973) and chronic (D.M. Thompson, 1974) administration of chlordiazepoxide and phenobarbital on the repeated acquisition of response sequences are examined. In the chronic drug test, behavior was more readily disrupted when the response chain alternated from session to session (learning condition) than when the response chain remained the same (performance condition). Also, chlordiazepoxide had a greater error-increasing effect than phenobarbital. The repeated acquisition of response sequences in the presence of external discriminative stimuli (chain-learning) has been shown to be a much more sensitive and stable baseline for assessing drug effects on learning than the repeated acquisition of similar sequences in the absence of discriminative stimuli (tandem-learning). This finding by D.M. Thompson (1975) is an interesting exception to the general finding that behavior under the control of external discriminative stimuli is less readily disrupted by drugs than behavior not under such control.

5.5.3. *Memory Processes*

There have been numerous studies examining the effects of pharmacological treatments on memory processes. In most of this research, memory is operationally defined in terms of performance on a delayed-response task or a delayed matching-to-sample task.* In the indirect method of delayed response, a discriminative stimulus signals which exit of a choice compartment leads to food. A delay is then introduced between the removal of the discriminative stimulus and the opportunity for the animal to make a response. In the direct method of delayed response, one of two containers is baited with food while the subject watches. After a delay period, both containers are presented to the subject, and it is allowed to choose one of them. Related to these paradigms is that of delayed alternation. In the delayed-alternation task, animals are trained in a two-choice apparatus until they consistently choose the side opposite the one chosen on the immediately preceding trial, i.e., L,R,L,R. A time delay between trials is then introduced to determine retention functions.

A number of animal species have been tested with these paradigms, including rats, dogs, monkeys, raccoons, pigeons, and cats. Fletcher (1965) has reviewed much of the literature on delayed-response performance and concludes that recall capacity is not involved as much as other intervening factors such as a response chain or response orientation that is maintained throughout the delay interval. However, other researchers have presented a different interpretation of delayed-response performance. W.A. Roberts

* It is noteworthy that simple test–retest designs are not even considered here. It should be clear from the preceding discussion, as well as from the section on aversively motivated behaviors (Section 6), that the effects of drug treatments in straight-alley tests, L-mazes, and passive avoidance are by no means indicative of either associative or memory changes. To be sure, these techniques have been used naïvely with the result that misleading conclusions have been drawn. The techniques we describe certainly are not foolproof, and only a sampling of methods is provided. Nevertheless, conclusions drawn from such experimentation have a greater degree of validity.

(1972*b*, 1974) maintains that both orientational cues and memory traces control delayed responding, with the weighting of each factor being determined by experimental conditions. B.A. Williams (1971*a*,*b*) interprets delayed-response performance as simply one form of discrimination learning in which retention deficits are believed to be due to incomplete learning and interference from previous learning. There is no easy solution to these controversies regarding the mechanisms of memory. At present, it appears that we must be satisfied with operational definitions of memory (Mello, 1971; Honig and James, 1971). The necessity of a concept of memory that is distinct from a concept of learning has been debated in both the animal (Spear, 1973) and human (Melton, 1963) literature.

To infer that a particular drug has disrupted the process of memory, it is necessary for *the same dose of the drug to have different effects on recall as a function of the delay interval.* That is, there must be an interaction between the length of the delay or retention interval and the drug-induced decrements in performance. Ksir (1974) failed to find such an interaction with scopolamine in the delayed-response performance of albino rats. A similar result was obtained by Heise and Conner (1972). Alpern and Marriott (1973) claimed that they demonstrated facilitation and disruption of short-term memory with physostigmine and scopolamine, respectively. They trained C57BL/6 mice in an electrified maze to develop a successive reversal-learning set and then delayed-response-tested them at intervals from 2 to 25 min. These delays, while not as long as those used by Deutsch (1971) in demonstrating the effects of cholinergic agents on long-term memory, are considerably longer than the 2.5- to 40-sec intervals employed by Ksir (1974). Tests of retention involving intervals of minutes, hours, or days are one aspect of the operational definition of long-term memory in the human literature (Glanzer, 1972; Baddeley, 1972; Craik and Lockhart, 1972). In fact, forgetting from the short-term store in humans is claimed to occur over a period ranging from 5 to 30 sec after stimulus presentation (Waugh and Norman, 1965; Shiffrin and Atkinson, 1969). Thus, it is possible that Alpern and Marriott (1973) were examining the effects of cholinergic agents on long-term memory rather than short-term memory.

An additional procedural problem encountered in the Alpern and Marriott (1973) report was that delay intervals were progressively lengthened and only one delay interval was tested each day. This meant that the drug dosage present at each delay interval may have varied due to differences in the metabolism of the drug. Methodologically, it would be preferable to present various delay intervals in a random order within a single session when assessing drug effects.

One of the most useful paradigms for investigating memory processes in monkeys, dolphins, and pigeons is delayed matching-to-sample. A matching-to-sample procedure typically involves a three-stimulus display as illustrated in Fig. 13, which is taken from a study involving rhesus monkeys (Mello, 1971). A correct response, in this situation, would be selection of the comparison stimulus that is identical to the sample stimulus. In delayed

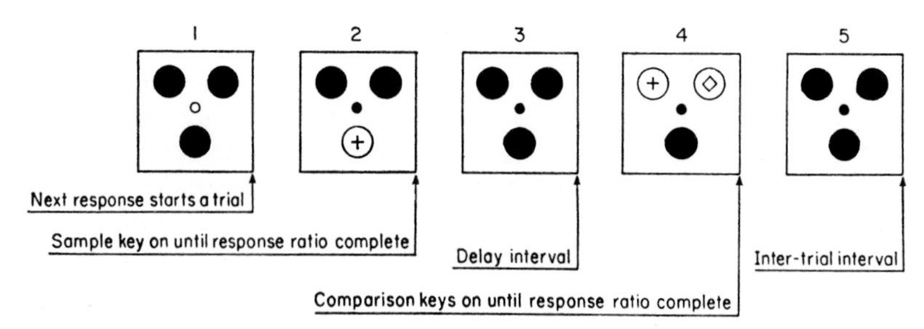

Fig. 13. Sequence of events in a single matching-to-sample trial. At the beginning of a trial, the monkey is confronted with three dark keys and a lighted ready light that indicates that the next response to the sample key will be effective in initiating a trial (1). Once the monkey presses the sample key, the ready light goes off and a stimulus appears on the sample key. The sample key remains lighted (2) until the monkey completes the response-ratio requirement. On completion of the response-ratio requirement, the entire panel becomes dark during the delay interval between the sample-stimulus offset and the comparison-stimuli onset (3). The two comparison keys are then illuminated with the sample key and the ready light dark (4). The comparison keys remain illuminated until the monkey has fulfilled the response-ratio requirement or until 1 min without a response has passed. During the intertrial interval (5), all keys and the ready light are dark. From Mello (1971); reprinted with permission; © Pergamon Press.

matching-to-sample, the delay between the sample-stimulus offset and comparison-stimuli onset can be of various durations. This paradigm has received considerable empirical attention (D'Amato, 1973; Jarrard and Moise, 1971; W.A. Roberts, 1972*a*; Herman, 1975). In addition, various theoretical models of the memory processes involved in delayed matching-to-sample have been proposed. In work with the pigeons, a simple trace-strength theory has received empirical support (W.A. Roberts, 1972*a*) and has been extended to situations in which proactive interference occurs (Grant and Roberts, 1973; Grant, 1975). On the other hand, several researchers working with monkeys and dolphins have emphasized the importance of temporal discriminations in accounting for retention in delayed matching-to-sample (D'Amato, 1973; Mason and Wilson, 1974; Herman, 1975). These theoretical models and the empirical data from which they are derived provide a valuable resource for the assessment of drug effects on memory.

An extensive series of experiments by Jarvik (1969) has shown that a variety of agents, such as pentobarbital, chlordiazepoxide, chlorpromazine, caffeine, and *d*-amphetamine, at various dosage levels reduced response rate without affecting matching accuracy at delays of 1 and 12 sec. However, scopolamine at doses of 0.05, 0.1, 0.2, and 0.4 mg/kg markedly reduced matching accuracy, especially at the 12-sec delay. In more recent studies, amphetamine has been found to reduce delayed-matching accuracy (Cook and Davidson, 1968). This effect was due to an increase in errors related to position preference and perseveration in normal rhesus monkeys (Glick *et al.*, 1970).

The effects of alcohol on the delayed-matching performance of rhesus monkeys has been examined by Mello (1971). The methodological details of her research were commendable, and she cogently discussed the relationship between the paradigm she employed and the theoretical processes in which she was interested. Procedures were incorporated to separate both attention and discriminative capacities from the analysis of drug effects on short-term memory. To ensure that the monkey was attending to the sample stimulus, an observing response was required to turn on the sample stimulus and 15 additional responses to the stimulus were required to ensure adequate exposure. To minimize the effects of discriminative capacities, very discriminable stimuli were employed. This was employed to minimize the possibility that perceptual impairment could account for performance decrements. The data indicated no specific impairment of short-term memory, although increasing doses of alcohol did reduce matching accuracy. Most of the errors occurred between 1- and 30-sec delays, while performance at longer delays was quite accurate despite blood alcohol levels above 200 mg/100 ml. In this study, Mello (1971) had employed a titrating delay procedure in which each correct response led to a 2-sec increase in delay and each incorrect response led to a 2-sec decrease in delay (Scheckel, 1965; Jarrard and Moise, 1971). In many studies, the particular delay-interval duration employed on any trial is randomly selected from a fixed set of delay durations. It may be worthwhile to compare both procedures (fixed and titrating delays) in terms of the drug effects obtained.

Research concerned with the effects of pharmacological agents on memory is especially exciting because of the possibility of an integrated account of these effects in both animals and humans. The relationship between animal and human memory has been discussed by Winograd (1971). In addition, research on how human memory is affected by agents such as alcohol (Talland, 1966; Tamerin *et al.*, 1971; Goodwin *et al.*, 1970; Carpenter and Ross, 1965) and marihuana (Abel, 1971) provides a base of comparative data. Zimmerberg *et al.* (1971) have indicated that cannabinoids have similar actions on the short-term memory of humans and animals. A knowledge of research on human memory may suggest guiding principles for the study of drug affects on animal memory.

5.5.4. Attentional Processes

The attentional processes involved in animal discrimination learning have attracted a great deal of theoretical and empirical interest (Riley, 1968; Gilbert and Sutherland, 1969; Sutherland and Mackintosh, 1971; Mackintosh, 1975; Riley and Leith, 1976). Attention can be empirically defined as the observation that some but not all elements of a stimulus complex will control responding when varied independently of each other. "Two-stage" theories of discrimination learning have assumed that orientation to stimuli (i.e., observing responses) precedes the development of stimulus control. However, attention appears to involve more than just an observing response.

Riley and Leith (1976) have outlined the problems inherent in the attempt to identify selective attention effects in animal discrimination learning. The main problem is that the study of how an organism learns to discriminate among stimuli confounds attentional processes with associative processes. As a method of controlling for associative processes, they recommended an investigation of the detection of stimulus attributes in divided and selective attention tasks. Blough (1969) had devised a steady-state procedure in which pigeons were trained to attend to both elements (tones and colors) of compound stimuli. He then alternated the birds between a divided-attention task, in which both dimensions were relevant, and a selective-attention task, in which only one dimension was relevant, followed by a return to the divided-attention task. While data suggestive of selective attentional processes were obtained with Blough's paradigm, there was still a problem in that selective attention was confounded with response competition (Leith and Maki, 1975).

The use of a matching-to-sample task to assess attentional mechanisms overcomes the problem inherent in Blough's paradigm (Riley and Leith, 1976). This technique was developed by Maki and Leuin (1972). On each trial, orientation to the response key at sample-stimulus onset was ensured by requiring that the pigeon peck the illuminated center key. This resulted in the sample stimulus being presented on this key for some period of time. Samples were either one-element stimuli or two-element stimuli (i.e., compounds of color and line orientation). Following termination of the sample, the two side keys were illuminated with elements from the sample dimension. A response to the side key illuminated with the element that formed all (element stimuli) or part (compound stimuli) of the sample presented on that trial was designated as correct. The purpose of developing this procedure was to determine whether sample durations necessary to maintain performance at 80% correct would be longer under conditions of divided attention (compound samples) than with nondivided attention (element samples). As Fig. 14 illustrates, compound samples require greater looking times than element samples for both color and angularity dimensions (Maki and Leuin, 1972).

Recent research has indicated that the performance differences between element and compound samples are not due to the psychophysical technique employed (Maki and Leith, 1973), to reduced perceptual clarity of compound samples (Leith and Maki, 1975), or to generalization decrement (Maki *et al.*, 1976). The role of short-term memory in producing element–compound differences is reportedly under investigation (Riley and Leith, 1976). This paradigm for investigating selective processes in attention offers the behavioral pharmacologist a promising assay for various pharmacological treatments. There is an indication in the pharmacological literature that drugs such as scopolamine may produce changes in habituation on the one hand (Bignami, 1976; Carlton, 1969) and attention deficits on the other (Brown and Warburton, 1971; Warburton and Brown, 1971, 1972). The paradigm discussed in this section may be worthwhile in pursuing these questions directly.

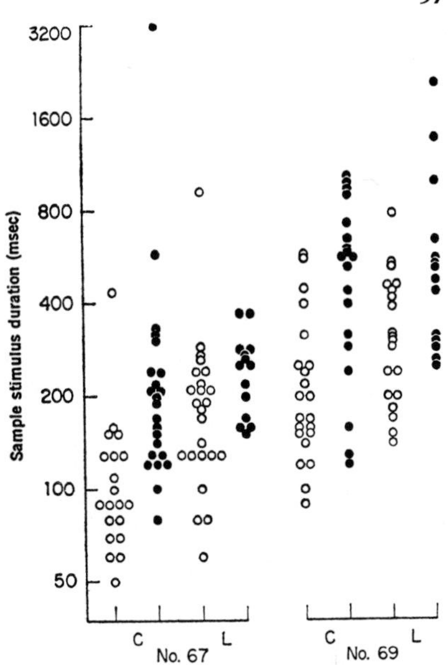

Fig. 14. Distributions of sample-stimulus durations obtained from two birds under conditions of divided and nondivided attention. Each point is the result of a matching-to-sample session in which the duration of the sample stimulus was varied to maintain performance constant at 80% correct. The durations for matching color and line orientation are plotted separately for each bird (C and L, respectively). (●) Divided-attention sessions; (○) nondivided-attention sessions. From Maki and Leuin (1972); reprinted with permission; © American Association for the Advancement of Science.

5.5.5. *Inhibition and Inhibitory Processes*

In Sections 3 and 4, considerable discussion surrounded the question of drug-induced inhibition and excitation. Yet for the behavioral pharmacologist to conclude that the effect of a drug is on inhibitory processes in a particular situation, there must be specific criteria for deciding when a stimulus is inhibitory. The multiplicity and vagueness in definitions of inhibition have plagued research (see footnote, p. 1). For example, the ability of an anticholinesterase, such as physostigmine, to facilitate discrimination performance is attributed to facilitation of inhibitory processes (Whitehouse, 1966; Leaton and Kreindler, 1972). This facilitation is reflected primarily in a reduced response rate to the negative discriminative stimulus in an operant situation. In other situations, such as maze-learning with the number of correct responses as the dependent variable, physostigmine has been found to facilitate both acquisition and performance (Cox and Tye, 1973, 1974). It is inferred from this work (and other work cited earlier) that the drug increased the inhibition of nonreinforced responses.

Isolation and quantitative measurement of inhibitory processes are necessary to provide a good empirical foundation for the assessment of pharmacological manipulations. Unfortunately, in behavioral pharmacology, the term inhibition has been so broadly defined that often any observed behavioral decrement has been taken as an indication of inhibition. In these situations, a number of alternative inferences are possible, such as a reduction of excitatory effects due to simple extinction, a failure to attend to a stimulus, and so on. Even in neuropsychology, the operation of inhibition is never inferred from simply observing a reduction in the ability of a stimulus to elicit neural firing (Diamond *et al.*, 1963; Thomas, 1972). Recently, a very thorough conceptual and empirical analysis of inhibition has been undertaken by a number of researchers interested in classic and instrumental conditioning (Rescorla, 1969; Hearst *et al.*, 1970; Hearst, 1972; Konorski, 1967).

Hearst (1972, pp. 6–7) defined an inhibitory stimulus as a "multidimensional environmental event that as a result of conditioning (in this case based on some negative correlation between presentation of the stimulus and subsequent occurrences of another event or outcome, such as 'reinforcement') develops the capacity to decrease performance below the level occurring when that stimulus is absent." There were some implications of this definition that Hearst regarded as important: (1) the inhibitory properties resulted from associative processes, and thus nonassociative effects were ruled out; (2) only when the conditions that maintained a response were kept constant and some discrete external stimulus that reduced responding was presented could one convincingly refer to inhibitory effects.

A number of assays for determining inhibitory properties of stimuli were presented by Hearst (1972). The most promising measures were the combined-cue (summation test) and new-learning (retardation test) techniques. In the combined-cue technique, the stimulus being tested for inhibitory properties must be combined for the first time with an excitatory stimulus (S^+), and reduce responding to the S^+ more than to the same stimulus in various (nonassociative) control groups. In the new-learning test, the presumed inhibitory stimulus must be harder to convert to an excitatory stimulus than the same stimulus in nonassociative control groups. Hearst (1972) indicated that both these tests must be used in combination to eliminate alternative accounts such as inattention or indifference. Similar analyses have been outlined in the case of Pavlovian conditioning (Rescorla, 1969). These recent developments in the conceptualization and measurement of inhibitory processes at the behavioral level merit careful consideration from behavioral pharmacologists interested in assessing drug effects on inhibition.

5.5.6. *Sensory Processes*

Assessment of drug effects on behavior requiring a discrimination requires careful control and measurement of stimuli. There is evidence indicating that drug–behavior relationships may not be independent of the selection of particular stimuli that control that behavior (McKearney, 1970). Stimuli varying in wavelength should be constant in luminance, and tones of varying intensity should not change with the subject's head position. Failure to properly control stimulus factors may allow subjects to make discriminations by using nonrelevant stimulus parameters (Krasnegor and Hodos, 1974). Even when the stimuli on the modality selected for study are properly controlled, there is the additional problem of response preferences that may be resistant to behavioral manipulation. Marked color preferences have been shown to exist for both the rhesus monkey (Sahgal *et al.*, 1975) and the pigeon (Delius, 1968; Sahgal and Iversen, 1975). These response preferences must be avoided, through counterbalanced designs, for example, if alternative interpretations of the data are to be eliminated. Even with these precautions, the problem must be faced that drug effects interact with stimulus modality (cf. Bignami, 1976; Frontali *et al.*, 1977).

For the behavioral pharmacologist interested in assessing the effects of drugs on sensory processes, a variety of methodologies can be found in Stebbins (1970*a*). One of the most interesting questions in this area is the modification of pain by analgesics. B. Weiss and Laties (1970) have employed a titration technique in which operant responses decreased the level of a continuously delivered shock. They found this technique to be more sensitive to the milder analgesics than traditional methods such as latency of tail-flick or the hot-plate test (Beecher, 1957).

Stebbins (1970*b*) has presented a technique that allows for the determination of auditory thresholds in monkeys via several psychophysical testing procedures. He has employed these auditory thresholds to demonstrate the frequency-specific hearing loss produced by antibiotic drugs such as ranamycin and neomycin. In addition, he has shown that a single, large intramuscular injection of sodium salicylate can produce a temporary hearing loss. Histological analysis of regions in the cochlea indicated a relationship between the anatomical and behavioral effects of these drugs.

5.6. Summary

This section on appetitively motivated operant behaviors began by examining the role that schedules of reinforcement can play in behavioral pharmacology. Schedules have been shown to significantly determine how a drug will affect behavior. Pharmacological agents can alter either the response rate or the response-patterning engendered by schedules, and these effects can be dissociated. In addition to having revealed the "rate-dependent" effects of drugs, schedules have been used to provide drug profiles, to examine the role of discriminative stimuli in modulating drug effects, and to assess the reinforcing properties of drugs.

Recent developments in the experimental analysis of behavioral processes have indicated the pitfalls and potential benefits of searching for specific drug effects on motivation, associative processes, memory, attention, inhibition, and sensory processes. There are many techniques that the experimental analysis of behavior makes available to the behavioral pharmacologist engaged in this search. This section has presented a representative but not exhaustive review of operant-appetitive paradigms and their use in behavioral pharmacology.

6. Aversively Motivated Behavior

The use of behavioral paradigms involving aversive motivation has become increasingly more frequent in the analysis of pharmacological effects.

The popularity of these paradigms likely stems from several factors: (1) The avoidance response can be learned within a single session, thus precluding the necessity of repeated testing, which may result in behavioral tolerance to the drug effects. (2) Some avoidance tasks can be acquired in a single trial, permitting specification of the *presumed* time at which the response-shock contingency was established. Consequently, the assessment of drug effects on memory consolidation is possible. (3) Finally, owing to the variety of avoidance tasks that are available, the judicious use of paradigms permits the evaluation of the nature of drug-induced changes in learning situations.

6.1. Components of Avoidance Behavior

Avoidance tasks can be separated into two broad categories—those tasks in which animals must emit a particular response to avoid shock (active avoidance) and those tasks in which animals must withhold responding to avoid shock (passive avoidance). The typical active-avoidance procedure consists of presenting the animal with a neutral stimulus (CS) such as a light or tone, followed subsequently by an aversive stimulus (US), such as shock, from which the subject can escape by performing a particular response. After several presentations of the CS followed by the US, the response occurs in anticipation to the shock, precluding the occurrence of the primary noxious stimulus. Of course, the response of the organism is not limited to these observable behaviors, but in addition there occur changes in respiration, heart rate, and GSR, as well as neurochemical events at both the central and peripheral levels (see Anisman, 1975*a*).

6.1.1. Drive–Expectancy

There exist a number of hypotheses that attempt to explain the mechanisms that subserve avoidance behavior. According to some theorists (e.g., Bolles, 1970, 1971), animals run to avoid shock, and the function of CS onset is that of a warning stimulus. In effect, the CS produces an *expectation* of shock, while CS offset acts as a feedback stimulus informing the animal as to the appropriateness of its response. Other theorists (e.g., Rescorla and Solomon, 1967) maintain that after the pairing of a CS with shock, fear is classically conditioned to the CS. Subsequently, the animal responds not so much to avoid the impending shock as to escape from the fear-provoking stimulus. A second version of this hypothesis does away with the concept of fear, and ascribes to the CS noxious properties simply because of its association with the shock (Dinsmoor, 1954). Unlike the position of Rescorla and Solomon (1967), this notion is not tied to fear as a hypothetical construct. Still other theorists (e.g., Herrnstein, 1969) maintain that avoidance responding is maintained by reductions in shock density. Again, this position assumes that fear is a gratuitous concept. Finally, Seligman and Johnston (1974) have proposed a model that, in effect, represents an amalgamation of previous models. Specifically, it is supposed that initially fear motivates responding.

Once the contingencies of the task are fairly well established, CS onset elicits the expectation of shock. Emission of an appropriate response similarly results in the expectation of no shock. Confirmation of the expectancies tends to increase response strength.

Although there is no consensus as to the most acceptable model for avoidance, the fact that so many contrary positions are available has important implications. Namely, it is still premature to consider the effects of pharmacological treatments in terms of modification of "fear" (e.g., see Konorski, 1967).

6.1.2. *Response Factors*

In an avoidance task, the animal must learn the appropriate response and where to direct this response, and be able to initiate the response. Bolles (1970) has indicated that one major determinant of the rate at which the response is acquired is based on the compatibility of the response requisite with the species-specific defense reaction (SSDR) of the organism under investigation. Presumably, there exists a hierarchy of response types within the organism's repertoire, with certain responses being prepotent over other responses. Among rodents, for example, the response of freezing is relatively potent after exposure to a stressor. Thus, if the response required of the animal is that it remain immobile, this response will be interpreted as being readily "learned." However, it is likely that the passive-avoidance response is a reflection of the animal's unconditioned response to stress, rather than an associative effect. If the response required for successful avoidance is one of running, then the immobility response needs to be suppressed first. If running is the next most dominant response, then this response will surface with the suppression of freezing. If this response leads to reduction of aversive stimulation, then it will be maintained. Conversely, if this response is not an appropriate one, then it too must be suppressed until an appropriate response emerges. In effect, the suggestion here is that responses are established by way of species-specific defensive behaviors, together with instrumental methods. If the response requisite is not an SSDR, then it should be acquired by instrumental methods alone—only after, however, the SSDRs have been suppressed. Accordingly, acquisition under these conditions should proceed relatively slowly. There currently exist a number of reports (e.g., Anisman and Wahlsten, 1974; Anisman and Waller, 1973; Bolles, 1971; Wahlsten, 1972) indicating that factors that modify response types will affect performance in a predictable manner.

6.1.3. *Stimulus Factors*

It has been suggested (Shettleworth, 1972; Testa, 1974) that just as in the case of certain response categories, the nature of the stimulus may also influence the rate at which an avoidance response is acquired. In particular, associability of the stimulus–shock contingency is related to the location and intensity of the stimuli. Moreover, certain stimuli may be more effective than others in serving in a warning capacity.

In the evaluation of pharmacological effects, the role of stimulus factors may become more complicated. That is, in addition to differences in associability, different CSs may also elicit varying unconditioned responses. Loud noises (as distinct from pure tones) may elicit greater motorigenic effects (i.e., less freezing) than a light CS (Frontali and Bignami, 1973; Rosic *et al.*, 1970; C.A. Smith *et al.*, 1961). Further, the response change induced by a drug may interact with different types of warning stimuli (see Bignami and Rosic, 1971).

6.1.4. Interaction of Associative and Nonassociative Factors

Although nonassociative factors may influence avoidance quite apart from involvement of associative factors, nonassociative factors may also aid in learning about response–shock contingencies. That is, by increasing the occurrence of a particular response being emitted, the likelihood of this response being associated with shock offset is increased. It is noteworthy that decreasing the occurrence of incompatible nonassociative responses does not assure that acquisition rate will be increased. For example, when the contingencies of the task are fairly difficult or ambiguous (e.g., Longo, 1966) or if strains of animals are used that suffer from apparent associative difficulties under certain conditions (Anisman, 1975*a–c*), reduction of shock-induced response suppression will not necessarily improve performance. For performance to be improved, it is necessary that reducing nonassociative factors permit the emergence of the response to be learned, and the response–shock relationship should not be compromised by ambiguous contingencies.

6.1.5. Deductions and Summary

The following basic formulations and deductions can be arrived at on the basis of the foregoing discussion.

1. The acquisition of an avoidance response requires the establishment of (1) a drive or expectancy of shock and (2) the acquisition of the correct instrumental response.
2. The rate at which the CS–shock contingency is acquired is based in part on the nature of the stimulus and its associability (in terms of the location and intensity of the CS) with the shock. Moreover, the effectiveness of the CS may be species-dependent. Of course, temporal factors such as intertrial interval, CS–shock interval, and intersession interval will affect the rate at which the contingencies are established (see review in McAllister and McAllister, 1971).
3. Offset of the CS serves to maintain behavior in its capacity to either (1) result in fear reduction or (2) serve as a feedback stimulus.
4. Given adequate associative abilities, or parameters of the task that

permit adequate acquisition of the avoidance contingencies, nonassociative factors may come to determine response rate. If the response requisite is compatible with the nonassociative effects elicited by the CS and shock, avoidance will progress readily. The occurrence of incompatible nonassociative factors will disrupt performance.

5. Although the nonassociative effects incurred in an avoidance situation are a consequence of shock, the nature of the CS may similarly induce nonassociative changes.

6. Other things being equal, elimination of incompatible nonassociative factors should enhance avoidance-response rate. Thus, elimination of inhibitory tendencies should augment active-avoidance performance but disrupt passive avoidance. Conversely, exacerbation of inhibitory tendencies should retard active but enhance passive avoidance.

7. In the *absence* of adequate associative abilities, elimination of incompatible nonassociative factors will not necessarily lead to enhanced avoidance; however, disruptive nonassociative factors may further retard performance. In effect, both the associative and nonassociative components may set rate limitations on avoidance acquisition.

6.2. Avoidance Techniques: Task Manipulations and Multiple-Testing Procedures

In the following sections, a variety of techniques will be described that permit the dissociation of drug effects on associative, nonassociative, and motivational factors. These involve multiple-testing procedures in which one component of avoidance is held constant while a second component is varied. Thus, through successive or concurrent testing, it may be possible to delineate the conditions necessary for drug effects to become apparent. For example, if drug effects are observed only when response suppression is strong, then the possibility must be considered that the drug affects nonassociative factors. Conversely, if the drug affects performance regardless of the level of response suppression, then the performance change is likely a consequence of effects on associative or motivational processes, or the drug has nonassociative effects beyond that of alleviating suppression.

6.2.1. *One-Way and Shuttle Avoidance*

Both these procedures involve a locomotor response to avoid shock. In the one-way task, animals consistently are required to run from one compartment to a second compartment to avoid shock (i.e., animals always run in the same direction). In contrast to the one-way task, where one compartment always serves as the safe area and the other as the danger area, in the shuttle task, the safe and dangerous compartments alternate from trial to trial. Acquisition in the one-way task proceeds considerably faster than in the shuttle task. Several factors contribute to the task differences. Although the running response is an SSDR, and thus should result in rapid acquisition in both tasks, the effectiveness of running as an SSDR is compromised in the shuttle task because animals are required to return to a compartment previously associated with shock. Similarly, because the running response is compromised, unconditioned freezing cannot be readily overcome in the shuttle task, whereas freezing is eliminated quickly in the one-way situation. The difficulties inherent in the shuttle task are further exaggerated by the fact that conflict associated with returning to cues previously associated with shock may induce response suppression, and in addition, the cue–shock contingencies may be ambiguous (cf. Anisman, 1975*a,b*; Anisman and Waller, 1973; Anisman and Wahlsten, 1974). Finally, because of the ambiguous nature of the shuttle response, the effectiveness of the feedback signal is reduced (Bolles, 1970), or the extent of fear reduction generated by CS offset may be minimal (McAllister and McAllister, 1971).

Performance in both the one-way and shuttle tasks is modifiable by a variety of manipulations. For example, if animals receive limited exposure to inescapable–unavoidable shock, in the presence of cues that subsequently signal danger, then subsequent performance is enhanced (see Fig. 15). Prior shock exposure presumably results in the establishment of the drive or expectancy of shock, thus requiring only the establishment of the running and place responses. In contrast, if animals receive preshock in the presence

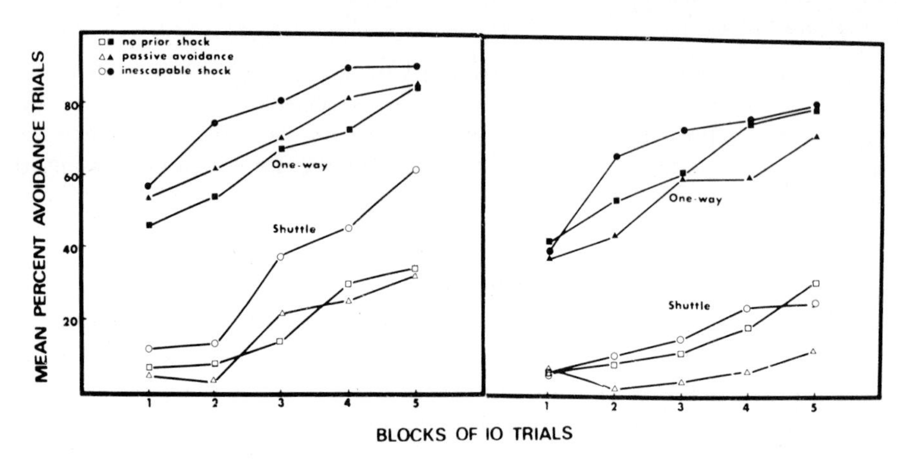

Fig. 15. Mean percentage of avoidance responses over blocks of ten trials in a one-way and shuttle task. *Left:* Holtzman rats received exposure either to a single inescapable–unavoidable shock (6 sec, 1.0 mA) in the presence of cues that subsequently signaled danger (light), to an equivalent amount of shock in the presence of the to-be-dangerous cues after animals made a response of crossing a hurdle, or to no prior shock. *Right:* Same procedure, but shock was delivered in the presence of the to-be-safe (dark) compartment. It will be noted that the facilitative effects of preshock in the to-be-dangerous compartments are eliminated when shock is contingent on a hurdle-crossing (the to-be-established response) or when shock is presented in the face of a stimulus that is subsequently associated with the safe compartment. These effects are additive in that both incompatible contingencies lead to disruption of performance relative to control groups. It is noteworthy that the effects described are true of both one-way and shuttle avoidance. From Anisman (1973*a*, Experiments 1 and 2); reprinted with permission; © American Psychological Association.

of cues that are associated with safety in the subsequent avoidance task, then performance may be disrupted (Anisman, 1973*b*; de Toledo and Black, 1967). The effects of preshock are considerably more complex than that indicated here, and more detailed descriptions are available in Anisman (1975*a*) and Anisman and Waller (1973). In any event, the important point for the present purpose is that manipulations of the cue–shock pairings during preshock exposure will modify later performance in a consistent fashion in both tasks. Whereas manipulations with associative consequences affect both one-way and shuttle avoidance, treatments that influence non-associative processes differentially affect performance in the two tasks. Increasing the intensity of the freezing response should disrupt avoidance when the required response is one of running. Conversely, decreasing the freezing response should augment avoidance. The shuttle task is exquisitely sensitive to these non-associative influences. Reduction of the immobility response by any one of a variety of techniques, e.g., *d*-amphetamine (Barrett *et al.*, 1972; Krieckhaus, 1965) or septal and hippocampal lesions (Fried, 1972; Schwartzbaum *et al.*, 1967; Kenyon and Krieckhaus, 1965; see also Isaacson, 1974), augments performance. Conversely, factors that increase freezing, such as intense shock (Anisman and Waller, 1972; Theios *et al.*, 1966; Moyer and Korn, 1964) or cingulate or mammillothalamic lesions (Lubar and Perachio, 1965; Krieckhaus, 1965), as well as treatment with drugs that increase ACh levels or result in catecholamine depletion (Anisman, 1973*b*; Bignami and Rosic, 1971; Ahlenius, 1973), tend to disrupt performance.

Unlike shuttle performance, avoidance in the one-way task is not greatly influenced by nonassociative manipulations. Neither increasing nor decreasing freezing tends to affect performance appreciably. In fact, numerous treatments that reduce freezing may come to disrupt one-way avoidance, e.g., scopolamine, *d*-amphetamine, and septal lesions (see review in Bignami, 1976). It seems that freezing can be overcome readily in the one-way task, and consequently, further reduction in response suppression will not augment performance. Indeed, if the treatment results in increased exploration, as is likely with anticholinergics, the exploratory response may compete with the appropriate defensive behavior necessary for successful avoidance (but, for a complete analysis of this and related problems, see Bignami, 1976).*

When drug effects are considered independently in the one-way and shuttle tasks, the conclusions derived are limited. However, when considered together, the nature of a particular drug effect can be clarified. Specifically, since both one-way and shuttle performance are sensitive to associative

* Most physiological and pharmacological investigations employing the one-way task have employed a procedure in which the animal is handled between trials. To say the least, this may prove to be a serious confounding variable. Witness, for example, the finding that the improvement in shuttle avoidance induced by *d*-amphetamine was eliminated if rats were handled during the intertrial interval (Griffiths and Wahlsten, 1974). To eliminate the handling factor, a number of apparatuses have been developed (e.g., see Anisman, 1973*b*; Anisman and Waller, 1973). Performance in these tasks proceeds more slowly than in the standard one-way task, and drugs that reduce freezing may elicit a transient improvement in performance.

manipulations, treatments that affect "learning" should modify performance in both tasks. In contrast, because only shuttle performance is sensitive to nonassociative manipulations, performance should be affected differentially in the two tasks after treatments that modify shock-induced suppression (see Suits and Isaacson, 1968; Anisman, 1973*b*). Although the combination of these tasks may be a sensitive index of nonassociative influences, it must be remembered that when the contingencies of the shuttle task are complex, reduction of suppression will have negligible effects on performance.

6.2.2. *Manipulation Avoidance*

The response necessary for escape–avoidance in such tasks involves pressing a lever or turning a treadmill. Not surprisingly, response acquisition, at least in the case of the lever-press task, proceeds very slowly. Bolles (1970, 1971) has argued convincingly that this is due to the fact that the lever-press response is not an SSDR. Further, the sequence of responses required of the organism is confusing. First, the animal is required to run, then approach and press the bar. The latter response involves one of not moving. This point is accentuated by the fact that animals typically remain immobile on the bar.

While reduction of immobility may augment performance in this task (cf. Anisman and Waller, 1973), drug effects may be obscured by the particularly slow rate at which the contingencies are established. Moreover, the use of this task does not elucidate whether changes in performance are due to associative or nonassociative factors.

6.2.3. *Passive Avoidance*

In contrast to the paradigms discussed thus far, the emission of a particular response results in punishment in the passive-avoidance situation. Subsequently, animals are reintroduced to the test chamber, and the latency of the previously punished response reoccurring is recorded. Predictably, response latency is increased considerably relative to nonshocked animals, or animals shocked in a different chamber. The source for the performance difference apparently is due to several factors: (1) inhibition of responding owing to the locomotor response being punished, (2) reluctance to return to the stimulus complex associated with shock, and (3) nonassociative effects of shock such as freezing (see Anisman, 1973*b*; Randall and Riccio, 1969). Theoretically, disruption of any of these components should reduce passive-avoidance behavior.

Owing to the various factors that contribute to successful passive-avoidance performance, the effects of drug treatments on associative vs. nonassociative factors cannot readily be divorced. Accordingly, when used alone, the passive-avoidance task generates a minimum of information regarding the nature of drug-induced performance changes. However, the effectiveness of this task in behavioral analyses is increased considerably when it is used in conjunction with an active-avoidance task. If a treatment affects associative or memory factors, then comparable behavioral changes should be observed in both tasks (i.e., if memory is enhanced, then improvement should be seen in both the active and passive avoidance tasks, whereas the converse should hold for agents that disrupt memory). If, on the other hand, nonassociative processes are affected by the drug treatment, then differential task effects should be observed (e.g., reduction in freezing should improve active but disrupt passive avoidance). Thus, in the case of drugs that apparently reduce response suppression (e.g., scopolamine or *d*-amphetamine), shuttle avoidance is enhanced while passive avoidance is disrupted (cf. Anisman, 1975*a,b*; Bignami and Rosic, 1971). It should be clear from the preceding discussion that the passive-avoidance task is an inadequate index of memory. In addition to the problems already delineated, it suffers because (1) of vulnerability to even mild nonassociative changes; (2) it results in excessive between-subject variance; and (3) it does not adequately assess drug effects as a function of task complexity.*

In view of the gross limitations of the passive-avoidance task, it is surprising to find that it still is used extensively as a unitary tool. This is likely a consequence of the exceptionally rapid rate at which the response is established. Under some conditions, only a single trial may be required for the avoidance response to be seen. It is assumed that because of this rapid learning, pharmacological treatments can be applied during the period of memory consolidation (i.e., injection immediately after shock) as compared to periods where memory has already been established (McGaugh and Petrinovich, 1965). Even in this respect, the technique is questionable, since the latency to peak drug effects may limit the effectiveness of posttrial drug injections.

6.2.4. *Conflict*

The conflict task, like the passive-avoidance situation, involves punishment of an emitted response. Animals are first trained to traverse an alley to obtain food or water reinforcement. Subsequently, shock is made contingent on the emission of this particular response. Whereas the passive-avoidance task involves only the suppression of responding, the approach tendency is maintained in the conflict task because of the food or water reinforcement. Since food–water is required from day to day, successive daily testing is possible without cessation of responding as might be seen in the passive-avoidance task. Moreover, in the conflict task, a measure can be obtained not only of response latency but also of the distance animals travel

* Because of the large number of different passive tasks in use (step-down, step-through, inhibitory response, Jarvik type, Calhoun type), it is not clear that these tasks would be similarly affected by drug treatments, since in some tasks the greatest source of variance involves nonassociative factors, whereas in others associative factors may contribute to the variance to a greater extent.

in the alley. Thus, the probability of floor effects masking behavioral changes is reduced. However, as in the case of the passive-avoidance task, it may be difficult to divorce associative from nonassociative factors. The varied aspects of the conflict paradigm are too numerous to be discussed in adequate detail; however, several excellent sources are available (e.g., Campbell and Church, 1969).

In any case, it is evident that the effects of drug treatments on conflict behavior may be due either to changes in the response to the aversive aspects of the task or to variation of the food-motivated response, or to both. It has been suggested that the opposing approach-avoidance tendencies should be affected equally by nonassociative factors involving changes in locomotor activity (Barry and Buckley, 1966). Thus, changes of a nonassociative sort induced by drugs should not represent a serious drawback. Moreover, because the approach and avoidance tendencies are established independently, the former being acquired first, it should be possible to assess the effects of drug treatments on the different components of the task. One of the more popular forms of the conflict paradigm is that developed by Geller and his associates (Geller, 1962; Geller and Seifter, 1960). Animals are trained on a multiple schedule in which intermittent reward is presented during one phase, while concurrent reward and punishment are presented during a second signaled phase. This paradigm permits the concurrent evaluation of drug effects during the nonconflict and conflict components, and thus allows for dissociation of inhibition and general depression of responding engendered by drug treatments.

6.2.5. *Conditioned Suppression*

A substantial amount of research has been concerned with changes in operant responding that occur when a conditioned stimulus (CS), established through Pavlovian conditioning, is superimposed on an operant baseline (Davis, 1968; Blackman, 1972). A decrease in operant responding, following presentation of a CS paired with shock, has been termed conditioned suppression. The term conditioned emotional response (CER) is sometimes used to refer to this empirical effect; however, it begs a particular theoretical position that presumes there is an underlying emotional state that is classically conditioned to the CS and interferes with the instrumentally conditioned behavior. There are other theoretical interpretations of the conditioned suppression effect, such as the punishment hypothesis, which maintains that the disruption of ongoing behavior is the result of an adventitious contingency between the operant response and punishment. It has also been suggested that conditioned suppression may be dependent on the degree of discriminative control exerted by the schedule of reinforcement, rather than competing responses or adventitious punishment effects (Blackman, 1972).

The effect of the CS on the operant baseline is typically expressed as a ratio of the response rate during the CS and the rate in the absence of the

CS. This is referred to as a suppression ratio, and it provides a control for individual differences in baseline response rate among subjects. There have been numerous methods used to compute these ratios, e.g., CS rate/control rate or (control rate − CS rate)/control rate, in which control rate is typically the rate of responding in the interval prior to CS presentation, which is of the same duration as the CS. In pharmacological experimentation, certain ratios (e.g., CS rate/control rate) are prone to misinterpretation. Assume that an animal, during the CS, has a response rate of 5 per minute and in the absence of the CS, a rate of 20 per minute. The suppression ratio would be 0.25. Next, assume an investigator reports a suppression ratio of 0.50 after administration of a drug to these same animals. While such a ratio indicates less suppression, it is not immediately evident whether this is attributable to changes in response rate during the CS or in the baseline, or in both. A ratio of 0.50 could have been obtained given any of the following ratios: 10/20, 5/10, 16/32. Interpretation of the effects of the drug would be dependent on the mechanism through which the suppression ratio changed. It has already been demonstrated that some drugs may differentially affect responding during CS and non-CS periods, while other drugs act in a nonselective manner (Barry and Buckley, 1966).

Reviewers of the effects of various pharmacological agents on conditioned suppression have uniformly noted the considerable variability in experimental findings (Davis, 1968; Kelleher and Morse, 1968). This has generally been attributed to the failure to take into account and precisely specify certain procedural details such as shock intensity and duration, nature and duration of the CS, and schedule of reinforcement maintaining the operant baseline. It is now clearly understood that a host of experimental parameters determine the degree of suppression obtained with this paradigm (Henton and Brady, 1970; Davis and Keruter, 1972; Blackman, 1972).

6.2.6. *Activity–Reactivity*

Since performance in learning tasks may be affected by factors such as response suppression, it would be advantageous to evaluate the effects of drugs directly on activity levels. However, if inferences are to be drawn to the avoidance situation, then activity should be measured after exposure to shock (i.e., reactivity). This factor takes on all the more importance when it is realized that general activity is a poor predictor of avoidance performance, whereas activity after shock correlates highly with avoidance behavior (cf. Anisman and Waller, 1973; Steranka and Barrett, 1974; J.M. Weiss *et al.*, 1968). This is not surprising, since in the absence of shock, animals will not necessarily exhibit individual defensive behaviors, whereas after shock, the prepotent response in the organism's repertoire should surface. Accordingly, if drug effects are to be evaluated, this should be done within the context of a shock paradigm. Indeed, it may well be advantageous to employ a repeated shock procedure in order to simulate the early trials of avoidance.

In addition to evaluating the effects of activity in the postshock situation, activity may also be measured during the presentation of a CS previously paired with shock, and with a CS presented on a random schedule within the shock situation. This may not only prove of value with respect to possible conditioned types of responses, but also may reveal whether the drug–CS combination elicits biases, e.g., locomotor excitation or locomotor suppression of an unconditioned nature.

If a particular agent improves avoidance performance, but has negligible effects on reactivity, it is likely that the observed drug effect on avoidance is not due to nonassociative factors. Conversely, if the drug eliminates shock-induced suppression, the possibility needs to be entertained that the drug modifies nonassociative processes. Thus, in the case of anticholinergics, for example, where response suppression is reduced and avoidance enhanced (Anisman, 1975*b*; Anisman *et al.*, 1975), the clear indication is that the drug modifies the defensive repertoire of the organism. Together with appropriate avoidance techniques (cf. Bignami, 1976), the case is strengthened considerably. With respect to effects of catecholaminergic stimulants, such as *d*-amphetamine, it is observed not only that shock-induced suppression is eliminated, but also that, in fact, after shock, activity levels may be increased relative to that observed in animals treated with amphetamine but not shocked (Anisman, 1976; Anisman and Cygan, 1975; Anisman *et al.*, 1975). This finding would suggest that unlike the effect of scopolamine, the effect of *d*-amphetamine is one of a *genuine* response excitation rather than simple disinhibition (for additional details, see Anisman, 1975*b*).

One example from our laborabory (Anisman, 1975*a*), although of a genetic rather than a pharmacological nature, may serve well as an illustrative example of how the activity–reactivity measure may be used. Mice of three inbred strains, together with the six reciprocal F_1's, were tested in activity, reactivity, inhibitory, one-way, and shuttle avoidance. As can be seen in Fig. 16, the performance of the hybrid crosses exceeded that of the parent strains in the shuttle task (overdominance). Yet, in the activity and reactivity situations, the performance of the F_1 mice generally was intermediate between that of the parent strains. Similarly, in the case of inhibitory avoidance, the F_1 performance was intermediate between that of the parent strains. Clearly, the high level of shuttle performance among the F_1 mice cannot be attributed to factors such as freezing. Rather, these data suggest that the hybrid superiority involves associative factors. Moreover, the fact that complete dominance, but not overdominance, was observed in a one-way task suggests that the hybrid superiority may be due to poorer abilities among inbred mice only when the task involves complex integrative abilities. In relatively simple tasks, where these integrative abilities are not essential, the inbred disadvantage is not apparent. This line of attack is by no means a novel one; however, it does permit differentiation of associative and nonassociative factors. With this capability, it represents a potent evaluative and predictive tool.

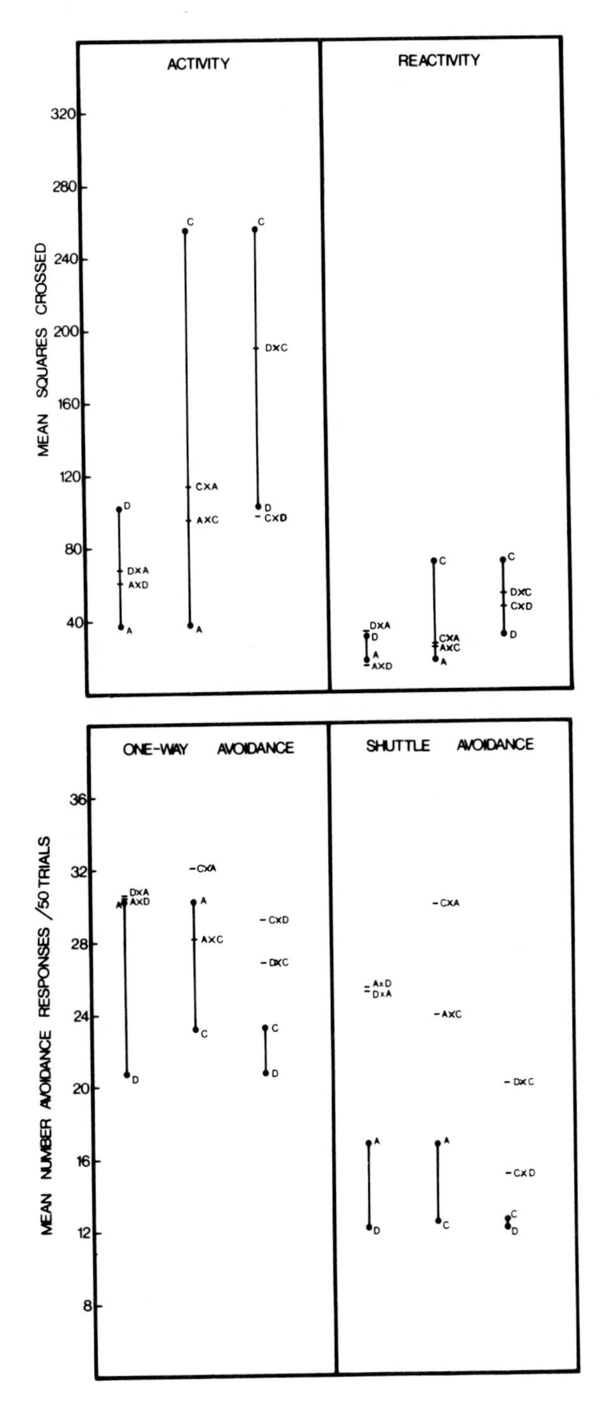

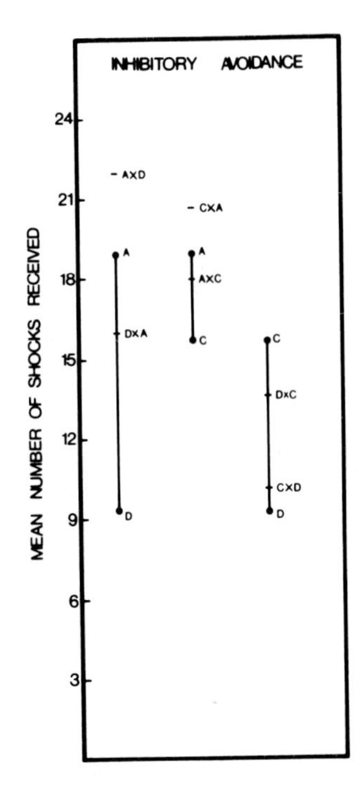

Fig. 16. Mean levels of activity before shock or after shock (reactivity), mean shocks received in an inhibitory task, and number of avoidances in a one-way and shuttle task in three inbred strains and the six reciprocal F_1's. (●) Inbred strains; (—) hybrid scores. From Anisman (1975*d*); reprinted with permission; © American Psychological Association.

6.2.7. *Shock Sensitivity*

In a one-way avoidance situation, acquisition is directly related to shock severity. In the shuttle task, an inverted U-shaped function is observed. With mild and severe shock, performance progresses poorly. Presumably, at low shock intensities, motivational levels are relatively low, thus eliciting poor performance. With high shock intensities, the unconditioned freezing response hampers shuttle performance, in which freezing cannot be suppressed quickly (cf. Anisman and Wahlsten, 1974; Theios *et al.*, 1966; Wahlsten and Sharp, 1969).

Inasmuch as shock sensitivity influences avoidance, it needs to be determined whether drug treatments influence the response to shock. There are, in fact, indications that pharmacological treatments will modify shock threshold. Houser and Van Hart (1973, 1974), for example, indicated that anticholinergics elevate shock threshold, while cholinergic agonists have the opposite effect. Similarly, Satinder and Hill (1974) and Satinder and Petryshyn (1974) reported that amphetamine may affect shock threshold, and further, these effects interact with the strain of rat employed. In these investigations, the independent measures were either flinch threshold or time spent on the shock grid. As might be imagined, low current intensities were employed, and it is therefore difficult to ascertain whether drug effects observed were reflective of higher intensities typically used in avoidance tasks. This notwithstanding, it is clear that information regarding shock sensitivity induced by drugs needs to be considered in conclusions drawn from avoidance studies.

In a genetic analysis Wahlsten (1972) employed an intriguing method that may well be applicable to pharmacological work. He ascertained responsivity, in terms of vocalization and jumping, in several strains of mice. In avoidance testing, he used not only an absolute shock intensity but also an equated shock intensity (i.e., the intensities at which strains exhibited equivalent jumping and squealing). Under these conditions, acquisition of stimulus–shock contingencies and reactivity to shock was found to influence avoidance performance. With respect to pharmacological analysis, it may prove useful to employ a similar technique. That is, drug effects on avoidance performance should be evaluated using drug doses that produce equipotent effects so far as aversive threshold is concerned.

6.3. *Discriminated Avoidance*

The methods described in previous sections indicate whether nonassociative factors influence performance; however, the possibility is not excluded that concomitantly associative changes are also produced. With the concurrent use of the active and passive tasks, arguments in favor of one hypothesis or another are greatly strengthened. However, the possibility exists that ceiling or floor effects preclude changes in the one-way and shuttle tasks, respectively. In effect, the use of this technique permits only indirect inference regarding associative factors.

6.3.1. *Y-Maze Discriminated Avoidance*

In the avoidance tasks described previously, the primary requisites for successful avoidance involved acquisition of the drive and instrumental response. With the additional requisite of appropriately directing the response, it may be possible to assess changes in associative processes. Barrett and his associates (Barrett *et al.*, 1973, 1974; Steranka and Barrett, 1974) have elegantly demonstrated the efficacy of the discrimination task using a symmetrical Y-maze. In the Y-maze task, each arm may serve either as the start (danger) or goal (safe) compartment. Shock is signaled by onset of a light in one of the arms, and to avoid shock, the animal is required to enter the illuminated arm within a set period of time. Thus, a measure can be obtained of correct avoidances or incorrect avoidances (i.e., leaving the start box and entering the correct or incorrect goal arm, respectively). In addition, correct vs. incorrect discriminations (as measured on both escape and avoidance trials) can also be assessed. Finally, activity levels can be measured concurrently. Accordingly, through this one task, it can be determined whether a drug treatment affects response initiation, as indicated by improved avoidance but no change in discrimination, or associative abilities, as indicated by improved avoidance and increased correct discriminations. Figure 17 shows the effects of scopolamine and *d*-amphetamine on avoidance and discrimination in two rat strains. It is clear from this figure that the two behaviors are differentially affected by drug treatments, thus suggesting that the drugs do not affect associative processes.

6.3.2. *Sidman Avoidance*

Unlike the other active-avoidance tasks described, the Sidman avoidance procedure (Sidman, 1953*a*,*b*) does not involve an explicit CS. In this task, a brief pulse of shock occurs at a set interval, say every 5 sec (S–S interval). By making an appropriate response such as running or pressing a bar, the subject can interrupt the sequence of shocks for a period of time, e.g., 20 sec (R–S interval). The rate at which the avoidance response is acquired is dependent on the length of each session and of the S–S and the R–S intervals. Since the shocks, once delivered, are inescapable, and since there is not an explicit CS onset or offset in this procedure, it creates an interesting problem with regard to the source of reinforcement. Some investigators have suggested that the reinforcement in the Sidman task is the avoidance of shock itself or the reduction in shock density (Herrnstein, 1969). Other investigators (Anger, 1963) have suggested that the shock or no shock following a response results in internal stimulus change. This stimulus "trace," in the case of a response, is paired with no shock; as time passes, the trace changes discriminably. Since shock is paired with an old trace, but not with a new trace, as the R–S interval approaches its end, the stimulus aspects of the trace become increasingly more aversive in their associative value, thus perpetuating the next response, and the concomitant trace. Whether these traces should be considered in

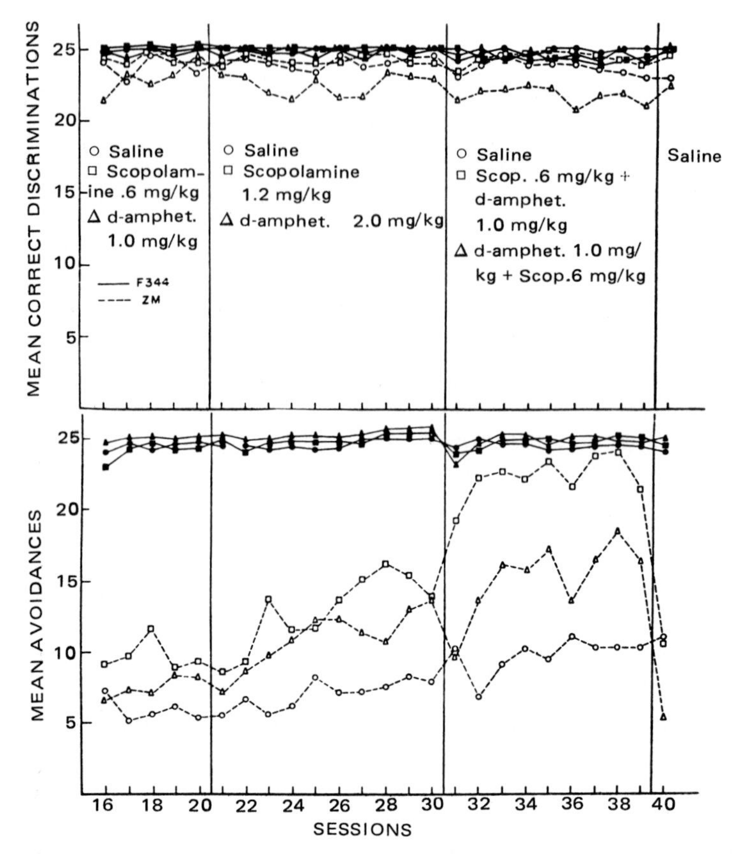

Fig. 17. Mean correct avoidances and correct discriminations in a Y-maze task among two rat strains as a function of scopolamine and *d*-amphetamine treatment. Note that the drug treatments augment avoidance in the low-avoidance strain (ZM), but not in the high-avoidance strain (F344). The strains do not differ in discrimination performance, nor do the drugs affect discrimination performance. Apparently, response inhibition, and not associative factors, is responsible for the strain difference. Similarly, the drugs augment performance as a result of response disinhibition. From Barrett *et al.* (1973); reprinted with permission; © American Psychological Association.

terms of "fear" is currently in dispute and probably will be for quite some time. One thing is eminently clear, however; when the response required is one that is part of the organism's repertoire in aversive situations (e.g., running), avoidance progresses more readily than when the response is not an SSDR (e.g., bar-pressing). It seems that the mode of responding is again a pertinent factor here.

Sidman avoidance has been used extensively in pharmacological work, particularly in evaluating neuroleptics. As with the other tasks, the drug treatment may be administered immediately prior to training to determine changes in acquisition. Alternatively, the drug may be administered after some training or after stable baseline performance has been established. The nature of a particular drug effect may be dependent on the rate of baseline responding. Kelleher and Morse (1968) and Dews (1958) indicated that the inhibitory effects of a drug treatment were greatest under high response rate, whereas the excitatory effect of a drug would be most pronounced under low baseline conditions.

An important aspect of the Sidman procedure is that it does not involve an explicit cue, such as a light or tone, thereby confounding drug effects with stimulus factors. Moreover, this procedure can be combined with Pavlovian techniques to evaluate discriminative functions (for appropriate control techniques, see Rescorla, 1967). That is, after a baseline level of responding is established, animals are subjected to inescapable shock in the presence of a particular cue (CS^+), and presented with a different cue (CS^-) in the absence of shock. Subsequently, animals are again placed in the Sidman task and baseline performance established. Intermittently, the CS^+ or CS^- is presented, to which animals exhibit predictable responding, i.e., increased response rate to the former and decreased response rate to the latter. Presumably, the Pavlovian procedures engender a conditioned excitatory response to CS^+, thereby increasing motivation to escape this stimulus, whereas CS^- is a signal of relative safety and consequently elicits decreased responding. If a particular drug improves associative abilities, then drugs administered during Pavlovian conditioning should augment the effects of the CS^+ and CS^- presented on the Sidman baseline. However, baseline performance should be unaffected, since the drug treatment was not associated with this procedure.

6.3.3. Go–No Go Avoidance

It was indicated earlier that the use of the active- and passive-avoidance paradigm represents a potentially useful evaluative technique. The go–no go paradigm represents the combination of these tasks in a within-subject design. Moreover, it may also be used to evaluate discrimination performance. In the go–no go task, animals are presented successively with two stimuli. In response to one CS (the go stimulus), the animal is required to make an active response, such as jumping a hurdle, to avoid shock. In response to the second, or no go, stimulus, the organism is required *not* to run in order to avoid shock.

As in the traditional active- and passive-avoidance tasks, those treatments that affect associative processes should elicit equivalent effects in both components of the go–no go task. That is, drugs that improve associative processes should decrease both errors of omission and errors of commission, whereas the converse should be true of drugs that disrupt associative processes. If, on the other hand, the drug treatment reduces the nonassociative effects of shock, then errors of omission should decrease and errors of commission should increase. Again, the converse should be true of drugs that increase the nonassociative effects of shock (see Bignami and Rosic,

1971; Carro-Ciampi and Bignami, 1968; Rosic and Bignami, 1970). As can be seen in Fig. 18, it is observed that physostigmine leads to decreased errors of omission (passive failures), but increased errors of commission (active failures). Evidently, the drug modifies response-modulating factors.

In addition to the role of nonassociative factors dealing with the response component, the go–no go task is amenable to the evaluation of stimulus factors in the effects of drug treatments. It appears to be the case that different CSs differentially affect behavior, either because these stimuli are more significant in an ethological sense, because they elicit motorigenic responses that augment response initiation, or because they affect subtle aspects of the learning process, e.g., feedback. Regardless of the source, it is clear that drug treatments may interact with stimulus factors (Bignami and Rosic, 1971; Frontali and Bignami, 1973; Frontali *et al.*, 1977). One variant of this paradigm is that of employing an asymmetrical arrangement; i.e., one signal is associated with shock, but in the presence of the other signal, shock is never presented (i.e., extinction). As with the symmetrical paradigm, extensive drug work has been carried out using this technique, and has served, among other things, to differentiate among various biases introduced

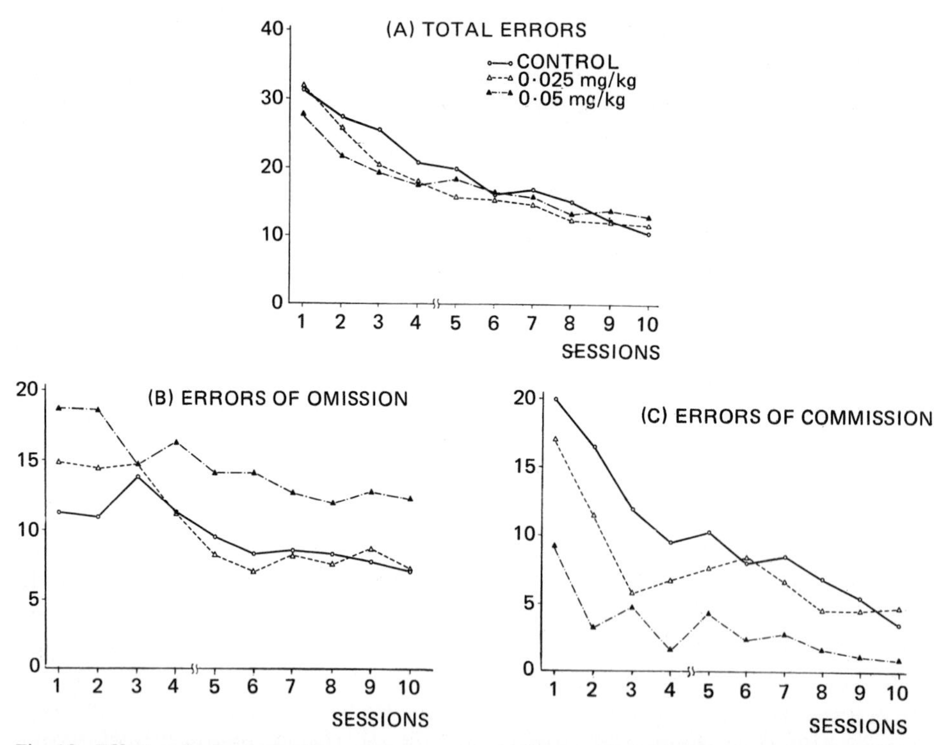

Fig. 18. Effects of physostigmine on go–no go avoidance discriminations over 10 training sessions of 50 trials each. Training commenced after a 70% correct criterion was achieved. Note the large errors of omission and the lessor errors of commission elicited by the drug. From Rosic and Bignami (1970); reprinted with permission; © Pergamon Press.

by drug treatments (see Bignami, 1977; Frontali *et al.*, 1976), as well as to determine whether comparable drug effects appear in both appetitive and aversive situations (see review in Bignami, 1977).

Given the various attributes of the go–no go task, e.g., differentiation of response factors, and evaluation of stimulus factors (remembering in addition that these are each treated as within-subject factors), this procedure seems most useful as regards pharmacological investigations.

6.4. Transfer Designs

Although the discrimination tasks are useful in attempting to differentiate between associative and nonassociative effects of drugs, the possibility exists that a treatment affects both these processes. One way in which this can be determined is that of training animals in the drug state, and subsequently retesting animals in the nondrug state. In the following sections, a variety of techniques are described that permit evaluation of transfer of training.

6.4.1. Reversal Learning

6.4.1a. Stimulus Factors. In a simultaneous-discrimination task, subjects initially are required to respond to one set of stimuli, and not to another, in order to avoid shock (e.g., run toward light but not toward dark). On the following day, the opposite arrangement is used. Adequate associative abilities should result in negative transfer in the reversal task, but positive transfer in a nonreversal task. In contrast, transfer effects should not be observed in either situation if a treatment hinders associative abilities (Anisman, 1975*e*). If good associative abilities are present, then errors should be increasingly less frequent with successive daily testing. Poor associative abilities should result in a slower between-days decline in error rate (Alpern and Marriott, 1972, 1973).

The transfer paradigm has recently been used in an attempt to differentiate between disruption of memory as opposed to response initiation induced by dopamine depletion or blockade (see Fibiger *et al.*, 1974, 1975). In particular, administration of the dopamine antagonist haloperidol or of 6-OHDA in high doses causes a large disruption of one-way avoidance. However, if animals are initially trained on the task in the absence of drug treatment, the disruptive influence of dopaminergic blockade is subsequently mitigated. Similarly, in a reversal task after pretraining, negative transfer was observed in 6-OHDA animals, suggesting that memory was not impaired. Apparently, DA antagonists affect performance by disrupting the animal's ability to initiate responses.

Although the transfer type of design may be very effective as an evaluative tool, certain precautions are necessary. For example, it must be ascertained that the observed results were not due to proactive drug effects or state-dependent learning. Even with the inclusion of the aforementioned control procedures, several problems may still persist. For example, it has been observed that performance in the nondrug state may be retarded, not because of associative effects, but likely because of stimulus factors other than state dependency *per se* (Barrett *et al.*, 1973, 1974; Barry and Krimmer, 1976; Bignami *et al.*, 1975). Morover, conditioned tolerance may also be observed as in the case of morphine or insulin (Siegel, 1975*a,b*). Finally, in a state-dependent type of design, Day 2 performance may be unduly biased, in that the effectiveness of the drug treatment may vary as a function of the extent to which the avoidance response was previously established (Bignami *et al.*, 1975; Fibiger *et al.*, 1974).

6.4.1b. Response Factors. Pharmacological treatments may affect the rate at which the response–shock contingency is established by virtue of the fact that the activity level of the animal is modified. That is, the acquisition of the cue–shock association may not be affected, but because the drug induces a particular response bias, the association between a response and the occurrence of shock is modified. Given this, it may be useful to evaluate reversal-learning involving responses as opposed to stimulus factors.

If animals are trained on a passive-avoidance task, subsequent active-avoidance performance is disrupted relative to the performance of naïve animals. Conversely, initial active-avoidance training will disrupt subsequent passive-avoidance performance (Klein and Spear, 1970*a,b*). It follows that if a drug treatment affects associative processes, then the negative transfer may be eliminated, and performance should be comparable to that of naïve animals. Of course, the contraindications presented for reversal learning of stimulus factors apply here as well. Together with stimulus-reversal, it should be possible to determine not only whether a pharmacological agent affects associative processes, but also whether they are indeed specific to stimulus or response variants of the behavior in question.

6.4.2. Acquired Drive

In the reversal-learning paradigms discussed, the initial phase of training involved the establishment of all the components of the avoidance task. Accordingly, it may not be possible to discern (1) selective effects on stimulus factors (2) selective changes in drive–expectation of shock, and (3) potential interaction of the drug treatment with the already-established instrumental response. It is possible to overcome these difficulties by evaluating acquisition of the stimulus–shock contingencies in the absence of adequate learning of the shock contingencies. Subjects are exposed to cue–shock pairings that are both inescapable and unavoidable. Subsequently, subjects are placed into the apparatus and permitted to escape the cues previously paired with shock. This paradigm follows closely the parameters of other classically conditioned responses (for greater detail, see McAllister and McAllister, 1965, 1971). The important point for the present purpose is that biases in terms of locomotor responding should not be formed during initial training. Moreover, the effects of the stress, in the absence of "learned contingencies," can be evaluated by initially presenting the CS and shock on independent-random schedules (see Rescorla, 1969).

This particular paradigm not only is useful from the standpoint of reducing locomotor biases, but also has been found to be relatively sensitive to measurement of generalization of the cues associated with shock. Moreover, the generalization gradient is dependent on temporal factors (see McAllister and McAllister, 1971), and thus is useful for the evaluation of time-dependent changes in performance.

There are several necessary control procedures that should accompany the use of the "acquired-drive" paradigm. These are, by and large, the same as those for other prior-shock-exposure paradigms that are presented in the next section.

6.4.3. *Prior Shock Exposure and Subsequent Avoidance*

As indicated previously, for the avoidance response to be established, the motivation to escape or avoid shock must be established, and the instrumental response is acquired subsequently. It follows that if the cue–shock contingency were established by preexposure to signaled-inescapable shock, then subsequent avoidance performance will be enhanced. Indeed, as can be seen in Fig. 19, in both the shuttle and the one-way task, preshock of moderate intensity (1 mA) enhanced subsequent performance (Anisman, 1973*a*; Anisman and Waller, 1972, 1973; de Toledo and Black, 1967, 1970). With intense preshock (2 mA), on the other hand, shuttle, but not one-way avoidance, was disrupted. From these results, it seems that the disruptive influence of inescapable shock involves the strong freezing response produced by the initial treatment, together with the slow rate at which freezing is suppressed in the shuttle task.

Since facilitation or disruption in performance can be produced by prior training, this technique may be ideally suited for pharmacological analysis. For example, under conditions wherein preshock disrupts performance, drug treatments that reduce freezing should not only eliminate the poor performance, but also, because of the prior CS–shock exposure, in fact lead to performance superior to that of naïve animals. In the one-way task, on the other hand, little effect should be observed.

If performance is compared among animals that initially received avoidance and those that received inescapable shock (in a yoked design), one can determine drug effects under conditions in which coping is or is not possible. Further, it should be possible to dissociate drug effects that influence response-learning from those that do not involve response factors but affect drive (expectancy). Accordingly, drug effects would not be contaminated by the presence of both components of the avoidance-learning process. More specifically, it has been suggested that certain drugs act on drive-modulating systems while other drugs act on response-modulating systems (see Bignami and Rosic, 1971). Thus, if an instrumental response has already been

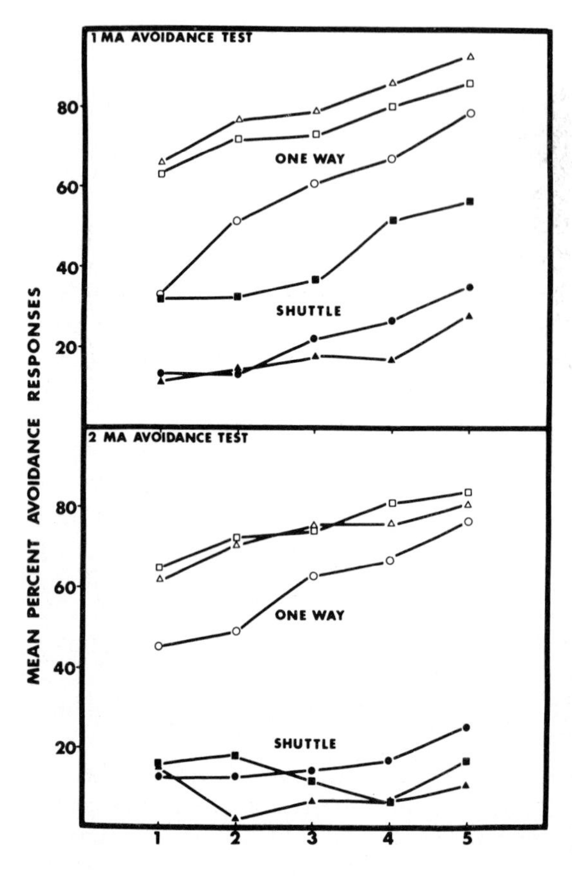

Fig. 19. Effects of ten signaled-inescapable shocks (0, 1, or 2 mA) on one-way and shuttle avoidance involving 1- or 2-mA shock. In the one-way task, preshock, regardless of intensity, enhanced avoidance. In contrast, in the shuttle task, preshock of 2 mA led to reduced response rate. Since effects are task-specific involving high or low freezing, nonassociative factors (freezing) are probably involved in the disruption of shuttle avoidance. Performance of nonpreshocked animals are denoted by circles, animals exposed to 1-mA shock by squares, and animals exposed to 2-mA shock by triangles. Adapted from Anisman and Waller (1972).

established, a drug that under other conditions improves avoidance performance by response modulation may have little or no effect (Oliviero, 1967) (but for a comprehensive analysis, see Bignami, 1977).

Several benefits can be derived from the preshock procedure: (1) Drugs may be administered during preshock to test for the effect of the drug on memory processes uncontaminated by instrumental process (Anisman, 1973*a*). (2) Drug dissociation can be determined by administering the drug during only one phase, during both phases, or during neither phase of training (see Anisman, 1975*e*). (3) The potential effects of the treatment on memory- vs. response-modulating systems can be determined by testing animals in a variety of tasks. For example, if a particular agent acts on memory processes, it should affect performance in different tasks in a comparable manner, i.e., improve one-way, shuttle, and passive-avoidance performance. In contrast, if the drug is acting on response-modulating sysyems, then the drug should differentially affect performance as a function of task (cf. Anisman, 1973*a*; Anisman and Waller, 1971). Parenthetically, in combination with the Y-maze discrimination task, associative factors can be assessed further (cf. Barrett *et al.*, 1973). (4) Finally, as a further test of memory processes, animals can be preshocked in the presence of specific cues and subsequently required to run toward or away from these cues.

Although this technique seems to be an effective one, several cautionary notes are needed. First, in terms of control procedures, it is not sufficient to use a CS–no-shock group, since latent inhibition may retard the performance of these animals, thereby increasing the difference in performance between preshocked and CS-only groups (see Lubow, 1973). Second, if a random CS–random shock paradigm is employed, there may be associative consequences of the shock–apparatus cues that need to be examined. Finally, the optimal delay between initial preshock and testing varies not only with shock intensity, but also with strain of rat or mouse used. Accordingly, care must be taken to employ adequate parameters in this respect (see Anisman, 1973*b*).

6.4.4. *Time-Dependent Variations of Transfer Effects*

Although change in performance is a monotonic function of time following initial training in the acquired-drive paradigm (McAllister and McAllister, 1971), this is not the case when the second phase of training involves avoidance-learning. Rather, in an active-avoidance task, the performance function is U-shaped, with the poorest performance occurring 1–6 hr after the initial stress session (Anisman, 1973*a*, 1975*a*; Kamin, 1963; Brush, 1971). While some investigators have attributed these changes to associative factors (Klein, 1972; Klein and Spear, 1970*a*,*b*), others have suggested that performance variations are due to nonassociative factors, possibly mediated by cholinergic and catecholaminergic mechanisms (Anisman, 1973*a*, 1975*a*; Barrett *et al.*, 1971; Pinel and Mucha, 1973*a*,*b*).

Support for the latter interpretation has been derived from various sources and is reviewed in Anisman (1975*a*). Briefly, although avoidance varies over the test intervals, discrimination performance remains unaffected (Barrett *et al.*, 1971; Steranka and Barrett, 1973). Similarly, in reversal-learning tasks, performance is disrupted at all intervals, suggesting adequate memory of the previously learned contingencies (Barrett *et al.*, 1971; de Toledo and Black, 1970). Finally, manipulations that affect nonassociative processes, e.g., shock intensity, task factors, and pharmacological treatments, have predictable effects on the U-shaped function. Figure 20 shows that the U-shaped function is observed in the shuttle task, wherein freezing is not suppressed readily, but not in the one-way task, wherein freezing is eliminated quickly. Furthermore, reduction in freezing behavior by scopolamine eliminates the U-shaped function, whereas increasing response-inhibition by physostigmine may, in fact, induce a U-shaped function, thus further attesting to the involvement of nonassociative factors.

If nonassociative changes are indeed responsible for the time-dependent changes in avoidance, then this might well constitute an ideal preparation for the evaluation of pharmacological effects. In the absence of floor and ceiling effects, elimination of the U-shaped function would implicate nonassociative factors. In contrast, upward or downward adjustment of the

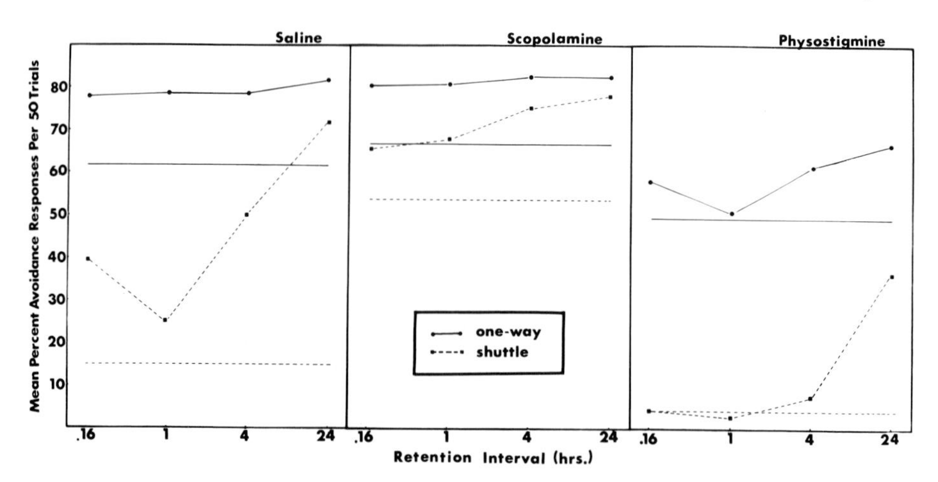

Fig. 20. One-way and shuttle avoidance at various intervals after exposure to ten signaled-inescapable shocks (1 mA). A U-shaped function was observed in the shuttle task, but not in the one-way task. Scopolamine hydrobromide (1.0 mg/kg) eliminated the U-shaped function. In the one-way task, physostigmine (0.5 mg/kg) promoted the U-shaped function. Horizontal dashed lines represent non-preshock controls. From Anisman (1973b); reprinted with permission; © American Psychological Association.

curve, without elimination of the U-shaped function, would suggest associative involvement.

6.4.5. *Drug Effects on Acquisition vs. Retention (Performance)*

It is important to differentiate between the effectiveness of a drug in modifying initial acquisition of an avoidance response and retention or retrieval of the response. If a drug improves acquisition, then this does not necessarily imply that comparable effects will be seen in the retention of the response (Singh *et al.*, 1974). During the acquisition phase, there may exist incompatible nonassociative responses that reduce performance. Once the response is established, these nonassociative effects should be minimal. Accordingly, even if the asymptomatic performance is not relatively high, performance in the retention phase should not be affected by drugs that modify nonassociative processes. Alternatively, drugs that affect retrieval need have little effect on acquisition, but should modify retention.

6.5. *Summary*

Avoidance behavior involves several components, e.g., associative, nonassociative, and motivational factors, as well as variations in the way an organism is prepared to deal with certain stimuli. Inasmuch as pharmacological treatments may affect performance by variation of any one of these factors, behavioral pharmacology needs to divorce or dissociate the various effects of drug treatments. In effect, techniques need to be used that may uncover the degree of variation attributable to changes in the subcomponents of avoidance. Basically, five methods are available that can be used to this end.

1. The use of tasks in which changes in one component will enhance performance in one task, but disrupt performance in the second task; e.g., reduction of freezing enhances active avoidance, but disrupts passive avoidance.
2. The concurrent use of tasks in which behavior in one task is modifiable by associative and nonassociative factors, while in the other task behavior is modifiable only by associative manipulations.
3. Tasks in which a discriminative component is included, such that a single manipulation can be used to determine the source of changes in avoidance as well as discriminated performance.
4. Transfer-of-training techniques that include a discriminated component (reversal learning), as well as possible variations in nonassociative factors.
5. Techniques in which associative ability is held constant, but involvement of nonassociative factors differs, or vice versa (i.e., strain manipulations).

Each of the techniques described may be indicative of the role played by associative as opposed to nonassociative factors. Needless to say, however, when several of these techniques are used concurrently, the strength of a particular position is considerably enhanced.

7. *Overview*

It is clear that in the evaluation of simple behaviors, great care must be taken to delineate the effects of treatments on individual attributes that compose the behavior. A wide variety of nonassociative behaviors, including biases of the response and stimulus class, and motivational and attentional factors, may affect performance in the same way that true associative changes might. It is not at all uncommon for nonassociative changes to mimic associative changes. We have presented a number of methods that may well be appropriate in evaluating drug effects on behavior. It should be clear, though, that the use of these techniques is not restricted to pharmacological analysis, but may well be appropriate for evaluation of other types of treatments.

While this chapter has been divided into broad behavioral categories, the overlap among these behaviors is apparent. Habituation, for example, was indicated to affect both appetitively and aversively motivated behaviors. Similarly, there may be considerable overlap among mechanisms subserving appetitively and aversively motivated behaviors. This being the case, the combination of various types of behaviors in pharmacological analysis is useful not only in the heuristic sense, but also as an analytic device.

A relatively restricted set of behaviors has been presented, while other behaviors such as aggression, self-stimulation, and feeding–drinking have not been touched on. Similarly, various independent variables, and their potential for interaction with pharmacological treatments (e.g., housing condition, genetic factors), either have received cursory attention or have not been dealt with. The failure to consider these variables in more detail does not reflect on their significance, but rather is a function of space limitations. It is noteworthy, though, that some of the techniques described may well be applicable to the examination of these behaviors.

ACKNOWLEDGMENTS. This research was supported by Grant A9845 from the National Research Council of Canada to H.A., and by a National Research Council Postgraduate Scholarship to A.J.S.

References

Abeelen, J.H.F.v., 1974, Genotype and the cholinergic control of exploratory behavior in mice, in: *The Genetics of Behavior* (J.H.F.v. Abeelen, ed.), pp. 347–374, Elsevier, New York.

Abeelen, J.H.F.v., and Strijbosch, H., 1969, Genotype-dependent effects of scopolamine and eserine on exploratory behavior in mice, *Psychopharmacologia* **16:**81–88.

Abeelen, J.H.F.v., Smits, A.J.M., and Raaijmakers, W.G.M., 1970, Central location of a genotype-dependent cholinergic mechanism controlling exploratory behavior in mice, *Psychopharmacologia* **19:**324–328.

Abel, E.L., 1971, Marihuana and memory: Acquisition or retrieval, *Science* **173:**1038–1040.

Ahlenius, S., 1973, Inhibition of catecholamine synthesis and conditioned avoidance acquisition, *Pharmacol. Biochem. Behav.* **1:**347–350.

Alpern, H.P. and Marriott, J.G., 1972, A direct measure of short-term memory in mice utilizing a successive reversal learning set, *Behav. Biol.* **7:**723–732.

Alpern, H.P., and Marriott, J.G., 1973, Short-term memory: Facilitation and disruption with cholinergic agents, *Physiol. Behav.* **11:**571–575.

Ando, K., 1975, Profile of drug effects on temporally spaced responding in cats, *Pharmacol. Biochem. Behav.* **3:**833–841.

Anger, D., 1963, The role of temporal discrimination in the reinforcement of Sidman avoidance behavior, *J. Exp. Anal. Behav.* **6:**477–506.

Anisman, H., 1973*a*, Effects of pretraining compatible and incompatible responses on subsequent one-way and shuttle-avoidance performance in rats, *J. Comp. Physiol. Psychol.* **83:**95–104.

Anisman, H., 1973*b*, Cholinergic mechanisms and alterations in behavioral suppression as factors producing time dependent changes in avoidance performance, *J. Comp. Physiol. Psychol.* **83:**465–477.

Anisman, H., 1975*a*, Time dependent variations in aversively motivated behavior: Nonassociative effects of cholinergic and catecholaminergic activity, *Psychol Rev.* **82:**359–385.

Anisman, H., 1975*b*, Differential effects of scopolamine and *d*-amphetamine on avoidance: Strain interactions, *Pharmacol. Biochem. Behav.* **3:**809–812.

Anisman, H., 1975*c*, Effects of scopolamine and *d*-amphetamine on one-way shuttle and inhibitory avoidance: A diallel analysis in mice, *Pharmacol. Biochem. Behav.* **3:**1037–1042.

Anisman, H., 1975*d*, Task complexity as a factor in eliciting heterosis in mice, *J. Comp. Physiol. Psychol.* **89:**976–984.

Anisman, H., 1975*e*, Acquisition and reversal learning of an active avoidance response in three strains of mice, *Behav. Biol.* **14:**51–58.

Anisman, H., 1976, Effects of scopolamine and *d*-amphetamine on locomotor activity before and after shock: A diallel analysis in mice, *Psychopharmacologia* **48:**165–173.

Anisman, H., 1978, Neurochemical changes elicited by stress: Behavioral correlates, in: *Psychopharmacology of Aversively Motivated Behavior* (H. Anisman and G. Bignami, eds.), pp. 1–62, Plenum Press, New York.

Anisman, H., and Cygan, D., 1975, Central effects of scopolamine and *d*-amphetamine on locomotor activity: Interaction with strain and stress variables, *Neuropharmacology* **14:**835–840.

Anisman, H., and Kokkinidis, L., 1975, Effects of scopolamine, *d*-amphetamine and other drugs affecting catecholamines on spontaneous alternation and locomotor activity in mice, *Psychopharmacologia* **45:**58–63.

Anisman, H., and Wahlsten, D., 1974, Response initiation and directionality as factors influencing avoidance performance, *J. Comp. Physiol. Psychol.* **87:**1119–1128.

Anisman, H., and Waller, T.G., 1972, Facilitative and disruptive effect of prior exposure to shock on subsequent avoidance performance, *J. Comp. Physiol. Psychol.* **78:**113–122.

Anisman, H., and Waller, T.G., 1973, Effects of inescapable shock on subsequent avoidance performance: Role of response repertoire changes, *Behav. Biol.* **9:**331–355.

Anisman, H., Wahlsten, D., and Kokkinidis, L., 1975, Effects of *d*-amphetamine and scopolamine on activity before and after shock in three mouse strains, *Pharmacol. Biochem. Behav.* **3:**819–824.

Anisman, H., Kokkinidis, L., Glazier, S., and Remington, G., 1976, Differentiation of response biases elicited by scopolamine and *d*-amphetamine: Effects on habituation, *Behav. Biol.* **18:**401–417.

Axelrod, J., 1970, Amphetamine: Metabolism, phsyiological disposition and its effects on catecholamine storage, in: *Amphetamines and Related Compounds* (E. Costa and S. Garattini, eds.), pp. 207–216, Raven Press, New York.

Baddeley, A.D., 1972, Retrieval rules and semantic coding in short-term memory, *Psychol. Bull.* **78:**379–385.

Bainbridge, J.G., 1968, The effect of psychotropic drugs on food reinforced behavior and on food consumption, *Psychopharmacologia* **12:**204–213.

Balster, R.L., and Schuster, C.R., 1973, Fixed-interval schedule of cocaine reinforcement: Effect of dose and infusion duration, *J. Exp. Anal. Behav.* **20:**119–129.

Barrett, J.E., 1974, Conjunctive schedules of reinforcement. I. Rate-dependent effects of pentobarbital and *d*-amphetamine, *J. Exp. Anal. Behav.* **22:**561–573.

Barrett, R.J., Leith N.J., and Ray, O.S., 1971, Kamin effect on rats: Index of memory or shock-induced inhibition, *J. Comp. Physiol. Psychol.* **71:**234–239.

Barrett, R.J., Leith, N.J., and Ray, O.S., 1972, Permanent facilitation of avoidance behavior by *d*-amphetamine and scopolamine, *Psychopharmacologia* **25:**321–331.

Barrett, R.J., Leith, N.J., and Ray, O.S., 1973, A behavioral and pharmacological analysis of the variables mediating active avoidance behavior in rats, *J. Comp. Physiol. Psychol.* **82:**489–500.

Barrett, R.J., Leith, N.J., and Ray, O.S., 1974, An analysis of the facilitation of avoidance acquisition produced by *d*-amphetamine and scopolamine, *Behav. Biol.* **11:**189–203.

Barry, H., 1978, Stimulus attributes of drugs, in: *Psychopharmacology of Adversely Motivated Behavior* (H. Anisman and G. Bignami, eds.), pp. 455–485, Plenum Press, New York.

Barry, H, and Buckley, J.P., 1966, Drug effect on animal performance and the stress syndrome, *J. Pharm. Sci.* **55:**1159–1183.

Bartholini, G., Blum, J.E., and Pletcher, A., 1969, Dopa-induced locomotor stimulation after inhibition of extracerebral decarboxylase, *J. Pharm. Pharmacol.* **21:**297–301.

Bartholini, G., Stadler, H., and Lloyd, K.G., 1973, Cholinergic–dopaminergic relations in different brain structures, in: *Frontiers in Catecholamine Research* (E. Usdin and S.H. Snyder, eds.), pp. 741–746, Pergamon Press, London.

Beecher, H.K., 1957, The measurement of pain: Prototype for the quantitative study of subject responses, *Pharmacol. Rev.* **9:**59–209.

Bignami, G, 1966, Anticholinergic agents as tools in the investigation of behavioral phenomena, *Excerpta Medica Int. Congr. Ser.* **129:**819–830.

Bignami, G., and Michalek, H., 1976, Nonassociative explanations of behavioral changes induced by central cholinergic drugs, *Acta Neurobiol. Exp.* **36:**5–90.

Bignami, G., and Michalek, H. 1978, Cholinergic mechanisms in aversively motivated behavior, in: *Psychopharmacology of Aversively Motivated Behavior* (H. Anisman and G. Bignami, eds.), pp. 173–255, Plenum Press, New York.

Bignami, G., and Gatti, G.L., 1969, Analysis of drug effects on multiple fixed ratio 33 fixed interval 5 min. in pigeons, *Psychopharmacologia* **15:**310–332.

Bignami, G., and Rosic, N., 1971, The nature of disinhibitory phenomena caused by central cholinergic (muscarinic) blockade, in: *Advances in Neuropharmacology* (O. Vinar, Z. Votava, and P.B. Bradley, eds.), pp. 481–495, North-Holland, Amsterdam.

Bignami, G., Rosic, N., Michalek, M., Milosevic, M., and Gatti, G.L., 1975, Behavioral toxicity of anticholinesterase agents: Methodological, neurochemical and neuropsychological aspects, in: *Behavioral Toxicology* (B. Weiss, and V.G. Latries, eds.), pp. 155–215, Plenum Press, New York.

Blackman, D., 1968, Response rate, reinforcement frequency and conditioned suppression, *J. Exp. Anal. Behav.* **11:**503–516.

Blackman, D., 1972, Conditioned anxiety and operant behavior, in: *Schedule Effects: Drugs, Drinking and Aggression* (R.M. Gilbert and J.D. Keehn, eds.), pp. 26–49, University of Toronto Press.

Blough, D.S., 1969, Attention shifts in a maintained discrimination, *Science* **166:**125–126.

Bolles, R.C., 1970, Species specific defense reactions and avoidance learning, *Psychol. Rev.* **77:**32–48.

Bolles, R.C., 1971, Species-specific defense reactions, in: *Aversive Conditioning and Learning* (F.R. Brush, ed.), pp. 183–233, Academic Press, New York.

Boren, J.J., 1966, The study of drugs with operant techniques, in: *Operant Behavior: Areas of Research and Application* (W.K. Honig, ed.), pp. 531–564, Appleton-Century-Crofts, New York.

Boren, J.J., and Devine, D.D., 1968, The repeated acquisition of behavioral chains, *J. Exp. Anal. Behav.* **11:**651–660.

Branch, M.N., 1975, Effects of chlorpromazine and amphetamine on observing responses during a fixed-interval schedule, *Psychopharmacologia* **42:**87–93.

Branch, M.N., and Gollub, L.R., 1974, A detailed analysis of the effects of *d*-amphetamine on behavior under fixed-interval schedules, *J. Exp. Anal. Behav.* **21:**519–539.

Breese, G.R., Cooper, B.R., and Hollister, A.S., 1975, Involvement of brain monoamines on the stimulant and paradoxical inhibitory effects of methylphenidate, *Psychopharmacologia* **44:**5–10.

Broadhurst, P.L., 1976, Pharmacogenetics, in: *Handbook of Psychopharmacology*, Vol. 7: *Behavioral Pharmacology in Animals* (L.L. Iversen, S.D. Iversen, and S.N. Snyder, eds.), pp. 265–320, Plenum Press, New York.

Brodie, B.B., Costa, E., Dlabac, A., Neff. H.H., and Smookler, H.H., 1966, Application of steady state kinetics to the estimation of synthesis rate and turnover time of catecholamines, *J. Pharm. Exp. Ther.* **154:**493.

Brodie, B.B., Cho, A.K., and Gessa, G.L., 1970, Possible role of *p*-hydroxynorephedrine in the depletion of norepinephrine induced by *d*-amphetamine and in tolerance to this drug, in: *Amphetamines and Related Compounds* (E. Costa and S. Garattini, eds.), pp. 217–230, Raven Press, New York.

Brody, J.F., 1970, Behavioral effects of serotonin depletion and *p*-chlorophenylalanine (a serotonin depletor) in rats, *Psychopharmacologia* **17:**14–33.

Brown, K., and Warburton, D.M., 1971, Attenuation of stimulus sensitivity by scopolamine, *Psychon. Sci.* **22:**297–298.

Bruner, J., and Tauc, L., 1966, Habituation at the synaptic level in Aplysia, *Nature (London)* **210:**37–39.

Brush, F.R., 1971, Retention of aversively motivated behavior, in: *Aversive Conditioning and Learning* (F.R. Brush, ed.), pp. 401–465, Academic Press, New York.

Bunney, B.S., Walters, J.R., Kuhar, M.J., Roth, R.H., and Aghajanian, G.K., 1975, D- and L-amphetamine stereoisomers: Comparative potencies in affecting the firing of central dopaminergic and noradrenergic neurons, *Psychopharm. Commun.* **1:**177–190.

Butcher, L.L., Engel, J., and Fuxe, K., 1970, L-Dopa-induced changes in central monoamine neurons after peripheral decarboxylase inhibition, *J. Pharm. Pharmacol.* **22:**313–316.

Byrd, L.D., 1973, Effects of *d*-amphetamine on schedule-controlled key pressing and drinking in the chimpanzee, *J. Pharmacol. Exp. Ther.* **185:**633–641.

Byrd, L.D., 1974, Modification of the effects of chlorpromazine on behavior in the chimpanzee, *J. Pharmacol. Exp. Ther.* **189:**24–32.

Callaway, E., III, 1973, Habituation of average evoked potentials in man, in: *Habituation* (H.V.S. Peeke, and M.J. Herz, eds.), pp. 153–174, Academic Press, London.

Campbell, B., and Church, R., 1969, *Punishment and Aversive Behavior*, Appleton-Century-Crofts, New York.

Campbell, B.A., and Mabry, P.D., 1973, The role of catecholamines in behavioral arousal during ontogenesis, *Psychopharmacologia* **31:**253–264.

Campbell, B.A., Lytle, L.D., and Fibiger, H.C., 1969, Ontogeny of adrenergic arousal and cholinergic inhibitory mechanisms in the rat, *Science* **166:**635–637.

Carey, R.S., and Kritkausky, R.P., 1972, Absence of a response-rate-dependent effect of *d*-amphetamine on a DRL schedule when reinforcement is signalled, *Psychon. Sci.* **26:**285–286.

Carlsson, A., 1970, Amphetamines and brain catecholamines, in: *Amphetamines and Related Compounds* (E. Costa, and S. Garattini, eds.), pp. 289–300, Raven Press, New York.

Carlton, P.L., 1961*a*, Augmentation of the behavioral effects of amphetamine by atropine, *J. Pharmacol. Exp. Ther.* **132:**91–96.

Carlton, P.L., 1961*b*, Augmentation of the behavioral effects of amphetamine by scopolamine, *Psychopharmacologia* **2:**377–380.

Carlton, P.L., 1963, Cholinergic mechanisms in the control of behavior by the brain, *Psychol Rev.* **70:**19–39.

Carlton, P., 1966, Scopolamine, amphetamine and light-reinforced responding, *Psychon. Sci.* **5:**347–348.

Carlton, P.L., 1968, Brain acetylcholine and habituation, *Prog. Brain Res.* **28:**48–60.

Carlton, P.L., 1969, Brain-acetylcholine and inhibition, in: *Reinforcement and Behavior* (J.T. Tapp, ed.), pp. 286–327, Academic Press, New York.

Carlton, P.L., and Advokat, C., 1973, Attenuated habituation due to parachlorophenylalanine, *Pharmacol. Biochem. Behav.* **1:**657–663.

Carlton, P.L., and Markiewicz, B., 1971, Behavioral effects of atropine and scopolamine, in: *Pharmacological and Biophysical Agents and Behavior* (E. Furchtgott, ed.), pp. 346–371, Academic Press, New York.

Carpenter, J.A., and Ross, B.M., 1965, Effect of alcohol on short-term memory, *Q. J. Stud. Alcohol* **26:**561.

Carro-Ciampi, G., and Bignami, G., 1968, Effects of scopolamine on shuttle-box avoidance and go–no go discrimination: Response–stimulus relationships, pretreatment baselines, and repeated exposure to drug, *Psychopharmacologia* **13:**89–105.

Catania, A.C., 1963, Concurrent performances: A baseline for the study of reinforcement magnitude, *J. Exp. Anal. Behav.* **6:**299–300.

Catania, A.C., 1973, The nature of learning, in: *The Study of Behavior: Learning Motivation, Emotion and Instinct* (J.A. Nevin, ed.), pp. 30–68, Scott, Foresman, Glenview, Ill.

Clark, F.C., 1962, Some observations on the adventitious reinforcement of drinking under food reinforcement, *J. Exp. Anal. Behav.* **5:**61–63.

Clark, R.B., 1965, The learning abilities of nereid polychaetes and the role of the supra-esophageal ganglion, *Anim. Behav. Suppl.*, pp. 89–100.

Cook, L., and Catania, A.C., 1964, Effects of drugs on avoidance and escape behavior, *Fed. Proc. Fed. Am. Soc. Exp. Biol.* **23:**818–835.

Cook, L., and Davidson, A.B., 1968, Effects of yeast RNA and other pharmacological agents on acquisition, retention and performance in animals, in: *Psychopharmacology: A Review of Progress 1957–1967* (D.H. Efron, ed.), PHS Publication No. 1836, U.S. Government Printing Office, Washington, D.C.

Cooper, J.R., Bloom, F.E., and Roth, R.H., 1974, *The Biochemical Basis of Neuropharmacology*, Oxford University Press, New York.

Corrodi, H., Fuxe, K., Hamberger, B., and Ljungdahl, A., 1970, Studies on central and peripheral noradrenaline neurons using a new dopamine-β-hydroxylase inhibitor, *Eur. J. Pharmacol.* **12:**145–155.

Costall, B., and Naylor, R.J., 1972, Possible involvement of noradrenergic area of the amygdala with stereotyped behavior, *Life Sci.* **11:**1135–1146.

Costall, B., and Naylor, R.J., 1974, The importance of the ascending dopaminergic systems to the extrapyramidal and mesolimbic brain areas for the cataleptic action of the neuroleptic and cholinergic agents, *Neuropharmacology* **13:**353–364.

Costall, B., and Olley, J.E., 1971, Cholinergic and neuroleptic induced catalepsy: Modification by lesions in the globus pallidus and substantia nigra, *Neuropharmacology* **10:**581–594.

Cox, T., and Tye, N., 1973, Effects of physostigmine on the acquisition of a position discrimination in rats, *Neuropharmacology* **12:**477–484.

Cox, T., and Tye, N., 1974, Effects of physostigmine on the maintenance of discrimination behavior in rats, *Neuropharmacology* **13:**205–210.

Craik, F.I.M., and Lockhart, R.S., 1972, Levels of processing: A framework for memory research, *J. Verb. Learn. Verb. Behav.* **11:**671–684.

Creese, I., and Iversen, S.D., 1973, Blockage of amphetamine induced motor stimulation and stereotypy in the adult rat following neonatal treatment with 6-hydroxydopamine, *Brain Res.* **55:**369–382.

Creese, I., and Iversen, S.D., 1975*a*, The pharmacological and anatomical substrates of the amphetamine response in the rat, *Brain Res.* **83:**419–436.

Creese, I., and Iversen, S.D., 1975*b*, Behavioral sequelae of dopaminergic degeneration: Post synaptic supersensitivity?, in: *Pre- and Postsynaptic Receptors* (E. Usdin and W.E. Bunney, eds.), pp. 171–190, Marcel Dekker, New York.

D'Amato, M.R., 1973, Delayed matching and short-term memory in monkeys, in: *The Psychology of Learning and Motivation: Advances in Research and Theory*, Vol. 7, Academic Press, New York.

Davis, H., 1968, Conditioned suppression: A survey of the literature, *Psychon. Monogr. Suppl.* **2:**283–291.

Davis, H., and Kreuter, C., 1972, Conditioned suppression of an avoidance response by a stimulus paired with food, *J. Exp. Anal. Behav.* **17:**277–285.

Davis, M., and Sheard, M.H., 1974, Habituation and sensitization of the rat startle response: Effects of raphe lesions, *Physiol. Behav.* **12:**425–433.

deFeudis, F.V., 1974, *Central Cholinergic Systems and Behavior*, Academic Press, New York.

Delius, J.D., 1968, Color preference shift in hungry and thirsty pigeons, *Psychon. Sci.* **13:**273–274.

De Toledo, L, and Black, A.H., 1967, Effects of preshock on subsequent avoidance conditioning, *J. Comp. Physiol. Psychol.* **63:**493–499.

De Toledo, L., and Black, A.H., 1970, Retention of aversively motivated responses in rats, *J. Comp. Physiol. Psychol.* **71:**276–282.

Deutsch, J.A., 1971, The cholinergic synapse and the site of memory, *Science* **174:**788–794.

Dews, P.B., 1955, Studies on behavior. II. The effects of pentobarbital, methamphetamine and scopolamine on performances in pigeons involving discriminations, *J. Pharmacol. Exp. Ther.* **115:**380–389.

Dews, P.B., 1958, Studies on behavior. IV. Stimulant actions of methamphetamine, *J. Pharmacol. Exp. Ther.* **122:**137–147.

Dews, P.B., 1962, A behavioral output enhancing effect of impramine in pigeons, *Int. J. Neuropharmacol.* **1:**265–272.

Dews, P.B., 1964, A behavioral effect of amobarbital, *Arch. Exp. Pathol. Pharmacol.* **248:**296–307.

Dews, P.B., 1970, Drugs in psychology: A commentary on Travis Thompson and Charles R. Schuster's *Behavioral Pharmacology*, *J. Exp. Anal. Behav.* **13:**395–406.

Dews, P.B., 1971, Drug–behavior interactions, in: *Behavioral Analysis of Drug Action* (J.A. Harvey, ed.), pp. 9–43, Scott, Foresman, Glenview, Illinois.

Diamond, S., Balvin, R.S., and Diamond, F.R., 1963, *Inhibition and Choice*, Harper and Row, New York.
Dinsmoor, J.A., 1954, Punishment. I. The avoidance hypothesis, *Psychol. Rev.* **61:**34–46.
Doteuchi, M., Wang, C., and Costa, E., 1974, Compartmentation of dopamine in rat striatum, *Mol. Pharmacol.* **10:**225–234.
Dougherty, J., and Pickens, R., 1973, Fixed-interval schedules of intravenous cocaine presentation in rats, *J. Exp. Anal. Behav.* **20:**111–118.
Douglas, R.J., 1966, Cues for spontaneous alternation, *J. Comp. Physiol. Psychol.* **62:**171–183.
Douglas, R.J., and Isaacson, R.L., 1965, Homogeneity of single trial response tendencies and spontaneous alternation in the T-maze, *Psychol Rep.* **16:**87–92.
Drew, W.G., Miller, L.L., and Baugh, E.L., 1973, Effects of Δ^9-THC, LSD-25 and scopolamine on continuous, spontaneous alternation in the Y-maze, *Psychopharmacologia* **32:**171–182.
Dudar, J.D., 1975, The effect of septal nuclei stimulation on the release of acetylcholine from the rabbit hippocampus, *Brain Res.* **83:**123–133.
Egger, G.J., Livesey, P.J., and Dawson, R.G., 1973, Ontogenetic aspects of central cholinergic involvement in spontaneous alternation behavior, *Dev. Psychobiol.* **6:**289–299.
Eisenstein, E.M., 1967, The use of invertebrate systems for the study of learning and memory, in: *The Neurosciences* (G.C. Quarton, T. Melnechuk, and F.O. Schmidt, eds.), pp. 653–665, Rockefeller University Press, New York.
Eisenstein, E.M., and Peretz, B., 1975, Comparative aspects of habituation in invertebrates, in: *Habituation* (H.V.S. Peeke and M.J. Herz, eds.), pp. 1–34, Academic Press, New York.
Falk, J.L., 1961, Production of polydipsia in normal rats by intermittent food schedule, *Science* **133:**195–196.
Falk, J.L., 1966, Schedule-induced polydipsia as a function of fixed-interval length, *J. Exp. Anal. Behav.* **9:**37–39.
Falk, J.L., 1972, The nature and determinants of adjunctive behavior, in: *Schedule Effects: Drugs, Drinking and Aggression* (R.M. Gilbert and J.D. Keehn, eds.), pp. 148–173, University of Toronto Press.
Fechter, L.D., 1974, Central serotonin involvement in the elaboration of the startle reaction in rats, *Pharmacol. Biochem. Behav.* **2:**161–173.
Fibiger, H.C., and Grewaal, D.S., 1974, Neurochemical evidence for denervation supersensitivity: The effect of unilateral substantial nigra lesions on apomorphine-induced increases in neostriatal acelylcholine levels, *Life Sci.* **15:**57–63.
Fibiger, H.C., Phillips, A.G., and Zis, P.S., 1974, Deficits in instrumental responding after 6-hydroxydopamine lesions of the nigro-neostriatal dopominergic projection, *Pharmacol. Biochem. Behav.* **2:**87–96.
Fibiger, H.C., Zis, A.P., and Phillips, G., 1975, Haloperidol-induced disruption of conditioned avoidance responding: Attenuation by prior training or by anticholinergic drugs, *Eur. J. Pharmacol.* **30:**309–314.
File, S.E., 1976, Are central cholinergic pathways involved in the habituation of exploration and distraction?, *Br. J. Pharmacol.* **56:**378–379.
File, S.E., and Plotkin, H.C., 1974, Habituation in the neonatal rat, *Dev. Psychobiol.* **7:**121–127.
File, S.E., and Wardell, A.G., 1975, Validity of head-dipping as a measure of exploration in a modified hole board, *Psychopharmacologia* **44:**53–59.
Fingl, E., and Woodbury, D.M., 1970, General principles, in: *The Pharmacological Basis of Therapeutics* (L.S. Goodman, and A. Gilman, eds.), pp. 1–35, Collier-Macmillan, Toronto.
Fletcher, H.J., 1965, The delayed response problem, in: *Behavior of Nonhuman Primates Modern Research Trends, I.* (A.M. Schier, H.F. Harlow, and F. Stollnitz, eds.), pp. 129–165, Academic Press, New York.
Franklin, K.B.J., and Quartermain, D., 1970, Comparison of the motivational properties of deprivation-induced drinking elicited by central carbachol stimulation, *J. Comp. Physiol. Psychol.* **71:**390–395.
Fried, P.A., 1972, Septum and behavior: A review, *Psychol. Bull.* **78:**292–310.
Frontalli, M., and Bignami, G., 1973, Go–no go avoidance discrimination in rats with simple

"go" compound and "no go" signals: Stimulus modality and stimulus intensity, *Anim. Learn. Behav.* **1:**21–24.

Frontali, M., Amorico, L., deAcetis, L., and Bignami, G., 1977, A pharmacological analysis of processes underlying differential responding: A review and further experiments with scopolamine, amphetamine, LSD-25, chlordiazepoxide, physostigmine, and chlorpromazine, *Behav. Biol.* **18:**1–74.

Geller, I., 1962, Use of approach avoidance behavior (conflict) for evaluating depressant drugs, in: *The First Hahnemann Symposium on Psychosomatic Medicine*, pp. 267–274, Lea and Febiger, New York.

Geller, I., and Seifter, J., 1960, The effects of meprobamate, barbiturates, *d*-amphetamine and promazine on experimentally induced conflict in the rat, *Psychopharmacologia* **1:**482–492.

Gilbert, R.M., and Sutherland, N.S., 1969, *Animal Discrimination Learning*, Academic Press, London.

Glanzer, M., 1953, Stimulus satiation: An explanation of spontaneous alternation and related phenomena, *Psychol. Rev.* **60:**257–268.

Glanzer, M., 1972, Storage mechanisms in recall, in: *The Psychology of Learning and Motivation: Advances in Research and Theory*, Vol. 5 (J.T. Spence and K.W. Spence, eds.), pp. 129–193, Academic Press, New York.

Glassman, E., 1969, The biochemistry of learning: An evaluation of the role of RNA and protein, *Annu. Rev. Chem.* **38:**605–646.

Glick, S.D., 1973, Enhancement of spatial preferences by (+)-amphetamine, *Neuropharmacology* **12:**43–47.

Glick, S.D., Levin, B., and Jarvik, M.E., 1970, Role of monkeys' spatial preferences in performance of a nonspatial task, *J. Comp. Physiol. Psychol.* **73:**56–61.

Glick, S.D., Jerussi, T.P., Waters, D.H., and Green, J.P., 1974, Amphetamine-induced changes in rat striatal dopamine and acetylcholine levels and relationship to rotation (circling behavior) in rats, *Biochem. Pharmacol.* **23:**3223–3225.

Glowinski, J., Kopin, I.J., and Axelrod, J., 1965, Metabolism of [^{3}H]norepinephrine in rat brain, *J. Neurochem.* **12:**25–30.

Goldberg, S.R., and Kelleher, R.T., 1976, Behavior controlled by scheduled injections of cocaine in squirrel and rhesus monkeys, *J. Exp. Anal. Behav.* **25:**93–104.

Goldstein, A., Aranow, L., and Kalman, S.M., 1969, *Principles of Drug Action*, Harper & Row, New York.

Goodwin, D.W., Othmer, E., Halikas, J.A., and Freeman, F., 1970, Short-term memory loss: Predictor of the alcoholic "blackout," *Nature (London)* **227:**201–202.

Grant, D.S., 1975, Proactive inference in pigeons short-term memory, *J. Exp. Psychol. Anim. Behav.* Process. **104:**207–220.

Grant, D.S., and Roberts, W.A., 1973, Trace interaction in pigeon short-term memory, *J. Exp. Psychol.* **101:**21–29.

Griffiths, D., and Wahlsten, D., 1974, Interacting effects of handling and *d*-amphetamine on avoidance learning, *Pharmacol. Biochem. Behav.* **2:**439–441.

Groves, P.M., and Lynch, G.S., 1972, Mechanisms of habituation in the brain stem, *Psychol. Rev.* **79:**237–244.

Groves, P.M., and Rebec, G.V., 1976, Biochemistry and behavior: Some central actions of amphetamine and antipsychotic drugs, *Annu. Rev. Psychol.* **27**(88):91–127.

Groves, P.M., and Thompson, R., 1970, Habituation: A dual process theory, *Psychol. Rev.* **77:**419–450.

Groves, P.M., and Thompson, R.F., 1973, A dual-process theory of habituation: Neural mechanisms, in: *Habituation, Physiological Substrates* (H.V.S. Peeke and M.J. Herz, eds.), pp. 175–213, Academic Press, New York and London.

Groves, P.M., Wilson, C.J., and Boyle, R.D., 1974, Brain stem pathways, cortical modulation, and habituation of the acoustic startle response, *Behav. Biol.* **10:**391–418.

Hamburg, D.A., Hamburg, B.A., and Barchas, J.D., 1975, Anger and depression in perspective of behavioral biology, in: *Emotions: Their Parameters and Measurement* (L. Levi, ed.), pp. 235–278, Raven Press, New York.

Harvey, J.A. (ed.), 1971, *Behavioral Analysis of Drug Action*, Scott, Foresman, Glenview, Illinois.
Hawkins, T.D., Schrot, J.F., Githens, S.U., and Everett, P.B., 1972, Schedule-induced polydipsia: An analysis of water and alcohol ingestion, in: *Schedule Effects: Drugs, Drinking and Aggression* (R.M. Gilbert and J.D. Keehn, eds.), pp. 95–128, University of Toronto Press.
Hearst, E., 1972, Some persistent problems in the analysis of conditioned inhibition, in: *Inhibition and Learning* (R.A. Boakes and M.S. Halliday, eds.), pp. 5–39, Academic Press, London.
Hearst, E., and Vane, J.R., 1967, Some effects of *d*-amphetamine on the behavior of pigeons under intermittent reinforcement, *Psychopharmacologia* **12**:58–67.
Hearst, E., Besley, S., and Farthing, G.W., 1970. Inhibition and the stimulus control of operant behavior, *J. Exp. Anal. Behav.* **14**:373–409.
Heise, G.A., and Conner, R., 1972, Variable ITI delayed spatial alternation in rats: Acquisition and drug effects, presented at the Behavioral Pharmacology Society Meeting, Chapel Hill, North Carolina.
Heise, G.A., and Lilie, N.L., 1970, Effects of scopolamine, atropine and *d*-amphetamine on internal and external control of responding on nonreinforced trials, *Psychopharmacologia* **18**:38–39.
Henton, W.W., and Brady, J.V., 1970, Operant acceleration during a pre-reward stimulus, *J. Exp. Anal. Behav.* **13**:205–209.
Herman, L.M., 1975, Interference and auditory short-term memory in the bottlenosed dolphin, *Anim. Learn. Behav.* **3**:43–48.
Herrnstein, R.J., 1969, Method and theory in the study of avoidance, *Psychol. Rev.* **76**:49–69.
Herrnstein, R.J., 1970, On the law of effect, *J. Exp. Anal. Behav.* **13**:243–266.
Hill, R.M., 1973, Characteristics of habituation by mammalian visual pathway units, in: *Habituation* (H.V.S. Peeke and M.J. Herz, eds.), pp. 139–152, Academic Press, London.
Hollard, V., and Davison, M.C., 1971, Preference for qualitatively different reinforcers, *J. Exp. Anal. Behav.* **16**:375–380.
Honig, W.K., and James, P.H.R., 1971, *Animal Memory*, Academic Press, New York.
Houser, V.P., and Van Hart, D.A., 1973, Changes in aversive threshold of the rat produced by adrenergic drugs, *Pharmacol. Biochem. Behav.* **1**:673–678.
Houser, V.P., and Van Hart, B.A., 1974, Modulation of cholinergic activity and the aversive threshold in the rat, *Pharmacol. Biochem. Behav.* **2**:631.
Iglauer, C., and Woods, J.H., 1974, Concurrent performances: Reinforcement by different doses of intravenous cocaine in rhesus monkeys, *J. Exp. Anal. Behav.* **22**:179–196.
Isaacson, R.L., 1974, *The Limbic System*, Plenum Press, New York.
Iversen, L.L., and Glowinski, J., 1966, Regional studies of catecholamines in the rat brain. II, *J. Neurochem.* **13**:671–682.
Iversen, S.D., and Iversen, L.L., 1975, *Behavioral Pharmacology*, Oxford University Press, New York.
Izquierdo, I., 1974, Possible peripheral adrenergic and cholinergic mechanisms in pseudoconditioning, *Psychopharmacologia* **35**:189–193.
Janowsky, D.S., El-Yousef, M.K., Davis, J.M., and Sekerke, H.J., 1972, Cholinergic antagonism of methylphenidate-induced stereotyped behavior, *Psychopharmacologia* **27**:295–303.
Jarrard, L.E., and Moise, S.L., 1971, Short-term memory in the monkey, in: *Cognitive Processes of Nonhuman Primates*, (L.E. Gerard, ed.), pp. 1–24, Academic Press, New York.
Jarvik, M.E., 1969, Effects of drugs on memory, in: *Drugs and the Brain* (P. Black, ed.), pp. 135–146, Johns Hopkins University Press, Baltimore.
Jarvik, M.E., 1972, Effects of chemical and physical treatments on learning and memory, *Annu. Rev. Psychol.* **23**:457–486.
Javoy, F., Thierry, A.M., Kety, S.S., and Glowinski, J., 1968, The effect of amphetamine on the turnover of brain norepinephrine in normal and stressed rats, *Commun. Behav. Biol.* **1A**:43–48.
Javoy, F., Agid, Y., Bouvet, D., and Glowinski, J., 1974, Changes in neostriatal DA metabolism after carbachol or atropine microinjections into the substantia nigra, *Brain Res.* **68**:253–260.
Jenkins, H.M., 1965, Measurement of stimulus control during discriminative operant conditioning, *Psychol. Bull.* **64**:365–376.

Jenkins, H.M., 1970, Sequential organization in schedules of reinforcement, in: *The Theory of Reinforcement Schedules* (W.N. Schoenfeld, ed.), pp. 63–109, Appleton-Century-Crofts, New York.

Jerussi, T.P., and Glick, S.D., 1975, Apomorphine-induced rotation in normal rats and interaction with unilateral caudate lesion, *Psychopharmacologia* **40:**329–334.

Kamin, L.J., 1963, Retention of an incompletely learned avoidance response: Some further analyses, *J. Comp. Physiol. Psychol.* **56:**713–718.

Karczmar, A.G., 1975, Cholinergic influences on behavior, in: *Cholinergic Mechanisms* (P.G. Waser, ed.), pp. 501–530, Raven Press, New York.

Keehn, J.D., 1972, Schedule-dependence, schedule-induction and the law of effect, in: *Schedule Effects: Drugs, Drinking and Aggression* (R.M. Gilbert and J.D. Keehn, eds.), pp. 65–94, University of Toronto Press.

Keehn, J.D., Coulson, G.E., and Kliebs, J., 1976, Effects of haloperidol on schedule-induced polydipsia, *J. Exp. Anal. Behav.* **25:**105–112.

Kelleher, R.T., and Morse, W.H., 1964, Escape behavior and punished behavior. *Fed. Proc. Fed. Am. Soc. Exp. Biol.* **23:**799–800.

Kelleher, R.T., and Morse, W.H., 1968, Determinants of the specificity of behavioral effects of drugs, *Ergeb. Physiol.* **60:**1–56.

Kendall, S.B., 1972, Some effects of response-dependent clock stimuli in a fixed-interval schedule, *J. Exp. Anal. Behav.* **14:**161–168.

Kenyon, J., and Krieckhaus, E.E., 1965, Enhanced avoidance behavior following septal lesions in the rat as a function of lesion size and spontaneous activity, *J. Comp. Physiol. Psychol.* **59:**466–468.

Kirkstone, B.J., and Levitt, R.A., 1970, Interactions between water deprivation and chemical brain stimulation, *J. Comp. Physiol. Psychol.* **2:**334–340.

Klein, S.B., 1972, Adrenal–pituitary influence in reactivation of avoidance learning memory in the rat after intermediate intervals, *J. Comp. Physiol. Psychol.* **79:**341–359.

Klein, S.B., and Spear, N.E., 1970*a*, Forgetting by the rat after intermediate intervals (Kamin effect) as retrieval failure, *J. Comp. Physiol. Psychol.* **71:**165–170.

Klein, S.B., and Spear, N.E., 1970*b*, Reactivation of avoidance learning memory in the rat after intermediate retention intervals, *J. Comp. Physiol. Psychol.* **74:**498–504.

Kokkinidis, L., and Anisman, H., 1976*a*, Interaction between cholinergic and catecholaminergic agents in a spontaneous alternation task, *Psychopharmacologia* **48:**261–270.

Kokkinidis, L, and Anisman, H., 1976*b*, Dissociation of the effects of scopolamine and *d*-amphetamine on spontaneous alternation: Genuine perseveration or dishabituation, *Pharmacol. Biochem. Behav.* **5:**293–297.

Kokkinidis, L., and Anisman, H., 1977, *d*-Amphetamine-induced perseveration in a Y-maze exploratory task: Differential effects of intraperitoneal and intraventricular administration, *Psychopharmacology* **52:**123–128.

Konorski, J., 1967, *Integrative Activity of the Brain*, University of Chicago Press.

Krasnegor, N.A., and Hodos, W., 1974, The evaluation and control of acoustical standing waves, *J. Exp. Anal. Behav.* **22:**243–249.

Krieckhaus, E.E., 1965, Decrements in avoidance behavior following mammilothalamic tractotomy in rats and subsequent recovery with *d*-amphetamine, *J. Comp. Physiol. Psychol.* **60:**31–35.

Ksir, C.J., Jr., 1974, Scopolamine effects on two-trial delayed-response performance in the rat, *Psychopharmacologia* **41:**127–134.

Ksir, C., 1975, Scopolamine and amphetamine effects on discrimination: Interaction with stimulus control, *Psychopharmacologia* **43:**37–41.

Laties, V.G., 1972, The modification of drug effects on behavior by external discriminative stimuli, *J. Pharmacol. Exp. Ther.* **183:**1–13.

Laties, V.G., 1975, The role of discriminative stimuli in modulating drug action, *Fed. Proc. Fed. Am. Soc. Exp. Biol.* **34:**1880–1888.

Laties, V.G., and Weiss, B., 1966, Influence of drugs on behavior controlled by internal and external stimuli, *J. Pharmacol. Exp. Ther.* **152:**388–396.

Laties, V.G., Weiss, B., Clark, R.L., and Reynolds, M.D., 1965, Overt "mediating" behavior during temporally spaced responding, *J. Exp. Anal. Behav.* **8:**107–116.

Leaton, R.N., 1968*a*, Effects of scopolamine on exploratory motivated behavior, *J. Comp. Physiol. Psychol.* **66:**524–527.

Leaton, R.N., 1968*b*, Effects of scopolamine and eserine on position discrimination learning with an exploratory incentive, *Psychon. Sci.* **12:**181–182.

Leaton, R.N., and Kreindler, M., 1972, Effects of physostigmine and scopolamine on operant brightness discrimination in the rat, *Physiol. Behav.* **9:**121–123.

Leith, C.R., and Maki, W.S., Jr., 1975, Attention shifts during matching-to-sample performance in pigeons, *Anim. Learn. Behav.* **3:**85–89.

Lints, C.E., and Harvey, J.A., 1969, Altered sensitivity to footshock and decreased content of serotonin following brain lesions in the rat, *J. Comp. Physiol. Psychol.* **67:**23–32.

Llewellyn N.E., Iglauer, C., and Woods, J.H., 1976, Relative reinforcer magnitude under a nonindependent concurrent schedule of cocaine reinforcement in rhesus monkeys, *J. Exp. Anal. Behav.* **25:**81–91.

Longo, V.G., 1966, Behavioral and electroencephalographic effects of atropine and related compounds, *Pharmacol. Rev.* **18:**965–996.

Lubar, J.F., and Perachio, A.A., 1965, One-way and two-way learning and transfer of an active avoidance response in normal and cingulectomized cats, *J. Comp. Physiol. Psychol.* **60:**46–52.

Lubow, R.E., 1973, Latent inhibition, *Psychol. Bull.* **79:**398–407.

Lyon, M., and Randrup, A., 1972, The dose–response effect of amphetamine upon avoidance behavior in the rat seen as a function of increasing stereotypy, *Psychopharmacologia* **23:**334–347.

Mackintosh, N.J., 1974, *The Psychology of Animal Learning*, Academic Press, London.

Mackintosh, N.J., 1975, A theory of attention: Variations in the associability of stimuli with reinforcement, *Psychol. Rev.* **82:**276–298.

Maki, W.S., Jr., and Leith, C.R., 1973, Shared attention in pigeons, *J. Exp. Anal. Behav.* **19:**345–349.

Maki, W.S., Jr., and Leuin, T.C., 1972, Information processing by pigeons, *Science* **176:**535–536.

Maki, W.S., Jr., Riley, D.A., and Leith, C.R., 1976, The role of test stimuli in matching to compound samples by pigeons, *Anim. Learn. Behav.* **4:**13–21.

Mallot, R.W., and Cumming, W.W., 1964, Schedules of interresponse time reinforcement, *Psychol. Rec.* **14:**211–252.

Mandel, P., and Ebel, A., 1974, Correlations between alteration in cholinergic system and behavior, in: *Neuro-chemistry of Cholinergic Receptors* (E. de Robertis and J. Schacht, eds.), pp. 131–140, Raven Press, New York.

Mandel, P., Ayad, G., Hermetet, J.C., and Ebel, A., 1974, Correlation between choline acetyltransferase activity and learning activity in different mice strains and their offspring, *Brain Res.* **72:**65–70.

Mason, M., and Wilson, M., 1974, Temporal differentiation and recognition memory for visual stimuli in rhesus monkeys, *J. Exp. Psychol.* **103:**383–390.

McAllister, W.R., and McAllister, D.E., 1965, Variables influencing the conditioning and the measurement of acquired fear, in: *Classical Conditioning: A Symposium* (W.F. Prokasy, ed.), pp. 172–191, Appleton-Century-Crofts, New York.

McAllister, W.R., and McAllister D.E., 1971, Behavioral measurement of conditioned fear, in: *Aversive Conditioning and Learning* (F.R. Brush, ed.), Academic Press, New York.

McGaugh, J.L., and Petrinovich, L.F., 1965, Effect of drugs on learning and memory, *Int. Rev. Neurobiol.* **8:**139–191.

McKearney, J.W., 1970, Rate-dependent effects of drugs: Modification by discriminative stimuli of the effects of amobarbital on schedule-controlled behavior, *J. Exp. Anal. Behav.* **14:**165–175.

McKearney, J.W., 1972, Schedule-dependent effects: Effects of drugs and maintenance of responding with response-produced electric shocks in: *Schedule Effects: Drugs, Drinking and Aggression* (R.M. Gilbert and J.D. Keehn, eds.), pp. 3–25, University of Toronto Press.

McKearney, J.W., 1973, Effects of methamphetamine and chlordiazepoxide on schedule-controlled and adjunctive licking in the rat, *Psychopharmacologia* **30:**375–384.

McLennan, H., 1970, *Synaptic Transmission*, Saunders, Toronto.
McMillan, D.E., and Campbell, R.J., 1970, Effects of *d*-amphetamine and chlordiazepoxide on spaced responding in pigeons, *J. Exp. Anal. Behav.* **14:**177–184.
Mellgren, S.I. and Srebro, B., 1973, Changes in acetylcholinesterase and distribution of degenerating fibers in the hippocampal region after septal lesions in the rat, *Brain Res.* **52:**19–36.
Mello, N.K., 1971, Alcohol effects on delayed matching to sample performance by rhesus monkey, *Physiol. Behav.* **7:**77–101.
Melton, A.W., 1963, Implications of short-term memory for a general theory of memory, *J. Verb. Learn. Verb. Behav.* **2:**1–21.
Meyers, B., and Domino, E.F., 1964, The effect of cholinergic blocking drugs on spontaneous alternation in rats, *Arch. Int. Pharmacodyn.* **150:**525–529.
Miller, N.E., and Barry, H., III., 1960, Motivational effects of drugs: Methods which illustrate some general problems in psychopharmacology, *Psychopharmacologia* **1:**169–199.
Modigh, K., 1974*a*, Functional aspects of 5-hydroxytryptamine turnover in the central nervous system, *Acta Physiol. Scand. (Suppl.)* **403:**1–56.
Modigh, K., 1974*b*, Studies on *dl*-5-hydroxytryptophan induced hyperactivity in mice, in: *Serotonin: New Vistas: Histochemistry and Pharmacology* (E. Costa, G.L. Gessa, and M. Sandler, eds.), pp. 213–218, Raven Press, New York.
Molander, L., and Randrup, A., 1974, Investigation of the mechanism by which L-dopa induces gnawing in mice, *Acta Pharmacol. Toxicol.* **34:**312–324.
Moorcroft, W.H., 1971, Ontogeny of forebrain inhibition of behavioral arousal in the rat, *Brain Res.* **35:**513–522.
Moore, K.E., and Thornburg, J.E., 1975, Drug-induced dopaminergic supersensitivity, in: *Advances in Neurology: Dopaminergic Mechanisms* (D. Calne, T.N. Chase, and A. Barbeau, eds.), pp. 93–104, Raven Press, New York.
Moyer, K.E., and Korn, J.H., 1964, Effect of UCS intensity on the acquisition and extinction of an avoidance response, *J. Exp. Psychol.* **67:**352–359.
Neff, N.H., and Costa, E., 1966, The influence of monoamine oxidase inhibition on catecholamine synthesis, *Life Sci.* **5:**951–959.
Neuringer, A.J., 1967, Effects of reinforcement magnitude on choice and rate of responding, *J. Exp. Anal. Behav.* **10:**417–424.
Nevin, J.A., 1973, The maintenance of behavior, in: *The Study of Behavior: Learning, Motivation, Emotion and Instinct* (J.A. Nevin, ed.), pp. 201–236, Scott, Foresman, Glenview, Illinois.
Nistri, A., Bartolini, A., Deffenu, G., and Pepeu, G., 1972, Investigations into the release of acetylcholine from the cerebral cortex of the cat: Effects of amphetamine, of scopolamine and of septal lesions, *Neuropharmacology* **11:**665–674.
Oliverio, A., 1967, Contrasting effects of scopolamine on mice trained simultaneously with two different schedules of avoidance conditioning, *Psychopharmacologia* **11:**39–51.
Oliverio, A., 1968, Effects of scopolamine on avoidance conditioning and habituation of mice, *Psychopharmacologia* **12:**214–226.
Oliverio, A., 1974, Genetic factors in the control of drug effects on the behavior of mice, in: *The Genetics of Behavior* (J.H.F.v. Abeelen, ed.), pp. 375–395, North-Holland, Amsterdam.
Oliverio, A., Eleftheriou, B.E., and Bailey, D.W., 1973, Exploratory activity: Genetic analysis of its modification by scopolamine and amphetamine, *Physiol. Behav.* **10:**893–899.
Pearson, V., 1973, Effect of scopolamine and atropine on habituation of the flexor withdrawal reflex, *Pharmacol. Biochem. Behav.* **1:**155–517.
Pedersen, V., 1968, Role of catecholamines in compulsive gnawing behavior in mice, *Br. J. Pharmacol.* **34:**219–220.
Pedersen, V., and Christensen, A.V., 1972, Antagonism of methylphenidate-induced stereotyped gnawing in mice, *Acta Pharmacol. Toxicol.* **31:**488–496.
Pepeu, G., and Bartolini, A., 1968, Effect of psychoactive drugs on the output of acetylcholine from the cerebral cortex of the cat, *Eur. J. Pharmacol.* **4:**254–263.
Pickens, R., and Thompson, T., 1968, Cocaine-reinforced behavior in rats: Effects of reinforcement magnitude and fixed-ratio size, *J. Pharmacol. Exp. Ther.* **161:**122–129.

Pinel, J.P.J., and Mucha, R.F., 1973*a*, Activity and reactivity in rats at various intervals after footshock, *Can. J. Psychol.* **27:**112–118.
Pinel, J.P.J., and Mucha, R.F., 1973*b*, Incubation and the Kamin effect in rats: Changes in activity and reactivity after footshock, *J. Comp. Physiol. Psychol.* **84:**661–667.
Platt, J.R., 1974, Are schedules of reinforcement necessary? A review of W.N. Schoenfeld and B.K. Cole's *Stimulus Schedules: The t-τ Systems*, *J. Exp. Anal. Behav.* **21:**383–388.
Pryor, G.T., Schlesinger, K., and Calhoun W.H., 1966, Differences in brain enzymes among five inbred strains of mice, *Life Sci.* **5:**2105–2111.
Randall, P.K., and Riccio, D.C., 1969, Fear and punishment as determinants of passive avoidance responding, *J. Comp. Physiol. Psychol.* **69:**550–553.
Randrup, A., and Munkvad, I., 1966, Dopa and other naturally occurring substances as causes of stereotypy and rage in rats, *Acta Psychiatr. Scand.* **191:**193–199.
Randrup, A., and Munkvad, I., 1970, Biochemical, anatomical and psychological investigations of stereotyped behavior induced by amphetamines, in: *Amphetamines and Related Compounds* (E. Costa and S. Garattini, eds.), pp. 695–714, Raven Press, New York.
Randrup, A., and Scheel-Kruger, J., 1966, Diethyldithiocarbamate and amphetamine stereotyped behavior, *J. Pharm. Pharmacol.* **18:**752.
Randrup, A., Munkvad, I., and Udsen, P., 1963, Adrenergic mechanisms and amphetamine induced abnormal behavior, *Acta Pharmacol.* **20:**145–157.
Rech, R.H., and Moore, K.E., 1971, *An Introduction to Psychopharmacology*, Raven Press, New York.
Reid, J.L., 1975, Dopamine supersensitivity in the hypothalamus?, in: *Advances in Neurology: Dopaminergic Mechanisms*, pp. 73–80, Raven Press, New York.
Rescorla, R.A., 1967, Pavlovian conditioning and its proper control procedures, *Psychol. Rev.* **74:**71–80.
Rescorla, R.A., 1969, Pavlovian conditioned inhibition, *Psychol. Bull.* **72:**77–94.
Rescorla, R.A., and Solomon, R.L., 1967, Two process learning theory: Relationships between Pavlovian and instrumental learning, *Psychol. Rev.* **74:**151–182.
Reynolds, G.S., 1968, *A Primer of Operant Conditioning*, Scott, Foresman, Glenview, Illinois.
Richardson, J.S., 1974, Basic concepts of psychopharmacological research as applied to the psychopharmacological analysis of the amygdala, *Acta Neurobiol. Exp.* **34:**543–562.
Riley, D.A., 1968, *Discrimination Learning*, Allyn and Bacon, Boston.
Riley, D.A., and Leith C.R., 1976, Multidimensional psychophysics and selective attention in animals, *Psychol. Bull.* **83:**138–160.
Roberts, D.C.S., Zis, A.P., and Fibiger, H.C., 1975, Ascending catecholamine pathways and amphetamine-induced locomotor activity: Importance of dopamine and apparent non-involvement of norepinephrine, *Brain Res.* **93:**441–454.
Roberts, W.A., 1972*a*, Short-term memory in the pigeon: The effects of repetition and spacing, *J. Exp. Psychol.* **94:**74–83.
Roberts, W.A., 1972*b*, A review of findings on the delayed-response problem as they pertain to the problem of short-term memory in animals, *Research Bulletin No. 228*, Department of Psychology, University of Western Ontario.
Roberts, W.A., 1974, Spaced repetition facilitates short-term retention in the rat, *J. Comp. Physiol. Psychol.* **84:**164–171.
Rosic, N., and Bignami, G., 1970, Scopolamine effects on go–no go avoidance discriminations: Influence of stimulus factors and primacy training, *Psychopharmacologia* **17:**203–215.
Rosic, N., Frontali, M., and Bignami, G., 1970, Stimulus factors affecting go–no go avoidance discrimination learning by rats, *Commun. Behav. Biol.* **4:**151–156.
Roth, R.H., Starjne, L., and Ueler, U.S. von, 1966, Acceleration of noradrenaline biosynthesis by nerve stimulation, *Life Sci.* **5:**1071.
Sahgal, A., and Iversen, S.D., 1975, Colour preferences in the pigeon: A behavioral and psychopharmacological study, *Psychopharmacologia* **43:**175:179.
Sahgal, A., Pratt, S.R., and Iversen, S.D., 1975, Response preferences of monkeys (*Macaca mulatta*) within wavelength and line-tilt dimensions, *J. Exp. Anal. Behav.* **24:**377–381.
Sanger, D.J., and Blackman, D.E., 1975, The effects of tranquilizing drugs on timing behavior in rats, *Psychopharmacologia* **44:**153–156.

Satinder, K.P., and Hill, K.D., 1974, Effects of genotype and postnatal experience on activity, avoidance, shock threshold, and open field behavior of rats, *J. Comp. Physiol. Psychol.* **86**:363–374.

Satinder, K.P., and Petryshyn, W.R., 1974, Interaction among genotype, unconditioned stimulus, *d*-amphetamine and one-way avoidance behavior of rats, *J. Comp. Physiol. Psychol.* **86**:1059–1073.

Sayers, A.C., Bürki, H.R., Asper, H., and Ruch, W., 1975, Neuroleptic-induced hypersensitivity of striatal dopamine receptors in the rat as a model of tardive dyskinesias: Effects of clozapine, haloperidol, loxapine and chlorpromazine, *Psychopharmacologia* **41**:97.

Scheckel, C.L., 1965, Self-adjustment of the interval in delayed matching: Limit of delay for the rhesus monkey, *J. Comp. Physiol. Psychol.* **59**:415–418.

Scheel-Kruger, J., 1970, Central effects of anticholinergic drugs measured by the apomorphine gnawing test in mice, *Acta Pharmacol. Toxicol.* **28**:1–16.

Scheel-Kruger, J., and Randrup, A., 1967, Stereotyped hyperactive behavior produced by dopamine in the absence of noradrenaline, *Life Sci.* **6**:1389–1398.

Schneider, B.A., 1969, A two-state analysis of fixed-interval responding in the pigeon, *J. Exp. Anal. Behav.* **12**:677–687.

Schwartz, B., and Williams, D.R., 1971, Discrete-trials spaced responding in the pigeon: The dependence of efficient performance on the availability of a stimulus for collateral pecking, *J. Exp. Anal. Behav.* **16**:155–160.

Schwartzbaum, J.S., Green, R.N., Beatty, W.W., and Thompson, J.B., 1967, Acquisition of avoidance behavior following septal lesions in the rat, *J. Comp. Physiol. Psychol.* **63**:93–104.

Segal, D.S., 1975, Behavioral and neurochemical correlates of repeated *d*-amphetamine administration, in: *Neurobiological Mechanisms of Adaptation and Behavior* (A.J. Mandell, ed.), pp. 247–262, Raven Press, New York.

Segal, D.S., and Mandell, A.J., 1974, Long-term administration of *d*-amphetamine: Progressive augmentation of motor activity and stereotypy, *Pharmacol. Biochem. Behav.* **2**:249–255.

Segal, E.F., 1962, Effects of *dl*-amphetamine under concurrent VI-DRL reinforcement, *J. Exp. Anal. Behav.* **5**:105–112.

Segal, E.F., and Holloway, S.M., 1963, Timing behavior in rats with water drinking as a mediator, *Science* **140**:888–889.

Seligman, M.E.P., and Johnston, J.C., 1973, A cognitive theory of avoidance learning, in: *Contemporary Approaches to Conditioning and Learning* (F.J. McGuigan and D.B. Lumsden, eds.), pp. 69–110, Winston, New York.

Shettleworth, S.J., 1972, Constraints on learning, in: *Advances in the Study of Behavior*, Vol. 4 (D.S. Lehrman *et al.*, eds.), Academic Press, New York.

Shiffrin, R.M., and Atkinson, R.C., 1969, Storage and retrieval processes in long-term memory, *Psychol. Rev.* **76**:179–193.

Shimp, C.P., 1967, The reinforcement of short interresponse times, *J. Exp. Anal. Behav.* **10**:425–434.

Sidman, M., 1953*a*, Avoidance conditioning with brief shock and no exteroceptive warning signal, *Science* **118**:157–158.

Sidman, M., 1953*b*, Two temporal parameters of the maintenance of avoidance behavior by the white rat, *J. Comp. Physiol. Psychol.* **46**:253–261.

Sidman, M., 1955, Technique for assessing the effects of drugs on timing behavior, *Science* **122**:925.

Sidman, M., 1956, Drug–behavior interaction, *Ann. N.Y. Acad. Sci.* **65**:282–302.

Sidman, M., 1960, *Tactics of Scientific Research: Evaluating Experimental Data in Psychology*, Basic Books, New York.

Siegel, S., 1975*a*, Conditioning insulin effects, *J. Comp. Physiol. Psychol.* **89**:189–199.

Siegel, S., 1975*b*, Evidence from rats that morphine tolerance is a learned response, *J. Comp. Physiol. Psychol.* **89**:498–506.

Singh, H.K., Ott, T., and Matthies, H., 1974, Effect of intrahippocampal injection of atropine on different phases of a learning experiment, *Psychopharmacologia* **38**:247–248.

Skinner, B.F., 1950, Are theories of learning necessary?, *Psychol. Rev.* **57**:193–216.

Smith, C.A., McFarland, W.L., and Taylor, E., 1961, Performance in a shock avoidance situation interpreted as pseudoconditioning, *J. Comp. Physiol. Psychol.* **54**:154–157.

Smith, C.B., 1964, Effects of *d*-amphetamine upon operant behavior of pigeons: Enhancement by reserpine, *J. Pharmacol. Exp. Ther.* **146:**167–174.

Smith, J.B., and Clark, E.C., 1975, Effects of *d*-amphetamine, chlorpromazine and chlordiazepoxide on intercurrent behavior during spaced-responding schedules, *J. Exp. Anal. Behav.* **24:**241–248.

Spear, N.E., 1973, Retrieval of memory in animals, *Psychol. Rev.* **80:**163–194.

Squire, L.R., 1969, Effects of pretrial and posttrial administration of cholinergic and anticholinergic drugs on spontaneous alternation, *J. Comp. Physiol. Psychol.* **1:**69–75.

Srebro, B., and Mellgren, S.I., 1974, Changes in postnatal development of acetylcholinesterase in the hippocampal region after early septal lesions in the rat, *Brain Res.* **79:**119–131.

Stadler, H., Lloyd, K.G., Gadea-Ciria, M., and Bartholini, G., 1973, Enhanced striatal acetylcholine release by chlorpromazine and its reversal by apomorphine, *Brain Res.* **55:**476–480.

Stebbins, W.C. (ed.), 1970*a*, *Animal Psychophysics: The Design and Conduct of Sensory Experiments*, Appleton-Century-Crofts, New York.

Stebbins, W.C., 1970*b*, Studies of hearing and hearing loss in the monkey, in: *Animal Psychophysics: The Design and Conduct of Sensory Experiments* (W.C. Stebbins, ed.), pp. 41–66, Appleton-Century-Crofts, New York.

Stein, L., 1964, Excessive drinking in the rat: Superstition or thirst?, *J. Comp. Physiol. Psychol.* **58:**237–242.

Stein, L., 1966, Habituation and stimulus novelty: A model based on classical conditioning, *Psychol. Rev.* **73**(4)**:**352–356.

Steranka, L.R., and Barrett, R.J., 1973, Kamin effect in rats: Differential retention or differential acquisition of an active-avoidance response, *J. Comp. Physiol. Psychol.* **85:**324–330.

Steranka, L.R., and Barrett, R.J., 1974, Facilitation of avoidance acquisition by lesion of the median raphe nucleus: Evidence for serotonin as a mediator of shock-induced suppression, *Behav. Biol.* **11:**205–213.

Stevens, R., and Cowey, A., 1973, Effects of dorsal and ventral hippocampal lesions on spontaneous alternation, learned alternation and probability learning in rats, *Brain Res.* **52:**203–224.

Stewart, W.J., 1975*a*, Environmental complexity does affect scopolamine-induced changes in activity, *Neurosci. Lett.* **1:**121–125.

Stewart, W.J., 1975*b*, Size of the environment as a determiner of effects of scopolamine, *Psychol. Rep.* **37:**175–178.

Stewart, W.J., and Blain, S., 1975, Dose–response effects of scopolamine on activity in an open field, *Psychopharmacologia* **44:**291–296.

Stone, E.A., 1975, Stress and catecholamines, in: *Catecholamines and Behavior*, pp. 31–72, Plenum Press, New York.

Stricker, E.M., and Adair, E.R., 1966, Bodily fluid balance, taste and postprandial factors in schedule-induced polydipsia, *J. Comp. Physiol. Psychol.* **62:**449–454.

Stromberg, U., 1970, DOPA effects on mobility in mice: Potentiation by MK-485 and dexchlorpheniramine, *Psychopharmacologia* **18:**58–67.

Suits, E., and Isaacson, R.L., 1968, The effects of scopolamine hydrobromide on one-way and two-way avoidance learning in rats, *Int. J. Neuropharmacol.* **7:**441–446.

Sutherland, N.S., and Mackintosh, N.J., 1971, *Mechanisms of Animal Discrimination Learning*, Academic Press, New York.

Swonger, A.K., and Rech, R.H., 1972, Serotonergic and cholinergic involvement in habituation of activity and spontaneous alternation of rats in a Y-maze, *J. Comp. Physiol. Psychol.* **81:**509–522.

Swonger, A.K., Chambers, W.F., and Rech, R.H., 1970, The effects of alterations in brain 5-HT on habituation of the cortical evoked response and the startle response in rats, *Pharmacologist* **12:**207.

Talland, G.A., 1966, Effects of alcohol on performance in continuous attention tasks, *Psychosom. Med.* **28:**596–604.

Tamerin, J.S., Weiner, S., Poppen, R., Steinglass, P., and Mendelson, J.H., 1971, Alcohol and memory: Amnesia and short-term function during experimentally induced intoxication, *Am. J. Psychiatry* **127:**1659–1664.

Tenen, S., 1967, The effects of *p*-chlorophenylalanine, a serotonin depletor, on avoidance acquisition, pain sensitivity and related behavior in the rat, *Psychopharmacologia* **10:**204–219.

Testa, T.J., 1974, Causal relationships and the acquisition of avoidance responses, *Psychol. Rev.* **81:**491–505.

Theios, J., Lynch, A.D., and Lowe, W.F., Jr., 1966, Differential effects of shock intensity on one-way and shuttle avoidance conditioning, *J. Exp. Psychol.* **72:**294–299.

Thierry, A.M., 1973, Effects of stress on the metabolism of serotonin and norepinephrine in the central nervous system of the rat, in: *Hormones, Metabolism and Stress: Recent Progress and Perspectives* (S. Nemeth, ed.), pp. 37–54, Publishing House of the Slovak Academy of Sciences, Bratislava.

Thierry, A.M., Fekete, M., and Glowinski, J., 1968, Effects of stress on the metabolism of noradrenaline, dopamine and serotonin (5-HT) in the central nervous system of the rat. I. Modifications of serotonin metabolism, *Eur. J. Pharmacol.* **4:**384–389.

Thierry, A.M., Blanc, G., and Glowinski, J., 1971, Effect of stress on the disposition of catecholamines localized in various intraneuronal storage forms in the brain stem of the rat, *J. Neurochem.* **18:**449–461.

Thomas, E., 1972, Excitatory and inhibitory processes in hypothalamic conditioning, in: *Inhibition and Learning* (R.A. Boakes and M.S. Halliday, eds.), pp. 359–380, Academic Press, London.

Thompson, D.M., 1973, Repeated acquisition as a behavioral baseline for studying drug effects, *J. Pharmacol. Exp. Ther.* **184:**506–516.

Thompson, D.M., 1974, Repeated acquisition of behavioral chains under chronic drug conditions, *J. Pharmacol. Exp. Ther.* **188:**700–713.

Thompson, D.M., 1975, Repeated acquisition of response sequences: Stimulus control and drugs, *J. Exp. Anal. Behav.* **23:**429–436.

Thompson, D.M., and Corr, P.B., 1974, Behavioral parameters of drug action: Signalled and response-independent reinforcement, *J. Exp. Anal. Behav.* **21:**151–158.

Thompson, R.F., and Spencer, W.A., 1966, Habituation: A model phenomenon for the study of neuronal substrates of behavior, *Psychol. Rev.* **73:**16–43.

Thompson, R.F., Groves, P.M., Teyler, T.J., and Roemer, R.A., 1973, A dual process theory of habituation: Theory and behavior, in: *Habituation: Behavioral Studies and Physiological Substrates* (H.V.S. Peeke and M.J. Herz, eds.), pp. 239–269, Academic Press, New York.

Thompson, T., and Pickens, R., 1972, Drugs as reinforcers: Schedule considerations, in: *Schedule Effects: Drugs, Drinking and Aggression* (R.M. Gilbert and J.D. Keehn, eds.), pp. 50–64, University of Toronto Press.

Thompson, T., and Schuster, C.R., 1968, *Behavioral Pharmacology*, Prentice-Hall, Englewood Cliffs, N.J.

Thornburg, J.E., and Moore, K.C., 1973, The relative importance of dopaminergic and noradrenergic neuronal systems for the stimulation of locomotor activity induced by amphetamine and other drugs, *Neuropharmacology* **12:**853–866.

Ungerstedt, U., 1971*a*, Stereotaxic mapping of the monoamine pathways in the rat brain, *Acta Physiol. Scand.* **367**(Suppl. 10)**:**1–48.

Ungerstedt, U., 1971*b*, Postsynaptic supersensitivity after 6-hydroxydopamine induced degeneration of the nigro-striatal dopamine system in the rat brain, *Acta Physiol. Scand.* **82**(Suppl. 367)**:**69–93.

Ungerstedt, U., Butcher, L., Butcher, S., Anden, N.E., and Fuxe, K., 1969, Direct chemical stimulation of dopaminergic mechanisms in the neostriatum of the rat, *Brain Res.* **11:**461–471.

Understedt, U., Ljungberg, T., Hoffer, B., and Siggins, G., 1975, Dopaminergic supersensitivity in the striatum, in: *Advances in Neurology: Dopaminergic Mechanisms* (D. Caine, T.N. Chase, and A. Barbeau, eds.), pp. 57–66, Raven Press, New York.

Vaillant, G.E., 1964, A comparison of chlorpromazine and imipramine on behavior in the pigeon, *J. Pharmacol. Exp. Ther.* **146:**377–384.

Van Voigtlander, P.F., and Moore, K.E., 1973, Turning behavior of mice with unilateral 6-hydroxydopamine lesions in the striatum: Effects of apomorphine, L-DOPA, amantadine, amphetamine and other psychomotor stimulants, *Neuropharmacology* **12:**451–462.

Wahlsten, D., 1972, Phenotypic and genetic relations between initial response to electric shock and rate of avoidance learning in mice, *Behav. Genet.* **2:**211–240.

Wahlsten, D., and Sharp, D., 1969, Improvement of shuttle-avoidance by handling during the intertrial interval, *J. Comp. Physiol. Psychol.* **67:**252–259.

Waller, M., and Waller, P.F., 1962, Effects of chlorpromazine on appetitive and aversive components of a multiple schedule, *J. Exp. Anal. Behav.* **5:**259–264.

Warburton, D.M., and Brown, K., 1971, Attenuation of stimulus sensitivity induced by scopolamine, *Nature (London)* **230:**126–127.

Warburton, D.M., and Brown, K., 1972, The facilitation of discrimination performance by physostigmine sulfate, *Psychopharmacologia* **27:**275–284.

Waugh, N.C., and Norman, D., 1965, Primary memory, *Psychol. Rev.* **72:**89–104.

Weiner, N., 1970, Regulation of norepinephrine biosynthesis, *Annu. Rev. Pharmacol.* **10:**273–290.

Weiner, N., 1974, Neurotransmitter systems in the central nervous system, in: *Drugs and the Developing Brain* (A. Vernadakis and N. Weiner, eds.), pp. 105–132, Plenum Press, New York.

Weisman, R.G., 1975, The compleat associationist: A review of N.J. Mackintosh's *The Psychology of Animal Learning*, *J. Exp. Anal. Behav.* **24:**383–389.

Weiss, B., and Heller, A., 1969, Methodological problems in evaluating the role of cholinergic mechanisms in behavior, *Fed. Proc. Fed. Am. Soc. Exp. Biol.* **28:**135–146.

Weiss, B., and Laties, V.G., 1963, Effects of amphetamine, chlorpromazine, and pentobarbital on behavioral thermoregulation, *J. Pharmacol. Exp. Ther.* **140:**1–7.

Weiss, B., and Laties, V.G., 1964, Effects of amphetamine, chlorpromazine, pentobarbital and ethanol on operant response duration, *J. Pharmacol. Exp. Ther.* **144:**17–23.

Weiss, B., and Laties, V.G., 1970, The psychophysics of pain and analgesia in animals, in: *Animal Psychophysics: The Design and Conduct of Sensory Experiments* (W.C. Stebbins, ed.), pp. 185–210, Appleton-Century-Croft, New York.

Weiss, J.M., Krieckhaus, E.E., and Conte, R., 1968, Effects of fear conditioning on subsequent avoidance behavior and movement, *J. Comp. Physiol. Psychol.* **65:**413–421.

Whitehouse, J.M., 1966, The effects of physostigmine on discrimination learning, *Psychopharmacologia* **9:**183–188.

Whitlow, J.W., Jr., 1975, Short-term memory in habituation and dishabituation, *J. Exp. Psychol.* **104:**189–206.

Williams, B.A., 1971*a*, Color alternation learning in the pigeon under fixed-ratio schedules of reinforcement, *J. Exp. Anal. Behav.* **15:**129–140.

Williams, B.A., 1971*b*, Non-spatial delayed alternation by the pigeon, *J. Exp. Anal. Behav.* **16:**15–21.

Williams, J.M., Hamilton, L.W., and Carlton, P.L., 1974, Pharmacological and anatomical dissociation of two types of habituation, *J. Comp. Physiol. Psychol.* **87:**724–732.

Williams, J.M., Hamilton, L.W., and Carlton, P.L., 1975, Ontogenetic dissociation of two classes of habituation, *J. Comp. Physiol. Psychol.* **89:**733–737.

Wimer, R.E., Reid, N., and Eleftheriou, E., 1973, Serotonin levels in hippocampus: Striking variations associated with mouse strain and treatment, *Brain Res.* **63:**397–401.

Winograd, E., 1971, Some issues relating animal memory to human memory, in: *Animal Memory* (W.K. Honig and P.H.R. James, eds.), pp. 259–278, Academic Press, New York.

Wuttke, W., 1970, The effects of *d*-amphetamine on schedule-controlled water licking in the squirrel monkey, *Psychopharmacologia* **17:**70–82.

Wuttke, W., and Innis, N.K., 1972, Drug effects upon behavior induced by second-order schedules of reinforcement: The relevance of ethological analyses, in: *Schedule Effects: Drugs, Drinking and Aggression* (R.M. Gilbert and J.D. Keehn, eds.), pp. 129–147, University of Toronto Press.

Zeiler, M.D., 1977, Schedules of reinforcement: The controlling variables, in: *Handbook of Operant Behavior* (W.K. Honig and J.E.R. Staddon, eds.), pp. 201–232, Prentice-Hall, Englewood Cliffs, New Jersey.

Zimmerberg, B., Glick, S.D., and Jarvik, M.E., 1971, Impairment of recent memory by marihuana and THC in rhesus monkeys, *Nature (London)* **233:**343–345.

2

Behavioral and Biochemical Methods to Study Brain Responses to Environment and Experience

Edward L. Bennett and Mark R. Rosenzweig

1. Introduction

Exposing laboratory rats to one or another laboratory environment was found in the early 1950's to affect their problem-solving ability (e.g., Forgays and Forgays, 1952; Hymovitch, 1952), and similar findings were later made with dogs (e.g., Melzack, 1962; Fuller, 1966), cats (Wilson *et al.*, 1965), and monkeys (e.g., Gluck *et al.*, 1973). Then, beginning in the late 1950's, experience in differential environments was discovered to affect numerous aspects of brain biochemistry and neuroanatomy of rodents (e.g., Krech *et al.*, 1960; Rosenzweig *et al.*, 1961, 1962; Bennett *et al.*, 1964; Ferchmin *et al.*, 1970; Greenough and Volkmar, 1973; Greenough, 1976; Walsh *et al.*, 1969). Bodily growth (Rosenzweig *et al.*, 1972*b*) and sleep–waking cycles (Tagney, 1973; McGinty, 1971; Lambert and Truong-Ngoc, 1976) were later found to be influenced by laboratory environments. Because not only these but also undoubtedly many other aspects of behavior and physiology are modified by environmental experience, it is clear that—at the minimum—investigators should describe the environments of their subjects in some detail in all research reports. Beyond this, selection of suitable environment(s) for the particular investigation should be undertaken with care. Either the environment in which animals are housed or specific experiences or training given during experimental sessions can be employed as independent variables to modulate many aspects of brain, body, and behavior.

Edward L. Bennett • Laboratory of Chemical Biodynamics, Lawrence Berkeley Laboratory, University of California, Berkeley, California 94720. *Mark R. Rosenzweig* • Department of Psychology, University of California, Berkeley, California 94720.

This chapter will describe some of the environmental and experimental manipulations that have been tested, and it will note some of the results that have been found; some of the methods used to assay cerebral effects of experience will also be described, and others will be cited. Specifically, this is how the content of the chapter will be apportioned: We will describe in some detail environmental and training situations and compare their effectiveness in producing significant changes in cerebral measures. We will then mention rather briefly some behavioral tests that have been used to characterize effects of prior experience and describe one procedure in detail. Finally, we will present in considerable detail the methods used to dissect brain samples and to obtain brain weights, and the analytical procedures employed to measure content of RNA and DNA and activities of acetylcholinesterase (AChE) and cholinesterase (ChE). The stablest and most statistically significant effects of environment on the brain that have been reported to date are changes in tissue weights (especially the cortical/subcortical weight ratio) and the RNA/DNA ratio of cortical tissue.

2. *Environmental and Training Techniques*

2.1. *Differential Environments*

2.1.1. *Basic Description*

Beginning in the late 1950s, our laboratory has done a large number of experiments in which we have compared the behavioral and cerebral consequences of assigning rodents for periods of time among three principal environments that we term "enriched condition" (EC), "standard colony" (SC), and "impoverished condition" (IC). The main characteristics of these environmental conditions can be seen in Fig. 1. Many variants of the enriched condition have been tested, as will be described below, but most appear to yield rather similar effects.

In the standard Berkeley EC, 10–12 same-sex animals are housed in a relatively large cage (76 × 76 × 45 cm) that is furnished with several varied stimulus objects. The exact size of the cage is not important, nor does it matter whether the floor is a metal grid or is covered with shavings. Variety in stimulation from the objects is secured in either of two ways, and these seem to yield equivalent results: (1) About 6 different objects are placed in the cage each day from a pool of about 25 objects. (2) When *N* groups are being given EC experience at the same time, different arrangements of objects are placed in *N* EC cages, and each group is moved from one cage to another each day; every *N*th day, all cages receive a new arrangement of objects. The exact nature of the objects does not seem to be important. They are chosen to induce exploratory behavior and to give the animals a variety of surfaces, visual forms, tastes, and smells. Objects that we have commonly used are illustrated in Fig. 2 (Rosenzweig and Bennett, 1969).

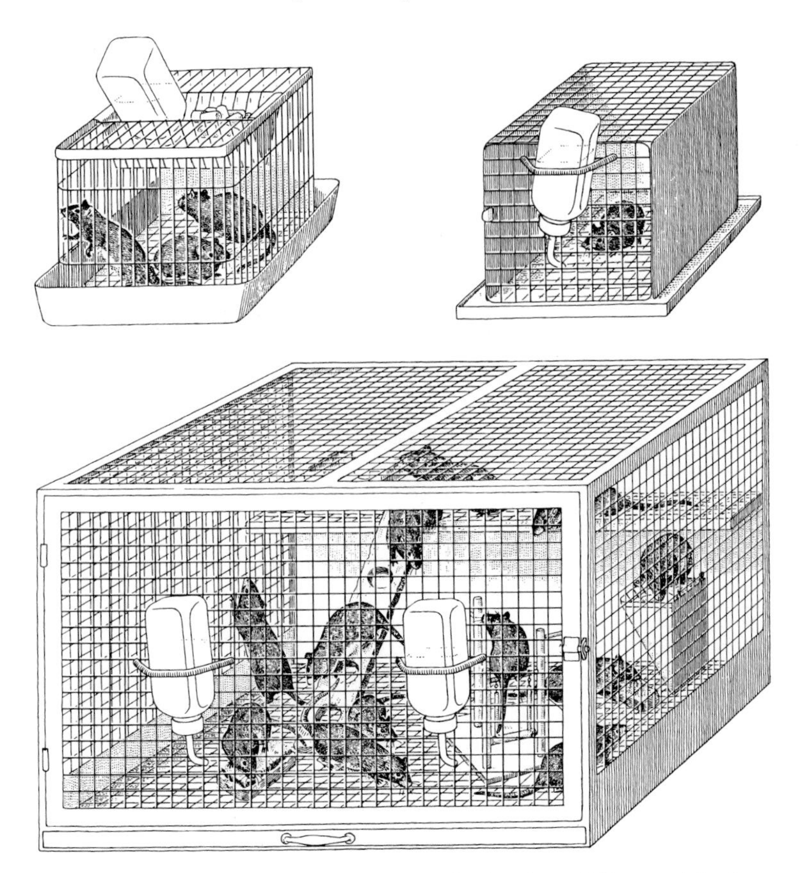

Fig. 1. Rats in three environments. *Top left:* standard colony (SC); *top right:* impoverished condition (IC); *bottom:* enriched condition (EC). Assigning rats to these environments for a period of days to weeks leads to significant changes in anatomy and chemistry of regions of the brain. From Rosenzweig *et al.* (1972*a*).

Standard colony (SC) animals are housed in a group of three in a conventional colony cage. Our colony cages for rats measure 21 × 34 × 20 cm, but any cage that meets standards for animal care will be adequate.

For the impoverished condition (IC), animals are housed singly in standard colony cages. In our early work, the IC subjects were housed in cages with solid walls (as illustrated in Fig. 1), and these were placed in quiet and dimly illuminated rooms. We later found that extracage stimulation is of little importance, so in many experiments we have housed the IC animals in SC cages and on the same cage racks as SC subjects.

All animals in all environmental conditions have food and water ad lib. Standard laboratory chow is provided as food.

In our earlier experiments, the enriched condition (EC) also included two additional features besides those described above: (1) a few trials of formal training per day and (2) daily exploration in a 75 × 75 cm open-field

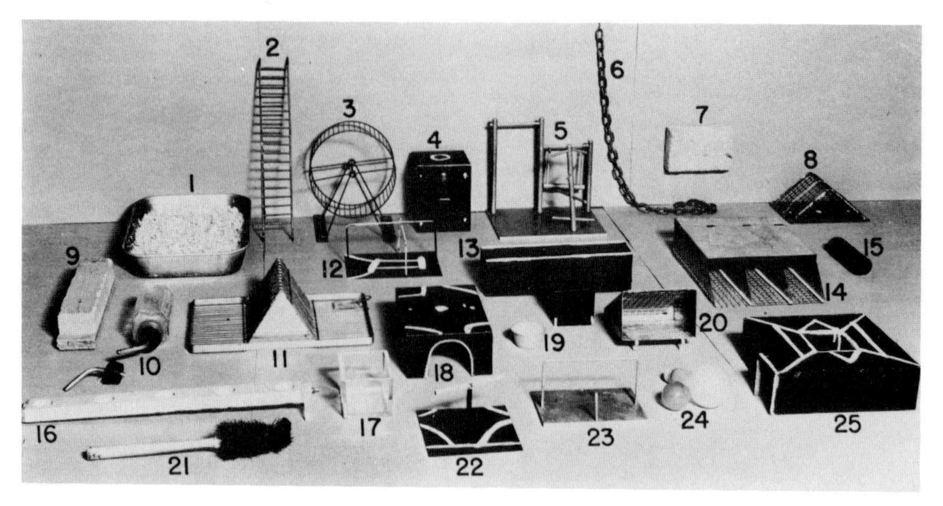

Fig. 2. Stimulus objects commonly used in the enriched condition. Typically, about six objects from this pool are placed in an EC cage. From Rosenzweig and Bennett (1969).

apparatus in which a different pattern of maze barriers was placed each day. Because of the inclusion of training (in the Krech Hypothesis Apparatus—a multiple-unit maze), we originally termed the enriched condition "environmental complexity and training" (ECT). In the mid-1960s, we tested whether the daily training trials contributed to the cerebral effects and found that they did not, so by 1968 (e.g., Rosenzweig *et al.*, 1968), we dropped the training and referred simply to the enriched condition. We also eliminated the sessions in the field with barriers to test whether the superiority of EC rats over IC rats in solutions of maze problems could be attributed to transfer of training from experience with barriers. The results showed that experience in the EC cage was sufficient to produce both superiority in maze solution and the cerebral differences with respect to IC animals; in fact, the field experience did not contribute measurably to either the behavioral or the cerebral effects, so we have not included it in experiments done since 1970.

2.1.2. *Dimensions of Differential Environments*

From the foregoing description, and from other environmental conditions that various investigators have tried, it appears that the enriched environment included both social and inanimate stimuli and that the latter can be divided into intracage and extracage stimuli. Some evidence is available as to the effectiveness of each of these sorts of stimulation in producing changes in behavior and in brain, and conclusions will be stated briefly here. [This topic is taken up in some detail and citations are given in Rosenzweig and Bennett (1977).]

Effects of social grouping have been studied by comparing animals housed in SC vs. IC environments. In some cases, the social group has been enlarged by housing 10 to 12 animals in an EC cage but without providing any inanimate stimulus objects [this has been called the group condition (GC)]. Social grouping vs. isolation aids subsequent learning and retention, but not as much as does the EC treatment. Social grouping also produces some cerebral effects, but these are significantly smaller than those caused by experience in EC (Rosenzweig *et al.*, 1978).

Inanimate stimulus objects within the EC cage aid subsequent learning, as can be seen by the superior performance of EC vs. GC animals. Larger cerebral effects are also produced by EC than by GC. Even single rats placed in EC cages with stimulus objects can develop brain values similar to those of EC rats and significantly different from those of IC rats, so social stimulation is not required. It should be noted, however, that single rats tend to be rather inactive and not to explore much, so production of EC cerebral effects in individual rats is facilitated if they are primed to interact with the stimulus objects, as described in the next section.

Extracage stimuli seem to be ineffective in producing either behavioral or cerebral effects in rats. We originally believed that ambient visual and auditory stimuli might be effective, so we placed the EC cages in a busy laboratory room and placed the IC cages in a quiet, dimly lighted room. Later, we found that typical cerebral differences between EC and IC rats were produced even when the EC cage was kept in a quiet, dimly lighted room and the IC cages were put in the busy laboratory room. We also found that rats housed in small individual cages within EC cages developed cerebral and behavioral characteristics like those of IC littermates, rather than like those of the EC rats (Ferchmin *et al.*, 1975). Some investigators have referred to an environment as "enriched" when rats were housed singly and were provided only extracage stimulation (Yeterian and Wilson, 1976), but we believe that this appellation is misleading.

2.1.3. *Enriched Experience for Individual Animals*

2.1.3a. Individual Rats in EC. To test effects of enriched experience that does not include the component of social stimulation, several conditions have been employed. The simplest has been just to place an individual rat in each EC cage. This was found not to work very well because individual laboratory rats tend to be rather inactive, remaining in a corner of the cage much of the time, resting or grooming. We found that the activity of each animal increases considerably when other animals are placed in the cage with it (Rosenzweig, 1971). We therefore undertook various methods to "prime" the activity of animals placed individually in EC. A method found to be effective with young rats placed into the experimental condition at weaning (about 25 days of age) or shortly thereafter was to give a small injection of methamphetamine (2 mg/kg) shortly before placing the animal into EC for a 2-hr period (Rosenzweig and Bennett, 1972). After an injection of the drug, the individual animal interacts vigorously with the varied stimulus objects, and during a 2-week or 4-week experiment it develops brain measures similar to those of the rat placed in EC in a group of 12. The methamphetamine is even more effective if given to rats during the dark phase of their daily cycle. This method cannot be used to prime the activity of older rats, because even low doses of methamphetamine induce stereotyped and abnormal behavior in adult rats.

2.1.3b. Individual Maze Training. Individual trials in mazes are another effective way of giving animals experience. On the basis of earlier experiments, we were doubtful about finding clear cerebral effects of maze training as compared with runway controls (Rosenzweig, 1971, pp. 335–336), but we later obtained significant effects (Bennett, 1976, p. 284), and Greenough (1976) has also reported effects of maze experience on brain measures.

Recently, we have been studying central effects of self-paced maze trials (Bennett *et al.*, 1979). To get rats to run self-paced trials, we separated their sources of food and water so that the animals had to traverse the maze to get from one to the other. The maxe was a box made of Plexiglas and measured 10 cm high × 74 cm wide × 74 cm deep; it was placed within an EC cage on flanges that supported it 15 cm above the cage floor (Fig. 3). Holes 7 cm in diameter were made at the four corners of the bottom and top of the

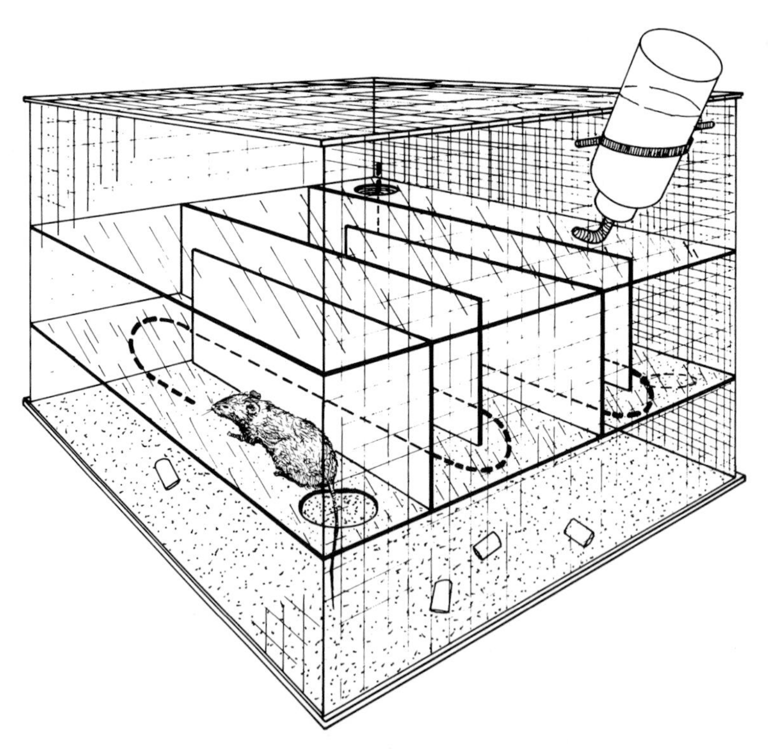

Fig. 3. Plastic maze placed within a large cage. The rat runs self-paced trials to get from the food station below to the water station above. Training in this situation leads to significant changes in brain anatomy and brain chemistry, when comparison is made with an appropriate cortical condition.

plastic box, so that the rats could crawl in and out of it, and any of these holes could be closed with a plastic door when desired. Plastic barriers could be placed within the box to provide a variety of maze patterns. Food pellets were made available, as usual, on the floor of the cage, but the water bottle was placed above the plastic box so that to get from food to water, the rat had to climb into the plastic box at an open corner in the bottom, traverse the box to an open corner at the top, climb out of the box, and stand on its top to reach the spout of the water bottle. Rats that eat dry food pellets like to drink frequently, so that they had to run up and down a number of times during each bout of feeding; the rats were rarely able to carry a food pellet up into the maze and never above it to the water station. In the experimental situation, the maze pattern was changed daily for 30 days. In a control condition, no barriers were placed in the plastic box, so that the animal had only to traverse the empty box and did not have to learn maze patterns.

Results have shown that the "empty-box" controls did not develop brain values different from those of littermate IC rats, whereas the animals that learned maze patterns differed significantly from IC and "empty-box" littermate controls. The maze-learning individuals developed brain values in the direction of EC rats, but somewhat below the EC level; the brain values of the maze rats were essentially the same as those of littermates placed in the GC, that is, placed in a group of 12 in a large cage but without access to varied stimulus objects. Thus, the maze situation yields significant cerebral effects without necessity of social stimulation and in an inanimate situation that is considerably simpler and easier to describe than that of the EC.

2.1.4. *Superenriched Environments*

In the last few years, some investigators have gone beyond the EC situation to provide animals with even more enriched experience. The methods used to enhance enrichment have included the following: (1) placing rats in succession in cages of a variety of sizes and shapes and placing naïve rats with experienced "guides" (Ferchmin *et al.*, 1970); (a) placing a large number of rats in two or three interlinked cages (Kuenzle and Knüsel, 1974; Davenport, 1976; Rosenzweig *et al.*, 1978); (2) placing rats in a large outdoor enclosure with a dirt floor (Rosenzweig *et al.*, 1972*a*, 1978; Bennett, 1976). Each of these methods will be described in some detail, and results will be noted.

2.1.4a. Ferchmin EC. Ferchmin *et al.* (1970), working in Argentina, reported that only 4 days in their enriched conditions sufficed to induce significant increases in weight of the cerebrum and changes in cerebral RNA and DNA. The environmental conditions were not spelled out in great detail in their report, but we were fortunate to have Drs. Ferchmin and Eterović in our laboratories during 1971–1974, and in discussion it became apparent that their enriched conditions were considerably more complex than our EC. They rotated rats twice a day among four cages of different sizes, two of the cages being considerably larger than ours and offering more opportunities for climbing and exploration. Also, since they had observed that the rats in their laboratory in Argentina seemed rather timid and slow to venture into the complex environment, they provided the naïve rats with more experienced "guides" that had already been in the environment for several days; in fact, the experimenters sometimes kept up a continuing production of enriched-experience rats in which the originally naïve rats then served as guides to a new group before being removed for brain analyses.

We set up conditions similar to those of the Argentinian laboratory and compared cerebral effects of Ferchmin EC (FEC) with those of our regular EC and IC conditions, using subjects of the Berkeley S_1 strain of rats. Results for weights of brain sections were reported by Bennett (1976, Table 17.3) for several experiments ranging in duration from 4 to 15 days. Both FEC and EC produced significant differences from IC in as little as 4 days; for occipital cortex, FEC vs. IC caused a difference of 5.2% ($P < 0.001$), EC vs. IC, 6.6% ($P < 0.001$); for total cortex, FEC vs. IC, 2.6% ($P < 0.001$), EC vs. IC, 2.2% ($P < 0.001$). The only apparent advantage of FEC over EC was seen in the rapidity of occurrence of differences in the ratio of weight of cortex to rest of brain. On this measure after 4 days, FEC vs. IC yielded a difference of 1.7% ($P < 0.001$), whereas EC vs. IC did not yet show a difference (0.2%, NS); after 8 days, the effects reached 2.6% ($P < 0.001$) and 1.7% ($P < 0.05$), respectively, but by 12 days the two environments yielded comparable effects on the cortical/subcortical ratio—2.6 and 2.3% respectively, both significant at beyond the 0.01 level of confidence. Thus, the more complex FEC environment was not more effective than the standard EC in altering most brain-weight measures.

2.1.4b. Large Social Group in Interlinked Cages. Kuenzle and Knüsel (1974) designed a "superenriched environment" in which a group of 70 rats lived in two large interconnected cages and had to shuttle back and forth across a bridge with changing gates and signals to find food and water. Over the 29-day experimental period, the rats were given successively more complicated problems to solve, and they had to perform athletic feats to survive. In two replications, groups were run simultaneously in the superenriched environment and in a reproduction of the Berkeley EC situation. The rats in the superenriched environment were reported to surpass the rats in EC in weight of occipital cortex, in length of cerebral hemispheres, and in ChE/AChE in occipital cortex, but not in RNA/DNA in occipital cortex. While these results are encouraging in showing even larger brain effects than had been found heretofore, it is clear that this experiment confounds the factors of social and inanimate enrichment, since the superenriched environment contained more stimulation along both dimensions than does the EC condition. Also, the stress on agility in the superenriched environment renders questionable the assertion of Kuenzel and Knüsel that the superenriched condition "is superior to the original [EC] in that it confronts the animal with true learning situations," whereas "the original setup mostly improves the animal's motor performance but does not provide for genuine learning situations."

Davenport (1976) used first the EC treatment and later a modified version of the Kuenzle and Knüsel superenriched environment in experiments on environmental therapy for neonatally hypothyroid rats. In an initial study in 1972 (Davenport *et al.*, 1976), the EC treatment was effective in helping to overcome behavioral deficits in hypothyroid rats. A further study in 1973 did not, however, yield these positive results (Davenport, 1976). Therefore, in later work, Davenport used a version of the superenriched environment in which 45 rats were housed in an apparatus constructed of three large cages interconnected by four tunnels with coded gates; the middle cage contained various objects to be explored. Davenport found the superenriched environment to be highly effective in counteracting the deficiencies caused by experimental cretinism. He did not, however, conduct a direct simultaneous comparison of the relative effectiveness of the standard Berkeley EC vs. the superenriched condition.

We have employed a version of the superenriched environment in our own laboratory; it was similar to that devised by Davenport. For this purpose, we placed three of our regular EC cages side by side on a table, spacing them 12 cm apart; two tunnels made out hardware cloth connected each pair of cages. The tunnels were 8 cm square in cross section; a swinging door could be placed in the middle of each tunnel, and the door could be set to swing in one direction only. On most days, food was placed in one of the end cages and water in the cage at the opposite end, so that the animals had to run back and forth between the cages to obtain food and water. Varied stimulus objects were placed in each of the cages. In our laboratory, the superenriched condition did not produce larger cerebral effects than were found in an EC group that was run simultaneously (Rosenzweig *et al.*, 1978).

2.1.4c. Seminatural Outdoor Environment. In a series of experiments, we placed laboratory rats in a seminatural outdoor environment (SNE) for 30 days and compared the cerebral effects with those found in littermates placed in EC or IC at the same time. The SNE was established in an outdoor "population pit" at the Field Station for Research in Animal Behavior above the Berkeley campus. The population pit is a 9 × 9 m concrete rectangle with a wire-mesh roof; earth was placed on the floor to a depth of about 30 cm. Some stones, branches, and pieces of wood lay on the surface of the dirt, and weeds grew in it. Four stations for food and water were placed in the pit, and food and water were available at at least one of the stations each day. For each experiment, 12 animals were placed together in the SNE.

We found that laboratory rats can thrive in an outdoor enclosure even during a wet winter when the temperature drops to the freezing point. When the ground was not too wet, the rats dug extensive burrows, something that their ancestors had no opportunity to do in the laboratory for over 100 generations. Even the enriched laboratory environment does not permit this kind of activity.

SNE was found to produce cerebral effects that were similar in their pattern of distribution over brain regions, but significantly larger in magnitude than those caused by EC. Some brain-weight data from these experiments were reported by Bennett (1976, Table 17.4). More extensive brain-weight data and also measures of RNA/DNA and AChE are given by Rosenzweig *et al.* (1978, Tables III and IV). In one set of four experiments with $N = 48$ per condition, SNE rats exceeded IC littermates by 11.2% ($p < 0.001$) in weight of occipital cortex and by 6.5% ($p < 0.001$) in the cortical/subcortical weight ratio, while the corresponding percentages for EC vs. IC littermates were 6.0 ($p < 0.001$) and 3.3 ($p < 0.001$).

2.1.5. *Assignment of Animals to Conditions*

To be able to make critical comparisons among animals that have been assigned to the differential environmental conditions, it is necessary to follow certain procedures in the assignment. The groups assigned to the conditions should be rather closely equivalent, but in order not to bias the outcome and to be able to make statistical tests of differences, each animal should have an equal chance of being placed in any of the conditions. To reduce variability and also the possible loss of animals during the course of an experiment, all runts or ill animals are excluded at the start. Insofar as possible, we take litters with at least as many littermates of the desired sex as there are experimental conditions, so that we can make littermate comparisons. As a further restriction on variability, we take only litters within which the range of body weights does not exceed 15%. The littermates are then assigned randomly so that each one goes to a different experimental condition. In other cases when littermates are not available, animals are weight-matched, and a weight-matched group is then treated like a litter.

Various further measures can be taken to restrict variability among subjects to be assigned to differential environments. Here are a few methods that may be considered for this purpose: One is the use of highly inbred strains of rats. We have used the Fischer inbred strain for this reason, but we did not find them to be less variable in brain weights or brain chemistry than rats of our S_1 breeding stock. Culling litters down to a standard number of pups (e.g., 4) provides larger and more uniform pups at weaning than does allowing the mothers to suckle numbers that range from 4 to 12 or more. Since mothers differ in their ability to care for young, Herman Epstein (personal communication) of Brandeis University rotates mothers daily among four litters to produce pups that are as uniform as possible. It is also possible to take the differences among litters into account by appropriate statistical techniques.

Animals of only a single sex are used in a given experiment for three reasons. First, since there are sex differences among rats in brain size and weight and in some measures of problem-solving behavior, the results must be analyzed separately for the two sexes. Second, including both sexes in group conditions can lead to territorial and aggressive behavior and, if the experiment lasts more than a few weeks, to the birth of young. These results of mixing the sexes may make conditions more "natural," but they greatly complicate the conditions. Also, pregnancy may alter brain values of rats (Diamond *et al.*, 1971; Hamilton *et al.*, 1977). Finally, female rats show smaller cerebral effects of environmental restriction than do males (Rosenzweig and Bennett, 1977), and female monkeys show less severe behavioral effects of isolation than do males (Sackett, 1972). Thus, while it has been worthwhile to investigate the effects of environment in each sex, it has been more convenient to perform parallel experiments with the sexes kept separate.

2.1.6. *The Question of a Baseline Environment*

As we have noted in a number of discussions (e.g., Rosenzweig, 1971), we use the terms "enriched environment" and "impoverished environment" only in relation to the baseline of the standard laboratory colony condition. This is a convenient baseline for many purposes, since it is the standard for much research in animal laboratories as well as in laboratories of biochemistry, pharmacology, nutrition, and other disciplines. It has been pointed out that even the "enriched condition" in the laboratory is undoubtedly impoverished compared to what rodents experience in a natural outdoor environment. As noted in Section 2.1.4c, we have found that housing a group of 12 rodents in a 9 × 9 m outdoor enclosure leads to somewhat greater development of brain values than does our indoor EC condition (Bennett, 1976; Rosenzweig *et al.*, 1978). But even this outdoor exposure is not a completely "normal" condition. A full comparison of the physiological and behavioral consequences of truly "normal" and laboratory conditions is difficult to undertake. Lessac and Solomon (1969) commented on this problem in the following way when describing their experiment on learning in "normally" reared and isolation-reared beagles:

> The definition of "normal" rearing is a difficult one. In a laboratory there is no "wild" environment, and so the ethologists stress how "abnormal" the laboratory is. But in the "wild" there are no psychologists teaching animals complex, abstract concepts! There seems to be no alternative, at the moment, to the purely arbitrary definition of normal rearing for each experiment. . . .

Granted that environment is an important variable and that any choice of environment brings some benefits and some liabilities, how should environments be chosen to study many aspects of brain and of behavior? We suggest that for laboratory rodents, an enriched environment of the sort described in Section 2.1.1 and illustrated in Fig. 1 is suitable for many purposes. It is practical to set up and to maintain, and it assures more complete development of both brain and behavior than occurs in restricted or standard colony environments. The trend toward automation of care in colonies that afford a minimum of stimulation runs counter to our recommendation and seems likely to produce inferior subjects for research. Whatever environment is employed in a given study, a full description of the environment and the handling should be a required part of each publication.

3. Behavioral Techniques to Test Effects of Prior Experience

As we mentioned in Section 1, it was found in the 1950's that exposing rats to an enriched environment led to better subsequent learning of a variety of tasks than did experience in an impoverished environment. Davenport (1976) has compiled reports of such experiments; he lists 32 papers giving positive results and 14 findings of no significant enhancement of learning in enriched-experience rats (see his Table 2, pp. 94–95). The positive results came mainly from studies employing relatively complex maze tests (e.g., Hebb–Williams maze, Lashley III maze), whereas lack of effect was mainly reported from simpler maze tests (e.g., multiple T-maze, Y-maze) or from tests of sensory discrimination, but this differentiation is not absolute. Even simpler situations, such as observation of behavior of an animal in an "open field" (e.g., an empty circular enclosure 1 m in diameter), have revealed differences among rats or mice from differential environments (e.g., Woods, *et al.,* 1960; Henderson, 1969). Since many investigators are now using operant or classic conditioning procedures for refined analysis of behavior, it is regrettable that [with the single exception of Ough *et al.* (1972)] these techniques have not been employed to characterize the effects of enriched or impoverished experience.

Description of procedures used in behavioral tests of prior experience would go beyond the scope of this chapter. Even the procedures employed in the apparently simple open-field test turn out to incorporate many variables, and a recent review of research with this instrument shows that seemingly inconsequential details of procedure can alter the behavior significantly (Walsh and Cummins, 1976). Some useful sources on formal observation and experimental testing of behavior of laboratory rodents are: Bureš *et al.* (1976) and Munn (1950).

Perhaps a few additional comments about behavioral procedures may not be amiss for investigators from other disciplines who may decide to incorporate behavioral tests in their programs of investigation.

1. As is true of other fields, published reports of behavioral research are not exhaustive, and they assume a background of knowledge and familiarity on the part of the reader. As we will note later under biochemical analyses (Section 4.4.1), certain chemical procedures require exacting standards of cleaning glassware, while others are less demanding, and we find analogies in the study of rodent behavior. In the case of a male rat placed in a maze for pretraining, the odor of a female rat that has traversed the area previously will be distracting—the male will try to locate the female. On the other hand, the presence of odor from other males of the same colony appears to be reassuring; a rat will begin to explore a maze more readily if it has been "ratted up" than if it is absolutely clean. Male mice, unlike male rats, are not distracted by the previous presence of females in the apparatus. Because of these characteristics of rats, we usually run experiments with only a single sex at a time.

2. Appropriate control of motivation of animal subjects during training and testing is of extreme importance. As Tolman demonstrated in the 1930's, a rat can know the shortest path through a maze but not run that path unless it is suitably motivated, e.g., by a food reward; this phenomenon is termed "latent learning." Furthermore, there may be competing motivations, such as the exploratory tendency in rats, that complicate the outcome.

3. Behavioral aspects of research on brain and behavior are at least as complicated as the cerebral aspects, and a great deal of progress has been made in laboratory studies of animal learning since Thorndike initiated them in 1898. It would be a waste of time and effort if neuroanatomists, neurochemists, and other biological scientists had to rediscover the same knowledge on their own and reinvent testing devices and techniques. For the present, the most promising approach seems to be interdisciplinary research in which investigators trained in behavioral sciences collaborate with investigators from the biological and chemical disciplines. Eventually, there may be scientists who can encompass the essential disciplines within a single skull.

3.1. Hebb–Williams Maze

The test that has most often been used to investigate behavioral differences as a consequence of previous enriched or impoverished experience is the Hebb–Williams (H-W) maze. It has also been used to study effects of postlesion enriched or impoverished experience on problem-solving behavior (e.g., Schwartz, 1964; Will *et al.*, 1976, 1977). We will describe here our own procedures with this test, which differ somewhat from those given in the original description by Rabinovitch and Rosvold (1951). The following description is adapted from that prepared for technicians and students in our laboratory.

The H-W maze test of problem-solving ability consists of a standard series of maze problems. These are set up by means of wooden barriers in a field 76 cm square and 9.8 cm high with a start box (SB) extending from one corner and a goal box (GB) extending from an opposite corner (see Fig. 4A). The apparatus is covered with a Plexiglas lid. The rats run 8 trials per day for food reward; a different problem is presented on each day.

Phase 1: Pretraining. Since rats without prior experience may not run at all, or may behave erratically when placed in a maze, several days of pretraining precede the series of actual test problems. During pretraining, rats are:

1. Deprived of food (except for the time they spend in the GB) and their weights brought down to 80% of predeprivation weight.
2. Introduced to eating a special kind of food (a mash made of ground-up rat chow mixed with water) in a strange situation (the GB of the maze)
3. Trained to run down a straight runway to find food in the GB at the end. (This also furnishes data on effects of prior experience on

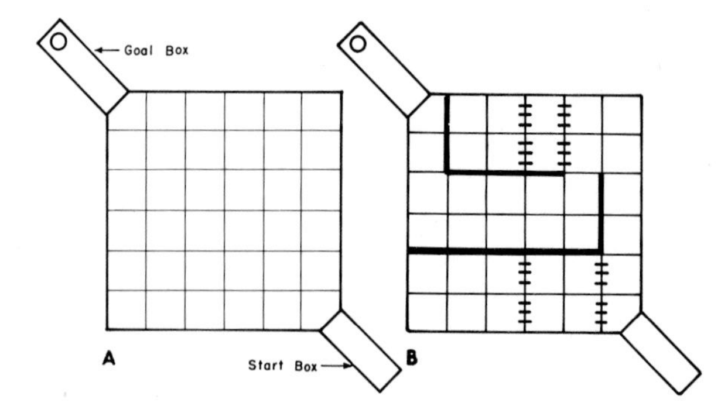

Fig. 4. (A) Hebb–Williams maze apparatus floorplan. (B) Barrier pattern and error zones for problem 3. The light solid lines are painted on the floor of the apparatus. The heavy lines indicate the positions of barriers. The hatched lines indicate error zones; these lines are not present in the maze, but they appear on the score sheets that the experimenters use for recording performance in the maze.

performance in a straightaway, a task that has been used frequently in other contexts.)

4. Introduced to several simple maze problems.

At the end of pretraining, most rats have reached a body weight at which they are optimally motivated and have learned to run through a maze to find food. They are now ready for the standard series of test problems.

Phase 2: Testing. Rats are tested on a new problem each day and are given eight successive trials for each problem: The rat is placed in the start box (SB), run through the maze to the goal box (GB), and allowed 30 sec to eat, and the procedure is repeated until the rat has run through the maze 8 times. The rat is then allowed extra time (about 5 min) to eat in a supplementary GB so it can get enough food to maintain its body weight over the next 24 hr of deprivation.

For each rat, body weight at the start and end of each day's testing is recorded. This way, the rat's weight can be monitored to make sure it stays close to 80% of its predeprivation weight; too much above this, and the rat may not be motivated to run; too much below, and it may be in poor health and even die. In addition, on each of the eight trials, the experimenter:

1. Records how long it takes the rat to come out of SB after the door is opened (SB time).
2. Draws on the record sheet the path the rat takes through the maze and counts any errors it makes.
3. Records how long it takes the rat to reach GB after leaving SB (running time).

The times and error scores are recorded first on a scoring sheet, then transferred to a face sheet that serves as a cumulative record of a single rat's performance throughout pretraining and testing.

On the score sheet for each problem, eight identical maze diagrams like the one shown in Fig. 4B are reproduced. The wooden barriers that define the problem are indicated by heavy solid lines. The hatched lines mark the error zones (these error-zone lines are not present in the maze, but are only on our scoring sheets). For each problem, there is *one* most direct path from SB to GB. If the rat deviates significantly from this path, it will cross one (or more) of the hatched lines. Each entry over such a line by the head and two front feet counts as an error.

The details of scoring start box time, errors, and running time are explained below. All other details of the procedure are best learned in practice.

Recording Errors and Time in the Hebb–Williams Maze

This is simple in principle, but there are a few special features to note.

Initial Errors. The first entry of a rat into an error zone on a given trial is an initial error. Therefore, initial errors for a trial equal the number of different error zones that the rat entered.

Repetitive Errors. Any entry after the first into the same error zone on a given trial is a repetitive error.

On most trials, a rat will make few errors, and one will have no difficulty in recording them. But occasionally a rat will make a large number of errors, and this can make recording difficult. From our experience, we have found that the following method works very well even when a rat makes so many errors that the tracing of its path is extremely complicated, so we recommend this procedure to you: As the rat runs through the maze, trace its path on the score test sheet and keep a running count of *all* errors that it makes, without distinguishing between initial and repetitive errors. When the rat reaches the goal box, that total is recorded. Now check to see how many error lines the rat crossed. This gives you the number of *initial errors.* Subtract initial errors from the total to obtain the number of repetitive errors.

Time Measures

Start Box (SB) time is the time the rat takes to emerge from the SB after the SB door has been opened. The stopwatch is started when the SB door is opened. As soon as the rat emerges into the apparatus and the door behind it is closed, SB time should be recorded. Continue timing with the stopwatch until the rat enters the goal box (GB), then subtract SB time from total time to obtain running time (R) and record R on the sheet.

Running time (R) starts when the rat has left the start box and the door behind it can be closed. Running time ends when the rat has entered the GB and the GB door can be closed.

Retracing. Sometimes a rat will enter the square just before the GB and almost go into the GB, then stop, turn away, and explore the maze further. When this happens, time taken to reach the GB initially is recorded as total time. The stopwatch is permitted to continue running to measure retracing time, that is, cumulative time minus the initial run.

Errors during retracing are tabulated separately. Typically, rats do not make many errors while retracing, but it will be helpful to score all the errors a rat makes on the way to GB. Then, if the rat starts to retrace, jot down the total errors made to the initial contact with GB. Start a new series of initial and repetitive errors for the retracing.

Exploration. Sometimes on later trials after a rat has already learned a pattern well, it will slow down on a particular trial and start to explore. When this happens, enter a big dot (●) beside both errors and total time. This is not done on early trials when most rats explore—record exploration only when a rat's performance on the problem has reached a high level and then the rat slows down and starts to nose around.

Detailed Schedule—Pretraining procedures for Hebb–Williams maze, using pretraining alley:

Day 1. Weigh each rat and put it in a clean individual colony cage with water but no food.

Day 2. Place a dish of mash in GB. Weigh rat and put it by itself in GB with door closed for 15 min. Weigh it again after taking it out. Starting this day and continuing through the experiment, weigh each rat just before putting it into the apparatus and just after removing it, and record these weights on the rat's record sheet.

Day 3. Same as Day 2.

Day 4. Same as Day 2 except only 10 min in GB.

Day 5. Place diagonal pretraining alley into the H-W apparatus. Open door to GB. Weigh rat and put it into SB. Open SB door and start stopwatch. When rat emerges from SB into pretraining alley, close SB door gently behind rat and record time to nearest second. This is SB time. When rat enters GB, close GB door gently and record time in alley. This is R (running) time. Give rat 30 sec in GB, then remove it, place it in SB, open door to GB and then to SB, and time both emergence from SB and alley-running during each trial. Allow rat a maximum of 5 runs in 15 min total in the apparatus. The last 5 min of the 15 is to be allowed for eating; 30 sec of this can be in the GB of the apparatus, and then the rat can be moved to an extra GB for the remaining $4\frac{1}{2}$ min. If the rat runs rapidly, it will not need all 15 min in the apparatus.

Some rats will not emerge readily from SB. It is best go give them a fair amount of time to emerge on their own, so follow this schedule: If rat has not left SB after 5 min, then guide it gently from SB into alley and close SB door behind it. If rat has not entered GB after 10 min in apparatus, guide it gently into GB and close GB door behind it. Record if the rat is guided.

Day 6. Same as Day 5.

Day 7. Remove diagonal alley from apparatus and set up barrier pattern D in maze. Errors are to be recorded from this day on. Detailed procedures are as follows.

1. Set up barrier pattern D in maze.
2. Open GB door; close SB door.
3. Weigh rat and record on face sheet.
4. Place rat in SB; open SB door when rat is facing door and start stopwatch.

5. After rat moves into field, close SB door gently, record SB time, and keep watch running for R time.
6. Record path of run on maze form and mark a stroke for each error.
7. When rat reaches food cup in GB, close GB door gently and stop watch.
8. After 30 sec in GB, remove rat to SB to begin next trial. Record path, errors, and time for each trial.

   Allow rat as many trials (maximum 8) as it can run in a total of 10 min from time it is first placed in SB. At end of last trial, allow rat 5 min in GB; thus, total time in apparatus may reach 15 min. Maximum time in GB will be 9 min.

   From this day on, the final period in GB can be adjusted to keep rat's weight between 80 and 85% of pretraining value. If weight at start of day is below 80%, increase final GB time by 1 min; if weight exceeds 85%, decrease final GB time by 1 min. Greater adjustments can be made, but it is best not to change feeding time more than 1 min from preceding day.

Day 8. Same as Day 7; give maximum of 8 trials with 5 min in GB after final trial. Thus, maximum time in GB is 8.5 min. Use pattern E.

Day 9. Use pattern F. Same as Day 7; give maximum of 8 trials with only 5 min in GB after last trial.

Day 10. Use pattern G. Give rat 8 trials with 30 sec in GB after trials 1–7 and 4 min after trial 8; total 8 min in GB.

Day 11. Use pattern H; otherwise, same as Day 10. The rat has now completed pretraining and is ready to run the standard maze patterns.

Day 12. Pattern 1. Same as Day 10.

Days 13–21. Procedure the same, for problems 2–5 and 7–11. (We have found that a series of 10 problems is sufficient to differentiate groups of rats, so we have shortened the series from 12 to 10 in the interests of economy. In addition, it should be noted that problem 12 of the standard series does not correlate well with results of the other problems and does not reliably separate groups on the basis of previous experience.)

4. Biochemical Techniques

4.1. General Considerations

One of the goals of the environmental manipulations and training that have been described in the previous sections has been to determine whether measureable differences in cerebral anatomy and biochemistry would result. At the initiation of this research, it was assumed (and it has been subsequently confirmed) that the differences produced by environmental manipulations would be relatively small. Therefore, the analytical techniques that have been chosen and developed are ones useful for processing relatively large numbers of samples in a standardized fashion and to high degrees of accuracy and reproducibility. Techniques for four biochemical measures are described.

These have been chosen from a large repertoire of measures that have been used on the basis of their being reliable, sensitive, convenient, and relatively easy indicators of prior environmental influences. These measures are (1) RNA content, (2) DNA content, (3) acetylcholinesterase (AChE) activity, and (4) cholinesterase (ChE) activity.

The brain-weight measures are compatible with and indeed represent a necessary first step in the measurement of either AChE and ChE activities or RNA and DNA content. Anatomical measures that have been used to test effects of differential experience on brain but that will not be described here include cortical thickness (Diamond *et al.*, 1966, 1967), counts of neurons and glia (Diamond *et al.*, 1964, 1966), dendritic spine counts (Globus *et al.*, 1973), dendritic branching (Greenough and Volkmar, 1973), and electron-microscopic measurement of synapses (West and Greenough, 1972; Diamond *et al.*, 1975).

4.2. *Weights of Brain Regions*

It has been found that environmental manipulations produce responses that differ among brain regions, and the most sensitive direct measures include weight of occipital cortex and weight of total cortex. The ratio of cortex to subcortex weight is an even more stable index and is the most reliable weight measure to use to detect the gross anatomical effects of environmental manipulations (Rosenzweig *et al.*, 1972*b*). Total brain weight is relatively independent of overall body growth after weaning and has not been found to be a sensitive indicator of environmental manipulations, other than ones that would be classified as extreme (such as prenatal or infantile malnutrition).

We describe in detail below the procedure employed for the dissection of rat brain into the six basic sections that we use routinely. This procedure can be readily adapted for use with other small rodents; it has been employed with minor modifications in studies with gerbils and mice (Rosenzweig and Bennett, 1969) and with ground squirrels (Rosenzweig *et al.*, 1979).

4.2.1. *Method of Dissection**

The unanesthetized rat is killed by decapitation, either by a Harvard guillotine (Harvard Apparatus Co., Dover, Massachusetts) or by inserting one blade of a pair of 8-inch double-sharp scissors through the neck and rapidly severing the spinal cord; subsequently, the head is completely removed. The calvarium is then removed, as described below, to expose the entire dorsal surface of the brain. [As an aid in visualizing the parts of the skull and brain, the reader may wish to refer to the excellent illustrations in Zeman and Innes (1963).] All dissection is carried out on brown waxed paper

* This description is taken from one distributed by our project starting in 1965, with subsequent revisions.

under a fluorescent light. The brown color provides good contrast with the tissue, and the waxed surface aids in removing bits of brain from the paper. A lamp with a magnifying glass attached (Luxo) is very useful for precise dissection.

After decapitation, the skin of the head is inverted and pushed forward from the neck toward the eyes so as to expose the muscles and the skull. At some points, the muscles can be freed from the underlying bone by cutting with a sharp scissors. Be careful not to cut through the skin with the scissors, because this will probably get hairs into the dissection. To remove the calvarium, first cut bilaterally through the dorsal surface of the medial orbit with a small bone forceps. (We use a Liston bone-cutting forceps, straight $5\frac{1}{2}$-inch.) This cut is at the anterior end of the olfactory bulbs; be careful not to damage the bulbs. With a scalpel, cut through the temporalis muscle 2–3 mm ventral to the attachment of the muscle. This provides an indication of the lateral extent to which the bone should be removed. Cut off the spinal column caudal to the supraoccipital bone. Clip away the calvarium, starting from the caudal extremity of the skull. Place the bone forceps in the foramen magnum and clip the sides of the foramen, enlarging it in an upward direction, again aiming toward the cut in the temporalis. Keep the point of the rongeurs turned outward toward the bone, in order not to damage the brain. Elevate the whole calvarium to the coronal suture. Continue cutting rostrally at an identical lateral level, and then elevate the frontal bone to the first cut in the medial orbit.

After removing the calvarium, cut the dura mater along the midline and reflect it back laterally, removing it from the dorsal surface of both hemispheres. Then position the plastic T-square on the brain as shown in Fig. 5.

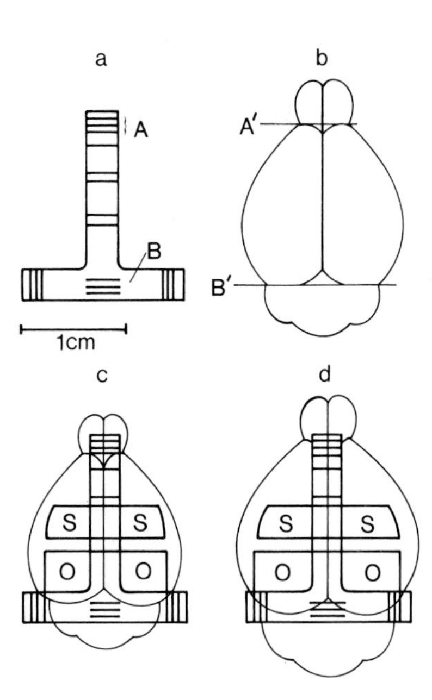

Fig. 5. Delimitation of cortical samples by means of a calibrated plastic T-square. (a) T-square with calibration; (b) dorsal surface of rat brain indicating anterior poles (A′) and posterior poles (B′) of the cerebral hemispheres; (c, d) small and large brains with locations of somesthetic (S) and occipital (O) samples demarcated by the T-square.

The stem of the T is flat on one surface and wedge-shaped on the other. Place the wedge-shaped surface of the stem downward so that it fits between the two hemispheres along the midline.

There are three transverse marks near the end of the stem of the T (as shown at A in Fig. 5a) and three transverse marks on the cross-bar just beyond the point where the stem joins the cross-bar (at B). These transverse lines are used as follows in positioning the T-square along the longitudinal axis of the brain: In Fig. 5b, the anterior extremity of the hemispheres is shown at A′ (the frontal poles) and the posterior extremity at B′ (the occipital poles). Move the T-square back and forth until these boundaries coincide with either the innermost, middle, or outermost pair of transverse lines (for a small, medium, or large brain, respectively). These placements are indicated for a small brain in Fig. 5c and for a large brain in Fig. 5d.

The length of the brain, as measured in this way, is used to determine which cross-marks near the end of the cross-bar to employ in determining the lateral extent of the samples. Thus, with a small brain, the cortical samples will be bounded laterally by the innermost pair of guidelines on the cross-bar of the T, as in Fig. 5c. With a large brain, the boundaries go out to the side lines nearest to the end of the cross-bar of the T, as in Fig. 5d. For the occipital sample, the lateral line should be an extension of the appropriate marking on the cross-bar. Beginning with the somesthetic sample, the lateral boundary should not be straight, but rather should follow the gentle curve of the lateral edge of the brain (see Figs. 5c and d).

Along the stem of the T, three sets of samples are provided for, as shown in Fig. 6. The occipital (O) sample has its anterior boundary indicated by the first cross-line on the stem of the T anterior to the cross-bar. There is then a small space (1.02 mm) between the occipital and somesthetic samples.

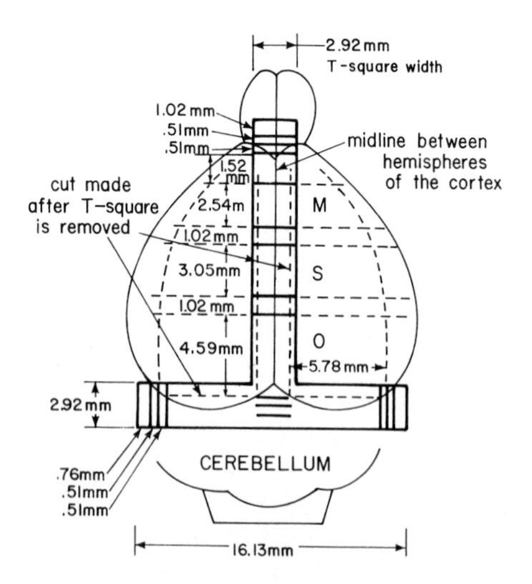

Fig. 6. T-square position on medium-sized adult rat brain. Detailed dimensions of the T-square are shown.

Table 1. Weights of Standard Brain Areas for Colony-Reared S_1 Male Rats

Brain area	Age at time of sacrifice (days)			
	30	60	100	250
	Mean weight (mg) ± S.D.			
Cortex				
Occipital	63 ± 4	66 ± 5	65 ± 5	74 ± 5
Somesthetic	50 ± 3	57 ± 4	56 ± 5	58 ± 4
Remaining dorsal	260 ± 14	299 ± 18	289 ± 17	286 ± 18
Ventral	210 ± 15	243 ± 15	273 ± 16	291 ± 19
TOTAL:	583 ± 30	665 ± 29	683 ± 29	709 ± 32
Subcortex				
Cerebellum	179 ± 12	224 ± 11	239 ± 12	249 ± 16
Medulla	120 ± 9	162 ± 8	180 ± 9	205 ± 12
Remaining subcortex	408 ± 20	477 ± 26	507 ± 32	550 ± 33
TOTAL:	708 ± 37	863 ± 40	925 ± 43	1004 ± 54
Total brain	1290 ± 65	1528 ± 65	1609 ± 63	1713 ± 83
Total cortex/subcortex	0.82 ± 0.02	0.77 ± 0.02	0.74 ± 0.03	0.71 ± 0.02
Body weight (g)	67 ± 12	200 ± 23	297 ± 29	410 ± 40

The next pair of lines bounds the somesthetic sample. Another space of 1.02 mm separates the somesthetic sample from the sample of motor cortex.

To take each sample of cortical tissue, circumscribe it with a scalpel, following the guidelines on the T-square. (We use a scalpel with a narrow blade, e.g., size 11 Bard-Parker.) The cuts made at the posterior end of the occipital sample and the cuts made adjacent to the midline between the cerebral hemispheres are to be made after the T-square has been removed from the brain. The medial boundaries of the sections are made 1 mm lateral to the midline; this is closer to the midline than the sides of the stem of the T would indicate, since the stem is 2.92 mm wide. The posterior border of the occipital sample is a function of the size of the brain. For a small brain (Fig. 5c), the posterior border of O should correspond to the anterior border of the bar of the T, but for larger brains (Figs. 5d and 6), the posterior limit of the occipital section is 0.5–0.8 mm behind the anterior border of the cross-bar of the T. In each case, the posterior corners of the occipital area reach practically to the boundaries of the brain, as shown in Fig. 6.

Before acquiring sufficient practice, a person tends to take occipital samples that are too small. As a guide in the first dissections, one can refer to Table 1, and the weight values should progressively approach the tabulated values as practice proceeds. Table 1 gives weight values at four ages; later tables will give RNA, DNA, AChE, and ChE for the same ages and brain regions. Since brains vary somewhat in size as a function of age, sex, and strain (and even among rats of the same age, sex, and strain), it is especially the relationships among the different samples and the cortical/subcortical ratio that should tend to approach the values indicated. Note that the cortical/subcortical ratio is higher in young rats than in adults; it tends to stabilize at around 0.70 as the animals reach the age of about 100 days. Not only should the means come to resemble the tabulated values, but also standard deviations should decrease for successive groups of dissections and approach the tabulated values, which are representative of those obtained from groups of 12 rats.

The desired cortical samples are thus completely circumscribed and can be peeled from the corpus callosum. To facilitate easy removal of the samples, it is advisable that none of the cuts penetrate the corpus callosum. We usually take the somesthetic sections from both hemispheres as one sample, and subsequently take the occipital sections from both hemispheres as another sample.

Following removal of the somesthetic and occipital areas (and motor area if desired), the remaining dorsal cortex is taken. This sample includes all the cerebral cortex remaining on the dorsal surface of the brain after removal of the preceding sections. We use the long, thin scalpel blade to dissect out the areas medial to the removed O and S samples, to proceed anteriorly to the frontal pole, then to clean off the left hemisphere anterior to the S area, and to continue laterally back to the occipital area and then medially to the original starting point. Take the dorsal cortex in segments to leave as much corpus callosum as possible. The same dissection is then performed on the right hemisphere. As one cuts laterally, the scalpel should rest on the inclined surface of the corpus callosum as it slopes downward and laterally, the point of the scalpel reaching the external surface of the cortex at the point where the brain shows its maximum width. The lateral boundary of this area is not clearly delineated, but with some practice it can be removed in a highly reliable way. By these procedures, we remove all the remaining dorsal cortex (as indicated by the dotted area in Fig. 7), thus leaving the cleaned white corpus callosum exposed.

At this point, the brain is removed from the skull and turned upside

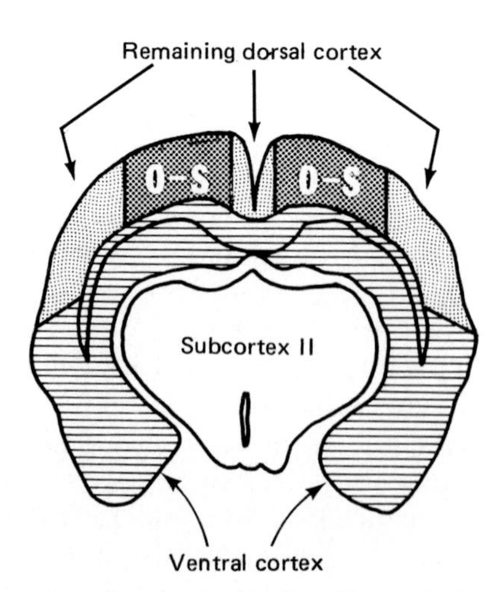

Fig. 7. Schematic transverse section of rat brain showing the occipital and somesthetic samples (O–S), remaining dorsal cortex, ventral cortex, and subcortex.

down. The olfactory bulbs and the olfactory tubercles are cut off and put aside to be used later as part of the "subcortex" sample. The next section, called "ventral cortex," also includes such structures as the corpus callosum, hippocampus, and amygdaloid complex. Working on the ventral surface, we gently free the ventral cortex from the adjacent hypothalamus with a size 15 Bard-Parker scalpel blade. The brain is then turned right side up again, and starting from the posterior aspect of the cortex, the choroid fissure is opened, exposing the internal capsule; this allows a clean separation of the ventral cortex from the underlying subcortex. Anteriorly, where the internal capsule becomes continuous with the corona radiata, the ventral cortical sample must be separated carefully from the underlying caudate nuclei, by cutting through the internal capsule and continuing around the frontal surface of the caudate nucleus. The presence of the lateral ventricle allows for a natural separation of the hippocampus from the underlying diencephalon.

The fifth section is called "rest of the brain" sample (RS). It includes the olfactory bulbs, the medulla and cerebellum, and all the core of the brain remaining after removal of the "ventral cortex." The medulla is separated from the spinal cord at the point where the medulla narrows to form the cord. When the medulla and cerebellum are desired as a separate sample, a cut is made at the cranial border of the pons while the brain is resting in an upside-down position. The cut is aimed toward the cerebellum, allowing for the natural separation between the cerebellum and caudal border of the inferior colliculus.

As each sample is removed, it is weighed and frozen. Rapid routine weighings accurate to 0.1 mg can be obtained with a semimicro projection balance capable of weighing to approximately 0.02 mg. Even more convenient is an electronic analytical balance such as a Mettler HE10, or a Sartorius 1602 or 2004 balance. These may be interfaced with data-printing units. For purposes of weighing, we prepare in advance small squares of waxed paper on which we have written the number of the animal and the section designation. These pieces of paper are weighed first and are then weighed again as soon as the tissue samples are placed on them. The order of removal and weighing is always the same, and the time of these operations is kept as uniform as possible to minimize errors due to drying of the tissues. When the five or six sections described above are removed, six or seven rats per hour can be sacrificed. The samples, on their pieces of paper, are placed on a block of dry ice and are stored in a deep freeze in petri dishes sealed with freezer tape until removal for chemical analysis.

The person performing the dissection and the person doing the weighing do not know the experimental condition to which any individual animal has been assigned, to guard against any possible bias in these procedures.

4.3. RNA and DNA Content

One of the most reliable biochemical indexes of differentiated environmental experience has been the content of RNA and the RNA/DNA ratio. Both these indexes increase significantly in the cortex with enriched experience. In over 600 paired comparisons of rats raised in enriched environments (EC) to littermates raised in impoverished environments (IC), the RNA/DNA ratio of the occipital cortex of the EC rats exceeded that of the IC rats by 7.6%; a difference in this direction was found in 90% of the pairs. For total cortex, the corresponding values were 3.7% increase and 89% of the comparisons.

The total RNA of the cortex of rats in EC exceeded that of rats in IC by 4.9%. This value approximated the percentage difference in cortical weight of the two groups (4.3%). We should note that just as for behavioral procedures and measurements of tissue weight, all chemical analyses are performed "blind" with animals identified only by code number and members of all experimental groups being interdigitated in an analytical series.

The following procedure for the determination of RNA and DNA in brain employs cetyltrimethylammonium bromide as the initial precipitant and represents a modification of the method of Schmidt and Thannhauser (1945), which, with various modifications, has frequently been employed for the determination of nucleic acids in brain. The method is based on the intrinsic UV absorption of the separated and hydrolyzed nucleic acids, and no corrections for interfering absorptions are required. Our method (Morimoto *et al.*, 1974) avoids problems of interference from contaminating materials that are particularly abundant in brain and problems of low recovery that are found with the other modified procedures.

Under routine conditions, 40 brain samples can be analyzed as a group for RNA and DNA over a 2-day period when the sample size permits duplicate analyses, or 80 samples if only single determinations are desired. The reliability of the assay is high; the standard deviation of 14 RNA analyses of one sample was found to be approximately 2%, and for DNA the standard deviation was approximately 3.5%. The standard deviation of the RNA/DNA ratio of the cortex in a sample of 12 rats from a given environmental condition is typically in the range of 2–3%.

In general, duplicate assays result in good agreement. We typically find 75% of our RNA assay values to agree within 3%, and 60% of our DNA duplicate analyses differ less than 4%. Occasionally, when samples are combined or are large (as in the case of the RS sample), triplicate analyses are performed. In these cases, RNA values that are more than 5% from the mean ($6\frac{1}{2}$% for DNA) are discarded.

Values of the nucleic acids for four ages and each of our standard brain regions are presented in Table 2 (RNA/weight); Table 3 (DNA/weight), and Table 4 (RNA/DNA). (A comparable presentation of AChE and ChE activities will be found in Tables 8–10.)

4.3.1. *Reagents for RNA and DNA Analyses*

Ethylenediaminetetraacetic acid buffer (0.11 M). Ethylenediaminetetraacetic acid, disodium salt, dihydrate (EDTA, Aldrich Chemical Co.), 41.8 g/liter, pH adjusted to 5.9 with KOH.

Table 2. RNA of Standard Brain Areas for Colony-Reared S_1 Male Rats

	Age at time of sacrifice (days)			
	30	60	100	250
Brain area	Mean specific RNA (μg/100 mg) ± S.D.			
Cortex				
Occipital	204 ± 7	171 ± 5	156 ± 5	152 ± 6
Somesthetic	199 ± 7	171 ± 6	146 ± 8	145 ± 6
Remaining dorsal	196 ± 6	165 ± 5	146 ± 5	144 ± 5
Ventral	192 ± 6	160 ± 3	145 ± 5	138 ± 4
TOTAL:	196 ± 6	164 ± 3	147 ± 4	143 ± 3
Subcortex				
Cerebellum	232 ± 8	187 ± 4	172 ± 3	167 ± 5
Medulla	195 ± 7	146 ± 3	129 ± 3	114 ± 4
Remaining subcortex	186 ± 5	150 ± 3	138 ± 3	133 ± 4
TOTAL:	200 ± 6	159 ± 3	146 ± 3	138 ± 3
Total brain	198 ± 5	161 ± 2	146 ± 2	140 ± 2
Total cortex/subcortex	0.98 ± 0.03	1.03 ± 0.02	1.01 ± 0.03	1.04 ± 0.03

Cetyltrimethylammonium bromide (0.08 M or 3%). An aqueous 30 g/liter solution of hexadecyltrimethylammonium bromide (CTAB). Technical grade, Eastman Kodak.

Deoxyribonucleic acid. Calf thymus DNA (A grade, Calbiochem) stock solution 2 mg/ml in 0.01 M Tris, pH 8.6. Several drops of chloroform are added to act as a preservative.

Scintillation solution. Forty milliliters Permafluor (Packard Instrument Co., Inc., Downers Grove, Illinois), 200 ml Bio-Solve (Beckman), and 33 ml butyric acid are diluted to 1 liter with toluene.

Ascorbic acid. Ascorbic acid (200 mg/ml) (Calbiochem) in H_2O.

Table 3. DNA of Standard Brain Areas for Colony-Reared S_1 Male Rats

	Age at time of sacrifice (days)			
	30	60	100	250
Brain area	Mean specific DNA (μg/100 mg) ± S.D.			
Cortex				
Occipital	97 ± 6	92 ± 5	97 ± 5	98 ± 4
Somesthetic	92 ± 5	96 ± 4	96 ± 5	96 ± 4
Remaining dorsal	93 ± 4	97 ± 5	98 ± 4	95 ± 5
Ventral	112 ± 5	104 ± 3	102 ± 6	95 ± 5
TOTAL:	100 ± 4	97 ± 3	99 ± 3	95 ± 3
Subcortex				
Cerebellum	757 ± 34	672 ± 30	607 ± 26	584 ± 22
Medulla	127 ± 6	104 ± 4	91 ± 4	80 ± 4
Remaining subcortex	148 ± 5	146 ± 9	143 ± 7	141 ± 9
TOTAL:	298 ± 12	275 ± 11	254 ± 11	239 ± 8
Total brain	208 ± 8	197 ± 7	188 ± 6	179 ± 6
Total cortex/subcortex	0.34 ± 0.02	0.35 ± 0.04	0.39 ± 0.02	0.40 ± 0.02

Table 4. RNA/DNA of Standard Brain Areas for Colony-Reared S_1 Male Rats

Brain area	Age at time of sacrifice (days)			
	30	60	100	250
	Mean specific RNA/DNA ± S.D.			
Cortex				
Occipital	2.10 ± 0.09	1.86 ± 0.06	1.63 ± 0.08	1.56 ± 0.06
Somesthetic	2.16 ± 0.08	1.78 ± 0.06	1.52 ± 0.08	1.51 ± 0.07
Remaining dorsal	2.11 ± 0.07	1.70 ± 0.09	1.49 ± 0.07	1.52 ± 0.09
Ventral	1.73 ± 0.04	1.54 ± 0.04	1.44 ± 0.08	1.46 ± 0.06
TOTAL:	1.96 ± 0.04	1.70 ± 0.05	1.49 ± 0.05	1.51 ± 0.05
Subcortex				
Cerebellum	0.31 ± 0.02	0.28 ± 0.01	0.28 ± 0.01	0.28 ± 0.01
Medulla	1.54 ± 0.03	1.40 ± 0.03	1.42 ± 0.04	1.42 ± 0.03
Remaining subcortex	1.29 ± 0.03	1.03 ± 0.03	0.97 ± 0.05	0.94 ± 0.06
TOTAL:	0.67 ± 0.02	0.58 ± 0.02	0.57 ± 0.03	0.58 ± 0.02
Total brain	0.95 ± 0.02	0.82 ± 0.02	0.78 ± 0.03	0.78 ± 0.03
Total cortex/subcortex	2.93 ± 0.07	2.99 ± 0.06	2.60 ± 0.13	2.62 ± 0.09

With the chloroform preservative, the DNA standard is stable for 4–6 weeks at room temperature; the other reagents are stable for 3–4 months.

4.3.2. *Instruments*

The instruments required are those commonly found in any well-equipped chemistry or biochemistry laboratory and include the following items: (1) A centrifuge such as a Sorvall RC-3 capable of 7000*g*. An HG-4 head and four adapters permitting the centrifugation of 40 samples is desirable. (2) Absorbance measurements are best made with a digital spectrophotometer equipped with a cell holder for four or five cuvettes. We have used a Beckman DU spectrophotometer updated with a Gilford Model 252-1 or a Gilford 220 unit. Other quality spectrophotometers such as a Cary 219 are also suitable. (3) A water-bath shaker maintains temperature at 37°C for RNA hydrolysis; DNA hydrolysis is performed in an oil bath maintained at 70°C. (4) A Vortex mixer is used for mixing the samples.

4.3.3. *Analytical Procedures*

4.3.3a. Sample Preparation. (See also description in Section 4.4.3a, p. 135.) All operations of the sample preparation are carried out at 0–5°C using cold solutions. If the sample size exceeds 100 mg, the sample is homogenized using a Potter-Elvehjem homogenizer in approximately 4 ml EDTA (0.11 M) buffer. Additional buffer is used to rinse the homogenizer and to bring the sample to a final concentration of 25.0 mg/ml. In those cases when the sample size is less than 100 mg (e.g., samples of occipital or somesthetic cortex), the sample is initially homogenized in 2.0 ml buffer, and two 1.0-ml rinses are used to transfer the entire sample to the culture tubes used for the analyses. A 2-ml quantity of 3% CTAB is added to 4 ml brain homogenate in 16 × 75 mm culture tubes, and the precipitate is allowed to form. After 1 hr, the precipitate is collected by centrifugation in a Sorvall RC-3 centrifuge at 7000*g* (5250 rpm) for 15 min. The supernatant is discarded, and the pellet is washed twice with 1 ml H_2O and once with 0.1 N potassium acetate in absolute ethyl alcohol. The pellet is thoroughly dispersed using a Vortex test tube mixer and then centrifuged between washings.

4.3.3b. Analysis of RNA. The pellet is dispersed with 100 μl H_2O and hydrolyzed in 1.1 ml 0.3 N KOH at 37°C for 1 hr in a shaker bath. During this 1-hr hydrolysis, samples are given two additional vigorous mixings using the Vortex mixer. After cooling, the alkaline digest is made 0.2 N in acid by the addition of 500 μl 1.3 N $HClO_4$, and allowed to stand for 15 min at 0°C. After centrifugation at 7000*g* for 15 min, the supernatant is recovered. The acid-insoluble fraction is washed twice with 500 μl 0.2 N $HClO_4$. The three supernatants comprising the RNA are pooled, and the volume is adjusted to 5.0 ml (final concentration $HClO_4$ 0.1 N).

The RNA content is assayed by absorbance at 260 nm, and calculated on the assumption that an absorbance of 1.00 at 260 nm is equivalent to 32 μg RNA/ml. This value has been reported for rat liver (Munro and Fleck, 1966). A value of 31.6 for rat brain RNA using the base composition data of Balazs and Cocks (1967) has been calculated, and values of 31.5 and 31.3 have been calculated for rat liver using data of Munro and Fleck (1966) and Mahler and Cordes (1966), respectively.

4.3.3c. Analysis of DNA. The acid-insoluble fraction is drained by inverting the tubes overnight over absorbant paper (e.g., Kimwipes). The pellet is thoroughly dispersed in the appropriate volume of 1 N $HClO_4$; 2.00 ml is used for cortical samples, 3.00 ml for rest of brain (RS) and medulla, and 6.00 ml for cerebellum. The DNA is heated with frequent mixing for 20 min at 70°C, cooled, and spun at 7000*g* for 15 min. The absorbance of the supernatant is determined at 266 nm. To control for minor variations in hydrolysis, for each group of samples, calf thymus DNA samples subjected to hydrolysis with 1 N $HClO_4$ are used as standards. An absorbance of 1.00 for 45 μg DNA/ml is typically found for calf thymus DNA; base analyses given for calf DNA and for rat tissues are very similar (Sorber, 1970).

4.3.3d. Determination of Radioactivity in the RNA Fraction. If desired, the radioactivity present in cerebral RNA of rats previously injected with [^{14}C]uridine or [^{3}H]uridine can be conveniently determined in the CTAB precipitate after hydrolysis with 0.3 N KOH. Aliquots are taken and further hydrolyzed overnight at 37°C; 500 μl of the hydroylsate is mixed in a glass scintillation vial with 18 ml scintillation solution; 50 μl 20% ascorbic acid is added to eliminate chemiluminescence. Samples are then counted in a liquid scintillation counter (such as a Packard Tricarb 3385 or 460C or a Beckman LS-9000).

4.4. AChE and ChE Activities

As a class, cholinesterases constitute a group of esterases that hydrolyze choline esters at a higher rate than other esters when hydrolysis rates are compared at optimum conditions with respect to substrate concentration, ionic strength, pH, and other parameters. Two classes of enzymes that hydrolyze choline esters exist in biological material—acetylcholinesterase (AChE) and cholinesterase (ChE). Unfortunately, since 1932, when an enzyme was prepared from horse serum by Stedman *et al.* (1932) and called "choline-esterase," considerable confusion has existed in the literature regarding the nomenclature of these two classes of enzymes. For example, AChE has been referred to as acetylcholinesterase, cholinesterase, e-type ChE, specific ChE, ChE I, and aceto ChE. ChE has been referred to as butyro ChE, propionic ChE, benzyl ChE, ChE, pseudo-ChE, s-type ChE, nonspecific ChE, ChE II, and X-ChE. To avoid further confusion, the Commission on Enzymes of the International Union of Biochemistry (Florkin and Stotz, 1965) has recommended that acetylcholine hydrolase (System No. 3.1.1.7) be the formal name and acetylcholinesterase (AChE) the trival name for the enzyme having the higher affinity for acetylcholine than for any other choline ester. In like manner, acylcholine acyl-hydrolase (System No. 3.1.1.8) or cholinesterase (ChE) is the term to be used for the other enzyme or enzymes that hydrolyze certain other esters of choline, i.e., butyrylcholine or propionylcholine, at a higher rate than acetylcholine. Despite these recommendations, it is still common to find confusion in the literature with respect to AChE and ChE.

AChE is widely distributed in the brain, and its principal function is to inactivate the neurotransmitter ACh after its release in the process of transmission of nerve impulses. The role of ChE in the mammalian CNS is still largely obscure. We have used ChE as an index of glial function, since it appears to be concentrated mainly in glial cells and in white fiber tracts. The ratio of AChE activity to ChE activity varies widely from one part of an organism to another, and also varies among species. Numerous studies have shown that in brain, retina, and erythrocytes, a high proportion of the total AChE–ChE activity is due to AChE. Unlike brain, intestine and blood serum are characterized by relatively high ratios of ChE to AChE. The properties of ChE vary more widely from species to species than do those of AChE. Isozymes of AChE have been demonstrated in numerous species.

Historically, acetylcholine, the natural substrate of AChE, or an analogue, acetyl-β-methylcholine (mechoyl), was the substrate of choice for AChE; butyrylcholine or benzoylcholine was the substrate of choice for ChE. Enzymatic activity was most frequently based on the amount of acid liberated on hydrolysis under standardized conditions of pH, substrate concentration, and other parameters. An autotitrator or "pH stat" was frequently employed to measure hydrolysis rates. While considerable precision could be obtained with this method under well-controlled and standardized conditions, it was relatively time-consuming and required considerable attention to details of technique to achieve highly reproducible answers.

In 1961, Ellman *et al.* (1961) described a colorimetric method for the determination of AChE utilizing acetylthiocholine (AcSCh) for substrate and the Ellman reagent, 5,5′-dithiobis-(2-nitrobenzoic acid) (DTNB), as the indicator of the extent of hydrolysis. In addition, by substituting butyrylthiocholine (BuSCh) for AcSCh, this method becomes a sensitive and convenient assay for ChE. The reactions involved are summarized below;

1a. $$H_2O + (CH_3)_3N^+CH_2CH_2SCOCH_3 \xrightarrow[\text{ChE (slow)}]{\text{AChE (fast) or}}$$

$$(CH_3)_3N^+CH_2CH_2S^- + CH_3COO^- + 2H^+$$

1b. $$H_2O + (CH_3)_3N^+CH_2CH_2SCOCH_2CH_2CH_3 \xrightarrow[\text{ChE (medium)}]{\text{AChE (slow)}}$$

$$(CH_3)_3N^+CH_2CH_2S^- + CH_3CH_2CH_2COO^- + 2H^+$$

2. $(CH_3)_3N^+CH_2CH_2S^-$ + O_2N–$C_6H_3(COO^-)$–S–S–$C_6H_3(COO^-)$–NO_2

$\xrightarrow[\text{Nonenzymatic}]{\text{Very fast}}$ $(CH_3)_3N^+CH_2CH_2$—S—S–$C_6H_3(COO^-)$–NO_2

+ O_2N–$C_6H_3(^-OOC)$–S^-

The product, 5-thio-2-nitrobenzoate, has an extinction coefficient of 13,600 at 412 nm; thus, measurement of its rate of formation provides a sensitive and highly specific measure of the hydrolysis rate of thiocholine esters. With appropriate modern spectrophotometers, a skilled analyst can readily perform analyses of 50 samples daily for both AChE and ChE in duplicate with an average difference between duplicate samples of less than 3%. The details of the procedures for AChE and ChE were developed after extensive investigation and comparison of the characteristics of hydrolysis of ACh, AcSCh, BuCh, and BuSCh as a function of pH, substrate concentration, source of enzyme and inhibitor type and concentration.

The hydrolysis rate of ACh (pH stat assay) and AcSCh (spectrophotometric assay) increases as a function of pH, but the pH dependence is less at pH 8.0 than at pH 7.0 (Fig. 8). However, above pH 8.0, the nonenzymatic hydrolysis rate increases rapidly. (The apparent high activity of AcSCh by the pH stat method can be attributed to a shift in the numbers of acid equivalents titrated/mole hydrolyzed as a function of pH.) The relative rates of hydrolysis of AcCh and BuCh (pH stat assay) and AcSCh and BuSCh (spectrophotometric assay) as a function of substrate concentration are compared in Fig. 9. The curves for the two acetyl substrates were very similar, with both substances having a maximum hydrolysis rate at about 10^{-3} M. Since the rate of hydrolysis of AcSCh is only slightly lower at 6×10^{-4} M, and the blank is reduced substantially, we recommend the lower concentration for routine assays. The best value—based on more than 200 analyses—for the ratio of rate of hydrolysis of 6×10^{-4} M AcSCh to 7.7×10^{-4} M ACh by rat brain tissue at pH 7.95 is 0.94. A similar ratio should be obtained with other sources of AChE that contain less than 10% ChE.

Although the enzyme–substrate relationships are highly specific, when one enzyme in a tissue (e.g., AChE) is present in much larger amounts than

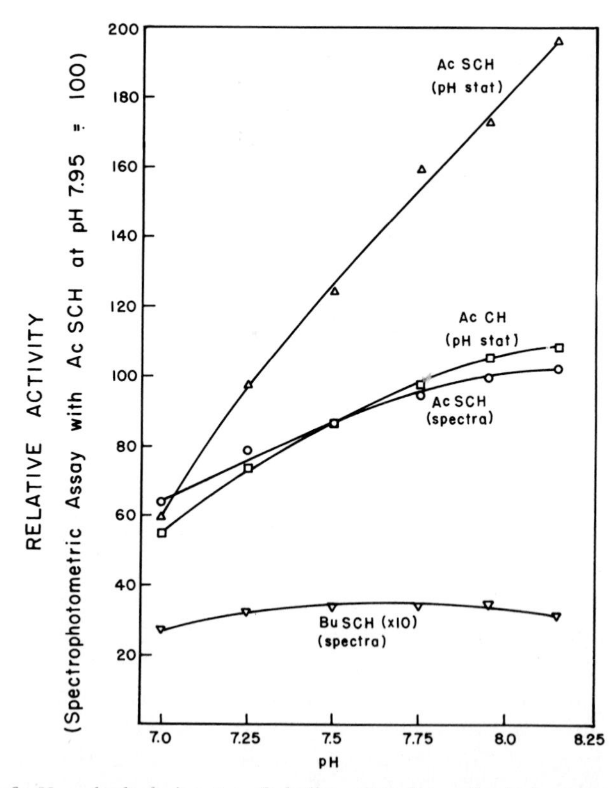

Fig. 8. Effect of pH on hydrolysis rates of choline esters by rat brain homogenate determined by rate of acid liberation (pH stat) and by DTNB method (spectra). The apparent high rate of hydrolysis of AcSCH as measured by the pH stat is discussed in the text.

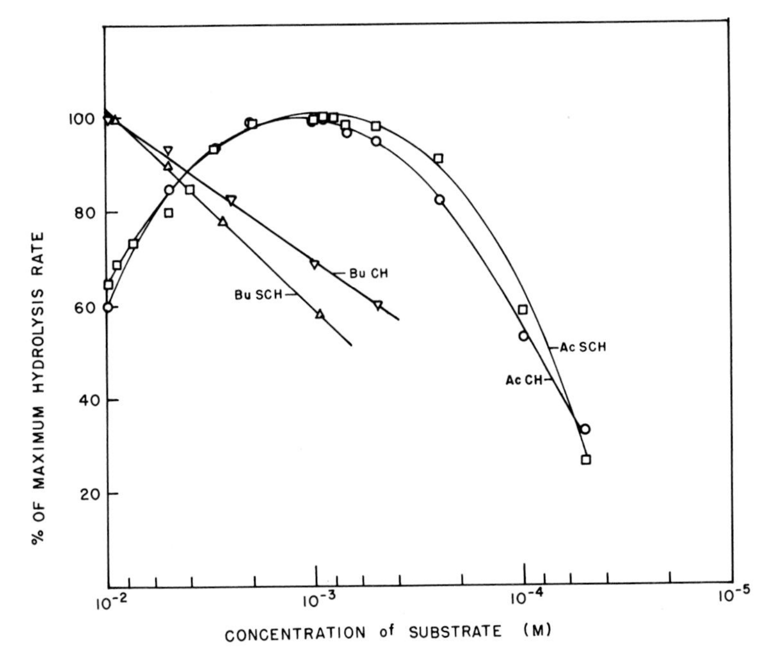

Fig. 9. Effect of substrate concentration on hydrolysis rates of choline esters by rat brain homogenate. The hydrolysis rates using the substrates for AChE are maximal at approximately 10^{-3} M; slightly lower concentrations are recommended to improve the ratio of net hydrolysis rate to blank hydrolysis rate. No concentration maximum is found with the substrates (BuCH, BuSCH) typically employed for ChE.

another related enzyme (e.g., ChE), it is desirable to have good estimates of the relative specificities under the conditions of assay. The availability of separate inhibitors of AChE and ChE in combination with several sources of AChE and ChE has permitted such estimates to be obtained. At the time of development of these assays, Bayless and Todrick (1956) had evaluated several selective inhibitors. These included 1:5-bis(4-trimethylammonium-phenyl)-pentan-3-one diiodide (BW62C47) and 1:5-bis-(4-allyldimethylam-moniumphenyl)-pentan-3-one diiodide (BW284C51), effective as inhibitors of AChE, and 10-(2-diethylaminopropyl)-phenothiazine (Lysivane, ethopro-pazine), effective as an inhibitor of ChE. Bayless and Todrick had used a number of substrates for AChE and ChE, but not the thiocholine esters. BW62C47 (5×10^{-6} M) and BW284C51 (1×10^{-7} M) each resulted in more than 95% inhibition of AcSCh hydrolysis by brain homogenates and less than 20% inhibition of the small amount of BuSCh hydrolysis. BW284C51 (5×10^{-7} M) reduced the hydrolysis of AcSCh by rat intestine by 10% and had no effect on the hydrolysis of BuSCh. Since BW284C51 is commercially available (Sigma Chemical Co.), it is the inhibitor of choice to determine ChE in the presence of AChE.

We were unable to obtain Lysivane, but were able to obtain a closely related compound, 10-(2-dimethylaminopropyl) phenothiazine (Promethazine). We found, using either rat brain or retina, that 2.5×10^{-5} M Promethazine inhibited hydrolysis of AcSCh (AChE activity) less than 5%, while the hydrolysis of BuSCh by intestine was inhibited more than 70%. This concentration is recommended to essentially eliminate any possible interference of ChE in assays for AChE in cerebral tissues of the rat.

Promethazine together with BW284C51 inhibited 99% of the hydrolysis of AcSCh by rat brain. In addition, eserine inhibited 98% of the hydrolysis of ACh, AcSCh, and BuSCh by brain or intestine, confirming that cholinesterases are being measured by the methods described.

Since rat retina contains AChE with little ChE, it provides a convenient source of tissue from the same species to further check the relative activities of AChE against AcSCh and BuSCh and the specificity of inhibitors of cholinesterases. Therefore, hydrolysis of AcSCh and BuSCh by homogenates of rat retina was investigated, and the effect of several concentrations of Promethazine and of 5×10^{-7} M BW284C51 and 10^{-5} M eserine on these hydrolysis rates was also studied. The results, summarized in Table 5, provide the best evidence that we have obtained to date of the high degree of specificity AChE has for AcSCh as compared to BuSCh. Thus, even in the absence of inhibitor, the hydrolysis of BuSCh by the retinal preparation is only 0.3 nmol/min per g, or 1.7% as rapid as the hydrolysis of AcSCh. When the effect of BW284C51 on the hydrolysis of BuSCh by the retinal preparation is considered, the hydrolysis of BuSCh by AChE is about 1% as rapid as the hydrolysis of AcSCh under our conditions of assay. The ChE inhibitor Promethazine inhibits less than 3% of the activity of the retinal homogenate against AcSCh at 4×10^{-5} M, a concentration that inhibits at least 80% of the activity of rat intestinal ChE against either BuSCh or AcSCh. These results confirm the selectivity of Promethazine as well as the high degree of "purity" of rat retina AChE. Over 98% of the activity against AcSCh is inhibited by BW284C51, and essentially all the activity is inhibited by 10^{-5} M eserine. Eserine inhibits the hydrolysis of BuSCh completely.

Table 5. Hydrolysis of Thiocholine Esters by Rat Retina[a]

Substrate	Concentration (M)	Inhibitor	Concentration (M)	Activity (nmol/min/mg)	Relative activity
AcSCh	7.7×10^{-4}	None	—	17.8	100
		Promethazine	4.6×10^{-6}	17.5	98
			4.6×10^{-5}	17.3	97
			9.2×10^{-5}	16.7	94
			2×10^{-4}	15.7	88
AcSCh	7.7×10^{-4}	284c51	5×10^{-7}	0.28	1.6
AcSCh	7.7×10^{-4}	Eserine	10^{-5}	0.28	0.5
BuSCh	1×10^{-3}	None	—	0.30	100
BuSCh	1×10^{-3}	284c51	5×10^{-7}	0.10	33
BuSCh	1×10^{-3}	Eserine	10^{-5}	0.00	0

[a] Rat retinas were homogenized in 0.1 M potassium phosphate buffer, pH 8.0; 1 mg retina in a total volume of 1 ml was used when AcSCh in the absence of inhibitor was used as substrate, and 10 mg tissue was used in all the other assays.

A limited number of studies with dog showed that while the ChE was approximately twice that of the rat, AChE activity was approximately one third that of the rat (Table 6). The fact that ChE represents an appreciable fraction of total cholinesterase activity in the dog but not in the rat emphasizes the desirability of determining the relative activities of AChE and ChE by appropriate inhibitor studies when applying these methods to previously little-studied sources of these enzymes.

4.4.1. *Reagents for AChE and ChE Analyses*

5,5'-Dithiobis-(2-nitrobenzoic acid (0.01 M). DTNB reagent without inhibitor is prepared by dissolving 400 mg DTNB in 100 ml 0.1 M sodium phosphate buffer, pH 7.0. This reagent is not used for routine assays, but is useful for studies of AChE and ChE activities in the absence of inhibitors. DTNB can be obtained from numerous suppliers, including Pierce Chemical Co., Calbiochem Corp., and Sigma Chemical Co.

DTNB-Promethazine (DTNB-I ChE). The stock solution of DTNB containing Promethazine, an inhibitor of ChE, is prepared by dissolving 48.2 mg of Promethazine hydrochloride [(*N*,*N*,α-trimethyl)-10H-phenothiazine-10-ethanamine; 10-(2-dimethylaminopropyl) phenothiazone] and 400 mg DTNB in 100 ml 0.1 M sodium phosphate buffer, pH 7.0. This solution is 1.5×10^{-3} M in Promethazine and 0.01 M in DTNB. Promethazine has been obtained from Wyeth Laboratories and Purpac Pharmaceutical Co.

For some studies, Promethazine may be omitted, since the hydrolysis of AcSCh by the ChE in mouse and rat brain is less than 5% of the total hydrolysis rate. However, as already noted in this section, relative ChE activity may be much higher in other tissues, species, or both.

DTNB-BW284C51 (DTNB-I AChE). For routine assays of ChE activity, a DTNB reagent containing a specific inhibitor of AChE is used. The specificity and commercial availability of 1,5-bis (*N*-allyl-*N*,*N*-dimethyl-4-ammonium-

Table 6. Comparison of Rates of Hydrolysis of Choline Esters by Rat and Dog Brain Tissues

	Substrate and concentration			
	AcSCh [7×10^{-4} M]		BuSCH [10^{-3} M]	
	AChE inhibitor		AChE inhibitor	
	−	+	−	+
Brain area	Hydrolysis rate (nmol/min/mg)			
Rat				
Cortex	8.68	0.53	0.5	0.32
Subcortex	18.4	1.1	0.9	0.57
Dog				
Cortex	2.91	.24	0.70	0.60
Subcortex	11.7	.99	2.20	1.88

phenyl)pentan-3-one dibromide (BW284C51) (Sigma Chemical Co.) make it the inhibitor of choice. A 3×10^{-3} M stock solution of BW284C51 is prepared by dissolving 20.1 mg BW284C51 in 10 ml distilled water; 1 ml of this is added to 400 mg DTNB dissolved in 100 ml water.

The effectiveness of BW284C51 as an inhibitor of AChE cannot be too strongly emphasized: Glassware used for the analysis of ChE in the presence of BW284C51 should be clearly marked and separated from glassware used for the analyses of AChE. Cells and glassware that have been in contact with BW284C51 must not be used for the analysis of AChE unless the articles have been scrupulously cleaned and checked against other known clean cells.

Acetylthiocholine (0.037 M). The AcSCh stock solution is prepared by dissolving 270 mg acetylthiocholine iodide in 25 ml distilled water. AcSChI can be obtained from numerous suppliers including Calbiochem, Sigma Chemical Co., Gallard-Schlesinger Chemical Mfg. Co., and Pfaltz and Bauer.

Butyrylthiocholine (0.063M). The BuSCh stock solution is prepared by dissolving 500 mg butyrylthiocholine iodide in 25 ml distilled water. BuSChI can be obtained from the suppliers of AcSChI.

Unidentified impurities in AcSCh and BuSCh have resulted in markedly low values for AChE and ChE using brain homogenates. New lots of any reagents should be checked against samples of known purity.

We routinely prepare sufficient quantities of these stock reagents for assay of the tissues from one or more behavioral experiments. Aliquot quantities of these stock reagents sufficient for several days' analyses are placed in individual small test tubes and stored frozen. Under these conditions, reagents are stable for at least 1 month.

4.4.2. Instruments

AChE and ChE activities are best determined with a recording spectrophotometer capable of determining the absorbance of four or five samples sequentially at frequent intervals. We have used a Beckman DU spectrophotometer equipped with a Gilford Model 220 absorbance detector, a Model 210 sample changer, and a Model 208 position offset. More modern, more suitable units include the Gilford Model 2400-2 or the Cary Model 219 spectrometer. These units are equipped with a thermostated cell compartment. We have used an auxillary thermostated block to prewarm the samples to 37°C. This permits groups of samples to be run consecutively with no loss of time between groups.

To maximize the number of analyses per day per analyst, it is necessary to have two spectrometers, one system used for AChE, the other set at twice the recorder sensitivity to analyze ChE. These instruments can be used simultaneously by an analyst who puts samples into them alternately. For a limited number of samples, the operator can change samples manually and read and record the absorbance at timed intervals over a 5- to 10-min period. However, these results will be less precise.

4.4.4. *Analytical Procedures*

4.4.3a. Sample Preparations. Since AChE activities are stable in frozen cerebral tissue, it is generally most convenient to use brain sections that have been dissected, weighed, frozen, and stored as described above. On the day of analysis, a petri dish of samples is removed from the deep freeze and placed on a slab of dry ice in an insulated box. To homogenize the samples, a commercial glass homogenizer with a Teflon pestle is used (A.H. Thomas No. 4288). The smaller size (A) is used for all samples weighing less than approximately 750 mg. The intermediate size (B) tissue grinder is used for the subcortical brain section when it includes the cerebellum and medulla. A Sunbeam Mixmaster motor, fitted with a Jacobs chuck into which the pestle can readily be locked, is recommended for homogenizing the samples. The motor is mounted on a heavy, stable supporting stand.

Frozen samples are readily transferred from the wax paper to the glass homogenizing tube, and any small trace of material that remains on the wax paper is wiped off on the Teflon pestle. A minimal size graduate (10, 25, 50, or 100 ml) is used for each sample. Samples are homogenized initially in a volume of the ice-cold 0.1 M sodium phosphate buffer, somewhat less than the final total volume that is desired, and then transferred to the chilled glass-stoppered graduate. The homogenizer and pestle are rinsed with several additional portions of buffer, and the grooves on the bottom of the pestle are inspected to make sure that no tissue is caught in them. These washings are added to the initial homogenate in the graduate. A plastic wash bottle containing chilled phosphate buffer is used for the rinses and to make the dilution to the desired final volume. The graduates containing the samples are stored in ice until the analyses are completed.

It is convenient to prepare a table with the desired final volumes before the samples are homogenized. The homogenate concentrations recommended for analyzing adult rat brain sections for AChE and ChE are summarized in Table 7 with the approximate aliquot used in each 3-ml assay.

Table 7. Recommended Tissue Concentrations and Aliquots for Analyses of Rat Brain AChE and ChE

	For AChE assay			For ChE assay	
Brain area	Tissue conc. (mg/ml)	Aliquot (μl)	pH 8.0 Phosphate buffer (ml)	Aliquot (ml)	pH 8.0 Phosphate buffer (ml)
Occipital cortex	3.0	1000	2.0	3.0	—
Somesthetic cortex	3.0	800	2.0	3.0	—
Remaining dorsal cortex	5.0	400	2.6	2.0	1.0
Ventral cortex	5.0	300	2.6	3.0	—
Subcortical brain	5.0	100	3.0	2.0	1.0
Cerebellum and medulla	5.0	400	2.6	2.0	1.0
Cerebellum	5.0	800	2.0	2.0	1.0
Medulla	3.0	300	2.6	3.0	—

These values are suitable for postweaning rats, and appropriate modifications of tissue homogenate concentrations are made when greatly different activities are expected, as from tissues from very young rats or other species.

4.4.3b. Analysis of AChE. To determine AChE activity, 0.1 M sodium phosphate buffer, pH 8.0, and the desired sample aliquot to make a known final volume of about 3 ml are pipetted into standard spectrophotometer cuvettes (Beckman or similar). "Lambda" pipettes are used to deliver the samples. Repipettes (Lab-Line Instruments, Inc.) provide a convenient and accurate method of adding the required amount of buffer. Alternatively, glass volumetric pipettes may be modified and individually calibrated to deliver 2.6 ml. A 100-μl quantity of DTNB reagent is added to each cell, and the contents are premixed by inverting several times. Small pieces of Parafilm provide convenient and leakproof closures for the cells. The solutions are prewarmed in incubator compartments (37°C) for 10–12 min, then removed, and 50 μl AcSCh reagent is added. Hamilton repeating syringes that dispense 50 μl/aliquot provide a convenient method for adding the DTNB and AcSCh rapidly and accurately. An alternative method is to prefill the required number of "Lambda" pipets with the appropriate reagent and place them horizontally on a grooved wooden or plastic pipette rack. The contents are again rapidly mixed and placed in the cell compartment (37°C) of the recording spectrophotometer. Optical density of 412 nm is recorded sequentially on each sample for 10–15 sec over a 10- to 12-min period. Recorder sensitivity is set for full scale of the recorder to be the equivalent of a change of 1 absorbance unit. Duplicate analyses are routinely made, normally in separate incubations and runs. These duplicate analyses should check within 3%. If they do not, additional analyses are made.

A set of 4 reagent blanks is run at least twice daily by substituting phosphate buffer for the brain homogenate. The AcSCh reagent blank should be about 0.0063 absorbance units/min, or about 8% of the sample

Table 8. AChE of Standard Brain Areas for Colony-Reared S_1 Male Rats

	Age at time of sacrifice (days)			
	30	60	100	250
Brain area	AChE (nmol AcSCh hyd./min/mg) ± S.D.			
Cortex				
Occipital	4.3 ± 0.5	5.8 ± 0.3	5.7 ± 0.4	5.5 ± 0.3
Somesthetic	5.5 ± 0.6	7.0 ± 0.4	7.0 ± 0.4	6.5 ± 0.4
Remaining dorsal	5.9 ± 0.6	7.6 ± 0.4	7.4 ± 0.4	7.0 ± 0.4
Ventral	8.5 ± 0.9	11.7 ± 0.6	11.6 ± 0.7	10.6 ± 0.6
TOTAL:	6.9 ± 0.6	9.1 ± 0.4	9.1 ± 0.4	8.5 ± 0.4
Subcortex				
Cerebellum and medulla	10.9 ± 0.5	9.8 ± 0.4	8.9 ± 0.3	8.4 ± 0.3
Remaining subcortex	24.0 ± 1.7	29.1 ± 1.4	28.0 ± 1.2	25.8 ± 1.1
TOTAL:	18.6 ± 1.3	20.5 ± 0.9	19.4 ± 0.9	18.0 ± 0.9
Total brain	13.3 ± 0.9	15.5 ± 0.6	15.1 ± 0.6	14.0 ± 0.6
Total cortex/subcortex	0.37 ± 0.03	0.44 ± 0.02	0.47 ± 0.02	0.48 ± 0.02

Table 9. ChE of Standard Brain Areas for Colony-Reared S_1 Male Rats

Brain area	Age at time of sacrifice (days)			
	30	60	100	250
	ChE (nmol BuSCh hyd./min/mg) ± S.D.			
Cortex				
Occipital	0.32 ± 0.04	0.34 ± 0.02	0.34 ± 0.03	0.35 ± 0.03
Somesthetic	0.35 ± 0.04	0.38 ± 0.02	0.38 ± 0.03	0.38 ± 0.02
Remaining dorsal	0.31 ± 0.04	0.34 ± 0.02	0.35 ± 0.03	0.36 ± 0.02
Ventral	0.33 ± 0.03	0.33 ± 0.02	0.31 ± 0.02	0.31 ± 0.01
TOTAL:	0.32 ± 0.03	0.34 ± 0.02	0.33 ± 0.02	0.34 ± 0.01
Subcortex				
Cerebellum and medulla	0.61 ± 0.05	0.55 ± 0.03	0.51 ± 0.02	0.50 ± 0.03
Remaining subcortex	0.64 ± 0.05	0.65 ± 0.03	0.63 ± 0.02	0.63 ± 0.03
TOTAL:	0.64 ± 0.04	0.59 ± 0.02	0.58 ± 0.02	0.57 ± 0.03
Total brain	0.50 ± 0.03	0.49 ± 0.02	0.47 ± 0.02	0.47 ± 0.02
Total cortex/subcortex	0.53 ± 0.05	0.56 ± 0.03	0.58 ± 0.03	0.60 ± 0.03

absorbance change. Reaction rates are linear for at least 10 min under the conditions described. High blanks may indicate an accumulation of tissue in the cells, and thorough cleaning is indicated.

4.4.3c. Analysis of ChE. The determination of ChE follows a procedure similar to that for AChE except that larger tissue aliquots and correspondingly less buffer diluent is required, 50 μl DTNB-I AChE is used and BuSCh is substituted for AcSCh, and recorder sensitivity is set at $\frac{1}{2}$ optical density unit per full scale deflection. The blank for BuSCh is approximately 0.0046 absorbance unit/min, about 20% of the rate of change observed with the brain samples. Duplicate ChE analyses should agree within 5%, and the reaction rate should be linear.

Table 10. ChE/AChE of Standard Brain Areas for Colony-Reared S_1 Male Rats

Brain area	Age at time of sacrifice (days)			
	30	60	100	250
	ChE/AChE × 100 ± S.D.			
Cortex				
Occipital	7.1 ± 0.7	6.1 ± 0.4	5.8 ± 0.6	6.3 ± 0.5
Somesthetic	6.0 ± 0.6	5.4 ± 0.4	5.4 ± 0.5	5.7 ± 0.4
Remaining dorsal	5.6 ± 0.5	4.4 ± 0.3	4.6 ± 0.4	5.1 ± 0.4
Ventral	4.0 ± 0.4	2.8 ± 0.2	2.6 ± 0.3	2.8 ± 0.2
TOTAL:	4.9 ± 0.5	3.7 ± 0.2	3.7 ± 0.3	3.9 ± 0.3
Subcortex				
Cerebellum and medulla	5.8 ± 0.5	5.6 ± 0.3	5.6 ± 0.3	6.0 ± 0.3
Remaining subcortex	2.7 ± 0.3	2.4 ± 0.2	2.3 ± 0.1	2.4 ± 0.2
TOTAL:	3.4 ± 0.4	2.9 ± 0.2	3.0 ± 0.2	3.2 ± 0.2
Total brain	3.7 ± 0.3	3.1 ± 0.2	3.2 ± 0.2	3.4 ± 0.2
Total cortex/subcortex	0.142 ± 0.012	0.125 ± 0.008	0.123 ± 0.009	0.124 ± 0.008

Calculation of AChE or ChE activity is based on the rate of change of absorbance (determined by fitting the best straight line to the recorder plot), the total assay volume, weight of sample, homogenate volume, aliquot of sample, and molar extinction of the 5-dinitrobenzoate anion. Answers are typically expressed in terms of nmoles of AcSCh or BuSCh hydrolyzed/min per mg wet weight tissue. As a guide in checking your procedure, typical values and standard deviations are given for the areas of brain that we usually analyze for rats of four different ages in Tables 8–10. AChE and weights of 15 brain areas from six lines of rats have been compared by Bennett *et al.* (1966). The 5-dinitrobenzoate anion has an extinction coefficient of 13,600 at 412 nm at pH 8.0.

NOTE ADDED IN PROOF. Since the writing of this paper, several other articles and reviews on effects of differential environments have appeared or are in press. A selection of these references is listed here: Diamond (1978), Floeter and Greenough (1979), Greenough and Juraska (1979), Greenough, *et al.* (1979), Jørgensen and Meier (1979), Rosenzweig (1979), Rosenzweig and Bennett (1978), Szeligo and Leblond (1977), and Walsh (1980*a*,*b*).

ACKNOWLEDGMENTS. The research of our project referred to in this chapter was supported by the Division of Biomedical and Environmental Research, Office of Environmental Research & Development, U.S. Department of Energy under Contract No. W-7405-ENG-48 (and still earlier by the Atomic Energy Commission) and by grants from the National Science Foundation and the National Institute of Mental Health (currently Grant 5 R01 MH 26704-5). Several of the papers referenced give specific acknowledgments; other material was newly compiled for this chapter. Marie Hebert Alberti and Hiromi Morimoto have been responsible for many years for devising and carrying out the dissection and biochemical procedures; Donald Dryden and Kenneth Chin were responsible for the behavioral procedures, and Jessie Langford maintained the records and carried out secretarial functions of the project.

References

Balazs, R., and Cocks, W.A., 1967, RNA metabolism in subcellular fractions of brain tissue, *J. Neurochem.* **14:**1035–1055.

Bayless, B.J., and Todrick, A., 1956. The use of a selective acetylcholinesterase inhibitor in the estimation of pseudocholinesterase activity in rat brain, *Biochem. J.* **62:**62–67.

Bennett, E.L., 1976, Cerebral effects of differential experience and training, in: *Neural Mechanisms of Learning and Memory* (M.R. Rosenzweig and E.L. Bennett, eds.), pp. 279–287, MIT Press, Cambridge, Massachusetts.

Bennett, E.L., Diamond, M.C., Krech, D., and Rosenzweig, M.R., 1964, Chemical and anatomical plasticity of brain, *Science* **146:**610–619.

Bennett, E.L., Diamond, M.C., Morimoto, H., and Hebert, M., 1966. Acetylcholinesterase activity and weight measure in fifteen brain areas from six lines of rats. *J. Neurochem.* **13:**563–572.

Bennett, E.L., Rosenzweig, M.R., Morimoto, H., and Hebert, M., 1979. Maze training alters brain weights and cortical RNA/DNA ratios, *Behav. Neurol. Biol.* **26:**1–22.

Bureš, J., Burešová, O., and Huston, J., 1976, *Techniques and Basic Experiments for the Study of Brain and Behavior,* Elsevier/North-Holland, Biomedical Press, Amsterdam.

Davenport, J.W., 1976, Environmental therapy in hypothyroid and other disadvantaged animal populations, in: *Environments as Therapy for Brain Dysfunction* (R.N. Walsh and W.T. Greenough, eds.), pp. 71–114, Plenum Press, New York.

Davenport, J.W., Gonzalez, L.M., Carey, J.C., Bishop, S.B., and Hagquist, W.W., 1976, Environmental stimulation reduces learning deficits in experimental cretinism, *Science* **191:**578–579.

Diamond, M.C., 1978, The aging brain: Some enlightening and optimistic results, *Am. Sci.* **66:**66–71.

Diamond, M.C., Krech, D., and Rosenzweig, M.R., 1964, The effects of an enriched environment on the histology of the rat cerebral cortex, *J. Comp. Neurol.* **123:**111–119.

Diamond, M.C., Law, F., Rhodes, H., Lindner, B., Rosenzweig, M.R., Krech, D., and Bennett, E.L., 1966, Increases in cortical depth and glia numbers in rats subjected to enriched environment, *J. Comp. Neurol.* **128:**117–125.

Diamond, M.C., Lindner, B., and Raymond, A., 1967, Extensive cortical depth measurements and neuron size increases in the cortex of environmentally enriched rats. *J. Comp. Neurol.* **131:**357–364.

Diamond, M.C., Johnson, R.E., and Ingham, C., 1971, Brain plasticity induced by environment and pregnancy, *Int. J. Neurosci.* **2:**171–178.

Diamond, M.C., Lindner, B., Johnson, R., Bennett, E.L., and Rosenzweig, M.R., 1975, Differences in occipital cortical synapses from environmentally enriched, impoverished, and standard colony rats, *J. Neurosci. Res.* **1:**109–119.

Ellman, G.L., Courtney, K.D., Andres, V.N., Jr., and Featherstone, R.M., 1961, A new and rapid determination of acetylcholinesterase activity, *Biochem. Pharmacol.* **7:**88–95.

Ferchmin, P.A., Eterović, V.A., and Caputto, R., 1970, Studies of brain weight and RNA content after short periods of exposure to environmental complexity, *Brain Res.* **20:**49–57.

Ferchmin, P., Bennett, E.L., and Rosenzweig, M.R., 1975, Direct contact with enriched environment is required to alter cerebral weight in rats, *J. Comp. Physiol. Psychol.* **88:**360–367.

Floeter, M.K., and Greenough, W.T., 1979, Cerebeller plasticity: Modification of Purkinje cell structure by differential rearing in monkeys, *Science* **206:**227–229.

Florkin, M., and Stotz, E.H., 1965, *Comprehensive Biochemistry,* Elsevier, Amsterdam.

Forgays, D.G., and Forgays, J.W., 1952, The nature of the effect of free-environmental experience on the rat, *J. Comp. Physiol. Psychol.* **45:**322–328.

Fuller, J.L., 1966, Transitory effects of experiential deprivation upon reversal learning in dogs, *Psychon. Sci.* **4:**273–274.

Globus, A., Rosenzweig, M.R., Bennett, E.L., and Diamond, M.C., 1973, Effects of differential experience on dendritic spine counts, *J. Comp. Physiol. Psychol.* **82:**175–181.

Gluck, J.P., Harlow, H.F., and Schiltz, K.A., 1973, Differential effect of early enrichment and deprivation on learning in the rhesus monkey (*Macaca mulatta*), *J. Comp. Physiol. Psychol.* **84:**598–604.

Greenough, W.T., 1976, Enduring brain effects of differential experience and training, in *Neural Mechanisms of Learning and Memory* (M.R. Rosenzweig and E.L. Bennett, eds.), pp. 255–278, MIT Press, Cambridge, Massachusetts.

Greenough, W.T. and Juraska, J.M., 1979, Experience-induced changes in brain fine structure: Their behavioral implications, in: *Development and Evolution of Brain Size: Behavioral Implications,* (M.E.Hahn, C. Jensen, and B.C. Dudek, eds.), pp. 295–320, Academic Press, New York.

Greenough, W,T., and Volkmar, F.R., 1973, Pattern of dendritic branching in occipital cortex of rats reared in complex environments, *Exp. Neurol.* **40:**491–504.

Greenough, W.T., Juraska, J.M. and Volkmar, F.R., 1979, Maze training effects on dendritic branching in occipital cortex of adult rats, *Behavior. Neurol. Biol.* **26:**287–297.

Hamilton, W.L., Diamond, M.C., Johnson, R.E., and Ingham, C.A., 1977, Effects of pregnancy and differential environments on rat cerebral cortical depth, *Behav. Biol.* **19:**333–340.
Henderson, N.D., 1969, Prior treatment effects on open-field emotionality: The need for representative design, *Ann. N.Y. Acad. Sci.* **159:**860–868.
Hymovitch, B., 1952, The effects of experimental variations on problem-solving in the rat, *J. Comp. Physiol. Psychol.* **45:**313–321.
Jørgensen, O.S., and Meier, E., 1979, Microtubular proteins in the occipital cortex of rats housed in enriched and in impoverished environments, *J. Neurochem.* **33:**381–382.
Krech, D., Rosenzweig, M.R., and Bennett, E.L., 1960, Effects of environmental complexity and training on brain chemistry, *J. Comp. Physiol. Psychol.* **53:**509–519.
Kuenzle, C.C., and Knüsel, A., 1974, Mass training of rats in a superenriched environment, *Physiol. Behav.* **13:**205–210.
Lambert, J.-F., and Truong-Ngoc, A., 1976, Influence de l'environnement instrumental et social sur la structure d'un échantillon du cycle veille-sommeil chez le rat Wistar male: Corrélations avec les modifications de l'excitabilité du système réticulo-cortical, *Agressologie* **17:**19–25.
Lessac, M.S., and Solomon, R.L., 1969, Effects of early isolation on the later adaptive behavior of beagles: A methodological demonstration, *Dev. Psychol.* **1:**14–25.
Mahler, H.R., and Cordes, E.H., 1966, *Biological Chemistry,* p. 176, Harper and Row, New York.
McGinty, D.J., 1971, Encephalization and the neural control of sleep, in: *Brain Development and Behavior* (D.J. McGinty and A.M. Adinolfi, eds.), pp. 335–357, Academic Press, New York.
Melzack, R., 1962. Effects of early perceptual restriction on simple visual discrimination, *Science* **137:**978–979.
Morimoto, H., Ferchmin, P.A., and Bennett, E.L., 1974, Spectrophotometric analysis of RNA and DNA using cetyltrimethylammonium bromide, *Anal. Biochem.* **62:**436–448.
Munn, N.L., 1950, *Handbook of Psychological Research on the Rat,* Houghton Mifflin, Boston.
Munro, H.N., and Fleck, A., 1966, The determination of nucleic acids, in: *Methods of Biochemical Analysis,* Vol. 14 (D. Glick, ed.), pp. 113–176, Interscience, New York.
Ough, B.R., Beatty, W.W., and Khalili, J., 1972, Effects of isolated and enriched rearing on response inhibition, *Psychon. Sci.* **27:**293–294.
Rabinovitch, M.S., and Rosvold, H.E., 1951, A closed-field intelligence test for rats, *Can. J. Psychol.* **5:**122–128.
Rosenzweig, M.R., 1971, Effects of environment on development of brain and of behavior, in: *The Biopsychology of Development* (E. Tobach, L.R. Aronson, and E. Shaw, eds.), pp. 303–342, Academic Press, New York.
Rosenzweig, M.R., 1979, Responsiveness of brain size to individual experience: Behavioral and evolutionary implications, in: *Development and Evolution of Brain Size: Behavioral Implications* (M.E. Hahn, C. Jensen, and B.C. Dudek, eds.), pp. 263–294, Academic Press, New York.
Rosenzweig, M.R., and Bennett, E.L., 1969, Effects of differential environments on brain weights and enzyme activities of gerbils, rats, and mice, *Dev. Psychobiol.* **2:**87–95.
Rosenzweig, M.R., and Bennett, E.L., 1972, Cerebral changes in rats exposed individually to an enriched environment, *J. Comp. Physiol. Psychol.* **80:**304–313.
Rosenzweig, M.R., and Bennett, E.L., 1977, Effects of environmental enrichment or impoverishment on learning and on brain values in rodents, in: *Genetics, Environment and Intelligence* (A. Oliverio, ed.), pp. 163–196, Elsevier/North-Holland, Amsterdam.
Rosenzweig, M.R., and Bennett, E.L., 1978, Experiential influences on brain anatomy and brain chemistry in rodents, in: *Studies on the Development of Behavior and the Nervous System. Early Influences* (G. Gottlieb, ed.), pp. 289–327, Academic Press, New York.
Rosenzweig, M.R., Bennett, E.L., and Sherman, P.W., 1979, Effects of field and laboratory environments on development of brain in ground squirrels: Evolution of brain plasticity. *Soc. Neurosci. Abstr.* **5:**634.
Rosenzweig, M.R., Krech, D., and Bennett, E.L., 1961, Heredity, environment, brain biochemistry, and learning, in: *Current Trends in Psychological Theory,* pp. 87–110, University of Pittsburgh Press.
Rosenzweig, M.R., Krech, D., Bennett, E.L., and Diamond, M.C., 1962, Effects of environmental

complexity and training on brain chemistry and anatomy: A replication and extension, *J. Comp. Physiol. Psychol.* **55**:429–437.

Rosenzweig, M.R., Love, W., and Bennett, E.L., 1968, Effects of a few hours a day of enriched experience on brain chemistry and brain weights, *Physiol. Behav.* **3**:819–825.

Rosenzweig, M.R., Bennett, E.L. and Diamond, M.C., 1972*a*, Brain changes in response to experience, *Sci. Am.* **226**(2):22–29.

Rosenzweig, M.R., Bennett, E.L., and Diamond, M.C., 1972*b*, Chemical and anatomical plasticity of brain: Replications and extensions, 1970, in: *Macromolecules and Behavior*, 2nd ed. (J. Gaito, ed.), pp. 205–277, Appleton-Century-Crofts, New York.

Rosenzweig, M.R., Bennett, E.L., Hebert, M., and Morimoto, H., 1978, Social grouping cannot account for cerebral or behavioral effects of enriched environments, *Brain Res.* **153**:563–576.

Sackett, G.P., 1972, Isolation rearing in monkeys: Diffuse and specific effects on later behavior, in: *Modèles Animaux du Comportement Humain,* Colloques Internationaux du C.N.R.S., Paris.

Schmidt, G., and Thannhauser, S.J., 1945, A method for the determination of desoxyribonucleic acid, ribonucleic acid, and phosphoproteins in animal tissues, *J. Biol. Chem.* **161**:83–89.

Schwartz, S., 1964, Effect of neonatal cortical lesions and early environmental factors on adult rat behavior, *J. Comp. Physiol. Psychol.* **57**:72–77.

Sorber, H.A., 1970, *Handbook of Biochemistry,* 2nd ed., pp. H-97–H-98, Chemical Rubber Co., Cleveland, Ohio.

Stedman, E., Stedman E., and Easson, L.H., 1932, Choline-esterase: An enzyme present in the blood-serum of the horse, *Biochem. J.* **26**:2056–2066.

Szeligo, F., and Leblond, C.P., 1977, Response of the three main types of glial cells of cortex and corpus callosum in rats handled during suckling or exposed to enriched, control and impoverished environments following weaning, *J. Comp. Neurol.* **172**:247–264.

Tagney, J., 1973, Sleep patterns related to rearing rats in enriched and impoverished environments, *Brain Res.* **53**:353–361.

Walsh, R.N., 1980*a*, Effects of environmental complexity and deprivation on brain chemistry and physiology: A review, *Intern. J. Neurosci.* **11**:77–89.

Walsh, R.N., 1980*b*, Effects of environmental complexity and deprivation on brain anatomy and histology: A review, *Intern. J. Neurosci.* (in press).

Walsh, R.N., and Cummins, R.A., 1976, The open-field test: A critical review, *Psychol. Bull.* **83**:482–504.

Walsh, R.N., Budtz-Olsen, O.E., Penny, J.E., and Cummins, R.A., 1969, The effects of environmental complexity on the histology of the rat hippocampus, *J. Comp. Neurol.* **137**:361–365.

West, R.W., and Greenough, W.T., 1972, Effect of environmental complexity on cortical synapses of rats: Preliminary results, *Behav. Biol.* **7**:279–284.

Will, B.E., Rosenzweig, M.R., and Bennett, E.L., 1976, Effects of differential environments on recovery from neonatal brain lesions, measured by problem-solving scores, *Physiol. Behav.* **16**:603–611.

Will, B.E., Rosenzweig, M.R., Bennett, E.L., Hebert, M., and Morimoto, H., 1977, Relatively brief environmental enrichment aids recovery of learning capacity and alters brain measures after postweaning brain lesions in rats, *J. Comp. Physiol. Psychol.* **91**:33–50.

Wilson, M., Warren, J.M., and Abbott, L., 1965, Infantile stimulation, activity and learning by cats, *Child. Dev.* **36**:843–853.

Woods, P.J., Ruckelshaus, S.I., and Bowling, D.M., 1960, Some effects of "free" and "restricted" environment rearing conditions upon adult behavior in the rat, *Psychol. Rep.* **6**:191–200.

Yeterian, E.H., and Wilson, W.A., 1976, Cross-modal transfer in rats following different early environments, *Bull. Psychonomic Soc.* **7**:551–553.

Zeman, W., and Innes, J.R.M., 1963, *Craigie's Neuroanatomy of the Rat,* Academic Press, New York.

3

Cell Fractionation

Robert Lahue

1. Introduction

For neurobiological work, and in general, the main aim of tissue-fractionation techniques is the study of the activity, composition, and morphology of subcellular components. An ultimate goal involves an understanding of the functional interrelationships of different cellular components as they are intergrated in the intact cell. In pursuit of these ends, two basic procedures are always followed. Initially, the membranes of the cells of interest are disrupted with the consequent release of subcellular components. This is usually accomplished by the application of shearing forces to the intact cells. Following this "homogenization" procedure, the "homogenate," suspended in some medium, is subjected to gravitation-like forces that cause the subcellular particles to sediment toward the bottom of the medium. Since the rate of this sedimentation is largely dependent on particle size and density, two parameters in which the particles do vary considerably, it is possible to achieve a separation of different particle populations. These separated subcellular particle populations then become the subject of further investigation—e.g., biochemical, morphological. Both homogenization and separation will be discussed below, but since the separation technique of centrifugation has played the primary role in the development of cell fractionation and is in fact its more critical aspect, it will be treated in greater detail.

1.1. Brief History of Centrifugal Fractionation

While particles that are dense enough will sediment through a medium under the force of gravity alone, this force is not great enough to sediment subcellular particles. The use of a centrifuge, however, does allow for the

Robert Lahue • Department of Psychology, Renison College, University of Waterloo, Waterloo, Ontario, Canada.

sedimentation of subcellular particles. Basically, a centrifuge is a device that produces a gravitation-like force, though of considerably greater magnitude than gravity, by rotating the medium and the particles contained therein through a circular orbit, at high velocities. The first systematic use of this possibility was initiated by Behrens (1932) and Bensley and Hoerr (1934), although the technical and theoretical developments of centrifugation by Svedberg and his colleagues as much as ten years earlier clearly laid the groundwork for the fractionation techniques (for their detailed treatment of ultracentrifugation, see Svedberg and Pedersen, 1940).

By the late 1940's, the technique of centrifugal fractionation of tissue homogenates had been regularized in a form that remains the basic format for current work. Claude (1946), Hogeboom *et al.* (1946, 1948) and Schneider (1946, 1948) developed a scheme for obtaining four fractions from rat or mouse liver. (Liver has remained perhaps the most extensively analyzed tissue in fractionation studies.) Following homogenization, originally in salt solutions but later in a sucrose medium (either 0.88 or 0.25 M), successive centrifugations yielded two fractions, one of which was rich in nuclei and the other in mitochondria. A third fraction, at that time of less obvious content, the microsomal, contained membrane fragments and was rich in endoplasmic reticulum. Finally, a supernatant fraction from the final centrifugation contained soluble, low-density material. An extension of this scheme by Novikoff *et al.* (1953), which is still often used in some form, resulted in the separation of six to ten successive fractions. As pointed out by deDuve (1971), this scheme was subsequently shown to have separated most of the cellular entities recognized at present, such as nuclei, mitochondria, lysosomes, peroxisomes, endoplasmic reticulum, and plasma membranes. Technical breakthroughs, including the commercial availability of high-speed centrifuges incorporating the latest design features, resulted in an ever-increasing amount of research on subcellular fractions, while modifications of technique such as the use of novel media produced greater refinement of results.

While the complex structure and function of neural tissue suggested for a time that subcellular fractionation might not be an appropriate technique for studying the brain, it did not take long for this misconception to be countered. Although presenting special problems, brain-tissue fractionations could be achieved with satisfactory resolution. Employing Claude's "four-fraction" scheme, Brody and Bain (1952) were able to study brain mitochondria, although the mitochondrial fraction from brain tissue is much more heterogeneous than that of, for instance, liver. A concentrated effort to analyze brain tissue originated with Whittaker and his associates (reviewed in Whittaker, 1965), who were interested in the distribution of acetylcholine and the enzymes involved in its metabolism. Thus, both acetylcholine and the enzyme that synthesizes it (choline acetyltransferase) were found to be largely bound in particulate fractions (Hebb and Smallman, 1956; Hebb and Whittaker, 1958), with the remainder free in the supernatant. Though acetylcholine and choline acetyltransferase were first localized in the mitochondrial fraction, a subsequent six-fraction scheme demonstrated that bound acetylcholine and choline acetyltransferase could be isolated in a fraction distinct from that containing the majority of the mitochondria (Whittaker, 1959). Later, it became apparent that the particles in this fraction consisted of pinched-off synaptic areas that were relatively insusceptible to disruption during homogenization. These synaptosomes, as they came to be called, contained small membrane-bounded synaptic vesicles that, in turn, were found to contain the bound neurotransmitter and synthetic enzyme.

2. Basic Information on Centrifugation

2.1. Centrifugal Force

When a particle moving around a circle of radius R at a constant speed v is at a point P, its velocity vector v may be drawn tangent to the circle at that point (Fig. 1a). Slightly later, when the particle has moved to point P', its velocity vector is represented by v'. The lengths (magnitudes) of the two vectors are the same, since the particle is assumed to be moving around the circle with a constant velocity. These two vectors do have slightly different directions, and thus, the particle having moved from P to P', the velocity of the particle has changed. To compute this change (Δv), the velocity vectors v and v' may be drawn from a common point (Fig. 1b). Thus, $v + \Delta v = v'$ or $\Delta v = v' - v$. Since the angle, $\Delta\theta$, between the two radii in Fig. 1a is the same as the angle between v and v' in Fig. 1b, and therefore the two isosceles triangles OPP' and LMN are similar, a proportion between the lengths of the corresponding sides may be written: $\Delta v/v = \Delta h/R$ and $\Delta v = v\Delta h/R$.

If the particle moves from point P to point P' in a time Δt, the equation for the magnitude of Δv may be written $\Delta v/\Delta t = v\Delta h/R\Delta t$. As Δt becomes very small, P and P' will be close together and as, in the limit, $\Delta t \rightarrow 0$, $\Delta h/\Delta t$ will equal v, the instantaneous speed. Thus, $\Delta v/\Delta t = v^2/R$. The left side of the equation becomes the instantaneous acceleration a_c of the moving particle, so $a_c = v^2/R$. Therefore, in traveling around the circle at constant speed, the particle is subject to an acceleration of constant magnitude a_c. As may be noted in Fig. 1b, as Δt (and therefore $\Delta\theta$) becomes very small, Δv becomes perpendicular to v. Since v is tangent to the circular path, a_c must be directed toward the center of the circle. This (a_c) is called the centripetal acceleration of the particle. The velocity vector of the particle moving at a constant speed around the circle does not change its magnitude, but it does change its direction at a constant rate, and this means that the particle experiences an acceleration of constant magnitude. The direction of the acceleration vector is, at any instant, perpendicular to the path and so has no component along the path; thus, the speed of the particle is unchanged by the centripetal acceleration. It may be said that the function of the centripetal acceleration is only to change the direction of motion of the particle, so that it rounds the circle instead of following a straight path (i.e., along the tangent) at constant speed, as it would if subject to no acceleration. Although the

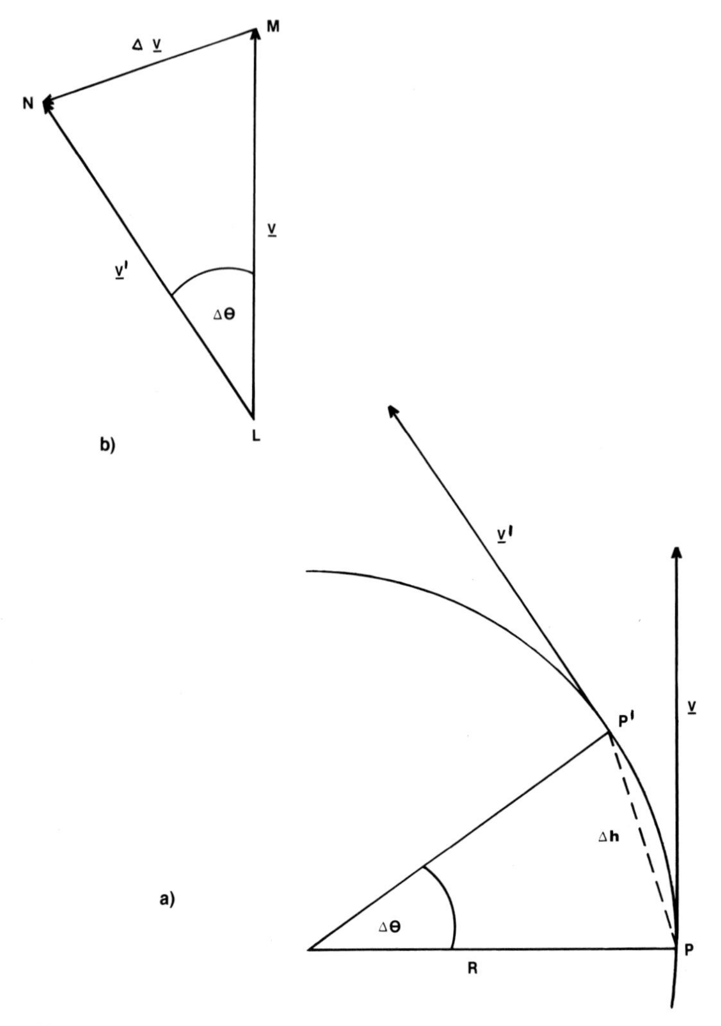

Fig. 1. Centrifugal force. (a) Depiction of the velocity of a particle moving in a circular orbit; (b) depiction of the change in the velocity of the particle necessary to maintain its circular orbit. Consult the text for further details.

particle remains at a constant distance from the center of rotation, it is always "falling" toward the center in the sense of veering inward from the momentary tangent direction.

For a particle of mass m, the magnitude of the centripetal force corresponding to the centripetal acceleration is $F_c = mv^2/R = ma_c$ and has the same direction as the acceleration vector. Centripetal force, then, is that force that constrains the particle to movement in a circular path. Were the particle attached to a string, the tension in the string would provide the centripetal force necessary to keep the particle in its circular path. The outward-pulling force of the particle on the string is referred to as centrifugal force. Strictly speaking, there is no such applied force. The particle, because of its inertia, is merely exhibiting its natural tendency to move in a straight line, as prescribed by Newton's first law. The centrifugal force is the reaction to the inward-directed centripetal force, which in this example is applied to the particle by the string, while the centrifugal reaction is the force exerted by the particle on the string. If the string constraining the particle to a circular path were replaced by a "massless" spring, calibrated to indicate force, the tension in the spring produced by the particle would register the centrifugal force. This centrifugal force, then, is equal in magnitude but opposite in direction to the centripetal force. Thus, the same formula as that which specifies the magnitude of the centripetal force may be used with regard to centrifugal force, $F = mv^2/R = ma_c$. However, for the purpose of centrifugation analysis, the component of the equation relating to the magnitude of the acceleration (i.e., v^2/R) is defined in terms of the angular velocity at which the particle is rotated (ω = angular velocity of rotation in radians per second; ω^2 = angular acceleration in radians per second per second) and its radial distance from the axis of rotation (r, measured in centimeters). The general formula for the determination of centrifugal force therefore becomes $F = m\omega^2 r$.

2.2. Centrifugal Force and Sedimentation

Now, if the particle being discussed is considered to be in a centrifuge tube filled with medium, an acceleration $\omega^2 r$ acts toward the center. The particle suspended in this medium in the tube is not connected to the axis of rotation, so that it is not accelerated toward the center. Rather, the inertia of the particle will tend to cause it to fly off at a tangent to the rotation, and the acceleration relative to the centrifuge tube and the medium will be $\omega^2 r$ in the opposite direction, i.e., away from the axis of rotation, toward the bottom of the tube. If no resistance were encountered, the particle would clearly reach the bottom very rapidly, but when suspended in the medium, it will travel much more slowly because of frictional forces.

This situation is equivalent to one in which the certrifuge tube and the medium are stationary and an acceleration analogous to an increased earth's gravitational field with a magnitude of $\omega^2 r$ acts on the particle. Or, alternatively, a force $m\omega^2 r$, where m is the effective mass of the particle in the medium, is causing it to move toward the bottom of the tube. Such a force, as already mentioned, does not really exist, though it is often convenient to assume that it does; this assumption may be justified in the context of centrifugation because in this situation the particle behaves as if this were so. The imaginary force is called "centrifugal force."

2.2.1. *Relative Centrifugal Force*

The force due to gravity is $f_g = mg$; g is the acceleration due to gravity. Centrifugal force is considered gravitation-like because both forces depend on mass and acceleration. It is customary to compare the magnitude of the centrifugal force with the gravitational force and express it as relative centrifugal force (RCF): $RCF \equiv \omega^2 r/g = 1.119(10^{-5})r(\text{rpm})^2$. In this expression, the symbol $(\text{rpm}) = (60/2\pi)\omega$. The RCF, or the g value, then, simply represents the accelerating field as a multiple of the earth's gravitational field.

3. Centrifuges

In practice, the production of centrifugal forces for use in neurobiological applications such as separations of mixtures of subcellular particles is achieved by the use of commercially available centrifuges. These instruments may be classified in various ways. One simple scheme is based on the magnitude of the relative centrifugal force (RCF) that a given instrument can produce. General-duty centrifuges are available in a wide range of models for both specialized and general applications in which forces of less than 10,000g are required. Superspeed centrifuges produce RCFs in the range of 10,000–50,000g, while ultracentrifuges produce forces in excess of 50,000g.

Basically, all these units consist of heavy-duty, series-wound motor capable of rotating a shaft at specified speeds either directly or through gears. In the simple, lower-speed models, some of which are small enough for benchtop use, a multiple-position switch is used to select one of several rotation speeds. Other models are capable of more nearly continuous variation in speed. Indication of the actual speeds of rotation obtained may derive from precalibration of settings or direct tachometer readout of actual shaft rotation or both. Many models, especially those capable of higher speeds, may have a flexible coupling between the motor and the rotation shaft to reduce vibration as well as motor-bearing wear occasioned by rotation of less than precisely balanced loads.

When the activities of sample materials are labile, as is often the case with samples of neurobiological interest, it is necessary to perform centrifugation as well as other procedures at reduced temperatures. Some of the general-duty models are small enough to be enclosed in a refrigerator during operation. Others may be run in a cold room to achieve temperature control. The higher-speed general-duty models as well as super-speed and ultracentrifuges can produce sample heating due to air friction encountered during rotation. To protect the samples, these units usually incorporate an internal refrigeration system capable of variable temperature control. The highest-speed models also include a means of evacuating the chamber in which the sample is rotated to further reduce the effects of air friction.

Most units, especially the larger, higher-speed models, are capable of relatively precise maintenance of rotation speeds even in the face of fluctuating line voltages. Times may be preset to maintain maximum desired rotation speeds for specified intervals. Most units also contain a braking mechanism to slow the rotation following the desired spin duration. Such brakes can reduce the stopping time by 50–75% compared to the time required for a coasting stop. Some brake mechanisms are calibrated to exert greater braking forces at higher speeds of rotation, with the braking force applied diminishing as the speed of rotation decreases. This provides a smooth, gradual rate of deceleration.

The sample is held in centrifuge tubes that may be made of glass for low-speed applications but that, because of the fragility of glass, are usually made of polycarbonate, polystyrene, or stainless steel. The tubes are usually capped to prevent spillage and, in the case of flexible tubes used in high-speed applications, to prevent deformation of the tube above the meniscus. These tubes are coupled to the axis of rotation by placing them in a centrifuge head or rotor that is attached to the drive shaft. Most instruments adapt to several different sizes and types of rotors. The size refers to the maximum volume the rotor will hold. A 6 × 50 ml rotor, for instance, will hold a maximum of 300 ml of sample. There is usually some trade-off in rotor design between the speed at which the rotor may be rotated and the volume of sample the rotor can hold. In general, the larger the capacity, the lower the maximum speed allowed. There are two main types of rotors of interest here. Swing-out rotors cause the sample tubes to be swung out during rotation so that their plane of rotation is perpendicular to the axis of rotation and the centrifugal force is in a direction parallel to a line through the vertical center of the tube. For complete sedimentation to occur, a particle initially near the top of the tube must travel down the full length of the tube. In angle rotors, the sample tube remains in a fixed position during centrifugation. The tube holes are usually inclined at some angle between 10 and 45° to the axis of rotation. For general purposes, the angle of the tubes is usually about 35°. Since the direction of the applied centrifugal force is essentially across the tubes in an angle rotor rather than down their length, the path the particles must travel is considerably shorter than with swing-out rotors. The nearer the tube is to being at right angles (i.e., parallel to the axis of rotation), the shorter is the effective path length and the more rapid is the completion of sedimentation. As will be discussed below, after the particles hit the outer wall of the tube, they tend to fall along it to the bottom of the tube.

3.1. Determination of Relative Centrifugal Force

As mentioned above, relative centrifugal force (RCF) depends not only on the angular acceleration but also on the radial distance of the accelerated object from the axis of rotation; i.e., $RCF \equiv \omega^2 r/g$. The magnitude of the force applied is usually given in the literature as an RCF with the unit g. However, since RCF is dependent on the radial distance, it is important to know how this distance was measured prior to the determination of the RCF. This distance may be measured in at least three different locations—at the top or the bottom of the centrifuge tube or in the middle. In an example from McCall and Potter (1973), when the tube is being rotated at 30,000 rpm in the swing-out rotor, the RCF at the top of the tube (R_{min} = 6.99 cm) is 70,000*g*; at the middle (R_{av} = 9.76 cm), 98,000*g*; and at the bottom of the tube (R_{max} = 12.54 cm), 126,000*g*. Actually, the minimum radius (R_{min}) should not necessarily be measured at the very top of the tube, but rather at the meniscus. Therefore, the reporting of RCF should indicate not only where the measurement was made but also the dimensions of the tube relative to the axis of rotation and the height to which the tube was filled. It should be clear that reporting only rpm is entirely inadequate.

The type of rotor also influences RCF. Thus, in angle heads, the position of the tube reduces the range from R_{min} to R_{max}, since the tube is not perpendicular to the axis of rotation. In another example (McCall and Potter, 1973), at 30,000 rpm at R_{min} (3.68 cm), the RCF is 37,000*g*, while at R_{av} (6.00 cm), it is 60,000*g*, and at R_{max} (7.87 cm), it is 80,000*g*. It is also important to note that in angle rotors that have positions for more than one row of tubes, the RCF values will differ for the individual rows, since their radial distances from the axis of rotation will differ. Since the RCF increases with radial distance and is greatest at the bottom of the tube, it is reasonable to report the force at R_{max}, since this is a maximal value and is independent of the level to which the tube is filled. Many reports, however, use the distance to the middle of the tube, thus reporting an average RCF for the tube (Allfrey, 1959).

A chart originally published by Dole and Cotzias (1951) has been widely reproduced. A straightedge intercepting a known radius and rpm on the chart will also cross at the RCF generated under those conditions. Values outside the ranges on the chart can be calculated by shifting decimals. Since RCF is directly proportional to the radial distance (r) from the axis of rotation, a change in r by a factor of 10 also changes RCF by a factor of 10. On the other hand, since RCF is related to the square of the angular velocity, a change in rpm by a factor of 10 alters the RCF by a factor of 100.

3.2. Analytical vs. Preparative Centrifugation

It is useful to distinguish between the terms analytical and preparative as they are likely to be encountered in the literature. Most simply, these refer to the type of centrifuge instrument being used. An analytical centrifuge is usually a very-high-speed ultracentrifuge. Many of these machines are able to produce RCFs in excess of 50,000*g*. More important, however, is the fact that they are typically designed to spin rotors that can hold only relatively small volumes of sample, contained in specially designed sample cells rather than in centrifuge tubes. Furthermore, these instruments are equipped with rather elaborate optical attachments of various types that make it possible to observe the movement of the sample during centrifugation and to quantify certain aspects of this movement. Usually, the sample is relatively pure or homogeneous, and the object of the centrifugation is less often the separation of impurities than it is the delineation of certain physical properties of the sample, such as its molecular weight. The more commonplace preparative centrifuges, which are essentially the types being discussed here, can in some cases be fitted with optical systems and adapted for analytical purposes. This eliminates, for some applications, the need for a very expensive analytical instrument and at the same time maximizes the versatility of the preparative model. Bowen (1966) discusses in some detail many of the features of both analytical and preparative machines. The theoretical formulations underlying the use of either type of machine are the same, although for most applications they can be less cumbersome for users of preparative instruments. Where relevant in this chapter, this theory will be discussed as it relates to preparative work. For a full review of centrifugal theory in general, but especially as it relates to analytical procedures, several sources are available (Bowen, 1970; McCall and Potter, 1973; Pickels, 1950; Schachman, 1959; Svedberg and Pedersen, 1940; Trautman, 1964; Williams, 1963).

In practice, various authors make other distinctions that are not based on the instrument alone. As mentioned above, the quantity and purity of the sample are often two criteria for analytical procedures basically requiring relatively small but homogeneous samples. In this respect, preparative centrifugation is frequently the method of choice for the separation of a homogeneous sample from a mixture prior to analytical work. Trautman (1964) suggests that if a sample is analyzed after centrifugation, then the experiment is preparative in nature. On the other hand, if the sample is observed optically during centrifugation, the experiment is analytical. Clearly, then, the analysis of a sample after a preparative experiment may entail an analytical-centrifugation experiment. Both procedures may be quantitative, and the references given above provide examples of analytical procedures that yield fractions for other types of analysis.

In discussing preparative centrifuges, Reid and Williamson (1974) (see also deDuve, 1964) make a useful distinction between experiments run under analytical and preparative conditions in the same type of instrument. Their distinction is not in terms of the quantity of sample being used, but in terms of the manner in which the sample is treated in light of the goal of the experiment. Thus, in an analytical procedure, the individual component populations of a mixture of subcellular particle types are separated from one another as completely as possible. All the fractions obtained, however, must be recovered and subjected to some further analysis with the goal of establishing how some product (e.g., an enzyme) is distributed in relation to its presence in the original mixture. As will become apparent subsequently such an experiment requires a number of compromises, since it is generally not possible to completely separate all the populations of subcellular components from one another. To establish the relationship of the product of interest to the starting material, it is necessary to produce a balance sheet that includes information from all the fractions obtained as they compare with the starting material. This strategy is discussed in more detail later (see Section 6.3). The preparative experiment, on the other hand, is seen as a means of isolating a particular subcellular component type for detailed study. It is desirable to obtain this component in as pure a state as possible, free from other components that in this case are considered as contaminants. In contrast to the analytical procedure, which greatly emphasizes total recoveries, the preparative strategy readily sacrifices yield for purity.

4. Homogenization

Before a tissue can be fractionated into its various subcellular components, cell integrity must be disrupted in some fashion. Surely, this homogenization step is of utmost importance, since the most careful centrifugation procedures can do no more than isolate subcellular particles that have been liberated from the tissue. Ideally, all cells in a tissue sample would be broken with the consequent release of intact subcellular organelles unaltered from their *in situ* state. With the aim of maximizing the intergrity of the subcellular components obtained after breaking cells, Claude (1946) used a very gentle mortar-and-pestle grinding on liver tissue. This, however, left a rather large proportion of the cells unbroken and, of course, frustrated one of the ideals. While this may be less than critical if a single homogeneous cell type is being homogenized, liver and most other tissues are heterogeneous regarding cell type, and were one population of cells to break more easily than another, the subcellular components of the homogenate could be unrepresentative of the whole tissue. Considerable experimentation and innovation have resulted in a variety of homogenization techniques that can satisfactorily disrupt cells from specific tissues. None of these techniques achieves the ideal, and in general, there is always a need to compromise quantitative cell disruption for the sake of organelle preservation, or vice versa. Two techniques most commonly used for cell disruption in a variety of tissues, including neuronal tissues, will be discussed here. Both these techniques disrupt cells in aqueous media by the application of shearing forces. Allfrey (1959) has reviewed many other techniques in the context of some of their special applications.

4.1. Coaxial Homogenizers

The most common homogenizers remain the Potter–Elvehjem type (as originally described by Potter and Elvehjem, 1936). These are available commercially (e.g., A.H. Thomas, Inc., Philadelphia, Pennsylvania) and, as described by the manufacturer, consist of a smooth-walled borosilicate glass grinding vessel and a piston-type Teflon pestle mounted on a stainless steel rod. The Teflon grinding heads have smooth, waxy surfaces that are resilient, practically inert, and not wetted by water. Additionally, the grinding heads may be autoclaved. The presence of powdered glass in homogenates that arises from the use of ground-glass pestles in some tissue grinders (homogenizers) as well as the generation of heat are both minimized with Teflon pestles. Futhermore, the clearance between the Teflon grinding head and the grinding vessel is maintained for extended periods, thus contributing to the reproducibility of results. The grinding vessels are made from precision-bore tubing with a 2-mm wall and are round-bottomed. A wider, top reservoir with a pouring spout accommodates the fluid displaced by the pestle. The pestle is rounded to conform to the bottom of the grinding chamber. The length of the pestle is approximately twice the diameter. The tip of the pestle may be either radially serrated to facilitate the passage of the tissue to the grinding surface or smooth. Various sizes are available, as indicated in Table 1.

4.1.1. *Theory of Operation*

The Potter–Elvehjem homogenizer and similar coaxial homogenizers disrupt cell membranes and, to a smaller extent, the membranes of cell organelles by shearing in a liquid velocity gradient. The gradient is produced by rotation of the pestle or moving it up and down relative to the grinding vessel, or both. In fact, most current techniques combine a rapidly rotating pestle with an up-and-down movement of the grinding vessel over the pestle. When this is done, the entire volume of medium and tissue is forced through the small space between the pestle and the tube, producing high shearing forces. The rate of shear is determined primarily by the radii of both the pestle and the tube, the rate of pestle rotation, and the speed of the up-and-down movement of the grinding vessel. When the speed of rotation is high, the tissue-containing medium is rapidly forced through the space between the pestle and the tube, resulting in high shearing forces. Relative to the speed of rotation of the pestle, the movement of the tube is small. The velocity of the liquid in the tube can be considered as at a minimum at the tube wall and at a maximum at the surface of the pestle, thus forming a velocity gradient. Since smaller distances between the pestle and the tube produce steeper gradients, the fit of the pestle within the tube can affect the homogenization (Allfrey, 1959).

However, maintenance of minimal distances between the pestle and the tube may limit cell rupture while damaging organelles of the cells that do rupture (Schneider and Kuff, 1964). While some other devices have been developed to circumvent this problem (e.g., Lang and Siebert, 1952; Dounce, 1953; Emanuel and Chaikoff, 1957), most workers continue to use the

Table 1. Dimensions of Potter–Elvehjem Homogenizers

	Model			
Size	AA	A	B	C
Chamber volume (ml)[a]	4	10	30	55
Vessel length (mm)	125	125	160	170
Grinding length (mm)	90	75	105	110
Chamber inside diameter (mm)	7.62	12.70	19.05	25.40
Pestle length (mm)	38	27	39	44

[a] The total capacity of the grinding vessel is slightly more than twice the capacity of the empty lower portion, which is given as the chamber volume.

Potter–Elvehjem-type homogenizer. Although commercially available homogenizers of this type are produced with a clearance of 0.100–0.150 mm, several workers (who bother to mention it) reduce the pestle diameter by turning it in a lathe. Actually, the early investigators in this area were quite aware of this and realized that the speed at which the tissue suspension could be forced between the wall of the tube and the rotating pestle was of more importance in producing a satisfactory homogenate than was the tightness of the pestle's fit (Hogeboom, 1955). In fact, the original design took this into account, employing a 0.23-mm clearance (Potter and Elvehjem, 1936). As Potter (1955) reported, the optimum clearance for different tissues varies. He suggests that the easiest way to standardize and select tubes and pestles that fit properly is to fill the tube with water and then insert the pestle to the bottom of the tube. If the tube can be lifted by raising the pestle, the clearance is appropriate for homogenization of liver and kidney. On the other hand, it should not be possible to lift the tube this way if muscle tissue is to be homogenized. The former condition is also generally appropriate for the homogenization of neuronal tissue. For the isolation of synaptosomes, Whittaker (1969) reports finding a clearance of 0.25 mm to provide satisfactory, reproducible results. Reviewing several techniques for the isolation of mitochondria from brain, Biesold (1974) reports the use of clearances ranging from 0.125 mm (Bellamy, 1959) to 0.25 mm (Vrba, 1967) and 0.3 mm (Jobsis, 1963). Most reported procedures for preparation of brain fractions use clearances in this range. Whittaker and Barker (1972) suggest, however, that clearances should not be thought of in absolute terms, but should be adjusted relative to the diameter of the pestle. Thus, in their article—which is, incidentally, one of the few that includes information on pestle diameter—a grinding vessel with an inside diameter of 30 mm was used with a pestle clearance of 0.25 mm, which represents an 8.33×10^{-3} ratio of clearance to pestle diameter. For a 20-mm-diameter tube, then, the clearance should be $20(8 \times 10^{-3})$, or 0.167 mm. Such adjustment is necessary to produce equivalent conditions in grinding vessels of different sizes, since shearing forces are related to pestle diameter and therefore to radial velocity at the surface of the pestle.

4.1.2. *Homogenizer Drive Units*

Dependent on the tissue being examined and, if relevant, the particular subcellular components of interest, it is possible to disrupt cells using a hand-rotated pestle. Best results are obtained with relatively soft tissues such as spleen, liver, and brain. However, as Whittaker (1969) and others have mentioned, hand homogenization generally leads to less complete disruption of cells than do other methods. The method of choice involves the use of a motor to turn the pestle. A major requirement of such a motor is that it produce sufficient torque to keep it from being temporarily slowed during the course of homogenization. However, it is desirable that some slippage be present in the system to prevent breakage in the event that the pestle binds in the grinding vessel, although this is not a serious problem when the Teflon pestles described in Section 4.1 are used. Even heavy-duty, series-wound motors (usually $\frac{1}{10}$ horsepower or less) commercially available as stirrers do not produce sufficient torque for this application, although they will usually stall before breakage of the pestle or grinding vessel occurs. Larger (about $\frac{1}{4}$ horsepower) induction motors are preferred, but such motors do not provide the necessary slippage if adapted to drive the pestle directly. This difficulty can be overcome if the motor is used to drive a mandrel through a pulley-and-belt arrangement. Provided the belt is not too tight, binding of the pestle in the grinding vessel will cause the belt to slip over the pulley, thus preventing breakage. This arrangement also provides a solution to another fairly basic requirement, variable speed (discussed in Section 4.1.3). If more than one set of pulley wheels are mounted on the motor and mandrel shaft, selection of various-size wheels can effect different drive ratios and, hence, speed of pestle rotation. The motor can be mounted on a bracket that provides some movement to adjust belt tension for different combinations of pulleys. To prevent loss of torque caused by using a larger pulley on the motor shaft than on the mandrel shaft to step up pestle speed, it is desirable to use a motor that rotates at least as fast as the maximum pestle speeds to be needed (probably in the 1000- to 2000-rpm range). The pulleys will then always be used to step down motor speed, with a concomitant gain rather than loss of mandrel torque. A variable-line-voltage transformer, when used with this sort of induction-motor arrangement, can provide some control over the torque applied to the mandrel without affecting speed. A tachometer can be mounted on the mandrel to indicate speed of rotation, or various pulley combinations can be calibrated with a stroboscopic tachometer. In all situations, it is desirable to use a foot switch to turn the motor on and off so that both hands may remain free. There are several ways of connecting the pestle shaft (usually $\frac{1}{4}$ inch diameter) to the mandrel. A piece of heavy-walled rubber tubing connecting both shafts is one option, and one that provides some potential slippage. A chuck of suitable size to accept the pestle shaft may be affixed to the mandrel. Alternatively, the end of the mandrel may contain a recess into which the pestle is either inserted and held with a setscrew or threaded. This treatment, though somewhat extended was deemed of value because suitable commercial drive apparatuses for this application are unavailable, so that most researchers must fabricate their own machines.

4.1.3. *Operational Principles*

In practice, a grinding vessel containing the tissue to be homogenized and a suitable medium (see Section 4.4) is raised and lowered over a rotating pestle. Care must be taken, of course, that the combined volume of the pestle and the tissue suspension does not exceed the capacity of the grinding vessel. One cycle of raising the vessel until the pestle reaches the bottom and lowering until all the tissue suspension is again below the pestle is referred to as a "stroke." Most of the cell breakage occurs as the cells pass through the liquid velocity gradient between the rotating pestle and the walls of the grinding vessel. Best results are obtained when the tissue suspension is rapidly transferred above and below the pestle during the stroke motion. This is one reason close-fitting homogenizers are less efficient, since they reduce the stroke speed and increase the danger of frictional heating. Of course, the tissue sample must be kept below the pestle until it has been sufficiently disrupted to pass above the pestle. In the complete transfer of the homogenate to the spaces above and below the pestle, it is best not to draw air into the homogenate. Thus, on the downward motion particularly, the movement should not proceed faster than the homogenate can fill the space below the pestle. Generally, the total number of strokes, or passes of the pestle, is mentioned in methodology sections. The range, reportedly, is anywhere from 2 or 3 to 10 or 15 total strokes. The optimum number must essentially be determined empirically unless a predetermined procedure is being followed. Not infrequently, no mention of the number of strokes is reported, although the duration of homogenization is sometimes given. Occasionally, authors include not only the number of strokes but also the length of time required for each stroke, thus allowing possible replication of stroke speed by other experimenters. Clearly, for the sake of reproducibility and interpretation, all procedural details including the number of strokes should be specified.

As mentioned above, the speed of pestle rotation is another important variable, but one that is rather easy to regulate. The speeds generally encountered in the literature range from 200 or 300 to 2000 rpm. Again, these values are usually determined empirically as those that provide optimum disruption for the recovery of subcellular components. Unfortunately, this information is not always included in the literature, although it is critically important in the interpretation of data that in some cases may very well be confounded by inadequate homogenization procedures.

4.1.4. Summary of Coaxial Homogenization

Thus, various physical aspects of the homogenization apparatus are of considerable importance. The drive assembly should be capable of providing a regulated speed with sufficient torque to minimize speed losses due to the load applied by the homogenizer itself. On the other hand, some form of slippage should be provided to reduce torque momentarily in the extreme case of the pestle's binding in the grinding vessel. The foot-operated on–off switch provides further controllability for this situation. When these rather basic criteria have been met, as mentioned above, four further variables have a large influence on the shearing forces produced in the homogenizer and, concomitantly, the quality and quantity of cell disruption. Three of these are relatively easy to control—pestle/grinding vessel clearance ratios, pestle diameter, and speed of pestle rotation. The clearance and speed of rotation are often reported in the literature, but the pestle diameter is rarely mentioned. It should be noted, however, that reports that indicate the size or capacity of the grinding vessel used allow deductions to be made about pestle diameter, at least in the case of commercially available homogenizers. Nevertheless, more frequent, explicit mention of this variable would prove useful in the standardization of at least comparable conditions among researchers.

The fourth important variable is the number of strokes applied to the tissue suspension. Many researchers report this, and it would be useful if all did. However, this is one aspect of the procedure that is less easy to control, since it requires human manipulation. Some variation in homogenization is to be expected due to the manner in which the grinding vessel is handled. The important factor here is the speed or duration of each stroke, although the use of maximum speeds should be fairly easy to replicate. In general, though, the course of each stroke is rather difficult to specify verbally, and conditions are most easily replicated following direct observation of the actual procedure in a given laboratory. Since this is usually impossible, it is obvious that there will always be some lack of precision in this area. While this has led to the description of homogenization as being an art and therefore not easily mastered, the situation is surely not so hopeless. In fact, rather good reproducibility can be obtained if careful attention is directed to all the factors discussed.

4.2. Blenders

An alternative to the coaxial homogenizer that is useful for some applications is the high-speed blender. There are two general types that are readily available at present. The Waring blender, based on the popular houshold appliance, has been refined for biological work. Basically, this consists of a high-speed motor that drives a stirrer mounted in the bottom of a cylindrical vessel. The walls of the vessel are indented vertically to form a cloverleaf cross section. These indentations serve to channel the tissue suspension onto the blades of the stirrer and at the same time produce considerable shearing forces as the suspension flows through the velocity gradient over the indented surfaces. The stirrer consists of four blades sharpened at right angles to one another and inclined at various angles to promote efficient mixing of the vessel contents. The high speeds at which the blades rotate (from 13,000 to 22,000 rpm) combined with the shape of the vessel produce complex shearing forces with considerable turbulence and cavitation. The vessel sits on top of the stirring base, and the leakproof bearing and shaft extend through the jar bottom. The jar–base joint is sealed by pressure on a neoprene gasket. This opens the possibility of leakage around the base, which may be especially pronounced if any organic solvents are used. Either fixed- or variable-speed models are available, some with control timers. A variety of models and accessories are available with vessel capacities ranging from 1 gallon down to 20 ml. Working volumes are usually about one half the capacity. Although the wide variety of vessel and stirrer sizes together with the variable-speed capability, which can be extended by the use of a variable-line-voltage transformer, permit the use of blenders for most tissue disruptions, certain inherent disadvantages tend to limit their use, in practice, to bulk homogenizations that cannot be carried out in coaxial homogenizers because of their smaller sizes or to the disruption of tissues that are particularly difficult to homogenize by other means. The most practical application of these blenders in neurobiological work is for bulk homogenization, since coaxial homogenizers are generally satisfactory for other applications.

Before the disadvantages of blenders are discussed, a variant of the blending principle will be considered. Vir Tis, for example, produces a homogenizer in which the shaft driving the stirring blades enters the vessel through a Teflon sealing cap on the top of a spherical homogenization vessel. This greatly reduces the problem of leakage. Flasks and blades are available in a large range of sizes for samples from 0.2 to 400 ml. The utility of this homogenizer for bulk work is, then limited. Stirring speeds are available from 100 to 45,000 rpm. The high speeds available, although combined with a simple round vessel, are capable of producing high shearing forces.

The major difficulties with these homogenizers when compared with the Potter–Elvehjem type center around the quality of cell disruption. Even at high speeds, many cells are not disrupted. This is not due to any gentleness of homogenization, since the particles of cells that are disrupted are often, in turn, disrupted themselves. Proteins may be denatured and enzymes inactivated. Overheating can occur, although Waring now provides a vessel with a cooling jacket, and the Vir Tis configuration is easily adapted for the use of a cooling bath. As mentioned in relation to coaxial homogenizers, the introduction of air into the suspension is undesirable, yet with blenders it is inevitable, and in fact, even undesirable frothing can occur. Generally, and in particular for neural tissue, operation of blenders at low speeds fails to resolve the difficulties mentioned. Thus, although they may be necessary for bulk homogenization and perhaps in some other applications, blenders do not provide a method of choice for the preparation of homogenates of neurobiological interest when these are to be subjected to subcellular-fractionation techniques.

4.3. Cellular Heterogeneity and Tissue Preparation

The heterogeneity of a tissue to be analyzed can be a source of both technical and theoretical problems for subcellular fractionation. Most neuronal tissue of interest (e.g., mammalian brain) is composed of a variety of neuron types that often differ markedly regarding function, morphology, and chemistry. Furthermore, glial cells comprise a large portion of the brain mass. There are about ten times as many glial cells as neurons, and of course, large areas of the brain are more or less heavily myelinated. Furthermore, considerable vascular and connective tissues are present, sometimes highly concentrated as in the choroid plexus. Finally, circulating blood cells will also be present.

The presence of so many types of cells greatly complicates the application of subcellular fractionation to neuronal tissue. Some procedures and interpretive cautions of value when considering such heterogeneity will be considered subsequently. However, some methods that are of value in reducing this heterogeneity will be mentioned here.

The problem of neuronal heterogeneity can sometimes be dealt with effectively, although not always without some loss of information. One common method is useful when a component of the brain (e.g., some nucleus) is relatively homogeneous regarding the types of neurons present. Such brain areas can be surgically isolated from the rest of the brain prior to homogenization. This method is even of some use when more heterogeneous brain areas are of interest. Thus, the cerebellum, cerebrum, hypothalamus, or other brain areas may be isolated by dissection prior to homogenization. Such dissection is frequently of considerable value in making possible the removal of large portions of myelinated fiber tracts, yielding tissue relatively rich in cell bodies, dendrites, and synapses. Each of these options, while reducing neuronal heterogeneity, does less to reduce heterogeneity arising from nonneuronal cell types. As should be expected, there are likely to be considerable shifts in the relative proportions of neuronal cell types, and comparisons of data obtained from initially isolated and intact tissues should be made very cautiously, if at all. Microdissection techniques have proved a quite useful solution to this problem. Individual cells can be isolated from neuronal tissue, resulting in preparations of high purity. Roots and Johnston (1972) have reviewed some techniques of value for such isolations in mammalian tissue, while Giacobini (1969) treats invertebrate preparations similarly and discusses some of the advantages of invertebrate preparations. Analysis of single cells presents some technical challenges that Giacobini (1969) has also considered (see also Volume 1, Chapter 2). Microdissected, single cells certainly resolve the problem of heterogeneity, although other considerations become important. The fact that these single-cell preparations may in many cases not be representative of a tissue as a whole must always be borne in mind. On the other hand, isolation of a virtually single-cell structure such as the crayfish stretch receptor (Giacobini, 1969) even eliminates most of the need for such caution. In general, the utility of such preparations in understanding basic processes in model systems should not be underestimated. Another series of model systems of high purity that provide greater amounts of tissue derive from cultured cell lines (see Volume 1, Chapter 4). The use of cloned cells has been reviewed by McMorris *et al.* (1973). Although problems of generalizability again arise due to the restricted number of cell types available and the fact that these cells do not derive from intact functional tissue, such systems are highly useful.

While the microdissection and clonal-system strategies provide cells that are neuronally homogeneous as well as free of nonneuronal contaminants, other techniques have been developed with much the same goal in mind. These techniques do not provide neuronal homogeneity, although nonneuronal contamination is considerably reduced. Extensive refinement has been achieved since the early work of Korey *et al.* (1958), and several recent articles have reviewed methods for the bulk isolation of neurons and glia (Podsulo and Norton, 1972; Rose, 1969; Rose and Sinha, 1974; Sellinger and Azeurra, 1974). These procedures provide populations of neurons and glia that are relatively pure although not nearly as homogeneous as may be obtained with the methods mentioned above. Neurons or glia or both can, of course, be subjected to analysis after such isolations. Furthermore, these bulk-separation techniques have the advantage of using initially intact, functional neuronal tissue as a starting point, although the heterogeneity of cell types obtained in either the neuronal or the glial population is still a source of analytical and interpretive importance. Finally, mention must be made of the fact that the neuronal population obtained by the various bulk-separation techniques is generally free of any axonal processes and for all practical purposes can be considered only as a neuronal soma population. Again, while these techniques provide some advantages, there are several limitations as well.

The presence of blood cells is another problem, but one that is readily and routinely circumvented. The simplest method involves decapitation, which has the advantage that anesthesia is generally not required when the decapitation is properly performed. This avoids the effects that anesthesia can have on sensitive brain metabolism (Brunner *et al*, 1971; Lindquist, Kehr and Carlsson, 1974). While decapitation greatly reduces the blood content of the brain, some blood cells do remain. Nevertheless, because of its simplicity, it is the most frequently used procedure. More complete blood removal can be obtained by intracardiac perfusion. McEwen and Zigmond (1972) describe one very thorough technique that does have the disadvantage of requiring anesthesia. The animal is rapidly anesthetized with a large dose of sodium pentobarbital (0.4 ml, 60 mg/ml for adult rats) that also includes 0.1 ml heparin (USP, 1000 U/ml) to reduce clotting. After the chest is opened, a perfusion cannula is inserted into the left ventrile while the right atrium is cut to allow blood to flow out. The chilled perfusion fluid is allowed to flow through the heart into the aorta. When the eyes, ears, and front paws become white, perfusion is complete. This usually takes less than 60 sec and 30 ml of perfusion fluid. When the brain is subsequently removed, it is a whitish-yellow color with no blood on its surface, in contrast to the unperfused brain, which is pink with surface blood vessels and blood clots very noticeable.

4.3.1. Minimizing Postmortem Deterioration

To reduce postmortem changes, it is always desirable to lower the temperature of isolated tissue as soon as possible after death (Alderman and Shellenberger, 1974). None of the methods available is entirely satisfactory, and so speed is essential. Decapitated or whole animals may be dropped into liquid N_2, but several seconds are required for complete freezing of the brain

(Swaab, 1971; Ferrendelli *et al.*, 1972). Freezing time can be shortened if the cranium is previously opened, but this requires anesthesia (Granholm *et al.*, 1968). The brains of decapitated or perfused animals can be dropped into liquid N_2 or placed on ice before or after removal. Microwave radiation can also be used to inactivate metabolism by heat (Balcom, Lenox, and Meyerhoff, 1975; Cheney, Racagni, Zsilla, and Costa, 1976; Guidott, Cheney, Trabucchi, Dotenchi, Wang, and Hawkins, 1974; Medina, Jones, Stavinoha, and Ross, 1975; Stavinoha, Weintraub, and Modek, 1973; Schmidt *et al.*, 1971), although this is also relatively slow and irreversibly inactivates enzymes, allowing only metabolites to be studied. A relatively new technique described as brain-blowing and detailed by Veech (1974) and Veech, Harris, Veloso, and Veech, (1973) allows the simultaneous removal and freezing of small portions of brain tissue. Again, none of the techniques available is entirely satisfactory, and speed must be emphasized. Furthermore, great care should be taken to standardize whatever procedure is used to maintain comparability among various brain isolations, and exact details should be reported to facilitate replication and interpretation.

4.3.2. Final Procedures

Since it is undesirable to force tissue past the homogenizing pestle until it has been disrupted into small pieces, a usual preliminary step involves mincing of the chilled tissue. Thus, scissors, a razor blade, or other suitable instrument is used to cut the tissue into small pieces. These pieces are then placed in the grinding vessel and homogenized. This procedure speeds homogenization and helps to reduce the role that connective tissues play in inhibiting thorough homogenization, though this is not as much of a problem with brain as with other tissues, even some other tissues of neurobiological interest. Although some organs contain so much connective tissue that it is necessary to press the organ through a wire mesh to remove some connective material prior to homogenization, this is not necessary with most brain samples. However, many investigators do follow a brief initial homogenization with a filtration of the homogenate through cheesecloth, silk, flannelette, or other material to remove connective tissues. The material passing through the cloth is then subjected to more complete homogenization.

4.4. Homogenization Media

Although the considerations outlined above can define fairly completely the degree of cell disruption achieved during homogenization, other factors also influence the characteristics of the particles in the homogenate and, concomitantly, the separation possible during centrifugation. The composition of the medium in which the tissue is suspended and homogenized is of particular importance. It is necessary to realize that when a cell is disrupted, its contents are released from the balanced internal environment of the cell. Osmotic, ionic, and pH balances become immediately susceptible to changes dependent on the composition of the medium. These factors, of course, can have dramatic influences on the structure and function of cell particles. Furthermore, the juxtaposition of particular organelles within the cell is disrupted and the concentrations of various soluble, cytoplasmic components dispersed in the medium. Any process spatially linked between organelles or organelles and cytoplasmic compartments is, of course, vulnerable to disruption following homogenization. A rather wide variety of physical and chemical artifacts may result from changes in subcellular relationships produced by homogenization. These artifacts may include leakage, absorption, adsorption, redistribution, and morphological alterations.

4.4.1. Artifacts

Since it was recognized quite early that homogenization would release the contents of the cell into an unnatural environment, the first suspension media consisted of what was thought to be a physiological salt solution that was isotonic and buffered at an approximately neutral pH. Though it has become clear that isotonicity and neutrality are generally desirable, the use of salt solutions to achieve these charateristics is rarely of any practical value because of the artifacts associated with their use. Salt solutions result in agglutination of particles of all sizes, loss of morphological integrity of some particles, and undesirable osmotic imbalances (deDuve and Berthet, 1954). Hogeboom *et al.* (1948) introduced the use of sucrose solutions in place of those containing mainly salts, and since these solutions produce much less artifactual difficulty, they remain the most common choice of suspension medium. The original recommendation by Hogeboom *et al.* (1948) was for hypertonic sucrose solutions (0.88 M), which maintained the rodlike appearance of mitochondria. Subsequent work made it apparent that better overall results could be obtained in isotonic (0.25 M) concentrations of sucrose (Hogeboom *et al.*, 1952). Sometimes the pH was lowered to 6.0 (Dounce, 1952), although this was found to produce artifacts such as agglutination (Hers *et al.*, 1951) and permeability increases in some particles (Cleland, 1952).

Various types of sucrose solutions are the media of choice for most homogenizations preliminary to subcellular fractionation. Some procedures call for the use of rather unusual sucrose solutions, while others do not use sucrose at all. These alternatives usually result in an overall sacrifice of the minimization of artifacts for the sake of increased yields or particular aspects of purity in specific subcellular fractions. Since sucrose solutions are most commonly employed in neurobiological work, they will be discussed below following consideration of the various artifacts associated with homogenization media.

4.4.1a. Evaluation of Artifacts. Perhaps the easiest or most straightforward evaluation of artifacts can be made at the morphological level. The development of electron microscopy and the refinement of various techniques associated with it, such as staining and sectioning techniques, have been most helpful in this regard. Thus, comparisons of sections from intact tissues and from isolated fractions have clearly demonstrated that dependent on the medium used, nuclei, mitochondria, Golgi material, synaptic vesicles, and synapses derived from isolated fractions can be obtained in a state that is essentially like that observed in intact tissues. Even the endoplasmic reticulum, drastically modified by the shearing forces involved in the homogenization procedure that results in the formation of vesicles, retains some of the morphological characteristics of undisrupted endoplasmic reticulum. Morphological examination, then, has revealed the qualitative reliability of organelle preservation in various media. Such examination is always required as a check on subcellular fractionations, especially if proven procedures are modified in any way. As will be mentioned below (see also Chapter 7), morphological procedures are also a valuable adjunct to procedures for estimating quantitative aspects of fractionation, for example, the purity of fractions obtained.

4.4.1b. Leakage. More serious and common artifacts are expected at the biochemical or molecular level, and these are also more difficult to analyze, since indirect methods must be employed. For example, the considerable research that has been devoted to mitochondrial isolation and function has revealed that the leakage of soluble components from these subcellular structures requires serious consideration. Clearly, many soluble components may be retained by mitochondria, but such retention requires careful medium control. Thus, the use of hypotonic media results in drastic losses of soluble materials. This is perhaps more of a problem for mitochondria than for other structures because of the semipermeable nature of the outer mitochondrial membrane. It is important to realize that disruption of the normal cytoplasmic environment of the mitochondria not only exposes them to unphysiological osmotic conditions, but also may remove substrates or cofactors, or both, that are necessary for the active and passive transport systems that normally maintain or contribute to the maintenance of the compartmentation of soluble materials in the mitochondria. Nuclei are subject to many of these considerations, and again, extensive research has delineated appropriate conditions for the retention of soluble components. Lysosomes, on the other hand, while extremely sensitive to osmotic conditions, probably do not lose their soluble components unless they have been ruptured either during homogenization or by osmotic forces. Synaptosomes (pinched-off nerve endings), which contain a variety of organelles such as mitochondria and synaptic vesicles, also contain cytoplasm, and the organelles are probably less influenced by unnatural medium conditions. On the other hand, the synaptosomes themselves are extremely sensitive to osmotic imbalances, which can lead to loss of soluble materials or actual structural disruption, thus exposing their contents to the homogenization medium. Leakage is a rather less important matter when considering endoplasmic reticulum. The vesicular structure of isolated endoplasmic reticulum is in itself an artifact of homogenization, and leakage of soluble materials from these vesicles seems rather less important than the disruption of the normal cytoplasmic–endoplasmic reticulum interface resulting from the dispersal of the cytoplasm into the homogenization medium.

In general, then, loss of soluble material is a serious problem. While retention of some soluble material has been clearly demonstrated, it is less easy to assess whether other components were lost. However, cross-validation with other techniques, such as isolation in nonaqueous media (see Allfrey, 1959), which might be quite unsatisfactory for general purposes, can help to establish the degree of loss. Such comparisons have revealed that in some few cases, appropriate medium composition can result in very good retention of soluble components. Since the results of isolation procedures can vary dependent on medium composition, tissue selection, and the subcellular particles of most interest, it is difficult to draw general conclusions. The best generalization, perhaps, is that the loss of soluble components must always be assumed, or at least considered, unless the procedures used are, or have previously been, specifically compared with independent standards (i.e., procedures) regarding leakage. Even here, there is often room for reasonable doubts about the cross-validation, and surely the first part of the generalization just noted is its most important aspect.

4.4.1c. Absorption. Absorption is a related problem. Soluble materials, in particular components of the homogenization medium, not normally associated with a specific subcellular structure may penetrate that structure. Similarly, unphysiological concentrations of normal soluble constituents may accumulate within various particles. Since absorption can be considered to be logically inverse to leakage, as just discussed, it will not be discussed further here. However, the cautions that apply to the loss of soluble material by subcellular particles may also be applied to their accumulation of soluble components.

4.4.1d. Adsorption. One artifact that has been less problematical is adsorption. Clearly, soluble enzymes, as well as other soluble elements, may be adsorbed on the surfaces of isolated particles. This would be most misleading in the case of enzymes normally compartmented in certain structures but released by homogenization and subsequently adsorbed on other structures. Soluble, cytoplasmic components can also cause problems, since disruption of cytoplasmic compartmentation might expose organelles to components that are not normally in contact with them. Even without this consideration, unphysiological pH characteristics of the medium or altered surface charges on subcellular structures could result in adsorption. Fortunately, adsorption does not seem to be a common occurrence and can be minimized by appropriate choice of medium. Furthermore, various quantitative procedures that are discussed below readily identify adsorption artifacts.

4.4.1e. Redistribution. Another artifact that can be readily indentified by quantitative procedures is redistribution. It should be apparent that both soluble material released or leaked from particles and insoluble or membrane-bounded material released by particle disruption may be absorbed or adsorbed by particles of a different density or sedimentation rate and appear in a particulate fraction with which they are not normally associated. Of course, in the absence of absorption or adsorption, these materials may be collected in the supernatant. In either case, this misleading redistribution may lead to faulty interpretation of the normal distribution of the materials. Even more misleading may be the redistribution of enzyme activators or inhibitors that alter the activity of enzymes. Although the enzymes may be obtained in the fraction with which they should be associated, they may be exposed to abnormal concentrations of inhibitors or activators. Of course, activators or inhibitors could be removed from particles that they normally contact, producing the inverse effect. Again, while redistribution is problematical, effective consideration of this artifact will reduce the risk of misinterpretation.

4.4.1f. Agglutination. Some media, especially those containing salts or those in which the pH is less than neutral, tend to promote clumping or agglutination of particles. The resulting agglutinated particles differ from their constituents and will appear in fractions based on their combined physical characteristics. A similar artifact may also result from certain aspects of the centrifugation procedures to be discussed. In either case, both morphological and biochemical determinations based on such fractions will be misrepresentative of the normal distribution of particles and their associated activities. This artifact can be reduced by appropriate choice of medium, although in some cases a medium that does promote agglutination may also produce some fraction that is desirable for a particular reason, such as purity of structural integrity of some particulate population.

4.4.2. *Effects of Sucrose*

Since sucrose is the medium of choice for most procedures, some of the effects of various solutions will be considered. Hypertonic solutions (0.88 M) retain the elongated shape of mitochondria as they are seen in intact tissue, but produce biochemical deficiencies such as a lack of oxidative phosphorylation. Sucrose solutions from 0.25 to 0.44 M do not have such a biochemical effect. Isolation of brain mitochondria is usually achieved in 0.25 M sucrose solutions (e.g., Jobsis, 1963), although 0.3 M (Tolani and Talwar, 1963) and 0.4 M (Stahl *et al.*, 1963) have also been successfully used. Isolation of nuclei and their subsequent functional integrity are also affected by sucrose concentration. Nuclei isolated in 0.25 M sucrose synthesize protein and RNA, while the use of 0.4 M sucrose results in a loss of these functions. Many procedures for the isolation of brain nuclei make use of 0.25 M sucrose (e.g., Bondy and Waelsch, 1965; Kato and Kurokawa, 1967), although 0.32 M sucrose has also produced satisfactory results (Dutton and Mahler, 1968). Synaptosomes are generally isolated in 0.32 M sucrose (Marchbanks, 1974; Whittaker, 1969; Whittaker and Barker, 1972). However, due to the osmotic sensitivity of synaptosomes, other concentrations of sucrose may sometimes be more appropriate. For example, the normal osmotic environment of cells in the *Octopus* brain requires that 0.8–1.0 M sucrose be used as the homogenization medium if satisfactory results are to be obtained (Jones, 1967).

4.4.3. Effects of Ions

The presence of ions in the sucrose medium also affects the morphological and functional states of subcellular particles. Thus, the presence of calcium inhibits oxidative phosphorylation of mitochondria, and some workers include ethylenediamine tetraacetic acid (EDTA) in the sucrose medium to chelate calcium (Stahl *et al.*, 1963; Tolani and Talwar, 1963). Ions, in general, promote particle agglutination or aggregation. Synaptosomal yields, for example, can be affected in this way (Gray and Whittaker, 1962). Divalent cations in concentrations as low as 0.5 mg-atom/liter (for calcium) and univalent cation concentrations as low as 20 mg-atoms/liter (for sodium) will cause agglutination (Whittaker, 1969). Both sucrose and water may contain calcium. Pure reagents should be used to reduce the calcium concentration. Whittaker (1969) reports that solutions of analytical-grade sucrose and double-distilled water contain about 12 μg-atoms calcium/liter. This is well below the threshold for agglutination. Passage of the sucrose solution over an ion-exchange resin reduced the calcium concentration to 2 μg-atoms/liter, but there was no improvement of synaptosomal yields with this as compared to the 12 μg-atoms/liter concentration. It is usually necessary to keep the use of buffers to a minimum to reduce agglutination. To bring sucrose solutions to neutrality, small amounts of 0.5 M tris base may be used. On the other hand, the absence of a divalent cation in the homogenization medium results in swelling and gelation of nuclei. Procedures specifically for the isolation of nuclei include calcium (e.g., as 3 mM $MgCl_2$), and buffering does not present any problem (McEwen and Zigmond, 1972). Relatively pure nuclei, free of cytoplasmic attachments, are achieved in this manner, but of course this is at the sacrifice of achieving satisfactory separations of other particle populations.

4.4.4. Effects of pH

The influence of ions causes another, technically related problem. It is usually desirable to maintain the pH of the medium at neutrality, but the addition of ionic buffers, which is the usual method of maintaining pH, can produce unwanted side effects, particularly agglutination. Although the tissue may initially be at a neutral pH, the release of autolytic enzymes during homogenization can lower the pH of the homogenate, and since many of these enzymes have pH optima in the acidic range, this effect is doubly serious. Agglutination is also promoted by lower pH. On the other hand, stability of the nuclear membrane is enhanced at an acidic pH. Alkaline pH tends to cause nuclei to swell and eventually dissolve as well as to cause agglutination even in the presence of divalent cations. Since ion concentration does not affect aggregation of nuclei, the use of phosphate buffer, for example, to maintain pH at 6.5 is permissible (McEwen and Zigmond, 1972). Mitochondrial isolations are usually achieved at pH 7.4, frequently, but not always, without buffering (Jobsis, 1963; Tolani and Talwar, 1963; Stahl *et al.*, 1963). When buffers are not used, the pH of the sucrose medium is adjusted initially to the desired pH, but of course there is no capacity to combat subsequent changes in pH. Synaptosomal isolations are effective at pH 6.5–7.0 (Whittaker and Barker, 1972). Much of the problem of autolytic changes in pH can be reduced by working rapidly at low temperatures. In fact, all procedures are usually carried out at temperatures just above 0°C.

4.4.5. *Effects of Tissue Concentration*

One further, related factor is the concentration of the tissue in the homogenization medium. Whittaker (1969) suggests that relatively concentrated homogenates are high in ionic strength, which may be the reason it is difficult to achieve satisfactory separations of such concentrated homogenates. In any case, the latter is a rather common finding, and almost all tissue concentrations are kept below 10% (wt./vol.) to eliminate this problem. Thus, each gram of brain weight is made up to a volume of 10 ml sucrose medium.

Homogenization conditions adopted for the isolation of specific subcellular components are summarized in Table 2.

5. *Differential Centrifugation*

Differential centrifugation is the general term used for the preparative centrifugation of more than one species of particles or macromolecules. It is based on the classic centrifugation technique called normal-rate separation and refers to the sedimentation of a suspension (or "solution") of particles or macromolecules under the influence of an RCF. Sedimentation is usually carried out in the presence of the "solvent" only, as distinct from the use of gradients, which is discussed in Section 5.3. The solvent may contain a buffer.

5.1. *Normal-Rate Separation*

For descriptive purposes, the normal-rate separation of a suspension of a single species of particle will be discussed first (as treated in McCall and Potter, 1973). Under initial conditions, the particles are homogeneously dispersed throughout the medium (solvent). Under ideal conditions (exceptions will be discussed below), the particles will sediment uniformly down the tube under the influence of the relative centrifugal force (RCF). At a given time, a boundary forms between the suspension and the portion of the medium that has been cleared of particles in the upper portion of the tube. Below the boundary region, an area of constant concentration of sedimenting material termed the "plateau" region borders a region in which there is an

Table 2. Homogenization Conditions for the Isolation of Specific Subcellular Components

Fraction of primary interest and species	Homogenizer clearance	Type[a] (size)	Speed (strokes)	Medium composition	Reference
Nuclear					
Rabbit, monkey, rat	0.009 inch	Glass + Teflon	500 rpm (4–6)	0.88 M sucrose; 1.5 mM $CaCl_2$	Bondy and Waelsch (1965)
Guinea pig	150 μm	Dounce plastic	1500 rpm (8 in 2 min)	0.32 M sucrose; 1.5 mM $CaCl_2$	Kato and Kurokawa (1967)
Rat	0.015–0.022 cm	Glass + Teflon (40–50 ml)	200 rpm (10)	0.32 M sucrose; 3 mM $MgCl_2$; 25% (wt./vol.) dilution	Dutton and Mahler (1968)
Rat	0.125 mm	A. H. Thomas P+E (Type A)	1000 rpm (20 slow)	0.32 M sucrose; 1 mM KH_2PO_4; pH 6.5; 3 mM $MgCl_2$	McEwen and Zigmond (1972)
Mitochondrial					
Rabbit, rat	0.3 mm	Glass + Teflon P+E	— (7)	0.25 M sucrose; 2 mM ATP; 5 mM citrate; 1 mM phosphate; 0.01% cysteine; pH 7.4; 7 volumes	Jobsis (1963)
Monkey	0.01 inch	Glass + Teflon P+E	1500 rpm (15)	0.3 M sucrose; 1 mM EDTA; pH 7.4; 9 volumes	Tolani and Talwar (1963)
Cow	0.10–0.15 mm	Glass + Teflon (200 ml)	— (10)	0.4 M sucrose; 1 mM EDTA; 0.02% polyethylene sulfonate; pH 6.8–7.0; 2 volumes	Stahl *et al.* (1963)
Rat	0.25 mm	Glass + Perspex Aldridge Type	500 rpm (5 min)	0.32 M sucrose	Vrba (1967)
Pigeon, rat	0.125 mm	P+E stainless steel (40 ml)	2000 rpm (8)	0.25 M sucrose; 4 volumes	Bellamy (1959)
Synaptosomes					
Guinea pig	0.25 mm	P+E (30 mm I.D.)	840 rpm (12)	0.32 M sucrose; pH 6.5–7.0	Whittaker and Barker (1971)
Octopus	0.25 mm	Aldridge Type	840 rpm (12)	0.7 M sucrose + 0.33 M urea; or 0.8 M sucrose; 10% (wt./vol.) dilution	Jones (1967)

[a] (P+E) Potter–Elvehjem.

increased concentration of sedimented material that eventually forms a pellet in the bottom of the tube. As sedimentation proceeds, the plateau region decreases progressively in size. At the same time, the increasing concentration differential between the clear solvent and the plateau region results in a net diffusion of particles into the clear region, thus decreasing the sharpness of the boundary.

For present purposes, the more common use of the normal-rate separation technique is for the fractionation of a mixture of subcellular particles, or differential centrifugation. Consider, as in Fig. 2a, a homogeneous mixture of three classes of particles. At relatively low centrifuge speeds, a large-particle fraction composed of whole cells, large cellular debris, nuclei, and mitochondria will sediment to the bottom of the tube. However, although each particle sediments at its own characteristic rate under the influence of

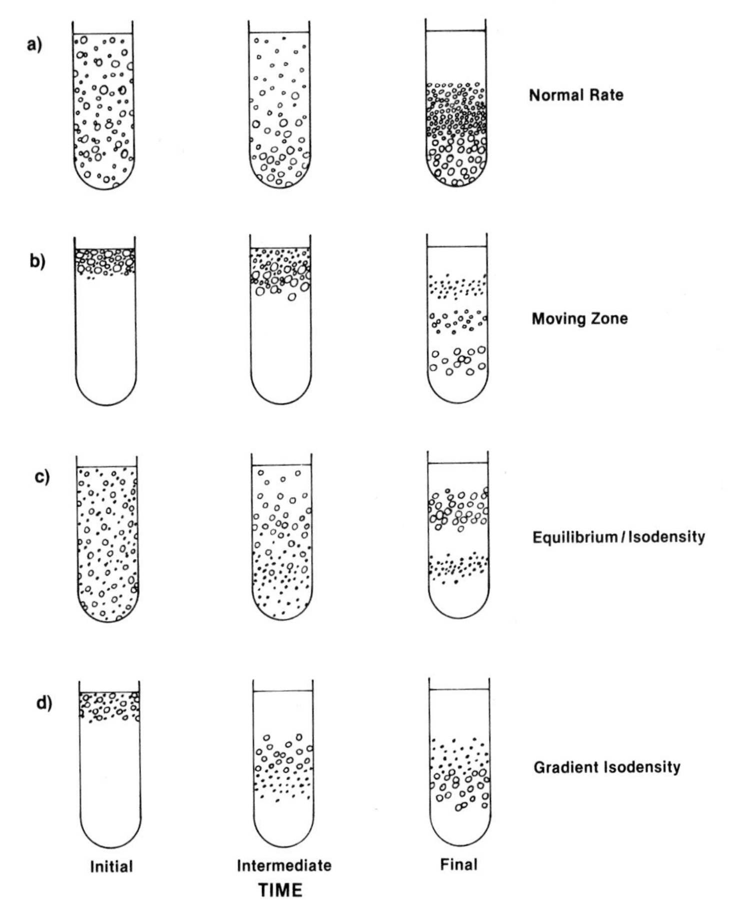

Fig. 2. Depiction of some of the centrifugation conditions in four types of separations. The conditions before application of the centrifugal force (initial time), part of the way through centrifugation (intermediate time), and at the end of the application of the centrifugal force (final time) are illustrated. Consult the text for further details.

an applied RCF, a pellet formed by sedimentation of large particles will be contaminated with smaller-size particles that were initially near the bottom of the tube. The medium-size particles that will sediment after the large particles will also be contaminated with smaller particles. The slowest-sedimenting particles, such as ribosomes or enzymes, may be obtained in a relatively pure fraction, although many of these particles will have been lost through contamination of previous fractions. The pellet obtained from any fraction, but in particular from the more repidly sedimenting fractions, may be resuspended in medium and centrifuged again to reduce the contamination of smaller particles. Several such "washings" may considerably reduce this contamination.

5.2. Factors That Influence Sedimentation

After a certain duration of centrifugation, a region of clear solvent appears from the meniscus down toward the bottom of the centrifuge tube. A boundary region may be seen between the solvent and the solution, then a plateau region of fairly constant concentration toward the bottom of the tube. Ideally, the boundary would remain a constant width and the plateau a constant concentration, and the clear solvent would increase in size by the amount that the plateau decreased as sedimentation progressed. However, sedimentation results in the formation of a region of clear solvent next to a region of more concentrated solution. This may be considered as a two-step gradient, and despite the applied centrifugal force, diffusion will occur back across the boundary. As sedimentation proceeds, the boundary region will become more diffuse and increase in width.

5.2.1. *Diffusion*

Particles that do not sediment from a solution under the force of gravity alone are kept in suspension by Brownian motion, i.e., forces acting on dispersed particles as a result of collisions between these particles and the molecules of the dispersing medium (as discussed by Trautman, 1964). The influence of gravity is not enough to overcome the collision forces of the dispersing medium on particles that show Brownian movement. When such random molecular motions result in transport of matter from one part of a system to another, the process is called diffusion. In a dilute solution, each particle behaves independently of the others, which it seldom meets, and as a result of collisions with solvent molecules, moves sometimes toward a region of higher and sometimes toward a region of lower concentration, with no preferred direction toward either. However, consider two, thin, equal elements of volume on either side of any horizontal section in a solution that has a lower concentration layered over a higher concentration. Although it is not possible to say which way any particular solute molecule will move in a given interval of time, it can be said that on the average, a definite fraction of particles initially in the lower element of volume will cross the section from below, and the same fraction of molecules in the upper element will cross from above. Thus, simply because there are more particles in the lower element than in the upper, there is a net transfer from the lower to the upper side of the section due to independent random molecular motions.

Fick's first law of diffusion states that the mass of solute diffusing across a plane per unit area per unit time is proportional to, and in a direction to diminish, the concentration gradient. The rate of diffusion, dm/dt, then, is proportional to the concentration gradient, dc/dx, and the cross-sectional areas, A, through which it occurs; i.e., ($dm/dt \propto -A(dc/dx)$. The negative sign arises because the material diffuses in the direction opposite to the concentration gradient. A proportionality constant, D, that becomes the diffusion coefficient may be added: $dm/dt = -DA(dc/dx)$; D is the rate of transfer of mass (m) of the material in a unit of time across a unit of area under the influence of a unit concentration gradient. The coefficient of proportionality, D, is the diffusion coefficient with the unit 10^{-7} cm^2/sec called a Fick and denoted F (McCall and Potter, 1973).

Fick's second law of diffusion relates the change of concentration with time to the rate at which the concentration gradient is changing with distance from the original boundary; i.e., $dc/dt = D(d^2c/dx^2)$. Thus, measurement of the change in concentration with time at a constant distance, x, as a function of time and distance from the original boundary should provide the necessary information for the determination of D.

5.2.2. *Friction*

Knowledge of D permits the determination of several other facts about the particle under examination. For example, as a particle moves through a medium, whether due to Brownian motion or to sedimentation, resistance occurs in the medium and a frictional force (F_f) is involved. This force is proportional to the velocity (dx/dt) and may be represented: $F_f = f\,(dx/dt)$, where f is a constant, known as the frictional coefficient. The frictional coefficient is related to the diffusion coefficient D as follows: $f = kT/D$, where k is the Boltzman constant and T is the temperature (Kelvin). The Boltzman constant is the proportionality constant from the ideal gas law and equals 1.38×10^{-16} erg/degree centigrade. The frictional coefficient (f) may be separated into three factors; the viscosity of the medium in which (D) is measured and the size and shape of the particle. This can be accomplished by defining an equivalent sphere of radius a_0 as denoted by f_0 and given by Stoke's law as $f_0 = 6\pi\eta a_0$, where η denotes the coefficient of viscosity of the medium. The frictional coefficient can then be rewritten as $f = 6\pi\eta a_0(f/f_0)$ where (f/f_0) is called the friction factor and represents the hypothetical velocity of the equivalent sphere compared with that of the actual particle.

5.2.3. *Effective Mass*

Thus, whether reference is being made to movement due to diffusion or sedimentation or both, the velocity of a particle is influenced by its size and density as well as the viscosity of the medium. For sedimentation to occur, there must be a density difference between the particle and the solvent. The effective mass (m_e) of the particle is the true mass (m) less the mass of the solvent it displaces (m_d); i.e., $m_e = m - m_d$. This may be rewritten as $m_e = m - m_d\rho = m(1 - \bar{v}\rho)$ where ρ is the density of the solvent and $\bar{v}$ is the partial specific volume of the particle. The partial specific volume ($\bar{v}$) is the increase in volume that occurs as a result of the addition of 1 kg of the particle to an infinite volume of water.

As described in Section 2, the acceleration due to an applied centrifugal field is equal to the product of the square of the angular velocity and the radial distance of the particle from the axis of rotation, or $\omega^2 r$. The centrifugal force exerted on the particle, which is equal to the product of its mass and the centrifugal acceleration, is calculated with the substitution of the effective mass of the particle for its true mass. Thus, $F = m(1 - \bar{v}\rho)\,\omega^2 r$. This force, which causes the particle to move toward the bottom of the centrifuge tube (i.e., away from the axis of rotation), is counteracted in part by the frictional force described above resulting from the resistance of the medium. After a certain time, assuming the centrifugal field to be constant and as particle velocity increases, these forces become equal in magnitude and the particle reaches a terminal velocity, moving through the medium at a uniform speed. Thus, $m(1 - \bar{v}\rho)\,\omega^2 r = (kT/D)(dx/dt)$. Both sides of the equation may be multiplied by Avogadro's number (N), $mN = M$ (M = molecular weight), and $kN = R$ (R = the gas constant, 8.314×10^7 erg/°C per mole). On rearrangement, then, $M = [RT/(1-\bar{v}\rho)]\,(dx/dt)(1/\omega^2 x)$. This is called the Svedberg equation, and it can be used, of course, to determine the molecular weight of macromolecules. However, such determinations require the use of an analytical centrifuge as fully detailed elsewhere (Bowen, 1970; McCall and Potter, 1973; Pickels, 1950; Schachman, 1959; Svedberg and Pedersen, 1940; Trautman, 1964; Williams, 1963). The equation is presented here as a demonstration of the various factors that influence particles during centrifugation, since these remain essentially the same whether analytical or preparative machines are used.

The factor $(dx/dt)\,(1/\omega^2 x)$ from the Svedberg equation is the rate of sedimentation in a unit field and is usually called the sedimentation coefficient (s) having the unit of seconds. Since the sedimentation coefficient usually has a value in the region of 10^{-13} sec, the Svedberg unit (S) was defined as S $= 1 \times 10^{-13}$ sec.

5.2.4. *Convection*

Under some circumstances, both the suspension medium and the suspended particles may move together in a process known as convection. Either temperature changes within the centrifuge tube or local concentration changes may cause density inversions leading to convection. Most centrifuges and rotors in use at present are capable of maintaining constant temperatures, and therefore thermal convection need not always be a serious consideration. Such convection may extend for considerable distances and is quite undesirable because of its uncontrolled effects on the pattern of particle sedimentation. The direction of sedimentation of particles with respect to the solvent in a cylindrical centrifuge tube in a swing-out rotor, for example, is not parallel to the walls of the tube because these walls are not parallel to the centrifugal field itself. Analytical procedures frequently make use of sector-shaped sample cells in which the walls are parallel to the field. However, in the preparative instance of interest here, the sedimenting particles collide with the tube walls and tend to accumulate there. This results in a region of greater density along the walls of the tube when compared to the partially vacated central area. As already mentioned, this process can have serious implications for the separation procedure as a whole. A relatively simple resolution of this problem, when sector-shaped analytical cells are not appropriate, is the use of a density gradient.

5.3. Density Gradients

Sucrose solutions are usually used to form density gradients. Appropriate combination of varying amounts of two sucrose solutions, one denser than the other, as described in Section 5.3.6, results in a gradient of density being formed in the centrifuge tube. The sucrose is least dense at the meniscus and becomes progressively denser toward the bottom of the tube. Density gradients formed with sucrose are relatively stable, and the gradient-containing tubes can be handled carefully without disturbing the gradient. Usually, the maximum density will be less than that of any of the sedimenting particles, and therefore, with sufficiently long application of centrifugal force, all the particles will sediment through the gradient and accumulate at the bottom of the tube. As the solution increases in density at any point due to the direction of sedimentation of the particles, it will sink only slightly until it reaches a level in the tube in which its density does not exceed the density of the gradient at that point. The sinking region of increased density is then turned toward the center of the tube back into the region where the density has decreased. The only systematic movement of the particles relative to the medium, then, is along the centrifugal-field vector in the center of the tube. Slight density gradients have also been shown to be beneficial in the maintenance of radial flow in angle rotors (Trautman, 1964).

5.3.1. *Normal-Rate Density-Gradient Separations*

There are a variety of techniques that employ density gradients. The first to be discussed is basically the same as the normal-rate separation that was used above to demonstrate the basic factors influencing centrifugal separations. This is simply a normal-rate separation on a shallow preformed sucrose density gradient. The particles of the homogenate are initially dispersed throughout the entire medium, and centrifugation proceeds in the same manner as described above. The shallow sucrose gradient reduces convectional forces, which, as mentioned above, are deleterious to systematic separations. The gradient also helps to reduce the back-diffusion of sedimenting particles from the boundary region into the clear solvent region.

Another practical aspect of centrifugal techniques is worth consideration at this point. With the exception of the use of zonal rotors (which are discussed in Section 5.3.7d), it is necessary, following a centrifugal run of sufficient duration to achieve a satisfactory separation, to decelerate the rotor completely in order to remove its contents. As the rotor slows down, the contents of the centrifuge tubes are subjected to forces that are capable of negatively affecting the separation that has already been achieved. The liquid near the walls of the centrifuge tubes is subjected to greater decelerative forces than the liquid in the center of the tubes. This results in a swirling movement of the tube contents in the direction of the rotor's motion. McCall and Potter (1973) mention that frictional interactions between the walls of the tube and their contents slows this swirling movement, and these interactions are proportional to the applied RCF at any instant, while the magnitude of the swirling force is proportional to the rate of deceleration. The obvious conclusion is that to minimize swirling, the rotor must be allowed to decelerate as slowly as possible. The greater the mass of the rotor, the slower it will decelerate. Although it is a rather prolonged process, it is usually desirable to allow the rotor to coast to a stop without the use of the brake, since this will provide the slowest rate of deceleration. Many centrifuges are known to exhibit considerable vibration at some speeds, especially lower speeds. This vibration in itself can reduce the degree of separation achieved, and it is best to decelerate past these particular speeds by using the brake, although some instruments may be adapted for auxiliary means of rotor stabilization at these critical speeds (e.g., deDuve *et al.*, 1959). In any case, the use of a density gradient adds considerable stability to the centrifugal separation and helps to maintain it even in the face of the sort of difficulties just mentioned. Finally, the maintenance of boundaries during handling after the rotor has been stopped is benefited by the use of a gradient. For these reasons, these rate- or moving-boundary-type separations have been called stabilized moving-boundary conditions when a density gradient is used (McCall and Potter, 1973).

5.3.2. *Moving-Zone Methods*

The use of a steeper preformed density gradient (i.e., one that spans a greater range of densities from the top of the tube to the bottom) provides the possibility of more complete separations than are achieved with the stabilized moving-boundary technique. The maximum gradient density is kept below that of the particles being centrifuged, although a small layer of very dense sucrose may be placed as a cushion in the bottom of the tube to prevent the formation of a pellet when this is not desirable. The gradient may be formed as in the earliest applications of this technique by layering small volumes of sucrose of decreasing densities (i.e., concentrations) on top of one another in the centrifuge tube, taking care to avoid mixing of the layers as they are applied. Such a stepwise or discontinuous gradient can be considered as a device for generating artificial separation bands, since different types of particles will tend to sediment to the interface between two sucrose layers even though they differ in sedimentation rate. While this may be a useful means of combining certain portions of the sedimentation distribution of particles for preparative purposes (i.e., reducing the number of fractions while increasing the particle concentrations within those fractions obtained), care should be taken that the illusion of a clear-cut separation that arises from the discontinuous gradient not be mistaken for reality (deDuve, 1971).

A technique that is less deceptive interpretively, although not always more desirable for preparative purposes, entails the use of a continuous gradient in which there is a smooth distribution of sucrose densities. The construction of such gradients is discussed in Section 5.3.6. In the words of deDuve (1971), "Density gradient centrifugation in a continuous gradient is the analytical method 'par excellence.' " When using it, it is possible to obtain an "entirely objective assessment of the frequency distribution curves of certain physical properties, such as density or sedimentation coefficient, from which in turn other characteristics of a population, including its size distribution, can be derived." In most cases, the sample to be fractionated is layered on top of the density gradient. Care must be taken to avoid mixing the sample with the upper portion of the gradient, and this may be achieved by introducing the sample through a needle or pipette bent at a 90° angle to allow the sample to be ejected in a direction parallel to the meniscus rather than being directed right into the gradient, which can result in mixing.

Techniques that are characterized by an initial concentration of the sample within a single layer are termed moving-zone techniques, since they make use of the movement of a zone or band of particles. Although, as mentioned, the initial layer is usually deposited on top of the gradient, when making use of flotation rather than sedimentation, the layer is placed below the gradient. It is necessary, in order to maintain the separations achieved during centrifugation, to terminate the RCF before the separated zones become combined in a pellet. The moving zone is capable of producing a density inversion as it moves into fresh solvent because the sedimenting particles increase the density of the solution. A density gradient is needed to stabilize this effect, and this has led to reference to these techniques under the general category of density-gradient centrifugation, although their primary feature is that of rate or velocity separation as in normal-rate separations, which do not employ gradients. More accurately, such separations should be and are referred to as rate-zonal or s-zonal to imply that the principal properties of the particles that influence their separation are their sedimentation velocities or coefficients. Each particle or class of particles will sediment at its own rate and will eventually form a band or zone in the gradient.

Moving-zone methods are of particular value in preparative work, since they allow the separation of faster-moving particles from slower-moving particles without pelleting and, possibly, in a single centrifugal run. Trautman (1964) has indicated that faster-moving particles, which, as mentioned previously, are initially obtained in a rather impure pellet following differential centrifugation, may be obtained in a relatively pure state when a moving-zone method is applied. This purity is equivalent to 10–20 cycles of washing and pelleting in differential centrifugation. On the other hand, the fractions collected are quite diluted compared to those obtained from pellets, and, as will be discussed further, considerably less starting material can be used. Zonal separations can be used to estimate the sedimentation coefficients of particles if a marker substance with a known coefficient is included in the sample for reference purposes. One difficulty with these techniques, which will also be discussed in relation to other methods, is that the moving zone requires the sample material to be moved through a changing solvent environment, which can alter some of the physical properties of some biological materials and may be due to irreversible osmotic and hydrostatic forces. This difficulty is not encountered when using moving-boundary methods. Figure 2 provides a comparison of moving-boundary and moving-zone techniques. Further discussion of the latter will be reserved for the general review of gradient methods below.

5.3.3. *Isopycnic Methods*

Other techniques are available, all of which make use of density gradients. However, while the moving-zone method, whether employing sedimentation or flotation, depends primarily on the sedimentation rate (s-rate) of the particles, isodensity or isopycnic techiques are more dependent on the buoyant density of particles and combine both sedimentation and flotation in the same separation. As mentioned above, the maximum density of the gradient in a rate-zonal separation is less than the density of the sample particles, and a sufficiently long application of the RCF will result in the formation of a pellet. Consequently, the duration of the centrifugal run is rather critical if optimal separations are to be achieved. Isodensity gradients, on the other hand, span the whole range of buoyant densities of the sample particles. The various zones of particles will sediment until they reach a level in the gradient at which the density of the solvent is equal to the buoyant density of the particles in the zone. The particles can then be considered to be floating at that level.

5.3.3a. Nonequilibrium Methods. Isopycnic techniques may be divided into two types depending on whether the gradient is formed before centrifugation in essentially the manner used in zonal separations or, as discussed later, during the course of the centrifugal run. Preformed isodensity gradients depend on the movement of particles, relatively rapidly, to a position in the gradient corresponding to their density. As pointed out by Trautman (1964), the density gradient is not permanent, since the material used to form it will eventually redistribute under the influence of the RCF, thus forcing the separated zones to the top or the bottom of the centrifuge tubes. However, the lag between the time needed to achieve optimal separation and that resulting in such redistribution is quite large and the duration of the run is much less critical than in the case of moving-zone methods. Furthermore, there is no formation of a pellet, since the densest particles will float before reaching the bottom of the tube. Also, in contrast to the moving-zone application, a purification of the sample material may be achieved, or, in other words, the fractions are not necessarily diluted in comparison with the original sample. It is also possible, of course, to determine the isodensity value of the particles if the density gradient itself can be calibrated. This is discussed in Section 5.3.7a. Although it is most common to apply the sample material as a layer on top of the gradient, it is also possible to distribute the sample homogeneously throughout the medium before the application of the RCF. Larger starting amounts of sample may be used and greater yields achieved, but the particles cannot move to a position in the gradient that was not previously occupied by slower-moving particles of the same or different density. Although this latter approach eliminates the problem of density inversion, it is of limited value for preparative work. Similar to the situation previously mentioned in regard to zonal separations, particles initially present in the lower portion of the tube may be irreversibly altered.

5.3.3b. Equilibrium Methods. The preformed gradients just described are, as mentioned, transient and have been characterized as nonequilibrium gradients. The second isodensity technique, in contrast, makes use of what is termed an equilibrium gradient and is therefore known as isodensity equilibrium. No layering or formation of the gradient is necessary before the RCF application. It is only necessary that the nominal density of the medium approximate the isodensity value of the sample. The gradient is formed under the influence of the RCF according to classic sedimentation equilibrium. That is, the particles or molecules that constitute the gradient medium redistribute in the centrifuge tube to the extent that either their sedimentation or their flotation is equal to their rate of diffusion. Thus, a stable gradient (i.e., in equilibrium) is formed. In contrast, diffusion of the sample material as separated is balanced by both sedimentation and flotation as already discussed for the general isodensity case. One limitation of the isodensity equilibrium method is that formation of the gradient is a rather lengthy matter, requiring 2 or 3 continuous days of RCF application, e.g., several hours for the gradient material to reach equilibrium and up to 30 hours more for the sample particles to form zones and reach an equilibrium. In contrast to other density-gradient methods, which most frequently employ sucrose or a similar substance as gradient material, isodensity equilibrium gradients are generally formed with the salt of a heavy alkali metal such as cesium chloride. The problem, again, of physical alterations of particles, medium composition, and long centrifugation times serves to somewhat limit the application of the isodensity equilibrium method for preparative purposes, although purification and concentration are achieved. Analytically, the technique is useful for the precise measurement of isodensity, homogeneity, and molecular weight. Both preformed-gradient-isodensity and isodensity-equilibrium methods are depicted in Fig. 2.

5.3.4. *Overview of Terminology*

The variety of methods available for use in preparative separation work is often quite confusing, and the lack of a truly standard terminology does little to simplify matters for the unsophisticated reader. It would be useful to review the previous sections at this point in order to arrange the various methods into a framework that will facilitate an understanding of their relationships. Figure 3 should be consulted in reference to this overview.

Much of the confusion arises from the fact that the centrifugation techniques of interest in preparative applications can be divided into more than one type of categorization, and the different categories are not always mutually exclusive. Thus, it is possible to consider those techniques that depend primarily on the sedimentation coefficients of the sample particles

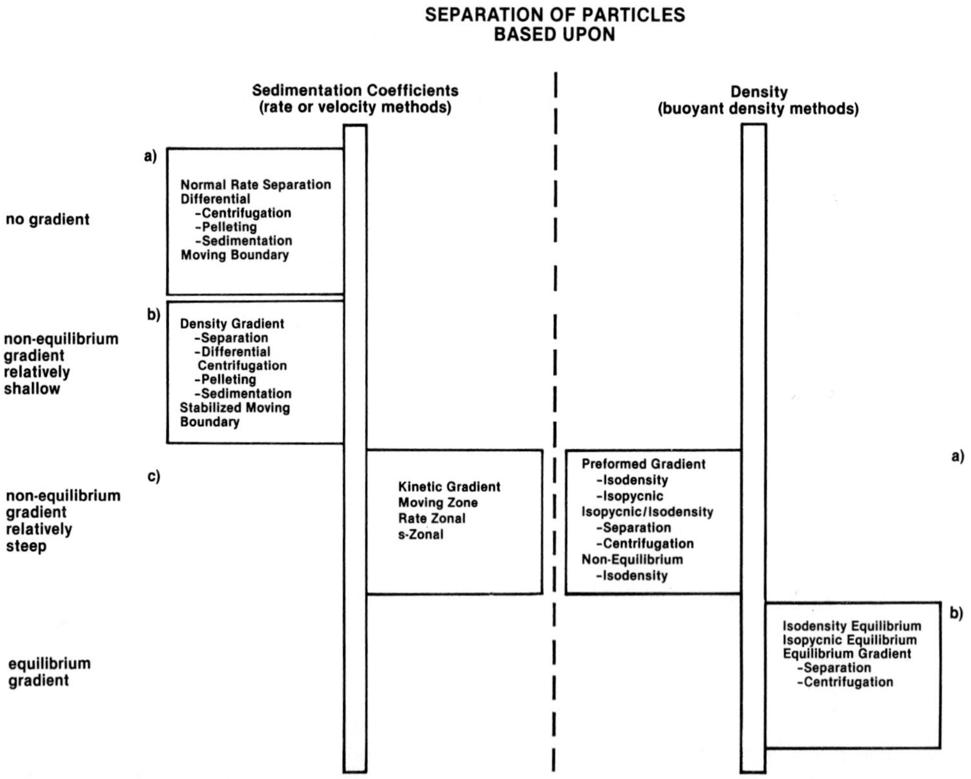

SEPARATION OF PARTICLES
BASED UPON

Sedimentation Coefficients
(rate or velocity methods)

Density
(buoyant density methods)

no gradient

a)
Normal Rate Separation
Differential
-Centrifugation
-Pelleting
-Sedimentation
Moving Boundary

non-equilibrium
gradient
relatively
shallow

b)
Density Gradient
-Separation
-Differential
Centrifugation
-Pelleting
-Sedimentation
Stabilized Moving
Boundary

non-equilibrium
gradient
relatively
steep

c)
Kinetic Gradient
Moving Zone
Rate Zonal
s-Zonal

a)
Preformed Gradient
-Isodensity
-Isopycnic
Isopycnic/Isodensity
-Separation
-Centrifugation
Non-Equilibrium
-Isodensity

equilibrium
gradient

b)
Isodensity Equilibrium
Isopycnic Equilibrium
Equilibrium Gradient
-Separation
-Centrifugation

Fig. 3. Classification of the various terminology used to refer to particle separations achieved by differing methods. Terms within a single box refer to the same separation conditions and are synonyms, while each box represents a different set of conditions as detailed in the text.

and may be categorized most simply as rate or velocity methods separately from those techniques that depend primarily on the buoyant densities of the sample particles to achieve their separation and may be categorized most simply as isodensity or isopycnic methods. The *rate* or *velocity* methods include: (a) normal-rate separation, also called differential centrifugation, differential pelleting, differential sedimentation, or moving-boundary separation; (b) stabilized moving boundary, also called density-gradient separation or density-gradient differential centrifugation, pelleting, or sedimentation; and (c) kinetic gradient, also called moving-zone, rate-zonal or s-zonal. The *buoyant-density* methods include: (a) preformed gradient isodensity, also called preformed isodensity gradient, preformed isopycnic gradient, nonequilibrium isodensity, or isopycnic separation (or centrifugation), or isopycnic zonal or *p*-zonal; and (b) isodensity equilibrium also called isopycnic-equilibrium or equilibrium-gradient centrifugation. Unfortunately, these lists of synonyms do not completely exhaust the possible references to be found in the literature, although the type of separation referred to by any synonym that has not been included here should be relatively easy to determine by comparison with the names that have been listed, since they should be essentially very similar. No effort has been made to include references for the initially rather bewildering array of terminology, since it was not deemed relevant and would have served only to complicate the presentation.

It is also possible to analyze methods on the basis of whether or not they make use of a density gradient. However, the only, rather simplistic, distinction in this regard is between moving-boundary and stabilized moving-boundary methods [i.e., types (a) and (b), respectively, under rate or velocity methods, above]. The distinction is rather superfluous, since the major function of the gradient in the stabilized moving-boundary method is the maintenance of a normal-rate separation both during and after the application of the RCF. It is more relevant to distinguish between methods that use a gradient in conjunction with rate or velocity and buoyant density, and this is implicit in the previous paragraph. All the preformed-density-gradient techniques may utilize either a continuous or a discontinuous gradient, but although this distinction is of some interpretive importance, it should not be a source of misunderstanding of types of separations as such. Finally, density-gradient techniques may be categorized according to whether or not the gradient is preformed, or non-equilibrium, or formed under the influence of the RCF, i.e., equilibrium. The only equilibrium techniques are listed above under (b) for buoyant-density methods.

One final source of categorization could be based on whether the sample is layered either above or below the medium or is incorporated homogeneously throughout the medium prior to centrifugation. This is, again, a rather arbitrary distinction, since most methods may use either procedure. However, in terms of the distinctions made in the first paragraph of this summary, the separations categorized as (a) and (b) under the rate or velocity methods and under (b) for the buoyant-density methods will always require the homogeneous incorporation of sample.

5.3.5. Gradient Materials

Several physical and chemical properties are relevant to the consideration of suspension media for gradient centrifugation (also discussed by deDuve *et al.*, 1959; Moore, 1969). An obvious initial criterion is that the basic component of the medium be soluble in sufficient quantity to produce densities that are appropriate for the range of particles to be fractionated and the particular centrifugation technique to be employed. Unfortunately, the higher concentrations of some of the most frequently used media (e.g., sucrose) are highly hypertonic and can irreversibly affect some subcellular components. Nevertheless, tonicity of the medium is an especially relevant criterion, and its effects should always be considered. Other criteria, however, are also important. The solute must, of course, be soluble in water, since most procedures, and all those considered here, make use of aqueous media. Additionally, the medium should exert little effect on pH or ionic strength. It is also desirable that the medium not denature any of the materials to be fractionated, nor should it be capable of inhibiting the activities of any of the biological systems to be studied. Finally, the medium should not possess any properties that interfere with any of the analytical procedures to be applied to the fractionated material. A variety of other effects of the medium on the isolation of various fractions have already been discussed with regard to homogenization media. The medium in which the sample is homogenized is, of course, often the medium in which the first centrifugal separation takes place. Since these aspects of gradient media will not be discussed further, please refer to Section 4.4.

5.3.5a. Sucrose. The first gradients, used simply for stabilization, in both angle rotors (Pickels, 1943) and horizontal or swing-out rotors (Brakke, 1951; Kahler and Lloyd, 1951), were made with sucrose, and sucrose has remained the most popular medium from its first use in subcellular fractionation (Hogeboom *et al.*, 1948) to the present. As noted by deDuve *et al.* (1959), pure sucrose gradients are far from ideal because, for example, in isopycnic sedimentation, very highly hypertonic solutions are required. Although a recommended search for other media has been ongoing, no completely satisfactory substitutes have been accepted. Sucrose solutions may be concentrated to yield densities of roughly 1.3 g/cm^3 without simultaneously being so viscous as to preclude their use in fractionation. Above this concentration, the viscosity increases very rapidly. The high densities of sucrose—and of other media discussed below—necessary in many separation procedures can result in the concentration of relatively large amounts of impurities in the medium that are present to a relatively small degree in the reagent as purchased. These impurities could either affect the activities of biological systems of interest or interfere with the analytical procedures used after fractionation. Thus, the lead concentration in a 50% solution of analytical, reagent-grade sucrose can be as high as 0.5 mg/liter, which would inactivate some enzyme systems (Reid and Williamson, 1974).On the other hand, some commercially available sucrose perparations have been reported to contain ribonuclease, which could interfere with ultraviolet absorption techniques because of its absorption in this region (Brakke, 1967). Hinton *et al.*, (1969) have reported cases of assays that must be corrected for the presence of sucrose itself. Treatment with decolorizing charcoal can remove some impurities (Moore, 1969), although more involved procedures may be required to remove contamination by bacteria that easily contaminate sucrose solutions or by plant ribonucleases. The latter are very resistant to heat inactivation, and since autoclaving of sucrose causes some hydrolysis and loss of viscosity, other means of inactivation are required (Reid and Williamson, 1974). Pretreatment with 0.1% diethylpyrocarbonate seems to be generally successful in the inactivation of these ribonucleases, but since it also denatures proteins and interacts with nucleic acids, it is necessary to allow it to decompose completely to ethanol and carbon dioxide before adding any sample material to the sucrose (Reid and Williamson, 1974). Glycerol, as discussed briefly by Moore (1969), has many of the same properties as sucrose, which renders it a less than satisfactory substitute, but it is volatile, which may be advantageous in certain applications.

5.3.5b. Ficoll. Some substitutes for sucrose have met with a degree of success. Specifically for use in density determination of cells and subcellular particles, Holter and Max Moller (1958) introduced a synthetic copolymer of sucrose and epichlorohydrin that has been called Ficoll (Pharmacia, Uppsala, Sweden; 200 Centennial Avenue, Piscataway, New Jersey). Concentrations of Ficoll higher than 50% are soluble in aqueous media due to the large number of hydroxyl groups on the polymer. Although it is vulnerable to oxidizing agents and unstable at low pH, it is stable at both neutral and alkaline pH. Ficoll may be autoclaved, but its acid hydrolysis is enhanced at high temperatures. The high average molecular weight of the polymer (400,000) results in very low osmotic activity, while its almost spherical shape results in a relatively low viscosity. Finally, even at high concentrations, it is non-toxic (Harwood, 1974). Thus, Ficoll is characterized by many qualities that are desirable in a gradient medium. However, some undesirable characteristics have also been reported (Marchbanks, 1974). Its quality varies, and dialysis is necessary to remove salt. Its viscosity is great enough to require longer centrifugation times than with sucrose. Finally, it is difficult to remove it from the fractionated material once it has been isolated. The high cost of Ficoll is sufficient to preclude its use in procedures requiring large volumes of medium such as those using zonal rotors, as discussed in Section 5.3.7d. Polyvinylpyrrolidone and polyglucose have some similar properties, but have been applied to cell-fractionation work much less frequently.

5.3.5c. Equilibrium-Gradient Materials. A third type of gradient material is also in common use, though essentially for isopycnic determinations only: the salts, usually chlorides, of alkali metals that can yield solutions that are much more dense yet less viscous than those obtainable with sucrose or Ficoll. To achieve sufficiently dense salt media, very high concentrations of these salts are necessary. These concentrations (as high as 8 M) have proven quite damaging to many fragile subcellular components, particularly in light of the long centrifugation times required to reach equilibrium (deDuve *et al.*, 1959; Moore, 1969). Many heavy-metal chloride salts available commercially seem to contain levels of impurities that when concentrated in a gradient medium may be harmful to subcellular components and their functioning. Decolorization charcoal treatment of salt solutions followed, if convenient, by recrystallization can result in satisfactory purification (Moore, 1969). A review by an early proponent of heavy salt media includes a section detailing methods of calculating salt concentrations necessary to achieve various desired densities (Vinograd, 1963).

Potassium citrate and potassium tartarate solutions have also seen some application. Problems, though, have been encountered. They are very active osmotically and at high concentrations may chelate metal ions, thus affecting the stability of some subcellular components (Moore, 1969). Bovine serum albumin gradients have also met with some success and are quite commonly used in the separation of whole cells. Some of their advantages and disadvantages in relation to cell separation have been discussed by Harwood (1974). As with other biological media such as sucrose, care must be taken when using serum albumin because of the possible presence of contaminants such as bacteria.

Moore (1969) presents several tables that compare the densities and viscosities of the various gradient media mentioned above.

In general, the buffering of various media is subject to considerable variation dependent on the techniques being used and the fraction or fractions that are of particular interest. Previous consideration has been given to the question of pH and buffer capacity in media (see Sections 4.4.3 and 4.4.4), these questions will not be reviewed again here.

5.3.6. Formation of Density Gradients

The earliest techniques used to form density gradients consisted of the manual layering of solutions of lower density on solutions of higher densities. Accurate layering could be achieved by forcing the medium from a syringe through a needle bent 90° at the end to allow carefully measured amounts of medium to flow in a direction parallel to the surface of the previous layer rather than directly into it. A smoother gradient could be produced by diffusion, if the layered gradient was allowed to stand for several hours. Alternatively, a brief stir of the entire contents of the tube with a wire also produced a relatively smooth distribution, although it is doubtful that the reproducibility of the gradients produced in this manner could equal that following diffusion alone. Early use of simple, two-layer gradients to improve resolution in differential centrifugation (Allfrey, 1959) led rapidly to multistep gradients useful for other techniques. While discontinuous gradients such as these are extremely simple technically and are also the gradients of choice in certain applications, continuous gradients are clearly of more practical value in most present-day applications.

Behrens (1938) described one of many simple devices that may be used to produce a continuous density gradient. Two burettes are connected in tandem. One burette contains a low-density or light solvent solution, while the other contains a high-density or heavy solvent solution. A wedge is fitted in the burette containing the light solvent, which produces a wedge-shaped column of solvent. Thus, increasingly larger amounts of the light solvent may be added to a fixed amount of the heavier solvent, thus allowing the precise construction of various gradients by the use of differing solvents or rates of flow or both.

"Gradient engines" provide more precise control over gradient parameters as well as greater, or more easily achieved, flexibility in types of gradients. The outflows of two individually piston-driven syringes, one syringe containing light solvent and the other containing heavy solvent, may be combined to produce a gradient (Anderson, 1955). The shape of the gradient may be controlled by the systematic variation of any one or a combination of syringe sizes, drive rates, and solvent densities. Other devices, similar to this one in principle, may also be used dependent on the type of gradient required (see the next section for more details on various types of gradients and the devices appropriate for the construction of each). Several engines have been described that, as in the one just mentioned, combine differential rates of flow of two solvent solutions, mix them in a separate chamber, and then add the mixture to the centrifuge tube. A second alternative involves the pumping of one solvent of a given concentration directly into a vessel containing the other solvent, where the mixing takes place. The mixed solvents are continuously added to the centrifuge tube. In both types of engine, the gradient shape produced may be influenced by the initial solvent densities, the rate of solvent mixing, and the rate of removal of the mixed solvents to the centrifuge tube. Some gradient engines are depicted in deDuve *et al.* (1959) and McCall and Potter (1973).

A final method of gradient formation has already been mentioned. Equilibrium isodensity gradients, usually consisting of a heavy-salt solution, are formed following prolonged centrifugation at high RCFs of an initially homogeneous solvent solution.

5.3.7. *Gradient Parameters and Functional Implications*

Several characteristics of density gradients may be systematically varied to obtain an almost limitless choice of gradients suited to any particular application. Generally speaking, in subcellular-fractionation work, density gradients facilitate the separation of various particle populations while helping to stabilize the boundaries of the separated fractions both during centrifugation and during handling after the run. Density-gradient techniques can eliminate or drastically reduce the need for successive pellet formations as in differential-centrifugation procedures. This is particularly advantageous in neurobiological work, since some structures of particular interest, e.g., synaptosomes, are adversely affected by pelleting. Reduction in the number of times a pellet is formed from the several washings necessary for purification in differential centrifugation to perhaps a single pelleting stage is quite successful in the retention of particle integrity. The gradient parameters chosen for a particular separation will depend on the desired end product of that separation. Procedures to resolve an entire homogenate into several fractions in which fraction purity may be sacrificed, at least temporarily, require different conditions than, for example, procedures designed to isolate a very pure particle fraction free of contamination by particles with similar sedimentation or density properties, or both. The latter will usually require a sacrifice in yield for the sake of purity. Of course, a given experiment may combine both procedures, with an initial separation followed by purification of the fractions obtained. In any case, the demands of different experiments necessitate the use of appropriate gradient characteristics.

5.3.7a. Parameters of Gradients. As mentioned in Section 5.3.4, nonequilibrium or preformed density gradients may be classified as either continuous or discontinuous. For the sake of clarity, the following discussion will be restricted to continuous gradients, although the same principles may be applied to discontinuous gradients, though, in general practice, in a less complex fashion. Density gradients may also be classified as either linear or exponential, with the latter further subdivided into convex and concave gradients. In linear gradients, the density of the medium increases linearly with distance from the axis of rotation or radius. Gradients in which the rate of change of increases in density decreases with increasing radius are convex. The formulas mentioned in the following discussion have been derived from McCall and Potter (1973).

In the most simply designed engines for producing exponential gradients, a volume of the gradient material of a given concentration (D_1) is continuously added at a fixed rate to a second solution of gradient material initially of a different density (D_2). The two solutions are thoroughly mixed as they combine, and the mixed solution is removed from its container and added to the centrifuge tube at the same rate as the first solution (D_1) is added to the mixing vessel. Thus, the volume (V_A) of solution in the mixing vessel always remains constant, and the centrifuge tube is filled at the same rate as D_1 is added to the mixing vessel. If the instantaneous density of the gradient leaving the mixing vessel is D_3 and the cumulative volume of gradient already added to the centrifuge tube is V_B at the instant that gradient of density D_3 is entering, the following equation holds:

$$2.303 \log \frac{D_1 - D_3}{D_1 - D_2} = -\frac{V_B}{V_A}$$

If the density of the first solution (D_1) is greater than the initial density of the medium in the mixing vessel (i.e., prior to mixing, thus, D_2), then the gradient formed will be convex in shape. If D_1 is less than D_2, the gradient formed will be concave. When convex gradients are being formed, solutions of lighter density are the first to leave the mixing vessel, since the vessel contains the lighter of the two solutions being combined. In practice, a tube through which the gradient mixture is flowing is lowered to the bottom of the centrifuge tube. Thus, as increasingly dense solution is added, the lighter portions of the gradient are lifted toward the top of the tube. In contrast, the material for concave gradients is allowed to simply run down the side of the centrifuge tube as the heaviest portion of the gradient leaves the mixing vessel first. Progressively lighter portions are layered on top.

With this in mind, reference may be made to Tables 3 and 4, which list the properties of two convex and two concave gradients as they leave the mixing chamber. The fixed parameters V_A, D_1, and D_2 are indicated for each of the four gradients. In the calculations for these tables, the equation above was rearranged as follows:

$$V_A \; 2.303 \log \frac{D_1 - D_3}{D_1 - D_2} = -V_B$$

Thus, the cumulative volume of gradient added to the centrifuge tube (V_B) was the dependent variable and D_3 was varied independently over a range of gradient concentrations from 5 to 30%. Obviously, for the sake of simplicity, a continuous gradient has been represented in discrete steps. In each case, the value of V_B is indicative of the amount of medium already added to the gradient at the instant that the concentration of the medium being added is precisely D_3. Thus, for example, in Table 3, as the concentration of medium is exactly 10%, a total of 1.666 cm^3 of medium has already been added to the convex gradient. For each gradient represented in Tables 3 and 4, the volumes of medium incorporated for certain ranges of concentration are also indicated. (ΔV_B). Thus, for the convex gradient in Table 3, the amount of medium of a concentration greater than 9% and less than or equal to 10% is 0.3610 cm^3.

Further rearrangement of the formula given above facilitated calculations when the cumulative volume (V_B) was allowed to vary independently from 0 to approximately 20 cm^3 (25 cm^3 for the shallow linear gradient). The instantaneous concentration of sucrose being delivered was determined as follows: $[\text{antilog} \; (-V_B/2.303 \; V_A)](D_1 - D_2) - D_1 = -D_3$. The same fixed parameters used in calculating the gradient characteristics in Tables 3 and 4 were used in the calculations that resulted in the data presented in Tables 5 and 6, respectively. Each of the latter two tables indicates the density of the gradient medium (D_3) entering at precisely the moment that the total volume of sucrose added to the centrifuge tube is equal to the value indicated as V_B. Thus, for the convex gradient in Table 5, when the volume of the added gradient material is exactly 10 cm^3, the sucrose concentration being added is 24.255%. Similar to the previous calculations, a difference column is also

Table 3. Steeper Gradients

D_3 (%)	Convex[a] V_B	Convex[a] ΔV_B	Linear[a] V_B	Linear[a] ΔV_B	Concave[a] V_B	Concave[a] ΔV_B
5	0.0000	—	0.0	—	19.8780	4.2290
6	0.3075	0.3075	0.8	0.8	15.6490	2.4738
7	0.6768	0.3193	1.6	0.8	13.1752	1.7551
8	0.9589	0.3321	2.4	0.8	11.4201	1.3615
9	1.3049	0.3460	3.2	0.8	10.0586	1.1123
10	1.6659	0.3610	4.0	0.8	8.9463	0.9405
11	2.0433	0.3774	4.8	0.8	8.0058	0.8147
12	2.4387	0.3954	5.6	0.8	7.1911	0.7186
13	2.8538	0.4151	6.4	0.8	6.4725	0.6428
14	3.2908	0.4370	7.2	0.8	5.8297	0.5815
15	3.7522	0.4614	8.0	0.8	5.2482	0.5309
16	4.2406	0.4884	8.8	0.8	4.7173	0.4883
17	4.7596	0.5190	9.6	0.8	4.2290	0.4522
18	5.3133	0.5537	10.4	0.8	3.7768	0.4209
19	5.9065	0.5932	11.2	0.8	3.3559	0.3938
20	6.5454	0.6389	12.0	0.8	2.9621	0.3698
21	7.2376	0.6922	12.8	0.8	2.5923	0.3488
22	7.9928	0.7552	13.6	0.8	2.2435	0.3298
23	8.8236	0.8308	14.4	0.8	1.9137	0.3130
24	9.7470	0.9234	15.2	0.8	1.6007	0.2977
25	10.7860	1.0390	16.0	0.8	1.3030	0.2838
26	11.9739	1.1879	16.8	0.8	1.0192	0.2712
27	13.3607	1.3868	17.6	0.8	0.7480	0.2597
28	15.0266	1.6659	18.4	0.8	0.4883	0.2490
29	17.1129	2.0863	19.2	0.8	0.2393	0.2393
30	19.9061	2.7932	20.0	0.8	0.0000	—

[a] Values used in construction of the gradients: Convex: V_A = 8.3, D_1 = 32.5%, D_2 = 5.0%; Linear: V_A = 20.0, D_1 = 55.0%, D_2 = 5.0%; Concave: V_A = 6.1, D_1 = 4.0%, D_2 = 30.0%.

included in Tables 5 and 6. In this instance, the values in the columns headed ΔD_3 indicate the difference in sucrose densities spanning a single cubic centimeter of gradient volume. As indicated for the convex gradient in Table 5, the change in density from the instantaneous concentration at precisely 9 cm^3 to that at 10 cm^3 is 1.055% (i.e., from 23.200 to 24.255%). To avoid confusion, note that since the order of addition of sucrose concentrations, from light to heavy, is inverted with concave gradients with respect to convex gradients, the values in Tables 3 and 4 are, essentially, read from top to bottom for the convex gradients and from bottom to top for the concave gradients. However, in Tables 5 and 6, although the density of the sucrose (D_3) changes in opposite directions for the convex in comparison to the concave gradients, the data for both gradients are read from top to bottom, since the independent variable was the cumulative volume (V_B) of the gradient in both instances. On the other hand, note that while ΔD_3 calculations for convex gradients always refer to the subtraction of a smaller volume

Table 4. Shallower Gradients

D_3 (%)	Convex[a]		Linear[a]		Concave	
	V_B	ΔV_B	V_B	ΔV_B	V_B	ΔV_B
5	0.0000	—	0.0	—	20.0354	2.1491
6	0.4175	0.4175	1.0	1.0	17.8863	1.7861
7	0.8475	0.4300	2.0	1.0	16.1002	1.5281
8	1.2906	0.4431	3.0	1.0	14.5721	1.3353
9	1.7479	0.4573	4.0	1.0	13.2368	1.1858
10	2.2202	0.4723	5.0	1.0	12.0510	1.0664
11	2.7084	0.4882	6.0	1.0	10.9846	0.9689
12	3.2138	0.5054	7.0	1.0	10.0157	0.8877
13	3.7376	0.5238	8.0	1.0	9.1280	0.8191
14	4.2812	0.5436	9.0	1.0	8.3089	0.7603
15	4.8461	0.5649	10.0	1.0	7.5486	0.7095
16	5.4340	0.5879	11.0	1.0	6.8391	0.6649
17	6.0470	0.6130	12.0	1.0	6.1742	0.6256
18	6.6872	0.6402	13.0	1.0	5.5486	0.5908
19	7.3572	0.6700	14.0	1.0	4.9578	0.5596
20	8.0599	0.7027	15.0	1.0	4.3982	0.5316
21	8.7987	0.7388	16.0	1.0	3.8666	0.5061
22	9.5774	0.7787	17.0	1.0	3.3605	0.4831
23	10.4006	0.8232	18.0	1.0	2.8774	0.4620
24	11.2738	0.8732	19.0	1.0	2.4154	0.4428
25	12.2033	0.9295	20.0	1.0	1.9726	0.4250
26	13.1970	0.9937	21.0	1.0	1.5476	0.4086
27	14.2643	1.0673	22.0	1.0	1.1390	0.3934
28	15.4171	1.1532	23.0	1.0	0.7456	0.3793
29	16.6703	1.2532	24.0	1.0	0.3663	0.3663
30	18.0436	1.3733	25.0	1.0	0.0000	—

[a] Values used in construction of the gradients: Convex: V_A = 14.4, D_1 = 40.0%, D_2 = 5.0%; Linear: V_A = 20.0, D_1 = 35.0%, D_2 = 5.0%; Concave: V_A = 10.6, D_1 = 0.55%, D_2 = 30.0%.

value from a larger [e.g., 24.555 (10 cm^3) − 23.200 (9 cm^3)], the reverse is true for the concave gradients. Thus, to obtain a ΔD_3 value of 0.899%, the instantaneous concentration at 10 cm^3 was subtracted from that at 9 cm^3 [i.e., 9.947 (9 cm^3) − 9.048 (10 cm^3) = 0.899].

Similar data are presented in Tables 3–6 for linear gradients. However, a different engine is used to produce a linear gradient. Although equal volumes of gradient material of concentration D_1 are added over equal amounts of time to a container of a second sucrose solution (D_2) as in the case of exponential gradients, the mixed gradient material is added to the centrifuge tube at twice the rate of the D_1 addition. For convenience, the concentration of D_1 is usually greater than that of D_2, and therefore lighter sucrose enters the centrifuge tube first and is displaced upward by the subsequent addition of heavier sucrose, as in convex gradients. The initial volume (V_A) of heavier sucrose (D_2) is equal to the final volume of the gradient. The general formula used to determine the characteristics of linear

Table 5. Steeper Gradients

V_B (cm³)	Convex		Linear		Concave	
	D_3	ΔD_3	D_3	ΔD_3	D_3	ΔD_3
0	5.000	—	5.00	—	30.000	3.931
1	8.121	3.121	6.25	1.25	26.069	3.336
2	10.888	2.767	7.50	1.25	22.733	2.832
3	13.340	2.452	8.75	1.25	19.901	2.404
4	15.515	2.175	10.00	1.25	17.497	2.040
5	17.442	1.927	11.25	1.25	15.457	1.732
6	19.151	1.709	12.50	1.25	13.725	1.470
7	20.666	1.515	13.75	1.25	12.255	1.248
8	22.009	1.343	15.00	1.25	11.007	1.060
9	23.200	1.191	16.25	1.25	9.947	0.899
10	24.255	1.055	17.50	1.25	9.048	0.763
11	25.191	0.936	18.75	1.25	8.285	0.648
12	26.020	0.829	20.00	1.25	7.637	0.550
13	26.756	0.736	21.25	1.25	7.087	0.466
14	27.408	0.652	22.50	1.25	6.621	0.397
15	27.986	0.578	23.75	1.25	6.224	0.336
16	28.498	0.512	25.00	1.25	5.888	0.285
17	28.957	0.459	26.25	1.25	5.603	0.243
18	29.355	0.398	27.50	1.25	5.360	0.205
19	29.712	0.357	28.75	1.25	5.155	0.135
20	30.028	0.316	30.00	1.25	4.980	—

Table 6. Shallower Gradients

V_B (cm³)	Convex		Linear		Concave	
	D_3	ΔD_3	D_3	ΔD_3	D_3	ΔD_3
0	5.00	—	5.00	—	30.000	2.651
1	7.347	2.347	6.00	1.0	27.349	2.412
2	9.537	2.190	7.00	1.0	24.937	2.196
3	11.581	2.044	8.00	1.0	22.742	1.998
4	13.487	1.906	9.00	1.0	20.744	1.817
5	15.266	1.779	10.00	1.0	18.927	1.654
6	16.925	1.659	11.00	1.0	17.273	1.497
7	18.473	1.548	12.00	1.0	15.767	1.369
8	19.917	1.444	13.00	1.0	14.398	1.247
9	21.264	1.347	14.00	1.0	13.151	1.134
10	22.521	1.257	15.00	1.0	12.017	1.032
11	23.693	1.172	16.00	1.0	10.985	0.939
12	24.787	1.094	17.00	1.0	10.046	0.855
13	25.807	1.020	18.00	1.0	9.191	0.778
14	26.759	0.952	19.00	1.0	8.413	0.708
15	27.647	0.888	20.00	1.0	7.705	0.644
16	28.476	0.823	21.00	1.0	7.061	0.586
17	29.249	0.773	22.00	1.0	6.475	0.534
18	29.970	0.721	23.00	1.0	5.941	0.484
19	30.643	0.673	24.00	1.0	5.457	0.442
20	31.270	0.627	25.00	1.0	5.015	—

gradients as detailed in Tables 3 and 4 is as follows:

$$D_3 = D_2 + (D_1 - D_2)\frac{V_B}{2V_A}$$

To solve for V_B as in Tables 5 and 6, this formula may be rearranged:

$$V_B = 2V_A\left(\frac{D_3 - D_2}{D_1 - D_2}\right)$$

In Table 3 then, the concentration of the medium entering the centrifuge tube is exactly 10% when 4.0 cm^3 has been delivered and the amount of medium greater than 9% and less than or equal to 10% is 0.8 cm^3. Again, Tables 5 and 6 indicate the concentration of sucrose entering the centrifuge tube (D_3) when exactly some specific volume (V_B) of sucrose has already been added. In Table 5, for example, when just 10 cm^3 has been added, D_3 equals 17.50%, and the change in density (ΔD_3) from that at 9 cm^3 to that at 10 cm^3 is 1.25% (i.e., from 16.25 to 17.50%).

While the data presented in Tables 3 and 4 are illustrative primarily of the progressive construction of continuous density gradients, Tables 5 and 6 are perhaps more useful for discussion of some of the characteristics of these gradients. For example, inspection of the ΔD_3 columns in the latter two tables demonstrates the fundamental nature of the three types of gradients. In each case, note the volume ΔD_3 as a function of increasing values of D_3 (NB: ΔD_3 values for the concave gradients must be read from bottom to top). Since these values are related to equal volumes of gradient (in cm^3) for each data point, they illustrate the rate of change of gradient density. For the convex gradients, the greatest rate of change (i.e., largest ΔD_3 values) occur in the initial portion of the gradient. The rate of change in density decreases exponentially throughout the length of the gradient. In general, the initial steepness of the gradient provides the potential for separating particles of widely varying properties, while the less steep subsequent portions of the gradient allow for the resolution and stabilization of larger zones of particles than are possible in the initial portions. Which particles are separated in the lower portions of the gradient instead of being retained in the initial portion is dependent on the density ranges of the gradient, the characteristics of the particles, and the magnitude and duration of the RCF application. Some related but more subtle factors may also influence the separation. These relate to the nature and amount of sample layered on the gradient. McCall and Potter (1973) suggest that a very approximate figure for a 20-cm^3 gradient, as used in the previous examples, would be 0.1–0.3 cm^3 solution containing 0.5–1.0 mg of each component to be separated. This is only a guide, and optimal conditions are best determined empirically. The sample overlay must, of course, be less dense than the lightest portion of the gradient to avoid uncontrolled sedimentation prior to centrifugation. However, within this limitation, it is also desirable that the concentration of particles in the overlay not be too dense. Concentrated samples may interact in a deleterious fashion with the initial portion of the gradient. If the initial portion of the gradient is too steep, large populations of particles may sediment through this region as a mass, forcing light particles that would be expected to float in that region into denser portions of the gradient. Alternatively, a large population of light particles may form a zone and effectively block the migration of larger or denser particles into the lower portions of the gradient. Such problems can largely be avoided by providing sufficient capacity in the initial stages of the gradient to dilute the sample before any boundaries are established. The concentration of the sample overlay should also be restricted to reduce mass movement of both large and small or light and heavy particles. Bowen (1970) reviews data obtained on the capacity of a density gradient to carry a zone of any size. Essentially, if the concentration of particles to be isolated in a single zone within a density gradient will result in the zone's being too large to be contained within a portion of the gradient of appropriate density, the zone will be unstable. If a zone is overloaded, sucrose may diffuse into the zone faster than macromolecules or particles can diffuse out. Droplets may then form within the zone that are greater in density than the rest of the gradient at that region, and these droplets may then fall through the gradient.

Concave gradients are rather different in character. In these gradients, the rate of change of density increases exponentially throughout the length of the gradient (see ΔD_3 in Tables 5 and 6). These gradients will allow large or heavy particles to move rapidly through the initial portions of the gradient, to be stopped eventually at lower portions of the gradient. The increasing rate of change of density in lower portions of the gradient allows particles reaching that region to be concentrated by the narrower density bands. Care must be taken that the concentrations of particles reaching these regions are not so great as to overload the zones formed. The upper portions of the gradient exhibit a more gradual change of density, and particles that sediment to a particular region of this upper gradient encounter areas of density appropriate to them. Overloading is less of a problem, and the gradual change in gradient density allows more room for particles to sediment to positions that are most appropriate. (In the lower portion of the gradient, particles differing slightly in size or density or both may overlap in the narrower zones.) The gradual density gradient, however, does not provide for a concentration of sample particles as compared to the steeper end of the gradient, but the shallower initial portions do not cause any difficulties with particle trapping as in the initial portions of convex gradients.

As noted, in Tables 5 and 6, the rate of change of density (ΔD_3) in linear gradients is not an exponential function, but rather a linear one. The capacity of zones in linear gradients, overloading, and concentration capability are all equal throughout the gradient, but do vary from one gradient to another dependent on the steepness, i.e., the magnitude of ΔD_3.

The data from the ΔD_3 columns in Tables 5 and 6 have been plotted in Fig. 4. These curves are, of course, suggestive of the profile in change of

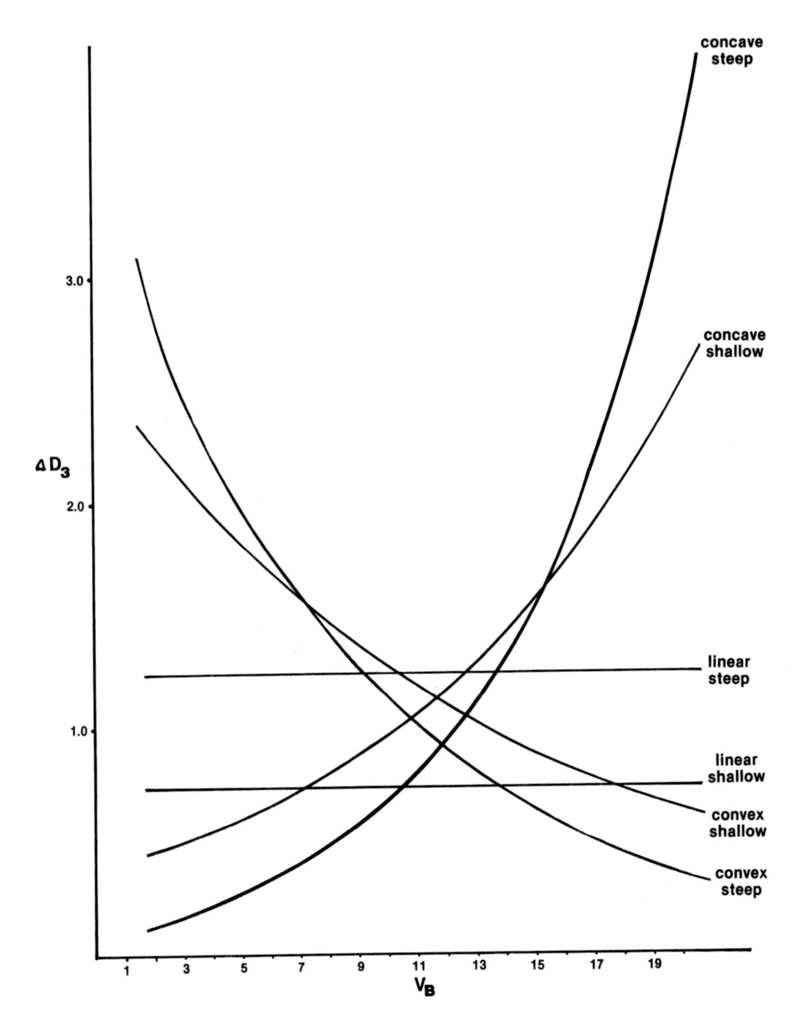

Fig. 4. Rates of change in the density (ΔD_3) of various density gradients plotted as a function of the level of the gradient in the centrifuge tube (V_B). The profiles for both steep and shallow versions of concave, linear, and convex gradients are plotted from the data presented in Tables 5 and 6.

density throughout the gradients. Convex gradients are characterized by a negatively exponential change from top to bottom of the gradient, while concave gradients change in a positively exponential fashion. Linear gradients demonstrate a constant rate of change of density throughout the entire length of the gradient. Figure 5 includes a similar presentation of the data from the D_3 columns in Tables 5 and 6 and depicts, simply, the change in density from the top to the bottom of the gradient expressed in sucrose concentration.

Another characteristic of density gradients can be noted in Tables 3–6 as well as in Figs. 4 and 5. Dependent on the parameters initially employed

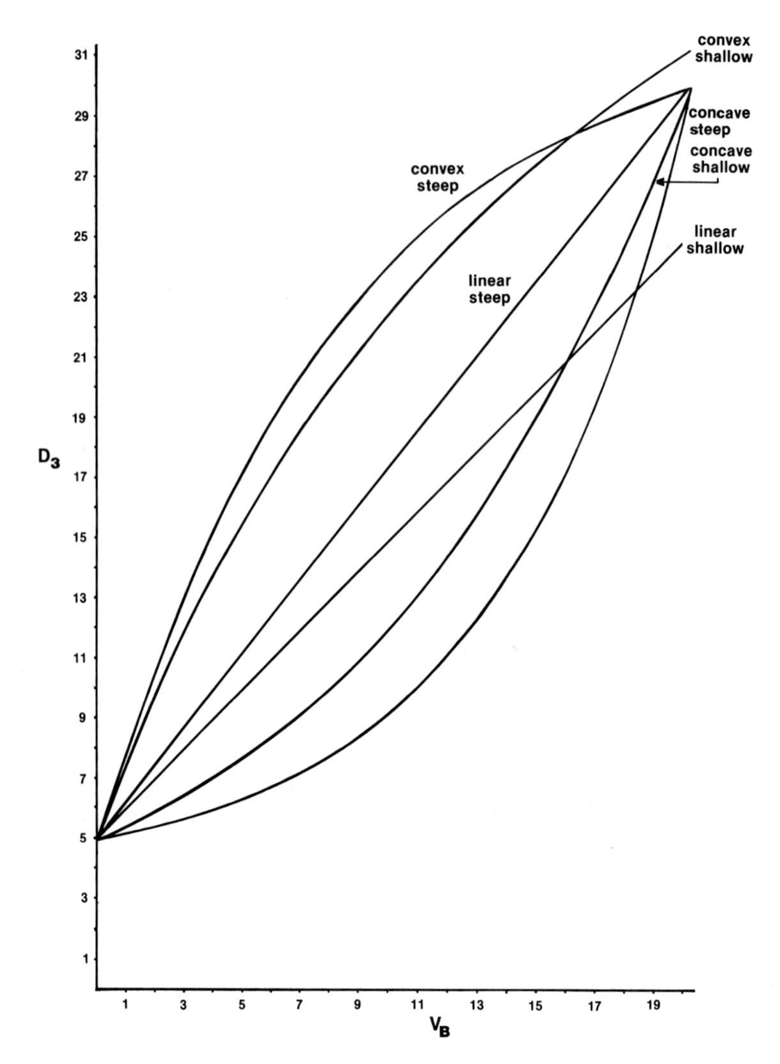

Fig. 5. Densities (D_3) of various density gradients plotted as a function of the level of the gradient in the centrifuge tube (V_B). The profiles for both steep and shallow versions of concave, linear, and convex gradients are plotted from the data presented in Tables 5 and 6.

in their construction (i.e., V_A, D_1, and D_2), gradients may be classified on a scale ranging from shallow to steep. This classification is noted most readily in Fig. 5 for the linear gradients. The steepness of the gradient refers to the magnitude of the rate of change of density and concomitantly for linear gradients both the range of densities contained in the gradient and the amount of gradient of a particular density. Thus, as in Tables 3 and 4, the volumes of gradient of a particular density, or more accurately within a narrow range of density, are smaller for the steeper of the two gradients (i.e., 0.8 cm^3 as compared with 1.0 cm^3). On the other hand, as in Tables 5 and 6, the rate of change of density (ΔD_3) is greater for the steeper of the two gradients (i.e., 1.25 as compared with 1.0%). Finally, the range of densities contained in the steeper gradient is larger than that in the shallower gradient (i.e., 5.0–30.0 as compared with 5.0–25.0 per cent). In a practical sense, steeper gradients will have the capacity to separate a greater range of particle sizes and densities because of the greater range of densities covered by the gradient. Shallower gradients, while useful for a more limited range of particle sizes and densities, will have greater zone capacity for particles with appropriate characteristics and will, in general, be able to carry stable zones of larger capacity. Of course, within the limits of zone capacity, steeper gradients will be more effective in concentrating particles in smaller zones because of the smaller bands of gradient of a given density.

The steepness of exponential gradients is also a direct function of ΔD_3, and the amount of gradient within a particular density range is related to this. However, the greater complexity of the mathematical description of exponential gradients makes it less easy to relate steepness to these characteristics. Nevertheless, reference to Tables 5 and 6 should demonstrate that the ΔD_3's for the steeper gradients are larger than those for the shallower gradients and that the inverse holds true for the ΔV_B's as noted in Tables 3 and 4. This is a rather crude description, since it does not fit the data when the lower portions of the gradients are considered. Figure 5 demonstrates, very generally, that the profile of exponential gradients will become more and more similar to that of linear gradients as they become more and more shallow. For the shallower gradients, the discrepancy in the relationships between the upper and lower portions of the gradient will be less than for steeper gradients and will therefore more closely resemble the relationships mentioned for linear gradients as steepness decreases.

Practical considerations resulting from gradient profile follow the previous discussion. As demonstrated by Figs. 4 and 5 as well as Tables 3–6, while exponential gradients may be constructed that cover essentially the same range of densities and are therefore capable of separating similar ranges of particle sizes and densities, the degree of separation of different particle populations, the stability of the separated zones, and the relative concentrations of the separated populations will depend not only on the steepness of the gradient but also on the portion of the gradient to which particular particle populations sediment. Thus, with convex gradients, the upper portions cover a wider range of densities per volume (i.e., the ΔD_3's will be smaller) than with shallow gradients. This allows greater separation of particle populations, though with a theoretical decrease in zone stability and concentrative ability. The inverse is true of the lower portions of the gradient.

As a final illustration, reference may be made to Figs. 6 and 7. In these figures, the hypothetical centrifuge tube containing the 20-ml gradients discussed above have been drawn as perfectly cylindrical with flat bottoms, rather than the round bottoms that the tubes would normally have. The tubes are drawn to scale with the initial dimensions of 16.1 mm in diameter and 114.3 mm in height. With a cross-sectional area of 2.04 cm^2, the height

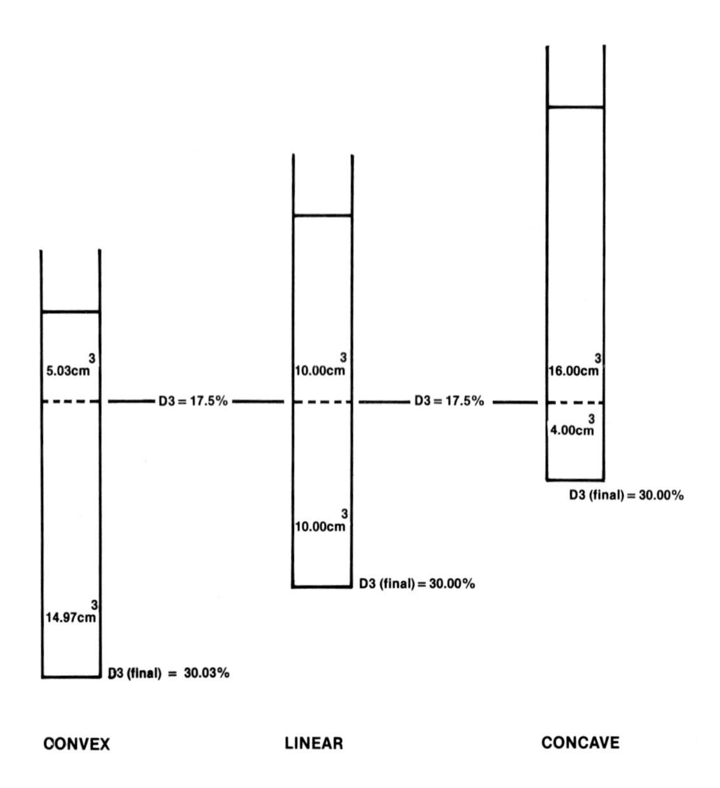

Fig. 6. Comparisons of the relative distributions of gradient density in steep convex, linear, and concave gradients that are depicted as filling centrifugation tubes. The tubes are drawn to scale, although flat rather than round bottoms have been drawn. As a reference point, the 17.5% density level in each tube is indicated, as are the volumes of the gradient above and below this level in each tube. This figure is based on the data in Table 3.

of the tube that will contain 1 cm^3 is 4.90 mm. In Fig. 6, all the gradients cover essentially the same range of densities (refer to Table 3). The middle of the linear gradient, which is at the 17.5% density level, was arbitrarily chosen for comparison with the same density level in the exponential gradients. Of course, one half the gradient is above this point and one half below for the linear gradient. However, the 17.5% level is near the top of the convex gradient but near the bottom of the concave gradient. Figure 7 includes a similar presentation for the shallower gradients of Table 4, but it must be noted that all the gradients do not cover the same range of densities in this case [i.e., the linear gradient density ranges only from 5.0 to 25.0% (see Table 6)].

When it is of interest to achieve more than simply a separation of particle populations, it is necessary to be able to estimate accurately the true density profile of the gradient on which a separation is obtained. One exemplary application would be for the estimation of sedimentation coefficients of particle populations that is possible with preparative centrifugation. One method for obtaining the necessary information entails the inclusion of particles or macromolecules of known sedimentation characteristics either in the sample or in a separate comparison centrifuge tube. Analysis of the recovery of the standard substance permits an estimation of the density of the gradient in the region of recovery. If this region also includes a particle population of interest, this information may be applied. This technique is, however, somewhat restrictive. Another method uses the same principle but requires less analysis and interpretation. As described by McCall and Potter (1973), the Beckman Instrument Company can supply a set of six density-gradient beads, each of which has a diameter of about 1 mm. They range in density from 1.1 to 1.6 g/cm^3 in steps of 0.1 g/cm^3. The different densities are color-coded. After reaching equilibrium during a preparative run, the beads, which are added prior to the run, indicate the density of the gradient at several levels. McCall and Potter (1973) have also described a more thorough technique, in common use, that makes use of the change in the refractive index of sucrose as it relates to density. After a centrifugal run, the refractive indexes of the fractions collected from the gradient are

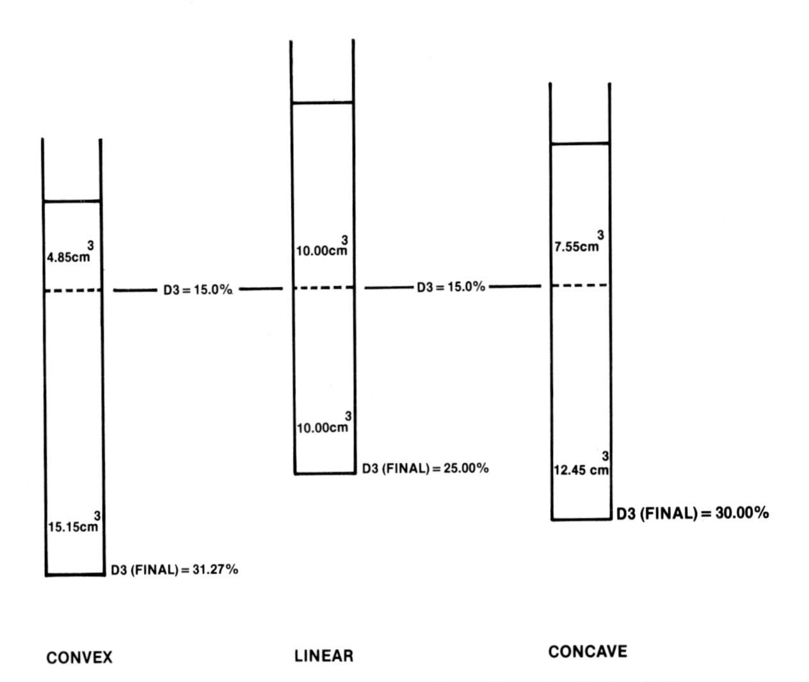

Fig. 7. Comparisons of the relative distributions of gradient density in shallow convex, linear, and concave gradients that are depicted as filling centrifugation tubes. The tubes are drawn to scale, although flat rather than round bottoms have been drawn. As a reference point, the 15.0% density level in each tube is indicated, as are the volumes of the gradient above and below this level in each tube. This figure is based on the data in Table 4.

determined with, for instance, an Abbe Refractometer. Reference to a calibration plot establishes the density of each fraction. The density figure obtained will be the average density for the fraction in question. In general, the smaller the fractions, the more accurate or representative will be the density profile obtained for the gradient. Since some fractions will contain separated particle populations that will alter the refractive index, this needs to be taken into consideration. McCall and Potter (1973) also briefly mention some other techniques that are not, as yet, of general interest.

5.3.7b. Selection of Density-Gradient Parameters: Practical Criteria. Traditionally, the shape of the gradient that is most appropriate for a particular separation has been determined through trial and error. Though the accumulation of empirical data for many types of separations has made this process somewhat less time-consuming, specifying conditions for a novel separation of a particle population or populations is still quite an involved matter. The densities and sedimentation rates of many of the subcellular particles released from neuronal tissue by homogenization are quite similar, which makes it rather difficult to determine satisfactory conditions for separation. However, since any differences in the size, shape, or density of particles may provide a basis for separation, effective use of any differences that exist among particle populations can lead to satisfactory separations. Since the physical similarity among subcellular particles of neurobiological interest greatly emphasizes the inefficiency of trial-and-error determinations of optimal gradient conditions, consideration of a more efficient, systematic, and theoretical treatment of this problem is needed. Cotman (1972) has presented such a treatment, which because of its intrinsic interest regarding gradient parameters in general, as well as its value in resolving the problems just mentioned, will be developed here.

For the sake of clarity, the separation of only two particle populations is considered, and since the separation of these particles can be determined by their sedimentation coefficients or densities or both, there are four possible cases to be considered. The first one would be the case in which the densities of the two particles are equal ($v_1 = v_2$) but their sedimentation coefficients differ (e.g., $S_1 > S_2$). The second case consists of the reverse situation (i.e., $v_1 > v_2$ and $S_1 = S_2$). A more complicated situation arises in case three, in which both parameters are larger for one of the particles (i.e., $v_1 > v_2$ and $S_1 > S_2$). In case four, both parameters also vary, but in opposite directions (e.g., $v_1 < v_2$ and $S_1 > S_2$). The efficiency of various gradients in achieving separations for each of the four cases was calculated in terms of resolution, i.e., the distance between the two particles as a percentage of the gradient path following the application of a particular total RCF. Resolution is a function of the distance traveled in a particular case by the faster-moving particles and the distance that the slower-moving particle is behind as a percentage of the gradient path. The larger the separation of the particles as a percentage of the gradient path, the larger is the resolution of a particular gradient. The data used for the discussion of the four cases were based on some of the approximate properties of subcellular particles, but were treated in a theoretical manner using computer simulation rather than actual empirical manipulation.

In the first case, since the particle densities are equivalent, no resolution can be obtained by isopycnic methods. Separation must be achieved by exploitation of the differential sedimentation coefficients of the particles. Maximum resolution is achieved in a homogeneous sucrose medium and is independent of the sucrose concentration of the medium. For particles of $v = 1.17$ g/cm^3 with $S_1 = 9.4 \times 10^3$ S and $S_2 = 5.6 \times 10^3$ S, resolution is 47% in both 10 and 25% sucrose media. Any change in the density of the gradient along the path of the particles decreases the resolution, and the decrease is greater for steeper gradients. This is because the faster-moving particle is affected more by the increasing gradient density relative to the slower-moving particle, since it reaches the denser regions of the gradient first. In a linear gradient with initial and final sucrose concentrations of 10 and 40% respectively, the maximum resolution possible is only 14%. Continued centrifugation after the point at which the faster particle is about half way down the gradient results in both particles eventually reaching their isopycnic points, since in this case their isodensity value of 38% sucrose is included in the gradient. Since there is no separation when particles of equivalent density reach their isopycnic level, it is necessary to determine the total RCF required that will result in maximum rate separation. Or, in other words, knowledge of the appropriate total RCF, which can be calculated, will indicate how maximum separation may be achieved. The same is true for the separation in homogeneous sucrose, since maximum resolution is achieved when the faster-moving particle has just reached the bottom of the tube. Further, centrifugation with the faster-moving particle stopped by the tube bottom will serve only to decrease resolution. The advantage offered by preliminary calculation of centrifugation parameters as opposed to trial-and-error determinations should be apparent.

In addition to the slope of the gradient, the initial sucrose concentration is also important. The closer the initial sucrose concentration is to the isodensity value of the particles, the greater will be the slowing of the particle with the higher sedimentation coefficient relative to the particle with the lower coefficient. Therefore, resolution will be reduced. Similarly, with a given gradient, the densities of the particles influence resolution except when a homogeneous medium is used. Separation in all sucrose media is also influenced by the sedimentation coefficients of the particles and is a function of their ratio rather than their absolute differences. S_1/S_2 ratios that are equivalent regardless of the absolute values of the sedimentation coefficients result in the same degree of resolution. Resolution increases as the ratio S_1/S_2 increases. Steepened gradient slope does not significantly affect separation if the S_1/S_2 ratio is large. However, this ratio is usually small for subcellular particles of interest, and in this situation, the steepness of the gradient slope is inversely related to resolution. Again, since the separations are achieved on the basis of rate sedimentation, it would seem apparent that greater differences in sedimentation coefficients will produce greater resolution.

In summary, then, for particles that differ in sedimentation coefficient but not in density, maximum resolution is achieved in homogeneous sucrose media. This resolution is not affected by the concentration of the homogeneous sucrose used unless the density is greater than the density of the particles. Increased steepness of gradient slope decreases separation unless the density of the particles is considerably greater than the initial or final density of the sucrose or if the ratio of the sedimentation coefficients is relatively large. Of course, it should be kept in mind that the use of a gradient is often of practical value for the reasons previously discussed. Therefore, it may be necessary to sacrifice the maximum resolution possible with homogeneous media. In such cases, knowledge of the effects of gradient parameters, as just discussed, is of particular value.

While the first case relies on rate separation because the particles differ only in sedimentation coefficient, when the particles do not differ in sedimentation coefficient but only in density, isopycnic techniques are most effective in achieving satisfactory separations. Since it is possible to construct a gradient such that the lighter particle bands near the top while the heavier particle does not reach its isodensity level until the bottom of the gradient path, maximum resolution is limited only by the length of the gradient path. Separations can also be achieved by rate methods, with best results obtained with conditions similar to those used for isopycnic separations. Maximum resolution with rate separation is achieved in a homogeneous sucrose medium in which the sucrose density is close to that of the lighter particle. Thus, relative to the heavier particle, the lighter particle will travel more slowly, since the density of the medium is closer to the density of the particle. The lower is the density of the medium in comparison to the density of the lighter particle, the poorer is the resolution, since, in essence, the separation technique approaches a pure rate method in which both particles have the same sedimentation coefficient and therefore sediment quite similarly. Resolution is only 7% in a 10% sucrose medium (though it is 78% in a 30% sucrose medium) as compared to the separation of 47% for the first case, in which the particles differed in sedimentation coefficient rather than their density.

Analogous to the first case, resolution is not dependent on the density difference between the two particles, but rather on the ratio of their density differences. Since resolution also depends on the interaction of particle density with the density of the sucrose medium, the ratio that influences resolution may be stated:

$$\left(\frac{v_2 - D_3}{v_1 - D_3}\right)\left(\frac{v_1 - D_{20,\mathrm{W}}}{v_2 - D_{20,\mathrm{W}}}\right)$$

where $D_{20,\mathrm{W}}$ is the density of water at 20°C. As long as this ratio remains constant, regardless of the absolute values of the variables in the equation, maximum resolution also remains constant. Increases in the value of this ratio lead to greater resolution. The resolution achieved by rate separation is independent of the sedimentation coefficients of the particles if the S_1/S_2 ratio remains equal to one.

When sucrose gradients are used, resolution depends on the slope of the gradient as well as the initial and final concentrations of the sucrose. Construction of a sucrose gradient by allowing the final concentration of the gradient to be equal to or close to the isopycnic banding point of the lighter particles results in a greater loss of resolution relative to that obtained in a homogeneous medium equal in concentration to that of the final portion of the gradient. The best resolution in gradients is obtained when they are very shallow and both the initial and final concentrations are close to the isopycnic value of the lighter particle. Resolution is poor in steep gradients or in homogeneous media in which the density of the medium is considerably different from the densities of the particles. Although the greatest resolution is obtained with isopycnic banding, there are other important factors that need to be considered. Maximum resolution is achieved by the use of a medium equal in density to the lightest particle if only two particle populations are to be separated. In this situation, the light particles are stopped at the sample zone, while the heavy particles can sediment to the bottom of the centrifuge tube. If the sample includes particles or populations with densities less than the lighter or greater than the heavier particles of interest, optimal results, overall, may be achieved by using a continuous gradient that includes the isopycnic densities of the particles of interest. Discontinuous gradients that include these densities within their range will help to concentrate a population of similar particles in a relatively small volume. Such gradients may also increase the resolution between two particle populations by providing a constant-density plateau of variable width between the steps at which the two particle populations are stopped. It is not necessary for the particles to reach equilibrium in isopycnic banding, since maximum resolution is reached prior to equilibrium. Full particle separation is achieved fairly rapidly, and subsequently the particles sediment essentially in parallel until they reach their isopycnic-banding densities. This would allow shorter centrifugation times than necessary for equilibrium without sacrifice of resolution when the sedimentation coefficients of the particles are equivalent. However, in some instances, an interaction between the particles and the gradient may lead to an increase in density. If this increase is specific for one class of particles, it may be desirable to use such a density increase to improve the separation of the particles. If this is desirable, it may be advisable to continue centrifugation beyond the point of maximum resolution (based on initial particle densities), since increases in particle density are most often noted after prolonged centrifugation.

In the third case, both the sedimentation coefficient and the density of one particle exceed those of the other particle. Thus, both factors will work in favor of separation, and the matter of resolution can be considered in terms of a combination of the discussions of the first and second cases. Rate separation is more efficient than in the previous cases, since both factors are effectual. Maximum resolution with rate methods is obtained if the lighter particle is held at its isopycnic point at the top of the gradient while the heavier particle is allowed to sediment to the bottom of the tube. If the differences between the two particles regarding one of these properties is large in relation to differences regarding the other property, then rate separations will be similar to those discussed for the first case (i.e., for large differences in S) and the second case (i.e., for large differences in v) above.

Maximum resolution in a homogeneous sucrose solution is obtained when the sucrose density is close to that of the lighter particle as in the second case. The differences in sedimentation coefficients considerably improve resolution in comparison to the second case, and this improvement is more profound, since the sucrose density and the density of the lighter particle are farther apart. The relationship that determines resolution in homogeneous sucrose media for this case is the product of the ratios detailed for the first and second cases, or:

$$\left(\frac{S_1}{S_2}\right)\left(\frac{v_2 - D_3}{v_1 - D_3}\right)\left(\frac{v_1 - D_{20,\mathrm{w}}}{v_2 - D_{20,\mathrm{w}}}\right)$$

Regardless of variations in the values actually entered into this equation, resolution will be equivalent whenever the product of the two ratios is the same. Since the value of the latter part of the formula will vary with the changing sucrose concentrations in a density gradient, the influence of the sedimentation coefficients of the particles on resolution cannot be determined as easily as was the case with homogeneous media. Without getting involved with the complexities necessary for such determinations, the following established relationships may be mentioned. The initial sucrose concentration of a density gradient affects separation. Maximal resolution is obtained when the initial sucrose density approximates the density of the lighter particle. When the initial sucrose density is lowered, resolution is progressively decreased. The ratio of the sedimentation coefficients of the particles can partially counteract the effects of lowered initial sucrose densities if the ratio is large. In general, separations of particles that differ in both density and sedimentation coefficient when one particle population is characterized by larger values in both variables may provide greater resolution than in either the first or second cases, in which the particles differ in only one variable. It should be noted, though, that centrifugation media that achieve greatest resolution when the particles differ in sedimentation coefficient only are quite different from those that are most useful when the particles differ in density only. Therefore, those gradients that are most efficient for separating the particles in the third case, when compared with either the first or the second case, are intermediate in character between those that are most effective in the first case and those that are most effective in the second case.

The fourth and final case is somewhat more complex, but important because it is not an uncommon one in neurobiological work. The complexity derives from the fact that while the particles differ in both density and sedimentation coefficient, the lighter particles have a larger sedimentation coefficient than the heavy particles. In contrast to the third case, the characteristics of the particles counteract one another, rather than work together to improve separation. In some gradients, as well as in homogeneous media, little or no separation is possible. For any pair of particles that fall into this category, there will be, for example, a homogeneous sucrose medium in which no separation will occur. Substitution of the appropriate values in the following equation, which is based on that presented for the third case, will allow a solution for D_3, the appropriate sucrose density:

$$\left(\frac{S_1}{S_2}\right)\left(\frac{v_2 - D_3}{v_1 - D_3}\right)\left(\frac{v_1 - D_{20,\mathrm{w}}}{v_2 - D_{20,\mathrm{w}}}\right) = 1$$

or

$$\frac{(v_2 - S_2v_2v_1 + v_1S_2D_{20,\mathrm{w}})\,(S_1v_1 - v_1D_{20,\mathrm{w}})}{(S_2D_{20,\mathrm{w}} - S_2v_2 - S_1v_1 - v_1D_{20,\mathrm{w}})} = D_3$$

The value of D_3, then, is that sucrose density in which the stated particles will sediment at the same rates. Separation may be achieved at values greater than or less than this zero-resolution concentration. At lower sucrose concentrations, the sedimentation coefficients of the particles influence separation positively, while density differences decrease resolution. Particles that have a sedimentation-coefficient ratio that is considerably greater than the ratio of the density differences will be separated most effectively in homogeneous media with a sucrose density that is lower than the zero-resolution concentration. Decreases in the density of the medium increase resolution, since they serve to diminish the effects of density differences. In a particular medium, resolution is determined, as noted above, by the product of the ratio of sedimentation coefficients and the ratio of the density differences. Increases in the sedimentation-coefficient ratio improve resolution if the ratio of density differences is held constant. When the latter ratio is not held constant, resolution is determined by the changes in this ratio relative to changes in the ratio of the sedimentation coefficients. If this ratio is large in comparison to the ratio of density differences, changes in the latter have a negligible effect on resolution. When the magnitude of the sedimentation-coefficient ratio is closer to the ratio of density differences, resolution is seriously diminished. The opposite is true when the density of the sucrose medium exceeds the zero-resolution concentration. In this instance, separation is based on density differences, and a relatively large density-difference ratio would indicate separations in this medium as most effective. In general, the greater is the density of the medium, the more will the density-difference ratio influence separation at the expense of a contribution from the ratio of sedimentation coefficients. When this ratio is held constant, increases in the ratio of density differences will lead to greater resolution. It should be noted that the ratio of density differences as determined by the formula given above will decrease as the density difference between the two particles increases. When the ratio of the sedimentation coefficients is not held constant, the effect of changes in the ratio of density differences will be relative to the ratio of the sedimentation coefficients as indicated in the formula. When the ratio of density differences remains large relative to the ratio of the sedimentation coefficients, changes in the latter will have little effect on resolution. When the two ratios are similar in magnitude, resolution will be greatly decreased.

The somewhat more involved case that occurs when density gradients are used may be considered in a similar fashion. The particle with the larger sedimentation coefficient and lower density will sediment more rapidly through the initial portion of the sucrose gradient than will the more dense particle with the lower sedimentation coefficient. If the gradient contains a sucrose density that is isopycnic with the density of the initially faster-sedimenting particle, this particle will sediment more slowly as it approaches its isodensity level and will be overtaken by the denser particle. After this intersection point, the denser particle will continue to sediment until it reaches its isodensity level or until it reaches the bottom of the centrifuge tube. Obviously, at the intersection point, resolution is at zero and is very poor anywhere near this point.

Prior to the intersection point, the particles separate on the basis of rate, i.e., their sedimentation coefficients, and conditions that enhance resolution based on dissimilarity of sedimentation coefficients will contribute to maximum separation in this portion of the gradient. The maximum separation possible is equal to the path length from the sample layer to the isodensity level of the lighter particle. Within this portion of the gradient, resolution will be enhanced, as discussed for the first case, by large sedimentation-coefficient ratios and by gradient shallowness. However, while separation in the first and third cases is aided when the density of the initial portion of the gradient approximates the density of either both particles (in the first case) or the lighter and slower-sedimenting particle (in the third case), such conditions are not useful for this fourth case. The lighter particle in this case is the one that sediments more rapidly by rate. Increased density early in the gradient will trap this particle and minimize the separation obtainable by rate. The best approach would be one in which the initial density of the gradient was well below the density of the lighter particle and the density of the gradient increased gradually with the isodensity level of the lighter particle near the bottom of the centrifuge tube.

After the particles intersect, separation is based on density differences, and considerations similar to those presented for the second case will increase resolution. Maximum resolution will be determined essentially by the length of the sedimentation path between the isodensity levels of the two particles or between the isodensity level of the lighter particle and the bottom of the tube. If the initial portion of the gradient approximates the density of the lighter particle, the lighter particle will float at this point, while the denser particle will sediment to its isopycnic point, which can be very close to or at the bottom of the path. If the denser particle is allowed to sediment the full length of the tube, then maximum resolution as well as the shortest centrifugation times are achieved in very shallow gradients in which both the initial and final gradient densities are close to the density of the lighter particle. As mentioned for the second case, this reduces the possibility of changes in particle properties resulting from prolonged exposure to a dense medium.

Since the sedimentation paths of the particles intersect in this fourth case, maximum resolution is obtained by emphasizing either a rate separation or an isopycnic separation and constructing a gradient accordingly. As indicated in the formula presented above, separation is influenced by the product of the ratio of the sedimentation coefficients and the ratio of the densities. Since the density ratio decreases as the difference in particle densities increases, resolution decreases as a function of density dissimilarity. Rate separation will be most effective if the ratio of the sedimentation coefficients is large and the density differential is small. Otherwise, isopycnic separation will prove more productive. Understanding this case is most important, since it is relevant for many particles of neurobiological interest. Thus, synaptosomal separation is complicated by the properties of lysosomes and small mitochondria, since they fall in this fourth category.

As already indicated, the foregoing topic has been more thoroughly treated in an article by Cotman (1972) that gives more attention to practical detail and numerous practical and theoretical examples. The accurate and reliable estimation of sedimentation coefficients of particles that is so important for the application of these principles is also treated in that work (see also Trautman, 1964).

5.3.7c. Unloading Fractions. Since the goal of preparative centrifugation in neurobiological work is rarely the simple separation of populations of subcellular particles alone, but rather the subsequent analysis of individual fractions by morphological, enzymatic, biochemical, or other methods, maintenance of the separations achieved during centrifugation while removing the contents of the centrifugation tube for further analysis is quite a critical matter. As previously mentioned, the use of dense gradient materials, such as sucrose, helps to stabilize the fractions during handling of the centrifuge tube. Nevertheless, it is always important to handle the tubes very carefully to minimize mechanical disturbances of the contents. Thermal convection can also lead to a loss of separation clarity. Since centrifugation is almost always performed at low temperatures in refrigerated instruments, collection of fractions should also be performed at low temperatures. Thus, centrifuge tubes are often moved rapidly to a cold room before unloading. Other methods such as cooling jackets are also of value in reducing thermal convection.

Although the differential-pelleting procedure may, as discussed previously, be less than ideal in many situations, it does allow a rather simple method of fraction recovery. After a particular run in the centrifuge, only one fraction is usually recovered, and this is confined within a pellet at the bottom of the centrifuge tube. The supernatant, in this case the nonpelleted material, may be decanted, resulting in two fractions—the decanted supernatant and the pellet remaining in the tube. One difficulty with this method is that the surface of the pellet may be resuspended to some extent and poured off with the supernatant, thus altering both fractions. The surfaces of pellets formed in angle rotors are not perpendicular to the walls of the centrifuge tube, which can result in an unavoidable amount of resuspension. The surface of the pellet is oriented vertically when the tube is in the rotor. Rotation of the tube about its vertical axis after removal from the head allows the pellet surface to be oriented horizontally during decanting. As the supernatant is poured, then, it flows across the surface of the pellet, rather than away from the surface.

A more useful means of removing the supernatant is aspiration. The suction of a Pasteur pipette with a rubber bulb, or a syringe, is used to remove the supernatant. Especially for use close to the surface of the pellet, it is advantageous to have the end of the pipette or needle bent at a 90° angle or in the shape of a J. Alternatively, the end of the pipette may be sealed and a hole opened in the side. The opening of the pipette is held at the supernatant–air interface, and both air and liquid are drawn into the aspirator. If necessary, a short, low-speed centrifugal run can be used to eliminate the air from the supernatant. Using this method, it is possible to remove the supernatant layer immediately above the pellet without resulting resuspension. Furthermore, this last portion of supernatant, which inevitably contains some incompletely pelleted material, due to diffusion, may be collected separately, thus avoiding contamination of the major portion of the supernatant.

Dependent on the pelleted material and how densely packed it is, the pellet may be removed intact or resuspended in a small volume of medium and then aspirated or decanted. Other particle populations may be recovered similarly in subsequent pellets formed following continued centrifugation. Density gradients may be unloaded by aspiration using good background illumination so that the edges of the zones can be clearly seen. Slightly more involved methods are usually employed to reduce the decreasing of the resolution of the fractions. Extension of a fine needle or piece of tubing to the bottom of the centrifuge tube allows the gradient to be unloaded from the bottom. Alternatively, addition of a very dense medium to the bottom of the centrifuge tube through the fine needle or tubing raises the contents of the tube, allowing successive fractions to be removed from the top of the tube. A tight-fitting stopper in the centrifuge tube, with one needle inserted through it to the bottom of the tube, and another needle inserted just into the tube, allows the contents of the gradient as they rise to be forced through tubing attached to the second needle. The narrower diameter of the tubing relative to the gradient diameter permits the boundaries between zones to be more finely distinguished than is the case with simple aspiration.

Several other methods require piercing of the centrifuge tube with a needle. The needle may enter the side of the tube at the lower boundary of a zone and the fraction be drawn off into a syringe. One or more fractions may be removed in this way. Piercing the bottom of the tube allows the contents to drip out. Addition of a tight-fitting stopper with a valve inserted on the top of the centrifuge tube permits control over the rate at which the gradient drips out of the hole in the bottom. Alternatively, a needle inserted through the bottom of the tube may be used to add very dense medium to the gradient from below, thus lifting the contents of the tube for removal from the top, as discussed above. Each of these techniques involving piercing eliminates the mixing that occurs when a needle or tube is passed through the gradient as in the previous methods. No matter which method is used, any movement of the gradient must proceed quite slowly, since the tendency of the material in the gradient to move more slowly near the edges of the centrifuge tube than in the center, due to frictional forces, will result in mixing unless the movement is quite gradual.

One further method of particular use when recovery is to be made of only a few fractions but from a large number of centrifuge tubes has also been developed. This method requires a device known as a tube slicer that in fact slices the tube at a particular level and separates the gradient above and below the level of the slice. Whittaker and Barker (1972) detail a tube slicer of one type. Other authors (deDuve *et al.*, 1959) present more information on this procedure.

5.3.7d. Zonal Rotors. While preparative centrifugation has made great progress using either angle-head or horizontal (swing-out) rotors and cylindrical centrifuge tubes, it has always been clear that these devices were less than ideal. As mentioned previously, the traditional cylindrical centrifuge tube does not allow all particles to sediment in the direction of the applied RCF without colliding with the wall of the tube, which is followed by sedimentation along the wall but no longer in parallel with the RCF. While some particles may be able to sediment through their full path in the same direction as the RCF, many particles will not behave this way. In fact, if the bottom of the centrifuge tube is at a distance from the axis of rotation that is twice that of the meniscus, then about 25% of the particles in the sample will collide with the walls during complete sedimentation. The convection that such collisions and wall sedimentation cause will also affect particles that do not contact the tube walls. In contrast, the walls of sector-shaped cells, which are standard in analytical centrifugation, are parallel to the force field. In such cells, macromolecules may sediment the length of the cell without colliding with the walls, and the direction of sedimentation is essentially always in the direction of the RCF. Resolution in such cells can be more precise, since a major factor (i.e., collisions with the wall) causing similar particles to sediment to different levels is eliminated. Of course, as discussed above, the density gradients used in preparative centrifugation are able to reduce sedimentation along the wall and consequent convection, but results are far from ideal. In general, analytical or sector-shaped cells have been of little use in preparative work because of their relatively small sizes and consequent small capacities.

Since it was clear that the separations obtainable with various populations of subcellular particles (or microorganisms, viruses, macromolecules, or other material of interest) are influenced not only by the properties of the particles, as has been discussed, but also by the capability of the separatory system, much effort has been devoted to refinement of centrifugation systems. Incorporating the inherent advantages of sector-shaped cells, various researchers, but particularly Norman G. Anderson, have developed a series of centrifuge rotors that meet the needs of preparative work. Limited analytical data can also be obtained with such rotors. These zonal rotors, as they are called, are essentially hollow cylinders of high aspect ratio. The latter term refers to the fact that either the height of the cylinder is significantly greater than its diameter or vice versa. These cylinders have been found to be more stable when rotated at high speeds than cylinders with low-aspect configurations. The hollow rotor is divided by internal radial septa oriented in parallel with the centrifugal field and thus giving rise to sector-shaped compartments. Anderson's rotors typically contain four septa (e.g., Anderson, 1966*a*), while one designed by Beaufay [Beaufay (1966), as discussed by Moore, 1969)] has only two septa. Theoretically, in a hollow cylindrical rotor, particles would always be free to sediment along the lines of force. However, rotation may cause turbulence due to swirling, especially during acceleration and deceleration. The inclusion of septa minimizes such disruptive influences yet does not interfere with sedimentation because of the orientation of the septa parallel to the centrifugal field. In addition to its two septa, the Beaufay rotor is also slightly eccentric in cross section to further reduce turbulence.

As will be discussed further, these zonal rotors are completely filled with gradient material and sample, and the latter can potentially sediment through the entire rotor in any direction. One disadvantage of sector-shaped centrifugal cells is that the particles undergo radial dilution as they sediment, since the width of the cell increases with distance from the axis of rotation. Therefore, zones of the same width but at different distances from the axis of rotation will vary considerably in volume. A population of particles confined initially to a zone of a certain width will be greatly diluted if confined to a zone of the same width but at a greater radial distance. This shortcoming can be overcome to a certain extent if zones at various distances are equivolumetric, that is, if the widths of the zones decrease with radius such that the total volumes of the zones remain equal. The latter arrangement is still not sufficient to concentrate particle populations under most circumstances, but does lead to one of the major advantages of zonal rotor systems, namely, their resolution capability. Since particle populations may be confined to ever-narrower zones as radius increases, without zone overloading, the zones can be separated from one another by larger volumes of gradient containing no particles than is possible in conventional systems. Thus, particle populations may be more clearly delineated from one another.

Zonal rotors, which come in various sizes, while they incorporate features of analytical cells, are capable of handling relatively large volumes of gradient and sample, ranging from 40 ml in the Beaufay rotor to 1670 ml in one of Anderson's more recent models (the B-XV rotor; an older model, the B-IV, could hold 1725 ml). Thus, these rotors compensate more than adequately for the small sizes of analytical cells. In fact, they are capable of handling much greater volumes than conventional angle or swing-out rotors.

All but the very earliest zonal-rotor models include features that overcome another problem inherent in the use of conventional equipment. Density gradients, which are an improvement over nongradient media, are nonetheless susceptible to mixing due to mechanical disturbances induced by handling both before and after centrifugation, as well as during acceleration and deceleration of the rotor head. Since the stability of such gradients is a function of the RCF to which they are subjected, zonal rotors have been designed so that gradients may be formed in them and sample applied while the rotor is spinning and the contents are under the influence of an RCF Similarly, the rotor may be unloaded while still under the influence of a centrifugal force. Since the gradients are greatly stabilized under these conditions, the greater resolution obtainable with zonal rotors, as already discussed, is extended further due to the maintenance of the separations achieved right through the unloading stage.

A major problem in the design of such dynamically loaded rotors was the fitting of the rotors with seals that could interface the rotor with the lines used to load and unload the gradient during rotation without leaking. This was particularly a problem at higher rotational speeds, and two means of approaching the problem have been used. The A series of Anderson rotors were quite broad relative to their height and had permanent seals, which limited their rotational speeds to about 6000 rpm in commercially available models (for a compendium of the characteristics of most commercially available models, see Cline and Ryel, 1971), although one model Anderson tested (A-IV) was operable at speeds as high as 18,000 rpm (Anderson, 1966*a*). These rotors are transparent on top, which allows sedimentation to be observed during the run, which in the commercially available models can generate an RCF of 10,000*g*. All the B-series rotors are larger in the height dimension relative to their breadth. The contents of these rotors cannot be observed during operation, and the early rotors in the series had permanent seals. Beginning with the B-X and B-XI rotors and including the commercially available B-XIV and B-XV rotors, the seals could be removed after gradient loading at a relatively low speed, thus allowing very high-speed operation. After deceleration, the seal may be reattached for dynamic unloading. These latter rotors can be spun at speeds up to 60,000 rpm, yielding forces of 247,000*g*. In practice, after a zonal rotor is brought to the temperature at which it will be operated, it is mounted in a refrigerated centrifuge and rotated to a standby velocity (between 2000 and 3000 rpm), and the seal assembly is attached (of course, in some rotors, the seal is permanently attached). A gradient pump is attached to an edge feed line that allows the gradient to enter the rotor at its maximum radius. The gradient, whatever its shape, is then pumped into the rotor, with the denser components constantly being added at the bottom of the gradient as the whole gradient is moved toward the center, or core, of the rotor. When the rotor has been filled with gradient, a dense cushion layer is added that forces the top of the gradient out of the rotor through a center feed connection. When the top of the gradient reaches a convenient point in the feed line, the sample may be injected with a syringe into the line so that no air is trapped. Less dense gradient material, as an overlay, may then be pumped into the center feed line, forcing the sample and the gradient back into position inside the rotor. Excess of the dense cushion layer leaves the rotor through the edge feed line. Accurate calibration of various volumes of gradient material and sample used allows the gradient to be precisely oriented within the rotor even though its contents are not visible in B-series rotors. After the feed lines are clamped off and the seal is removed, in models with which this is possible, the rotor may be accelerated to the desired operating speed. When the centrifugal run is complete, the rotor is slowed to standby speed, and after the seal is in place and the feed lines freed, the gradient is forced out of the center feed line by pumping the dense cushion layer back into the rotor through the edge feed line. The gradient is collected in fractions either manually or automatically, the sucrose density of each fraction is calibrated, and subsequent analyses may then be performed.

The shape of the gradient used in a zonal rotor has an important influence on the resolution that can be obtained. Density gradients that are linear with volume are the easiest to prepare. A gradient profile that is linear with volume in a cylindrical centrifuge tube will also be linear with regard to radial distance from the axis of rotation, as discussed above. However, because of the sector-shaped configuration of zonal rotors, zones containing equal volumes of gradient vary in their radial width. Thus, in the B-XIV rotor, the first 100 ml of gradient extends from the 0.0 radius point to a distance of 3.126 cm, while the second 100-ml zone extends only from 3.126 to 4.024 cm in radial distance (Cline and Ryel, 1971). It is also possible to construct gradients that are linear with radius in zonal rotors. The use of this type of linear gradient facilitates the calculation of sedimentation coefficients following separation. More complex, concave or convex, gradients can also be produced, although it is difficult to calculate the manner in which these should be constructed for zonal rotors (Birnie *et al.*, 1972).

In addition to the zonal rotors just described, many of which can be used in conventional centrifuges, special models have been constructed that can achieve higher rotational velocities as well as continuous flow and separation of sample during centrifugation. Several reviews have treated both practical and theoretical applications of zonal rotors in more detail (Anderson, 1966*b*; Birnie *et al.*, 1972; Bowen, 1966; Cline and Ryel, 1971; Cotman, 1972; McCall and Potter, 1973; Moore, 1969; Reid and Williamson, 1974).

6. Analytical vs. Preparative Strategies

When reference is made to the type of centrifuge and technique used, the amount of sample analyzed, and the timing of fraction analysis, virtually all tissue fractionation can be considered to be preparative in nature. However, as mentioned earlier, centrifugal subcellular fractionation, which in the aforementioned sense is always preparative, can be pursued either analytically or preparatively when these terms are used in a somewhat different sense (deDuve, 1964). When the emphasis of a fractionation procedure is placed on the recovery of one pure fraction or a few pure fractions at the expense of quantitative recovery of the particles associated with the fraction(s) of interest as well as knowledge of other fractions, the run may then be termed preparative in this second sense. Ideally, it would be desirable to recover all fractions in a pure state, but this has not been, and no doubt never will be, possible. Analytical procedures, though, do emphasize recovery of all the starting material in the fractions produced during centrifugation.

The design and execution of preparative procedures are fairly straightforward, since, in general, it is necessary only to isolate one particle population from others without any interest in the state of the latter. Such procedures are not uncommon and are often justifiable in terms of the degree to which they simplify the design and execution of both the separation technique and subsequent analysis. The shortcomings of strictly preparative procedures, however, are many, though they all derive from the lack of quantitative recovery. It is obvious that no information is obtained concerning the extent to which the particular fraction of interest is recovered from the original sample. A fraction that is obtained in a rather pure, or homogeneous, state (i.e., not contaminated by particles from any other particle population) may contain only a small percentage of the particular particle population initially present in the homogenate. The particles collected in the fraction may not be representative of the particle population in the homogenate. This becomes a problem when it is recalled that any sample, and most surely the majority of samples of neurobiological interest, are not likely to be homogeneous regarding the types of cells they incorporate, therefore it is not unreasonable to expect a rather significant degree of heterogeneity among the particles of even a single particle population in the homogenate. The problem is compounded by the fact that in designing a fractionation procedure to isolate a pure fraction, it may be necessary to collect only a portion of a particular fraction. Thus, for instance, when particle populations sediment with similar rates, it is often possible to achieve purity if only part of a fraction, i.e., its leading or trailing edge, is collected and the other parts of the fraction that may overlap with another particle population are discarded. In summary, then, preparative procedures may or may not lead to a biased sampling of a particle population, but in themselves provide no means of assessing the degree of bias. Many of the artifacts previously discussed with reference to homogenization media, such as leakage, adsorption, absorption, and redistribution, may also influence the properties of the fraction collected, and again, a strictly preparative strategy will not be able to identify these artifacts.

Obviously, data obtained following preparative fractionation must be interpreted very cautiously. However, as mentioned above, preparative techniques are of some value in that they may considerably reduce the time and effort involved in analyzing a particular fraction and do have the advantage of providing greater particle homogeneity than might otherwise be obtained. There is, then, some justification for a preparative approach, and the validity of the data obtained may be enhanced if certain precautions are taken. As will be discussed below, the analytical approach to preparative centrifugation allows a quantitative evaluation of the artifacts and sampling biases just mentioned. If analytical work has been used to validate the procedures used to obtain a pure fraction, subsequent experiments using the same isolation procedure may omit the analytical-validation steps. This will be acceptable only if the preparative isolation procedure used is identical in all aspects with that which has been analytically verified. This applies to all aspects of the procedure, from the initial sampling of tissue to the collection of the appropriate fraction. Clearly, a procedure validated for one species or strain of animal cannot be applied to another species or strain without further validation. Furthermore, even when the same strain or species is used, care must be taken that the animals used in subsequent experiments have had essentially the same experience as those used initially. Factors such as age, size, feeding schedules, handling, and others can significantly affect the properties of subcellular particles and must be controlled. Even when these factors have been rather scrupulously considered, it is probably a good idea to periodically recheck the validity of a preparative procedure, since subtle changes in it that may not be apparent could affect experimental outcomes. It also seems clear that the preparative approach is likely to be of most value to the investigator who is interested in studying only the properties of some particle population and is able to use a standard source of experimental material and carefully follow the validated procedure. Preparative techniques are less easily justified when the experimental question deals, as experimental questions often do, with any sort of comparison between groups, such as experimentally treated vs. control animals. Validation of a technique in control animals does not necessarily justify its use with experimental animals even if only one variable has been manipulated. Further validation is necessary, and in fact it may be best to simply plan on the routine use of analytical procedures in such work.

6.1. Analytical Approach

The goal of the analytical approach is, essentially, to make use of all relevant information that may be obtained from subcellular fractionation in order to avoid the shortcomings of a strictly preparative approach, as already discussed. After an early preoccupation with a preparative approach to subcellular fractionation, analytical techniques were developed that emphasized quantitative recovery but also allowed the collection of relatively homogeneous fractions (Claude, 1946; Hogeboom *et al.*, 1948; Schneider, 1948). Thus, a compromise was reached that to a certain extent satisfied the aims of the preparative strategy without sacrificing the validation possible through the analytical approach.

This early analytical work resulted in a fractionation scheme that is still frequently used and that clearly underlies most, if not all, fractionation schemes in use at present. In this classic scheme, a tissue homogenate is separated into four fractions. Each fraction contains a major subcellular component and is as free as possible of contamination by other subcellular components. This scheme made sense in terms of what was known about subcellular organelles in those days before the widespread use of the electron microscope. The obvious physical characteristics of the fractions, such as their color and texture, added further to the adoption of this scheme. Subsequent morphological and biochemical analyses of the various fractions provided further justification. The major subcellular component of the first fraction, the fraction that sediments most rapidly, is the nucleus. The second fraction contains large granules, subsequently identified as mitochondria. The third contains the small granules or microsomes, and the fourth or soluble fraction includes the cytoplasm and any material that does not sediment under the conditions of the experiment.

DeDuve (1964) has discussed the need for accurate use of terminology in relation to the fractions obtained with any technique. Each of the four fractions enumerated above is identified with its major subcellular component. Thus, the first fraction, which contains the nuclei, is called the nuclear fraction. The others are referred to as the mitochondrial, microsomal, and soluble fractions. These names had a strictly operational meaning initially. The mitochondrial fraction, for instance, was simply that fraction obtained under certain conditions that also happened to be rich in mitochondria. It became clear that each of the particulate fractions contained subcellular components other than the major component for which the fraction was named. Nevertheless, it is common in the literature to see the mitochondrial fraction referred to as the mitochondria, for example, and the nuclear fraction referred to as the nuclei. As deDuve (1964) has pointed out, such usage is inappropriate. The use of the name of an organelle to identify a fraction has led easily to the situation in which the fraction is dealt with as though it were a homogeneous sample of that organelle. The history of subcellular fractionation has been plagued with erroneous interpretations of data that have arisen from this semantic and logical inconsistency.

DeDuve (1971) has reviewed the compositions of the four fractions as they are known at present, and although he has particularly considered liver tissue, essentially the same considerations hold for brain tissue or in fact almost any other sample. In addition to nuclei, the nuclear fraction contains a significant concentration of plasma-membrane fragments. Furthermore, this fraction will also contain blood cells if any remain in the tissue at the time of homogenization. In addition, whole cells and large cellular debris, both of which result from incomplete homogenization, also appear in this fraction. Finally, other subcellular particles, especially mitochondria, also occur commonly in this fraction. Several factors probably account for this latter contamination, including trapping of sedimenting particles and agglutination. Careful control of centrifugation conditions or washing of the nuclear fraction, or both, can reduce this contamination. The nuclear fraction has also been successfully subfractionated, allowing the separation of chromatin, nucleoli, and nuclear ribosomes and polysomes. The plasma-membrane fragments typical of the nuclear fraction have also been isolated and subjected to quite intensive investigation.

The mitochondrial fraction includes, in addition to mitochondria, lysosomes and peroxisomes. In isolations from neuronal tissues, synaptosomes also sediment in this fraction. It is quite difficult to resolve this fraction into its components without significant cross-contamination. The most successful procedure for collection of uncontaminated mitochondria requires that only a portion of the mitochondrial fraction, at the leading edge of the zone, be collected after rate-zonal separation. One shortcoming of this method is that the sample is biased in favor of more rapidly sedimenting mitochondria. A variety of techniques have also been used to subfractionate the mitochondria themselves after they have been disrupted. The outer and inner mitochondrial membranes have been separated from both the inner matrix and the soluble components of the mitochondria.

Lysosomes are also quite difficult to purify in any quantity because of the heterogeneity of their physical characteristics. Pretreatment of experimental animals with Triton WR-1339 reduces the density of lysosomes relative to mitochondria and has been used extensively to facilitate their separation. One problem with this technique is, of course, that the lysosomes are loaded with an artificial substance that could alter their properties. Separation of peroxisomes, also known as microbodies, from lysosomes is facilitated by pretreatment with Triton WR-1339.

Synaptosomes, which sediment in much the same manner as mitochondria and in fact usually include mitochondria within their membranes, may be isolated with some difficulty. A variety of approaches are possible and will be discussed below. One strategy, though, is analogous to that mentioned for mitochondria in that the trailing edge of the sedimenting zone is collected. Again, this leads to somewhat biased sampling. After isolation, synaptosomes can be subfractionated into their various subcomponents including mitochondria, membranes, vesicles, soluble material, and others.

The microsomal fraction was originally called the small-granule fraction, and the term microsomes has come to be used in place of small granules. In contrast to the names of other fractions, which are based on well-known organelles, the identifier microsomes has, or at least had, a purely operational significance. The microsomal fraction was simply that fraction that included a variety of subcellular particles the complete sedimentation of which required a relatively high centrifugal force. For a time, evidence seemed to indicate that microsomes were in fact a distinct subcellular structure, and thus they were considered to be the major component of the microsomal fraction. It is now clear that microsomes are not a distinct entity, but rather an operationally defined class of subcellular particles. It would be most appropriate to always refer to the microsomal fraction rather than to microsomes.

The matter of logical inconsistency in this case is most problematical because of the implicit reference to nonexistent organelles.

The microsomal fraction consists largely of vesicles formed from endoplasmic reticulum disrupted during homogenization. It has been thought that the rough and smooth endoplasmic reticula would each be represented by a population of particles, the first containing ribosomes. Attempts to separate two distinct populations have not been particularly successful, and deDuve (1971) has presented evidence suggesting that there is only one population of vesicles that vary on a continuum of ribosomal inclusion from vesicles very heavily loaded with ribosomal particles to vesicles free of such inclusions. Small plasma-membrane fragments also sediment in this fraction, amounting to as much as one half the total plasma membrane. The plasma membrane obtained in this fraction would seem to be biochemically identical to that obtained in the nuclear fraction. When low concentrations of digitonin are added to this fraction, the digitonin seems to be bound to the cholesterol of the plasma membrane, resulting in a "digitonin shift." The latter is an increase in the equilibrium density of the plasma-membrane fragments that allows them to be separated from the other major components of the fraction.

Membranes of the Golgi apparatus appear in this fraction and can be purified, especially when digitonin has been added to the fraction. The digitonin-shift phenomenon also aids the separation of another class of membrane fragments that seems to be made up of mitochondrial outer membranes that are sheared from the mitochondria during homogenization. Special treatments also allow the separation of ribosomal particles from the endoplasmic-reticulum vesicles. Small mitochondria and about one fifth of the total lysosomes and peroxisomes also sediment with this fraction. Many biochemical components of the soluble or supernatant fraction that seem to be adsorbed to ribosomes may be encountered in this fraction, as well as free synaptic vesicles that are of special neurobiological interest.

The final fraction, the supernatant, consists essentially of the cytoplasm released from the homogenized cells and any material that is soluble therein and not sedimented under the centrifugal forces used to sediment the microsomal fraction. This includes soluble materials released from damaged or leaking organelles. Not all soluble material will end up in this fraction, since absorption, adsorption, or other forms of trapping result in sedimenting of some soluble materials in the particulate fractions. Free ribosomes and polysomes may also fail to sediment with other particulate matter, and thus are found in the supernatant. Larger particles may also contaminate this fraction, though usually as a result of mixing during decanting or aspiration.

As is clear from the proceding discussion, the fractions obtained in the classic four-fraction scheme are far from pure. However, each fraction may be further fractionated to achieve greater resolution and purity of its several components. With the addition of density-gradient techniques to the original differential-pelleting procedure used to establish the scheme, the separation of many fractions has been greatly facilitated. Examples of these more elaborate procedures as they have been applied in neurobiological work will be given below. For the interested reader, a selected bibliography of fractionation literature is presented at the end of this chapter. But at this point, it is necessary to consider the techniques used to establish the validity of fractionation procedures and to interpret their outcomes.

6.2. *Validation of Fractionation Schemes*

To assess the results of a fractionation experiment, it is necessary to identify the components of each fraction and to estimate the homogeneity or purity of the fraction. In general, there are two ways of approaching these questions, and thorough analytical techniques employ both. The first approach is morphological and attempts to relate the structure of isolated subcellular components with the more or less classic structures observed in sections from intact cells. Thus, the nuclear fraction was so named, in part, because light-microscopic examination of the fraction revealed a preponderance of components that were structurally analogous to the classic cytological descriptor. Continuing problems with this approach are technical in nature. The earlier work was restricted to the resolving power of the light microscope and thus of little use for fractions that consisted of particles below this level of resolution. The advent of the electron microscope has extended the resolution of morphological techniques to include all subcellular components down to the level of large macromolecules. However, the manner in which the fractions are sampled also presents some interpretational difficulties at both the light-microscopic and, especially, the electron-microscopic level. At lower levels of resolution, of course, smaller particles were not sampled at all. On the other hand, while electron-microscopic techniques will resolve smaller particles, the sections viewed are extremely small and of necessity can represent only a small portion of the fraction. If the constituents of the fraction were equally distributed in the sections examined, this would present little problem. But this is not usually the case, since constituents that differ in density or sedimentation coefficient are likely to be distributed in the fraction according to these variables, with the lower portions of the fraction including relatively more of the more rapidly sedimenting or denser particles than the rest of the fraction. As will be discussed below, it is possible to circumvent this difficulty (e.g., see Cotman, 1972), but it remains an important consideration. Another shortcoming of morphological techniques, at least as they are most frequently applied, is that they are essentially qualitative in nature and do not provide sufficient data to satisfy the needs of analytical validation and interpretation. Although the usual procedures are satisfactory in indicating whether or not certain subcellular particles are present in a fraction, they are much less satisfactory as indicators of the extent to which a particular components contributes to, for example, the total volume or mass of a fraction. Morphometric techniques have been adapted to the examination of subcellular fractions, and quantitative, statistically valid analyses of fractions are possible, that yield data on the number, size, and shape of the particles contained in various fractions (Baudhuin *et al.*, 1967; Baudhuin, 1968; Elias *et al.*, 1971; Vrensen and De Groot, 1973). Expanded use of such procedures, especially when coupled with various staining or labeling techniques (see Chapters 7 and 9), will certainly lead to a greater contribution of morphological analysis to the analytical interpretation of fractionation data.

The second approach to the validation of fractionation schemes is essentially biochemical. The biochemical approach is conceptually similar to the morphological approach in that the defined characteristics of various subcellular components are assumed to be associated with the presence of these particles in a given fraction. The difference, of course, is that the characteristics in this case are biochemical rather than morphological. Although the assay of various biochemical entities may be suitable for analytical validation, the most fruitful work has centered on enzyme assays. Two postulates have provided the framework within which analytical validation has proceeded. The first postulate is the assumption that "a given enzyme belongs to a single intracellular component in the living cell" (deDuve and Berthet, 1954). Clearly, if an enzyme is to be used as an indicator or marker for the presence of a certain particle, then it must be associated only with that particle. In fact, it is possible to make valid use of an enzyme marker even though its distribution may extend to more than one population of particles (deDuve, 1964). Such cases, which emphasize the nature of the postulate as a working hypothesis, are nevertheless still most easily conceptualized in terms of the single-location postulate as extended to other than unimodal distributions. The second postulate assumes that particles "of a given population are enzymically homogeneous or, at least, cannot be separated by centrifuging into subgroups differing significantly in relative enzymic content" (deDuve *et al.*, 1955). This postulate of enzymatic or biochemical homogeneity is also critical, of course, if an enzyme is to be a valid indicator or marker for a particle population. Were an enzyme associated with only a portion of the particles constituting a population, interpretation of the distribution of the particle population within various fractions could be erroneous, and certainly estimation of the size of the population would be distorted.

Although both these postulates are critical for the use of marker enzymes in validation of fractionation schemes, there is some flexibility possible regarding their application. As already mentioned, it is possible to work with an enzyme that exhibits a bi- or multimodal distribution, and there are many examples of enzymes distributed in this manner. However, in most cases of enzymes exhibiting complex distributions, it has been possible to demonstrate that the enzymes localized in different sites were actually composed of isozymes. The isozymes were found to have unimodal distributions. Thus, when appropriately sensitive analyses are performed, the postulate of single location is not as vulnerable as it may seem (deDuve *et al.*, 1962).

It should also be made clear that the postulate of single location need not hold for all enzymes to be of value. It is only necessary that it hold reasonably well for some enzymes that can then be used as markers. The distribution of other enzymes that are not used as markers is, in general, quite irrelevant for validation purposes. Of course, if the measurable product used to quantify a particular enzyme actually requires the joint action of another enzyme or system of enzymes, the distribution of these other enzymes could influence observations concerning the marker. Thus, a given quantity of a marker enzyme could exhibit more activity in one fraction that contained an excess of the related enzymes from the multienzyme system than a greater quantity of the marker would exhibit in another fraction that had a limited content of the related enzymes. The use of markers that are components of multienzyme systems should be avoided whenever possible. Similar confusion can result if a separate, unrelated enzyme is capable of reducing the substrate to the same end product as is being measured for the marker enzyme.

The postulate of biochemical homogeneity is necessary for quantitative evaluation of data, since it defines the relationship between the marker enzyme and the particle with which it is associated. If the postulate is taken as true in a particular case, then the distribution of the enzyme in the fractionated material may also be taken to represent the distribution of the particle associated with it. The postulate need only be true in a broad sense, since it is unlikely to hold if more narrowly defined. Thus, it is not reasonable to expect that all the particles of a population are of the same size, density, and other characteristics, or that each may be characterized by the same amount of marker-enzyme activity. However, just as there is an expectation that the physical properties of a population of particles will be normally distributed about some mean values, there is also, reasonably an expectation that the biochemical characteristics of the particles will be distributed in a similar fashion.

Finally, note should be made of the fact that some marker enzymes are specific for certain tissues and will not be useful in the analysis of other types of tissues. It would not be surprising to find that some marker enzymes are also specific to certain phyla, species, or even strains of experimental animals.

6.3. Balance Sheets

The key to the validation of fractionation schemes lies in the collection of sufficient data to produce a balance sheet for each separation experiment. This requires that none of the materials employed in the fractionation be discarded. Accurate records of each step of the procedure, including particularly the factors by which various fractions have been diluted from the original starting material, must be maintained. The goal of this balance-sheet strategy is to allow the investigator to compare the activities of various enzymes in the fractionated material with the activities of these enzymes in the unfractionated homogenate. The activity of a particular enzyme is assayed not only in the fraction for which it is assumed to be a marker but also in all the fractions. The total activity of the enzyme in all fractions may then be compared with the activity of the homogenate to determine its recovery in the fractionated material. If the total activities associated with the fractions are equal to that in the homogenate, then the recovery is said to be 100%. Any significant difference from 100% recovery may indicate some complicating factor either in the assay procedure or in the fractionation scheme. Recoveries less than 100% could indicate that the fractionation procedure itself had caused an inhibition or inactivation of some enzyme, by osmotic forces, for instance. Alternatively, the fractionation procedure may have dissociated the enzyme from some factors that served to activate it in the homogenate. Conversely, if recoveries exceed 100% fractionation may have freed the enzyme from some factor that inhibited its activity in the homogenate. It could also be possible that the stresses of fractionation exposed to the assay some enzyme that had previously been isolated in the homogenate. A first step toward resolving some of these possibilities involves the use of a reconstituted homogenate. Samples of each of the isolated fractions are combined in the same proportions as they initially represented in the homogenate. The reconstituted homogenate then contains a representative sample of all the material present in the original homogenate. However, any irreversible changes that occurred during fractionation, such as disruption of membranes or vesicles, will be retained in the reconstituted homogenate. In situations in which the recoveries obtained from the fractions differed significantly from the homogenate and the recovery in the reconstituted homogenate was more similar to the fractional recoveries, the differences may be attributable to irreversible changes occurring during fractionation. On the other hand, if recoveries in the reconstituted homogenate resemble those of the original homogenate more than those of the fractions, then it might be assumed that fractionation had removed enzyme from either inhibition or activation, as the case may be, and that reconstitution had resulted in a return to the original state as encountered in the homogenate.

Assaying each fraction for all the marker enzymes of interest in a particular experiment and accurate recording of all the dilution factors involved allow the experimenter to relate all the enzyme activities to some reference figure such as the original wet weight of the tissue or the initial volume of the homogenate. Thus, activity measures may be given as units of enzyme activity per gram of original tissue or per milliliter of homogenate. It is more common practice now to assay the protein content of the homogenate and all the fractions and use these figures as reference points. Thus, enzyme activities may be expressed in terms of units of activity per milligram of protein in the homogenate.

Table 7 demonstrates the dilution factors involved in a simple fractionation of 500 mg mouse brain into four fractions. Since these data are used here only as an example, the details of the fractionation procedure are not included. Dilution factors are included both for comparison with the original tissue and for comparison of the concentration of that tissue in the homogenate. In considering the dilution factor, note the values indicated for the nuclear fraction. This fraction can be operationally defined most easily as consisting of those particles that sediment under the particular centrifugation

Table 7. Simple Fractionation Scheme for Mouse Brain

Fraction or component	Protocol	Concentration relative to Tissue	Concentration relative to Homogenate
Starting material	500 mg mouse brain	1:1	9:1
Homogenate	Homogenize in 9 volumes (i.e., 9 ml medium/g starting materials).	1:9	1:1
	Centrifuge; save supernatant.		
	Resuspend pellet in 9 ml/g starting material.		
	Centrifuge; save supernatant.		
Nuclear	Resuspend pellet in 5 ml/g starting material; save fraction.	1:5	1:8:1
S-1	Combine supernatants.	1:18	1:2
	Centrifuge S-1; save supernatant.		
	Resuspend pellet in 18 ml/g starting material.		
	Centrifuge; save supernatant.		
Mitochondrial	Resuspend pellet in 2 ml/g starting material; save fraction.	1:2	4.5:1
S-2	Combine Supernatants.	1:36	1:4
	Centrifuge S-2; save supernatant.		
	Resuspend pellet in 9 ml/g starting material.		
	Centrifuge; save supernatant.		
Microsomal	Resuspend pellet in 3 ml/g starting material; save fraction.	1:3	2.99:1
Soluble (S-3)	Combine supernatants.	1:45	1:5

conditions applied to the homogenate. By definition, then, a quantitative recovery of these particles is assumed for the pellet of this fraction. Since this pellet contains all the particles originally present in the tissue that sediment under given conditions and this pellet is then diluted with 5 ml medium for each gram of original tissue, the contents of this pellet are diluted by a factor of 5 in comparison with their original concentration in the tissue (i.e., dilution of 1:5). However, since the homogenate is already diluted by a factor of 9 in comparison with the original tissue, the concentration of the particles in the nuclear fraction is greater than in the homogenate (i.e., 9/5:1 or 1.8:1). The same discussion applies, in general, to each fraction. Table 7 also indicates one of the problems encountered in differential centrifugation. The final supernatant, or soluble fraction, is greatly diluted in comparison with all the other fractions as well as in comparison with the original tissue. Washing of fractions increases their purity, but also increases the dilution of the supernatant. Thus, if each particulate fraction had been washed twice instead of once, the dilution of the supernatant compared to the original tissue would be 1:81 instead of 1:45.

Marker enzymes are assayed under various conditions dependent on the procedures required by the assay used. In the present case, for example, one assay requires 50 μl (0.05 ml) of sample and yields a figure, X, in units of enzyme activity for that amount of sample. Since the sample assayed represents a portion of the material available in a given fraction, multiplication by an appropriate factor will yield the total activity in that fraction. For example, a 50-μl sample of the nuclear fraction (which totals 2.5 ml in volume) is equal to 0.02 of that fraction. If the 50-μl sample yields an activity of X_N, then the total activity of the fraction (X_N) is equal to 50 X_N (i.e., the activity recorded equaled 2.0% of the fraction; therefore, multiplication by 50 yields the total activity of the fraction), which represents the total contribution by the nuclear fraction to the activity of that particular enzyme in the original tissue. Similar calculations may be performed for each of the other fractions, yielding values for X_{M-1}, X_{M-2}, and X_{S-3} as well as for the homogenate, X_H. The sum of the fractional activities, X_T, is expected to approximate the total activity as estimated in the homogenate. The amount of activity recovered in each fraction is indicated in percentage by, for example, in the case of the nuclear fraction, $(X_N/X_H) \times 100$. The total recovery is calculated similarly, $(X_T/X_H) \times 100$.

The calculations described above yield a value for the activity of a given enzyme in the particular amount of tissue originally homogenized. Of more value, frequently, is a relative representation of the data that readily allows comparisons among experiments that differ, for instance, in terms of the amount of starting material. When activities are expressed in relation to the weight of the original tissue, the relative units used are usually per gram of tissue. Since the original tissue in this example weighed 0.5 g, the total activity would be $2X_T$ per gram of tissue. Each of the fractional activities could be expressed similarly. Thus, the activity of the nuclear fraction would be $2X_N$ per gram of tissue.

An alternative method for presenting results may also be encountered. Thus, the activity of the nuclear fraction may be expressed as activity per milliliter rather than as the total activity of the fraction. Since the sample that yields X_N is 50 μl, $20X_N$ would represent the activity per milliliter in the nuclear fraction. Since the nuclear fraction has been diluted by a factor of 5 for each gram of original tissue, the activity due to the components of the nuclear fraction present in the original tissue would be $5 \times 20X_N$ units per gram, which results in the same figure as in the previous method of calculation. Similar calculations can be performed in which activity is based relative to the volume of the homogenate, taking account of the different dilution factors involved.

To present results that are more immediately meaningful in themselves, researchers often relate enzyme activity, not to volume or wet weight, but to some biochemical factor that can be measured precisely in each fraction. The most common and in fact the more or less standard measure for this purpose is protein content. Using a suitable assay, the protein content is measured in samples from each fraction as well as from the homogenate. Calculations are performed as above relating protein content to tissue weight. This is usually expressed as milligrams of protein per gram of tissue. Recoveries are also calculated. The upper portion of Table 8 presents some sample data based

Table 8. Sample Balance

	Protein		Acetylcholinesterase (AChE)				
Fraction	Content	Protein recovered (%)	Activity[b]	Activity recovered (%)	Specific activity	Purification	Relative specific activity
Homogenate	163.1	—	751.89	—	4.61	—	1.00
Nuclear	49.2	31.95	174.66	29.56	3.55	−1.06	0.93
Mitochondrial	32.9	21.36	131.27	22.22	3.99	−0.62	1.04
Microsomal	28.4	18.44	163.58	27.68	5.76	1.15	1.50
Soluble	43.5	28.25	121.37	20.54	2.79	−1.82	0.73
TOTAL:	154.0	100.00	590.88	100.00	—	—	—
Recovery (%)	94.42						
Mitochondrial	32.9	—	131.27	—	3.99	—	1.00
A	4.8	16.49	18.19	16.27	3.79	−0.20	0.99
B	18.2	62.54	87.36	78.16	4.80	0.81	1.25
C	6.1	20.96	6.22	5.56	1.02	−2.97	0.27
TOTAL:	29.1	100.00	111.77	100.00	—	—	—
Recovery (%)	88.45		85.15				

[a] Adapted from Stahl and Swanson (1975).
[b] Micromoles hydrolyzed per hour.
[c] Micromoles oxidized per hour.
[d] Moles oxidized per hour.

on the four-fraction separation discussed above. The values obtained in the actual protein assays have been adjusted relative to tissue weight and dilution factors. In a given experiment in which all the dilution factors and assay conditions are known, it would be possible to work from the protein-content figures given and estimate directly, for instance, the amount of protein assayed in the samples. The same would hold true for any of the enzyme-marker assays.

The first column of Table 8 presents the actual protein contents of the homogenate and each of the four fractions expressed in milligrams of protein per gram wet weight of original tissue. The protein assayed for the four fractions is summed (ΣP) and then expressed as a percentage of the protein found in the homogenate. This figure (94.42%) represents the recovery of protein in the fractions. Additionally, with the total amount of protein in the fractions treated as 100%, the percentage of recovered protein attributable to each fraction was calculated. These figures are presented in the second column of the table.

Exemplary data for enzymes assayed are also included in the table. For each of three enzymes, acetylcholinesterase, monoamine oxidase, and succinate dehydrogenase, the data are presented in the same format. The first column presents the total activity found in the homogenate and each fraction expressed in appropriate units (i.e., micromoles of substrate hydrolyzed or micromoles of substrate oxidized per hour per gram wet weight of the original tissue). As with the protein assays, the recovery of the combined

Sheet: Extended Version[a]

Monoamine oxidase (MAO)					Succinate dehydrogenase (SD)				
Activity[c]	Activity recovered (%)	Specific activity	Purification	Relative specific activity	Activity[d]	Activity recovered (%)	Specific activity	Purification	Relative specific activity
6083.63	—	37.3	—	1.00	2038.75	—	12.5	—	1.00
1840.08	39.75	37.4	0.1	1.24	674.04	40.15	13.7	1.2	1.26
2148.37	46.40	65.3	28.0	2.17	950.81	56.63	28.9	16.4	2.65
397.60	8.59	14.0	−23.3	0.47	45.44	2.71	1.6	−10.9	0.15
243.60	5.26	5.6	−31.7	0.19	8.70	0.52	0.2	−12.3	0.01
4629.65	100.00	—	—	—	1678.99	100.00	—	—	—
76.10					82.35				
2148.37	—	65.3	—	1.00	950.81	—	28.9	—	1.00
18.24	1.11	3.8	−61.5	0.07	0.48	0.07	0.1	−28.8	0.01
889.98	54.32	48.9	−16.4	0.87	343.98	52.08	18.9	−10.0	0.83
720.17	44.57	119.7	54.4	5.71	315.98	47.84	51.8	22.9	2.28
1638.39	100.00	—	—	—	660.44	100.00	—	—	—
76.26					69.47				

fractions is compared with the homogenate. The proportion of recovered activity contributed by each fraction is also presented. Recovery data are important, since they indicate the yield of the centrifugation procedure. The overall-recovery figure indicates how much protein, or enzyme activity, is lost (or gained) in the separated material compared with the unfractionated homogenate. Of course, for practical purposes, it is desirable to achieve the highest yields possible simply to produce sufficient fractionated material for analysis. More important, high yields contribute significantly to the validity of fractionation experiments. When yields are not satisfactory, it is likely that material has been lost, but it is impossible to determine whether the recovered material is truly representative of the material in the homogenate. It is possible that some cellular component may be selectively lost. Although losses due to the handling of materials (e.g., during fraction collection) are inevitable, great care must be taken to keep these at a minimum. Beaufay *et al.* (1964) have provided a technique that allows for correction of known handling or sampling losses. Yields could also be reduced due to inhibition or removal from sources of activation of enzymes, but it is extremely unlikely that all the markers examined would be affected similarly. In any event, such occurrences can usually be clarified if fractionations are run under different conditions in an effort to isolate the cause of the reduced yields.

The third column of figures for each enzyme in Table 8 represents the specific activities of the homogenate and each of the fractions. The specific activity of a fraction is simply the ratio between the activity measured and the protein content of the fraction or homogenate. Thus, the ratio between acetylcholinesterase activity in the homogenate and its protein content (751.89/163.1 = 4.61) is the specific activity of the homogenate, and its units are micromoles of acetylcholine hydrolyzed per hour per milligram protein. Specific activities are a measure of the degree of purification of the markers being examined. The activity recovered in one fraction may greatly exceed that recovered in another fraction and may seem, therefore, to be concentrated in that fraction. However, if the fraction containing the greater amount of activity also exhibits a high protein recovery, then the concentration of the marker in that fraction may be lower than in a fraction exhibiting lower activity but also less protein. Protein content is taken as a convenient indicator of the amount of material recovered in a fraction. Since specific activities indicate the amount of enzyme recovered in terms of the amount of the material from which it was recovered, they also indicate the degree of concentration, or even purity, of the fractions. As can be seen in Table 8, specific activities can take on a rather wide range of values. This occurs because different marker enzymes vary in activity, and this variation greatly influences the magnitude of the specific activities. The reference point for interpretation of the specific activities of fractions is the specific activity of the homogenate. If the specific activity of a given fraction is less than that of the homogenate, then the marker is less concentrated in the fraction than it is in the homogenate. Although not usually expressed as such, this is an indication of purification, since it means that the concentration of some marker that is presumably a contaminant for that fraction has been reduced, thus leading to greater purity of the fraction in terms of some other marker. Of course, a fractional specific activity that exceeds that of the homogenate indicates a purification of the marker in that fraction relative to the homogenate. The difference between the specific activity of a fraction and the specific activity of the homogenate is a direct measure of purification. These values have been calculated and appear in the fourth column for each enzyme. Note, for instance, the succinate dehydrogenase data. Although over 40% of the enzyme activity is recovered in the nuclear fraction, the purification obtained is negligible, while a recovery of 57% in the mitochondrial fraction is associated with a purification that is 16 times greater than in the homogenate.

Although specific activities are the most frequently used means of data presentation, they do have at least two minor weaknesses. The magnitudes of the range of specific activities vary from marker to marker, as does the baseline value of the homogenate. Thus, it is not as easy as it might be to compare the purification of different markers or even of different experiments. The use of purification values in addition to the specific activities does simplify the discussion somewhat. The other weakness results from the fact that the specific-activity calculations for the fractions take no account of the recovery obtained with the fractions in comparison with the homogenate. Dependent on the overall recovery, specific-activity figures will contain some degree of inaccuracy when purification from the homogenate is considered.

In the first of a series of classic papers on the fractionation of neuronal tissue (DeRobertis *et al.*, 1962, 1963; Rodriguez de Lores Arnaiz and DeRobertis, 1962), DeRobertis and his colleagues advocated the use of a different measure, which was called relative specific activity. Use of this measure eliminates both the problems just mentioned. Relative specific activity is defined as the ratio between the percentage of enzyme activity recovered and the percentage of protein recovered in a given fraction or in the homogenate. Since, by convention, the recoveries in the homogenate are always 100%, the relative specific activity of the homogenate is always 1.00 (i.e., 100/100), no matter which marker is being considered. Relative specific activities in the fractions that are less than 1.00 indicate a dilution of the marker in comparison with the homogenate. Values greater than 1.00 indicate a concentration of the marker in the fraction in comparison with the homogenate. The last column of figures under each enzyme in Table 8 lists the relative specific activities.

The lower portion of Table 8 represents, in a similar fashion, the results obtained following the subfractionation on a density gradient of the original mitochondrial fraction. As is obvious, the data for the original mitochondrial fraction are repeated from above. Otherwise, all values for the three mitochondrial subfractions (A, B, and C) are calculated essentially independently of the data in the upper portion of the table. Recoveries are calculated with reference to the original mitochondrial fraction, not with reference to the homogenate as previously. In fact, all values are calculated in the same manner as previously, except that the original mitochondrial values are used in place of homogenate values. The reasoning behind such treatment is that the subfractionation of the mitochondrial fraction is logically analogous to the original fractionation of the homogenate.

The figures in Table 8 were included to demonstrate the principles involved in the calculation, presentation, and interpretation of balance sheets. No table encountered in the literature would ever be so complete. Table 9 is included as an example of what might be expected in the literature. It presents the same data as Table 8, but in a more economical form. Much of the information in Table 8 is actually redundant in the sense that every value in Table 8 may be calculated directly from the values included in Table 9.

Balance-sheet data may also be presented graphically, and frequently both graphs and tables are included in the literature. Generally, one graph is required for each marker enzyme as well as for protein, although the latter may be combined in a marker-enzyme graph, in which case the concept of specific activity is implied. DeDuve (1967) has outlined generally accepted graphic formats that he and his associates have developed.

DeDuve first stresses the importance of achieving satisfactory recoveries before making any attempts to present data. The absolute amounts of an enzyme or other biochemical component found in each subcellular fraction (Q) are summed for all fractions (ΣQ). If the sum is not too different from the amount found in the homogenate, then $(Q \times 100)/\Sigma Q$ may be taken as the fractional amount (expressed as a percentage) of the component present in a given fraction. This is, of course, the procedure used to calculate the percentage of recovered activity in Tables 8 and 9. As already mentioned, unsatisfactory recoveries can lead to biased results and should be avoided.

Table 9. Sample Balance Sheet: Simplified Version

Fraction	Protein		AChe		MAO		SD	
	Content	Protein recovered (%)	Activity	Activity recovered (%)	Activity	Activity recovered (%)	Activity	Activity recovered (%)
Homogenate	163.1	—	4.61	—	37.3	—	12.5	—
Nuclear	49.2	31.95	3.55	29.56	37.4	39.15	13.7	40.15
Mitochondrial	32.9	21.36	3.99	22.22	65.3	46.40	28.9	56.63
Microsomal	28.4	18.44	5.76	27.68	14.0	8.59	1.6	2.71
Soluble	43.5	28.25	2.79	20.54	5.6	5.26	0.2	0.52
Recovery (%)	94.42		78.59		76.10		82.35	
Mitochondrial	32.9	—	3.99	—	65.3	—	28.9	—
A	4.8	16.49	3.79	16.27	3.8	1.11	0.1	0.07
B	18.2	62.54	4.80	78.16	48.9	54.32	18.9	52.08
C	6.1	20.96	1.02	5.56	119.7	44.57	51.8	47.84
Recovery (%)	88.45		85.15		76.26		69.47	

Given satisfactory recovery, graphic representation is usually achieved best in histogram form. Rectangles of width Δx are aligned along the abcissa. Each rectangle represents a fraction, and the ordering of the rectangles is dependent on the experimental conditions. The height of the rectangles is proportioned to the contribution of each fraction to the recovered activity (i.e., $[(Q \times 100)/\Sigma Q] \times \Delta x$, or some equivalent). The surface area of each rectangle is equal to the percentage amount of the marker contained in a particular fraction. If all fractions are represented, the surface area of the whole histogram is equal to 100%.

In its simplest form, the histogram will include rectangles, all of an equal, arbitrary width (Δx). Though not very informative, such a representation may be necessitated by a lack of information concerning protein content or some other physical measure derived from the centrifugation conditions. When protein content is known, then Δx may assume a value for each rectangle that is a function of the percentage of recovered protein in the fraction represented by that rectangle, i.e., $(P \times 100)/\Sigma P$. Substitution of this term for Δx in the formula for the height of the rectangle, above, yields $[(Q \times 100)/\Sigma Q] \times [(P \times 100)/\Sigma P]$, which on rearrangement results in $(Q \times \Sigma P)/(\Sigma Q \times P)$, which is in fact the general formula for relative specific activity. Thus, the fractions are represented by rectangles the widths of which are a linear function of the recovered protein in the fraction and the heights of which are a function of the relative specific activity of the fraction with respect to the marker being considered. The fractions are aligned from left to right along the abcissa in the order in which they are isolated, which is to say in order of decreasing sedimentation coefficients. One of the early examples of this technique may be seen in deDuve *et al.* (1955). The data from the example given above have been graphed in Fig. 8 for the four main fractions and the three enzymes assayed.

Following density-gradient centrifugation, more information is available concerning the physical properties of the fractionated material. The widths of the rectangles that represent the fractions can be made a function of the volume of the fractions as a percentage of the total volume of the gradient. The heights of the rectangles are based on the degree of concentration of the marker in the fraction compared with its concentration if the initial homogenate were evenly distributed throughout the gradient. Beaufay *et al.* (1959, 1964) elaborate on the methodology involved in the presentation of such data.

Beaufay *et al.* (1964) have also detailed procedures that allow the density profile of a gradient to be estimated. If density equilibrium has been achieved prior to the end of centrifugation, then the data can be plotted with Δx being equivalent to density intervals, yielding a density-distribution histogram. If the conditions of centrifugation are known accurately and fully enough, it is possible to compute a value of Δs, the interval of sedimentation coefficients spanned by the fraction, which can be substituted for Δx (general principles have been discussed above and in deDuve *et al.*, 1959). In differential-sedimentation experiments, the sedimentation coefficient of the lightest

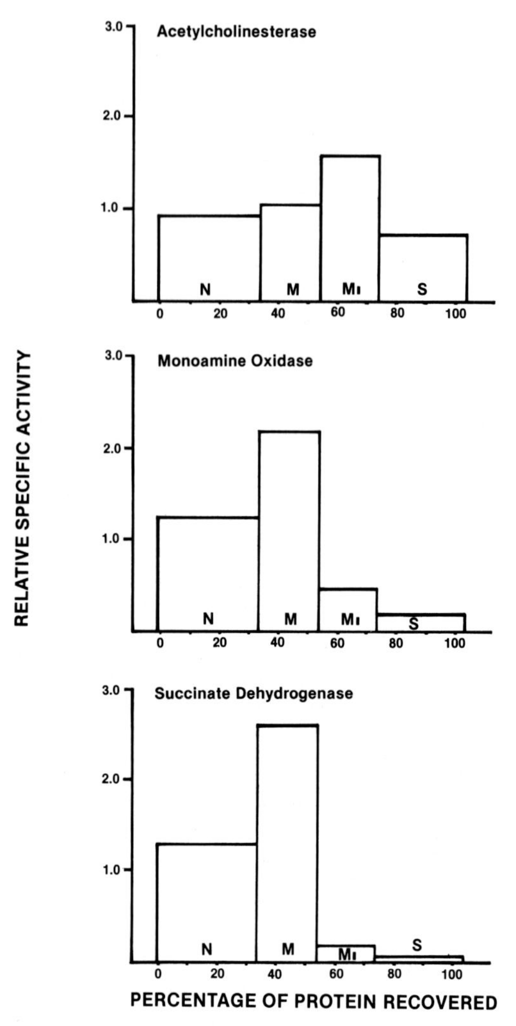

Fig. 8. Relative specific activities of three enzymes in four major fractions depicted as a function of the percentage of protein recovered in each fraction. This figure is based on the data in Table 8.

particle to be fully sedimented can be calculated if the conditions of sedimentation are known (deDuve and Berthet, 1953; Applemans *et al.*, 1955). Finally, deDuve (1967) also discusses the conditions under which a particle-size distribution may be calculated and plotted along the abcissa.

To summarize regarding the validation of fractionation schemes, one of the early workers in this area will be paraphrased (Schneider, 1972; also Schneider and Hogeboom, 1951). Several important principles must be followed. Morphological and cytological examinations of each fraction must be undertaken in an attempt to closely correlate the cytological and morphological properties of the particles in the fractions with the properties of those particles in the intact cell. Ideally, each fraction should consist of a single type of particle. It is also necessary that the yield of the isolated particles encompass a relatively large proportion of those particles contained in the whole tissue. The physical properties of particles in a given type of cell are likely to vary considerably, and furthermore, most tissues examined are comprised of several cell types, which contributes further to particle heterogeneity. Quantitative yields are important, since particles that are not recovered for any reason may in fact be classifiable as some specific subcategory of the particle class. Failure to recover these particles will lead to a nonrepresentativeness in the recovered fraction. As an ideal, a fractionation procedure should provide the isolated cell components not only in a highly purified state but also in high yield.

The enzymatic activity of each fraction must be determined. The summed activity of all the fractions must equal the activity of the unfractionated homogenate. As already mentioned, it is necessary to consider the possibilities that there may be enzyme inhibitors or activators present in the homogenate and that the distribution, after fractionation, of the enzyme and the inhibitor or activator may differ significantly. In such a case, the summed fractional activities will either exceed or fall short of the activity of the homogenate. To clarify the reasons for such results, the activity of the enzyme can be assayed in various combinations of fractions as well as in the individual fractions in an attempt to localize the inhibitor or activator. If such procedures do not resolve the poor recovery and if even a completely reconstituted homogenate is incapable of exhibiting full recovery, then the possibility of irreversible damage during fractionation must be considered. As has been repeatedly emphasized, examination of an isolated subcellular component in the absence of concomitant examination of all the other components of the homogenate cannot be considered to extend knowledge of cells or tissues as a whole. Such studies must be considered as only suggestive until more thorough examination of all the components of the homogenate have been completed.

After these criteria have been met, the localization of an enzyme in a particular cellular component is indicated when a large percentage of the total activity of the homogenate is recovered in the fraction, the specific activity of the fraction is several times greater than that of the homogenate, and the specific activity of the fraction is unaffected by repeated sedimentation.

References

Abood, L.G., 1969, Brain mitochondria, in: *Handbook of Neurochemistry* (A. Lajtha, ed.), Vol. II, pp. 303–326, Plenum Press, New York.

Acs, G., Neidle, A., and Waelsch, H., 1961, Brain ribosomes and amino acid incorporation, *Biochim. Biophys. Acta* **50:**403–404.

Alderman, J.L., and Shellenberger, M.K., 1974, 7'-Aminobutyric acid (GABA) in the rat brain: Re-evaluation of sampling procedures and the post mortem increase, *J. Neurochem.* **22:**937–940.

Aldridge, W.N., 1957, Liver and brain mitochondria, *Biochem. J.* **67:**423–431.

Allfrey, V., 1959, The isolation of subcellular components, in: *The Cell*, Vol. I (J. Brachet and A.E. Mirsky, eds.), pp. 193–290, Academic Press, New York.

Allfrey, V., and Mirsky, A.E., 1959, Biochemical properties of the isolated nucleus, in: *Subcellular Particles* (T. Hayashi, ed.), pp. 136–207, Ronald Press, New York.

Anderson, N.G., 1955, Mechanical device for producing density gradients in liquids, *Rev. Sci. Instrum.* **26:**891–892.

Anderson, N.G., 1966*a*, An introduction to particle separations in zonal centrifuges, *Natl. Cancer Inst. Monogr.* **21:**9–39.

Anderson, N.G. (ed.), 1966*b*, *The Development of Zonal Centrifuges, Natl. Cancer Inst. Monogr.* **21**.

Anderson, N.G., and Green, J.G., 1967, The soluble phase of the cell, in: *Enzyme Cytology* (D.B. Roodyn, ed.), pp. 475–509, Academic Press, New York.

Applemans, F., Wattiaux, R., and deDuve, C., 1955, Tissue fractionation studies. 5. The association of acid phosphatase with a special class of cytoplasmic granules in rat liver, *Biochem. J.* **59:**438–445.

Autilio, L.A., Norton, W.T., and Terry, R.D., 1964, The preparation and some properties of purified myelin from the central nervous system, *J. Neurochem.* **11:**17–27.

Autilio, L.A., Appel, S.H., Pettis, P., and Gambetti, P.L., 1968, Biochemical studies of synapses *in vitro.* I. Protein synthesis, *Biochemistry* **7:**2615–2622.

Babitch, J.A., Breithaupt, T.B., Chiu, T.-C., Garadi, R., and Helseth, D.L., 1976, Preparation of chick brain synaptosomes and synaptosomal membranes, *Biochim. Biophys. Acta* **433:**75–89.

Balcom, G.J., Lenox, R.H., and Meyerhoff, J.L., 1975, Regional gamma-aminobutyric acid levels in rat brain determined after microwave fixation, *J. Neurochem.* **24:**608–613.

Baudhuin, P., 1968, L'Analyse morphologique quantitative de fractions subcellulaires, Thesis, University of Louvain, Louvain, Belgium.

Baudhuin, P., 1974, Morphometry of subcellular fractions, in: *Methods in Enzymology*, Vol. XXXII (J. Fleischer and L. Packer, eds.), pp. 3–20, Academic Press, New York.

Baudhuin, P., Evrard, P., and Berthet, J., 1967, Electron microscopic examination of subcellular fractions. 1. The preparation of representative samples from suspensions of particles, *J. Cell Biol.* **32:**181–191.

Beaufay, H., 1966, *Thèse d'Agrégation de l'Enseignment Superieur*, Université Catholique de Louvain, Louvain, Belgium, Centerick S.P., Louvain, Belgium.

Beaufay, H., Bendall, Baudhuin, P., Wattiaux, R., and deDuve, C., 1959, Tissue fractionation studies. 13. Analysis of mitochondrial fractions from rat liver by density-gradient centrifuging, *Biochem. J.* **73:**628–637.

Beaufay, H., Jacques, P., Baudhuin, P., Sellinger, O.Z., Berthet, J., and deDuve, C., 1964, Tissue fractionation studies. 18. Resolution of mitochondrial fractions from rat liver into three distinct populations of cytoplasmic particles by means of density equilibration in various gradients, *Biochem. J.* **92:**184–205.

Behrens, M., 1932, Untersuchungen an isolierten Zell and Gewebebestand. I. Mitteilung. Isolierung von Zellkernen des Kalbsherzmuskels, *Z. Physiol. Chem.* **209:**59–74.

Behrens, M., 1938, Zell- und Gewebetrennung, in: *Handbuch der biologischen Arbeitsmethoden* (E. Abderhalden, ed.), Vol. X, Part 10, II, pp. 1363–1392, Urban and Schwarzenberg, Berlin.

Bellamy, D., 1959, The distribution of bound acetylcholine and choline acetylase in rat and pigeon brain, *Biochem. J.* **72:**165–168.

Bensley, R.R., and Hoerr, N.L., 1934, Studies on cell structure by the freezine–drying method. VI. The preparation and properties of mitochondria, *Anat. Rec.* **60:**449–455.

Biesold, D., 1974, Isolation of brain mitochondria, in: *Research Methods in Neurochemistry*, Vol. 2 (N. Marks and R. Rodnight, eds.), pp. 39–52, Plenum Press, New York.

Birnie, G.B., Fox, S.M., and Harvey, D.R., 1972, Separation of polysomes, ribosomes and ribosomal subunits in zonal rotors, in: *Subcellular Components: Preparation and Fractionation* (G.D. Birnie, ed.), pp. 235–250, Butterworths, London.

Bloom, F.E., 1970, Correlating structure and function of synaptic ultrastructure, in: *The Neurosciences: Second Study Program* (F.O. Schmitt, ed.), pp. 729–747, Rockefeller University Press, New York.

Bodian, D., 1967, Neurons, circuits and neuroglia, in: *The Neurosciences: A Study Program* (G.C. Quarton, T. Melnechuk, and F.O. Schmitt, eds.), pp. 6–23, Rockefeller University Press, New York.

Bonanou-Tzedaki, S.A., and Arnstein, H.R.V., 1972, Isolation of animal polysomes and ribosomes, in: *Subcellular Components: Preparation and Fractionation* (G.D. Birnie, ed.), pp. 215–234, Butterworths, London.

Bondy, S.C., and Waelsch, H., 1965, Nuclear RNA polymerase in brain and liver, *J. Neurochem.* **12:**751–756.

Bourne, G.H., and Tewari, H.B., 1964, Mitochondria and the Golgi complex, in: *Cytology and Cell Physiology* (G.H. Bourne, ed.), pp. 377–421, Academic Press, New York.

Bowen, T.J., 1966, Laboratory centrifuges, in: *Instrumentation in Biochemistry* (T.W. Goodwin, ed.), Biochemical Society Symposium No. 26, pp. 1–24, Academic Press, New York.

Bowen, T.J., 1970, *An Introduction to Ultracentrifugation,* Wiley-Interscience, London.

Bradford, H.F., 1972, Cerebral cortex slices and synaptosomes: *In vitro* approaches to brain metabolism, in: *Methods of Neurochemistry*, Vol. 3 (R. Fried, ed.), pp. 155–202, Marcel Dekker, New York.

Brakke, M.K., 1951, Density gradient centrifugation: A new separation technique, *J. Am. Chem. Soc.* **73:**1847–1848.

Brakke, M.K., 1967, Density-gradient centrifugation, in: *Methods in Virology*, Vol. II (K. Maramorusch and H. Koprowski, eds.), pp. 93–118, Academic Press, New York.

Brody, T.M., and Bain, J.A., 1952, A mitochondrial preparation from mammalian brain, *J. Biol. Chem.* **195:**685–696.

Brown, F., and Danielli, J.F., 1964, The cell surface and cell physiology, in: *Cytology and Cell Physiology* (G.H. Bourne, ed.), pp. 239–310, Academic Press, New York.

Brunner, E.A., Passonneau, J.V., and Molstad, C., 1971, The effect of volatile anaesthetics on levels of metabolites and on metabolic rate in brain, *J. Neurochem.* **8:**2301–2316.

Bunge, R.P., 1970, Structure and function of neuroglia: Some recent observations, in: *The Neurosciences: Second Study Program* (F.O. Schmitt, ed.), pp. 782–797, Rockefeller University Press, New York.

Campagoni, A.T., and Mahler, H.R., 1967, Isolation and properties of polyribosomes from cerebral cortex, *Biochemistry* **6:**956–967.

Chappel, J.B., and Hansford, R.G., 1972, Preparation of mitochondria from animal tissues and yeast, in: *Subcellular Components: Preparation and Fractionation* (G.D. Birnie, ed.), pp. 77–91, Butterworths, London.

Cheney, D.L., Racagni, G., Zsilla, G., and Costa, E., 1976, Differences in the action of various drugs on striatal acetylcholine and choline content in rats killed by decapitation or microwave radiation, *J. Pharm. Pharmacol.* **28:**75–77.

Claude, A., 1946, Fractionation of mammalian liver cells by differential centrifugation. I. Problems, methods and preparation of extract, *J. Exp. Med.* **84:**51–58.

Cleland, K.W., 1952, Permeability of isolated rat heart sarcosomes, *Nature (London)* **170:**497–499.

Cline, G.B., and Ryel, R.B., 1971, Zonal centrifugation, in: *Methods in Enzymology*, Vol. 22 (W.B. Jakoloy, ed.), pp. 168–204, Academic Press, New York.

Cotman, C.W., 1968, Doctoral dissertation, Department of Chemistry, Indiana University, Bloomington, Indiana.

Cotman, C.W., 1972, Principles for the optimization of centrifugation conditions for the fractionation of brain tissue, in: *Research Methods in Neurochemistry*, Vol. 1 (N. Marks and R. Rodnight, eds.), pp. 45–93, Plenum Press, New York.

Cotman, C.W., 1974, Isolation of synaptosomal and synaptic plasma membrane fractions, in: *Methods in Enzymology*, Vol. XXXI (S. Fleischer and L. Packer, eds.), pp. 445–452, Academic Press, New York.

Cotman, C.W., Mahler, H.R., and Anderson, N.G., 1968, Isolation of a membrane fraction enriched in nerve-end membranes from rat brain by zonal centrifugation, *Biochim. Biophys. Acta* **163:**272–275.

Cotman, C.W., Brown, D.H., Harrell, B.W., and Anderson, N.G., 1970, Analytical differential centrifugation: An analysis of the sedimentation properties of some synaptosomes, mitochondria and lysosomes from rat brain homogenates, *Arch. Biochem. Biophys.* **136:**436–447.

Cotman, C.W., Herschman, H., and Taylor, D., 1971, Subcellular fractionation of cultured glial cells, *J. Neurobiol.* **2:**169–180.

Davis, B.D., 1970, Recent advances in understanding ribosomal action, in: *The Neurosciences: Second Study Program* (F.O. Schmitt, ed.), pp. 920–927, Rockefeller University Press, New York.

deDuve, C., 1959, Lysosomes, a new group of cytoplasmic particles, in: *Subcellular Particles* (T. Hayashi, ed.), pp. 128–159, Ronald Press, New York.

deDuve, C., 1964, Principles of tissue fractionation, *J. Theor. Exp. Biol.* **6:**33–59.

deDuve, C., 1967, General principles, in: *Enzyme Cytology* (D.B. Roodyn, ed.), pp. 1–26, Academic Press, New York.

deDuve, C. 1971, Tissue fractionation: Past and present, *J. Cell Biol.* **50:**200–550.

deDuve, C., and Berthet, J., 1953, Reproducibility of differential centrifugation experiments in tissue fractionation, *Nature (London)* **192:**1142.

deDuve, C., and Berthet, J., 1954, The use of differential centrifugation in the study of tissue enzymes, *Int. Rev. Cytol.* **3:**225–275.

deDuve, C., Pressman, B.C., Gianetto, R., Wattiaux, R., and Applemans, F., 1955, Tissue fractionation studies. 6. Intracellular distribution patterns of enzymes in rat-liver tissue, *Biochem. J.* **60:**604–617.

deDuve, C., Berthet, J., and Beaufay, H., 1959, Gradient centrifugation of cell particles: Theory and applications, in: *Progress in Biophysics and Biophysical Chemistry*, Vol. 9 (J.A.V. Butler and B. Katz, eds.), pp. 325–369, Pergamon Press, London.

deDuve, C., Wattiaux, R., and Baudhuin, P., 1962, Distribution of enzymes between subcellular fractions in animal tissues, in: *Advances in Enzymology*, Vol. 24 (F.F. Nord, ed.), pp. 291–358, Interscience, New York.

DeRobertis, E., and Rodriguez de Lores Arnaiz, G., 1969, Structural components of the synaptic region, in: *Handbook of Neurochemistry*, Vol. II (A. Lajtha, ed.), pp. 365–392, Plenum Press, New York.

DeRobertis, E., Pellegrino, de Iraldi, A., Rodriguez de Lores Arnaiz, G., and Salganicoff, L., 1962, Cholinergic and non-cholinergic nerve endings in rat brain. I. Isolation and subcellular distribution of acetylcholine and acetylcholinesterase, *J. Neurochem.* **9:**23–35.

DeRobertis, E., Rodriguez de Lores Arnaiz, G., Salganicoff, L., Pellegrino de Iraldi, A., and Zieher, L.M., 1963, Isolation of synaptic vesicles and structural organization of the acetylcholine system within brain nerve endings, *J. Neurochem.* **10:**225–235.

Detering, N.K., and Wells, M.A., 1976, Detection of myelin in the optic nerve of young rats by sedimentation equilibrium in a CsC1 gradient, *J. Neurochem.* **26:**247–252.

Dole, V.P., and Cotzias, G.C., 1951, A nomogram for the calibration of relative centrifugal force, *Science* **113:**532–553.

Dounce, A.L., 1952, The interpretation of chemical analyses and enzyme determinations on isolated cell components, *J. Cell. Comp. Physiol.* **39**(Suppl. 2)**:**43–74.

Dounce, A.L., 1953, The isolation and composition of cell nuclei and nucleoli, in: *The Nucleic Acids*, Vol. II (E. Chargaff and J.N. Davidson, eds.), pp. 93–153, Academic Press, New York.

Dutton, G.R., and Mahler, H.R., 1968, *In vitro* RNA synthesis by intact rat brain nuclei, *J. Neurochem.* **15:**765–780.

Elias, H., Hennig, A., and Schwartz, D.E., 1971, Stereology: Applications to biomedical research, *Physiol. Rev.* **51:**158–200.

Elson, D., 1967, Ribosomal enzymes, in: *Enzyme Cytology* (D.B. Roodyn, ed.), pp. 407–473, Academic Press, New York.

Emanuel, C.F., and Chaikoff, I.L., 1957, An hydraulic homogenizer for the controlled release of cellular components from various tissues, *Biochim. Biophys. Acta* **24:**254–261.

Feit, H., and Barondes, S.H., 1970, Colchicine-binding activity in particulate fractions of mouse brain, *J. Neurochem.* **17:**1355–1364.

Feit, H., Dutton, G., Barondes, S.H., and Shelanski, M.L., 1971, Microtubule protein identification and transport to nerve endings, *J. Cell Biol.* **51:**138–147.

Fernandez-Moran, H., 1967, Membrane ultrastructure in nerve cells, in: *The Neurosciences: A Study Program* (G.C. Quarton, T. Melnechuk, and F.O. Schmitt, eds.), pp. 281–304, Rockefeller University Press, New York.

Ferrendelli, J.A., Gay, M.H., Sedgewick, W.G., and Chang, M.M., 1972, Quick freezing of the murine CNS: Comparison of regional cooling rates and metabolic levels when using liquid nitrogen or Freon-12, *J. Neurochem.* **19:**979–987.

Georgiev, G.P., 1967, The nucleus, in: *Enzyme Cytology* (D.B. Roodyn, ed.), pp. 27–100, Academic Press, New York.

Giacobini, E., 1969, Chemistry of isolated invertebrate neurons, in: *Handbook of Neurochemistry*, Vol. II (A. Lajtha, ed.), pp. 195–239, Plenum Press, New York.

Glees, P., and Meller, K., 1968, Morphology of neuroglia, in: *The Structure and Function of Nervous Tissue*, Vol. I (G.H. Bourne, ed.), pp. 301–323, Academic Press, New York.

Glowinski, J., and Iversen, L.L., 1966, Regional studies of catecholamines in the rat brain. I. The disposition of (^{3}H) norepinephrine, (^{3}H) dopamine and (^{3}H) dopa in various regions of the brain, *J. Neurochem.* **13:**655–669.

Granholm, L., Kaasik, A.E., Nilsson, L., and Siesjo, B.K., 1968, The lactate/pyruvate ratios of cerebrospinal fluid of rats and cats related to the lactate/pyruvate, the ATP/ADP, and the phosphocreatine/creatine ratios of brain tissues, *Acta Physiol. Scand.* **74:**398–409.

Gray, E.G., and Whittaker, V.D., 1962, The isolation of nerve endings from brain: An electron microscopic study of cell fragments derived by homogenization and centrifugation, *J. Anat. (London)* **96:**79–88.

Green, D.E., 1959, Mitochondrial structure and function, in: *Subcellular Particles* (T. Hayashi, ed.), pp. 84–103, Ronald Press, New York.

Guidotti, A., Cheney, D.L., Trabucchi, M., Dotenchi, M., Wang, C., and Hawkins, R.A., 1974, Focused microwave radiation: A technique to minimize post-mortem changes of cyclic nucleotides, DOPA and choline and to preserve brain morphology, *Neuropharmacology* **13:**1115–1122.

Gurd, J.W., Jones, L.R., Mahler, H.R., and Moore, W.J., 1974, Isolation and partial characterization of rat brain synaptic plasma membranes, *J. Neurochem.* **22:**281–290.

Hamberger, A., Blomstrand, C., and Lehninger, A.L., 1970, Comparative studies on mitochondria isolated from neuron-enriched and glia-enriched fractions of rabbit and beef brain, *J. Cell Biol.* **45:**221–234.

Harwood, R., 1974, Cell separation by gradient centrifugation, in: *International Review of Cytology*, Vol. 38 (G.H. Bourne and J.F. Danielli, eds.), pp. 369–403, Academic Press, New York.

Hebb, C.O., and Smallman, B.N., 1956, Intracellular distribution of choline acetylase, *J. Physiol.* **134:**385–392.

Hebb, C.O., and Whittaker, V.P., 1958, Intracellular distributions of acetylcholine and choline acetylase, *J. Physiol.* **142:**187–196.

Hers, H.G., Berthet, J., Berthet, L., and deDuve, C., 1951, Le systeme hexose-phosphatosique. III. Localisation intracellulaire des ferments par centrifugation fractionée, *Bull. Soc. Chim. Biol.* **33:**21–44.

Hinton, R.H., 1969, Purification of plasma-membrane fragments, in: *Subcellular Components: Preparation and Fractionation* (G.D. Birnie, ed.), pp. 119–156, Butterworths, London.

Hinton, R.H., Burge, M.L.E., and Hartman, G.C., 1969, Sucrose interference in the assay of enzymes and proteins, *Anal. Biochem.* **29:**248–256.

Hogeboom, G.H., 1955, Fractionation of cell components of animal tissues, in: *Methods in Enzymology*, Vol. I (S.P. Colowick and N.O. Kaplan, eds.), pp. 16–19, Academic Press, New York.

Hogeboom, G.H., Claude, A., and Hotchkiss, R.D., 1946, The distribution of cytochrome oxidase and succinoxidase in the cytoplasm of the mammalian liver cell, *J. Biol. Chem.* **165:**615–629.

Hogeboom, G.H., Schneider, W.C., and Palade, G.E., 1948, Cytochemical studies of mammalian tissues. I. Isolation of intact mitochondria from rat liver: Some biochemical properties of mitochondria and submicroscopic particulate material, *J. Biol. Chem.* **172:**619–635.

Hogeboom, G.H., Schneider, W.C., and Striebich, M.J., 1952, Cytochemical studies. V. On the isolation and biochemical properties of liver cell nuclei, *J. Biol. Chem.* **196:**111–120.

Holter, H., and Max Moller, K., 1958, A substance for aqueous density gradients, *Exp. Cell Res.* **15:**631–632.

Holtzman, S.G., 1974, Interactions of pentazocine and naloxone on the monoamine content of discrete regions of the rat brain, *Biochem. Pharmacol.* **23:**3029–3035.

Hydén, M., 1960, The neuron, in: *The Cell*, Vol. IV (J. Brachet and A.E. Mirsky, eds.), pp. 215–323, Academic Press, New York.

Jobsis, F.F., 1963, A study of preparative procedures for brain mitochondria, *Biochim. Biophys. Acta* **74:**60–68.

Johnston, I.R., and Mathias, A.P., 1969, The biochemical properties of nuclei fractionated by zonal centrifugation, in: *Subcellular Components: Preparation and Fractionation* (G.D. Birnie, ed.), pp. 53–75, Butterworths, London.

Jones, D.G., 1967, An electron microscope study of subcellular fractions from *Octopus* brain, *J. Cell Sci.* **2:**573–586.

Jones, D.G., 1972, On the ultrastructure of the synapse: The synaptosome as a morphological tool, in: *The Structure and Function of Nervous Tissue*, Vol. VI (G.H. Bourne, ed.), pp. 81–129, Academic Press, New York.

Kahler, H., and Lloyd, B.J., Jr., 1951, Sedimentation of polystyrene latex in a swinging-tube rotor, *J. Phys. Colloid Chem.* **55:**1344–1350.

Kato, T., and Kurokawa, M., 1967, Isolation of cell nuclei from the mammalian cerebral cortex and their assortment on a morphological basis, *J. Cell Biol.* **32:**649–662.

Kennedy, E.P., 1967, Some recent developments in the biochemistry of membranes, in: *The Neurosciences: A Study Program* (G.C. Quarton, T. Melnechuk, and F.O. Schmitt, eds.), pp. 271–280, Rockefeller University Press, New York.

Kirkpatrick, J.B., Hyams, L., Thomas, V.L., and Howley, P.M., 1970, Purification of intact microtubules from brain, *J. Cell Biol.* **47:**334–394.

Koenig, E., 1972, Ribonucleic acid of nervous tissue, in: *The Structure and Function of Nervous Tissue*, Vol. IV (G.H. Bourne, ed.), pp. 179–214, Academic Press, New York.

Koenig, H., 1969, Lysosomes, in: *Handbook of Neurochemistry*, Vol. II (A. Lajtha, ed.), pp. 255–301, Plenum Press, New York.

Koenig, H., 1974, The isolation of lysosomes from brain, in: *Methods in Enzymology*, Vol. XXXI (S. Fleischer and L. Packer, eds.), pp. 457–477, Academic Press, New York.

Koenig, H., Gaines, D., McDonald, T., Gray, R., and Scott, I., 1964, Studies of brain lysosomes. I. Subcellular distribution of five acid hydrolases, succinate dehydrogenase and gangliosides in rat brain, *J. Neurochem.* **11:**729–743.

Korey, S.R., Orchan, M., and Brotz, M., 1958, Studies of white matter. I. Chemical constitution and respiration of neuroglial and myelin enriched fractions of white matter, *J. Neuropathol. Exp. Neurol.* **17:**430–438.

Koslow, S.H., Racagni, G., and Costa, E., 1974, Mass fragmentographic measurement of norepinephrine, dopamine, serotonin and acetylcholine in seven discrete nuclei of the rat tel-diencephalon, *Neuropharmacology* **13:**1123–1130.

Kuczenski, R., Segal, D.S., and Mandell, R.J., 1975, Regional and subcellular distribution and kinetic properties of rat brain choline acetyltransferase—some functional considerations, *J. Neurochem.* **24:**39–45.

Lang, K., and Siebert, G., 1952, Untersuchungen über Stoffwechselprozesse in isolierten Zellkernen: Methodik der Gewinnung reiner intakter Zellkernen in beliebigen Massstab, *Biochemie* **2:**322–360.

Lehninger, A.L., 1967, Cell organelles: The mitochondrion, in: *The Neurosciences: A Study Program* (G.C. Quarton, T. Melnechuk, and F.O. Schmitt, eds.), pp. 91–100, Rockefeller University Press, New York.

Lehninger, A.L., 1970, Mitochondria and their neurofunction, in: *The Neurosciences: Second Study Program* (F.O. Schmitt, ed.), pp. 827–839, Rockefeller University Press, New York.

Levi, G., Bertollini, A., Chen, J., and Raiteri, M., 1974, Regional differences in the synaptosomal uptake of ^{3}H-gamma-aminobutyric acid and ^{14}C-glutamate and possible role of exchange processes, *J. Pharmacol. Exp. Ther.* **188:**429–438.

Lindquist, M., Kehr, W., and Carlsson, A., 1974, Effect of pentobarbitone and diethyl ether on the synthesis of monoamines in rat brain, *Naunyn-Schmiedeberg's Arch. Pharmacol.* **284:**263–277.

Mahler, H.R., and Brown, B.J., 1968, Protein synthesis by cerebral cortex polysomes: Characterization of the system, *Arch. Biochem. Biophys.* **125:**387–400.

Marchbanks, R.M., 1974, Isolation and study of synaptic vesicles, in: *Research Methods in Neurochemistry*, Vol. 2 (N. Marks and R. Rodnight, eds.), pp. 79–98, Plenum Press, New York.

Marks, N., 1974, Preparation of brain mitochondrial membranes, in: *Research Methods in Neurochemistry*, Vol. 2 (N. Marks and R. Rodnight, eds.), pp. 54–77, Plenum Press, New York.

Matthies, H., Rauca, C., and Liebmann, H., 1974, Changes in the acetylcholine contents of different brain regions of the rat during a learning experiment, *J. Neurochem.* **23:**1109–1113.

McCall, J.S., and Potter, B.J., 1973, *Ultracentrifugation*, Bailliere Tindall, London.

McEwen, B.S., and Zigmond, R.E., 1972, Isolation of brain cell nuclei, in: *Research Methods in Neurochemistry*, Vol. 1 (N. Marks and R. Rodnight, eds.), pp. 134–161, Plenum Press, New York.

McMorris, F.A., Nelson, P.G., and Ruddle, F.H., 1973, Contributions of clonal systems to neurobiology, *Neurosci. Res. Program Bull.* **11:**411–536.

Medina, M.A., Jones, D.J., Stavinoha, W.B., and Ross, D.H., 1975, The levels of labile intermediary metabolites in mouse brain following rapid tissue fixation with microwave irradiation, *J. Neurochem.* **24:**223–227.

Michaelis, E.K., Michaelis, M.L., and Boyarsky, L.L., 1974, High-affinity glutamic acid binding to brain synaptic membranes, *Biochim. Biophys. Acta* **367:**338–348.

Moore, D.H., 1969, Gradient centrifugation, in: *Physical Techniques in Biological Research*, Vol. II (D.H. Moore, ed.), Part B, pp. 285–314, Academic Press, New York.

Moses, M.J., 1964, The nucleus and chromosomes: A cytological perspective, in: *Cytology and Cell Physiology* (G.H. Bourne, ed.), pp. 424–558, Academic Press, New York.

Nagata, Y., Mikoshiba, K., and Tsukada, Y., 1974, Neuronal cell body enriched and glial cell enriched fractions from young and adult rat brains: Preparation and morphological and biochemical properties, *J. Neurochem.* **22:**493–503.

Newkirk, R.F., Ballou, E.W., Vickers, G., and Whittaker, V.P., 1976, Comparative studies in synaptosome formation: Preparation of synaptosomes from the ventral nerve cord of the lobster (*Homarus americanus*), *Brain Res.* **101:**103–111.

Nomura, M., 1970, Assembly of ribosomes, in: *The Neurosciences: Second Study Program* (F.O. Schmitt, ed.), pp. 913–920, Rockefeller University Press, New York.

Norton, W.T., 1974, Isolation of myelin from nerve tissue, in: *Methods in Enzymology*, Vol. XXXI (S. Fleischer and L. Packer, eds.), pp. 435–474, Academic Press, New York.

Novikoff, A.B., Podber, E., Ryan, J., and Noe, E., 1953, Biochemical heterogeneity of the cytoplasmic particles isolated from rat liver homogenate, *J. Histochem. Cytochem.* **1:**27–46.

Oestreicher, A.B., and vanLeeuwen, C., 1975, Isolation and partial characterization of fractions enriched in synaptosomes from chick brain, *J. Neurochem.* **24:**251–259.

Pickels, E.G., 1943, Sedimentation in the angle centrifuge, *J. Gen. Physiol.* **26:**341–360.

Pickels, E.G., 1950, Centrifugation, in: *Biophysical Research Methods* (F.M. Uber, ed.), pp. 67–105, Interscience, New York.

Podsulo, S.E., and Norton, W.T., 1972, The bulk separation of neuroglia and neuronal perikarya, in: *Research Methods in Neurochemistry*, Vol. 1 (N. Marks and R. Rodnight, eds.), pp. 19–32, Plenum Press, New York.

Pollard, H.B., Barker, J.L., Bohr, W.A., and Dowdall, M.J., 1975, Chlorpromazine: Specific inhibition of l-noradrenaline and 5-hydroxytryptamine uptake in synaptosomes from squid brain, *Brain Res.* **85:**23–31.

Potter, V.R., 1955, Tissue homogenates, in: *Methods in Enzymology*, Vol. I (S.P. Colowick and N.O. Kaplan, eds.), pp. 10–15, Academic Press, New York.

Potter, V.R., and Elvehjem, C.A., 1936, A modified method for the study of tissue oxidations, *J. Biol. Chem.* **114:**495–504.

Rappoport, D.A., Maxcy, P., Jr., and Daginawala, H.F., 1969, Nuclei, in: *Handbook of Neurochemistry*, Vol. II (A. Lajtha, ed.), pp. 241–254, Plenum Press, New York.

Raskin, N.H., and Sokoloff, L., 1970, Alcohol dehydrogenase activity in rat brain and liver, *J. Neurochem.* **17:**1677–1687.

Reid, E., 1967, Membrane systems, in: *Enzyme Cytology* (D.B. Roodyn, ed.), pp. 321–405, Academic Press, New York.

Reid, E., 1972, Preparation of lysosome rich fractions with or without peroxisomes, in: *Subcellular Components: Preparation and Fractionation* (G.D. Birnie, ed.), pp. 93–118, Butterworths, London.

Reid, E., and Williamson, R., 1974, Centrifugation, in: *Methods in Enzymology*, Vol. XXXI (S. Fleischer and L. Packer, eds.), pp. 713–733, Academic Press, New York.

Renaud, F.L., Rowe, A.J., and Gibbons, I.R., 1968, Some properties of the protein forming the outer fibers of cilia, *J. Cell Biol.* **36:**79–90.

Rich, A., 1967, The ribosome—a biological information transducer, in: *The Neurosciences: A Study Program* (G.C. Quarton, T. Melnechuk, and F.O. Schmitt, eds.), pp. 101–112, Rockefeller University Press, New York.

Robertson, J.D., 1970, The ultrastructure of synapses, in: *The Neurosciences: Second Study Program* (F.O. Schmitt, ed.), pp. 715–728, Rockefeller University Press, New York.

Rodriguez de Lores Arnaiz, G., and DeRobertis, E.D.P., 1962, Cholinergic and non-cholinergic nerve endings in rat brain. II. Subcellular localization of monoamine oxidase and succinate dehydrogenase, *J. Neurochem.* **9:**503–508.

Rodriguez de Lores Arnaiz, G., Alberici, M., and DeRobertis, E., 1967, Ultrastructural and enzymic studies of cholinergic and non-cholinergic synaptic membranes isolated from brain cortex, *J. Neurochem.* **14:**215–225.

Roodyn, D.B., 1959, A survey of metabolic studies on isolated mammalian nuclei, *Int. Rev. Cytol.* **8:**279–344.

Roodyn, D.B., 1967, The mitochondrion, in: *Enzyme Cytology* (D.B. Roodyn, ed.), pp. 103–178, Academic Press, New York.

Roodyn, D.B., 1972, Sonic methods for isolating nuclei, in: *Subcellular Components: Preparation and Fractionation* (G.D. Birnie, ed.), pp. 15–51, Butterworths, London.

Roots, B.I., and Johnston, P.V., 1972, Nervous system cell preparations: Microdissection and micromanipulation, in: *Research Methods in Neurochemistry*, Vol. I (N. Marks and R. Rodnight, eds.), pp. 3–17, Plenum Press, New York.

Rose, S.P.R., 1969, Neurons and glia: Separation techniques and biochemical interrelationships, in: *Handbook of Neurochemistry*, Vol. II (A. Lajtha, ed.), pp. 183–193, Plenum Press, New York.

Rose, S.P.R., and Sinha, A.K., 1974, Incorporation of amino acids into proteins in neuronal and neuropil fractions of rat cerebral cortex: Presence of a rapidly labelled neuronal fraction, *J. Neurochem.* **23:**1065–1076.

Schachman, H.K., 1959, *Ultracentrifugation in Biochemistry*, Academic Press, New York.

Schmidt, M.J., Schmidt, D.E., and Robison, G.A., 1971, Cyclic adenosine monophosphate in brain areas: Microwave irradiation as a means of tissue fixation, *Science* **173:**1142–1143.

Schneider, W.C., 1946, Intracellular distribution of enzymes. 1. The distribution of succinic dehydrogenase, cytochrome oxidase, adenosinetriphosphatase, and phosphorus compounds in normal rat tissues, *J. Biol. Chem.* **165:**585–593.

Schneider, W.C., 1948, Intracellular distribution of enzymes. III. The oxidation of octanoic acid by rat liver fractions, *J. Biol. Chem.* **176:**259–266.

Schneider, W.C., 1972, Methods for the isolation of particulate components of the cell, in: *Manometric and Biochemical Techniques* (W.W. Umbrait, R.H. Burris, and J.F. Stauffer, eds.), pp. 196–212. Burgess, Minneapolis.

Schneider, W.C., and Hogeboom, G.H., 1951, Cytochemical studies of mammalian tissues: The isolation of cell components by differential centrifugation: A review, *Cancer Res.* **11:**1–22.

Schneider, W.C., and Kuff, E.L., 1964, Centrifugal isolation of subcellular components, in: *Cytology and Cell Physiology* (G.H. Bourne, ed.), pp. 19–89, Academic Press, New York.

Schochet, S.S., Jr., 1972, Neuronal inclusions, in: *The Structure and Function of Nervous Tissue*, Vol. IV (G.H. Bourne, ed.), pp. 129–177, Academic Press, New York.

Sellinger, O.Z., and Azeurra, J.M., 1974, Bulk separation of neuronal cell bodies and glial cells in the absence of added digestive enzymes, in: *Research Methods in Neurochemistry*, Vol. 2 (N. Marks and R. Rodnight, eds.), pp. 3–38, Plenum Press, New York.

Shantha, T.R., Manocha, S.L., Bourne, G.H., and Kappers, J.A., 1969, The morphology and cytology of neurons, in: *The Structure and Function of Nervous Tissue*, Vol. II (G.H. Bourne, ed.), pp. 1–67, Academic Press, New York.

Shelanski, M.L., 1974, Methods for the neurochemical study of mictotubules, in: *Research Methods in Neurochemistry*, Vol. 2 (N. Marks and R. Rodnight, eds.), pp. 281–300, Academic Press, New York.

Shelanski, M.L., and Feit, H., 1972, Filaments and tubules in the nervous system, in: *The Structure and Function of Nervous Tissue*, Vol. VI (G.H. Bourne, ed.), pp. 47–80, Academic Press, New York.

Shelanski, M.L., and Taylor, E.W., 1967, Isolation of a protein subunit from microtubules, *J. Cell Biol.* **34:**549–554.

Shelanski, M.L., and Taylor, E.W., 1968, Properties of the protein subunit of central-pair and outer-doublet microtubules of sea urchin flagella, *J. Cell Biol.* **38:**304–315.

Shelanski, M.L., Albert, S., DeVries, G.H., and Norton, W.T., 1971, Isolation of filaments from brain, *Science* **174:**1242–1245.

Siakotos, A.N., 1974*a*, The isolation of nuclei from normal human and bovine brain, in: *Methods in Enzymology*, Vol. XXI (S. Fleischer and L. Packer, eds.), pp. 452–457, Academic Press, New York.

Siakotos, A.N., 1974*b*, Procedures for the isolation of brain lipopigments: Ceroid and lipofuscin, in: *Methods in Enzymology*, Vol. XXXI (S. Fleischer and L. Packer, eds.), pp. 478–485, Academic Press, New York.

Sjöstrand, F.S., 1964, The endoplasmic reticulum, in: *Cytology and Cell Physiology* (G.H. Bourne, ed.), pp. 311–375, Academic Press, New York.

Smith, S.J., McLaughlin, P.J., and Zagon, I.S., 1976, Granule neurons and their significance in preparations of isolated brain cell nuclei, *Brain Res.* **103:**345–349.

Spohn, M., and Davison, A.N., 1972, Separation of myelin fragments from the central nervous system, in: *Research Methods in Neurochemistry*, Vol. 1 (N. Marks and R. Rodnight, eds.), pp. 33–43, Plenum Press, New York.

Stahl, W.L., and Swanson, P.D., 1975, Effects of freezing and storage on subcellular fractionation of guinea pig and human brain, *Neurobiology* **5:**393–400.

Stahl, W.L., Smith, J.C., Napolitano, Z.M., and Basford, R.E., 1963, Brain mitochondria. I. Isolation of bovine brain mitochondria, *J. Cell Biol.* **19:**293–307.

Stavinoha, W.B., Weintraub, S.T., and Modak, A.T., 1973, The use of microwave heating to inactivate cholinesterase in the rat brain prior to analysis for acetylcholine, *J. Neurochem.* **20:**361–371.

Strauss, W., 1967, Lysosomes and related particles, in: *Enzyme Cytology* (D.B. Roodyn, ed.), pp. 239–319, Academic Press, New York.

Svedberg, T., and Pedersen, K.O., 1940, *The Ultracentrifuge*, Oxford University Press, London.

Swaab, D.F., 1971, Pitfalls in the use of rapid freezing for stopping brain and spinal cord metabolism in rat and mouse, *J. Neurochem.* **18:**2085–2092.

Tamir, H., Rapport, M.M., and Roizon, L., 1974, Preparation of synaptosomes and vesicles with sodium diatrizoate, *J. Neurochem.* **23:**943–949.

Tata, J.R., 1969, Preparation and properties of microsomal and sub-microsomal fractions from secretory and non-secretory tissues, in: *Subcellular Components: Preparation and Fractionation* (G.D. Birnie, ed.), pp. 185–213, Butterworths, London.

Tolani, A.J., and Talwar, G.P., 1963, Differential metabolism of various brain regions (biochemical heterogeneity of mitochondria), *Biochem. J.* **88:**357–362.

Trautman, R., 1964, Ultracentrifugation, in: *Instrumental Methods of Experimental Biology* (D.W. Newman, ed.), pp. 211–297, Maxmillan, New York.

Veech, R.L., and Hawkins, R.A., 1974, Brain blowing: A technique for *in vivo* study of brain metabolism, in: *Research Methods in Neurochemistry*, Vol. 2 (N. Marks and R. Rodnight, eds.), pp. 171–182, Plenum Press, New York.

Veech, R.L., Harris, R.L., Veloso, D., and Veech, E.H., 1973, Freeze–blowing: A new technique for the study of brain *in vivo*, *J. Neurochem.* **20:**183–188.

Vrba, R., 1967, Assimilation of glucose carbon in subcellular rat brain particles *in vivo* and the problems of axoplasmic flow, *Biochem. J.* **105:**927–936.

Vinograd, J., 1963, Sedimentation equilibrium in a buoyant density gradient, in: *Methods in Enzymology*, Vol. VI (S.P. Colowick and N.O. Kaplan, eds.), pp. 854–870, Academic Press, New York.

Vrensen, G., and De Groot, D., 1973, Quantitative stereology of synapses: A critical investigation, *Brain Res.* **58**:25–35.
Weisenberg, R.C., Borisy, G.G., and Taylor, E.W., 1968, The colchicine-binding protein of mammalian brain and its relation to microtubules, *Biochemistry* **7**:4466–4479.
Whittaker, V.P., 1959, The isolation and characterization of acetylcholine-containing particles from brain, *Biochem. J.* **72**:694–706.
Whittaker, V.P., 1965, The application of subcellular fractionation techniques to the study of brain function, *Prog. Biophys. Mol. Biol.* **15**:39–95.
Whittaker, V.P., 1968, The morphology of fractions of rat forebrain synaptosomes separated on continuous sucrose density gradients, *Biochem. J.* **106**:412–417.
Whittaker, V.P., 1969, The synaptosome, in: *Handbook of Neurochemistry*, Vol. II (A. Lajtha, ed.), pp. 327–364, Plenum Press, New York.
Whittaker, V.P., 1970, The investigation of synaptic function by means of subcellular fractionation techniques, in: *The Neurosciences: Second Study Program* (F.O. Schmitt, ed.), pp. 761–768, Rockefeller University Press, New York.
Whittaker, V.P., and Barker, L.A., 1971, The subcellular fractionation of brain tissue with special reference to the preparation of synaptosomes and their component organelles, in: *Methods of Neurochemistry*, Vol. 2 (R. Fried, ed.), pp. 1–52, Marcel Dekker, New York.
Whittaker, V.P., Michaelson, I.A., and Kirkland, P.J.A., 1964, The separation of synaptic vesicles from nerve-ending particles (synaptosomes), *Biochem. J.* **90**:293–303.
Williams, J.W. (ed.), 1963, *Ultracentrifugal Analysis in Theory and Experiment*, Academic Press, New York.
Wilson, W.S., Schulz, R.A., and Cooper, J.R., 1973, The isolation of cholinergic synaptic vesicles from bovine superior cervical ganglion and estimation of their acetylcholine content, *J. Neurochem.* **20**:659–667.
Yamamura, H.I., and Snyder, S.H., 1974, Postsynaptic localization of muscarinic cholinergic receptor binding in rat hippocampus, *Brain Res.* **78**:320–326.
Zomzely, C.E., Roberts, S., and Rapaport, D., 1964, Regulation of amino acid incorporation into protein of microsomal and ribosomal preparations of rat cerebral cortex, *J. Neurochem.* **11**:567–582.
Zomzely, C.E., Roberts, S., Gruber, C.P., and Brown, D.M., 1968, Cerebral protein synthesis. II. Instability of cerebral messenger ribonucleic acid–ribosome complexes, *J. Biol. Chem.* **243**:5396–5409.
Zomzely, C.E., Roberts, S., Peache, S., and Brown, D.M., 1971, Cerebral protein synthesis. III. Developmental alterations in the stability of cerebral messenger ribonucleic acid–ribosome complexes, *J. Biol. Chem.* **246**:2097–2103.
Zomzely-Neurath, C., and Roberts, S., 1972, Brain ribosomes, in: *Research Methods in Neurochemistry*, Vol. 1 (N. Marks and R. Rodnight, eds.), pp. 95–137, Plenum Press, New York.

Selected Bibliography of Fractionation Literature

Whole Cells: Neurons and Glia

Neurobiological Reviews

Bodian, D., 1967, Neurons, circuits and neuroglia, in: *The Neurosciences: A Study Program* (G.C. Quarton, T. Melnechuk, and F.O. Schmitt, eds.), pp. 6–23, Rockefeller University Press, New York.
Bunge, R.P., 1970, Structure and function of neuroglia: Some recent observations, in: *The Neurosciences: Second Study Program* (F.O. Schmitt, ed.), pp. 782–797, Rockefeller University Press, New York.
Glees, P., and Meller, K., 1968, Morphology of neuroglia, in: *The Structure and Function of Nervous Tissue*, Vol. I (G.H. Bourne, ed.), pp. 301–323, Academic Press, New York.

Hydén, M., 1960, The neuron, in: *The Cell*, Vol. IV (J. Brachat and A.E. Mirsky, eds.), pp. 215–323, Academic Press, New York.

Podsulo, S.E., and Norton, W.T., 1972, The bulk separation of neuroglia and neuronal perikarya, in: *Research Methods in Neurochemistry*, Vol 1 (N. Marks and R. Rodnight, eds.), pp. 19–32, Plenum Press, New York.

Sellinger, O.Z., and Azeurra, J.M., 1974, Bulk separation of neuronal cell bodies and glial cells in the absence of added digestive enzymes, in: *Research Methods in Neurochemistry*, Vol 2 (N. Marks and R. Rodnight, eds.), pp. 3–38, Plenum Press, New York.

Shantha, T.R., Manocha, S.L., Bourne, G.H., and Kappers, J.A., 1969, The morphology and cytology of neurons, in: *The Structure and Function of Nervous Tissue*, Vol. II (G.H. Bourne, ed.), pp. 1–67, Academic Press, New York.

Original Neurobiological Papers

Cotman, C.W., Herschman, H., and Taylor, D., 1971, Subcellular fractionation of cultured glial cells, *J. Neurobiol.* **2**:169–180.

Hamberger, A., Blomstrand, C., and Lehninger, A.L., 1970, Comparative studies on mitochondria isolated from neuron-enriched and glia-enriched fractions of rabbit and beef brain, *J. Cell Biol.* **45**:221–234.

Nagata, Y., Mikoshiba, K., and Tsukada, Y., 1974, Neuronal cell body enriched and glial cell enriched fractions from young and adult rat brains: Preparation and morphological and biochemical properties, *J. Neurochem.* **22**:493–503.

Nuclei

General Reviews

Allfrey, V., and Mirsky, A.E., 1959, Biochemical properties of the isolated nucleus, in: *Subcellular Particles* (T. Hayashi, ed.), pp. 136–207, Ronald Press, New York.

Georgiev, G.P., 1967, The nucleus, in: *Enzyme Cytology* (D.B. Roodyn, ed.), pp. 27–100, Academic Press, New York.

Johnston, I.R., and Mathias, A.P., 1969, The biochemical properties of nuclei fractionated by zonal centrifugation, in: *Subcellular Components: Preparation and Fractionation* (G.D. Birnie, ed.), pp. 53–75, Butterworths, London.

Moses, M.J., 1964, The nucleus and chromosomes: A cytological perspective, in: *Cytology and Cell Physiology* (G.H. Bourne, ed.), pp. 424–558, Academic Press, New York.

Roodyn, D.B., 1959, A survey of metabolic studies on isolated mammalian nuclei, *Int. Rev. Cytol.* **8**:279–344.

Roodyn, D.B., 1972, Sonic methods for isolating nuclei, in: *Subcellular Components: Preparation and Fractionation* (G.D. Birnie, ed.), pp. 15–51, Butterworths, London.

Neurobiological Reviews

McEwen, B.S., and Zigmond, R.E., 1972, Isolation of brain cell nuclei, in: *Research Methods in Neurochemistry*, Vol. 1 (N. Marks and R. Rodnight, eds.), pp. 134–161, Plenum Press, New York.

Rappoport, D.A., Maxcy, P., Jr., and Daginawala, H.F., 1969, Nuclei, in: *Handbook of Neurochemistry*, Vol. II (A. Lajtha, ed.), pp. 241–254, Plenum Press, New York.

Siakotos, A.N., 1974*a*, The isolation of nuclei from normal human and bovine brain, in: *Methods in Enzymology*, Vol. XXI (S. Fleischer and L. Packer, eds.), pp. 452–457, Academic Press, New York.

Siakotos, A.N., 1974*b*, Procedures for the isolation of brain lipopigments: Ceroid and lipofuscin, in: *Methods in Enzymology*, Vol. XXXI (S. Fleischer and L. Packer, eds.), pp. 478–485, Academic Press, New York.

Original Neurobiological Papers

Smith, S.J., McLaughlin, P.J., and Zagon, I.S., 1976, Granule neurons and their significance in preparations of isolated brain cell nuclei, *Brain Res.* **103:**345–349.

Mitochondria

General Reviews

Bourne, G.H., and Tewari, H.B., 1964, Mitochondria and the Golgi complex, in: *Cytology and Cell Physiology* (G.H. Bourne, ed.), pp. 377–421, Academic Press, New York.
Chappel, J.B., and Hansford, R.G., 1972, Preparation of mitochondria from animal tissues and yeast, in: *Subcellular Components: Preparation and Fractionation* (G.D. Birnie, ed.), pp. 77–91, Butterworths, London.
Green, D.E., 1959, Mitochondrial structure and function, in: *Subcellular Particles* (T. Hayashi, ed.), pp. 84–103, Ronald Press, New York.
Lehninger, A.L., 1967, Cell organelles: The mitochondrion, in: *The Neurosciences: A Study Program* (G.C. Quarton, T. Melnechuk, and F.O. Schmitt, eds.), pp. 91–100, Rockefeller University Press, New York.
Roodyn, D.B., 1967, The mitochondrion, in: *Enzyme Cytology* (D.B. Roodyn, ed.), pp. 103–178, Academic Press, New York.

Neurobiological Reviews

Abood, L.G., 1969, Brain mitochondria, in: *Handbook of Neurochemistry* (A. Lajtha, ed.), Vol. II, pp. 303–326, Plenum Press, New York.
Aldridge, W.N., 1957, Liver and brain mitochondria, *Biochem. J.* **67:**423–431.
Biesold, D., 1974, Isolation of brain mitochondria, in: *Research Methods in Neurochemistry*, Vol. 2 (N. Marks and R. Rodnight, eds.), pp. 39–52, Plenum Press, New York.
Lehninger, A.L., 1970, Mitochondria and their neurofunction, in: *The Neurosciences: Second Study Program* (F.O. Schmitt, ed.), pp. 827–839, Rockefeller University Press, New York.
Marks, N., 1974, Preparation of brain mitochondrial membranes, in: *Research Methods in Neurochemistry*, Vol. 2 (N. Marks and R. Rodnight, eds.), pp. 54–77, Plenum Press, New York.

Original Neurobiological Papers

Cotman, C.W., 1968, Doctoral dissertation, Department of Chemistry, Indiana University, Bloomington, Indiana.
Hamberger, A., Blomstrand, C., and Lehninger, A.L., 1970, Comparative studies on mitochondria isolated from neuron-enriched and glia-enriched fractions of rabbit and beef brain, *J. Cell Biol.* **45:**221–234.
Whittaker, V.P., 1968, The morphology of fractions of rat forebrain synaptosomes separated on continuous sucrose density gradients, *Biochem. J.* **106:**412–417.

Lysosomes

General Reviews

deDuve, C., 1959, Lysosomes, a new group of cytoplasmic particles, in: *Subcellular Particles* (T. Hayashi, ed.), pp. 128–159, Ronald Press, New York.

Reid, E., 1972, Preparation of lysosome rich fractions with or without peroxisomes, in: *Subcellular Components: Preparation and Fractionation* (G.D. Birnie, ed.), pp. 93–118, Butterworths, London.

Strauss, W., 1967, Lysosomes and related particles, in: *Enzyme Cytology* (D.B. Roodyn, ed.), pp. 239–319, Academic Press, New York.

Neurobiological Reviews

Koenig, H., 1969, Lysosomes, in: *Handbook of Neurochemistry*, Vol. II (A. Lajtha, ed.), pp. 255–301, Plenum Press, New York.

Koenig, H., 1974, The isolation of lysosomes from brain, in: *Methods in Enzymology*, Vol XXXI (S. Fleischer and L. Packer, eds.), pp. 457–477, Academic Press, New York.

Original Neurobiological Papers

Koenig, H., Gaines, D., McDonald, T., Gray, R., and Scott, I., 1964, Studies of brain lysosomes. I. Subcellular distribution of five acid hydrolases, succinate dehydrogenase and gangliosides in rat brain, *J. Neurochem.* **11**:729–743.

Synaptosomes

Neurobiological Reviews

Bloom, F.E., 1970, Correlating structure and function of synaptic ultrastructure, in: *The Neurosciences: Second Study Program* (F.O. Schmitt, ed.), pp. 729–747, Rockefeller University Press, New York.

Bradford, H.F., 1972, Cerebral cortex slices and synaptosomes: *In vitro* approaches to brain metabolism, in: *Methods of Neurochemistry*. Vol. 3 (R. Fried, ed.), pp. 155–202, Marcel Dekker, New York.

DeRobertis, E., and Rodriguez de Lores Arnaiz, G., 1969, Structural components of the synaptic region, in: *Handbook of Neurochemistry*, Vol. II (A. Lajtha, ed.), pp. 365–392, Plenum Press, New York.

Jones, D.G., 1972, On the ultrasructure of the synapse: The synaptosome as a morphological tool, in: *The Structure and Function of Nervous Tissue*, Vol. VI (G.H. Bourne, ed.), pp. 81–129, Academic Press, New York.

Robertson, J.D., 1970, The ultrastructure of synapses, in: *The Neurosciences: Second Study Program* (F.O. Schmitt, ed.), pp. 715–728, Rockefeller University Press, New York.

Whittaker, V.P., 1969, The synaptosome, in: *Handbook of Neurochemistry*, Vol. II (A. Lajtha, ed.), pp. 327–364, Plenum Press, New York.

Whittaker, V.P., 1970, The investigation of synaptic function by means of subcellular fractionation techniques, in: *The Neurosciences: Second Study Program* (F.O. Schmitt, ed.), pp. 761–768, Rockefeller University Press, New York.

Whittaker, V.P., and Barker, L.A., 1971, The subcellular fractionation of brain tissue with special reference to the preparation of synaptosomes and their component organelles, in: *Methods of Neurochemistry*, Vol. 2 (R. Fried, ed.), pp. 1–52, Marcel Dekker, New York.

Original Neurobiological Papers

Autilio, L.A., Appel, S.H., Pettis, P., and Gambetti, P.L., 1968, Biochemical studies of synapses *in vitro*. I. Protein synthesis, *Biochemistry* **7**:2615–2622.

Babitch, J.A., Breithaupt, T.B., Chiu, T.-C., Garadi, R., and Helseth, D.L., 1976, Preparation of chick brain synaptosomes and synaptosomal membranes, *Biochim. Biophys. Acta* **433**:75–89.

Cotman, C.W., 1968, Doctoral dissertation, Department of Chemistry, Indiana University, Bloomington, Indiana.

Cotman, C.W., 1974, Isolation of synaptosomal and synaptic plasma membrane fractions, in: *Methods in Enzymology*, Vol. XXXI (S. Fleischer and L. Packer, eds.), pp. 445–452, Academic Press, New York.

Newkirk, R.F., Ballou, E.W., Vickers, G., and Whittaker, V.P., 1976, Comparative studies in synaptosome formation: Preparation of synaptosomes from the ventral nerve cord of the lobster (*Homarus americanus*), *Brain Res.* **101:**103–111.

Oestreicher, A.B., and vanLeeuwen, C., 1975, Isolation and partial characterization of fractions enriched in synaptosomes from chick brain, *J. Neurochem.* **24:**251–259.

Pollard, H.B., Barker, J.L., Bohr, W.A., and Dowdall, M.J., 1975, Chlorpromazine: Specific inhibition of 1-noradrenaline and 5-hydroxytryptamine uptake in synaptosomes from squid brain, *Brain Res.* **85:**23–31.

Tamir, H., Rapport, M.M., and Roizin, L., 1974, Preparation of synaptosomes and vesicles with sodium diatrizoate, *J. Neurochem.* **23:**943–949.

Whittaker, V.P., 1968, The morphology of fractions of rat forebrain synaptosomes separated on continuous sucrose density gradients, *Biochem. J.* **106:**412–417.

Yamamura, H.I., and Snyder, S.H., 1974, Postsynaptic localization of muscarinic cholinergic receptor binding in rat hippocampus, *Brain Res.* **78:**320–326.

Plasma Membranes: Synaptic and Glial Membranes

General Reviews

Brown, F., and Danielli, J.F., 1964, The cell surface and cell physiology, in: *Cytology and Cell Physiology* (G.H. Bourne, ed.), pp. 239–310, Academic Press, New York.

Hinton, R.H., 1969, Purification of plasma-membrane fragments, in: *Subcellular Components: Preparation and Fractionation* (G.D. Birnie, ed.), pp. 119–156, Butterworths, London.

Kennedy, E.P., 1967, Some recent developments in the biochemistry of membranes, in: *The Neurosciences: A Study Program* (G.C. Quarton, T. Melnechuk, and F.O. Schmitt, eds.), pp. 271–280, Rockefeller University Press, New York.

Reid, E., 1967, Membrane systems, in: *Enzyme Cytology* (D.B. Roodyn, ed.), pp. 321–405, Academic Press, New York.

Neurobiological Reviews

Cotman, C.W., 1974, Isolation of synaptosomal and synaptic plasma membrane fractions, in: *Methods in Enzymology*, Vol. XXXI (S. Fleischer and L. Packer, eds.), pp. 445–452, Academic Press, New York.

DeRobertis, E., and Rodriguez de Lores Arnaiz, G., 1969, Structural components of the synaptic region, in: *Handbook of Neurochemistry*, Vol. II (A. Lajtha, ed.), pp. 365–392, Plenum Press, New York.

Fernandez-Moran, H., 1967, Membrane ultrastructure in nerve cells, in: *The Neurosciences: A Study Program* (G.C. Quarton, T. Melnechuk, and F.O. Schmitt, eds.), pp. 281–304, Rockefeller University Press, New York.

Original Neurobiological Papers

Babitch, J.A., Breithaupt, T.B., Chiu, T.-C., Garadi, R., and Helseth, D.L., 1976, Preparation of chick brain synaptosomes and synaptosomal membranes, *Biochim. Biophys. Acta* **433:**75–89.

Cotman, C.W., 1968, Doctoral dissertation, Department of Chemistry, Indiana University, Bloomington, Indiana.

Cotman, C.W., Mahler, H.R., and Anderson, N.G., 1968, Isolation of a membrane fraction enriched in nerve-end membranes from rat brain by zonal centrifugation, *Biochim. Biophys. Acta* **163:**272–275.

Cotman, C.W., Brown, D.H., Harrell, B.W., and Anderson, N.G., 1970, Analytical differential

centrifugation: An analysis of the sedimentation properties of some synaptosomes, mitochondria and lysosomes from rat brain homogenates, *Arch. Biochem. Biophys.* **136:**436–447.
Gurd, J.W., Jones, L.R., Mahler, H.R., and Moore, W.J., 1974, Isolation and partial characterization of rat brain synaptic plasma membranes, *J. Neurochem.* **22:**281–290.
Michaelis, E.K., Michaelis, M.L., and Boyarsky, L.L., 1974, High-affinity glutamic acid binding to brain synaptic membranes, *Biochim. Biophys. Acta* **367:**338–348.
Rodriguez de Lores Arnaiz, G., Alberici, M., and DeRobertis, E., 1967, Ultrastructural and enzymic studies of cholinergic and non-cholinergic synaptic membranes isolated from brain cortex, *J. Neurochem.* **14:**215–225.

Synaptic Vesicles

Neurobiological Reviews

Bloom, F.E., 1970, Correlating structure and function of synaptic ultrastructure, in: *The Neurosciences: Second Study Program* (F.O. Schmitt, ed.), pp. 729–747, Rockefeller University Press, New York.
DeRobertis, E., and Rodriguez de Lores Arnaiz, G., 1969, Structural components of the synaptic region, in: *Handbook of Neurochemistry*, Vol. II (A. Lajtha, ed.), pp. 365–392, Plenum Press, New York.
Marchbanks, R.M., 1974, Isolation and study of synaptic vesicles, in: *Research Methods in Neurochemistry*, Vol. 2 (N. Marks and R. Rodnight, eds.), pp. 79–98, Plenum Press, New York.
Whittaker, V.P., 1970, The investigation of synaptic function by means of subcellular fractionation techniques, in: *The Neurosciences: Second Study Program* (F.O. Schmitt, ed.), pp. 761–768, Rockefeller University Press, New York.

Original Neurobiological Papers

Cotman, C.W., 1968, Doctoral dissertation, Department of Chemistry, Indiana University, Bloomington, Indiana.
Tamir, H., Rapport, M.M., and Roizin, L., 1974, Preparation of synaptosomes and vesicles with sodium diatrizoate, *J. Neurochem.* **23:**943–949.
Whittaker, V.P., Michaelson, I.A., and Kirkland, P.J.A., 1964, The separation of synaptic vesicles from nerve-ending particles (synaptosomes), *Biochem. J.* **90:**293–303.
Wilson, W.S., Schulz, R.A., and Cooper, J.R., 1973, The isolation of cholinergic synaptic vesicles from bovine superior cervical ganglion and estimation of their acetylcholine content, *J. Neurochem.* **20:**659–667.

Endoplasmic Reticulum; Ribosomes, Polysomes

General Reviews

Birnie, G.B., Fox, S.M., and Harvey, D.R., 1972, Separation of polysomes, ribosomes and ribosomal subunits in zonal rotors, in: *Subcellular Components: Preparation and Fractionation* (G.D. Birnie, ed.), pp. 235–250, Butterworths, London.
Bonanou-Tzedaki, S.A., and Arnstein, H.R.V., 1972, Isolation of animal polysomes and ribosomes, in: *Subcellular Components: Preparation and Fractionation* (G.D. Birnie, ed.), pp. 215–234, Butterworths, London.
Elson, D., 1967, Ribosomal enzymes, in: *Enzyme Cytology* (D.B. Roodyn, ed.), pp. 407–473, Academic Press, New York.
Rich, A., 1967, The ribosome—a biological information transducer, in: *The Neurosciences: A Study Program* (G.C. Quarton, T. Melnechuk, and F.O. Schmitt, eds.), pp. 101–112, Rockefeller University Press, New York.

Sjöstrand, F.S., 1964, The endoplasmic reticulum, in: *Cytology and Cell Physiology* (G.H. Bourne, ed.), pp. 311–375, Academic Press, New York.

Neurobiological Reviews

Davis, B.D., 1970, Recent advances in understanding ribosomal action, in: *The Neurosciences: Second Study Program* (F.O. Schmitt, ed.), pp. 920–927, Rockefeller University Press, New York.

Nomura, M., 1970, Assembly of ribosomes, in: *The Neurosciences: Second Study Program* (F.O. Schmitt, ed.), pp. 913–920, Rockefeller University Press, New York.

Zomzely-Neurath, C., and Roberts, S., 1972, Brain ribosomes, in: *Research Methods in Neurochemistry*, Vol. 1 (N. Marks and R. Rodnight, eds.), pp. 95–137, Plenum Press, New York.

Original Neurobiological Papers

Acs, G., Neidle, A., and Waelsch, H., 1961, Brain ribosomes and amino acid incorporation, *Biochim. Biophys. Acta* **50:**403–404.

Campagoni, A.T., and Mahler, H.R., 1967, Isolation and properties of polyribosomes from cerebral cortex, *Biochemistry* **6:**956–967.

Koenig, E., 1972, Ribonucleic acid of nervous tissue, in: *The Structure and Function of Nervous Tissue*, Vol. IV (G.H. Bourne, ed.), pp. 179–214, Academic Press, New York.

Mahler, H.R., and Brown, B.J., 1968, Protein synthesis by cerebral cortex polysomes: Characterization of the system, *Arch. Biochem. Biophys.* **125:**387–400.

Zomzely, C.E., Roberts, S., and Rapaport, D., 1964, Regulation of amino acid incorporation into protein of microsomal and ribosomal preparations of rat cerebral cortex, *J. Neurochem.* **11:**567–582.

Zomzely, C.E., Roberts, S., Gruber, C.P., and Brown, D.M., 1968, Cerebral protein synthesis. II. Instability of cerebral messenger ribonucleic acid–ribosome complexes, *J. Biol. Chem.* **243:**5396–5409.

Zomzely, C.E., Roberts, S., Peache, S., and Brown, D.M., 1971, Cerebral protein synthesis. III. Developmental alterations in the stability of cerebral messenger ribonucleic acid–ribosome complexes, *J. Biol. Chem.* **246:**2097–2103.

Filaments, Tubules, and Other Neuronal Inclusions

Neurobiological Reviews

Schochet, S.S., Jr., 1972, Neuronal inclusions, in: *The Structure and Function of Nervous Tissue*, Vol. IV (G.H. Bourne, ed.), pp. 129–177, Academic Press, New York.

Shelanski, M.L., 1974, Methods for the neurochemical study of microtubules, in: *Research Methods in Neurochemistry*, Vol. 2 (N. Marks and R. Rodnight, eds.), pp. 281–300, Academic Press, New York.

Shelanski, M.L., and Feit, H., 1972, Filaments and tubules in the nervous system, in: *The Structure and Fundtion of Nervous Tissue*, Vol. VI (G.H. Bourne, ed.), pp. 47–80, Academic Press, New York.

Siakotos, A.N., 1974*a*, The isolation of nuclei from normal human and bovine brain, in: *Methods in Enzymology*, Vol. XXI (S. Fleischer and L. Packer, eds.), pp. 452–457, Academic Press, New York.

Siakotos, A.N., 1974*b*, Procedures for the isolation of brain lipopigments: Ceroid and lipofuscin, in: *Methods in Enzymology*, Vol. XXXI (S. Fleischer and L. Packer, eds.), pp. 478–485, Academic Press, New York.

Original Neurobiological Papers

Feit, H., and Barondes, S.H., 1970, Colchicine-binding activity in particulate fractions of mouse brain, *J. Neurochem.* **17:**1355–1364.

Feit, H., Dutton, G., Barondes, S.H., and Shelanski, M.L., 1971, Microtubule protein identification and transport to nerve endings, *J. Cell Biol.* **51:**138–147.

Kirkpatrick, J.B., Hyams, L., Thomas, V.L., and Howley, P.M., 1970, Purification of intact microtubules from brain, *J. Cell Biol.* **47:**334–394.

Renaud, F.L., Rowe, A.J., and Gibbons, I.R., 1968, Some properties of the protein forming the outer fibers of cilia, *J. Cell Biol.* **36:**79–90.

Shelanski, M.L., and Taylor, E.W., 1967, Isolation of a protein subunit from microtubules, *J. Cell Biol.* **34:**549–554.

Shelanski, M.L., Albert, S., DeVries, G.H., and Norton, W.T., 1971, Isolation of filaments from brain, *Science* **174:**1242–1245.

Weisenberg, R.C., Borisy, G.G., and Taylor, E.W., 1968, The colchicine-binding protein of mammalian brain and its relation to microtubules, *Biochemistry* **7:**4466–4479.

Myelin

Neurobiological Reviews

Norton, W.T., 1974, Isolation of myelin from nerve tissue, in: *Methods in Enzymology*, Vol. XXXI (S. Fleischer and L. Packer, eds.), pp. 435–474, Academic Press, New York.

Spohn, M., and Davison, A.N., 1972, Separation of myelin fragments from the central nervous system, in: *Research Methods in Neurochemistry*, Vol. 1 (N. Marks and R. Rodnight, eds.), pp. 33–43, Plenum Press, New York.

Original Neurobiological Papers

Autilio, L.A., Norton, W.T., and Terry, R.D., 1964, The preparation and some properties of purified myelin from the central nervous system, *J. Neurochem.* **11:**17–27.

Cotman, C.W., 1968, Doctoral dissertation, Department of Chemistry, Indiana University, Bloomington, Indiana.

Detering, N.K., and Wells, M.A., 1976, Detection of myelin in the optic nerve of young rats by sedimentation equilibrium in a CsCl gradient, *J. Neurochem.* **26:**247–252.

Soluble Phase

General Reviews

Anderson, N.G., and Green, J.G., 1967, The soluble phase of the cell, in: *Enzyme Cytology* (D.B. Roodyn, ed.), pp. 475–509, Academic Press, New York.

Tata, J.R., 1969, Preparation and properties of microsomal and sub-microsomal fractions from secretory and non-secretory tissues, in: *Subcellular Components: Preparation and Fractionation* (G.D. Birnie, ed.), pp. 185–213, Butterworths, London.

4

Polyacrylamide Gel Electrophoresis: Principles, Techniques, and Micromethods

Reinhard Rüchel, Y. Peng Loh, and Harold Gainer

1. Introduction

Ever since the discovery that electrophoretic techniques were valuable for the fractionation of mixtures of colloids (for historical commentary see Tiselius, 1937; Cann, 1968), various investigators have attempted to modify and improve these techniques for their specific purposes. The subsequent evolution of these techniques has been very rapid, and electrophoresis probably represents the most popular method for the fractionation of macromolecules used by modern-day biologists. The reason for its popularity is due to many factors. It represents a very high-resolution analytical method (i.e., provides information about the size, conformation, and charge of molecules) as well as a fractionation procedure that may be preparative. It is relatively easy to learn to use and is relatively inexpensive to introduce into the laboratory. Electrophoresis is a versatile technique, and can be used both in parallel and in series with a wide variety of other methods employed by biologists. In view of these considerations and the current interest in neurobiology concerning the roles of proteins in the structure and function of the nervous system, many neurobiologists have made use of polyacrylamide gel electrophoresis (PAGE). These methods have been particularly valuable in the analysis of structural proteins in axons (e.g., tubulin, neurofilament

---

Reinhard Rüchel • Biophysics Branch, Armed Forces Institute of Pathology, Washington, D.C. 20306. ***Y. Peng Loh*** and ***Harold Gainer*** • Section on Functional Neurochemistry, Laboratory of Developmental Neurobiology, National Institute of Child Health and Human Development, National Institutes of Health, Bethesda, Maryland 20205. Dr. Rüchel's present address is: Department of Microbiology, Institute of Hygiene, Kreuzbergring 57, D-3400 Göttingen, West Germany.

protein), the immunocytochemistry of neurons, the isolation of nerve-specific proteins, the analysis of secretory-granule-membrane and membrane-receptor proteins, and the study of the axonal transport of proteins. An elegant example of the use of PAGE in the analysis of proteins transported intraaxonally in the visual system is the work of Willard *et al.* (1974).

The major purpose of this chapter is to introduce to the naïve as well as to the more experienced neurobiologist a variety of PAGE techniques that are used relatively infrequently, but that we believe are particularly suited for studies on the nervous system. Although conventional PAGE methods are very valuable (and will be reviewed here as well), particularly for preparative needs, one of the difficulties in studying the nervous system is the necessity to fractionate and detect proteins extracted from small samples of tissue. Since the nervous system is unique in its morphological and functional heterogeneity, and correlative biochemical analysis must often be done on material ranging from single neurons to small clusters of neurons (e.g., specific ganglia and brain nuclei), we have devoted a substantial portion of this chapter to relevant micromethods.

The chapter is roughly divided into five sections. The first deals with the general concepts of electrophoresis. This is followed by a discussion of the various types of electrophoresis and the principles that underlie these different methods. The various forms of PAGE (e.g., continuous and discontinuous PAGE, isotachophoresis, isoelectric focusing, the use of nonionic and ionic detergents, and two-dimensional techniques) are particularly stressed in this section. The next section contains practical information as regards the technology of gel electrophoresis. Although this information is largely presented in the context of microelectrophoretic technology, most of it applies equally well to conventional techniques. Many of the examples that are illustrated come from our studies on single, identified neurons from the abdominal ganglion of *Aplysia californica* (Frazier *et al.*, 1967). However, application of these techniques to more complex vertebrate nervous tissues is effective as well. The final two sections are concerned with an evaluation of the limitations and caveats associated with the use of PAGE techniques and with a discussion of other methods that are easily used in conjunction with PAGE.

2. Basic Concepts

Electrophoresis refers to the migration of ions in an electrical field. Because of their different migration velocities, different ionic species in a mixture will separate and hence can be detected in different locations following an electrophoretic run. The electrophoretic velocity of an ion is determined by the applied electrical field strength (E) (i.e., potential gradient, voltage gradient, or volts/cm), which is the driving force; by the viscosity and sieving properties of the medium in which electrophoresis is taking place; and by the electrophoretic mobility of the ion itself. The free electrophoretic mobility

m_0 is considered a characteristic constant for a given ion and is defined as $m_0 = d/(t \cdot E) = V/E = Q/f$, where d/t is the migration distance in time (which is the velocity, v), Q is the effective charge of the ion, and f is the frictional resistance of the migrating ion. According to Stoke's law, the frictional resistance is defined as $f = 6\pi \cdot r \cdot v \cdot \eta$ (where r is the particle radius, v is the velocity, and η is the viscosity of the medium). Thus, the effective charge Q and the Stoke's radius r are inherent parameters of any given ionic species that determine its free mobility (in water) and its effective mobility (i.e., in a gel). The migration velocity v is further defined as $v = E \cdot m_0 = (m_0 \cdot J)/x$ where J is the current density (amperes/cm^2) and x is the specific conductivity of the medium ($\text{ohms}^{-1} \cdot \text{cm}^{-1}$). If the medium has higher viscosity than water, the free mobility m_0 is reduced to the effective mobility (m).

For electrophoretic separations performed in a gel, the relationship of m_0 and m is defined by $\log m = \log m_0 - K \cdot T$, where K is the retardation constant of the ion and T is the total gel concentration. Since K is a function of the molecular weight of the migrating ion, this constant has been employed in the evaluation of the molecular weights of ionic species by the electrophoretic technique (Ferguson, 1964; Rodbard and Chrambach, 1971; Thorun, 1971) (see also Section 5.2). For a more thorough development of these concepts, see Maurer (1971).

In biology, electrophoretic techniques have been used primarily for the separation of peptides, proteins, and nucleic acids. All these macromolecules contain charged groups and are therefore referred to as macroions. Proteins and peptides are amphoteric (amphoprotic) ions due to the ionizable groups ($-NH_3^+$, $-COO^-$) of their amino acids. Thus, the net charge of a protein is dependent on the pH of the medium and may even change its polarity by the supression of dissociation of one type of charged group and by the enhancement of the dissociation of the other type.

The polarity of a macroion is extremely important in electrophoresis, since it determines the migration direction. All proteins and peptides have isoelectric points, the isoelectric point being the pH at which the number of positive and negative charges on the molecule are equal. At the isoelectric point, an amphoteric ion has zero net charge and will not migrate in an electrical field. The isoelectric point is an important constant and can be measured by isoelectric focusing, which is a special case of electrophoresis (Vesterberg, 1973*a*; Righetti and Drysdale, 1974).

All other types of electrophoresis are performed in buffer solutions that provide a defined pH throughout the electrophoretic run. The ionic strength of the buffer reduces the net charge of the protein to the effective charge. Thus, in buffers of high ionic strength, the migration velocity will be low (Maurer, 1971), which is an important factor in discontinuous buffer systems (see Section 3.8). The only nonaqueous medium of importance is formamide, which is specifically used in electrophoresis of RNA (Staynov *et al.*, 1972). In general, a higher electrical field strength yields faster and better separation of macroions. However, this is limited, since the electrical power (volts $\times$ amperes = watts) passing through the system produces heat (watts $\times$

mechanical equivalent of heat), thereby increasing diffusion, which limits the concentration of the macroions in separated peaks. To overcome this drawback, cooling devices are often employed in electrophoresis. However, cooling will improve the separation only if heat exchange is rapid and homogeneous throughout the cross section of electrophoresis carrier (i.e., the gel); otherwise, thermal gradients will occur and cause distortion of the protein peaks (see below).

The electrophoretic mobility of a macroion is also dependent on its size (i.e., equivalent to the Stokes radius). This factor allows for calculation of molecular weight of the macroion using electrophoretic techniques (see Section 5.2). However, particles that have a pronounced ellipsoid shape and an asymmetrical distribution of surface charge tend to align with their long axis perpendicular to the electrical field (electrical birefringence) and migrate more rapidly than their molecular weight would predict (Felgenhauer, 1974). The sizes and shapes of macroions are particularly important when electrophoresis is performed in media such as starch gels or polyacrylamide gels with restrictive sieving properties.

Proteins also differ in their solubility properties, thereby influencing the choice of the electrophoretic technique to be used. Water-soluble proteins such as serum proteins do not cause problems in any type of electrophoresis. Other proteins, however, are bound to membrane structures by hydrogen (van der Waals) bonds and ionic bonds and have to be solubilized by various specific treatments (e.g., high salt concentrations, urea, or detergents). Strong chaotropic substances such as guanidine chloride or guanidine cyanate that have to be used at high concentrations are charged, and therefore cannot be used in electrophoresis. There are hydrophobic proteins the surfaces of which are dominated by hydrophobic amino acids such as phenylalanine, tyrosine, and tryptophan, or by associated lipids (e.g., the proteolipid protein of myelin). These proteins can be solubilized only in alcohol [e.g., 75% butanol (Maddy *et al.*, 1972)] or by detergents (Helenius and Simons, 1975). Detergent molecules have a hydrophilic end that is either strongly polar (e.g., Triton R, Tween R, Nonidet R, Brij R) or ionic [e.g., sodium dodecyl sulfate (SDS), cetyltrimethyl-ammoniumbromide] (see Helenius and Simons, 1975) and have a hydrophobic chain that interacts with appropriate sites of the protein by hydrophobic bonds (Tanford, 1973). Thus, detergents form micelles around hydrophobic proteins and render them water-soluble. Nonionic detergents do not affect the charge of a protein significantly, whereas ionic detergents break all noncovalent bonds in the protein and, if combined with the reduction of inter- and intramolecular disulfide bonds, destroy its tertiary structure totally. Disulfide bonds can be broken up by reducing agents such as 2-mercaptoethanol, thioglycolic acid, or dithiothreitol. By formation of micelles around the unfolded peptide chain, the native charged groups of the protein are hidden under a cover of ionic detergent. These detergent–protein micelles have ellipsoid shapes, and the long axis of the aggregate varies with the molecular weight of the incorporated peptide chain, while the charge is considered to be constant per unit length (Pitt-Rivers and

Impiombato, 1968; Reynolds and Tanford, 1970*a,b*; Jones, 1975; Takagi *et al.*, 1975). Thus, under these conditions, the charge density and sign of the protein surfaces become uniform, and only size or molecular weight influences the mobility of the protein–detergent micelle (Shapiro *et al.*, 1967; Weber and Osborne, 1969; Dunker and Rueckert, 1969). Protein–detergent interactions have been intensively investigated with SDS, and additional features of other detergents are reviewed by Helenius and Simons (1975). Limitations of the SDS method are described in Section 3.9.

3. Outline of Electrophoretic Techniques

3.1. General Terminology

When one enters the field of electrophoresis, one encounters a wide variety of techniques and terms describing different variations of the method. For example: *Free electrophoresis* (carrier-free electrophoresis) is performed in buffer solution without supporting matrix and can be used for determination of mobility, m_0. *Carrier electrophoresis* is performed in a supporting matrix to reduce convection and diffusion. Paper, cellulose acetate foil, agar gel, starch gel, and polyacrylamide gel are used as carriers. The latter two types of gels provide molecular-sieve properties, since the pore sizes of their matrix structure can interfere with the migration of the macroions (Smithies, 1964; Fawcett and Morris, 1966; Richards and Lecanidou, 1974). This type of sieving should not be confused with the sieving properties of dextran beads (e.g., Sephadex gel chromatography). In columns containing Sephadex beads, the smaller particles are supposed to penetrate the meshwork of the bead core, while the larger particles are excluded and elute first at the end of the column. In gel electrophoresis, the sequence is reversed; i.e., smaller particles migrate faster than larger particles. A good general introduction to gel electrophoresis has been written by Maurer and Allen (1972*a*).

The pore size of the polyacrylamide gel, which is critical for molecular sieving, can be controlled over a certain range by variation of the gel mixture. Smaller pores are produced by increased acrylamide concentration as well as by increased cross-linker (bisacrylamide) concentration. The latter causes maximal retardation of migrating particles at about 5%C [C is the part of the comonomer (acrylamide + bisacrylamide) concentration, T (%, wt./vol.), that is bisacrylamide] (Hjertén *et al.*, 1969). Higher degrees of cross-linking diminish the sieving effect of the gel. The structural properties of polyacrylamide gel with respect to molecular sieving have recently been investigated using electron microscopy (Rüchel and Brager, 1975; Rüchel *et al.*, 1977).

A *continuous* carrier exerts the same influence on migrating particles at any point in space. Such a continuous carrier can be a rectangular paper strip, a gel of cylindrical shape, or a rectangular gel slab, made of one gel mixture, and thus provides a constant pore size throughout the gel. *Discontinuous* carriers are, for instance, wedge-shaped paper strips (used to achieve concentration of the sample during the run) or gels made up of layers of gel with different constituent concentrations, e.g., stepped gradients (Been and Rasch, 1972; Shaw and Hennen, 1975). A continuous-gradient gel has a continuously decreasing pore size provided by a continuously increasing constituent concentration in the running direction of the electrophoresis.

3.2. Moving-Boundary Technique

The *moving boundary* was the first important electrophoretic technique that was used as a scientific tool (Tiselius, 1937) (for a historical review, see Cann, 1968). It is a carrier-free electrophoresis and is usually performed in a U-shaped glass tube. The cell is filled with running buffer and the sample is overlayered at one end (see Fig. 1). The leading ion, in this case an anion, must have a higher mobility than any other anion in the sample. All the macroanions of the sample (i.e., proteins) arrange behind the leading ion in a sequence of decreasing mobility when the voltage is applied. Thus, the moving-boundary technique does not produce total separation of the different macroionic species of the sample. Instead, it provides a segregation of the sample, and the interfaces between the various migrating constituents can be traced by schlieren optics (Tanford, 1961).

3.3. Zone Electrophoresis

Zone electrophoresis uses one common buffer throughout the whole system (including in the sample). Under such conditions, different macroions migrate with different velocities, and separation takes place as a function of time and migration distance. Hence, zone electrophoresis has been compared with elution chromatography (Everaerts, 1973). After the run is started, all the migrating ionic constituents are susceptible to diffusion, since there is no concentrating force stabilizing the bands, as in the moving-boundary technique or in disk electrophoresis. Therefore, high starting concentrations of the sample are required, and can be achieved by reduction of the conductivity in the sample compartment [i.e., by diluting the buffer in the sample compartment (see Hjertén *et al.*, 1965)]. The lower conductivity in the sample raises the local electrical field strength, and therefore the sample ions are accelerated in the migration direction and concentrate at the interface to the running buffer in front of the sample. Thus, conductivity in the sample segment is raised by concentration of the macroions and buffer ions, and the displacement volume of the sample is decreased. If the sample is applied in the middle of the system, migration to both electrodes according to the polarity of the sample ions is possible. This provides information about the presence of both acidic and basic proteins in the sample in one separation run.

To reduce convection and diffusion, zone electrophoresis is usually performed in a paper or gel supporting carrier. Since the latter may provide sieving properties, high separation power can be achieved if the common

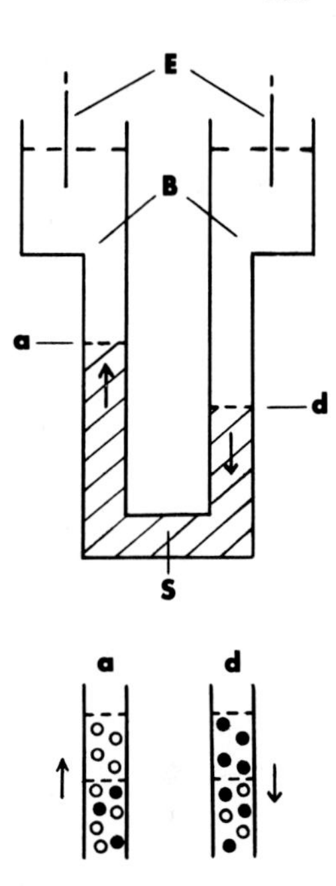

Fig. 1. Scheme of moving-boundary electrophoresis. (E) Electrodes; (B) buffer; (S) sample; (a) ascending boundary; (d) descending boundary. The stabilized sample is placed in a U-shaped glass vessel, and the sample is carefully overlayered with buffer to provide undisturbed interfaces. The common buffer contains an ionic species with high electrophoretic mobility that may be compared with the leading ion of a discontinuous buffer system. Only partial separation can be achieved by this technique, which is useful for determination of free electrophoretic mobilities. Such partial separation as described for two ionic species (i.e., proteins) in the area of the boundaries (a, d) can be traced either by measurement of the refractive index, representing protein concentration, or by schlieren optics, sensing the developing interfaces. Since this technique is a carrier-free electrophoresis, convection due to heat development is a serious problem that limits electrical field strength.

buffer is made up of low-mobility ions (e.g., tris–glycine or tris–borate buffer at basic pH). Zone electrophoresis in phosphate buffers of neutral pH is widely used with addition of SDS (Shapiro *et al.*, 1967; Weber and Osborn, 1969; Dunker and Rueckert, 1969). Good separation of high-molecular-weight RNA has been achieved using this electrophoretic technique at slightly basic pH (Wolfrum *et al.*, 1974). Electrophoresis of nucleic acids in nonaqueous formamide (Staynov *et al.*, 1972), microelectrophoresis of RNA-bases in sulfuric acid (Edström, 1964), and separation of proteins in acetic acid (Maddy and Kelly, 1971) or concentrated formic acid (Mokrasch, 1975) are also effective continuous electrophoretic methods.

3.4. Disk Electrophoresis

Disk electrophoresis (also called multiphasic zone electrophoresis or discontinuous electrophoresis) combines the moving-boundary technique with zone electrophoresis. Its name is related to the sharp protein and marker dye disks that develop during the run and to the use of discontinuous buffer systems and discontinuous gels (as carriers) as well (Ornstein, 1964; Jovin, 1973).

For the separation of macroionic species, this technique requires a sieving carrier (starch- or polyacrylamide-gel). The gel buffer in front of the sample contains the leading ion, which has to have the highest mobility of all ions in the system migrating in the direction of separation. The sample is applied on top of a spacer or stacking gel of low acrylamide concentration. The spacer gel may also contain a different buffer at a pH somewhat different from the pH of the system, to decrease the effective charge of the trailing ion of the running buffer when entering this gel (Ornstein, 1964). The spacer gel sits atop the separation gel, which has a higher acrylamide concentration, to provide molecular sieving. The sample is overlayered by the running buffer, the ionic species of which (migrating in the direction of the separation) has to have the lowest mobility in the whole system. This ion

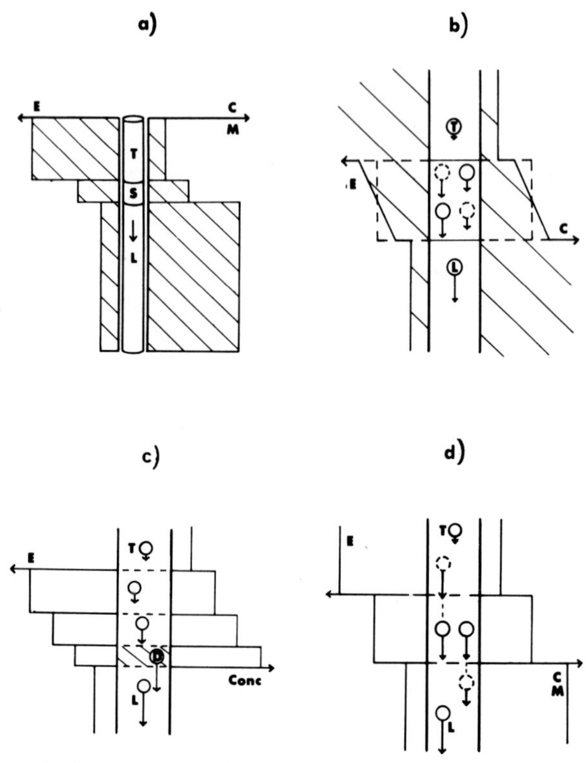

Fig. 2. (**a**) Scheme of a discontinuous buffer system with sample segment representing conditions of a discontinuous electrophoresis at the start of the run. The leading ion (L) has high electrophoretic mobility providing high electric conductivity, while the trailing ion [(T) terminator] has low electrophoretic mobility (M) and provides only low electrical conductivity (C). Thus, the electrical field strength [voltage gradient (E)] in the area of the trailing ion (contained in the running buffer) is high, while the field strength in the area of the leading ion (contained in the gel buffer) is low. The sample (S) is a mixture of various ionic species averaging intermediate conductivity and field strength. (**b**) Scheme of the changing ion distribution in the sample compartment right after the start of the electrophoretic run. (S) Sample segment; (T) trailing ion; (L) leading ion. The length of the vectors indicates the electrophoretic mobility of the ion. Assuming only two ionic species in the sample compartment that migrate in the running direction, ions with higher mobility concentrate at the front of the sample compartment (left side), while ions of lower mobility concentrate at the rear end of the sample segment (right side). Such demixing of the sample causes a shift in conductivity within the sample segment that is compensated by a shift in electrical field strength as indicated. This provides for constant current density throughout the electrophoretic system. (**c**) Scheme of the sample compartment of a discontinuous electrophoresis after demixing of the sample. Every ionic species has formed its

is called the terminator or the trailing ion. The free mobilities of all macroionic species in the sample must be intermediate between the mobilities of the leading and trailing ions. When a voltage is applied, a steady-state stack is formed due to the regulating function (Kohlrausch, 1897), which is comparable to the moving boundary described in Section 3.2. The regulating-function concept states that the sum of the ratios of concentration vs. mobility of all ionic species present in any compartment of an electrophoretic system is constant. Thus, in a simple case: $(\text{conc.}^{+}/m^{+}) + (\text{conc.}^{-}/m^{-}) = \text{const.}$ Hence, the concentration of an ionic species in the system is dependent on its electrophoretic mobility (the latter being a constant under given conditions). Discontinuous electrophoresis is always unidirectional and may be either cathodic or anodic. Therefore, only cations or anions, respectively, have to be considered further, and the equivalent presence of counterions is provided for by the law of electroneutrality. If the pH of the whole system is kept constant (Allen, 1971), the influence of the counterion may be neglected, even if conditions in different compartments of the electrophoresis system are compared, since the ratio of the counterion is constant in any compartment.

At the start of the run, the compartment containing the leading ion with the highest mobility (and therefore the highest conductivity), but the lowest voltage gradient (and resistance), is in front of the sample, and the compartment containing the trailing ion with lowest mobility (and conductivity), but the highest voltage gradient (and resistance), is in the rear of the sample. The sample segment itself has intermediate conductivity and an intermediate voltage gradient, ruled by the average of the mobilities of the sample ions (Fig. 2a). After the voltage has been applied, the local voltage gradient in the sample accelerates the sample ions with high mobilities to the front of the

---

own subcompartment in which they all migrate with identical speed. The sequence of those compartments is ruled by electrophoretic mobilities as indicated by vectors. Assuming that all ionic species of the sample have been applied in equal amounts, the displacement volume of the subcompartments appears as indicated. According to the regulating function, high-mobility ions are concentrated more than those of lower mobility. Therefore, the displacement volume of each ionic compartment is related to the specific electrophoretic mobility and is ruled by the concentration of the leading ion. This feature of the regulating function has been called "retrograde regulation" (Jovin, 1973). The arrangement of ion compartments is called "steady-state stack" (Ornstein, 1964). Any marker dye (D) added to the sample should have the highest electrophoretic mobility of all sample ions. The dye thus travels directly behind the leading ion. (**d**) Scheme of a steady-state stack containing only one ionic species between the leading (L) and trailing (T) ions. An ion of the sample that falls back into the area of the trailing ion (left side) becomes exposed to higher electrical field strength. Due to its higher mobility, the particle becomes faster than the (constant) speed of the whole steady-state stack, and therefore the ion is pushed out of the compartment of the trailing ion back into the area of lower electrical field strength, where the particle's velocity is decelerated to the system speed. An ion of the sample that gets into the area of the leading ion (right side) is exposed to lower electrical field strength. Thus, the particle cannot maintain the speed of the system and falls back into the area of appropriate electrical field strength. This correcting feature of the regulating function provides self-stabilization of the steady-state-stack arrangement by opposing diffusion, and it provides for the high separation power of discontinuous electrophoresis and isotachophoresis.

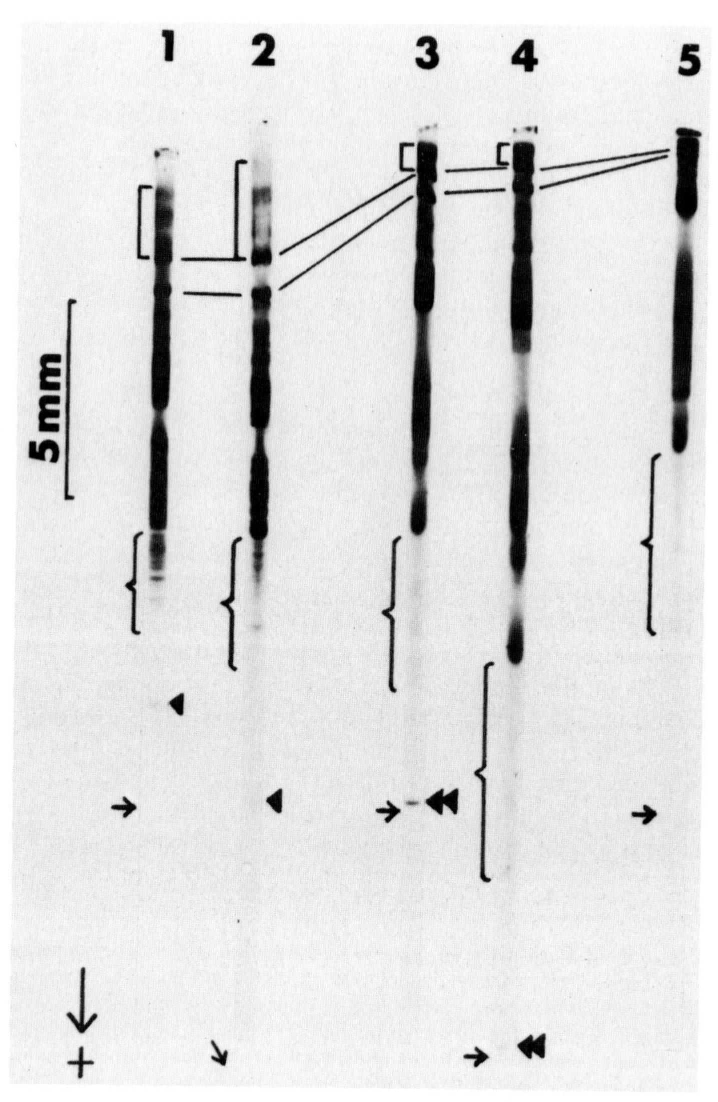

Fig. 3. Comparison of separation patterns of fish serum proteins [from striped bass (*Morone saxatilis*)] in different polyacrylamide gels. Electrophoresis was performed in parallel runs. Gels 1 and 2 are continuous-gradient gels (T_{max}, 40%; *C*, 5%). In these gels, electrophoresis was performed in the tris-glycine/tris-sulfate discontinuous buffer system at pH 8.4. The marker dye was still present in gel 1 (→), but it had already passed gel 2 when the run was stopped. Gels 3 and 4 are constant-pore-size gels (*T*, 10%; *C*, 4.5%). Prior to polymerization, the gel mixture had been overlayered with water, which produces a very short pore gradient. Thus, the entry of larger proteins into the gel is improved. The buffer system used is identical to the one described above. The final position of the marker dye is indicated (→). Gel 5 has a constant pore size (*T*, 12.5%; *C*, 4.5%). In this run, a continuous tris-glycine buffer of pH 8.4 was used throughout the system. This buffer is identical to the running buffer in gels 1–4. The final position of the marker dye is indicated (→). Continuous-gradient gels provide a wider separation of high-molecular-weight proteins, as indicated by the position of two protein peaks in the upper region of the gels. The top region of the protein pattern ([) is separated significantly only in the gradient gels, while gel 5 shows precipitation of protein even in the top of the gel. Low-molecular-weight

compartment, while the sample ions with low mobilities fall back to the interface of the trailing ion in the running buffer (Fig. 2b). Thus, the sample demixes and new subcompartments develop due to the different mobilities. Each subcompartment maintains a defined voltage gradient. All ionic species of the sample are finally arranged in the "stack" in sharp "disks" in a sequence of decreasing mobilities (Fig. 2c). The stack moves with constant velocity, and no mixing of ionic species within the stack will occur. Mixing is prevented by the stepped voltage gradient, which provides a self-stabilizing power, forcing the macroions to stay in the disk area of the appropriate field strength (Fig. 2d). The self-stabilization of the disk arrangement is described by the term "steady-state stack" (Ornstein, 1964).

The concentration of the various ionic species in the stack is strictly ruled by the concentration of the leading ion, since the ratio of concentration vs. mobility can be altered only by change of the concentration of the leading ion. Therefore, the regulating function (see above) can be rewritten neglecting the counterion: $\text{conc.}_L/m_L = \text{conc.}_1/m_1 = \text{conc.}_2/m_2 = \cdots = \text{conc.}_T/m_T$, where $m_L > m_1 > m_2 > \cdots > m_T$, L is the leading ion, and T is the trailing ion. Thus, a twofold increase in the concentration of the leading ion causes a twofold increase in all other concentrations in the stack. This implies that the displacement volume of each disk will be half. This "retrograde regulation" (Jovin, 1973) causes the very sharp separations achieved by disk electrophoresis and is limited only by heat development and by the solubility properties of the proteins involved. The protein disks of the stack are separated when the stack encounters the sieving separation gel, where the effective mobility of the macroions is lowered beyond the mobility of the trailing ion mainly due to the sieving effects of the lower gel. If a spacer gel containing buffer of lower pH (in the anionic discontinuous system) has been used, the entry into the separation gel is coincident with an increase of the mobility of the trailing ion (Ornstein, 1964). This may favor the separation of the steady-state stack, but it has to be kept in mind that the mobilities of the sample ions are also increased. When the trailing ions have passed the protein disks, the latter migrate under conditions of zone electrophoresis and are susceptible to diffusion, although diffusion is largely restricted by the gel. Therefore, in regular disk electrophoresis, the steady-state-stack arrangement is used only as a means of sample concentration and preseparation (Hjertén *et al.*, 1965).

---

proteins are spread over a wide range ({) in the gels of constant pore size, but the peaks are relatively blurred in comparison with those in the gradient gels. The gels that had a discontinuous buffer (1–4) show a sharp front peak (◀, ◀◀). In the gradient gels, the front peak (◀) is widely separated from the marker dye and yields only one peak on refractionation in a 60% gradient gel. The front peak in gels 3 and 4 (◀◀) migrates directly behind the dye marker. By refractionation in a 60% gradient gel, this peak is separated into two peaks. Therefore, the front peak of gels 3 and 4 proved to be a remainder of the steady-state stack of the discontinuous buffer system used. Due to the continuous buffer used, no such front peak occurs in gel 5. Lacking the concentrating power of a steady-state stack, the low-molecular-weight proteins appear even more blurred than in gels 3 and 4 despite the higher acrylamide concentration in gel 5.

In *gradient gels*, the pore size of which is continually decreasing, the separation of the steady-state-stack arrangement of protein disks does not take place at the moment of entry into the gel; rather, different protein disks are sieved out of the stack at various places in the gel where deceleration due to sieving reduces the mobility of the protein below the mobility of the trailing ion (Cox and Teven, 1976). Thus, the phase of steady-state stack (moving boundary) is prolonged and the phase of zone electrophoresis is comparatively shorter than in regular disk electrophoresis. This yields a significant improvement in the sharpness of the separated protein bands (Maurer and Dati, 1972). This phenomenon is illustrated in Fig. 3 using fish serum proteins. Note the improved resolution of the low-molecular-weight proteins in particular.

Another unique feature of the gradient gel is the possibility of separation of macroionic mixtures with a much broader molecular-weight range than is allowed by disk electrophoresis in constant-pore-size gels [i.e., continuous gels (Rodbard *et al.*, 1971; Lambin *et al.*, 1976)]. The latter type of gel will separate macroions of about one order of magnitude difference in molecular weight. Any larger species cannot enter the gel, while any smaller species may not be sieved out of the stack and is thus not separated. Steep-gradient

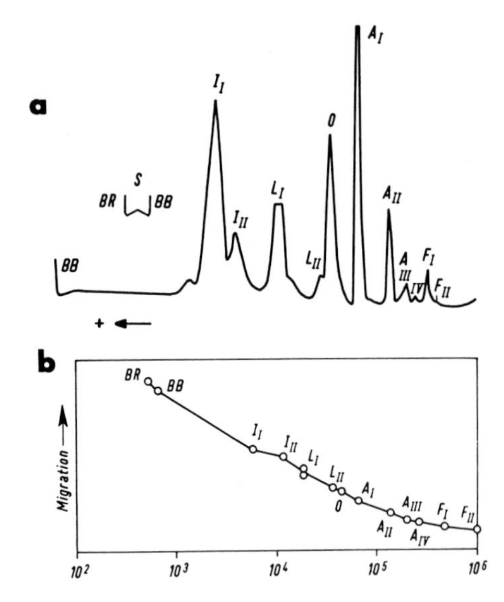

Fig. 4. (**a**) Pherogram of an electrophoretic separation of standard proteins and marker dyes in a 5-μl gradient gel (T_{max}, 52%; C, 3.8%) in a detergent-free discontinuous buffer system. Before being stained, the gel was cut at the peaks of the dyes. Thus, the gel segment (S) permits the localization of these two molecular species. (BB) Bromophenol blue; (BR) bromophenol red; (I) insulin; (L) β-lactoglobulin; (O) ovalbumin; (A) bovine serum albumin; (F) ferritin. The roman-numeral subscripts denote oligomers. (**b**) Plot of the relative migration distances against the logarithm of molecular weights. Only the insulin dimer migrates considerably too far, suggesting an aggregate of two insulin monomers. Due to electrical birefringence, the aggregate migrates through the gel by adjusting its long axis to the axis of the gel. Reproduced by permission from Rüchel *et al.* (1974).

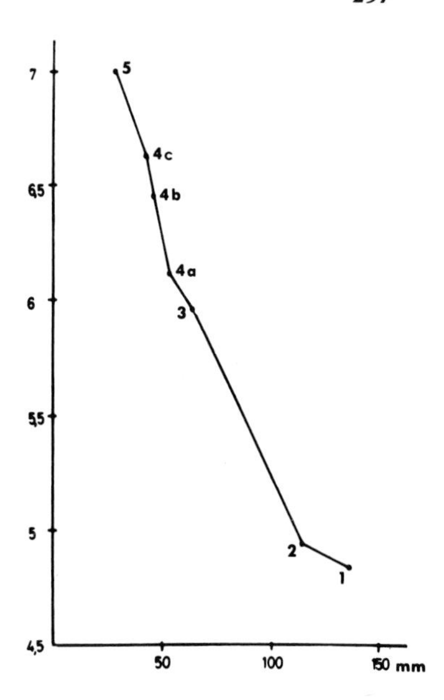

Fig. 5. Separation of high-molecular-weight particles in a 5-μl gradient gel (T_{max}, 40%; C, 2%). The running conditions were as described in the Fig. 4 caption. (1) Human serum albumin, molecular weight (MW) 69,000; (2) transferrin, MW 88,000; (3) α_2-macroglobulin, MW 900,000; (4a–c) 40 S, 60 S, and 80 S ribosomal particles of rabbit reticulocytes, MWs ≈ 1.3×10^6 and 2.8×10^6, and 4.2×10^6. (5) Friend leukemia virus particles, MW 1×10^7. The logarithm of the molecular weight was plotted against the relative migration distance as recorded by the microdensitometer (transmission 1:10).

gels, however, can separate mixtures covering more than three orders of magnitude in molecular weight (see Figs 4 and 5). Another interesting feature of gradient-gel electrophoresis is the possibility of achieving a "dead run" of macroions into a "pore limit" of the gel (Felgenhauer, 1974; Margolis and Kenrick, 1968; Slater, 1969; Rüchel *et al.*, 1973*b*). Such an approach is comparable to isoelectric focusing methods, since both electrophoretic techniques approach a final equilibrium. In this regard, the "dead run" can also be compared with isopycnic centrifugation in a density gradient. One principal drawback of such gradient-gel electrophoresis (i.e., the dead run) must be considered. Its driving force is primarily a function of the electrical field strength and of the effective charge of the macroion under consideration (Nee, 1975), and the charge is just that parameter the influence of which has to be overcome in favor of size and shape. Therefore, ideal results cannot be obtained. The dead-run approach yields practical results only after very long running times, and with macroions with charges that do not differ too much from one another (Rüchel, 1977). The experiment shown in Fig. 15 seems to be the best that can be achieved in such a "dead run."

3.5. Isotachophoresis

Isotachophoresis is used as a synonym for steady-state stack, if the stack arrangement is used on its own as a separation technique (Everaerts, 1972; Arlinger, 1971). It has been proposed that the term "isotachophoresis" should be used for such types of steady-state stacks only when the protein disks are separated by spacers (Chrambach *et al.*, 1972). The spacers used are the ampholytes used in isoelectric focusing (carrier ampholytes), which arrange in the stack between the protein species according to their mobilities and thus separate the proteins in the stack (Catsimpoolas, 1973; Arlinger, 1975).

Isotachophoresis is usually performed as a carrier-free electrophoresis in long glass capillaries and has been used for the separation of ionic species ranging from small inorganic compounds to proteins (Arlinger, 1976; Everaerts, *et al.*, 1976). Other workers have used large-pore polyacrylamide gels so that larger column diameters can be employed (Chrambach *et al.*, 1972; Griffith and Catsimpoolas, 1972; Baumann and Chrambach, 1976*b*). To render the technique stationary, counterflow of buffer has been tried. However, this attempt has been hampered by disturbance of the separation pattern due to convection (Everaerts *et al.*, 1976). The separation patterns are detected at interfaces between ionic species using potential, conductivity, and UV detectors (Everaerts *et al.*, 1976). Isotachophoresis has been developed primarily as a preparative tool, and no real microversion is known to us at present.

3.6. Isoelectric Focusing

Isoelectric focusing is the electrophoresis of amphoteric macroions in a pH gradient and their separation based on their isoelectric points. The pH gradient is established either in a density gradient or in a large-pore polyacrylamide gel by carrier ampholytes that arrange in an electrical field due to their isoelectric point and cause the development of a "natural" pH gradient that is stable in its steady-state phase (Rilbe, 1973). Carrier ampholytes are amphoteric molecules of intermediate size that were first synthesized for electrophoretic purposes by Vesterberg (1973*b*). By variation of the ratios of basic and acidic charges in the numerous ampholyte species, the isoelectric points of the final mixtures cover, at present, a maximum pH range of 2–11. Carrier ampholytes of pH 3.5–10 (or smaller ranges) are distributed by LKB (Bromma, Sweden) under the trademark "Ampholine." Other brands are "Bio-Lyte" (pH 3–10) from Bio-Rad (Richmond, California), "Servalyt" (pH 2–11) from Serva (Heidelberg, West Germany), and "Pharmalyte" from Pharmacia (Uppsala, Sweden). The carrier ampholytes not only make the pH gradient, but also provide conductivity and buffering capacity and act as spacers between different protein species, as described in Section 3.5. The pH gradient, which is a large steady-state stack of zero mobility, is not really continuous but is a stepped gradient, each amphoteric species contributing one plateau to the whole arrangement. The pH gradient can be established either in an electrophoretic prerun (Gainer, 1973) or as a one-step procedure including the separation of the sample since the separation of amphoteric protein species is fundamentally the same process as the establishment of the pH gradient. The arrangement of the pH gradient is caused by H^+ ions coming from the acid, which is used as a terminating electrolyte at the anode, and from OH^- ions entering the gel from the cathode, where a base is used as terminating electrolyte. The choice of acid and base is related to the pH range of the carrier ampholytes used; thus, a broad pH range requires strong terminating electrolytes such as sulfuric acid and sodium hydroxide (Vesterberg, 1973*a*). When the electrodes (platinum wire or carbon rods) are applied directly on top of the gel, terminating electrolytes can be avoided and the H^+ and OH^- ions can be provided by electrolysis of water (Vesterberg, 1973*a*; Righetti and Drysdale, 1974).

An amphoteric ion (either a protein or a carrier ampholyte ion) with positive charge will travel to the cathode until it encounters so many OH^- ions, and thus a basic pH, that the dissociation of basic groups on the particle is decreased. Its positive charge is diminished while negative charges are freed by dissociation of H^+ ions until both basic and acidic dissociation are in equilibrium. In this state, the amphoteric particle has zero net charge and stops migration at its isoelectric point in the developing pH gradient (Fig. 6). In contrast to the other electrophoretic techniques mentioned (with the exception of dead runs in gradient gels), -isoelectric focusing is a technique that reaches a quasi-equilibrium wherein the steady state maintained by the electrical field stabilizes the pH gradient and thus the separation pattern

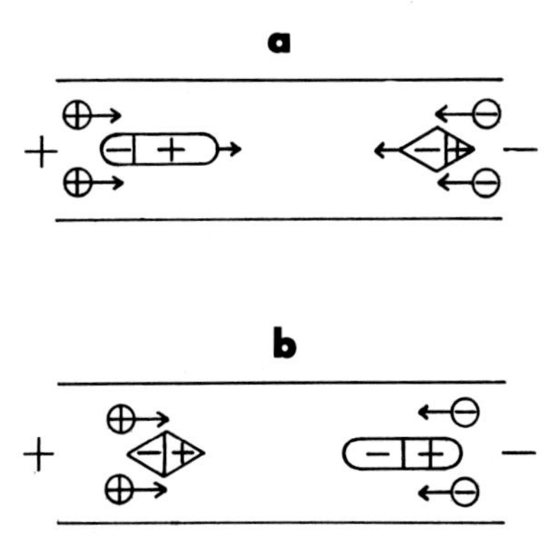

Fig. 6. (**a**) Two amphoteric ions (carrier ampholytes or proteins) randomly distributed at the start of an isoelectric focusing run. Since their net charge at the constant pH of the system is either positive (basic position, left) or negative (acidic position, right), the two amphoteric ions migrate to the counterpole until they encounter increasing amounts of H^+ and OH^- ions, respectively. This pH shift decreases the acidic dissociation of the acidic ampholyte and the basic dissociation of the basic ampholyte. At the same time, the basic dissociation of the primarily acidic ampholyte is enhanced, as is the acidic dissociation of the primarily basic ampholyte. Thus, the net charge of both particles is continuously decreased; their migration velocities are slowing down to zero when the dissociation of both charged groups on the same amphoteric particle approaches an equilibrium (**b**). With all amphoteric species thus asymptotically approaching their isoelectric points, a pH gradient is established, and the conductivity of the system decreases sharply. At constant voltage, therefore, the drop of current can be used as an indicator for the establishment of the pH gradient. Although called "continuous," the pH gradient is "stepped." Each amphoteric species in the system makes up one "step." Thus, it is desirable to have carrier ampholyte mixtures with many species to produce a smooth gradient. The carrier ampholytes used to create the gradient provide a residual conductivity even at their isoelectric point, providing an electrical flow that maintains the pH gradient against diffusion. The carrier ampholytes have buffering capacity to stabilize the local pH over a period of time. OH^- ions and H^+ ions form water, which is also an ampholyte, and establishes a "peak" at pH 7. The development of such a water peak produces an expansion of the pH gradient after a prolonged running time. Therefore, the determination of the minimum focusing time is important. (See the discussion of the use of a marker dye in the text.)

over a period of time (for details, see Vesterberg, 1973*b*; Righetti and Drysdale, 1974; Catsimpoolas, 1973).

In general, isoelectric focusing is restricted to proteins that can be solubilized in the absence of ionic detergents. Recently, methods for the isoelectric focusing of membrane proteins have been reported (Garewal and Wasserman, 1974; Novak-Hoffer and Siegenthaler, 1975; Ames and Nikaido, 1976). Large proteins are best electrofocused in a highly cross-linked carrier (Baumann and Chrambach, 1976*b*), and electrofocusing in Sephadex gels (Radola, 1975; O'Brien *et al.*, 1976) allows for rapid recovery of the proteins for further analysis.

3.7. Two-Dimensional Techniques

Recent extensions of electrophoretic methodologies include two-dimensional techniques. These have been developed to obtain fingerprint patterns of macroions. Such two-dimensional mapping techxiques are extremely useful if highly complex mixtures are to be separated (O'Farrell, 1975; Ames and Nikaido, 1976; Kaltschmidt and Wittmann, 1970). A great deal of information can be obtained by the combination of isoelectric focusing (which is a separation procedure depending exclusively on differences in charge) with a separation step dependent exclusively on differences in size. The latter step is provided by SDS electrophoresis if the size of single peptide chains is of interest (O'Farrell, 1975; Ames and Nikaido, 1976; Iborra and Buhler, 1976). To circumvent denaturation of protein by detergents, either cationic electrophoresis in acidic buffers (Wrigley and Shepherd, 1973) or gradient-gel electrophoresis can be used in the second dimension (Emes *et al.*, 1975; Iborra and Buhler, 1976).

3.8. Buffer Systems

With the exception of isoelectric focusing, all types of electrophoresis are performed in specific buffer solutions to maintain constant pH throughout the run. In zone electrophoresis, the buffer is *continuous*; i.e., there is only one buffer throughout the whole system. Phosphate buffers of pH 7–8 are most widely used. In disk electrophoresis, *discontinuous* buffer systems are used. These are combinations of at least two different buffers having the counterion in common (Ornstein, 1964; Allen, 1971). The combination of *tris-chloride* (as the leading ion) and *tris-glycine* (as the trailing ion) is most widely used as an anionic discontinuous buffer system. Many other discontinuous buffer systems have been calculated (Jovin, 1973), and some preferentially used cathodic and anodic discontinuous buffer systems have been listed elsewhere (Mauer and Allen, 1972*b*). Discontinuous buffer systems yield separation power in only one direction; hence, the polarity has to be kept in mind. The ionic strength of a discontinuous buffer system determines the disk displacement volume and the sharpness of the separated peaks to a certain extent, as described in Section 3.4. In general, the migration velocity is inversely related to the ionic strength of the buffer system used. To maintain sufficiently high velocities, the voltage can be raised only to the limits of heat development.

Various additives to electrophoresis buffers are used for special purposes. The most common is the addition of 0.1% SDS for the separation of SDS proteins. One must use caution in the use of additives in discontinuous buffer systems. They may disturb the formation of the steady-state stack either if they have high electrophoretic mobility (as, for instance, with SDS) or if they do not keep pace with the sample (in the case of low mobility), and may thus be useless (e.g., see Rüchel *et al.*, 1975).

3.9. Electrophoresis in Sodium Dodecyl Sulfate

The use of the anionic detergent SDS in electrophoresis has two purposes: (1) to solubulize membrane and other highly hydrophobic proteins from biological tissue (i.e., if other solubilizing agents are not effective) and (2) to obtain an estimate of the molecular weight of the protein directly from the electrophoretic run. For the latter purpose, the anionic detergent SDS is reacted with the proteins, and electrophoresis is performed in SDS-continaing buffers. Since the noncovalent SDS–protein bonds are ionic and primarily hydrophobic, the micellelike SDS–proteins have to be stabilized during the run by a steady supply of SDS from the upper (cathodic) buffer reservoir. Free SDS migrates much faster than any SDS–protein complex in the gel; thus, SDS should not be incorporated into the separation gel (which keeps the options for use of the gel open).

According to Laemmli (1970), SDS, due to its high electrophoretic mobility, interferes with the formation of the steady-state stack in disk electrophoresis and can be tolerated at only about 0.1% maximum concentration in the running buffer. This SDS concentration is not always sufficient to stabilize the SDS–protein complexes if they are exposed to severe physical stress in the sieving gel. Disintegration of SDS–protein complexes leads to artifactual peaks representing various amounts (ratios) of SDS bound to the same protein. In such cases, SDS electrophoresis in continuous buffer systems is to be preferred, since the stress exerted on the SDS–protein complexes is less than in the discontinuous systems, though at the expense of separation power (Rüchel *et al.*, 1974; Camacho *et al.*, 1975).

Molecular-weight-dependent migration of proteins in the presence of SDS can be expected only if the proteins (peptide subunits) have been totally unfolded. Therefore, all inter- and intramolecular disulfide bonds in the protein should be reduced by treatment with reducing agents (e.g., 2-mercaptoethanol, dithiothreithol, thioglycolic acid) (for details, see Dunker and Rueckert, 1969; Shapiro *et al.*, 1967; Weber and Osborn, 1969). Molecular-weight-dependent migration in SDS is best for proteins larger than 15,000 daltons in molecular weight (Weber and Osborn, 1975; Neville, 1971). Smaller peptides show erratic behavior, since they cannot acquire the ellipsoid shape typical for SDS–proteins (Reynolds and Tanford, 1970*b*). Highly charged proteins (either acidic, such as pepsin, or basic, such as lysozyme) also behave abnormally, for their native charge is not sufficiently overcome by the SDS. Glycoproteins have also been reported to behave abnormally in SDS electrophoresis (Williams and Gratzer, 1971; Tung and Knight, 1972; Russ and Polakova, 1973; Banker and Cotman, 1972).

The SDS protein ratio is crucial in the preparation of the sample. At least 2 times more SDS than protein (by weight) has to be applied, but if the SDS surplus exceeds a ratio of about 5:1, the mass of SDS influences the separation pattern of the sample proteins by displacement of the peaks due to the SDS–micelles present in high amounts at such SDS concentrations (Rüchel *et al.*, 1974; Richter-Landsberg *et al.*, 1974; Stoklosa and Latz, 1975). Impurities in commercially available SDS can be a problem (see Birdi, 1976).

In some cases, it is desirable to solubilize the proteins (e.g., membrane proteins) in SDS, but to maintain the charge of the proteins in the electrophoretic separation (e.g., as in isoelectric focusing). This can be done by first removing SDS from the protein before the electrophoretic run (Tuszynski and Warren, 1975) or by including high concentrations of competing nonionic detergents (e.g., Nonidet P-40) in the sample during the run (Ames and Nikaido, 1976).

4. Practical Aspects of Electrophoresis and Micromethods

4.1. The Case for Miniaturization

Because of the heterogeneity of the nervous system and the necessity to study small regions of the brain, electrophoretic techniques in neurobiology are even more useful if they can be scaled down in size to increase the sensitivity of the system. Miniaturization has been performed with zone electrophoresis, disk electrophoresis, gradient-gel electrophoresis, and isoelectric focusing and two-dimensional combination techniques. With the employment of Coomassie blue R 250 as a general protein stain, 10^{-8} to 10^{-10} g protein can be detected in a separated peak. With radioactive labels, the sensitivity can be increased still more, depending on the specific activity of the protein, by using either autoradiography of the gel or scintillation counting of gel slices (discussed in Section 6.2).

The general goal of analytical electrophoretic techniques is the separation of macroionic mixtures and the identification and comparison of the macroionic species in terms of electrophoretic parameters. This can be achieved with microelectrophoretic techniques, although handling of these methods requires a slightly greater level of skill. Microgels offer many additional advantages: (1) Only very small amounts of sample are required. (2) Very high concentration power can be achieved when high electrical field strengths are applied. This allows separation of proteins out of very dilute mixtures (e.g., Rüchel, 1976). (3) Only very short running times are required (with the exception of gradient-gel dead-run experiments), and staining and destaining of the gels are fast as well. (4) The use of capillaries to cast microgels is facilitated by capillary forces, which allows filling of the tube and the production of gel gradients in an elegant manner. (5) The favorable heat-dissipation properties of microgels are a unique feature. Since the ratio of surface to volume changes in favor of the surface if the gel size is decreased, cooling devices are rarely required and heat dissipation from the gel is improved. Therefore, higher ionic strength and electrical field strength can be applied to the run, which results in improvement of both separation and peak sharpness.

4.2. *Zone Microelectrophoresis*

Virtually any electrophoretic system that is used at the conventional scale can be adapted to the microscale. The various techniques specifically associated with microelectrophoresis (such as filling the capillary tubes with the gel, the appropriate apparatus for electrophoresis, and the removal of the gels for staining) are discussed extensively by Neuhoff (1973*a*).

One zone-microelectrophoresis system that we have used with success is an acid–urea gel system (Davis *et al.*, 1972) adapted to the microscale (Loh and Gainer, 1975). The running pH of the system is 2.7, and Triton X-100 can be included in the sample (2%) and the gel (0.1%). As with all such systems, separation of proteins is based on charge and size, and can provide very high resolution. At this pH, most if not all proteins are positively charged and migrate toward the cathode. It has been calculated that for a protein with a molecular weight of 15,000 daltons, a change of one positively charged residue would produce a 3% change in electrophoretic mobility in such a system (Panyim and Chalkley, 1969). The use of such an acid–urea gel system in the electrophoresis of proteins from identified neurons in *Aplysia* is illustrated in Fig. 7

Zone electrophoresis in microgradient gels can yield good results if buffers of low conductivity are used, thereby providing fast migration of macroionic species if a high voltage is applied. The separation power and sharpness of peaks in such a system are due primarily to the action of the gradient and are thus size-dependent. This is especially useful in "dead-run" experiments (see Fig. 15). Continuous (zone) electrophoresis of SDS–proteins has also been performed at the microscale (see Section 4.4), and zone microelectrophoresis of certain types of RNA and of RNA-bases is the method of choice (see Section 4.8).

4.3. *Disk Microelectrophoresis*

This microelectrophoretic technique offers the superior separation properties of disk electrophoresis and has been used by various authors in neurobiology (Grossbach, 1965; Hydén *et al.*, 1966; Neuhoff, 1968, 1973*a*). Due to the relatively large surface of the gels cast in capillaries, their removal

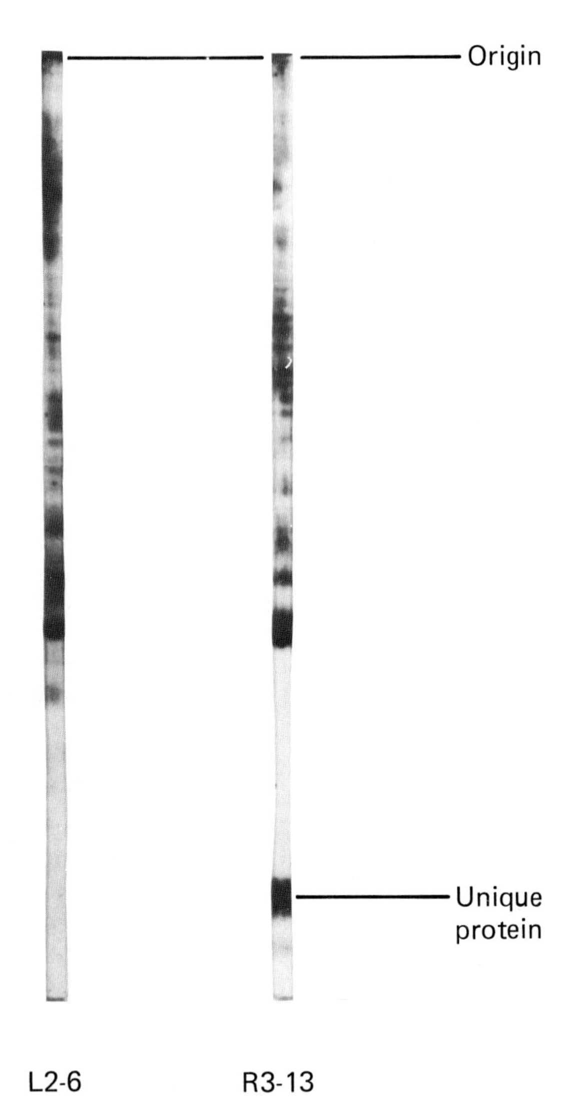

Fig. 7. Acid–urea gel (minigel) electrophoresis of proteins from two identified groups of molluscan neurons. Stained by Coomassie blue. Note the high concentration of unique protein in R3–13 neurons not found in L2–6 neurons. From Loh and Gainer (1975).

from the capillary is difficult, particularly if gels of constant pore size are employed. This has been overcome by the addition of nonionic detergent to the gel mixture (Hydén *et al.*, 1966). Such gels cannot be used in SDS electrophoresis, since ionic and nonionic detergents at the required concentrations are incompatible during electrophoresis. Alternatively, frozen capillary gels can be pushed out of the capillaries by applying pressure in the course of thawing, even if the nonionic detergent has been omitted (Gainer, 1971). Therefore, the capillaries should be siliconized prior to use.

The production of disk-microelectrophoresis gels is somewhat elaborate, since it involves several steps of pipetting into the narrow capillary tube, and one has to deal with polymerizing gel solutions that have to be processed quickly. Disk electrophoresis requires an additional spacer gel to be layered between separation gel and sample. The spacer gel may be replaced by a layer of high-density sucrose or Sephadex (Hydén *et al.*, 1966), but since the spacer requires a different buffer, such completed gels for disk electrophoresis can be stored for only about 2 days prior to use. Disk microelectrophoresis will yield good results if separation of proteins in a narrow molecular-weight range is attempted. More complex mixtures covering several orders of magnitude in molecular weight require several runs in separation gels of different acrylamide concentrations.

4.4. Gradient Microelectrophoresis

Continuous microgradient gels of cylindrical shape have been produced either by a gradient-making apparatus (Smeds and Björkman, 1972; Dames and Maurer, 1974) or by use of capillary force (Rüchel *et al.*, 1973*b*; Rüchel, 1976). The former procedure yields a batch of almost identical gels in one production step, but the addition of nonionic detergent is still needed to allow the gels to be pushed out of the capillaries. In addition, a minimal gel concentration of 2% is obtained at the top of the gel. The procedure that utilizes capillary forces can be performed without any special equipment, and gradients ranging from a polymerization minimum of about 1% up to 60% acrylamide have been produced by this technique (see Fig. 8). Gradient gels of this type do not require the addition of nonionic detergent and need not be cast in siliconized capillaries. They can be pushed out of the capillaries with a tight-fitting steel wire, and separation of SDS–proteins can be performed (Rüchel *et al.*, 1974). However, the gradients created by capillary forces are not as uniform as those that come from a single batch made in a gradient mixer. The latter type may therefore be preferred if experimental standard conditions are to be strictly maintained, whereas the gradient gels created by capillary force are superior if experimental conditions are changed often and versatility is required. The very high maximum acrylamide concentrations in capillary gradient gels cannot be reproduced in larger-scale tubes, since the heat development due to the polymerization causes internal disruption of such gels. Such very highly concentrated gels are especially suitable for dead-run experiments, in which macroions are run into narrowing pores until their mobility is slowed down and approaches zero. Under such pore-limit conditions, interpretation of the separation pattern in terms of molecular weight is possible without the presence of ionic detergent. Steep-gradient gels provide favorable conditions for the separation of peptides smaller than the insulin monomer (see Fig. 9 and Rüchel, 1976). Even the separation of relatively small ions such as synthetic dyes is possible due to sieving in a very narrow-pore gel (see Fig. 4.)

For discontinuous-buffer systems used with gradient gels, a spacer gel

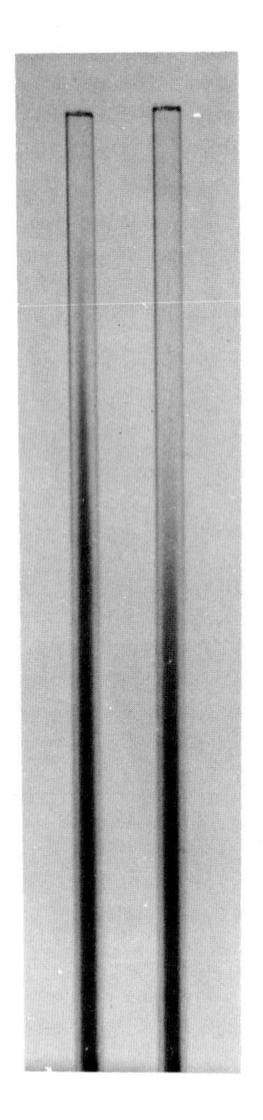

Fig. 8. Capillaries (5 μl) right after suction of the gel gradient (*left*) and after another 2 min (*right*). Marker dye was added to the second gel solution to visualize the development of the gradient. The necessity for a lag period between suction and polymerization of the gel gradient to allow mixing of the two gel solutions is demonstrated. Reproduced with permission from Rüchel *et al.* (1973*b*).

does not have to be layered between the separation gel (which is the gradient) and the sample, since the formation of the steady-state stack takes place in the large-pore top of the gradient gel. Thus, even with the use of discontinuous buffers, electrophoretic preruns can be performed to remove ionic radicals from the gel. This may be important if enzyme activities are to be measured in the gel. Microgradient gels can be used with continuous buffers as well. Separation and sharpness of the protein bands under such conditions are closely related to the steepness of the gradient gel (see Figs. 4 and 5). Continuous buffers are preferable if high-molecular-weight nucleic acids are to be separated (see below). The choice of the slope of the gradient depends on the separation problem: If the sample contains a large variety of particles and molecular weights, a steep gradient may be preferable to screen the overall pattern. If there are only small differences in molecular weight among

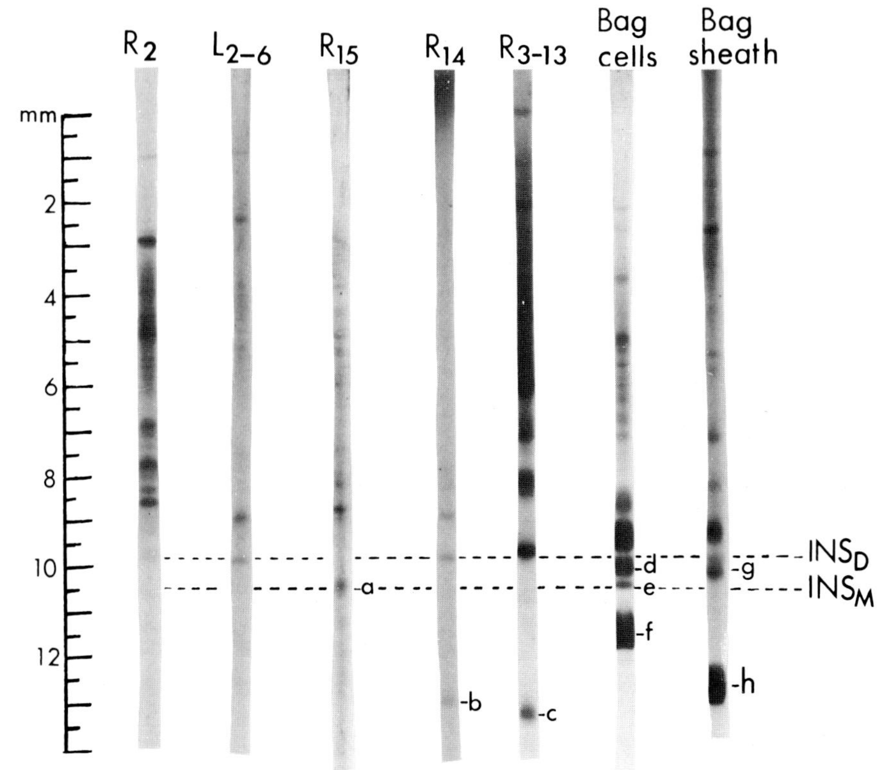

Fig. 9. Coomassie-blue-stained patterns of *Aplysia* neuronal, water-soluble proteins separated on gradient (1–30%) microgels. Proteins were water-extracted from different neurons (indicated above the gels) that were treated with 100% ethylene glycol and stored at −70°C. The mobilities (on a separate gel) of two standards, insulin dimer (INS_D, MW 12,000) and insulin monomer (INS_M, MW 6000), are indicated. The bands that are specific to the neurons are labeled *a–h*. From Loh *et al.* (1977).

the proteins, a gradient of low slope should be used to gain the best separation in the area of interest.

Gradient gels in capillaries can be stored for a long time in the refrigerator if they are kept in gel buffer and if contamination with microorganisms is prevented (i.e., by the addition of 0.02% sodium azide to the holding solution); thus, a large variety of gradient gels can be kept on the shelf for immediate use, making this technique very flexible. Continuous-slab-gradient gels have been described of the size of microscope slides by Maurer and Dati (1972) and at less than the size of a postage stamp by Rüchel (1977).

The use of microgradient gels for the separation of water-soluble proteins in various identified *Aplysia* neurons is illustrated in Fig. 9. Note the relatively sharp peaks of the low-molecular-weight proteins [i.e., smaller than the insulin dimer (INS_D)] and the characteristic staining pattern for each cell

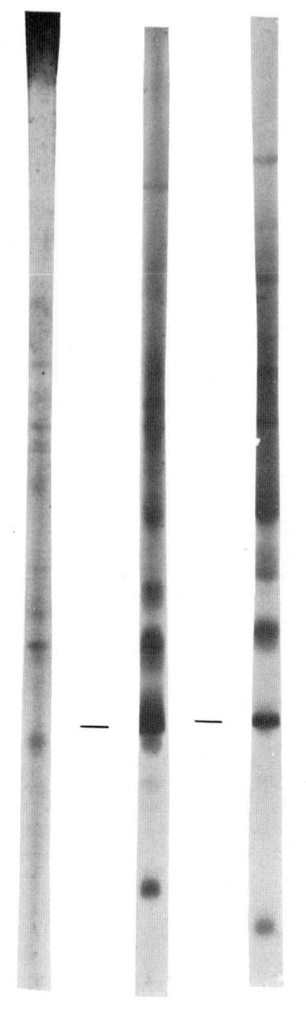

Fig. 10. Coomassie-blue-stained pattern of proteins separated on gradient (1–30%) microgels. The left and right gels show the distribution of water-soluble proteins extracted from an ethylene-glycol-treated R_{15} neuron and R_{3-13} neurons, respectively. The middle gel shows the pattern of a mixture of proteins from R_{15} and R_{3-13}. Note that bands R_{15} and R_{3-13}, which appear to co-run, are in fact different, since they do not overlap when the proteins of both cells are mixed and electrophoresed. From Loh *et al.* (1977).

type. Since these gels (shown in Fig. 9) were cast by the capillary-force method, which does not always produce identical gradients, "mixing" experiments (such as that shown in Fig. 10) are useful to demonstrate that a particular protein peak is specific to a particular cell type.

4.5. *Microisoelectric Focusing*

Isoelectric focusing in capillary gels is performed essentially as in the conventional scale (Gainer, 1973; Quentin and Neuhoff, 1972; Grossbach, 1972). The capillaries have to be filled quickly by capillary force, and the gel mixture (containing the ampholytes) should be kept on ice to delay polymerization until all the capillary tubes are filled. The sample is applied directly on top of the gel and has to be carefully stabilized with sucrose or glycerol to prevent mixing with the terminating electrolyte. Nonionic additives such as Triton or urea do not interfere with the focusing process, but buffers, ionic detergents, and higher amounts of salt must be avoided (Vesterberg, 1973*a*). The sensitivity of protein detection by staining on a microisoelectric-focusing gel appears to be somewhat lower than on a gradient gel of comparable size. The carrier ampholytes have to be washed out of the gel with trichloroacetic acid (TCA) prior to being stained. The apparent protein loss may be due to the contact with the alcohol that is present in the staining process, since TCA has been described as making certain proteins alcohol-soluble (Michael, 1962). In this instance, the aforementioned surface volume ratio of capillary gels may be a disadvantage, since it would facilitate diffusion and the consequent loss of proteins solubilized by alcohol. For the same reason, the staining procedure should not be performed at high temperatures, as has been recommended for gels of conventional size (Söderholm *et al.*, 1972).

The choice of polarity is critical in microisoelectric focusing. An extension of the pH gradient directed to the lower electrode, which is not polarity-dependent but is more pronounced with the anode at the bottom, has been observed (Rüchel, 1976). Therefore, separation of acidic proteins is enhanced if the anode is in the lower position and that of basic proteins if the cathode is in the lower position (Graesslin *et al.*, 1971; Rüchel, 1977). Microelectrofocusing is performed at about 100 volts/cm and takes about 20 min in gels of 2.5-cm length. The focusing stage of the run can be determined by evaluating when the anionic marker dye xylene cyanole FF reaches its isoelectric point (Rüchel, 1977). The fragile electrofocusing gels can easily be pushed out of the tubes by water pressure if the capillaries have been siliconized.

Examples of the use of isoelectric focusing on the microscale are shown in Figs. 11 and 12. Figure 11 illustrates the electrofocusing of various standard proteins [i.e., lactic dehydrogenase (LDH) isozymes and various markers with known isoelectric points] and the pH gradient found on such a microgel at the end of a run. The staining patterns of proteins from specific, identified neurons of *Aplysia* after electrofocusing are shown in Fig. 12. Note the specific patterns for each individual neuron and the large amount of very basic protein (pI > 10) in cells R3–13 and R14.

4.6. *Two-Dimensional Microelectrophoresis*

Two versions of this technique have been described. One involves cathodic electrophoresis in acidic buffer and SDS electrophoresis in the second dimension (Bosselman and Kaulenas, 1976). In this case, both separation steps are dependent primarily on size differences of the separated macroions (in the case of SDS, entirely) as opposed to charge differences. A

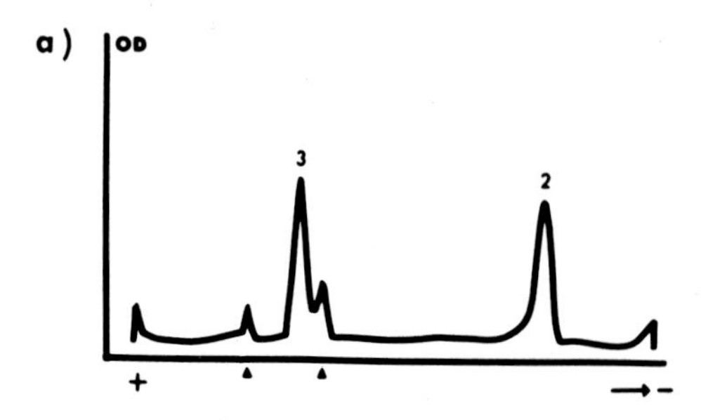

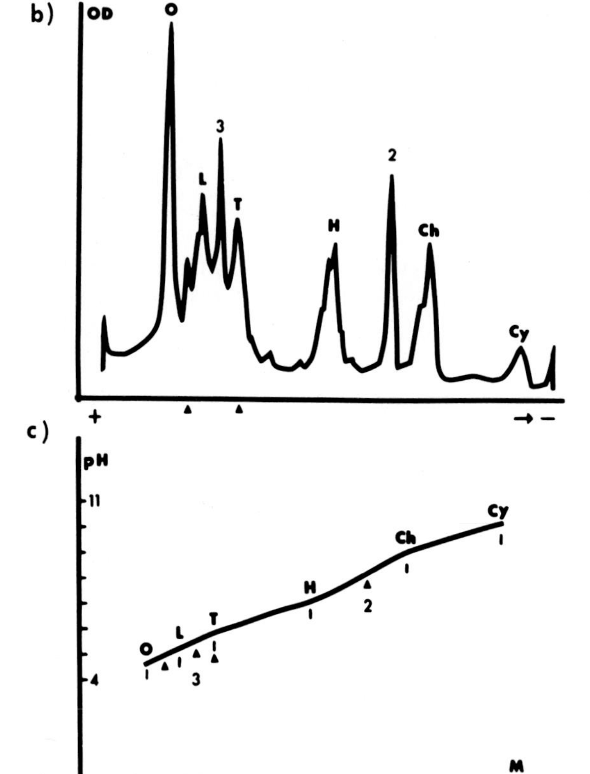

Fig. 11. (**a**) Optical-density (OD) scan of electrofocused lactic dehydrogenase (LDH) from rabbit muscle [Sigma type 2 (2)] with subunits M_4 and from beef heart [Sigma type 3 (3)] with subunit patterns H_4 (*left*), H_3M (*middle*), and H_2M_2 (*right*). The enzymes were detected by the tetrazolium system (cf. Fig. 15). (**b**) OD pattern of electrofocused LDHs (2, 3) and marker proteins of known isoelectric points: (O) ovalbumin, pl 4.6; (L) β-lactoglobulin, pI 5.2; (T) transferrin, pI 5.8; (H) hemoglobin, pI 7.0; (Ch) chymotrypsinogen-A, pI 8.9; (Cy) cytochrome *c*, pI 10.0. The enzymes were localized by the tetrazolium assay prior to staining with Coomassie blue. (**c**) Plot of the isoelectric points of the proteins, separated as shown in (b), against their linear migration distance (M). According to repeated experiments, the isoelectric point of LDH 2 has been determined at 8.1, while the species of LDH 3 have isoelectric points of 4.9, 5.5, and 5.8, respectively.

diagonal array of protein spots representing the separation pattern is found in the second-dimension gel, and hence, charge differences can hardly be identified. The second version combines microelectrofocusing and gradient microelectrophoresis (Rüchel, 1977). The small slab gradient gels employed in this mapping technique can be produced rather simply with an automatic pipette, and maximum acrylamide concentrations of about 40% can be reached. Higher concentrated gel may burst the flat chamber in which the gel is cast. The procedures for producing the microslab gradient gels and a simple electrophoresis setup are illustrated in Fig. 13.

When proteins are driven, during the electrophoretic separation, against

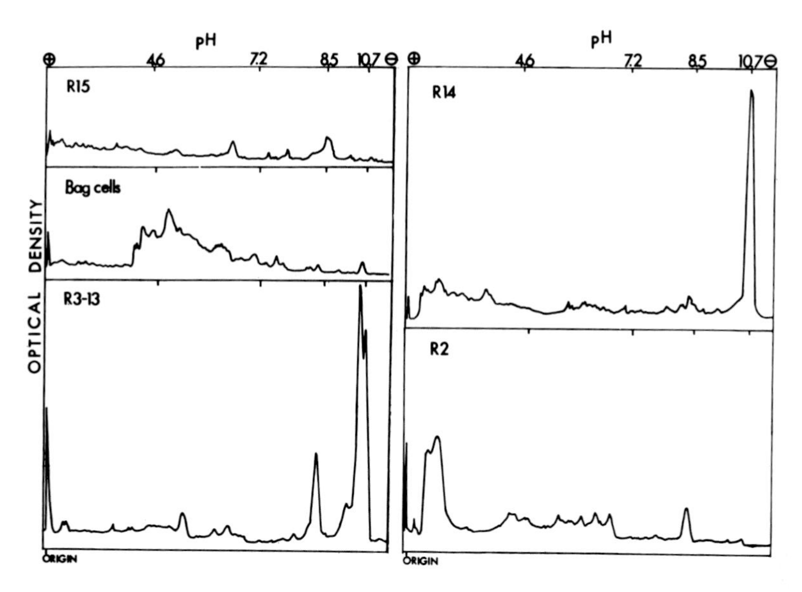

Fig. 12. Optical-density scans of proteins from different neurons (indicated in top left corner), extracted by urea–Triton X-100 and separated on isoelectric focusing gels, pH 2–11. *Upper abscissa:* pH gradient determined from marker proteins: ovalbumin (pI 4.6), hemoglobin (pI 7.2), chymotrypsin A (pI 8.5), and cytochrome *c* (pI 10.7); *lower abscissa:* mobility from cathode to anode; *ordinate:* optical density. Note that only the R_{14} and R_{3-13} neurons have a dominance of basic proteins with isoelectric points close to 10. From Loh *et al.* (1977).

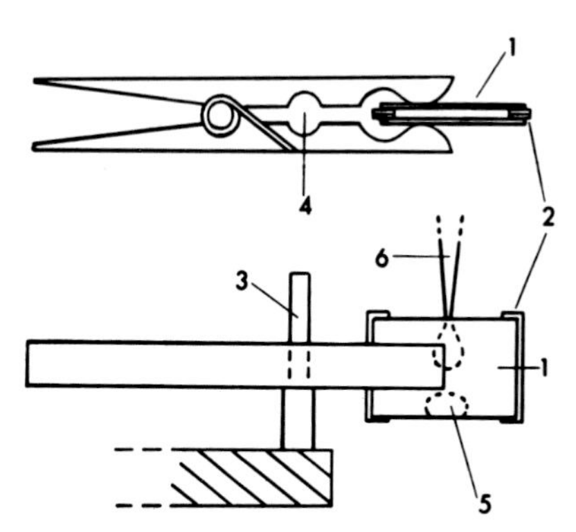

Fig. 13. Electrophoresis setup. Small slab gradient gels are cast in a chamber (1) made of two pieces of slide glass, which are held apart by spacers (2). Double-layered pieces of cover slips (e.g., thickness No. 2, Fisher Scientific Co.) can be used as spacers, providing a distance of approximately 0.5 mm, which is sufficient to apply the cylindrical gel of the first-dimension run on top of the slab gel. The gel chamber is held together by a clothespin with a second hole (4) by which the chamber can be mounted on a rack (3). Prior to filling the chamber with the gel solution in the tip of the pipette (6), a sucrose solution (10 μl, 60%) is placed in the chamber to seal its lower end (5).

their pore limit in the gradient slab gel, they tend to spread and form frontal (or horizontal) disks (Rüchel, 1977; Margolis and Wrigley, 1975). This phenomenon (illustrated in the two-dimensional runs in Figs. 14–17), while a drawback in terms of fingerprint-type mapping, can also serve as an indicator that such patterns may be appropriate for molecular-weight estimation. Several examples are shown for this very powerful technique. Figure 14 shows the results of two-dimensional separation of serum albumin. The pattern of human serum albumin oligomers (Fig. 14A) reveals a vertical distribution of proteins (representing the monomer, dimer, and trimer) that have the same isoelectric points but different sizes. A similar separation of bovine serum albumin (BSA) (Fig. 14B) also reveals the vertical array of BSA oligomers, but in addition shows the charge isomers of BSA monomer and dimer. Figure 15 shows the two-dimensional electrophoresis of LDH isozymes, which were run on an isoelectric-focusing gel in the first dimension (a typical isoelectric-point pattern is shown in Fig. 11a). The second dimension, a continuous slab gradient, required a long running time for the "dead-run" condition. Note the presence of two major stained isozymes [LDH_2, pI = 8.1; LDH_3, pI = 5.5 (seen in Fig. 11c)], as well as two minor stained disks at pH 5.0 and 5.8.

Two-dimensional electrophoresis on the microscale of proteins extracted from biological tissues is illustrated for rat serum in Fig. 16 and for the bag cells (identified neurosecretory-cell population) in *Aplysia* (Fig. 17). The bag cells contain a large number of low-molecular-weight proteins in the pH range 4–5.

A more elaborate but very sensitive approach is the use of cylindrical gradient gels in the second dimension. In this method, proteins have to be detected in the first-dimension gel, and the appropriate gel slides have to be cut out for elution of the protein into the gradient gels. This procedure does not produce peak-spreading; the results of such experiments on specific proteins in the identified neurons from *Aplysia* are shown in Figs. 18 and 19.

4.7. *Microelectrophoresis in Sodium Dodecyl Sulfate*

SDS electrophoresis on the microscale can be performed essentially as has been described for conventional systems (Gainer, 1971; Rüchel *et al.*, 1974). In addition to the various potential artifacts discussed earlier, the formation of artifactual bands due to the breakdown of the SDS–proteins should be considered (Shapiro *et al.*, 1967; Rüchel *et al.*, 1975). The electrical field strength applied in SDS microgel runs should not exceed 40 volts/cm in discontinuous buffer systems so as to reduce SDS-stripping as far as possible (Rüchel *et al.*, 1974). Since the ratio of protein to SDS in the sample is critical, as has been described above, the SDS treatment of very small samples is problematical. If the protein concentration of the sample is in question, as for instance when dealing with microliter volumes of extracts from single cells, it is advisable to start with an undersaturation of SDS. Therefore, an aliquot of the sample should be run without addition of SDS to

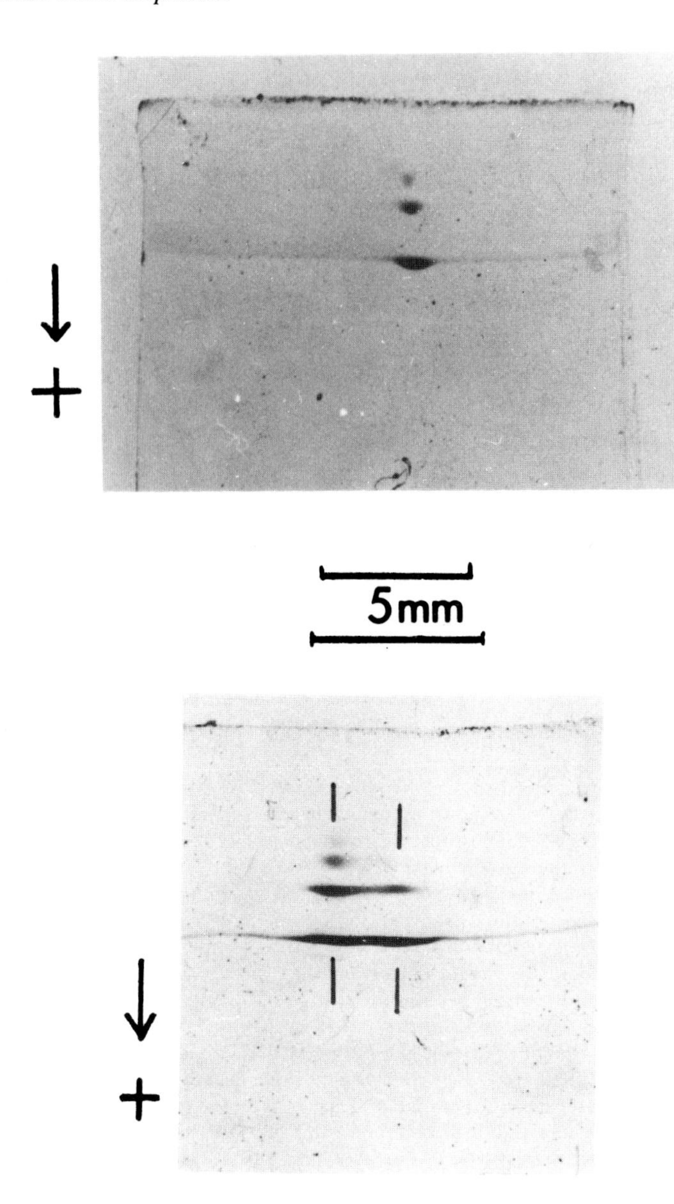

Fig. 14. (Top) Two-dimensional electrophoresis of human serum albumin (HSA) after Coomassie blue staining. Following isoelectric focusing, the HSA was eluted into a continuous slab gradient gel (T_{max}, 40%; C, 5%) by means of the discontinuous buffer system of pH 8.4. The pattern of HSA oligomers reveals a vertical line of disks, typical for the size of isomeric particles that have focused in the same place in the first-dimension run. (Bottom) Two-dimensional electrophoresis of bovine serum albumin (BSA) after Coomassie blue staining. After isoelectric focusing, the albumin was eluted into a continuous slab gradient gel (T_{max}, 30%; C, 5%) by means of the discontinuous buffer system of pH 8.4. The pattern shows not only the vertical array of BSA oligomers, but also the charge isomers of BSA monomer and dimer. The difference in isoelectric points is about 0.3 pH unit. The apparent spreading of the protein disks indicates the approach to their pore limit in the gel.

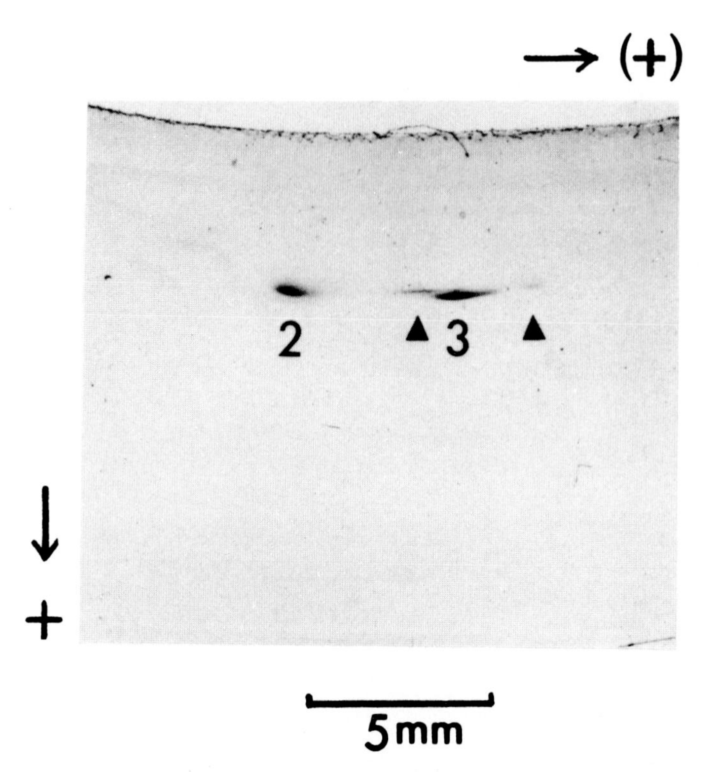

Fig. 15. Two dimensional, "dead-run" electrophoresis of LDH types 2 (left) and 3 (right). The enzymes were detected by the tetrazolium assay. In the first dimension, the enzymes were separated according to their charge properties, as shown in Fig. 11a. The second-dimension run was performed in a continuous gradient gel (T_{max}, 40%; C, 5%). Tris-borate buffer (pH 9.5) was used throughout the system. The buffer had a maximum concentration of 150 mM tris and was distributed in the gel in a continuous concentration gradient, as well as in the acrylamide. The buffer gradient is provided by the omission of buffer from the persulfate solution. This buffer gradient provides low conductivity in the gel top and the supernatant, thus allowing rapid elution of protein from the first-dimension gel into the slab gel. The terminal stage of electrophoresis shown here was attained only after a 10-hr run at 100 volts/cm gel. In earlier stages of the run, the basic LDH 2 trails, as a blurred spot, behind the favorably migrating acidic LDH 3. At the terminal stage, all proteins have closely approached their pore limit in the gel and form an almost horizontal line in the gel. Such migration behavior is typical for charge isomers of proteins in a steep-gradient gel. Since the difference between the pI of the basic LDH 2 (8.1) and the separation pH in the gel gradient (9.5) is smaller than the difference between the pI of the separated LDH species, this experiment outlines the power of pore-limit (dead-run) electrophoresis that is used to identify charge-isomeric proteins.

the sample. SDS provided by the running buffer will cause all protein to migrate down to the anode, and the protein can be checked for intensity and precipitations after Coomassie blue staining. Comparing the results of such pilot runs with separations of standard proteins, one will be able to correct the SDS content of the sample. Small samples that have once been oversaturated with SDS are almost lost, since removal of SDS by dialysis or

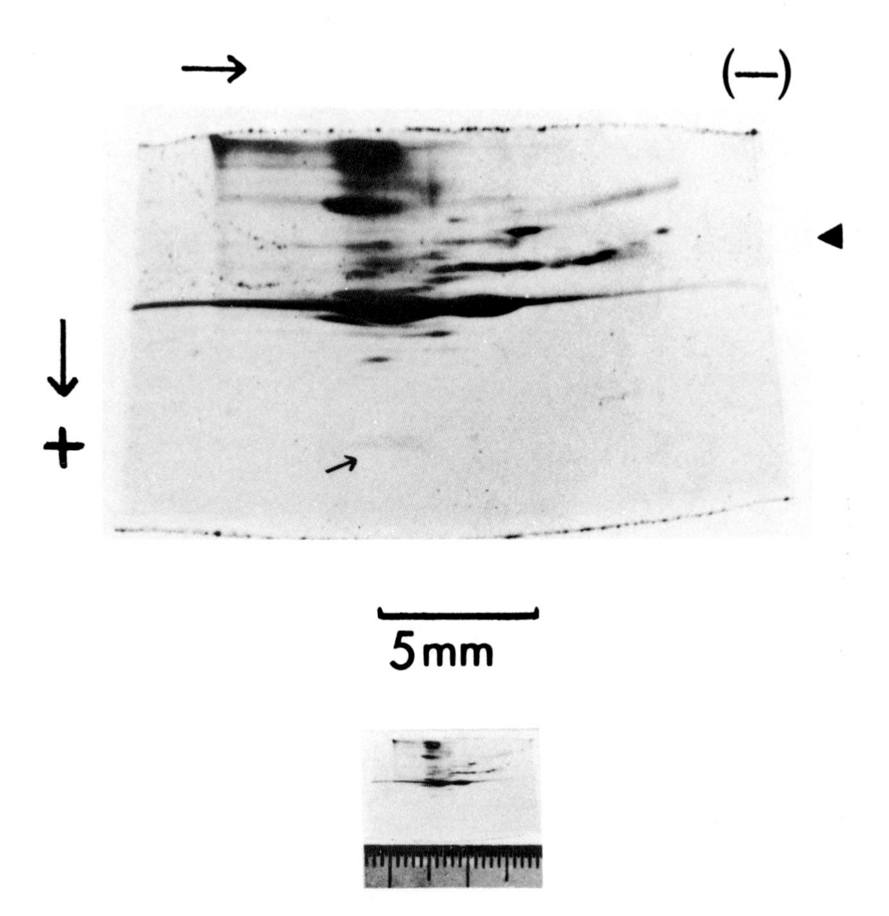

Fig. 16. Two-dimensional electrophoresis of rat serum proteins after staining with Coomassie blue. In the first dimension, electrofocusing was performed in the indicated direction. The section of the cylindrical electrofocusing gel covering approximately pH 3–8 was cut out and applied on top of a continuous-gradient gel (T_{max}, 40%; C, 5%). The discontinuous buffer system of pH 8.4 was used at 100 volts/cm gel for 30 min. The run was stopped when bromophenol blue had just passed the gel and xylene cyanole was still short of the gel edge by about one fifth of the gel's length. The pattern shows two charge-isomeric albumins that have spread over the whole width of the gel. Such spreading is due partly to the elution of protein into the supernatant during the transfer of the cylindrical gel on top of the slab gel. Spreading also indicates the approach of a protein to its pore limit in the gradient gel. A series of protein spots (◀) that stretch from the acidic range to the neutral range and form a slightly bent curve can be seen. Depending on their apparent molecular weight (>100,000) and charge properties, these proteins may be immunoglobulins of the A and G types. At prolonged running times, these proteins form an almost horizontal line in the gel that indicates their closely related molecular weights. Peptides such as those in front of the albumin (→), however, have passed through the gel under such conditions. A picture of the slab gel in its original size has been added for comparison.

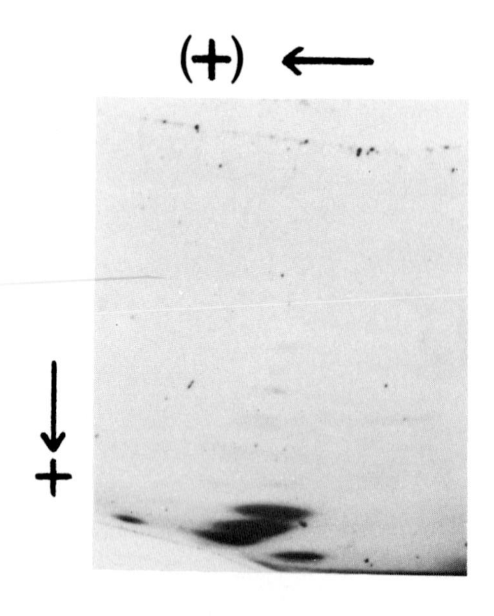

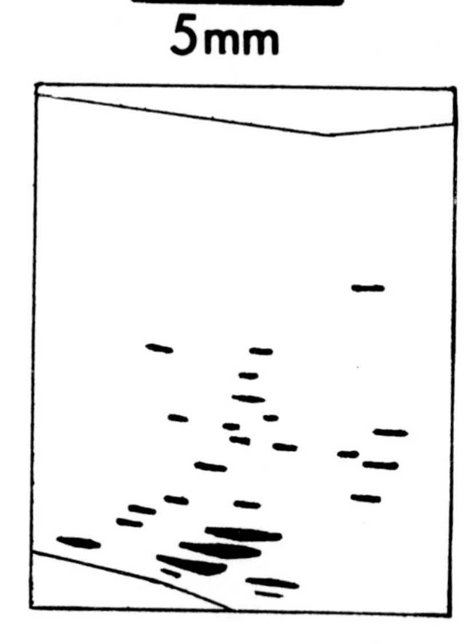

Fig. 17. Two-dimensional electrophoresis of water-soluble proteins stained with Coomassie blue that have been extracted from the bag cells of the abdominal ganglion of *Aplysia californica* and their sheath by means of ethyleneglycol passage and osmotic shock. After isoelectric focusing in the indicated direction, the proteins were eluted into a continuous slab gradient gel (T_{max}, 30%; C, 5%) by means of the discontinuous buffer system of pH 8.4. The run was stopped when bromophenol blue had just reached the lower end of the gel. The area of the slab gel shown here covers the pH range of 4–5 and demonstrates the variety of proteins with low molecular weights that are contained in such neurosecretory cells. With the use of radioactive-labeled protein and autoradiography of the gel, the sensitivity of such separation can be further increased.

precipitation of the protein in cold acetone (Weber and Osborn, 1975) may not be possible.

Continuous SDS electrophoresis in capillary tubes according to the large-scale method of Weber and Osborn (1969), employing phosphate buffer, does not yield consistent results with capillary gels of about 2-cm length used in disk and gradient microelectrophoresis. Only longer capillary gels will yield reasonable results. However, such gels have to be pushed out of the capillary tube according to the method described by Gainer (1971) or by breaking the glass tube surrounding the gel (Maizel, 1971). Good separations

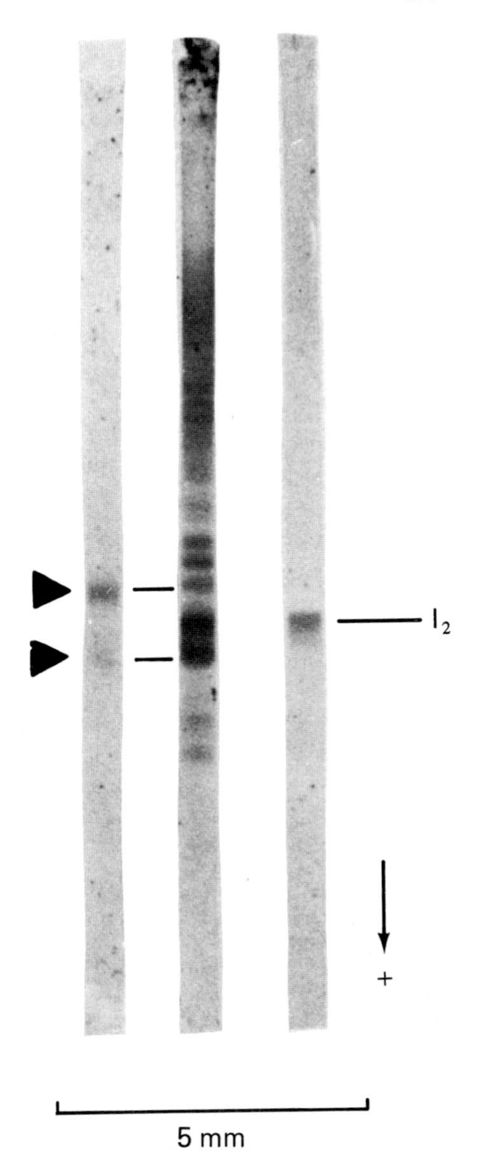

Fig. 18. Selective, two-dimensional electrophoresis of proteins from neurons of *Aplysia californica* after they were stained with Coomassie blue. Proteins extracted from the bag cells and the sheath of the abdominal ganglion (compare Fig. 12) were submitted to electrofocusing, and a peak identified by its white appearance in 10% TCA was cut out of the pH 5.5 area. The gel segment was washed for 2 min in H_2O prior to elution of the protein into a steep, cylindrical gel gradient (T_{max}, 60%; C, 4%). The discontinuous buffer system of pH 8.4 was used. The run was performed at 120 volts/cm gel until bromophenol blue reached the lower fifth of the gel and xylene cyanole migrated only two thirds of the gel length. After staining, two peaks can be identified (▶) that correspond with the pattern of all water-soluble proteins of the same sample run under comparable conditions (middle gel). Since insulin dimer has been added to the heterogeneous sample, an estimation of the molecular-weight range is possible. The position of insulin dimer (I_2) alone is shown in the right gel. Such a selective, two-dimensional technique, though more elaborate, is also more sensitive to faint protein peaks, since no peak-spreading hampers the pattern as in the slab gels.

of SDS–proteins in regular microgradient gels can be achieved by employing continuous buffers with anions of lower mobility than phosphate [i.e., glycine or borate (see Fig. 20)]. Continuous SDS electrophoresis is important as a means of comparison if patterns of SDS disk electrophoresis are suspected of artifacts (Rüchel *et al.*, 1974; Camacho *et al.*, 1975).

4.8. Microelectrophoresis of Nucleic Acids

Extraction of nucleic acids from single cells or small tissue samples has been done, employing different combinations of phenol, heat treatment,

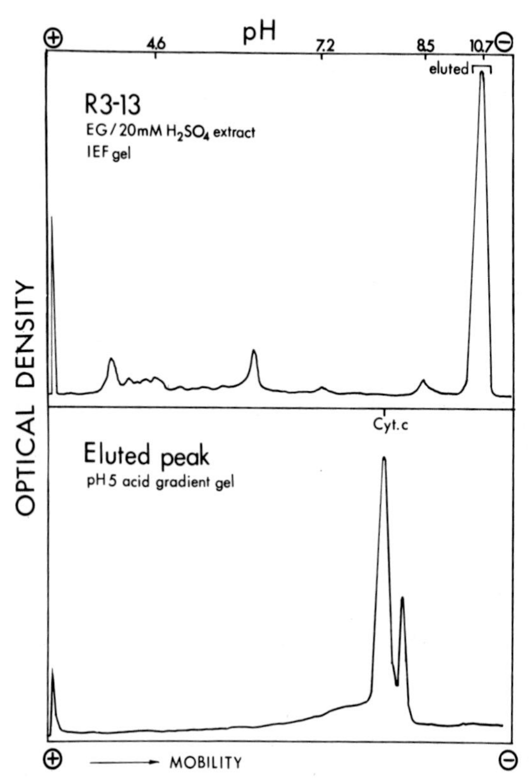

Fig. 19. *Top:* Optical-density scans of proteins from R3–13 separated on isoelectric focusing (IEF) gels, pH 2–11. The proteins were extracted with 20 mM H_2SO_4 from neurons that were pretreated with 100% ethylene glycol. The figure shows that the very basic proteins of pI 10 or greater can be extracted with mild acid. *Bottom:* The basic peak of R3–13 (see the top panel) was eluted from the IEF gel and rerun on an acid gracient (1–40%) microgel, pH 5, under conditions approaching the "dead run." The optical-density scan of the stained patterns on the acid gel shows that the basic peak was separated into two peaks with molecular weights equal to and less than cytochrome *c*. From Loh *et al.* (1977).

protease, and SDS (Peterson, 1970) or diethyl–oxidiformate–SDS and protease–SDS–phenol or SDS–mercaptoethanol–diethyl oxidiformate–pronase (Wolfrum *et al.*, 1974). Such treatment is used to destroy nucleases that would otherwise cause excessive lysis of high-molecular-weight nucleic acids. Thus far, none of these extraction procedures has yielded ideal results, and further improvement is desirable. Since both types of nucleic acids (i.e., RNA and DNA) are extracted, differentiation can be achieved by treatment with either RNAase or DNAase prior to separation.

Gradient-gel microelectrophoresis has been used for the separation of nucleic acids. In such gels, low-molecular-weight RNA can be separated readily in discontinuous buffer systems, but larger RNA tends to precipitate. In continuous buffer systems of higher ionic strength (and therefore higher conductivity), separation of the high-molecular-weight RNA is improved, but separation of the small RNA species is poor. Such problems are due in part to the high net charge of nucleic acids. To improve the separation of large RNA species without deterioration of the separation of small RNAs, the effective charge of the acids has been lowered by use of creatinine buffers at pH 5 (Wolfrum *et al.*, 1974). High-molecular-weight RNA separation is improved by lowering the bis concentration in the gel (Shaaya, 1976). The limit of sensitivity in the microgradient gel is 0.1 μg of a mixture of RNAs

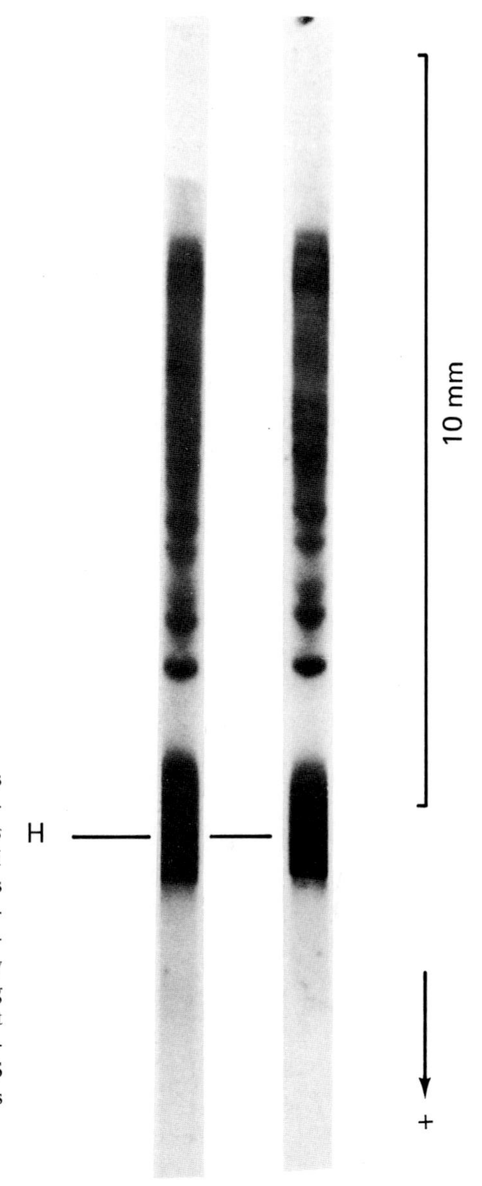

Fig. 20. Separation of unreduced SDS–proteins from human erythrocyte ghosts in 5-μl gradient gels (T_{max}, 35%; C, 5%). A *continuous* buffer system (50 mM tris + glycine to pH 8.4) containing 0.1% SDS was used. The runs were performed at 40 volts/cm gel until bromophenol blue and xylene cyanole were separated by about 2 mm. At this relatively early stage of separation, no considerable stripping of SDS has been observed. The imposing front peak (H) is due to heavy hemoglobin contamination. Such imposing front peaks in SDS runs occur in discontinuous buffer systems as well and are likely to be heterogeneous.

that can be separated and detected by their optical-density patterns. RNAs with molecular-weight differences of less than 1% have been determined using this electrophoretic technique (Sprinzl *et al.*, 1975). The separation of RNAs extracted from whole rat brain (Fig. 21) and from identified neurons of *Aplysia* (Fig. 22) has been possible with this technique.

Radioactive-labeled RNA has been extracted from salivary glands of insects (Egyháti *et al.*, 1969) and from large molluscan neurons (Peterson, 1970) and separated in agarose slabs using a continuous buffer system. In

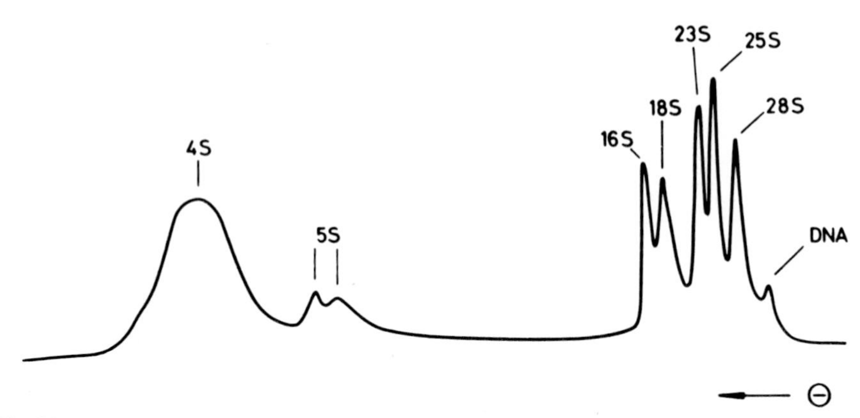

Fig. 21. Separation of various nucleic acids in a 5-μl gradient gel (T_{max}, 40%; C, 2%) employing a continuous tris–sodium phosphate buffer of pH 7.8 in the presence of 1 mM EDTA; 0.1% SDS was added to the sample only. The sample contained 4 S, 5 S, 18 S, and 28 S RNA from rat brain; 4S, 5S, 18 S, and 25 S RNA from root tips of pea; and 4 S, 5 S, 16 S, and 23 S RNA from *E. coli.* The DNA peak originates from the pea preparation and has been identified by its violet image after staining with toluidine blue. From Wolfrum *et al.* (1974).

these studies, RNA from nuclei and cytoplasm has been analyzed separately, yielding information about the RNA turnover in these cells. The base composition of 4 S RNA after gel electrophoresis has been investigated at the microgram level employing tritium label, and chromatography in the second-dimension run (Chia *et al.*, 1973). Agarose–polyacrylamide composite gel has been used in capillaries for separation of RNA that has been detected by staining with methylene blue (Kuzmin *et al.*, 1975). Such composite gels (of agarose and acrylamide) have been cast in microslab size and used for separation of RNA at the 10^{-8} to 10^{-10} level. The evaluation of such separations has been performed by scanning in UV light at about 260 nm wavelength (Ringborg *et al.*, 1968). Since this technique requires micromanipulation and special equipment, its use has been limited. This is also true for the highly sensitive continuous high-voltage electrophoresis technique designed by Edström (Edström, 1964; Edström and Neuhoff, 1973), which allows the detection of separated RNA-bases at the nanogram level in UV light.

It should be mentioned that isoelectric focusing of tRNAs has been performed successfully, although at the regular scale, yielding isoelectric points in the area of pH 4 (Righetti and Drysdale, 1973), and that two-dimensional separation of low-molecular-weight RNAs has been accomplished (Varricchio and Ernst, 1975).

4.9. *Equipment for Microelectrophoresis*

Any type of commercially available glass tube providing an inner diameter of about 0.3–0.5 mm and a length of about 30 mm may be used for the production of cylindrical microgels. Brands that provide stable capillaries

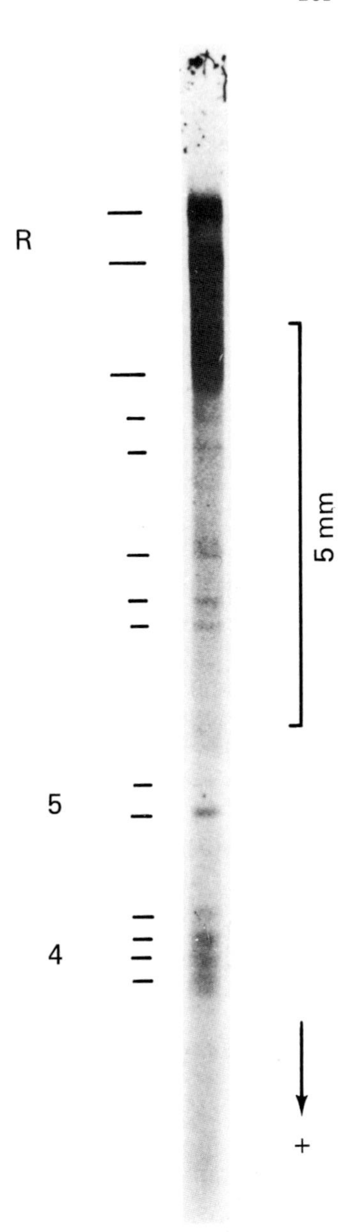

Fig. 22. Separation of RNAs extracted from R3–13 neurons in *Aplysia* on 1–30% *T*/5% *C* 5-μl gradients, pH 8.4 discontinuous buffer. Several 4 S peaks and two 5 S peaks as well as higher-molecular-weight RNA (presumably ribosomal RNA) can be seen. The faint intermediate peaks may be breakdown products of the ribosomal RNA.

with thick glass walls are to be preferred (i.e., Brand D6980, Wertheim, West Germany; Corning, Corning, New York; or Clay-Adams, Parsippany, New Jersey; the latter two have narrow ends and have to be cut into the appropriate length with a diamond, but they are considerably cheaper). New capillaries may be used as they come, but cleaning with bleach, as has been described (Rüchel and Wolfrum, 1974; Rüchel, 1976), enhances the suction properties of the capillaries and allows the "recycling" of used capillaries. Glass surfaces that have been treated with strong base may expose fixed negatively charged groups that cause an electroendosmotic flow to the cathode. This is especially harmful in SDS electrophoresis and causes smearing of protein over the gel surface. Such endosmotic flow can be detected by the development of flat gas bubbles squeezed between the glass wall and the gel during the run.

Microelectrophoresis can be performed with any type of D.C. power supply yielding about 40–250 volts in a controlled range. A modest and inexpensive version employs only 9-volt dry-cell batteries and homemade electrophoresis racks (see Fig. 23). Any work with microelectrophoretic techniques also necessitates the use of thin glass pipettes. Such pipettes can be pulled freehand out of Pyrex glass tubing (4 mm O.D., 2 mm I.D.) in an oxygen-supported flame. Since the outer diameter of the pipette tip should be less than half the inner diameter of the capillary tube that contains the gel, the pipette tips have to be thin (i.e., 0.1–0.2 mm). Therefore, the Pyrex tubing is pulled to about 2-mm outer diameter in a first step and to the final shape in a second step. The glass pipettes can be fitted with a piece of flexible tubing and a mouthpiece for pipetting. Since pipetting on such a small scale requires a certain degree of skill, the application of samples can be facilitated at the expense of running time. Capillaries containing gels are immersed in gel buffer, and the gas bubbles on top of the gel are removed by a vacuum. Afterward, the capillary can be fitted with a short piece of flexible tubing (with a wider inner diameter) into which the sample can be applied more easily than into the capillary itself.

4.10. Recent Advances

Recently, gel electrophoretic "mini" techniques have been described which are suitable for protein separation at the 10^{-7} gram scale (Tulchin *et al.*, 1976; Condeelis, 1977). General problems of the SDS–protein interaction

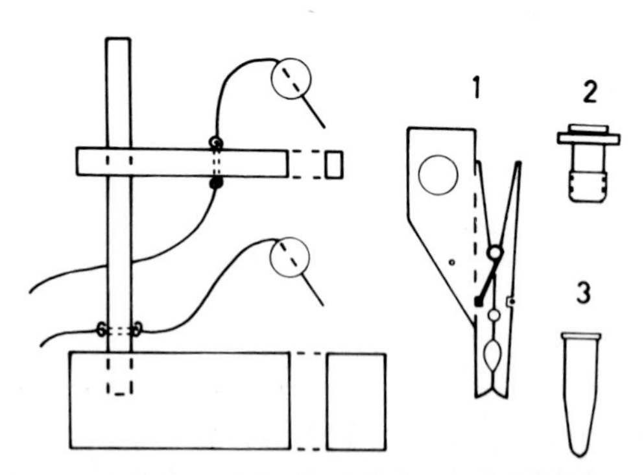

Fig. 23. Simplified electrophoresis equipment providing continuous height adjustment. (1) Holder for the upper buffer reservoir. (2) The lower buffer reservoir (3) fits into the hole of the stand. Both buffer reservoirs are made of commercial reaction tubes. The upper buffer reservoir is fitted with a rubber cap and a rubber ring (slice of rubber tubing). Both electrodes are platinum wire. For general use, stainless steel wire will perform as well. Reproduced with permission from Rüchel and Wolfrum (1974).

have been treated by Tanford and Reynolds (1976) and the state of SDS-electrophoretic techniques has been reviewed by Wyckoff *et al.* (1977) and by Stegemann (1979).

The microgradient gel technique has been further applied to identified neurons of *Aplysia californica*; a specific pattern of such cells has been demonstrated which is likely to be due to neurosecretory peptides (Rüchel *et al.*, 1977; Loh *et al.*, 1977). Immune precipitation of protein antigens (after separation in microgradient gels) by incubation of whole gels or gel slices in agarose-containing antisera has been described by Neuhoff and Mesecke (1977).

A staining procedure has been developed that allows for the investigation of charge properties of SDS–proteins after separation in polyacrylamide gels (Rüchel *et al.*, 1978); this procedure may be applied to proteins which, due to limited solubility, are not accessible to electrofocusing.

Microelectrofocusing has been updated further by Bispink and Neuhoff (1976), and the problem of staining proteins after electrofocusing has been reinvestigated by Vesterberg *et al.* (1977). The whole field of isoelectric focusing, which is still very much in development, has been reviewed by Radola and Graesslin (1977) and Radola (1980). Two-dimensional microelectrophoresis has been treated by Rüchel (1977); such stamp-sized slab gradient gels can also be polymerized with slots allowing for parallel runs of samples in one dimension. The slots are formed by a comb that is cut out of X-ray film and which is inserted into the top of the gel chamber after filling with the gradient mixture. Single-layer spacers for the gel chamber can simply be cut from cover slips used with hemocytometers since they have a thickness of approximately 0.5 mm.

Microgradient slab gels of approximately one square inch in size have recently been developed by Poehling and Neuhoff (1980). Such gels allow for high-resolution separation of proteins in either one- or two-dimensional mode. Other improvements of the microgel techniques have been reviewed by Grossbach and Kasch (1977) and by Neuhoff (1980).

Trace amounts of protein in gels can be localized by new silver-staining techniques (Oakely *et al.*, 1980; Merril *et al.*, 1979; Switzer *et al.*, 1979).

5. Related Techniques and Analysis of Gel Patterns

5.1. Qualitative and Quantitative Analysis of Gel Patterns

Macromolecules that have been separated on polyacrylamide microgels can be detected by staining with the same dyes that are typically used for conventional-size gels (for a discussion of the various staining and detection techniques that are available, see Maurer, 1971). Dyes that are commonly used in microelectrophoresis include Coomassie blue R250 and Amido black 10B for proteins, toluidine blue and methylene blue for nucleic acids, and periodic acid–Schiff (PAS) stain for glycoproteins (Rüchel *et al.*, 1974; Hydén *et al.*, 1966; Zacharius *et al.*, 1969). A less commonly used but potentially valuable approach is that of "charge-staining" [i.e., the use of acidic or basic dyes to "fix," via salt linkages, highly basic or acidic proteins, respectively (Rüchel, 1976)]. Thus, one can obtain information about the charge properties of proteins by the staining procedure itself (e.g., strong binding of crystal violet suggests a highly basic protein, whereas strong binding of methylene blue indicates that the protein is highly acidic). In addition, "charge-staining" is reversible, providing an aid in the detection of a protein on a gel for the purpose of elution (e.g., for two-dimensional electrophoresis).

Stained microgels should be evaluated as soon as possible, since they tend to pick up dye and dust particles on their surfaces, which produces difficulty if they have to be photographed and enlarged. Some dirt can be removed prior to photography or densitometry by washing the gel rapidly (for seconds *only*) in absolute methanol. The latter washing must be done with continuous motion to prevent the gel from sticking to the wall of the vessel. Photography of microgels can be done using a single-lens reflex camera fitted with bellows, and by keeping the gels in transmitted light. A light flash is very efficient as a light source (e.g., the Illumitran® equipment, Bowens, London). More detail of the separation pattern can often be revealed by densitometry using a microdensitometer (e.g., the Joyce-Loebl Microdensitometer). If such expensive equipment is not available, the microgel can be magnified optically to fit into a conventional gel scanner (see Fig. 24).

The component patterns on polyacrylamide gels can be quantitated by various optical methods (see Maurer, 1971). For microgels, the areas of the densitometric electropherograms of the stained peaks can be measured either by planimetry or by weighing the cut-out paper of the peak area. The calculation of the protein present in the peak can then be determined from standard calibration curves, in which peak areas of standard proteins are plotted vs. amount of protein (Fenner *et al.*, 1975; Kruski and Narayan, 1974; Bertolini *et al.*, 1976; Hsieh and Anderson, 1975).

5.2. Determination of Molecular Weight

Two methods are available for the estimation of a protein's molecular weight by PAGE: (1) electrophoresis in SDS(discussed in Sections 3.9 and 4.7) and (2) by "Ferguson-plot" analysis (Ferguson, 1964), which is based on the effective electrophoretic mobility of a macroion in a defined gel system. In the latter method, the proteins need not be denatured, and information is also obtained about the charge properties of the protein. To determine the effective electrophoretic mobility of a macroion, the migration distance of the ion is divided by the migration distance of a marker dye, which migrates freely with the buffer front. Typical marker dyes that are used are bromophenol blue or bromophenol red in anodic buffer systems and methyl green in cathodic buffer systems. In steep-gradient gels, the gel top may be used as the point of reference, since even synthetic dyes are sieved (retarded) in such gels (Rüchel *et al.*, 1974).

The free electrophoretic mobility m_0, which is a physical constant for a

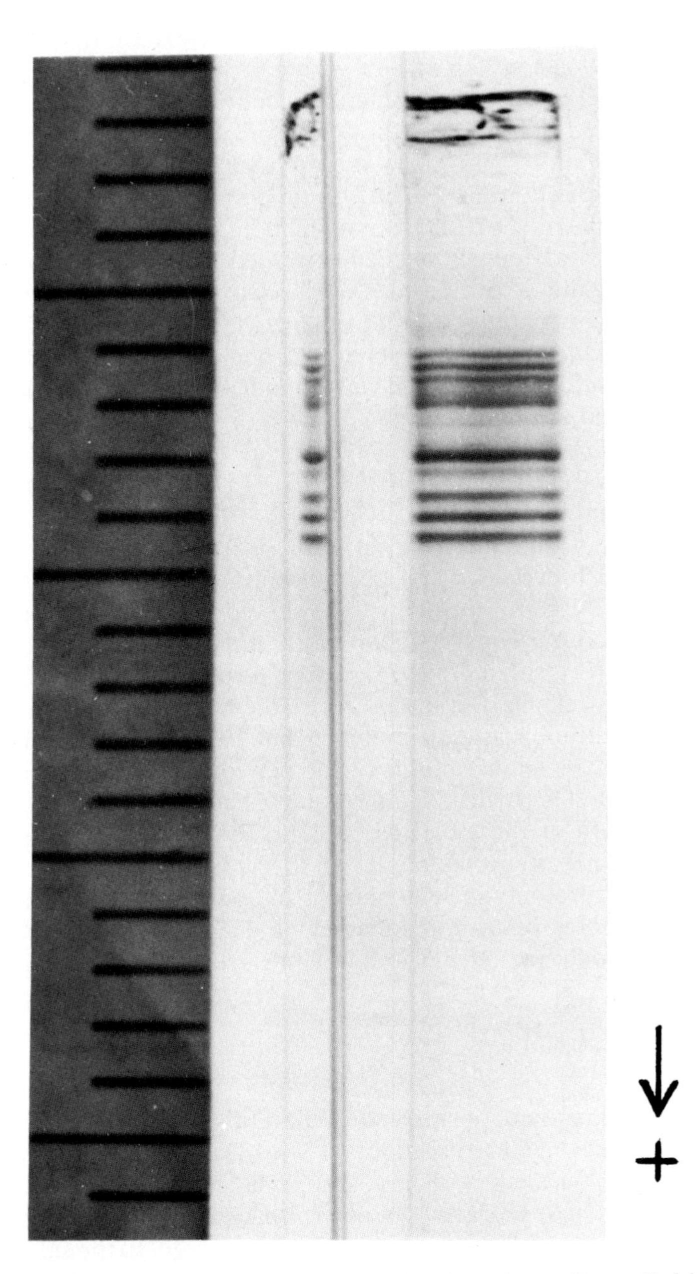

Fig. 24. A cylindrical glass rod placed in the top of a microgel to act like a cylindrical magnifying lens. The image of the same gel held in the same cuvette (made of slide glass) but covered with a coverslip is shown at the left. With the use of such a simple arrangement, an expensive microdensitometer may not be needed for evaluation of small gels. The separated pattern shows myelin proteins from rat brain and various standard proteins (cytochrome *c*, catalase, BSA, and conalbumin). The discontinuous buffer system of pH 8.4 was used in the presence of 0.1% SDS. From Rüchel (1976).

given ionic species, can be calculated according to an equation first described by Ferguson (1964): $\log m = \log m_0 - K \cdot T$, where m is the effective mobility of the macroion in a gel of comonomer concentration T. The retardation coefficient K has been defined as $K = C \cdot \mathrm{MW} + m_0$, where C is the cross-linkage of the gel (Maurer, 1971) [see also the alternative definition of K by Chrambach and Rodbard (1971)]. Thus, the molecular weight (MW) of the particle can be calculated if m_0 (the free mobility of the macroion in a gel of zero concentration) and K have been determined. When $\log m$ is plotted against T, a straight line is produced, the slope of which is defined by the retardation coefficient K and the intercept of which with the $\log m$ scale defines the $\log m_0$, which is related to charge properties of the macroion. To draw a Ferguson plot, electrophoretic runs in at least three gels of different and exactly controlled comonomer concentration (T) at constant cross-linking (C) are required. Examples of the plots have been shown by various authors (e.g., see Thorun, 1971; Banker and Cotman, 1972; Hedrick and Smith, 1968; Chrambach and Rodbard, 1971; Ugel *et al.*, 1971; Hearing *et al.*, 1976).

Certain failings of the Ferguson-plot analysis have been pointed out by Felgenhauer (1974), especially when gel concentrations above 20% T were used. In such gels, migrating proteins were slowed down to approximately zero mobility (dead run), which is theoretically impossible according to the Ferguson equation (Felgenhauer, 1974). Using zero mobility as revealed by pore-limit electrophoresis in continuous gel gradients, an exclusion limit T_{EL} (comonomer concentration at zero mobility) has been proposed for determination of the hydrodynamic volume of the macroion under consideration (Felgenhauer, 1974). The hydrodynamic volumes as defined by the Stoke's radii allow calculation of the molecular weight, assuming that the average partial specific volume of proteins is close to 0.73 g/ml (Andrews, 1965). Rodlike and fibrous proteins such as fibrinogen and myosin as well as large bulky structures such as IgM and α_2-macroglobulin need special consideration, since they cannot be treated like weight-equivalent spheres (Felgenhauer, 1974).

5.3. *Sources of Artifacts*

In sieving gels of constant pore size, the separation pattern of a macroionic mixture remains unchanged throughout the run, whereas in gradient gels, changes in position may occur. For example, in gradient gels, a smaller particle of low charge may be sieved out of the steady stack of a discontinuous buffer system earlier than a larger particle of higher charge. In the subsequent stages of zone electrophoresis, the velocity of the larger particle will slow down faster due to sieving in the gradient gel, and the smaller particle may pass the larger particle or even pass out of the gel. Such phenomena can be investigated by a series of runs in identical gradient gels with different running times. Such an experiment is illustrated in Fig. 25.

In any type of electrophoresis, certain typical *sources of artifacts* that may influence the separation pattern must be considered (see Ressler, 1973). A

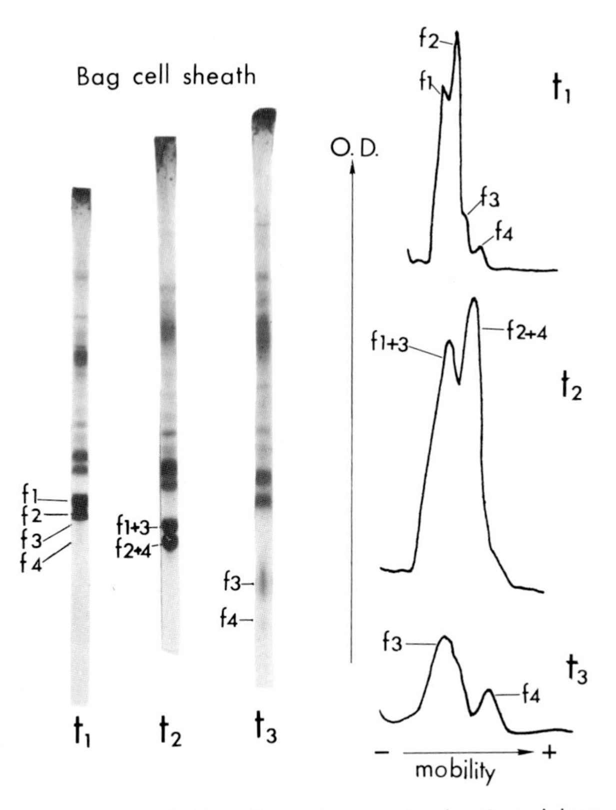

Fig. 25. Photographs of three 1–30% gradient microgels showing the staining pattern of water-soluble, bag-cell-sheath proteins after different running times ($t_1 < t_2 < t_3$). The optical-density (O.D.) patterns of the rapidly moving proteins (f_1–f_4) on the three gels are shown at the right of the figure. At t_1, the mobility of the four bands is $f_4 > f_3 > f_2 > f_1$; at t_2, $f_2 = f_4 > f_1 = f_3$; at t_3, $f_1 + f_2$ have moved off the gel and $f_4 > f_3$ remain, indicating that the molecular weights are $f_3 > f_4 > f_1, f_2$.

general requirement for electrophoresis is that a charge of the sample ions sufficient to allow migration be present. Therefore, the isoelectric points of amphoteric ions in the sample should differ more than 1 pH unit from the pH of the running buffer (Leaback, 1975). Lack of effective charge is indicated if a protein migrates slowly and displays a tail. In discontinuous buffer systems, such a lack of charge may prevent a protein from entering the stack at the very beginning of the run. Certain macroions such as high-molecular-weight RNA or lipoproteins tend to precipitate when they are forced into high concentration, for instance, in a steady-state stack. Precipitate peaks also occur frequently in discontinuous SDS electrophoresis and can be identified by their distorted images. To prevent tailing in continuous electrophoresis of RNA, a continuous buffer system has to be used with elevated conductivity or the pH of the buffer system used has to be made more acidic. In discontinuous SDS electrophoresis, precipitation indicates a lack of SDS if the precipitate occurs anywhere in the pattern. However, a precipitate in

Fig. 26. Staining pattern of electrophoretically separated (water-soluble) bag-cell proteins on 1–30% gradient microgels in the presence (+) and absence (−) of SDS. Note that SDS interferes with resolution of low-molecular-weight proteins (<12,000) on these gels.

front of the pattern is often due to low-molecular-weight proteins (peptides) that are not separated by the SDS gel system but are displaced by SDS–micelles. SDS–micelles can be detected by staining with toluidine blue (Rüchel *et al.*, 1974). SDS interference with the resolution of low-molecular-weight proteins is illustrated in Fig. 26. If peaks are bent in the direction of migration, the applied electrical field strength has been too high, and a thermal gradient has developed in the cross section of the gel. Since raised temperature increases dissociation and thus charge of migrating macroions, particles in the center of the gel travel faster and produce bent peaks. Smearing of proteins on the outer surface of the gel is due to electroendosmosis, which is discussed in Section 4.9.

In disk electrophoresis, the use of constant-pore-size gel confines the range of macroions to be separated to about one order of magnitude in molecular weight. Thus, the appearance of a fast-migrating protein peak just behind the dye marker is likely to represent unsieved peptides, and therefore a steady-state stack. Separation of the sample in a gel of significantly higher concentration will usually allow separation of such peptides.

A major source of artifacts in isoelectric focusing is the contamination of the sample with terminating electrolyte. Not only may the protein be denatured by the contact with base or acid, but also the electrolyte adds so much conductivity to the sample compartment that the relatively low-charged protein ions will practically not migrate any more and are lost as a precipitate in the top of the gel. When using polyacrylamide as supporting medium for isoelectric focusing, the pore size has to be large enough to avoid sieving. This requirement is fulfilled in most cases by use of 5% gel (see Baumann and Chrambach, 1976*a*). Considerable sieving prevents a protein from reaching its isoelectric point. Proteins smeared in the center of the gel are probably being retarded by sieving. Their position in the gel is variable when parallel runs are compared.

As has been mentioned above, several groups of proteins behave abnormally when SDS electrophoresis patterns are evaluated in terms of molecular weight. Some additional sources of artifacts found with SDS electrophoresis include the following: Disulfide bonds that have been cleaved by reduction may be reoxidized in the presence of SDS and produce apparent high-molecular-weight peaks in the pattern (Rüchel *et al.*, 1974; Richter-Landsberg *et al.*, 1974). Disulfide formation during SDS electrophoresis can be prevented either by the presence of an electrophoretically mobile reducing agent (Rüchel *et al.*, 1975) or by alkylation of the exposed sulfhydryl groups immediately after reduction (Maizel, 1971). Since SDS is linked to proteins by noncovalent bonds only, severe sieving during electrophoresis can cause breakdown of the SDS–protein micelle. Such stripping of the SDS may result in additional peaks, since SDS binds to proteins in at least two stoichiometric ratios, each of which can be represented by a separate peak in the gel (Rüchel *et al.*, 1974). To avoid the stripping of SDS in a gradient gel, the applied field strength may be lowered and the run has to be stopped early; as an indicator for timing, the separation of the two anionic marker dyes bromophenol blue and xylene cyanole FF can be used in the discontinuous buffer system.

Blurred patterns and precipitations of protein are also caused by the interference of nonionic lipids (including nonionic detergents) with the ionic SDS. Therefore, the presence of Triton has to be avoided in SDS microgels, and extracts of membrane proteins yield better separations if they are washed and precipitated in ether–ethanol (3:2) prior to SDS treatment (Morell *et al.*, 1975). Proteolysis remains a potential problem when treating proteins with SDS, since several proteases (e.g., pronase and proteinase K) survive SDS alone at higher temperatures for quite a while and may even show enhanced activity (Wolfrum *et al.*, 1974). Therefore, proteins should be loaded with SDS in the presence of reducing agent and boiled for 1 min if possible. In addition, protease blockers may be used (Weber *et al.*, 1972). If mercapto-ethanol concentrations significantly higher than 1% (vol./vol.) are employed for reduction of the SDS–protein complex, the boiling time is critical, since protein aggregation may occur under these conditions (Morell *et al.*, 1975).

Proteins have to be stained to full saturation at once with Amido black or Coomassie blue, since superficially stained peaks will not take up more dye if staining is repeated. Only SDS–proteins will pick up dye to full saturation out of dilute solutions over an extended period of time. Nucleic acids bind with appropriate dyes primarily via salt (ionic) bonds. Therefore, staining of nucleic acids in the presence of acid is very inefficient. Destaining of nucleic acids has to be performed in deionized water only. However, addition of some alcohol and dye is helpful to prevent fading of the peaks (Wolfrum *et al.*, 1974).

6. Interfacing Electrophoresis with Other Techniques

Electrophoresis on a large scale has been transformed into a preparative tool, but preparative electrophoresis has to rely on bulk preparations (e.g., whole brain homogenates) and is therefore confined to limited applications in neurobiology (e.g., for the investigation of myelin proteins or synaptosomal proteins). Electrophoresis as an analytical tool is used mostly as an intermediate or final step in an experimental sequence to demonstrate results at the macromolecular level.

6.1. Elution and in Situ Histochemistry

To make separated macroions accessible for further investigation, they have to be eluted from the gel. Elution is most effectively achieved by diffusion only if large pore gel or other nonsieving media are used. If the gel is polymerized with cross-linkers other than bisacrylamide, the protein can be liberated by dissolving the gel (Maurer, 1971). Electrophoretic elution from microgels has been described various times and is a requirement for two-dimensional electrophoresis (Neuhoff and Schill, 1968; Dames *et al.*, 1972). The special features of elution of SDS–proteins from gel have recently been reviewed (Weber and Osborn, 1975).

Since elution from the gel is always accompanied by a loss of protein, techniques dealing with the protein in place in the gel (*in situ*) are often preferred. Enzymatic activities of various dehydrogenases have been measured in microgels employing the tetrazolium system for detection. The sensitivity of such measurements was thus increased by three orders of magnitude as compared with the usual optical tests (Cremer *et al.*, 1972; Quentin and Neuhoff, 1972). In addition, peroxidase activity has been measured in microgels, after separation, by the diaminobenzidine reaction (Phillips-Burnett *et al.*, 1978), as well as glycosidases (Marz *et al.*, 1976). The activities of proteases have been determined in gels containing gelatin as a substrate, and the proteases have been located in the gel by negatively stained peaks (Klockow, 1976; Ward, 1976). Activity of DNA-dependent DNA-polymerase and of DNA-dependent RNA-polymerase has been demonstrated in microgels by both indirect binding tests and direct polymerization tests (Neuhoff and Lezius, 1968; Neuhoff *et al.*, 1968, 1970*a*) (for a review, see Neuhoff, 1973*a*).

Various enzymes have been recovered after SDS electrophoresis and subsequent removal of the detergent (for a review, see Weber and Osborn, 1975). Since these methods involve elution of the enzyme from the gel (with dialysis and ion exchange in the presence of urea), they are not suited for the microscale. However, SDS has been stripped from proteins by electrical forces in microgradient gels under pore-limit conditions (Rüchel *et al.*, 1974). Recovery of enzymatic activity by this method has been shown with β-galactosidase, and the investigation of membrane-bound enzyme activity after SDS electrophoresis in the gel seems possible. SDS has also been removed from an enzyme by incubation of the gel in SDS-free buffer containing substrate (Strauss, 1975). Awdeh (1976) has recently described a simple dialysis device that allows the handling of volumes down to 5 μl and may be very useful in the procedures discussed above.

6.2. *Immunological and Autoradiographic Techniques*

Several immunological techniques have been adapted to microelectrophoresis. A combination of disk microelectrophoresis with radial immune diffusion in agar gel has been described (Neuhoff and Schill, 1968). Crossed immune electrophoresis of the Laurell type has been combined with disk microelectrophoresis and yields a sensitivity of 3 μg for the antigen if 40 μl antiserum is employed (Dames *et al.*, 1972; Giebel, 1974). In addition, a multiphasic, one-dimensional immune electrophoresis has been described in polyacrylamide gels utilizing the mobility difference of antigen and antibody in a sieving gel (Fitschen, 1964; Louis-Ferdinand and Blatt, 1967). A microversion of this technique employing the dead-run property of gradient gels has been outlined (Rüchel *et al.*, 1973*a*, 1974). Further immunoelectrophoretic methods have been reviewed recently (Verbruggen, 1975).

SDS–proteins eluted from the gel have been used successfully for immunization. Immunization has also been achieved employing SDS–protein in a gel homogenate, although the latter approach yielded antibodies against gel components as well (Weber and Osborn, 1975).

A major use of PAGE is in the separation of radioactive proteins. In recent years, the trend is away from gel slicing and counting (i.e., by liquid scintillation or gamma counting) and toward autoradiographic analysis using X-ray films. Autoradiography has several advantages, which include less tedium in obtaining the data and better resolution of the individual radioactive peaks (mainly the minor peaks). Furthermore, only autoradiography is appropriate for two-dimensional electrophoresis. The disadvantages are greater difficulty in quantitation and limitations regarding double-label experiments. The recent development of scintillation autoradiography (also known as fluorography) allows for more rapid development of autoradiograms from gels and detection of tritium-labeled proteins as well (Bonner and Laskey, 1974). Preexposure of the X-ray films to a brief flash of light greatly increases the sensitivity of fluorography and also permits quantitative interpretation of the autoradiogram (Laskey and Mills, 1975). The improvement in quantitation is due to the correction, on the sensitized film, of the nonlinear relationship between radioactivity and absorbance of the film image. Autoradiography has also been performed with whole microgels using radiation-monitoring films (Neuhoff *et al.*, 1973).

6.3. *Peptide Mapping and Amino Acid Analysis of Proteins Following Electrophoresis*

Peptide mapping and amino acid sequencing following electrophoresis have been described, to date, only on a large scale (Bray and Brownlee, 1973; Bridges, 1976). Amino acid analysis of proteins after SDS electrophoresis using [^{14}C]dansyl has also been described (Weiner *et al.*, 1972), by employing microchromatography on polyamide sheets (for a review, see Neuhoff, 1973*b*). This method detects 10^{-12} mol amino acid, and therefore, analysis of the composition of the major protein peaks in microgels should be possible. However, this has not yet been reported. Determination of the C and N termini of peptide chains in the presence of SDS has been performed often (Neuhoff *et al.*, 1970*b*; Weiner *et al.*, 1972). Since at least 2.5×10^{-10} mol protein is needed for each cycle of the Edman-dansyl method employed for determination of the N terminus (Weiner *et al.*, 1972), this approach is not yet sensitive enough for the analysis of proteins separated in microelectrophoresis. This is also true for the determination of the C terminus employing carboxypeptidase A, which requires at least 0.2 mg protein (Neuhoff *et al.*, 1970*b*). A method described for the amino acid analysis of stained protein bands in gels may also be able to be scaled further down to fit with microelectrophoresis (Houston, 1971).

7. *Concluding Remarks*

In less than 40 years, the technique of electrophoresis has become one of the most widely used analytical methods in the biological sciences. Although protein biochemistry has made an impact on neurobiology only relatively recently, a very large literature on proteins and peptides in the nervous system already exists. PAGE plays an important role in this literature, and its usage is expected to increase. In this chapter, we have focused not only on the conventional uses of PAGE but also on nonconventional approaches. As with all techniques, PAGE is afflicted with artifacts, and it is hoped that the diagnostic tests and remedies discussed in these pages will assist the investigator in avoiding them.

Our preoccupation with microelectrophoretic techniques comes largely from our interests in protein specificity in the nervous system. Not all neurobiologists are dedicated to this approach, and many will still prefer to work (effectively) on the larger scale. However, micromethods can be useful for these investigators as well, simply because screening for appropriate gel systems is considerably more rapid by these techniques. In addition, it should be pointed out that micromethods have certain virtues not shared by the conventional-scale systems. For example, the easy production of microgradient gels ranging from 1 to 60% acrylamide is not possible at all in larger gels. The lower gel concentrations allow for the fractionation of ribosomal and virus particles as well as proteins at the 10^6 molecular-weight level (see

Fig. 4). This ability of microgradient gels to separate particles should, with some effort, provide a method for the fractionation of subcellular organelles by PAGE. Manipulation of the cross-linkage parameter (Baumann and Chrambach, 1976*b*) could conceivably extend the present range of particle separation.

Many technological advances with PAGE are currently under way, and as this chapter indicates, new uses for the technique will continue to emerge. At the present time and state of the art, PAGE provides a most versatile and essential biochemical tool for the neurobiologist.

References

Allen, R.C.,1971, Polyacrylamide gel electrophoresis with discontinuous voltage gradients, in: *Disc Electrophoresis* (H.R. Maurer, ed.), pp. 29–31, Walter de Gruyter, Berlin.

Ames, G.F.L.,and Nikaido, K., 1976, Two-dimensional gel electrophoresis of membrane proteins, *Biochemistry* **15:**616.

Andrews, P., 1965, Gel-filtration behavior of proteins related to their molecular weights over a wide range, *Biochem. J.* **96:**595.

Arlinger, L., 1971, Isotachophoresis in capillary tubes, in: *Protides of Biological Fluids* (H. Peeters, ed.), pp. 513–519, Pergamon Press, Oxford.

Arlinger, L.,1975, Analytical isotachophoresis in capillary tubes: Analysis of hemoglobin, hemiglobin cyanide and isoelectric fractions of hemoglobin cyanide, *Biochim. Biophys. Acta* **393:**396.

Arlinger, L., 1976, Preparative capillary isotachophoresis: Principle and some applications, *J Chromatogr.* **119:**9.

Awdeh, Z.L., 1976, Dialysis of micro samples, *Anal. Biochem.* **71:**601.

Banker, G.A., and Cotman, C.W.,1972, Measurement of free electrophoretic mobility and retardation coefficient of protein–sodium dodecyl sulfate complexes by gel electrophoresis, *J. Biol. Chem.* **247:**5856.

Baumann, G. and Chrambach, A., 1976*a*, Gram-preparative protein fractionation by isotachophoresis: Isolation of human growth hormone isohormones, *Proc. Natl. Acad. Sci. U.S.A.* **73:**732.

Baumann, G. and Chrambach, A., 1976*b*, A highly cross-linked, transparent polyacrylamide gel with improved mechanical stability for use in isoelectric focusing and isotachophoresis, *Anal. Biochem.* **70:**32.

Been, A.C., and Rasch, E.M., 1972, A vertical micro system for discontinuous electrophoresis of insect tissue proteins using thin sheets of polyacrylamide gel, *J. Histochem. Cytochem.* **20:**368.

Bertolini, M.J., Tankersley, D.L., and Schroeder, D.D., 1976, Staining and destaining polyacrylamide gels: A comparison of Coomassie blue and Fast green protein dyes, *Anal. Biochem.* **71:**6.

Birdi, K.S., 1976, Comparative purity of commercially available sodium dodecyl sulfates: A simple criterion, *Anal. Biochem.* **74:**620.

Bispink, G., and Neuhoff, V., 1976, Isoelektrische Fokussierung in Mikrogelen zur Fraktionierung komplexer Proteingemische im Nanogrammbereich, *Hoppe-Seyler's Z. Physiol. Chem.* **357:**991.

Bonner, W.M., and Laskey, R.A., 1974, A film detection method for tritium-labelled proteins and nucleic acids in polyacrylamide gels, *Eur. J. Biochem.* **46:**83.

Bosselman, R.A., and Kaulenas, M.S.,1976, A rapid two-dimensional micro electrophoretic procedure for the analysis of ribosomal proteins, *Anal. Biochem.* **70:**281.

Bray, D., and Brownlee, S.M., 1973, Peptide mapping of proteins from acrylamide gels, *Anal. Biochem.* **55:**213.

Bridges, J., 1976, High sensitivity amino acid sequence determination: Application to proteins eluted from polyacrylamide gels, *Biochemistry* **15:**3600.

Camacho, A., Carrascosa, J.L., Vinuela, E., and Salas, M., 1975, Discrepancy in the mobility of a protein of phage ϕ 29 in two different SDS polyacrylamide gel systems, *Anal. Biochem.* **69:**395.

Cann, J.R., 1968, Recent advances in the theory and practice of electrophoresis, *Immunochemistry* **5:**107.

Catsimpoolas, N., 1973, Isoelectric focusing and isotachophoresis of proteins, *Sep. Sci.* **8:**71.

Chia, L.-L.S.Y., Randerath, K., and Randerath, E., 1973, Base analysis of ribopolynucleotides by tritium incorporation following analytical polyacrylamide gel electrophoresis, *Anal. Biochem.* **55:**102.

Chrambach, A., and Rodbard, D., 1971, Polyacrylamide gel electrophoresis, *Science* **172:**440.

Chrambach, A., Kapadia, G., and Cantz, M., 1972, Isotachophoresis in polyacrylamide gel, *Sep. Sci.* **7:**785.

Condeelis, J.S., 1977, A sodium dodecyl sulfate micro gel electrophoresis technique suitable for routine laboratory analysis, *Anal. Biochem.* **77:**195.

Cox, H.D., and Teven, J.M.G., 1976, On the mechanism of the molecular-sieve effect in polyacrylamide gel electrophoresis, *J. Chromatogr.* **123:**261.

Cremer, T., Dames, W., and Neuhoff, V., 1972, Micro disk electrophoresis and quantitative assay of glucose-6-phosphate dehydrogenase at the cellular level, *Hoppe-Seyler's Z. Physiol. Chem.* **353:**1317.

Dames, W., and Maurer, H.R., 1974, Simultaneous preparation for electrophoresis of a large number of micro polyacrylamide gels with continuous concentration gradients, in: *Electrophoresis and Isoelectric Focusing in Polyacrylamide Gel* (R.C. Allen and H.R. Maurer, eds.), pp. 221–231, Walter de Gruyter, Berlin.

Dames, W., Maurer, H.R., and Neuhoff, V., 1972, Micro antigen–antibody crossed electrophoresis in vertical agarose gels following micro disk electrophoresis, *Hoppe-Seyler's Z. Physiol. Chem.* **353:**554.

Davis, R.H., Copenhaver, J.H., and Carver, M.J., 1972, Characterization of acidic proteins in cell nuclei from rat brain by high-resolution acrylamide gel electrophoresis, *J. Neurochem.* **19:**473.

Dunker, A.K., and Rueckert, R.R., 1969, Observations on molecular weight determinations on polyacrylamide gel, *J. Biol. Chem.* **244:**5074.

Edström, J.E., 1964, Micro extraction and micro electrophoresis for determination and analysis of nucleic acids in isolated cellular units, in: *Methods in Cell Biology*, Vol. 1 (D.M. Prescott, ed.), pp. 417–447, Academic Press, New York.

Edström, J.E., and Neuhoff, V., 1973, Micro electrophoresis for RNA and DNA base analysis, in: *Micro Methods in Molecular Biology* (V. Neuhoff, ed.), pp. 215–256, Springer, Berlin.

Egyházi, E., Daneholt, B., Edström, J.E., Lambert, B., and Ringborg, U., 1969, Low molecular weight RNA in cell components of *Chironomus tentans* salivary glands, *J. Mol. Biol.* **44:**517.

Emes, A.V., Latner, A.L., and Martin, J.A., 1975, Electro focusing followed by gradient electrophoresis: A two-dimensional polyacrylamide gel technique for the separation of proteins and its application to the immunoglobulins, *Clin. Chim. Acta* **64:**69.

Everaerts, F.M., 1972, Isotachophoresis, *J. Chromatogr.* **65:**3.

Everaerts, F.M., 1973, Panel discussion, in: *Towards a Unified Electrophoretic Viewpoint*, (D. Rodbard, moderator), *Ann. N. Y. Acad. Sci.* **209:**515.

Everaerts, F.M., Geurts, M., Mikkers, F.E.P., and Verheggen, T.P.E.M., 1976, Analytical isotachophoresis, *J.Chromatogr.* **119:**129.

Fawcett, J.S., and Morris, C.J.O.R., 1966, Molecular sieve chromatography of proteins on granulated polyacrylamide gels, *Sep. Sci.* **1:**9.

Felgenhauer, K., 1974, Evaluation of molecular size by gel electrophoretic techniques, *Hoppe-Seyler's Z. Physiol. Chem.* **355:**1281.

Fenner, C., Traut, R.R., Mason, D.T., and Wikman-Coffelt, J., 1975, Quantification of Coomassie blue stained proteins in polyacrylamide gels based on analyses of eluted dye, *Anal. Biochem.* **63:**595.

Ferguson, K.A., 1964, Starch-gel electrophoresis—application to the classification of pituitary proteins and polypeptides, *Metabolism* **13**:985.

Fitschen, W., 1964, A quantitative study of antigen–antibody combination during disk electrophoresis in acrylamide gel using iodine-131 labelled human growth hormone, *Immunology* **7**:307.

Frazier, W.T., Kandel, E.R., Kupfermann, I., Waziri, R., and Coggeshall, R.E., 1967, Morphological and functional properties of identified neurons in the abdominal ganglion of *Aplysia californica*, *J. Neurophysiol.* **30**:1288.

Gainer, H., 1971, Micro disk electrophoresis in sodium dodecyl sulfate, *Anal. Biochem.* **44**:589.

Gainer, H., 1973, Isoelectric focusing of proteins at the 10^{-10} to 10^{-9} g level, *Anal. Biochem.* **51**:646.

Garewal, H.S., and Wasserman, A.R., 1974, Triton X-100–4 M urea as an extraction medium for membrane proteins, *Biochemistry* **13**:4063.

Giebel, W., 1974, Identification of serum proteins in a combined micro-disk and micro-antigen–antibody crossed electrophoresis, in: *Electrophoresis and Isoelectric Focusing in Polyacrylamide Gel* (R.C. Allen and H.R. Maurer, eds.), pp. 231–233, Walter de Gruyter, Berlin.

Graesslin, D., Trautwein, A., and Bettendorf, G., 1971, Gel isoelectric focusing of glycoprotein hormones, *J. Chromatogr.* **63**:475.

Griffith, A., and Catsimpoolas, N., 1972, General aspects of analytical isotachophoresis of proteins in polyacrylamide gel, *Anal. Biochem.* **45**:192.

Grossbach, U., 1965, Acrylamide gel electrophoresis in capillary columns, *Biochim. Biophys. Acta* **107**:180.

Grossbach, U., 1972, Microelectrofocusing of proteins in capillary gels, *Biochem. Biophys. Res. Commun.* **49**:667.

Grossbach, U., and Kasch, E., 1977. Microelectrophoresis at the cellular level, in: *Techniques of Biochemical and Biophysical Morphology*, Vol. 3 (D. Glick and R.M. Rosenbaum, eds.), pp. 81–101, Wiley, New York.

Hearing, V.J., Klingler, W.G., Ekel, T.M., and Montague, P.M., 1976, Molecular weight estimation of Triton X-100 solubilized proteins by polyacrylamide gel electrophoresis, *Anal. Biochem.* **72**:113.

Hedrick, J.L., and Smith, A.J., 1968, Size and charge isomer separation and estimation of molecular weights of proteins by disk electrophoresis, *Arch. Biochem. Biophys.* **126**:155.

Helenius, A., and Simons, K.,1975, Solubilization of membranes by detergents, *Biochim. Biophys. Acta* **415**:29.

Hjertén, S., Jerstedt, S., and Tiselius, A., 1965, Some aspects of the use of continuous and discontinuous buffer systems in polyacrylamide gel electrophoresis, *Anal. Biochem.* **11**:219.

Hjertén, S., Jerstedt, S., and Tiselius, A., 1969, Apparatus for large scale preparative polyacrylamide gel electrophoresis, *Anal. Biochem.* **27**:108.

Houston, L.L., 1971, Amino acid analysis of stained bands from polyacrylamide gels, *Anal. Biochem.* **44**:81.

Hsieh, W.C., and Anderson, R.E., 1975, Quantitation of stained proteins in SDS polyacrylamide gels with lysozyme as internal standard, *Anal. Biochem.* **69**:331.

Hydén, H., Bjurstam, K., and McEwen, B., 1966, Protein separation at the cellular level by micro disk electrophoresis, *Anal. Biochem.* **17**:1.

Iborra, F.,and Buhler, J.M., 1976, Protein subunit mapping: A sensitive high resolution method, *Anal. Biochem.* **74**:503.

Jones, M.N., 1975, A theoretical approach to the binding of amphipathic molecules to globular proteins, *Biochem. J.* **151**:109.

Jovin, T.M., 1973, Multiphasic zone electrophoresis: I, II, III, *Biochemistry* **12**:871.

Kaltschmidt, E., and Wittmann, H.G., 1970, Ribosomal proteins, VII. Two-dimensional polyacrylamide gel electrophoresis for finger printing of ribosomal proteins, *Anal. Biochem.* **36**:401.

Klockow, M., 1976, Lokalisierung von Proteinasen nach der Disk Elektrophorese in Polyacrylamid Gel, *Kontakte (Merck)* **1**(76):39.

Kohlrausch, F., 1897, Ueber Concentrations-Verschiebungen durch Electrolyse im Inneren von Lösungen und Lösungsgemischen, *Ann. Phys. Chem. N.F.* **62**:209.

Kruski, A.W., and Narayan, K.A., 1974, Some quantitative aspects of the disk electrophoresis of ovalbumin using Amido black 10 B stain, *Anal. Biochem.* **60**:431.

Kuzmin, S.V., Mikichur, N.I., Naumova, L.P., and Sandakhchiev, L.S., 1975, Micro electrophoresis of RNA in 10^{-8}–10^{-9} g amount, *Anal. Biochem.* **65**:405.

Laemmli, U.K., 1970, Cleavage of structural proteins during the assembly of the head of bacteriophage T_4, *Nature (London)* **227**:680.

Lambin, P., Rochu, D., and Fine, J.M., 1976, A new method for determination of molecular weights of proteins by electrophoresis across a sodium dodecyl sulfate (SDS)–polyacrylamide gradient gel, *Anal. Biochem.* **74**:567.

Laskey, R.A., and Mills, A.D., 1975, Quantitative film detection of ^3H and ^{14}C in polyacrylamide gels, *Eur. J. Biochem.* **56**:335.

Leaback, D.H., 1975, Isoelectric focusing as an aid to other techniques for the separation and characterization of proteins, in: *Isoelectric Focusing* (J.P. Arbuthnott and J.A. Beeley, eds.), pp. 201–211, Butterworths, London.

Loh, Y.P., and Gainer, H., 1975, Low molecular weight specific proteins in identified molluscan neurons: I, *Brain Res.* **92**:181.

Loh, Y.P., Rüchel, R., and Gainer, H.,1977, Specific, water soluble polypeptides in identified neurons of *Aplysia californica, Hoppe-Seyler's Z. Physiol. Chem.* **358**:667.

Louis-Ferdinand, R., and Blatt, W.F., 1967, Quantitative immunoassay by disk electrophoresis, *Clin. Chim. Acta* **16**:259.

Maddy, A.H., and Kelly, P.G., 1971, Acetic acid as a solvent for erythrocyte membrane proteins, *Biochim. Biophys. Acta* **241**:290.

Maddy, A.H., Dunn, M.J., and Kelly, P.G., 1972, The characterization of membrane proteins by centrifugation and gel electrophoresis: A comparison of proteins prepared by different methods, *Biochim. Biophys. Acta* **288**:263.

Maizel, J.V., 1971, Polyacrylamide gel electrophoresis of viral proteins, in: *Methods in Virology*, Vol. 5 (K. Maramorosch and H. Koprowski, eds.), pp. 179–246, Academic Press, New York.

Margolis, J., and Kenrick, K.G., 1968, Polyacrylamide gel electrophoresis in a continuous molecular sieve gradient, *Anal. Biochem.* **25**:347.

Margolis, J., and Wrigley, C.W., 1975, Improvement of pore gradient electrophoresis by increasing the degree of cross-linking at high acrylamide concentrations, *J. Chromatogr.* **106**:204.

Marz, L., Barna, J., and Ebermann, R., 1976, Localization of glycosidases on polyacrylamide gel, *J. Chromatogr.* **123**:495.

Maurer, H.R., 1971, *Disk Electrophoresis*, pp. 1–29, Walter de Gruyter, Berlin.

Maurer, H.R., and Allen, R.C., 1972*a*, Polyacrylamide gel electrophoresis in clinical chemistry: Problems of standardization and performance, *Clin. Chim. Acta* **40**:359.

Maurer, H.R., and Allen, R.C., 1972*b*, Useful buffer and gel systems for polyacrylamide gel electrophoresis, *Z. Klin. Chem. Klin. Biochem.* **10**:220.

Maurer, H.R., and Dati, F.A., 1972, Micro polyacrylamide flat gel electrophoresis on microscope slides, *Anal. Biochem.* **46**:19.

Merril, C.R., Switzer, R.C., and Van Keuren, M.L., 1979, Trace polypeptides in cellular extracts and human body fluids detected by two-dimensional electrophoresis and a highly sensitive silver stain, *Proc. Natl. Acad. Sci. U.S.A.* **76**:4335.

Michael, S.E., 1962, The isolation of albumin from blood serum or plasma by means of organic solvents, *Biochem. J.* **82**:212.

Mokrasch, L.C., 1975, Separation of water-insoluble membrane proteins from mouse brain subcellular fractions by gel electrophoresis in concentrated formic acid, *Fed. Proc. Fed. Am. Soc. Exp. Biol.* **34**:308.

Morell, P., Wiggins, R.C., and Jones–Gray, M., 1975, Polyacrylamide gel electrophoresis of myelin proteins, a caution, *Anal. Biochem.* **68**:148.

Nee, T.W., 1975, Electrophoretic mobility, *J. Chromatogr.* **105**:251.

Neuhoff, V., 1968, Micro Disk Electrophorese von Hirnproteinen, *Arzneimittelforschung* **18**:35.

Neuhoff, V., 1973*a*, Micro electrophoresis on polyacrylamide gels in: *Micromethods in Molecular Biology* (V. Neuhoff, ed.), pp.1–83, Springer-Verlag, Berlin.

Neuhoff, V., 1973*b*, Micro determination of amino acids and related compounds with dansyl chloride, in: *Micromethods in Molecular Biology* (V. Neuhoff, ed.), pp. 85–147, Springer-Verlag, Berlin.

Neuhoff, V., 1980, Recent advances in microelectrophoresis, in: *Electrophoresis—79* (B.J. Radola, ed.), pp. 203–218, Walter de Gruyter, Berlin.

Neuhoff, V., and Lezius, A., 1968, Nachweis und Charakterisierung von DNS-Polymerasen durch Micro Disc Elektrophorese, *Z. Naturforsch. Teil B.* **23:**812.

Neuhoff, V., and Mesecke, S., 1977, Direkte immunologische Identifizierung von Proteinen nach elektrophoretischer Fraktionierung in Polyacrylamid Mikrogradientengelen und Einfluss von Detergentien auf die Immunpräzipitation, *Hoppe-Seyler's Z. Physiol. Chem.* **358:**1623.

Neuhoff, V., and Schill, W.B., 1968, Kombinierte Mikro Disk Elektrophorese und Mikro Immunpräzipitation von Proteinen, *Hoppe-Seyler's Z. Physiol. Chem.* **349:**795.

Neuhoff, V., Schill, W.-B., and Sternbach, H., 1968, Mikro Disk elektrophoretische Analyse reiner DNA-abhängiger RNA-Polymerase aus *E. coli*: I, *Hoppe-Seyler's Z. Physiol. Chem.* **349:**1126.

Neuhoff, V., Schill, W.-B., and Jacherts, D.,1970*a*, Nachweis einer RNA-abhängigen RNA-Replikase aus immunologisch kompetenten Zellen durch Mikro Disk Elektrophorese, *Hoppe-Seyler's Z. Physiol. Chem.* **351:**157.

Neuhoff, V., Weise, M., and Sternbach, H., 1970*b*, Micro analysis of pure DNA-dependent RNA-polymerase from *Escherichia coli*, IV. Determination of the amino acid composition, *Hoppe-Seyler's Z. Physiol. Chem.* **351:**1395.

Neuhoff, V., Wolfrum, D.I., and Rüchel, R., 1973, Autoradiography of micro gels, *Naturwissenschaften* **60:**476.

Neville, D., 1971, Molecular weight determination of protein dodecyl sulfate complexes by gel electrophoresis in a discontinuous buffer system, *J. Biol. Chem.* **246:**6328.

Novak-Hofer, I., and Siegenthaler, P.A., 1975, Isoelectric focusing of membrane proteins from spinach chloroplasts, *FEBS Lett.* **60:**47.

Oakley, B.R., Kirsch, D.R., and Morris, N.R., 1980, A simplified ultrasensitive silver stain for detecting proteins in polyacrylamide gels, *Anal. Biochem.* **105:**361.

O'Brien, T.J., Liebke, H.H., Cheung, H.S., and Johnson, L.K., 1976, Isoelectric focusing in a Sephadex column, *Anal. Biochem.* **72:**38.

O'Farrell, P.H., 1975, High resolution two-dimensional electrophoresis of proteins, *J. Biol. Chem.* **250:**4007.

Ornstein, L., 1964, Disc electrophoresis. I, Background and theory, *Ann. N. Y. Acad. Sci.* **121:**321.

Panyim, S., and Chalkley, R., 1969, High resolution acrylamide gel electrophoresis of histones, *Arch. Biochem. Biophys.* **130:**337.

Peterson, R.P., 1970, RNA in single identified neurons of *Aplysia*, *J. Neurochem.* **17:**325.

Phillips-Burnett, C.C., Anderson, W.A., and Rüchel, R., 1978, Characterization of the estrogen-induced uterine peroxidase by microelectrophoresis, *J. Histochem. Cytochem.* **26:**382.

Pitt-Rivers, R., and Impiombato, F.S.A., 1968, The binding of sodium dodecyl sulfate to various proteins, *Biochem. J.* **109:**825.

Poehling, M.P.,and Neuhoff, V., 1980, One- and two-dimensional electrophoresis in micro-slab gels, *Electrophoresis* **1**(in press).

Quentin, C.-D., and Neuhoff, V., 1972, Micro isoelectric focusing for the detection of LDH isoenzymes in different brain regions of rabbit, *Int. J. Neurosci.* **4:**17.

Radola, B.J., 1975, Some aspects of preparative isoelectric focusing in layers of granulated gels, in: *Isoelectric Focusing* (J.P. Arbuthnott and J.A. Beeley, eds.), pp. 182–197, Butterworths, London.

Radola, B.J. (ed.), 1980, *Electrophoresis—79*, Walter de Gruyter, Berlin.

Radola, B.J., and Graesslin, D. (eds.), 1977, *Electrofocusing and Isotachophoresis*, Walter de Gruyter, Berlin.

Ressler, N., 1973, A systematic procedure for the determination of the heterogeneity and nature of multiple electrophoretic bands, *Anal. Biochem.* **51:**589.

Reynolds, J.A., and Tanford, C., 1970*a*, Binding of dodecyl sulfate to proteins at high binding ratios: Possible implications for the state of proteins in biological membranes, *Proc. Natl. Acad. Sci. U.S.A.* **66:**1002.

Reynolds, J.A., and Tanford, C., 1970*b*, The gross conformation of protein–sodium dodecyl sulfate complexes, *J. Biol. Chem.* **245:**5161.

Richards, E.G., and Lecanidou, R., 1974, Polymerization kinetics and properties of polyacrylamide gels, in: *Electrophoresis and Isoelectric Focusing in Polyacrylamide Gel* (R.C. Allen and H.R. Maurer, eds.), pp. 16–21, Walter de Gruyter, Berlin.

Richter-Landsberg, C., Rüchel, R., and Waehneldt, T.V., 1974, Discontinuous gel electrophoresis of reduced membrane proteins, *Biochem. Biophys. Res. Commun.* **59:**781.

Righetti, P.G., and Drysdale, J.W., 1973, Small-scale fractionation of proteins and nucleic acids by isoelectric focusing in polyacrylamide gels, *Ann. N. Y. Acad. Sci.* **209:**163.

Righetti, P.G., and Drysdale, J.W., 1974, Isoelectric focusing in gels, *J. Chromatogr.* **98:**271.

Rilbe, H., 1973, Historical and theoretical aspects of isoelectric focusing, *Ann. N. Y. Acad. Sci.* **209:**11.

Ringborg, U., Egyházi, E., Daneholt, B., and Lambert, B., 1968, Agarose–acrylamide composite gels for micro fractionation of RNA, *Nature (London)* **220:**1037.

Rodbard, D.,and Chrambach, A., 1971, Estimation of molecular radius, free mobility, and valence using polyacrylamide gel electrophoresis, *Anal. Biochem.* **40:**95.

Rodbard, D., Kapadia, G.,and Chrambach, A., 1971, Pore gradient electrophoresis, *Anal. Biochem.* **40:**135.

Rüchel, R., 1976, Sequential protein analysis from single identified neurons of *Aplysia californica*: A micro electrophoretic technique involving polyacrylamide gradient gels and isoelectric focusing, *J. Histochem. Cytochem.* **24:**773.

Rüchel, R., 1977, Two-dimensional micro-separation technique for proteins and peptides, combining isoelectric focusing and gel gradient electrophoresis, *J. Chromatogr.* **132:**451.

Rüchel, R., and Brager, M.D., 1975, Scanning electron microscopic observations of polyacrylamide gels, *Anal. Biochem.* **68:**415.

Rüchel, R., and Wolfrum, D.I., 1974, A simple outfit for micro electrophoresis in polyacrylamide gels, *GIT Fachz. Lab.* **18:**1098.

Rüchel, R., Mesecke, S., Wolfrum, D.I., and Neuhoff, V., 1973*a*, in: *Micromethods in Molecular Biology* (V. Neuhoff, ed.), pp. 200–202, Springer-Verlag, Berlin.

Rüchel, R., Mesecke, S., Wolfrum, D.I., and Neuhoff, V., 1973*b*, Mikroelektrophorese an kontinuierlichen Polyacrylamid Gradientengelen: I, *Hoppe-Seyler's Z. Physiol. Chem.* **354:**1351.

Rüchel, R., Mesecke, S., Wolfrum, D.I., and Neuhoff, V., 1974, Micro electrophoresis in continuous polyacrylamide gradient gels: II, *Hoppe-Seyler's Z. Physiol. Chem.* **355:**997.

Rüchel, R., Richter-Landsberg, C., and Neuhoff, V., 1975, Mikroelektrophorese in kontinuierlichen Polyacrylamid Gradientengelen: IV, *Hoppe-Seyler's Z. Physiol. Chem.* **356:**1283.

Rüchel, R., Loh, Y.P., and Gainer, H., 1977, A technique for the selective extraction of water-soluble polypeptides from identified neurons of *Aplysia californica, Hoppe-Seyler's Z.Physiol. Chem.* **358:**659.

Rüchel, R., Retief, A.E., and Richter-Landsberg, C., 1978*a*, Charge-dependent staining of proteins after electrophoretic separation in polyacrylamide gels, *Anal. Biochem.* **90:**451.

Rüchel, R., Steere, R.L., and Erbe, E.F., 1978*b*, Transmission-electron microscopic observations of freeze–etched polyacrylamide gels, *J. Chromatogr.* **166:**563.

Russ, G., and Polakova, K., 1973, The molecular weight determination of proteins and glycoproteins of RNA enveloped viruses by polyacrylamide gel electrophoresis in SDS, *Biochem. Biophys. Res. Commun.* **55:**666.

Shaaya, E., 1976, Separation of high molecular weight heterodisperse RNA and mRNA by polyacrylamide gels, *Anal. Biochem.* **75:**325.

Shapiro, A.L., Vinuela, E., and Maizel, J.V., 1967, Molecular weight estimation of polypeptide chains in SDS polyacrylamide gels, *Biochem. Biophys. Res. Commun.* **28:**815.

Shaw, P.H., and Hennen, S., 1975, A high-resolution micro gel system for analyzing small amounts of protein with a wide diversity of molecular weights, *Microchem. J.* **20:**183.

Slater, G.G., 1969, Stable Pattern formation and determination of molecular size by pore-limit electrophoresis, *Anal. Chem.* **41:**1039.
Smeds, S., and Björkman, U., 1972, Micro-scale protein separation by electrophoresis in continuous polyacrylamide concentration gradients, *J. Chromatogr.* **71:**499.
Smithies, O., 1964, Starch gel electrophoresis, *Metabolism* **13:**974.
Söderholm. J., Allestam, P., and Wadström, T., 1972, A rapid method for isoelectric focusing in polyacrylamide gel, *FEBS Lett.* **24:**89.
Sprinzl, M., Wolfrum, D.I., and Neuhoff, V., 1975, Separation of Aminoacyl-tRNA^{Phe} and tRNAs^{Phe} with partially hydrolysed 3′- end by polyacrylamide gradient micro electrophoresis, *FEBS Lett.* **50:**54.
Staynov, D.Z., Pinder, J.C.,and Gratzer, W.B., 1972, Molecular weight determination of nucleic acids by gel electrophoresis in non-aqueous solution, *Nature (London) New Biol.* **235:**108.
Stegemann, H., 1979, SDS-gel electrophoresis in polyacrylamide, merits and limits, in: *Electrokinetic Separation Methods* (P.G. Righetti, C.J. van Oss, and J.W. Vanderhoff, eds.), pp. 313–336, Elsevier, Amsterdam.
Stoklosa, J.T., and Latz, H.W., 1975, Electrophoretic behavior of protein dodecyl sulfate complexes in the presence of various amounts of sodium dodecyl sulfate, *Anal. Biochem.* **68:**358.
Strauss, M., 1975, Determination of the subunit molecular weight of hypoxanthine-guanine phosphoribosyltransferase from human erythrocytes by recovery of enzyme activity from sodium dodecyl sulphate gels, *Biochim. Biophys. Acta* **410:**426.
Switzer, R.C., Merril, C.R., and Shifrin, S., 1979, A highly sensitive silver stain for detecting proteins and peptides in polyacrylamide gels, *Anal. Biochem.* **98:**231.
Takagi, T., Tsujii, K., and Shirahama, K., 1975, Binding isotherms of sodium dodecyl sulfate to protein polypeptides with special reference to SDS-polyacrylamide gel electrophoresis, *J. Biochem.* **77:**939.
Tanford, C., 1961, *Physical Chemistry of Macromolecules*, John Wiley & Sons, New York.
Tanford, C., 1973, *The Hydrophobic Effect*, John Wiley &Sons, New York.
Tanford, C., and Reynolds, J.A., 1976, Characterization of membrane proteins in detergent solutions, *Biochim. Biophys. Acta* **457:**133.
Thorun, W., 1971, Der Molekularsiebeffekt bei der Polyacrylamidgel Elektrophorese als Funktion von Molekülgrösse, Gelkonzentration und Gelvernetzung sowie seine Anwendung zur Molekulargewichtsbestimmung, *Z. Klin. Chem. Klin. Biochem.* **9:**3.
Tiselius, A., 1937, A new apparatus for electrophoretic analysis of colloidal mixtures, *Trans. Faraday Soc.* **33:**524.
Tulchin, N., Ornstein, L.,and Davis, B.J., 1976, A Microgel system for disc electrophoresis, *Anal. Biochem.* **72:**485.
Tung, J.S., and Knight, C.A., 1972, Relative importance of some factors affecting the electrophoretic migration of proteins in sodium dodecyl sulfate polyacrylamide gels, *Anal. Biochem.* **48:**153.
Tuszynski, G.P., and Warren, L., 1975, Removal of sodium dodecyl sulfate from proteins, *Anal. Biochem.* **67:**55.
Ugel, A.R., Chrambach, A., and Rodbard, D., 1971, Fractionation and characterization of an oligomeric series of bovine keratohyalin by polyacrylamide gel electrophoresis, *Anal. Biochem.* **43:**410.
Varricchio, F.,and Ernst, H.J., 1975, Separation of low molecular weight RNA by polyacrylamide gel electrophoresis, *Anal. Biochem.* **58:**485.
Verbruggen, 1975, Quantitative immunoelectrophoretic methods: A literature survey, *Clin. Chem.* **21:**5.
Vesterberg, O., 1973*a*, Isoelectric focusing of proteins in thin layers of polyacrylamide gel, *Sci. Tools* **20:**22.
Vesterberg, O., 1973*b*, Physicochemical properties of the carrier ampholytes and some biochemical applications, *Ann. N. Y. Acad. Sci.* **209:**23.

Vesterberg, O., Hansen, L., and Sjösten, A., 1977, Staining of proteins after isoelectric focusing in gels by new procedures, *Biochim. Biophys. Acta* **491:**160.

Ward, C.W., 1976, Detection of proteolytic enzymes in polyacrylamide gels, *Anal. Biochem.* **74:**242.

Weber, K., and Osborn, M.,1969, The reliability of molecular weight determinations by dodecyl sulfate–polyacrylamide gel electrophoresis, *J. Biol. Chem.* **244:**4406.

Weber, K., and Osborn, M., 1975, Proteins and sodium dodecyl sulfate: Molecular weight determination on polyacrylamide gels and related procedures, in: *The Proteins*, 3. ed., Vol. 1 (H. Neurath and R.L. Hill, eds.), pp. 179–223, Academic Press, New York.

Weber, K., Pringle, J.R., and Osborn, M., 1972, Measurement of molecular weights by electrophoresis on SDS-acrylamide gel, in: *Methods in Enzymology*, Vol. 26 (C.H.W. Hirs and S.N. Timasheff, eds.), pp. 3–27, Academic Press, New York.

Weiner, A.M., Platt, T., and Weber, K., 1972, Amino-terminal sequence analysis of protein purified on a nanomole scale by gel electrophoresis, *J. Biol. Chem.* **247:**3242.

Willard, M., Cowan, W.M., and Vagelos, P.R., 1974, The polypeptide composition of intra-axonally transported proteins: Evidence for four transport velocities, *Proc. Natl. Acad. Sci. U.S.A.* **71:**2183.

Williams, J.G., and Gratzer, W.B., 1971, Limitation of the detergent–polyacrylamide gel electrophoresis method for molecular weight determination of proteins, *J. Chromatogr.* **57:**121.

Wolfrum, D.I., Rüchel, R., Mesecke, S., and Neuhoff, V., 1974, Mikroelektrophorese in kontinuierlichen Polyacrylamid Gradientengelen: III, *Hoppe-Seyler's Z. Physiol. Chem.* **355:**1415.

Wrigley, C.W., and Shepherd, K.W., 1973, Electrofocusing of grain proteins from wheat genotypes, *Ann. N. Y. Acad. Sci.* **209:**154.

Wyckoff, M., Rodbard, D., and Chrambach, A., 1977, Polyacrylamide gel electrophoresis in sodium dodecyl sulfate-containing buffers using multiphasic buffer systems: Properties of the stack, valid R_f-measurement, and optimized procedure, *Anal. Biochem.* **78:**459.

Zacharius, M.R., Zell, T.E., Morrison, J.H., and Woodlock, J.J., 1969, Glycoprotein staining following electrophoresis on acrylamide gel *Anal. Biochem.* **30:**148.

5

Classic Methods in Neuroanatomy

J. Voogd and H.K.P. Feirabend

1. Introduction

Classic methods in neuroanatomy accentuate certain characteristic structural features of the nervous system. The problems in applying these methods in neurobiology usually do not concern the histotechnical procedure but are inherent in the complexity of the nervous system itself. In the choice of a method, one should realize that the value of a given method is determined as much by its ability to accentuate certain histological features as by its selectivity in omitting other structural components.

Neuroanatomy begins with gross dissection of the hardened brain and the study of the topography of the white and the gray matter in myelin- and Nissl-stained sections. Myelin stains derive their use from the accentuation of one of the main properties of the vertebrate brain, namely, the concentration of myelinated axons in what we call tracts, fascicles, capsules, or funiculi. But for the myelin sheath, these methods omit everything including the nerve cells themselves. Nissl stains reduce the gray matter to the stained cell bodies of the nerve cells and the nuclei of glial cells. Dendrites, unmyelinated axons, and their terminal boutons, which account for most of the volume and determine the structural identity of the gray matter, remain unstained. Still, the Nissl method is an indispensable tool in the study of cytoarchitectonics, i.e., the identification of groups of nerve cells as nuclei or cortical layers by their size, shape, and packing density. Perhaps the greatest drawback inherent in these histological procedures is the loss of the three-dimensional structure of the brain. Reconstruction methods, therefore, are a necessary complement to histological technique. Essentially, three-dimensional reconstruction takes place in the mind of the investigator, but it may

---

J. Voogd and *H.K.P. Feirabend* • Laboratory of Anatomy and Embryology, University of Leiden, Leiden, The Netherlands.

be greatly assisted by a judicious use of different planes of sectioning, graphic and spatial reconstruction, and a continuing recourse to gross dissection.

Topographical methods such as the myelin and Nissl stains also bridge the gap between gross anatomy and histology of the brain. Histological stains, such as the neurofibrillar and the Golgi methods, impregnate nerve cells with their axons and dendritic processes more completely, but often lack a clear differentiation into nuclei and fiber tracts as seen in Nissl- and myelin-stained sections.

Prior to gross dissection, the brain should be hardened. Initially, ethyl alcohol or potassium dichromate [solution of Müller (1859): potassium dichromate, 2.5%; sodium sulfate, 1%] solutions were used for this purpose. Formaldehyde was not introduced in histology until the end of the 19th century (Blum, 1893). Dichromate and alcohol have been retained as fixatives in histology—dichromate as a mordant in many myelin stains, ethyl alcohol as one of the fixatives of choice for the Nissl method. These and other histological methods were developed in anatomical and neuropathological laboratories in Central and Southern Europe, along with the art of microtomy (microtomes were developed by Jung, Reichert, Minot, and Rutherford in the 1870's), tissue-embedding methods [paraffin (Klebs, 1869), celloidin (Duval, 1879)], and the histological dyes, both natural and synthetic.

Application of the myelin, Nissl, and neurofibrillar staining methods in the adult vertebrate nervous system has produced a wealth of descriptive and topographical detail. In addition, these methods were used in the study of the developing brain—Nissl methods in studies of cytological development and differentiation, neurofibrillar stains for tracing the outgrowth of nerve tracts, and myelin stains for establishing the characteristic temporal and spatial patterns in which nerve tracts acquire their myelin sheaths (Flechsig's myelogenetic method). Finally, these methods found an application in the experimental investigation of neuronal degeneration, which until recently was the only way to establish the polarity, course, and termination of long-fiber connections in the nervous system. The ultimate disappearance of the myelin sheaths of the distal, transected part of a nerve fiber in so-called orthograde or wallerian degeneration is visible with myelin stains. The degeneration products of the sheath of a degenerated axon can be displayed with the Marchi method (Marchi and Algeri, 1885). Later, neurofibrillar methods (Apáthy, 1892; Bielschowsky, 1902, 1904, 1935; Ramón y Cajal, 1903) were introduced, modified (Glees, 1946; Nauta and Ryan, 1952), and extensively used to impregnate the degenerating axons and their terminals. The Nissl method can be used to study the profound changes that occur in the nerve-cell body on transection of the axon. This "retrograde reaction" (Nissl's "primäre Reizung") was often used to establish the origin of nervous pathways.

Although these methods are largely responsible for the present state of descriptive and topographical anatomy of the brain and of its connectivity, they have not contributed greatly to the basic histological concepts that determine our thinking about the fundamental structure and function of the nervous system. Often, the methods used to elucidate the histology of the nerve cell were, and are, less orthodox (Van der Loos, 1967) and less suitable to apply as routine methods in descriptive and experimental investigations. Thus, the first demonstration of the general structure of the multipolar nerve cell, its single axon, and its multiple dendritic processes was made by Deiters (1865) by microdissection of cells from the spinal cord of the ox (Fig. 1). The same exclusivity surrounds the "reazione nera" of Golgi (1873), which impregnates complete nerve cells with the finest ramifications of their processes and which, by virtue of its staining only a few of the nerve cells present in any one region and the use of thick sections, performs a kind of histological microdissection of the brain. Although it was and is extensively used and modified, it remains a tool for the few who followed in the footsteps of Ramón y Cajal and shaped our present-day concepts in neurohistology.

Some of these histological methods, such as Weigert's myelin stain, are no longer in general use, though they still provide most of the illustration material for most modern textbooks. Degeneration methods to trace fiber connections are in the process of being replaced by axonal-transport techniques. Some of the methods used in descriptive light microscopy have been replaced by scanning electron microscopy or the use of semithin, plastic-embedded sections. However, classic methods have left ineffaceable traces in neuroanatomy. A judicious application of modern methods is not possible without a knowledge of these older methods, and in many cases the use of the older methods may produce results that cannot be reached in any other way.

2. *Dissection and Gross Anatomy of the Brain*

In lower vertebrates and small mammals, dissection generally has to be limited to the exploration of the external morphology and the ventricular system. In larger brains, dissection of the main tracts and grisea becomes feasible. Immersion or perfusion fixation in 10% formalin gives suitable material. In small brains, dissection (Fig. 2) or a combination of dissection with making casts of the ventricles (Fig. 3) may lead to unexpected results. Methods for making ventricular casts of small brains with plastics were described by Böhme and Franz (1967), Badawi (1967), and Böhme (1969).

Dissection of fiber tracts in larger brains, a method perfected for the human brain by Ludwig and Klingler (1956), is facilitated by repeated freezing and thawing of the fixed brain prior to dissection. In material treated in this way, the white matter is disrupted by the formation of ice crystals, and fiber tracts can be traced by blunt dissection (Fig. 4). Generally, freeze artifacts also break up the gray matter, but the resulting unnatural appearance is associated with a much darker aspect of the gray matter, which then stands out clearly from the surrounding fiber tracts.

Dissection should be accompanied by the study of thick sections cut in the three major planes. Macrotomes have been developed for this purpose,

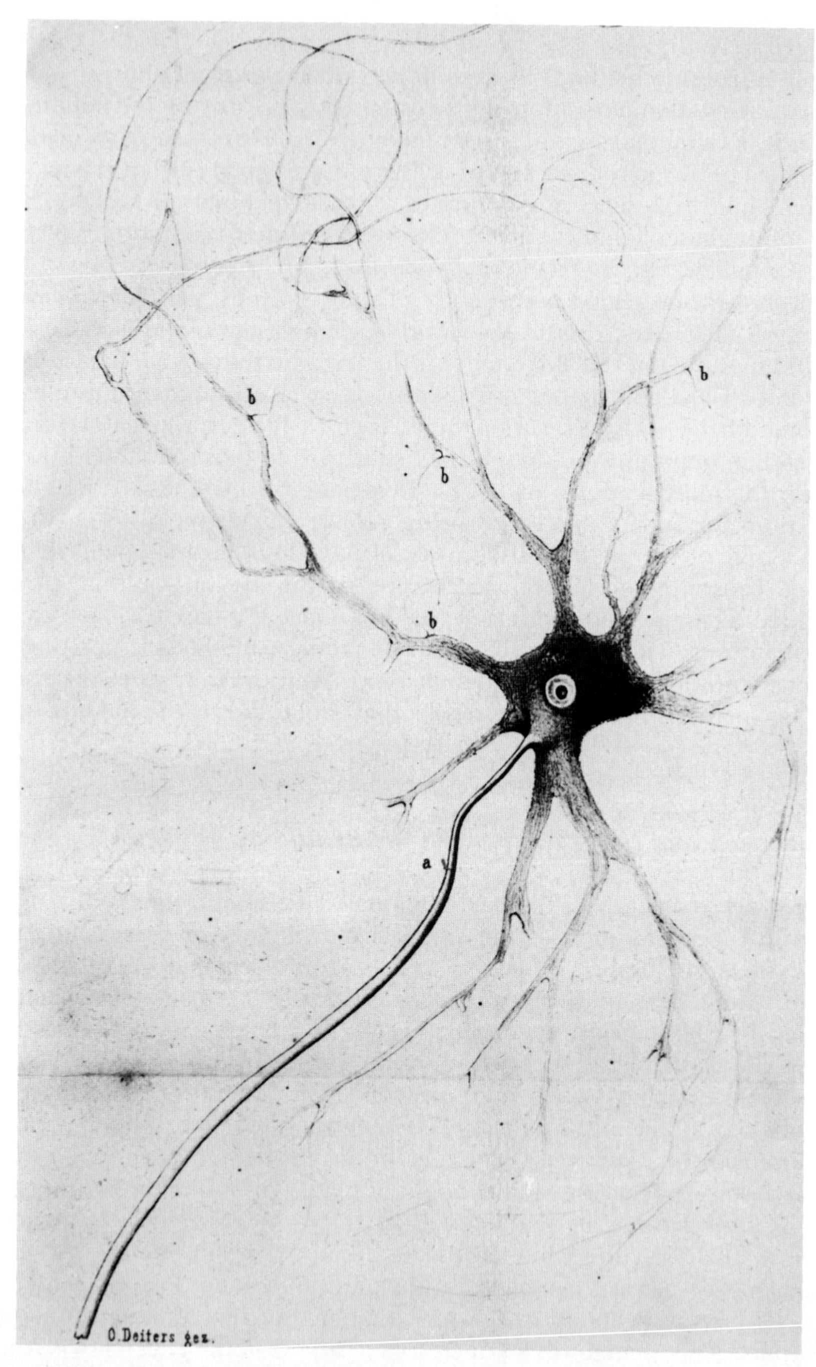

Fig. 1. Multipolar nerve cell dissected from the spinal cord (Deiters, 1865). One of the first illustrations in which the processes of the nerve cell are correctly indicated. The axon is indicated by (a). Deiters thought that a second system of axons takes its origin from the small protrusions visible on the dendrites (b).

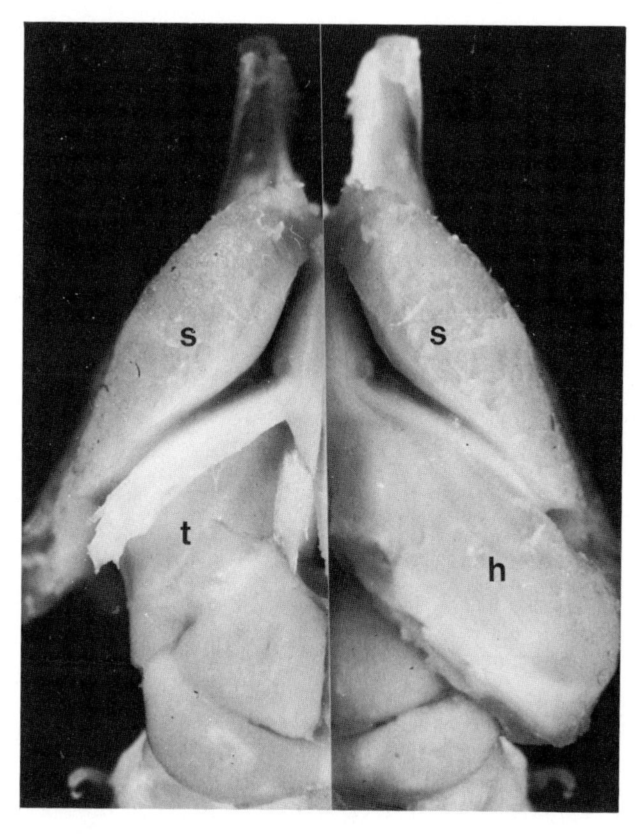

Fig. 2. Dissection of the rat brain showing the striatum (s), the hippocampus [(h) right], and after its removal the fornix and the thalamus [(t) left].

but the job can be done as well or better with a commercial rotary-blade meat slicer. To reduce distortion, the brain should be embedded in polyurethane foam or in gelatin in a styrofoam box. Box and embedded brain are sectioned as a whole. In this way, continuous series of 2- to 3-mm-thick sections can be obtained. In such sections, the contrast between gray and white matter can be increased by superficial Berlin blue staining of the gray matter with the method of Mulligan (1931) (Tompsett, 1955, 1970).

Few dissection guides for the brains of lower mammals are available; the laboratory guides for the sheep brain by Ranson and Clarke (1959) and Northcut (1966) and Craigie's *Neuroanatomy of the Rat,* edited by Zeman and Innes in 1963, remain exceptions. Information on the gross anatomy of the vertebrate brain remains scattered through the comparative-anatomical literature. Useful entries can be found in Ariëns Kappers *et al.* (1960), Brauer and Schober (1970), Smith (1902), Haller von Hallenstein (1934), Llinás and Precht (1976), Papez (1929), and Pearson (1972).

In many experiments, it will be necessary to have a knowledge of the vascular system of the brain (Kier, 1974) and gross dissection generally does

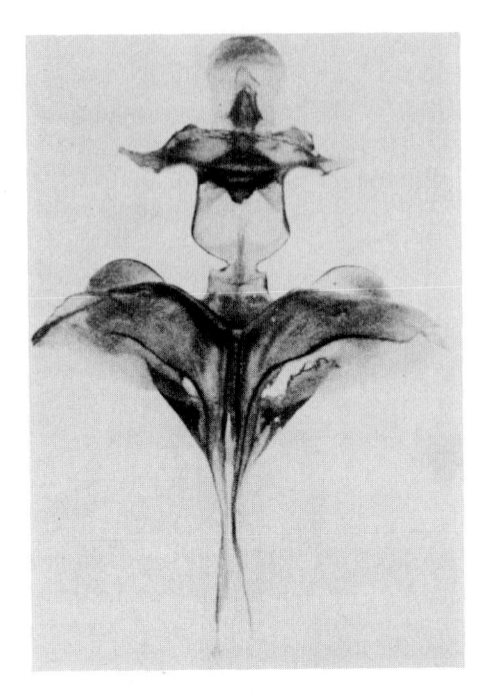

Fig. 3. Cast in Tensol® cement of the ventricular system of the chicken. Reproduced from Böhme and Franz (1967) with the permission of S. Karger, Basel and New York.

not suffice to display it completely. To this purpose, injection techniques with polyester resins, followed by dissection or maceration of the tissue (Tompsett, 1970), or with radioopaque substances (Kaplan, 1953; Wollschlaeger and Wollschlaeger, 1974; du Boulay, 1974) have been developed. Good results may be obtained by injection with India ink in gelatin followed by sectioning and clearing in agents such as oil of aniseed (Fig. 5).

The topographical relationships between brain and skull and the position of the spinal cord in the vertebral canal are of great practical importance. References to segmentation of the spinal cord, its ascensus, and the formation

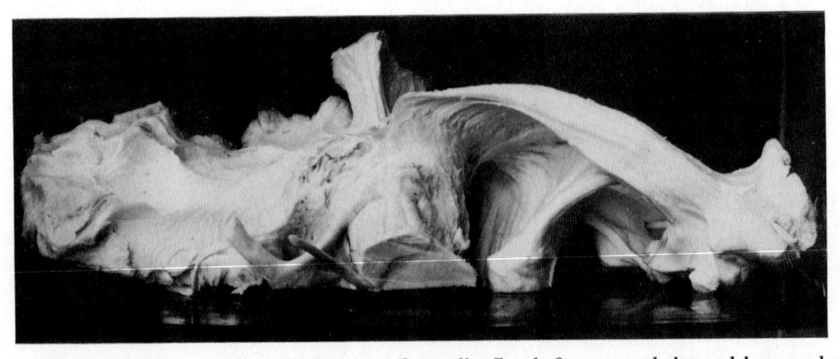

Fig. 4. Dissection of the optic radiation in a formalin-fixed, frozen-and-thawed human brain.

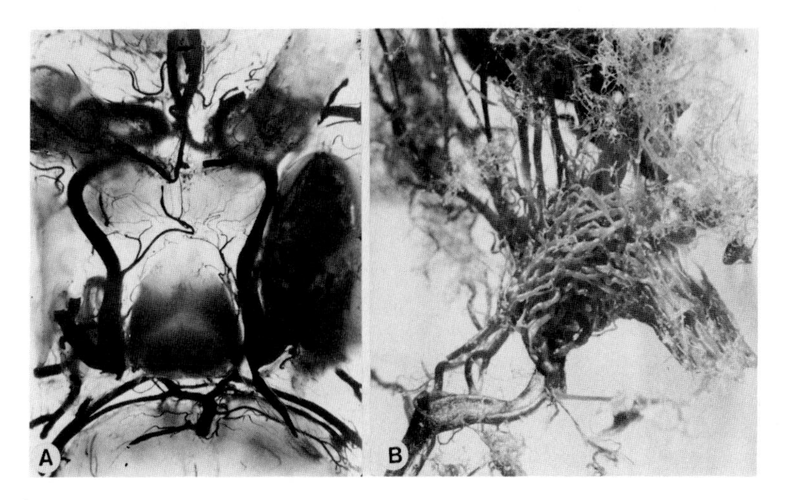

Fig. 5. (A) India ink injection of the arterial circle of Willis at the base of the brain of the cat. (B) Plastoid cast of the rete mirabile, which surrounds the maxillary artery in the cat (Martinéz Martinéz, 1967).

of the limb plexus may be found in the works cited above and in Nieuwenhuys (1964) and Van den Akker (1970).

Relationships between bony landmarks of the skull and internal structures of the brain are especially important in stereotactic surgery, which was introduced by Horsley and Clarke (1908) (see also Clarke, 1920; Jasper and Ajmone-Marsan, 1954). For humans, the variability in these relationships, and methods to compensate for them, have been extensively discussed (Delmas and Pertuiset, 1959; Van Manen, 1967). For smaller experimental animals, fewer data are available. In this respect, the study by Zweers (1971) of stereotactic surgery in the duck, in which he measured the variability of head structures and of the relationships among the outside of the head, the base of the skull, and certain brain elements, may serve as a model. Useful information on the relationships between skull and brain and on the construction of stereotactic apparatuses can be found in many stereotactic atlases, lists of which were published by the Brain Information Service Publication Office of the University of California, David Kopf Instruments, and Heimer and Lohman (1975). The relationships of the brainstem and the base of the skull in birds were discussed in a paper on a ventral surgical approach by Vielvoye *et al.* (1977).

3. Reconstruction Methods

The process of transforming two-dimensional into three-dimensional structures generally occurs in the mind of anyone studying serial sections through the brain. The composition of spatial or graphic reconstructions from serial sections may greatly facilitate this process or contribute to the presentation of the ultimate result. Spatial reconstructions can be made from wax, cardboard, styrofoam, glass, or perspex. In all methods, an accurate positioning of the consecutive sections is essential. To this purpose, reference lines can be drawn on the sides of the blocks containing the embedded brain prior to sectioning. This method is especially useful when the embedding medium is still present on the finished slide, as will be the case with plastic- or celloidin-embedded specimens. In the case of paraffin embedding, the medium is dissolved prior to staining and the reference lines are no longer visible. In this case, and also when frozen sections are employed, needle tracks through the brain, marked with India ink, can be used instead. Generally, one track, in combination with the midline, is sufficient to position transverse or horizontal sections; for sagittal sections, two parallel tracks are necessary. As an alternative, the outline of the brain, derived from photographs of the brain made prior to sectioning, can be used to stack sections in a correct position (Gribnau and Lammers, 1976).

Generally, drawings from sections, rather than microphotographs, are used in a reconstruction. Accurate drawings may be obtained with microprojectors, with photographic prints of the actual sections in which the desired structures can be marked, or by the use of an x–y plotter coupled to the stage of a microscope (Glaser and Van der Loos, 1965; Grant and Boivie, 1970).

The classic reconstruction method using wax plates (Born, 1883; Blechschmidt, 1955; Gaunt, 1971) is rather cumbersome. Cardboard and especially styrofoam are easier to utilize (Gribnau and Lammers, 1976). The styrofoam method can be improved by coating the plates with stripped or liquid photographic emulsion on which the drawings of the sections can be photographed (Fig. 6). Often, it may be necessary to make several reconstructions from sections in different planes because surface relief parallel to the sections is smoothed in spatial reconstructions.

Transparent-glass-plate reconstructions, on which the sections can be drawn, can be prepared rapidly, but can be viewed only from the front or the back. The advantage of transparency may be combined, by the use of perspex, with accurate reproduction of the external form of the region of the central nervous system to be reconstructed (Fig. 7).

Graphic reconstruction also depends on the presence of reference lines in the sections that are to be reconstructed. Simple, ortholinear projections of cells or terminals on the surface of the brain are easy to prepare (Fig. 8). In some instances, natural coordinates in the brain, instead of the Cartesian system of coordinates, may be used to prepare surface projections. The topological maps of the brain of lower vertebrates (Fig. 9) constructed by Nieuwenhuys (1974) were obtained in this way.

Graphic reconstruction is markedly improved when perspective is introduced by compressing the serial sections with the aid of a pantograph as though they were viewed at a certain angle (Tinkelenberg, 1979, 1980) (Fig. 10). With a modification of this pantograph, it is even possible to prepare pairs of graphic reconstructions that can be viewed with a stereoscope.

Fig. 6. Reconstruction in styrofoam plates coated with an emulsion on which the drawings of the sections of the cerebellum of the fowl were photographed.

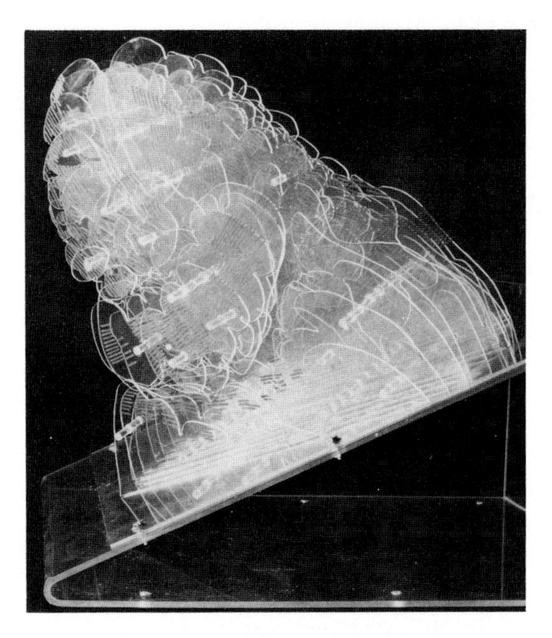

Fig. 7. Reconstruction in perspex of the cerebellum of the cat (posterior view).

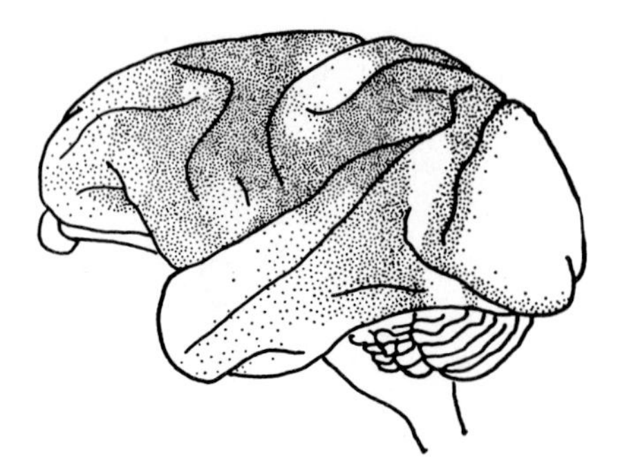

Fig. 8. Graphic reconstruction of the termination of the fibers of the corpus callosum on the lateral surface of the brain of *Macaca mulatta* (after Ebner and Myers, 1962). Reproduced from Ettlinger (1965) with the permission of J. & A. Churchill, Ltd., London.

Computer-generated reconstructions are useful in studying large numbers of individual cells (Wann *et al.*, 1973; Hillman *et al.*, 1976), but cannot replace simple methods available to any investigator.

4. Fixation and General Histological Procedure

4.1. Fixation

Fixation, embedding, and sectioning are necessary prerequisites for histological staining and largely determine which staining methods can be applied. Formalin fixation is most widely used and offers a wide choice of staining methods (Table 1), including the Golgi method. When ultrastructural criteria for good fixation are applied, glutaraldehyde offers certain advantages. However, for light microscopy, formalin fixation generally fulfills most

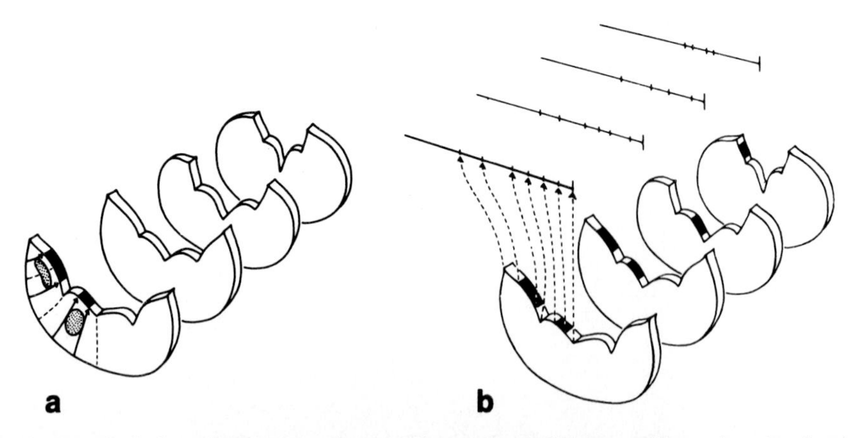

Fig. 9. Topological projection according to Nieuwenhuys (1974). (a) Illustration showing how structures are projected on the ventricular surface using the "natural system of coordinates" in the brain. (b) Illustration showing the ventricular surface flattened out. (c) A topological projection of the brain of the frog (*Rana esculenta*). (c) Reproduced from Opdam *et al.* (1976) with the permission of the Wistar Press, Philadelphia.

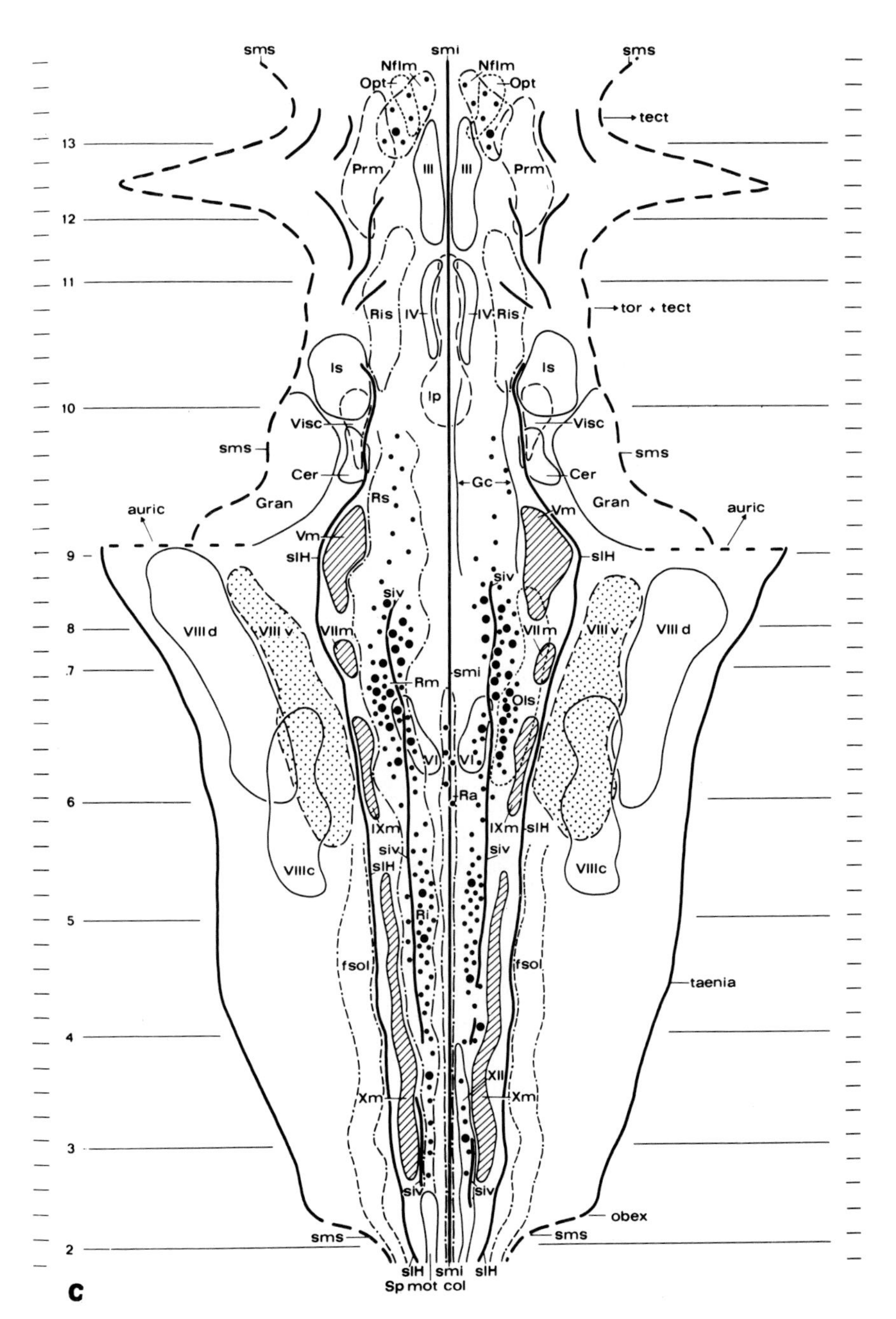

Fig. 9. (*Continued*)

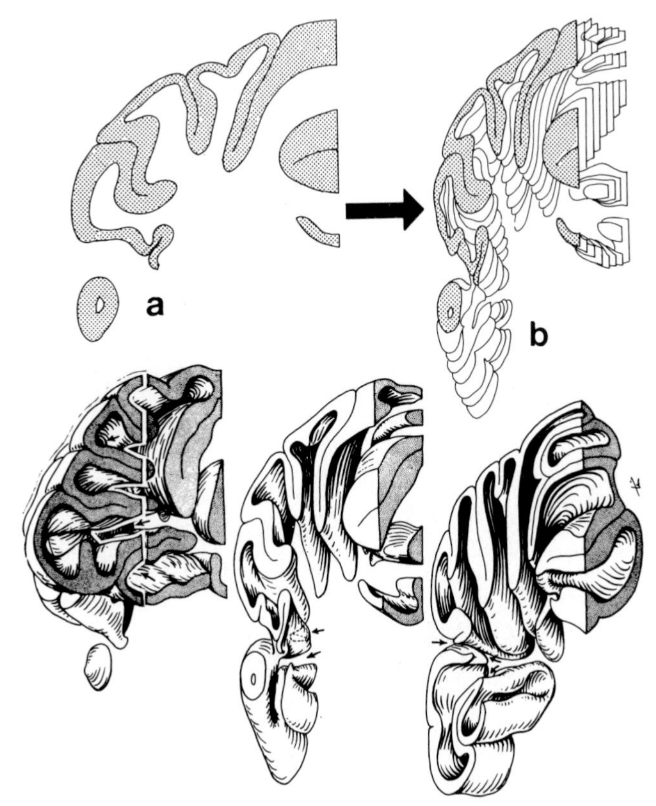

Fig. 10. Introduction of perspective in graphic reconstructions according to Tinkelenberg. Sections, one of which is depicted in (a), are compressed with a pantograph, as though they were viewed at a certain angle, and placed in serial order (b). The final drawing of a midsagitally sectioned cerebellum of the mouse was prepared by studying pairs of drawings under a stereoscope.

Table 1. Some Stains That Can Be Applied After Formalin or Ethanol Fixation, Mordanting, and Different Embedding and Sectioning Procedures

Prestaining procedure	Formalin					Ethanol	
Mordanting in dichromate	+	+					
Embedding in paraffin	+		+			+	
Embedding in celloidin		+		+			+
Frozen sections					+		
Nissl stain	+		+	+	+	+	+
Myelin stain							
Häggqvist					+[a]		
Weigert	+	+			+[a]		
Klüver	+		+	+	+		
Silver impregnation							
Bodian			+				
Nauta					+		

[a] For these stains, the sections should be mordanted on the slide.

requirements. Moreover, storage of the brains in formalin is possible for an indefinite period; in glutaraldehyde, nervous tissue deteriorates rapidly.

Generally, perfusion fixation is preferable to immersion fixation. When immersion fixation is used, an ample amount of fixative [usually 10% formalin (= formaldehyde 4% in distilled water)] should be used. If possible, the brain should not rest on the bottom of the bottle, but should be suspended or allowed to rest on a perforated scale. Fixative should be replaced several times. After a fixation of 1–2 weeks, large immersion-fixed brains should be trimmed to ultimate size for further fixation and processing.

In perfusion fixation, the fixative can be administered through the heart or through one of the great arteries. In the former, artificial respiration may be necessary. In the case of perfusion through the abdominal aorta, the thoracic cavity remains intact. Perfusion should be preceded by the intravenous administration of heparin to prevent clotting and sodium nitrite to dilate the cerebral vessels. To prevent cerebral edema, which may occur when the fixative contains dichromate, acetazolamide should be administered intravenously.

Perfusion is started with physiological saline (0.9% sodium chloride) until the blood is washed out, followed by 10% formalin (1 liter/kg body weight); sufficient pressure is obtained by placing the bottles at a height of 1 m above the animals. When more complete preservation of cytological structure is desired, the fixatives employed in ultrastructural studies can be used (Ross, 1972) (see also Chapter 7).

To prevent artifacts, especially the occurrence of dark, shrunken neurons, the brain should be left in the skull for at least 3 hr after the perfusion. For several stains, such as the silver-impregnation methods for degenerated axons, the best results are obtained when the brains are stored for at least 4 weeks in 10% formalin before being processed. For the fixation of lipids in myelin staining, the brain should be stored in Baker formalin (4% formaldehyde, 1% calcium chloride in distilled water). When neutralized formalin is used, a much-disputed prescription in fixation for many silver-impregnation methods, the formalin should be buffered with an acetate buffer.

A general survey of embedding methods and section-cutting can be found in Steedman (1960).

4.2. Frozen Sections

As shown in Table 1, formalin fixation followed by making frozen sections offers the widest choice of staining methods. Disadvantages of the frozen-section method are the difficulty in handling the friable sections, which may fall apart and get lost; the thickness of the sections, which is usually 20 μm or more; and the presence of freeze artifacts. Fewer freeze artifacts are seen when the blocks are left in sucrose–formalin (dissolve 300 g sucrose in 725 cc distilled water, add 100 cc formaldehyde 40%) until they sink to the bottom and are ready for sectioning. However, frozen sections never attain the histological quality of sections of paraffin- or celloidin-embedded material. Freezing can be circumvented by the use of a Vibratome.

With this instrument, thin, 20-μm sections can be cut with a vibrating knife from soft material. The maximum width of the sections is limited to 3 mm. To ease the handling of frozen sections, embedding methods in egg albumen (Snodgress and Dorsey, 1963) or gelatin have been developed, and are consistent with most staining methods. To keep sections from falling apart, they can also be attached to a piece of paper that is frozen to the top of the block (Collewijn and Noorduin, 1969). To transfer whole, frozen sections from unembedded material to the slide is easier in a cryostat, but this procedure is restricted to the sectioning of unfixed tissue. Storage of frozen sections in formalin is possible over long periods when they are kept in closed boxes at 4°C. For staining, the sections are assembled in perforated porcelain beakers that are fitted in a frame and passed through glass staining troughs containing the staining solutions (Fig. 11). Frozen sections are mounted on slides from phenol–carbol, which is easy, or from alcohol–gelatin (Albrecht, 1954), which is time-consuming but avoids shrinkage.

4.3. Paraffin Embedding

Before embedding of tissue in paraffin or one of its modern substitutes is possible, the tissue has to be dehydrated and cleared in solvents that mix with paraffin. Strict dehydration and infiltration schedules have to be observed to obtain good material. This is especially true for the embedding of material that has been mordanted in dichromate after initial fixation in formalin. In this case, dehydration should be long enough to remove alcohol-soluble material—which interferes with later myelin staining—from the block, but not so long that the blocks become brittle and fracture during infiltration of the heated paraffin or during sectioning. Paraffin embedding always involves shrinkage, but offers the advantages of a firm support of friable tissue, the possibility of cutting thin serial sections that may be stored

Fig. 11. Sieve beakers placed in a frame, used for staining frozen sections.

for an indefinite time, and good histological quality. Paraffin sections are mounted on slides with gelatin, albumen, or filtered serum, and are deparaffinized before being stained.

4.4. Celloidin and Plastic Embedding

Celloidin embedding is time-consuming, but there is less chance of the tissue's becoming too hard. Mordanted material for Weigert myelin staining, therefore, is embedded in celloidin. Celloidin blocks and sections are stored in 70% ethyl alcohol, and during sectioning, the blocks are kept wet with it. The sections can be marked with a number in India ink on a dry corner of the block; they can be kept indefinitely when stacked with interleaved sheets of rice paper in serial order in closed bottles with 70% ethyl alcohol. The thickness of the sections, 30 μm or more, makes celloidin the method of choice for Golgi impregnation and the production of serial reference sections of large brains. Celloidin sections are stained loose. After being stained, they are attached on slides. Cover glasses are weighted to flatten the sections on the slide.

Plastic embedding is used in electron microscopy (see Chapter 7). For light microscopy, the use of 1-μm "semithin" sections made from this material and stained with toluidine blue are often helpful as an intermediate step toward ultrastructural studies or for their fine cytological detail. The small size of these sections is often the limiting factor. Methods permitting the production of large sections from material embedded in plastics, and appropriate stains, have been developed for special purposes (Cathey, 1962; Anker *et al.*, 1974) in which the hardness of the embedding medium (cutting sections from heads without previous decalcification), its resistance to deformation (the production of sections for a stereotactic atlas), or the thinness of the sections (fiber-counting) is important.

5. Nissl Stains

5.1. Nissl's Original Method and Its Modifications

In Nissl's original method, which he introduced in 1894 (Nissl, 1894) and discussed at length in 1910 (Nissl, 1910), alcohol-fixed tissue is sectioned unembedded, stained by heating the sections in an aqueous solution of methylene blue and venetian soap, differentiated in aniline–alcohol, and passed through cajaputi oil and benzine to be cover-slipped with xylene–colophonium. In nerve cells, the Nissl bodies in the cytoplasm and in the large dendrites, the nuclear membrane, and the nucleolus are stained (Fig. 12). In smaller nerve cells and in the nuclei of glial and other cells, the nuclear chromatin can be more prominent. The stained structures stand out against a clear background. The main part of the dendrites, the axon, and the glial cytoplasm remain unstained. According to Nissl, strict adherence to

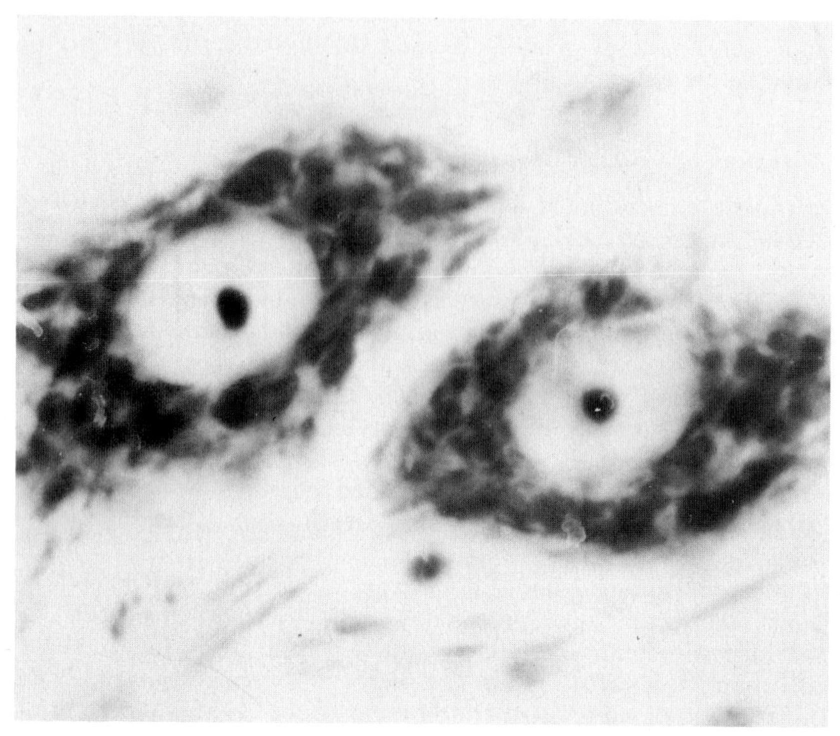

Fig. 12. Nissl-stained motoneurons of the facial nucleus of the rabbit. Cresyl violet, celloidin embedding, 20-μm section.

the original procedure yields images that are equivalent though not identical to the living nerve cells. Only under these conditions do changes in the appearance of the nerve cells reflect changes in the living cell. Artifacts are introduced and the results are made more difficult to reproduce by the use of other fixatives, such as formalin; embedding the tissue in celloidin or paraffin; and the use of other dyes and different differentiation and mounting agents. According to Nissl, these modifications therefore do not produce an image equivalent to the living cell.

The term "Nissl stain" is now applied to the staining of nervous tissue with basic dyes with the object of obtaining preparations in which only the perikarya of nerve cells and the nuclei of all cells are stained. With the establishment of criteria for good fixation and the demonstration that Nissl bodies in the cytoplasm of nerve cells are the equivalent of the granular endoplasmic reticulum (Palay and Palade, 1955), Nissl's scruples about the introduction of modifications of his original method have lost much of their validity.

With immersion fixation in many changes of 96% ethyl alcohol, followed by celloidin embedding, staining with cresyl violet, and differentiation with aniline–alcohol, thick, 30-μm preparations with high contrast that are excellent material for cytoarchitectonic studies can be obtained. In cresyl-violet-stained paraffin-embedded or frozen sections from material fixed in formaldehyde or glutaraldehyde or both, the contrast is generally less, but is sufficient for localization purposes and for the counterstaining of myelin stains, silver impregnations, and autoradiographs. For optimal results, fixation methods for ultrastructural studies should be employed. The most common artifacts are dark, shrunken neurons (Fig. 13), which are more frequent when the brain is removed from the skull immediately after perfusion or when it is not handled with care (Cammermeyer, 1962; Stensaas *et al.*, 1972). In frozen sections, the cells often appear to be blown up (Fig. 14). This can be prevented by passage of the blocks through sucrose–formalin prior to sectioning (see Section 4.2) and the avoidance of large temperature shifts during the freezing. In paraffin and frozen sections, nucleoli may get lost (Cammermeyer, 1967). Uneven staining in sections from alcohol-fixed material can be due to differences in penetration of the fixative.

5.2. Cytoarchitectonics

To quote Olszewski and Baxter (1954):

> The term cytoarchitectonics is applied to a method of anatomical investigation which is primarily concerned with patterns of arrangement and morphological details of nerve cells as revealed by magnifications within the range of the ordinary light microscope. The primary objective of the cytoarchitectonic method is the subdivision of the cellular masses of the nervous system into regions with distinctive morphological characteristics. Such regions are referred to as "areas" in the cortical grey matter, and as "nuclei" in the subcortical white matter. The value of the method rests on the hypothesis that the criteria used for cytoarchitectonic subdivisions are of biological significance.

Different properties of the Nissl-stained nerve cell such as size, shape, orientation, packing density, and staining quality each contribute alone or in combination to the parcellation of the gray matter as it is seen in thick, Nissl-

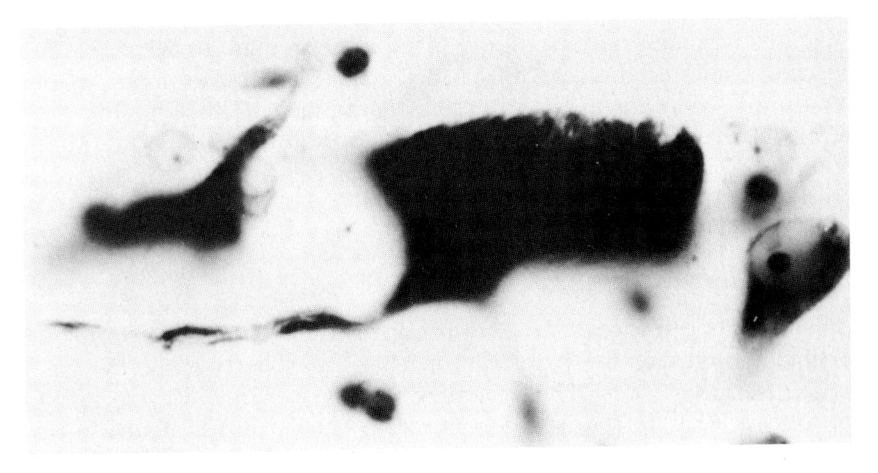

Fig. 13. "Dark neuron" from the same section as in Fig. 12.

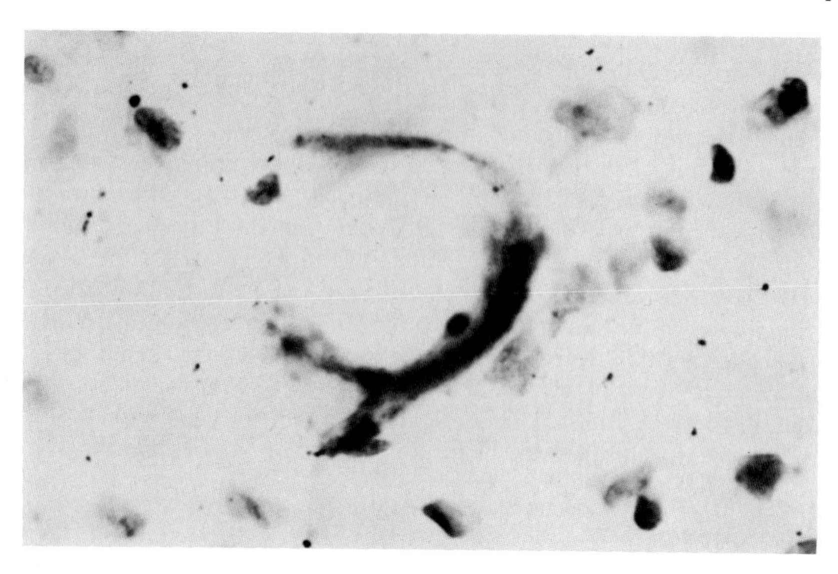

Fig. 14. Nissl-stained cell, "blown up" during freeze-sectioning.

stained sections (Fig. 15). Quantification of most of these cellular parameters is possible, but subjective rather than objective criteria still prevail, and many of the cytoarchitectonic subdivisions remain arbitrary. A good example of this is the atlas of the human thalamus by Dewulf (1971), in which different investigators applied their subdivision on the same histological material. Cytoarchitectonics, the taxonomy of the gray matter, is indispensable, but great care should be taken in applying it to new material. Borders should not be copied indiscriminately from the literature, but criteria for the subdivision of nuclei should be checked because slight differences in staining quality, orientation of the sections, and age of the animal may cause large deviations from the original description.

Quantitative descriptions of a number of neurological parameters in the cytoarchitectonics of the cerebral cortex, expressed as a function of depth below the pia, were given by Bok (1959) and Smit (1968) (Fig. 16). Similar data for the cerebellar cortex were reported by Palkovits *et al.* (1971). Methods for the measuring and counting of nerve cells can be found in the volume on stereological methods edited by Elias (1967) and in Konigsmark (1970).

5.3. Axonal Reaction of the Perikaryon

Nerve cells often react on transection of their axon with complex but reversible changes in their perikaryon that are collectively known as the "primäre Reizung" (Nissl, 1894), as the "axonal reaction" (Meyer, 1901), or, inappropriately, as retrograde degeneration. The principal features of the axonal reaction and its metabolic significance were extensively reviewed by Lieberman (1971), and its application for establishing the origin of fiber pathways in the brain was treated by Cowan (1970).

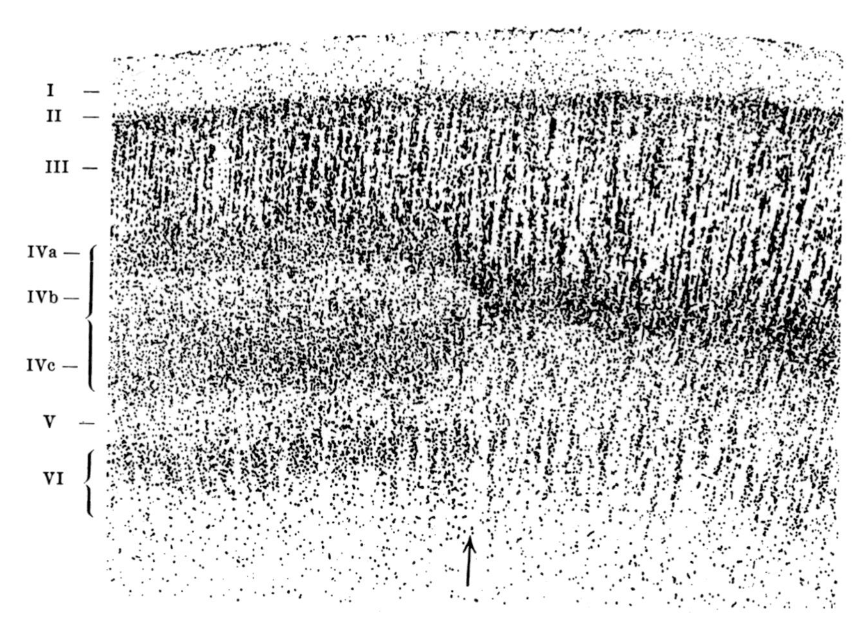

Fig. 15. Cytoarchitecture of the border region of the striate cortex in an 8-month human fetus. On the left side, the characteristic tripartitioning of Brodmann's area 17 is visible. Nissl stain. Reproduced from Brodmann (1909).

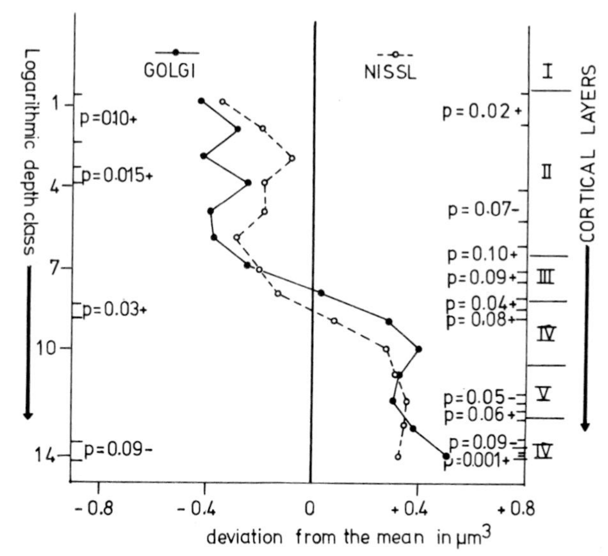

Fig. 16. Deviation from the mean volume of the cell body in Nissl-stained sections (mean volume 7.5347 μm^3) and Golgi–Cox-stained sections (mean volume 7.4229 μm^3) from the visual cortex of the rabbit. The data are collected from 14 logarithmic depth classes (*left ordinate*). Their correspondence with the cortical layers is shown on the *right ordinate*. On the *left* and *right ordinates*, the depth levels with a significant increase or decrease in volume according to the test of Kendall are indicated. The author concludes that the Golgi-impregnated neurons are a representative sample from the neurons at different depths of the cortex. Redrawn from Smit (1968) with the permission of Dr. G. J. Smit.

The axonal reaction includes swelling of the cell body, disintegration and redistribution of the Nissl substance toward the periphery of the cell [chromatolysis (Marinesco, 1896)], and migration of the nucleus to the cellular periphery (Fig. 17). These changes are usually observed in motor neurons when their axons are interrupted in the periphery. Chromatolysis reaches its peak in 1 or 2 weeks, but in the course of several months, the cells may regain their normal appearance concomitant with the establishment of new functional connections of the axons. In the central nervous system, such a regeneration of the axons is not the rule, and the chromatolytic phase is usually short and followed by rapid degeneration and disappearance or atrophy of the cell. In the central nervous system, this type of reaction not only is present after axonal transection but also is seen after removal of afferent connections (transneuronal reaction). Factors that influence the development of the axonal reaction include the age of the animal and the type and the site of the lesion. In young animals, retrograde (Brodal, 1939) and transneuronal changes (Torvik, 1956) develop more rapidly and more completely than in adults. The type of lesion influences the occurrence of chromatolysis in motor neurons because sharp transections of the nerve are less effective in this respect than repeated sections of the nerve, crush lesions, or tearing of the nerve. Moreover, the efficacy of lesions increases with a decrease of the distance between the cell and the lesion. This, and the resistance of some adult central nerve cells to transection of their axons, is often explained by the presence of collaterals of the axon, proximal to the lesion, that protect the perikaryon from the effects of a more distal axonal transection. Important differences exist in the time course of the axonal

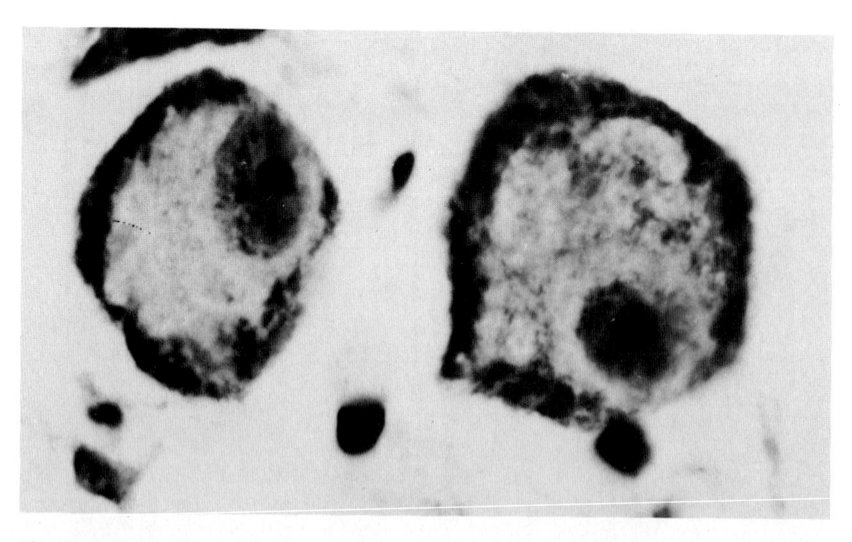

Fig. 17. Chromatolytic neurons from the inferior olive of the cat after ablation of the contralateral cerebellum. Note central chromatolysis and peripheral displacement of nucleus and Nissl substance.

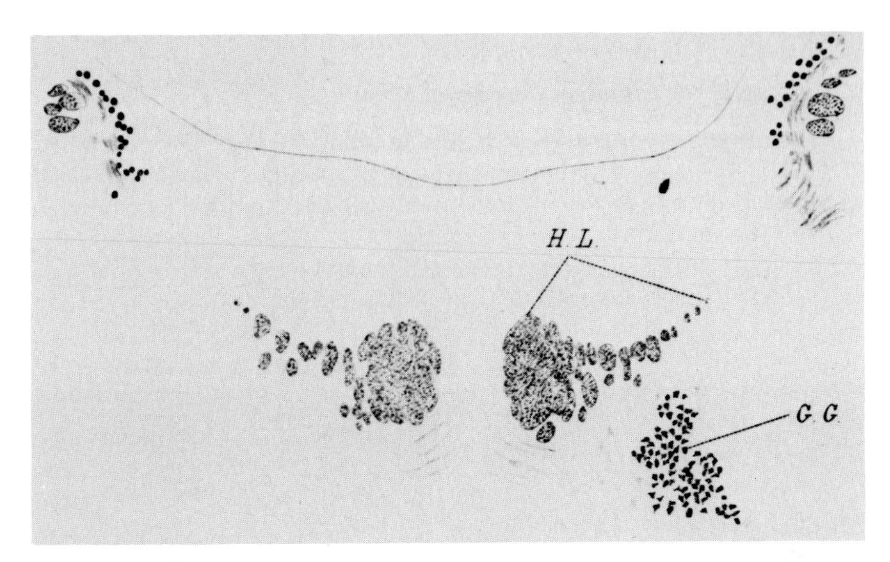

Fig. 18. Disappearance of the cells of the left ventral tegmental nucleus [(G.G.) Gudden's ganglion, as he himself called it] after a lesion of the ipsilateral dorsal nucleus of the mammillary body in a young cat. (H.L.) Medial longitudinal fascicle. From Von Gudden (1889).

reaction in different functional systems of the brain. Retrograde or transneuronal changes or both in the thalamus on removal of the cerebral cortex develop extremely rapidly, whereas other nuclei, such as the inferior olive, are much more resistant to removal of the target of their axons, in this case the cerebellum. However, changes do occur in certain parts of the inferior olive after cerebellectomy. In these parts, chromatolytic neurons swell excessively (neuronal hypertrophy), but this reaction does not subside, and hypertrophic neurons remain present for a year or more after cerebellectomy (Voogd and Boesten, 1976).

The rapid disappearance of nerve cells after axonal transection in the central nervous system of young animals was used by Von Gudden (1870, 1889) as a means of establishing the origin of certain fiber connections in the brain (Fig. 18). Von Gudden's method, the first experimental method of its kind, was modified by Brodal (1939), who did not wait until the cells had disappeared but made a positive identification of the cells of origin in his experiments by looking for chromatolysis in nerve cells of young animals (modified Von Gudden method). The main problems in using the axonal reaction in hodological investigations have been outlined above. Difficulties inherent in all experimentation on newborn animals, the possibility of confusing between axonal and transneuronal reactions, and the difficulty in establishing changes in small neurons have also limited the application of the modified Von Gudden method. At present, it is being replaced by retrograde-transport methods (Kristensson and Olsson, 1971; La Vail and La Vail, 1972).

6. Myelin Stains

6.1. Staining of the Lipids in the Myelin Sheath

Myelin becomes visible when viewed in polarized light (Fig. 19) in frozen sections from formalin-fixed material that are mounted, without dehydration, in glycerin–gelatin. For the anisotropy of myelin, the optic axis is oriented radially to the main direction of the fiber.

The lipids in myelin from fresh or aldehyde-fixed material that has not been passed through fat solvents can be stained with osmium tetroxide or

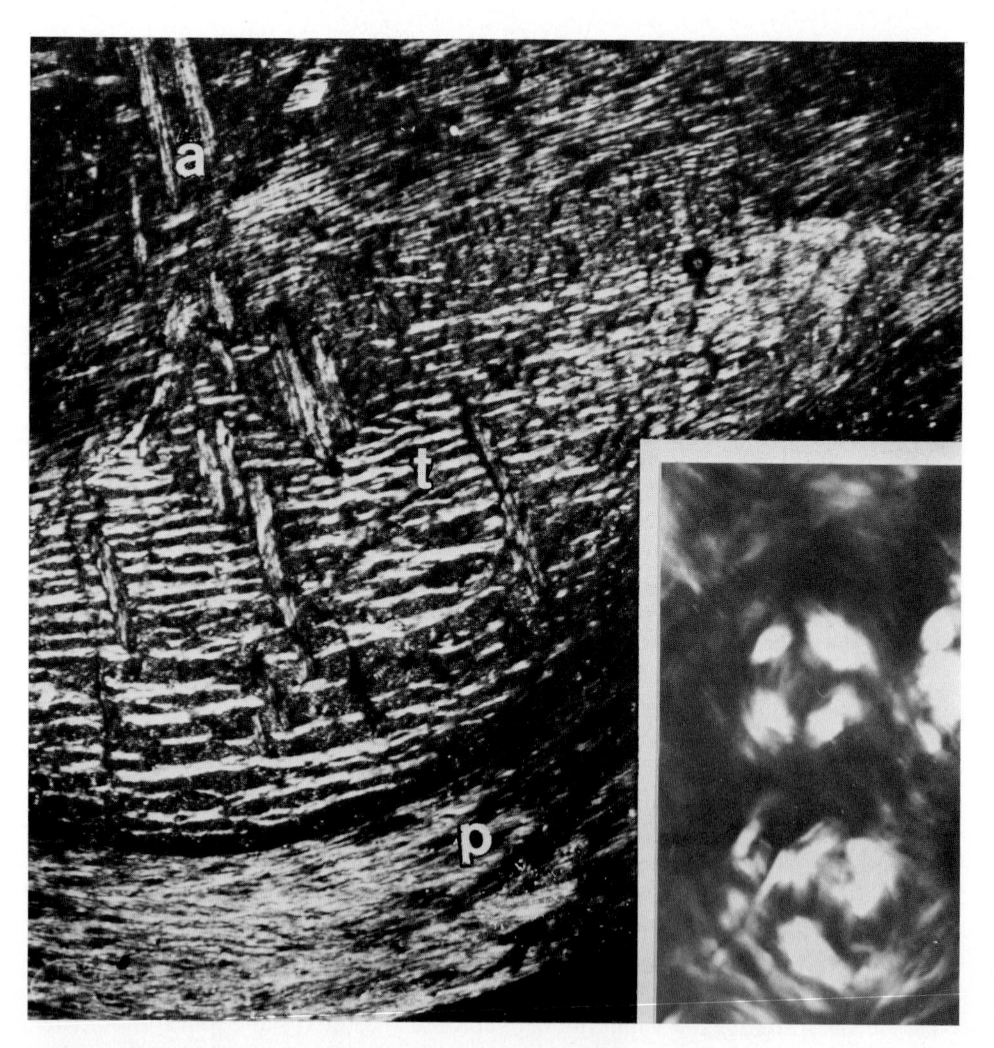

Fig. 19. Formalin-fixed, frozen parasagittal section mounted in glycerin–gelatin and viewed in polarized light. Part of the trapezoid body (t), the abducens nerve (a), and the pyramid (p) are shown. *Inset:* Crosslike image of transversely sectioned sheath in polarized light.

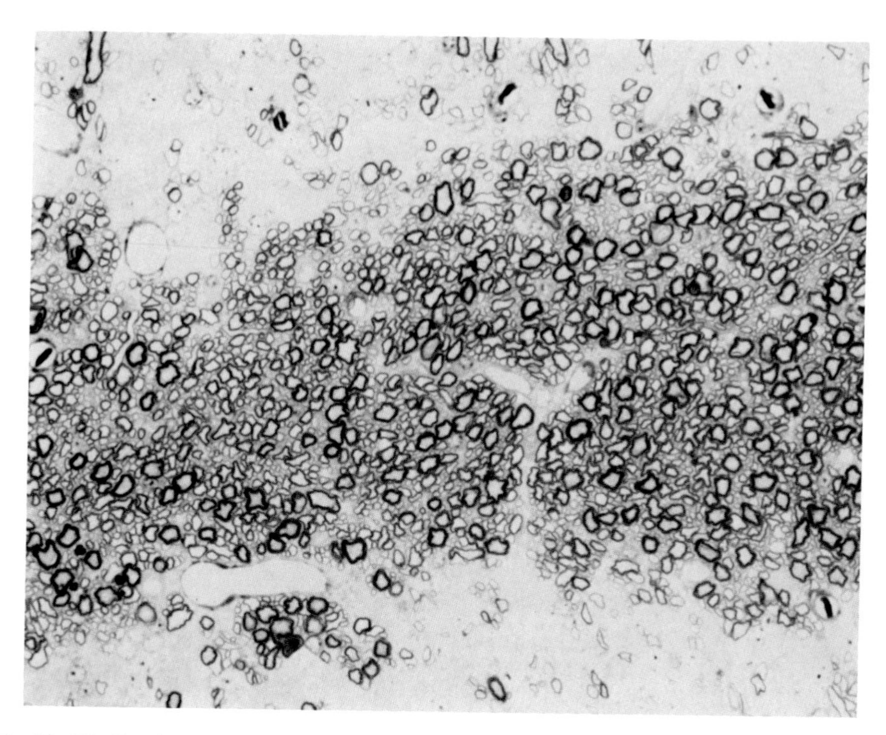

Fig. 20. Myelin sheaths in semithin 1-μm, toluidine-blue-stained section from chicken fixed tissue postfixed with osmium tetroxide and embedded in epon (Epikote 812). The white matter of the cerebellum is shown.

Sudan black. In the first myelin stain, introduced by Exner (1881) in 1881, small tissue fragments were fixed for 6–8 days in osmic acid. Sections, made without embedding, were studied in alkalified glycerin. Myelin stains black against a clear background. The limitation of the method is the very slow penetration rate of osmium into the tissue. In a modern version of this method, Vibratome sections from aldehyde-fixed material are postfixed with osmium tetroxide (for details, see Chapter 7). In semithin, toluidine-blue-stained sections from the plastic-embedded material (Figs. 20 and 21), the myelin sheaths are distinct. It is the method of choice for cytological and quantitative investigations. Occasionally, staining of sheaths with Sudan black has been used (König and Klippel, 1963); however, we have no personal experience with this method.

6.2. *Mordanting of Myelin: Weigert and Häggqvist Methods*

Before formalin came into general use, brain tissue was fixed in solutions of potassium dichromate [solution of Müller (1859): potassium dichromate, 12.5%; sodium sulfate, 1%]. Unlike formalin, dichromate is not a good general fixative. Its action on lipids (mordanting), which are made insoluble in fat solvents, is used to obtain selective staining of myelin in the Weigert

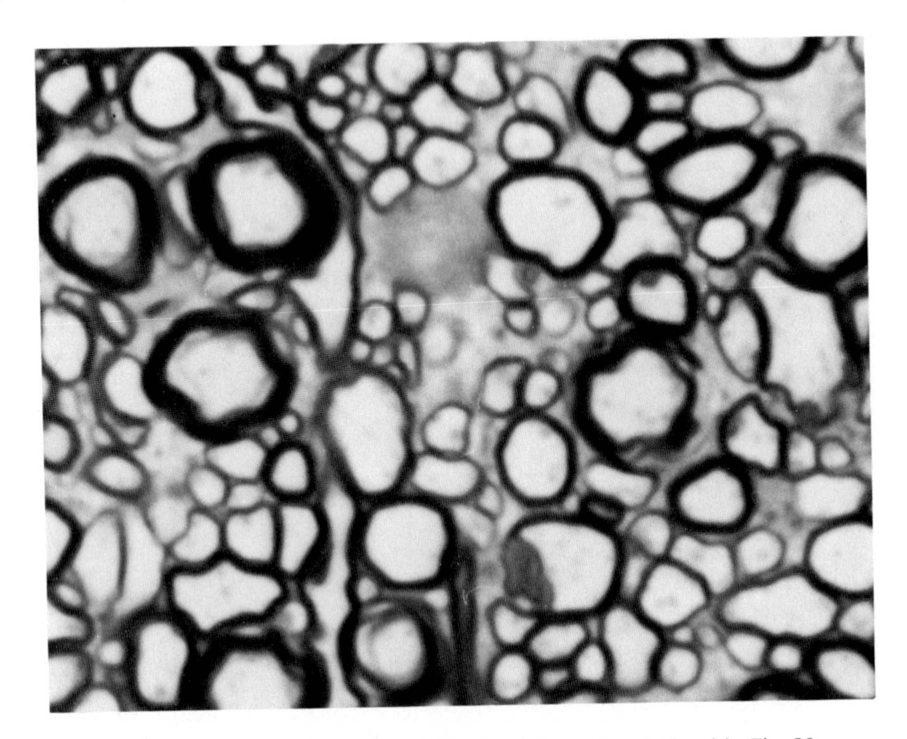

Fig. 21. High magnification of myelin sheath in section depicted in Fig. 20.

methods and to increase contrast among different tissue constituents in the Häggqvist–Alzheimer–Mann stain. Originally, brains fixed in Müller's solution were stained with carmine, introduced by Gerlach (1858). In carmine-stained sections, myelin, nucleus, cytoplasm and Nissl substance, axon and dendrites, and glial cell nuclei and cytoplasm are all stained in different shades of red (Fig. 22). The completeness of the staining is due, not to the carmine, but to the mordanting with dichromate, which exerts its action not only on the myelin but also on other cell components. Contrast, which is very faint in the carmine method, could be increased by the use of Mann's staining solution containing eosin and methyl blue on formalin-fixed and dichromate-mordanted material (Alzheimer, 1910). In the original Alzheimer–Mann method, frozen sections were made of the mordanted blocks. Häggqvist (1936) modified the method for paraffin. Frozen sections can be mordanted on the slide.

With the Häggqvist method, the myelin stains red and axons, dendrites, cytoplasm, and nuclei in different shades of blue. The method is especially useful to study tracts of myelinated fibers because the individual axons and their myelin sheath are stained in contrasting colors. At lower magnification, tracts consisting of small fibers appear as darker areas contrasting with lighter-staining coarse fibers and with the heavily stained neuropile of the nuclei (Figs. 23 and 24). The method greatly facilitates the topographical analysis of the white matter because it gives an accurate impression of the caliber specter of certain tracts, an impression that can be further quantified. Häggqvist's method has been used extensively in comparative-anatomical investigations of the vertebrate brain (Verhaart, 1970; 1976; Nieuwenhuys and Opdam, 1976).

Weigert (1884) tried to improve on the carmine method by making it more selective for myelin. In the Weigert methods, metal derivatives in the mordanted myelin combine with a dye into an insoluble compound, a "lacquer." After formalin fixation and mordanting, the tissue is embedded in celloidin. Depending on the mordant used, a chrome, copper, or iron

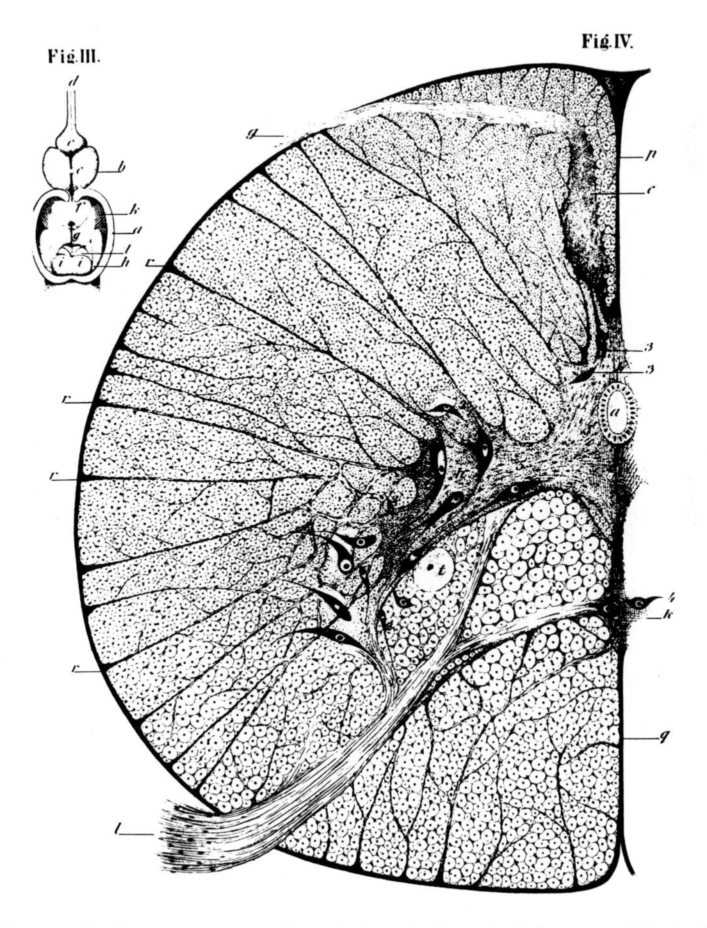

Fig. 22. Lithograph of a transverse section of the spinal cord of the trout (Stieda, 1861). The section was prepared by chromic acid fixation, sectioning unembedded, and staining with carmine 3 years after Gerlach introduced this staining method in neurohistology. Note the staining of all tissue components. Compare with Figs. 23 and 24, which show Häggqvist-stained material.

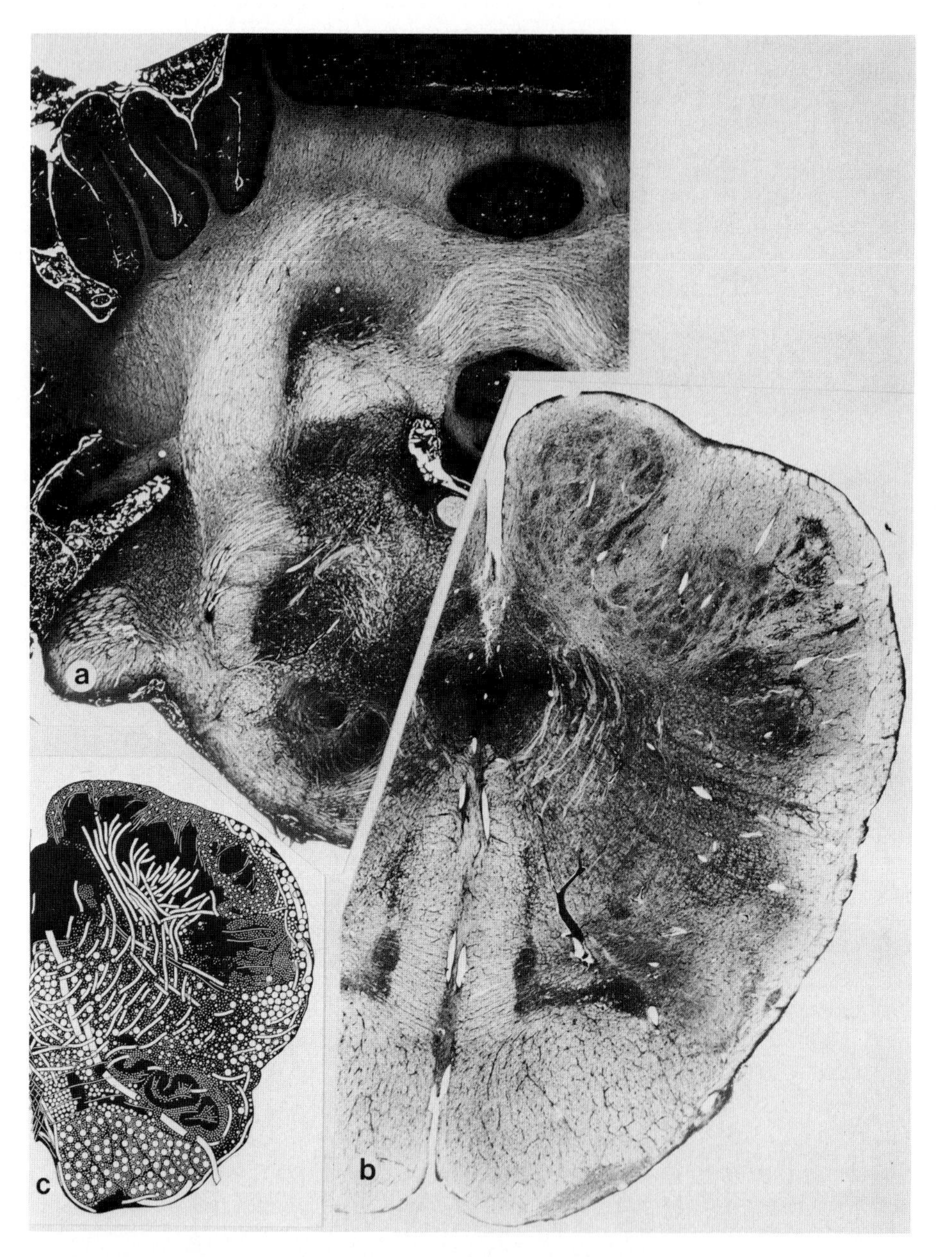

Fig. 23. Frozen 24-μm sections stained with the Häggqvist modification of the Alzheimer–Mann stain. (a) Transverse section through the cerebellum and brainstem of the cat. (b) Transverse section through the caudal medulla oblongata of man. (c) Diagram of approximately the same level as illustrated in (b). White circles indicate the size of the myelinated fibers that characterize the different tracts.

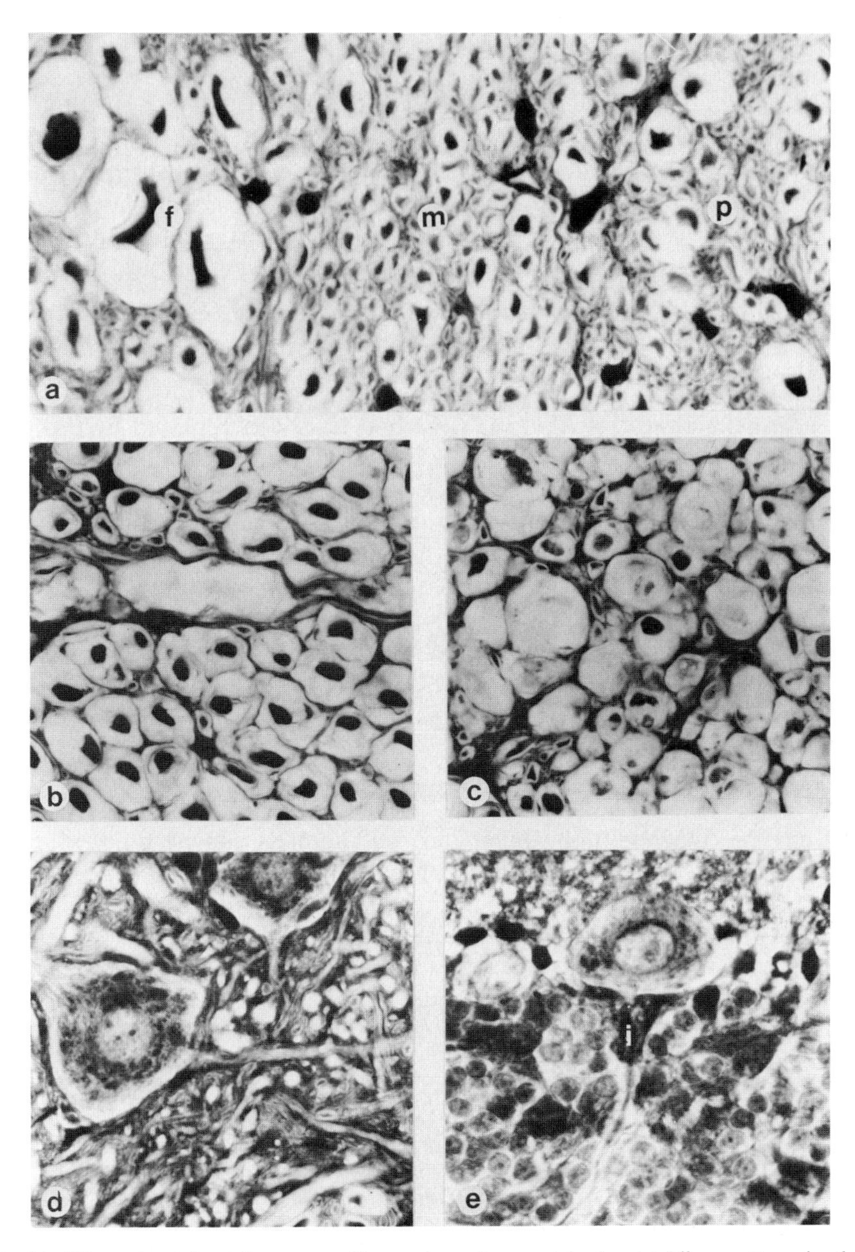

Fig. 24. Häggqvist-stained 5-μm paraffin sections from cat brain. (a) Fiber pattern in three adjoining tracts: (f) medial longitudinal fascicle; (m) medial lemniscus; (p) pyramid. (b) Transversely sectioned large myelinated fibers in one of the reticulospinal tracts in the medial longitudinal fascicle. Note the longitudinally sectioned degenerated fiber. (c) Degenerated and normal fibers in the same tract as in (b). (d) Portion of the motor nucleus of the facial nerve. Dendrites stain as hollow tubes. (e) Purkinje cell with initial segment (i) and beginning of myelin sheath.

lacquer is produced with hematoxylin as a dye. Unbound dye is removed by differentiating the sections with potassium ferricyanide–lithium carbonate solutions or potassium permanganate. In well-stained preparations, myelin stands out bluish-black against a yellow or unstained background (Fig. 25). Numerous modifications of the method for different embedding and sectioning methods are available. As a method to enhance contrast between white and gray matter in topographical studies of the brain, it is unsurpassed. Weigert-stained sections are still used extensively in teaching and provide the majority of the illustrations in current atlases and textbooks on neuroanatomy.

6.3. *Klüver–Barrera Method*

In most laboratories, Weigert's method has been replaced by the stain of Klüver and Barrera (1953), which is simpler to execute because no mordanting is necessary. Formalin-fixed tissue is dehydrated, embedded in either paraffin or celloidin, sectioned, and stained with Luxol fast blue, a dye that binds to porphyrins present in myelin. After being differentiated in lithium carbonate, the sections are counterstained with cresyl violet. The cresyl violet enhances the staining of the myelin and produces a combined Nissl stain of the perikarya. The appearance of the greenish-blue myelin in Klüver–Barrera-stained sections (Fig. 26) is similar to, but never reaches the finish of, the Weigert myelin stain.

6.4. *Myeloarchitecture: Counting and Measuring of Nerve Fibers*

Both the Weigert and Klüver-Barrera methods can be used in "myeloarchitectonic" studies (Vogt and Vogt, 1902) to subdivide the white matter and, especially, the gray matter on the basis of the amount and the disposition of the myelinated fibers. Myelin, unlike the perikarya of nerve cells in cytoarchitectonic investigations, cannot be described and measured in simple terms. Myelinated fibers can be afferent, efferent, or intrinsic. Moreover, the lower limit of visibility for thin myelin sheaths depends on the staining quality, and unmyelinated fibers and terminals cannot be seen at all. The subjective element in myeloarchitecture is much greater than in cytoarchitecture, and the value of myeloarchitectonic parcellations of the gray matter often remains uncertain.

The Weigert and Klüver–Barrera stains were not meant as cytological methods and should not be used as such. In Weigert sections, the individual sheath is often difficult to identify, owing to clotting and overlayering in the thick sections. Most of the myelin in Klüver–Barrera-stained sections is dissolved, and the stained components represent the neurokeratin skeleton (Fig. 27), which little resembles the sheath as it is known from ultrastructural studies.

For cytological and quantitative purposes, semithin sections of osmium-fixed material (see Chapter 7) or, in certain cases, Häggqvist-stained sections

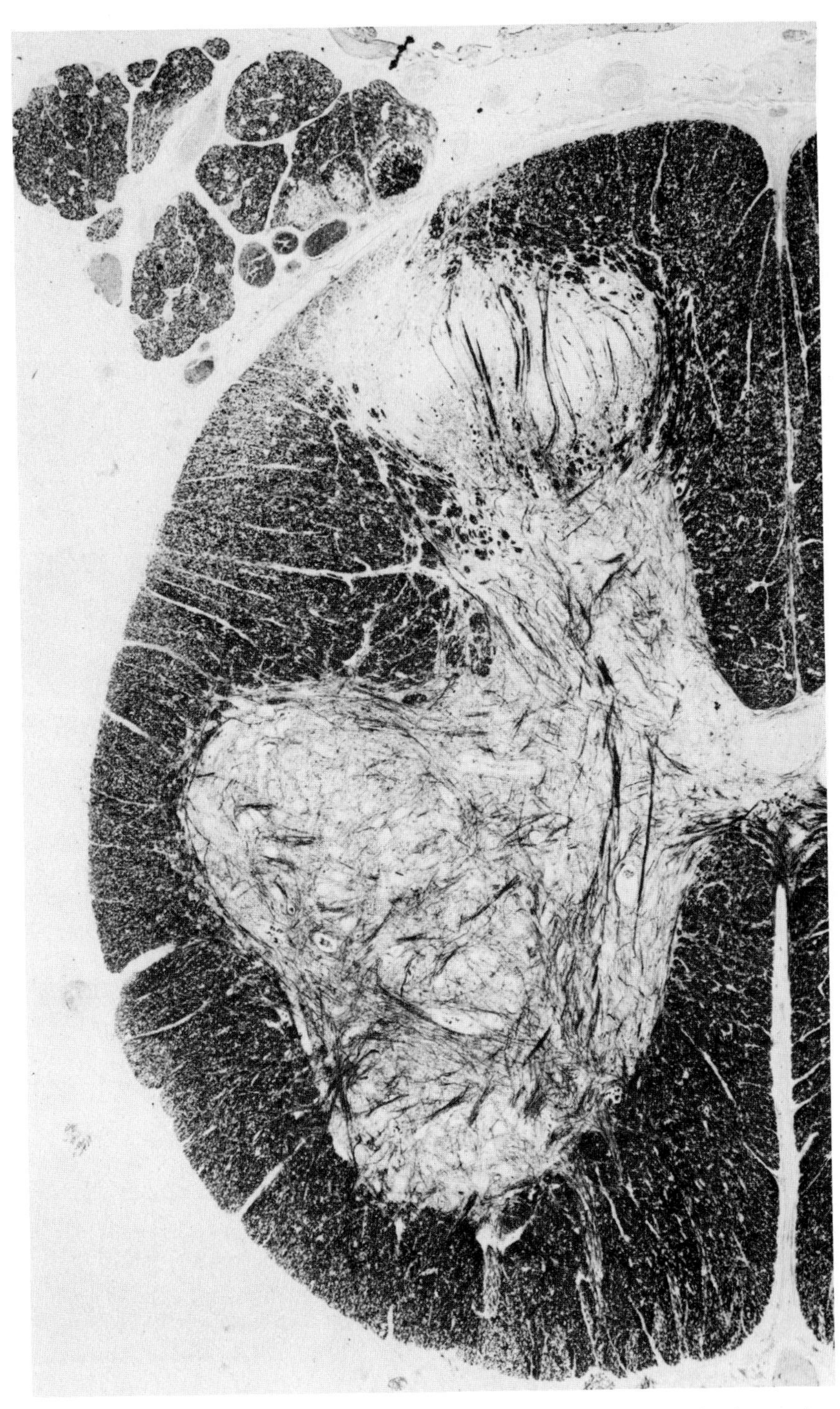

Fig. 25. Weigert-stained 30-μm celloidin section through the human lumbar cord.

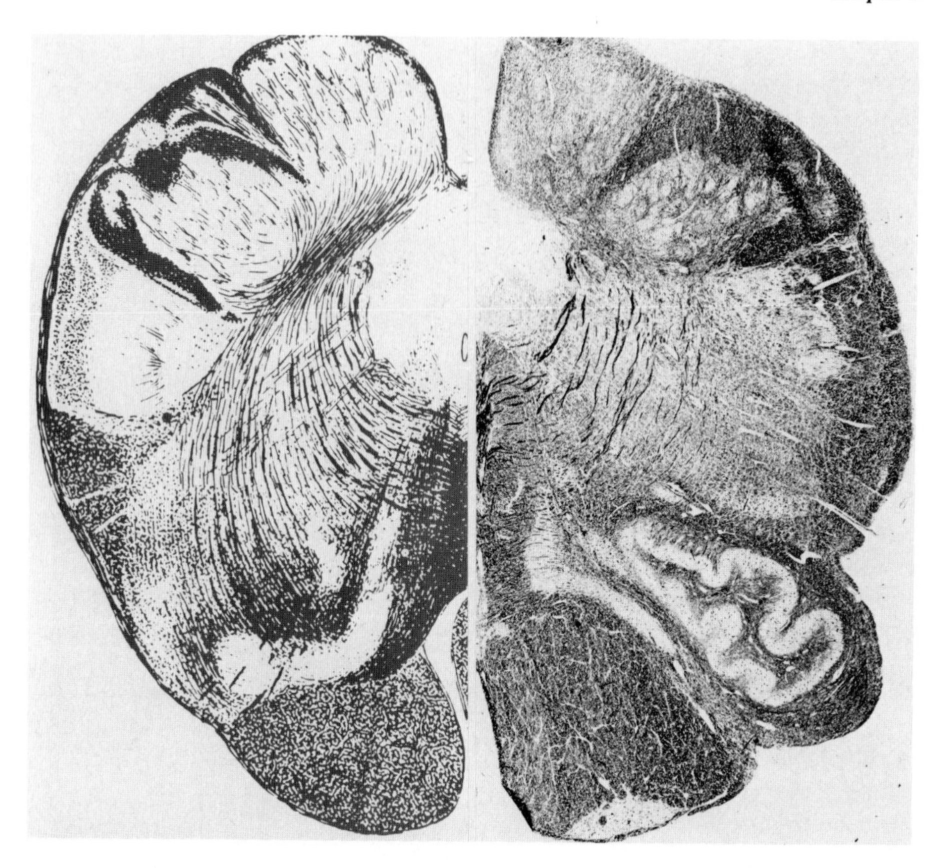

Fig. 26. Klüver–Barrera-stained 10-μm paraffin section of the human medulla oblongata. At left, a diagram of a similar, Weigert-stained section is depicted. To assess the relative merit of the Klüver–Barrera and Weigert methods and the Häggqvist method in delineating tracts in the white matter, this diagram should be compared to the one in Fig. 23c.

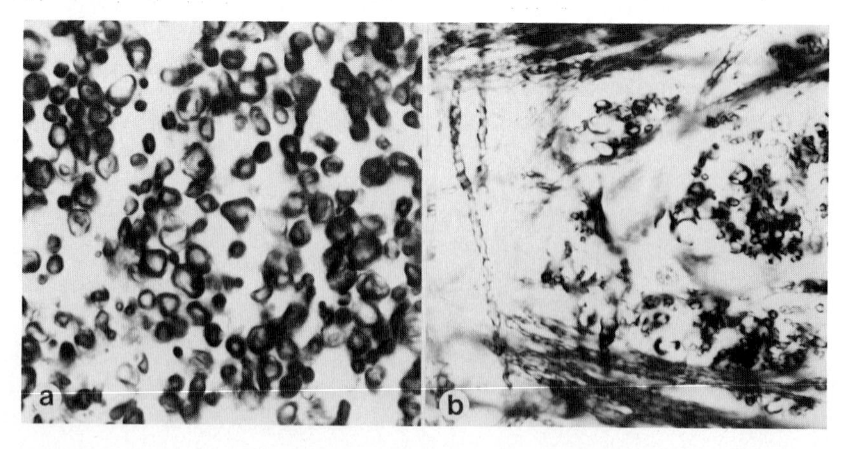

Fig. 27. Staining of myelin with the Weigert method [(a) transversely sectioned sheaths] and the Klüver–Barrera method [(b) both transversely and longitudinally sectioned sheaths, from which only the "neurokeratin" skeleton is left after dehydration].

should be used. With Häggqvist staining, large areas of white matter are available; with semithin sections, the surface area that can be covered is more limited. However, fiber-counting from semithin sections (P. Schnepp and G. Schnepp, 1971; G. Schnepp *et al.*, 1971) is more accurate, especially in the lower range of myelinated fibers, and it can be combined with ultrastructural visualization of the unmyelinated fibers. Measuring and counting fibers from Häggqvist-stained sections were described by Van Crevel and Verhaart (1963). These authors described an elegant method to quantify components of nerve tracts with a different origin by taking into account the element of degeneration rate of myelinated fibers.

6.5. Myelogenesis, Myelin Degeneration, and Demyelinization

Tracts acquire their myelin sheaths in a definite sequence that is repeated in the successive outgrowth of nerve fibers from the neuroblasts as well as in the sequence of phylogenetic development of the central nervous system (Flechsig, 1927). Observation of the development of myelinization provides an excellent method for studying localization in white matter if Weigert-stained sections from celloidin-embedded material are used. Because the Weigert-stained sections do not reveal the polarity of nerve fibers, interest in the method had already waned when Marchi's method for tracing degenerated myelinated fibers came into general use. In the hands of Von Bechterew (1899) and others, the Weigert method revealed the human nervous system in unsurpassed detail.

The first systematic investigation of secondary degeneration of nerve fibers was made by Waller (1850, 1851). With the crude methods at his disposal, he described the changes in a nerve fiber after severance, i.e., breakdown of the distal portion with the proximal portion remaining intact. It remained for Marchi (Marchi and Algeri, 1885) to develop a stain for degenerated myelin. In the Marchi method, the tissue is first fixed in Müller's dichromate solution and subsequently in a mixture containing 8% potassium dichromate and 0.3% osmium tetroxide. In celloidin sections of the thin blocks, the degenerated myelin appears as black, often vacuolated, droplets against a light brown background (Fig. 28). The Marchi reaction is not immediately positive after transection. Usually, these "vital" phenomena of myelin degeneration are strongest from 8 until 32 days after the lesion. However, the degeneration rate of myelinated fibers depends on species, age, body temperature, fiber system (in peripheral nerve fibers the process is faster), and fiber size [large fibers degenerate faster than those of small caliber, but resorption of the debris occurs more rapidly in thin fibers (Van Crevel and Verhaart, 1963)]. In Marchi staining, artifacts are common. Often, black deposits are observed in otherwise normal nerve roots and tracts. The interpretation of Marchi material is also complicated by the occurrence of retrograde degeneration of the proximal axon, which may occur with long survival times. Establishing the termination of fibers is difficult with the Marchi impregnation because the distal portions of axons are usually not myelinated. As a method for tracing degenerated fibers, it

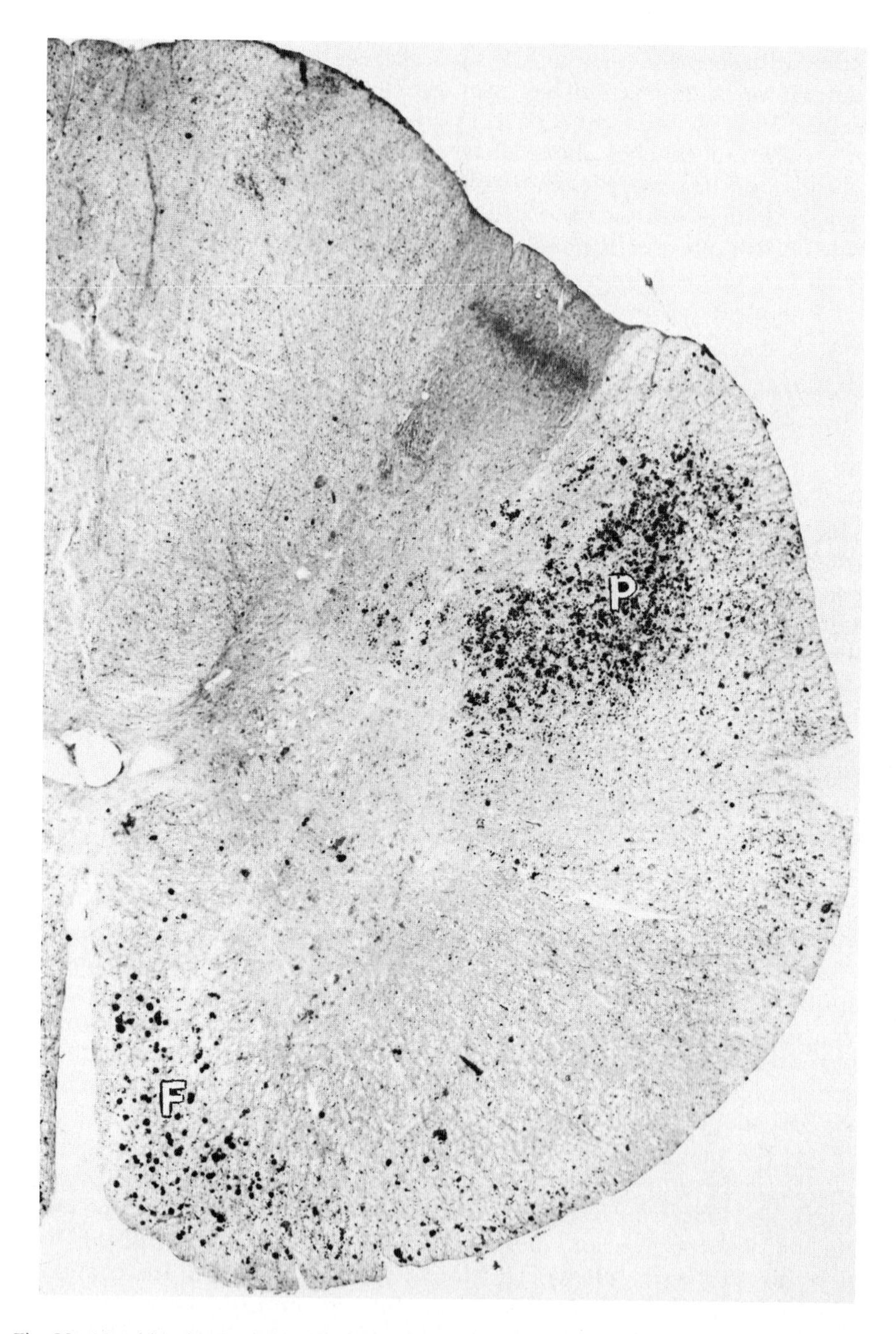

Fig. 28. Marchi-stained section through the spinal cord of the cat showing droplets of degenerated myelin in the area of the lateral pyramidal tract (P) and the medial longitudinal fascicle (F).

has become obsolete since the introduction of the suppressive silver methods in the 1950's. Recently, Marchi osmification, in dichromate–osmium solutions, was reintroduced for ultrastructural investigations of peroxidase-injected cells (Cullheim and Kellerth, 1976).

Similar degenerative changes in myelinated fibers can be observed with the Häggqvist–Alzheimer–Mann stain. Here, the degenerated axons soon disappear, but the degenerated, vacuolated myelin stains bright red (See Fig. 24) (Van Crevel and Verhaart, 1963). Ultimately, the degeneration of myelinated fibers results in demyelinization, which can be studied with the Weigert or the Klüver–Barrera method.

7. *Reduced-Silver Methods*

7.1. *Neurofibrillar Stains*

Neurofibrils are argyrophilic components of the nerve cell that precipitate metallic silver from solutions of silver salts or ammoniacal silver solutions after reduction. The fibrillary character of nerve-cell processes had been observed with different stains before Bielschowsky (1902, 1904) and Ramón y Cajal (1903) introduced their reduced-silver methods, which showed the neurofibrils with much greater clarity. In nerve cells, neurofibrils appear as bundles of parallel filaments that can be traced from one process into another. Sometimes the neurofibrils interconnect. In the soma, they often combine into a reticulum, while in terminal boutons, they are often arranged in loops (Fig. 29). During the first half of this century, prior to the advance of electron microscopy, the nature of neurofibrils remained a matter of

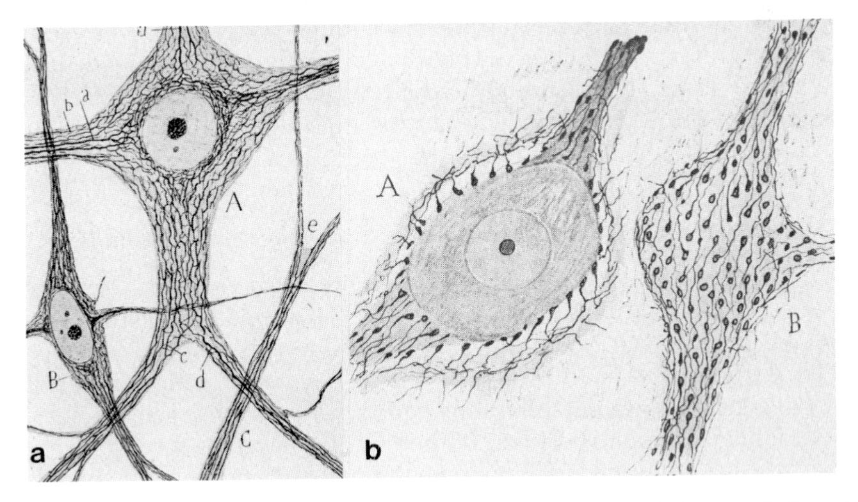

Fig. 29. Neurofibrils (**a**) and axonal endfeet (**b**) of cells of the spinal gray of the rabbit impregnated with the reduced-silver method of Cajal (1903).

conjecture. In this period, neurofibrillar stains played an important part in the discussions between supporters and opponents of the neuron theory, which is the assumption that all processes of nerve cells end free. Observations of silver-impregnated fibrils, which seemed to be continuous from one nerve cell into another or into periterminal networks that merged with the cytoplasm of effector cells, were used as arguments against the contiguity of nerve cells or were interpreted as artifacts. From our present knowledge of the ultrastructure of nerve cells, it seems likely that the neurofibrils of light microscopy correspond to coalesced neurofilaments and microtubules. But it is also known that argyrophilia of nerve cells is not necessarily limited to these cell organelles (Walberg, 1964).

Present silver techniques are based on the procedures introduced by Bielschowsky and Ramón y Cajal in the beginning of this century. Cajal's reduced-silver methods (Ramón y Cajal, 1903, 1907) are mainly block stains. Small blocks of fresh or fixed tissue are impregnated in solutions of silver nitrate and reduced with pyrogallol, hydroquinone, or formaldehyde. The method of Bielschowsky (1902, 1904) can be applied to blocks as well as to frozen sections. In this method, the silver nitrate is followed by immersion in an ammoniacal silver solution. For the reduction, formalin is used. Often, silver impregnations are intensified by gold toning in diluted solutions of gold chloride, which tinges the background a light purple.

The neurofibrillar methods and their modifications have been applied to many problems in neurohistology. To appreciate their potential use, one should consult the chapters of Bielschowsky (1935) and Ramón y Cajal (1935). At present, their main applications seem to be in neuroembryology, in problems of axonal regeneration and degeneration, and as a general stain for nervous tissue. In neuroembryology, block impregnations according to Ramón y Cajal or Ranson (1911) are often used. Excessive shrinkage and uneven impregnation—the surface of the block is usually overimpregnated—are important disadvantages of these block stains. For the study of axons and axon terminals, the Bielschowsky method and its modifications (Glees, 1946; Nauta and Gygax, 1951) should be used. A good neurofibrillar method for general use on formalin-fixed, paraffin-embedded tissue (Fig. 30) was developed by Bodian (1936).

7.2. *Silver Impregnation of Degenerating Axons and Axon Terminals*

The establishment of nervous connections by the tracing of degenerated axons is possible with the Bielschowsky silver-impregnation method, but it is hindered by the large amount of impregnated normal fibers. In 1946, Glees (1946) introduced a modification of the Bielschowsky method that impregnates degenerated terminal boutons and thin fibers more selectively. When terminal degeneration is studied in the inferior olive (Blackstad *et al.*, 1951), it is characterized by a general increase in number of the ring- or club-shaped boutons that were already known from earlier studies (Figs. 29 and 31), a shift from weakly impregnated ringlike terminals to the solid club-shaped

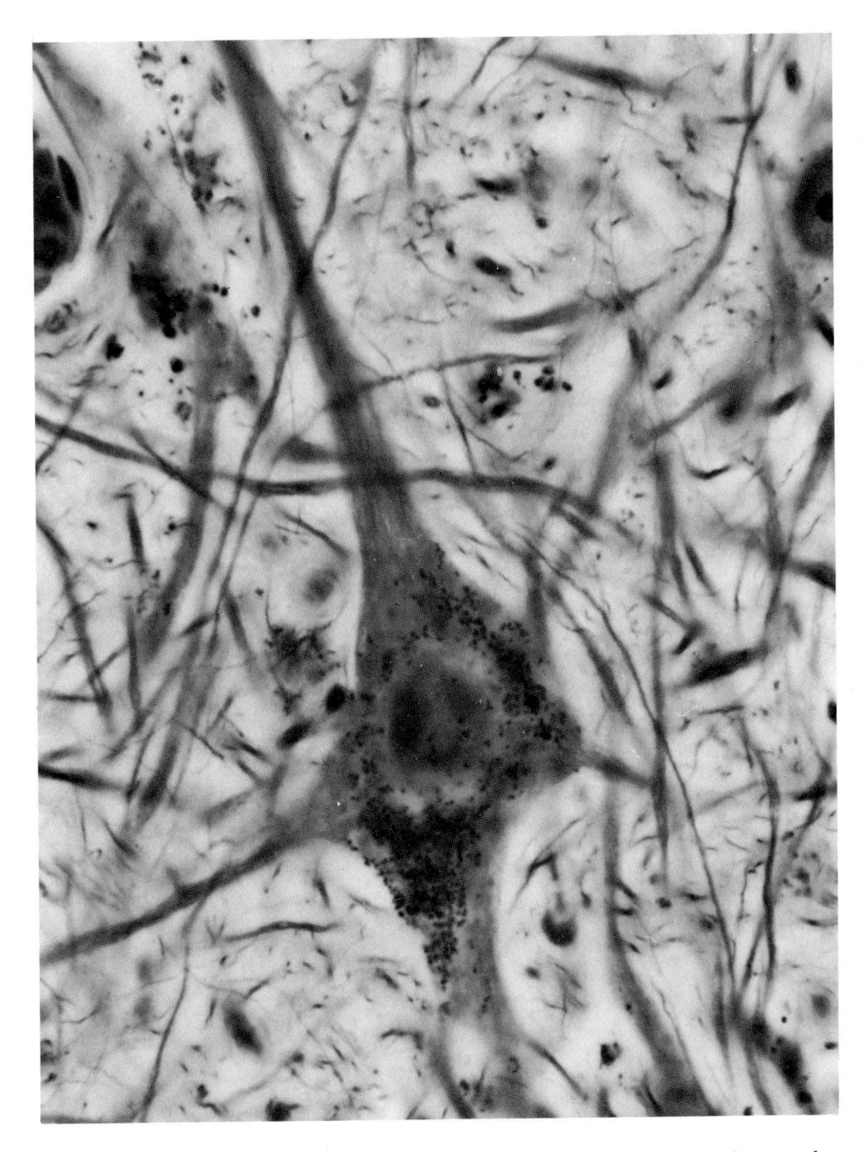

Fig. 30. Bodian silver impregnation of a 15-μm paraffin section showing a pigmented neuron of the human substantia nigra.

ones, and argyrophilia, fragmentation, and vacuolation of the thin preterminal axons. That the normal appearance, distribution, and number of boutons have to be known before the presence of terminal degeneration can be inferred from experimental material was stressed by Cowan and Powell (1956), who described a selectively distributed "pseudo degeneration" in Glees's preparations from the hypothalamic region of the normal monkey. More recently, Walberg (1971) found silver-impregnated boutons to be

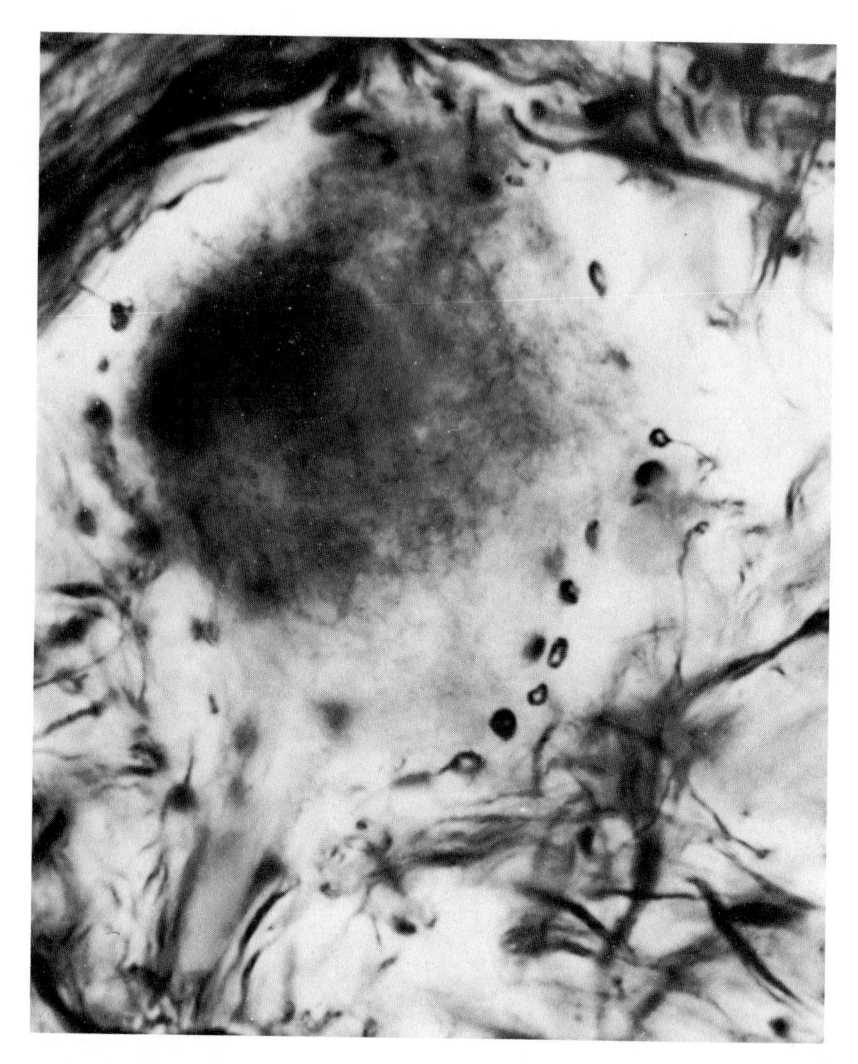

Fig. 31. Glees silver impregnation of a 20-μm frozen section showing compact and ringlike boutons on a motor neuron in the spinal cord of the cat. Compare Fig. 29b.

absent in an ultrastructural study of Glees-stained material of the inferior olive. According to him, the rings and solid "boutons" of light microscopy correspond to the silver-impregnated axoplasm of transversely sectioned fibers with or without central unimpregnated mitochondria.

Soon after Glees's terminal degeneration method, Nauta's suppressive silver impregnation for degenerated axons became available. His first method (Nauta and Gygax, 1951) was a modification of the Glees method, but in 1952, Nauta and Ryan (1952) showed that the staining of normal fibers is suppressed when tissue sections are oxidized with permanganate prior to the modified Nauta–Glees–Bielschowsky silver impregnation. The Nauta–Ryan method has been modified by many of the numerous investigators who used it, namely, Chambers *et al.* (1956) (introduction of Laidlaw's silver carbonate solution instead of the ammoniacal silver solution), Fink and Heimer (1967) (use of uranyl acetate alone or in combination with silver nitrate), and De Olmos (1969) (cupric–silver method). The Nauta method and its modifications were reviewed in *Contemporary Research Methods in Neuroanatomy,* edited by Nauta and Ebbesson (1970).

Suppression of argyrophilia in the Nauta technique affects primarily the staining of normal axons but does not spare the degenerated ones. The relative proportions of normal and degenerated fibers in the sections can be adjusted, by the choice of a suitable time and concentration of the permanganate, in such a way that the impregnated degenerated fibers can be traced under low-power magnification (Fig. 32). Argyrophilia alone is not a certain sign of axon degeneration; it should occur together with fragmentation and vacuolation of the axon (Fig. 33). Terminal, i.e., bouton, degeneration usually cannot be distinguished with certainty because argyrophilic boutons have the same appearance as impregnated fragments of passing axons. The term "preterminal degeneration," therefore, was applied to the impregnation of a fabric of thin degenerated axon collaterals in putative terminal nuclei of

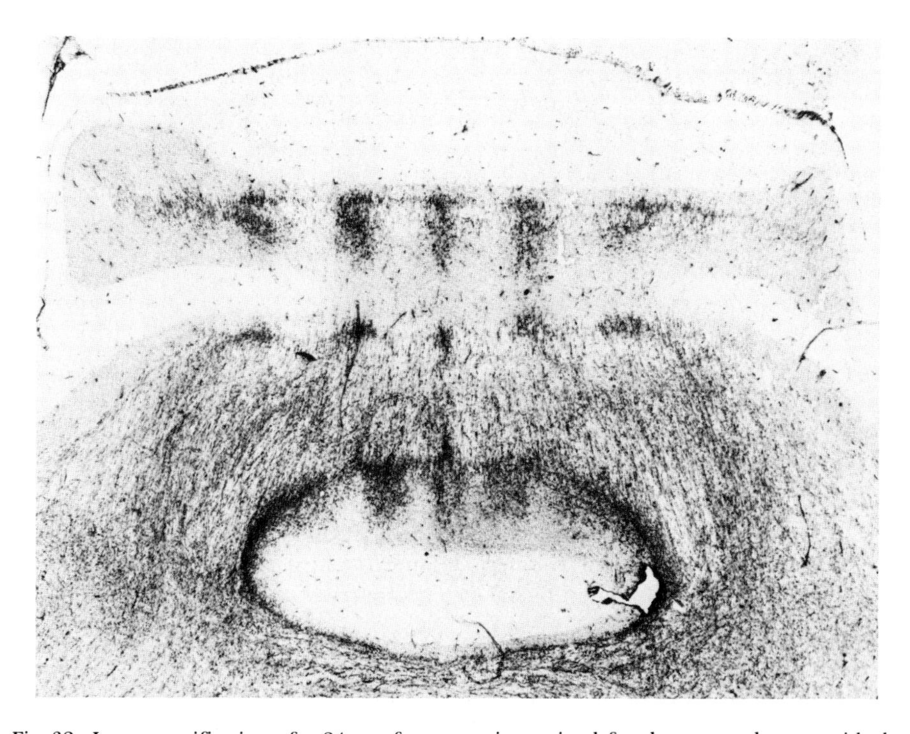

Fig. 32. Low magnification of a 24-μm frozen section stained for degenerated axons with the Nauta–Ryan silver-impregnation method. Degenerated spinocerebellar fibers in the anterior lobe of the cerebellum of tupaia, which distribute in a number of parasagittal concentrations, are shown. Survival time 7 days.

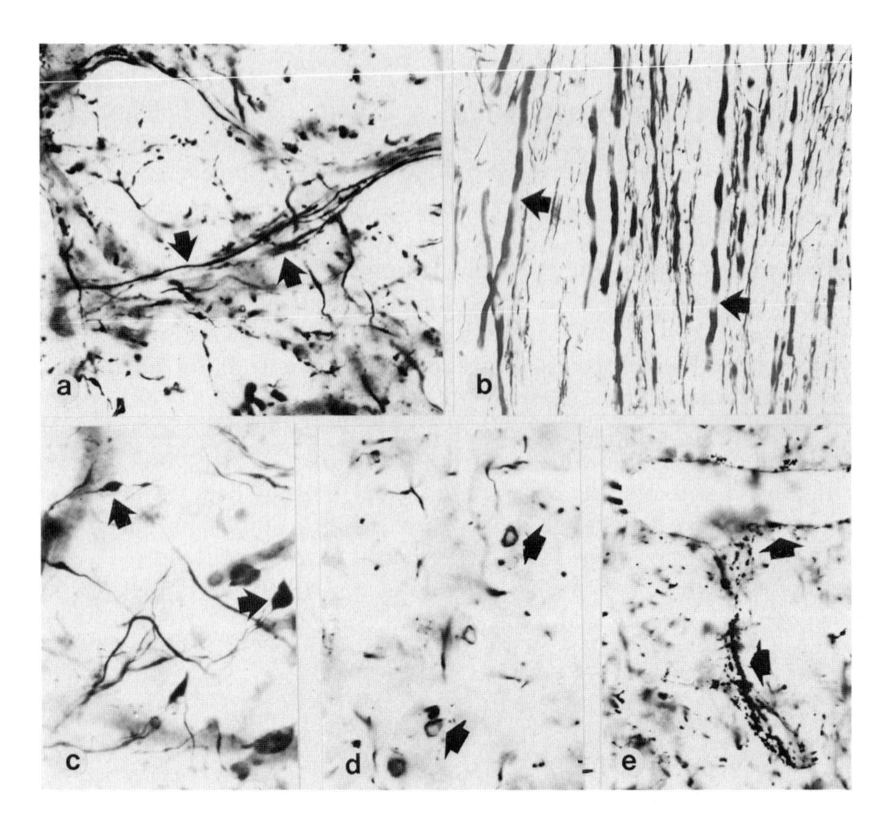

Fig. 33. Degenerated axons and artifacts in the Nauta–Ryan silver-impregnation method. (a) Fragmented, degenerated fibers and normal impregnated axon (↓); (b) uneven impregnation of normal axons; (c) fusiform enlargements in normal axons; (d) impregnated myelin cuffs; (e) impregnated reticulin fibers of vessel walls.

a tract. It was contended that the Fink and Heimer (1967) uranyl modification does impregnate degenerated terminals. With this method, degeneration in terminal nuclei often appears as a dust of small impregnated particles. Some of these particles were found by Heimer and Peters (1968) to correspond to impregnated boutons. In general, subsequent electron-microscopic verification is necessary to establish the presence of terminal degeneration. In some cases, however, the size or the shape of the terminals distinguishes them sufficiently to be recognized in an early, argyrophilic state of degeneration. This is the case in the so-called mossy-fiber rosettes (Fig. 34) and the climbing-fiber terminals (Fig. 35) in the cerebellum.

For the interpretation of axonal degeneration, its time course should be taken into account. In warm-blooded animals, terminal-bouton degeneration can be observed as early as 1 day after the lesion of the axon, but may have disappeared a day or so later. Degeneration of climbing-fiber terminals in mammals, for instance, disappears completely after 3 days' survival (Desclin, 1974; Groenewegen and Voogd, 1977). Degeneration of larger terminals and preterminal axons generally remains visible for 3 or 4 weeks (Vielvoye and Voogd, 1977). Silver impregnation of the degenerated parent axon becomes possible 1 or 2 days after the degenerated terminals can be stained, but impregnated remnants of the axon remain present for as long as 12 months after the original intervention (Knook, 1965). Possible differences in the time course of silver impregnation of degenerating axons of different caliber have not been systematically studied. For this reason and other reasons, the degeneration method cannot be used for quantitative studies. The conclusions on the degeneration rate of myelinated fibers (Van Crevel and Verhaart, 1963) quoted in Section 6.5 cannot be applied because they were taken from Häggqvist-stained material.

As a metabolic process, axonal degeneration is subject to the effects of body temperature in poikilothermic animals. Therefore, longer survival times should be used (Ebbesson, 1970). Systematic differences in degeneration rate are certainly present. In peripheral nerves, the rate is higher than in central systems; in unmyelinated fibers, degeneration proceeds more rapidly

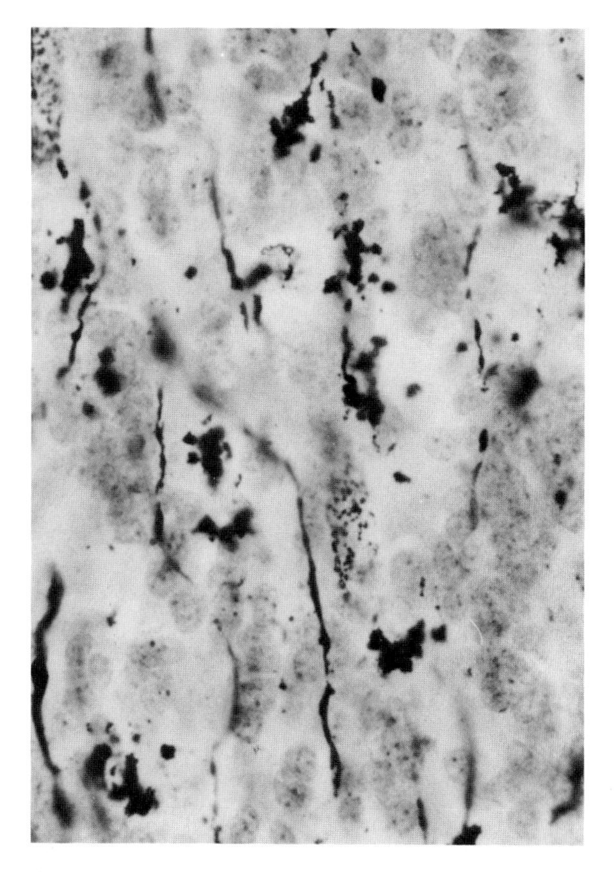

Fig. 34. Degeneration of the terminals (rosettes) of pontocerebellar mossy fibers in the cerebellar cortex of the chicken. Frozen section, 24 μm, Fink–Heimer method, survival time 4 days.

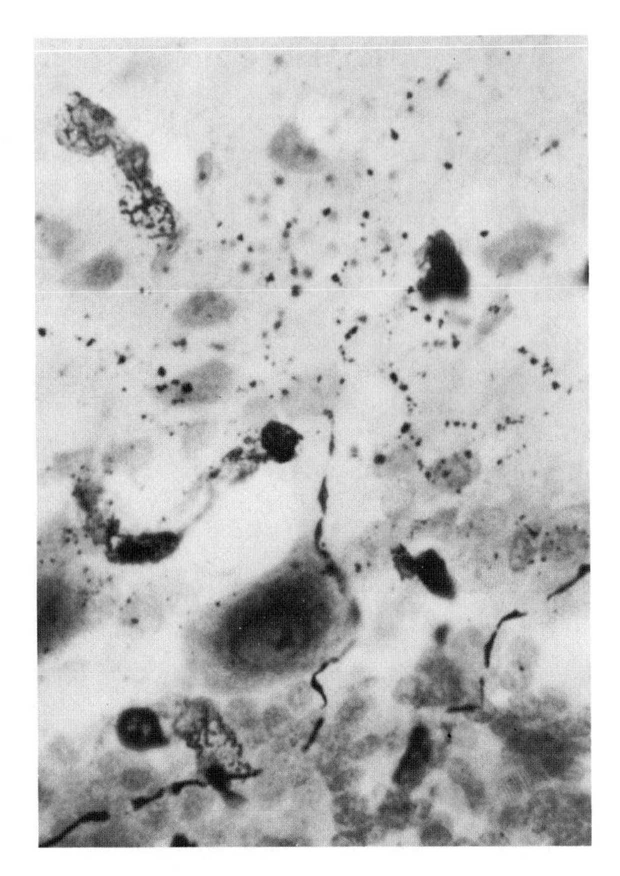

Fig. 35. Degeneration of climbing fiber in the cerebellar cortex of the chicken, 2 days after a lesion of the contralateral inferior olive. Frozen section, 24 μm, Fink–Heimer method. Reproduced from Freedman *et al.*, 1977, *J. Comp. Neurol.* **175:**243 with the permission of The Wistar Press, Philadelphia.

(Heimer and Wall, 1968). As a general rule, terminal degeneration is best studied with survival times of 2–3 days and degeneration in preterminal collaterals and parent fibers with 7–10 days' survival. However, the optimal survival time should be established for each individual experiment.

The limitations of the Nauta method are inherent in the staining technique and the process of axon degeneration. Furthermore, they became apparent when more recent axonal-tracing techniques were applied. When an ammoniacal silver solution is used in the Nauta method or its modifications, both the "dark" and the "filamentous" ultrastructural types of axonal degeneration are stained (Walberg, 1972). The uranyl–(Fink and Heimer, 1967) and cupric–silver modifications (De Olmos, 1969) impregnate more preterminal axons and boutons, but principally they do not differ from the original procedure of Nauta and Ryan (1952). When Laidlaw's silver carbonate solution is used instead, the impregnation is less complete and limited to

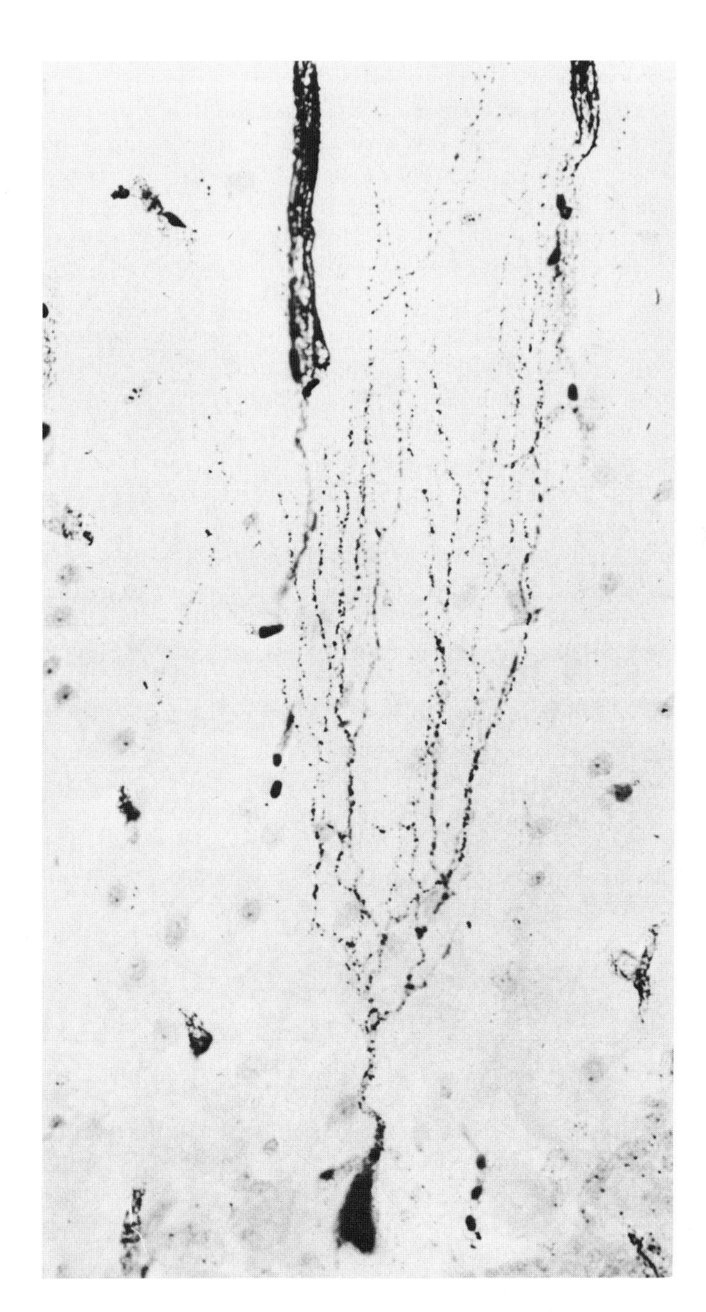

Fig. 36. Silver impregnation of a degenerated Purkinje cell in the cerebellar cortex of the chicken. Frozen section, 24 μm, Fink–Heimer-impregnated. Compare Fig. 38.

the dark type of axon degeneration (Walberg, 1972). Contrary to the Nauta method, the Glees method can still be applied after alcohol extraction. According to Walberg (1972), the Glees method impregnates mainly the filamentous type of axonal degeneration.

The full extent of some fiber connections, such as the monoaminergic paths in the brain that were discovered with histochemical or transport techniques, or both, cannot be demonstrated with the Nauta method. The main advantage of the anterograde-transport technique with tritiated amino acids over the degeneration method lies in the selective uptake of the precursor by the cell body. The perennial problem in the degeneration techniques, namely, the interruption of passing axons when a certain focus is destroyed, therefore does not arise when transport techniques are used.

Artifacts are not uncommon in the Nauta technique (Glees and Nauta, 1955). Fusiform enlargements or irregular impregnation of normal fibers (see Fig. 33b) should be distinguished from axonal degeneration. Irregular, dustlike impregnation of the background and myelin cuffs (see Fig. 33d) can often be prevented when the blocks are stored for some weeks before processing, by the use of clean glassware and fresh solutions, and by adjusting the ammoniacal silver solution with sodium hydroxide and citric acid. Apart from the axon, the cell body, and dendrites, and the reticular fibers in vessel walls can be impregnated. Silver impregnation of cell body and dendrites is seen when the cells degenerate in vascular disturbances (Fig. 36) and was used in retrograde-degeneration studies in young mammals (Grant and Aldskogius, 1967). The impregnation of reticular fibers (see Fig. 33e) cannot be prevented when an ammoniacal silver solution is used.

Degeneration methods are also used in electron microscopy. To bridge the gap between light and electron microscopy, the silver-impregnation methods and other stains were modified for plastic-embedded material (Holländer and Vaaland, 1968; Heimer, 1969; Chan-Palay, 1977). However, they have only found limited application.

8. Golgi Method

The method of Golgi (1873) is the oldest neurohistological method still in use. In sections from Golgi-impregnated material, staining is not limited to certain cell components, but reveals the complete contours of the cells and their processes in their full complexity. The nerve cells are visible against a clear background because only a small proportion (2%, according to Van der Loos, 1959) of the cells are impregnated. The reason for this selectivity is not known.

In the original Golgi method, fresh tissue is hardened in Müller's dichromate solution and impregnated in 1% silver nitrate. In the rapid Golgi method, fresh and small pieces are hardened in 4 parts 3.5% potassium dichromate and 1 part 1% osmic acid for 2–7 days and transferred to 0.75% silver nitrate for 1–2 days. In double and triple impregnations, the pieces are returned to the osmium–dichromate and further impregnated in the same silver nitrate solution. Generally, the pieces are embedded in celloidin and sectioned at 70–100 μm. When dehydration is complete, the sections can be coverslipped, but formerly they were only covered with balsam and no coverslips were applied. With the rapid Golgi method, the impregnated cells and axons are reddish against a yellow background (Fig. 37).

A reliable modification of the Golgi method was introduced by Cox (1891). In the Golgi–Cox method, hardening and impregnation take place in a single bath of potassium dichromate and potassium chromate to which sublimate is added. In certain cells, there is formed a yellow compound of divalent mercury that is transformed into a black substance by alkalization of the blocks or the sections (Ramón-Moliner, 1970). The Golgi–Cox method is simple and reliable, but the impregnation is often rather coarse and axons are never stained. Counterstaining with a Nissl stain has been successfully attempted by Van der Loos (1956) and Turcotte and Ramón-Moliner (1965).

Golgi impregnation is facilitated when it is preceded by perfusion fixation. The rapid Golgi method can be applied to aldehyde-fixed tissue slices (Palay and Chan-Palay, 1974). Direct perfusion with chrome–osmium

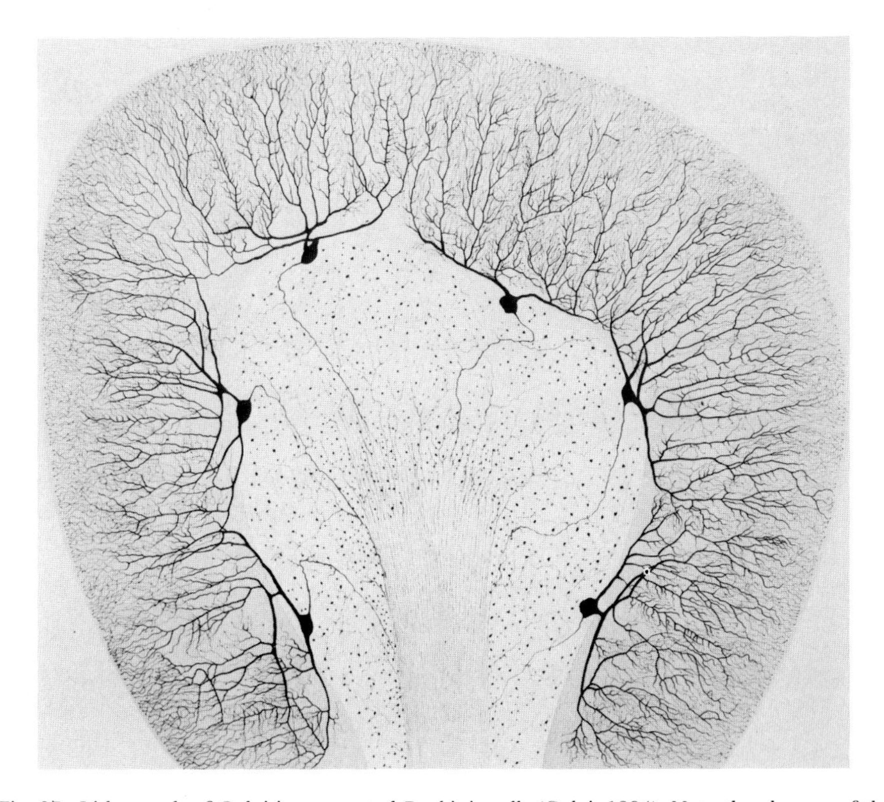

Fig. 37. Lithograph of Golgi-impregnated Purkinje cells (Golgi, 1894). Note the absence of the impregnation of spines. Collaterals of the axons get lost in a continuous reticulum in the granular layer.

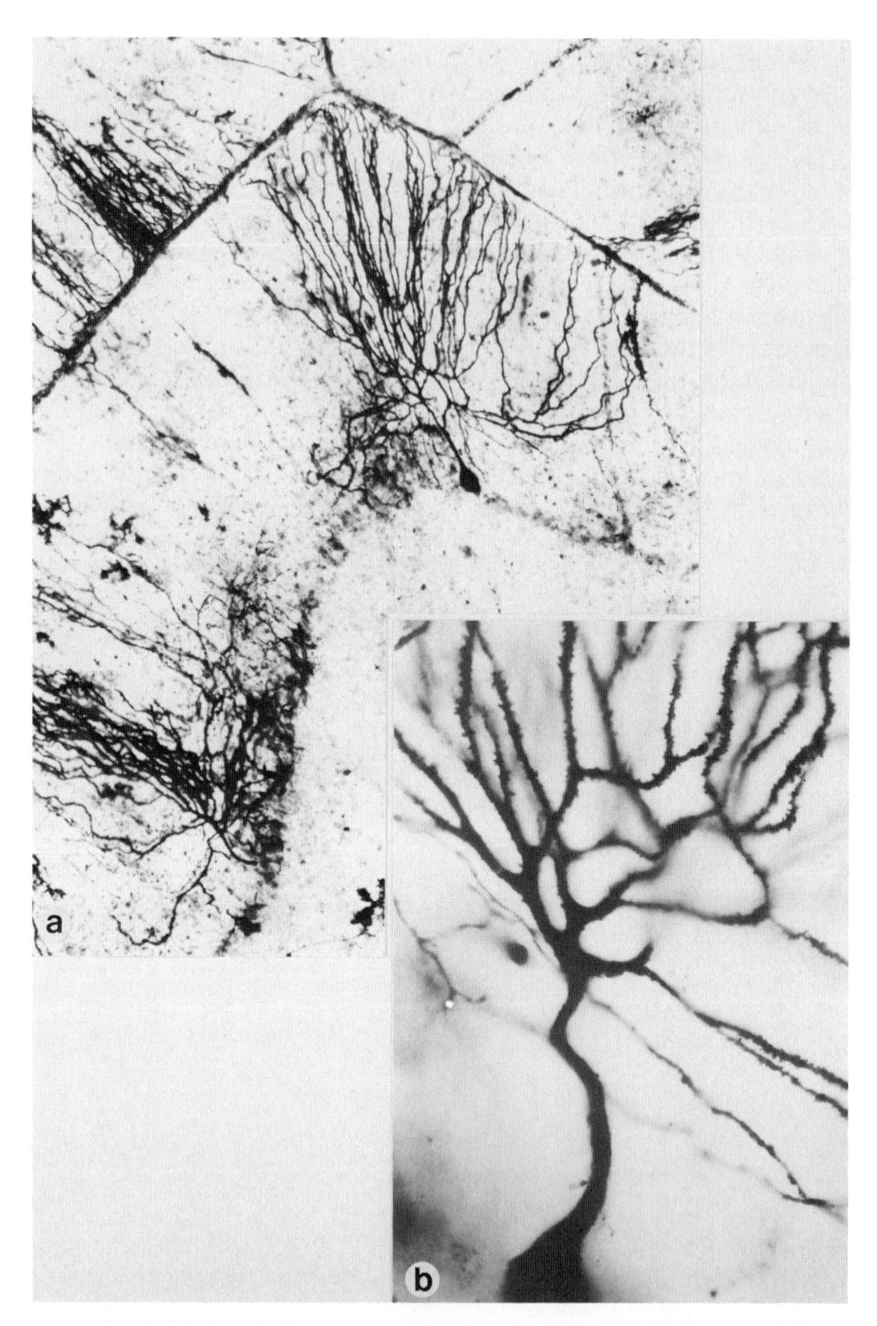

Fig. 38. Impregnated Purkinje cells of the chicken. Del Rio Hortega method. (a) Low magnification. (b) Higher magnification. Note the impregnation of spines on distal parts of the dendrites.

solutions for the rapid Golgi method was described by Morest and Morest (1966). According to these authors, the method demonstrates fine morphological detail more consistently, but the costs are prohibitive. In the Golgi–Kopsch procedure (Tömböl, 1968), the animals are perfused with a freshly prepared mixture of potassium dichromate and formalin (or glutaraldehyde). The hardened brain is sliced and impregnated in 0.75% silver nitrate. A similar procedure, after del Rio Hortega (1928), is followed in our laboratory (Fig. 38). In most Golgi modifications, the best results are obtained with young animals. Myelinated axons or cells located in the white matter remain the most difficult to impregnate.

Our experience with the Golgi method is rather limited. In investigations of neuronal development and in the classification of neurons according to shape, dendritic morphology, and origin and collateralization of the axon, the method is indispensable. Such classifications are possible only from a large sample of material, impregnated with different modifications of the Golgi method. Such classifications can be refined by quantification of the dendritic arborizations (Sholl, 1956; Uylings and Smit, 1975). However, the impression of orderliness in the complex relationships among nerve cells in Golgi-impregnated material is conveyed most effectively in the line drawings of authors who combine scientific observation and artistic sense in the great tradition of Ramón y Cajal.

9. Histological Methods

The preceding survey of classic methods in neuroanatomy is incomplete, and the choice of histological methods in this section was guided mainly by our personal experience. Important methods, such as Ehrlich's supravital methylene blue method, glial stains, and the chrome–silver endfeet techniques on mordanted material (Sterling and Kuypers, 1966), are not mentioned. More complete instructions for these and other methods can be found in the works listed in the bibliography at the end of the chapter.

9.1. Gelatin Embedding

The fixed brain is washed in running tapwater for 2–3 days to remove formaldehyde. Next, it is immersed in a 5% gelatin solution (Oxoid, bacteriological gelatin), which is kept in an oven at a temperature of 37°C. After 12–24 hr, the brain is transferred to a 10% gelatin solution held at the same temperature, in which it remains for about 5 hr.

The embedding is carried out using a freshly prepared 10% gelatin solution at 37°C. The brain can be either embedded in a block of gelatin or covered by a thin mantle of gelatin. The latter method is usually sufficient and is more convenient in mounting the sections on slides.

For block-embedding, a plastic box (Lipshaw®) of the appropriate size can be used as a mold. The brain is placed in the box. After being filled with the freshly prepared 10% gelatin, the box is placed in the freezer for quick hardening of the gelatin. Freezing of the block should be avoided. As soon as the gelatin sets, the block can be removed from the box to be hardened in a 10% solution of buffered formalin. After 1–2 weeks, the block is ready for sectioning on a freezing microtome. The thickness of the sections is usually 20–30 μm.

As already stated, embedding in a thin mantle of gelatin is sufficient and more convenient. For this method, the brain is placed on a petri dish and poured with the 10% gelatin solution. Next, the dish is placed in the freezer until the gelatin sets; freezing should be avoided. This procedure must be repeated several times to give the gelatin mantle its appropriate thickness. Finally, the gelatin-coated brain can be removed from the dish and processed in the usual way.

9.2. *Paraffin Embedding*

Paraffin with a melting temperature between 54 and 58°C is most commonly used. To improve the embedding and sectioning properties, substances can be added to the paraffin. These additives are already present in many commercial products. Beeswax (25 g in 1000 ml paraffin) will prevent formation of air-filled spaces during crystallization, plastic polymers will soften the consistency of the paraffin, and addition of dimethylsulfoxide (DMSO) will improve the penetration into the tissue. We have obtained satisfactory results with a 1 : 1 mixture of Paraplast® and Histowax®.

Procedure

1. Washing: The fixed brain is washed thoroughly for some days in running tapwater.
2. Dehydration: The success of paraffin embedding depends very much on a thorough dehydration of the brain in an ascending series of ethanol, consisting of 50, 60, 70, 90, 96, and 100% ethanol. The 96% and the 100% ethanol are changed once. Dehydration of brain slices smaller than 10 mm will take 1 day, whereas 2 days are recommended for larger pieces. The tissue should not stay longer than 24 hr in 100% ethanol. Table 2 presents two complete examples of dehydration schedules.
3. Methylbenzoate: To remove the ethanol, the brain is passed through three baths of fresh methylbenzoate (I, II, and III). This part of the procedure takes 1 day (baths I and II, 4 hr each; bath III, overnight). Initially, the brain floats; after 1 hr, it sinks to the bottom. There is no maximum time for the stay in bath III; 12 hr is sufficient.
4. Toluene: Methylbenzoate, which cannot be mixed with paraffin, has to be removed by passing the brain through toluene, in which it should stay no longer than 20 min or the tissue will become too hard. During this time, the bath is kept in an oven at 60°C to raise the temperature of the brain to the level of the warmed paraffin to which it is transferred next.

Table 2. Two Examples of Dehydration Schedules in the Paraffin-Embedding Procedure

Bath	Size of tissue pieces	
	Smaller than 10 mm	Larger than 10 mm
	Time	
Ethanol		
50%	9:00	9:00
60%	10:00	10:00
70%	11:00	11:00
80%	12:00	14:00
90%	13:00	17:00 (Overnight)
96%	14:00	9:00
96%	15:00	12:00
100%	16:00	17:00 (Overnight)
100%	17:00 (Overnight)	9:00
Methylbenzoate I	9:00	11:00

5. Paraffin: Prior to the embedding, the brain is passed through two freshly prepared baths of melted paraffin of the chosen composition, which should be kept in an oven at 60°C. In the first bath of paraffin, it remains for 3 hr. Next, it is transferred to the second bath, in which it stays overnight for at least 24 hr.
6. Embedding: For the embedding, freshly prepared paraffin is used, which can be kept outside the oven in a Lipshaw® electric paraffin pitcher. Plastic boxes (Lipshaw®) are convenient to use as embedding molds. At first, a thin layer of paraffin is poured into the box. When this begins to set, the brain can be placed in its proper position. Realize that the bottom of the block will be the plane of sectioning. Next, the box is filled to the brim. As long as the paraffin is still soft, the brain can be positioned using heated forceps. A Bunsen burner is indispensable not only for heating the forceps but also for repeated melting of the solidified paraffin in the upper part of the box to avoid formation of a deep central pit in the block. This is done by waving the flame over the top of the box from time to time during the first hour. Then the box is left until the next day, when the block can be removed.

9.3. Celloidin Embedding

Fewer staining methods are available for celloidin-embedded material as compared with paraffin- or gelatin-embedded tissue. However, celloidin can be considered when (1) thick sections (40–120 μm) are required (e.g., Golgi methods) or (2) shrinkage of the brain has to be reduced to a minimum (e.g., preparation of a stereotaxic atlas), or in the combination of both requirements. Shrinkage is restricted, since there is no heating during the procedure.

Procedure

1. Washing: Formalin-fixed material is washed overnight in running tap-water. Golgi-stained brain slices are shaken with ethanol 50% to remove silver chromate precipitates.
2. Dehydration: Brain slices of 3–4 mm (see Section 9.7)] are carefully dehydrated using an ascending series of ethanol: 50% (1 hr), 70% (2 hr), 96% (1 hr), and 100% (1 hr); shake continuously. For the Golgi–del Rio Hortega modification, 0.5% silver nitrate is added to the 50% ethanol. Whole brains are dehydrated according to the instructions given in Section 9.2.
3. Ether–ethanol: The brain or the slices are transferred from 100% ethanol to a 1 : 1 (vol./vol.) mixture of diethylether (water-free) and 100% ethanol in which the tissue remains for 24 hr. The mixture has to be renewed twice.
4. Celloidin infiltration: For embedding, nitrocellulose [(LVN) low viscosity nitrocellulose, Gurr] is recommended instead of celloidin because it has a lower viscosity and therefore penetrates better into the tissue. The whole or the sliced brain is passed through an ascending series of nitrocellulose dissolved in ether–ethanol [see (3) above]: 5% (24 hr), 10% (24 hr), 20% (24 hr), and 30% (48 hr). Stored nitrocellulose should be kept moist with *n*-butyl alcohol; when dry, it is explosive. In preparing the nitrocellulose solutions, nitrocellulose is first dissolved in 100% ethanol; then the same volume of diethylether is added. The different concentrations can either be prepared separately or diluted from a 30% solution with ether–ethanol.
5. Embedding: The brain is placed in a cardboard box that is filled with freshly prepared 30% nitrocellulose solution.
6. Hardening: For this purpose, the box is placed in an exsiccator in chloroform vapor. After 24–28 hr, the block is hard enough to be removed from the box. Next, the block is put in 70% ethanol for 2–3 days, which completes the hardening. For longer periods, the blocks can be stored in 96% ethanol in the refrigerator.
7. Sectioning: Celloidin blocks are cut with a concave, so-called B-knife. During sectioning, block and knife must be moistened continuously with 70% ethanol. The sections are collected in 70% ethanol.
8. Further processing: Different procedures are employed for (a) brains that are already impregnated (see Section 9.7) and do not need counterstaining and (b) brains that have to be stained or counterstained.
   a. The sections are passed through 70% ethanol, 96% ethanol, a 2 : 1 (vol./vol.) mixture of isopropyl alcohol and 100% ethanol (twice), and toluene (twice). The sections are then ready for mounting. Slides are covered with a thick layer of a neutral mounting medium such as Malinol®. The sections are put on top with another thick layer of Malinol. The cover glasses are weighted with small weights, and ample drying time is allowed.

b. The sections are referred to water passing through a descending series of ethanol: 70, 50, 25, and 15%. After two changes of distilled water, they are ready for staining.

9.4. Nissl Stains

Procedures are given for the Nissl stain on (9.4.1) paraffin-, (9.4.2) gelatin-, and (9.4.3) celloidin-embedded material.

9.4.1. Nissl Stain on Paraffin-Embedded Material

Use material that is sufficiently fixed in 10% formalin or Baker's formalin. For embedding, see Section 9.2. The section thickness is 8–10 μm.

Procedure

1. Mounting: The sections are mounted on clean slides covered with a thin layer of filtered serum or albumen–glycerin. A small amount of gelatin–water (a knife-point of gelatin dissolved in 100 ml distilled water) is put on the glass while it lies on a heating plate. When the section is placed on the water, it should stretch, but the paraffin should not melt. Before it overstretches, the glass should be removed from the plate and the water poured off. As soon as the section sticks to the glass, it is put in an oven at 37°C for further drying. This will take at least 24 hr.
2. Removal of paraffin: For this purpose, the slides are passed through two changes of xylene for 3 min each.
3. Hydration: The sections are passed through a descending series of ethanol: 100% (twice), 96, 70, and 40%, and distilled water. Take at least 2 min for each step.
4. Cresylechtviolet: The slides are transferred to a 0.5% solution of cresylechtviolet (Chroma) with 5 ml glacial acetic acid added for each 1000 ml. Staining is started at room temperature, but the bath is put immediately into an oven at 60°C for 10–15 min and left to cool for 5 min outside the oven.
5. Washing: Rinse quickly in distilled water.
6. Differentiation: The slides stay for some seconds in 96% ethanol; test slides should be made to establish the optimal time. As an alternative, differentiation can be carried out in aniline–ethanol (see Section 9.4.2).
7. Ethanol 100%: Twice.
8. Xylene: Two or three changes.
9. Covering: The sections are coverslipped with a mounting medium such as Malinol®.

9.4.2. Nissl Stain on Gelatin-Embedded Material

The gelatin blocks are cut on a freezing microtome into 20- to 30-μm-thick sections, which are usually stored in citrate-buffered 10% formalin.

Procedure

1. Mounting: The sections are mounted from 70% ethanol on slides covered with a thin layer of filtered serum. Next, they are dried for 24 hr at 37°C.
2. Distilled water: Wash for a few seconds.
3. Cresylechtviolet: The slides are transferred to a heated (60°C) 0.5% solution of cresylechtviolet to which 5 ml glacial acetic acid is added for each 1000 ml and kept in an oven at 60°C for 15 min.
4. Distilled water: Wash for a few seconds.
5. Differentiation: Subsequently, the slides are rapidly passed through:
   a. Ethanol 96%.
   b. Aniline–ethanol 96% (preparation: fill 100 ml aniline, Riedel-de Haën, up to 1000 ml with ethanol 96%).
   c. Aniline–ethanol 96%–HCl. In this solution, excess of cresylechtviolet is removed; test slides should be made to determine the optimal time (preparation: fill 100 ml aniline up to 1000 ml with ethanol 96% and add 5 drops hydrochloric acid 36%).
   d. Ethanol 96%: to remove excess aniline.
6. Ethanol 100%: Two rapid changes to complete the dehydration.
7. Xylene: Two changes.
8. Covering: Coverslips are mounted with Malinol®.

9.4.3. *Nissl Stain on Celloidin-Embedded Material*

The following procedure is intended for thick sections (40–120 μm). It should be noted that this staining is carried out on unmounted sections.

1. Distilled water: Two changes.
2. Cresylechtviolet: The sections are transferred to a 0.5% solution of cresylechtviolet with an addition of 5 ml glacial acetic acid for each 1000 ml in which they stay for 15 min at room temperature.
3. Distilled water: Wash.
4. Differentiation: Use either the method described in Section 9.4.1 or that described in Section 9.4.2. In either case, the differentiation is ended in ethanol 96%.
5. Further dehydration: In a 2:1 mixture of isopropylalcohol and ethanol 100%; two changes.
6. Toluene: Two changes.
7. Mounting: See Section 9.3, step 8.

9.5. Myelin Stains

9.5.1. *Häggqvist Stain*

This method can be carried out either on paraffin sections of mordanted brains or on frozen sections that are mordanted afterward. The paraffin sections are the most appropriate for a combined study of myelo- and cytoarchitecture. The modification on frozen sections is especially recommended as a parallel stain to experimental series when the experimental results have to be considered in comparison with the myeloarchitecture. Since these two procedures differ in many aspects, they are described separately.

9.5.1a. Häggqvist Stain on Paraffin Sections

1. Fixation: Baker's formalin is the fixative of choice for this method. It is composed of 100 ml formaldehyde 40%, 100 ml of a 10% solution of calcium chloride, and 800 ml distilled water. Preferably, the brain is perfused with this fixative, although material that is perfused with formalin and even human autopsy material can be used. A prolonged stay (2–4 weeks or longer) in Baker's fixative is required for good staining results. During this period, the material is kept alternately in an oven at 37° C (overnight) and at room temperature. The fixative is renewed every 2 weeks.
2. Washing: Rinse for at least 48 hr in running tapwater.
3. Mordanting: The fixed brain is mordanted in a 5% and a 10% Baker–potassium dichromate solution (preparation: dissolve 50 and 100 g potassium dichromate, respectively, in 1000 ml of a 1% solution of calcium chloride). The time depends on the size of the brain. Insufficient mordanting gives less satisfactory staining results. When the time is too long, the tissue becomes too hard to cut. For each species and for each part of the CNS (spinal cord, brainstem, and cerebrum), the right mordanting time has to be established empirically. In general, the tissue stays for 1–2 weeks in each of the two Baker–dichromate solutions. During the mordanting period, the material is kept alternately in an oven at 37°C (overnight) and at room temperature.
4. Washing: Rinse for 48 hr in running tapwater.
5. Dehydration and paraffin embedding: Follow the instructions in Section 9.2.
6. Sectioning: Ribbons of 5- to 6-μm-thick sections are cut with a bi-plane C-knife.
7. Mounting (see Section 9.4.1, step 1): When sections work loose, they can be attached to the glass by heating the slide in a gas flame until the paraffin is evaporated.
8. Removal of paraffin: The slides are passed through two changes of xylene (preferably fresh, especially the second one) for 3 min each.
9. Hydration: The sections are referred to water in a descending series of ethanol: 100% (2 min), 96% (1 min), 70% (1 min), and 50% (1 min). Finally, three changes of distilled water.
10. Phosphomolybdic acid: The slides remain for $1\frac{1}{2}$ hr in a 10% aqueous solution at room temperature.
11. Washing: Rinse in distilled water (three changes) or, alternatively, in running tapwater for $1\frac{1}{2}$ hr.

12. Mann's solution: The slides are put in a solution of 2.630 g methyl blue (BDH) and 0.530 g eosin (Gurr) in 1000 ml distilled water to which 15 drops of glacial acetic acid is added. Some thymol crystals are added to prevent the formation of mildew. The slides are left in this solution for 3 hr, for 1½ hr in an oven at 37°C, and for 1½ hr at room temperature.

It should be noted that the amounts of methyl blue and eosin differ from those given in the original prescription of Mann's solution. The aforementioned methyl blue can be substituted with the same dye of other manufacture (Gurr, Ral, Fluka) or with brilliantwollblau (BASF). As a result, the intensity of the blue staining may differ from dye to dye. This also holds for the red staining with the use of a different brand of eosin.

Instead of the addition of glacial acetic acid, an acetic acid–sodium acetate buffer at a pH of 5.4 can be used in Mann's solution (preparation of the buffer: add 145 ml 0.1 M glacial acetic acid to 855 ml 0.1 M sodium acetate, and correct the pH, if necessary, to 5.4). In this buffer, the given amounts of methyl blue and eosin are dissolved.

13. Differentiation: The slides are passed quickly (several seconds) through ethanol 96% (which dissolves eosin) and ethanol 100% (which dissolves methyl blue).
14. Xylene: Use two subsequent baths of xylene. In the first bath, which removes ethanol, the slides should stay no longer than 2 min; otherwise, the differentiation will proceed.
15. Coverslips are mounted with Malinol®.

9.5.1b. Häggqvist Stain on Frozen Sections

1. Frozen sections are mounted and dried on slides as indicated in Section 9.4.2. The coating of the slides with serum should be thin to avoid staining. When the blocks are embedded in gelatin, the gelatin adhering to the sections will stain blue without interfering with the staining of the sections.
2. The mounted sections are mordanted in Baker's 10% potassium dichromate solution (see Section 9.5.1a). The sections are placed in the warmed solution (60°C) and left in the mordant for 3 or 4 days at room temperature.
3. The slides are washed in running tapwater for 6 hr, rinsed in one change of distilled water, and left in 96% ethanol overnight.
4. The next morning, the slides are rinsed in three changes of distilled water and then left in a 10% solution of phosphomolybdic acid for 1½ hr at room temperature.
5. After being rinsed in three changes of distilled water, the sections are stained in Mann's solution (see Section 9.5.1a) at 40°C for 1½ hr and subsequently allowed to cool at room temperature for 1½–24 hr. A longer staining period at room temperature results in a darker stain.
6. The slides are differentiated for 10–30 sec in 96% ethanol and absolute ethanol and cleared in two changes of xylene. Cover glasses are mounted with Malinol®.

9.5.2. *Klüver–Barrera Stain*

This method is a combined Nissl and myelin stain. Fixation is in 10% formalin or Baker formalin. The method can be carried out on paraffin, frozen (gelatin), or celloidin sections.

9.5.2a. *Klüver–Barrera Stain on Paraffin Sections*

1. Fixation and embedding: After at least 3 weeks' fixation, the brain is embedded in paraffin (see Section 9.2). Human autopsy material is preferably fixed in Baker's fixative for at least 2 months.
2. Sectioning: Cut 10- to 15-μm-thick sections.
3. Mounting: See Section 9.4.1.
4. Removal of paraffin: Two changes of xylene—the first for several minutes, the second for at least 1 hr. This helps to prevent a patchy distribution of Luxol fast blue.
5. Pass the slides through 100% ethanol and several changes of 96% ethanol.
6. Luxol fast blue: The sections are stained overnight (16–24 hr) in an oven at 60°C in a 0.1% solution of Luxol fast blue (Du Pont) in 96% ethanol to which 5 ml glacial acetic acid per 1000 ml is added.
7. Washing: After cooling, excess stain is removed in 96% ethanol, followed by two quick changes of distilled water.
8. Differentiation: Myelin should retain the blue dye, whereas the dye should be removed from the gray matter. For this purpose, the slides are immersed alternately for 3–5 sec in a 0.05–0.1% (aqueous) solution of lithium carbonate and in 70% ethanol; the differentiation can be slowed down by using a lower concentration of lithium carbonate. Finish with two changes of 70% ethanol.
9. Distilled water: Two changes.
10. Cresylechtviolet: The sections are now stained for 6 min in a warmed 0.1% solution (60°C) of cresylechtviolet (Chroma) to which 5 ml glacial acetic acid per 1000 ml is added. Higher concentrations of cresylechtviolet may be used (up to 0.25%), as may longer staining times.
11. Differentiate quickly in 96% ethanol.
12. Ethanol 100%.
13. Xylene: Two changes.
14. The sections are coverslipped with Malinol®.

9.5.2b. *Klüver–Barrera Stain on Frozen Sections*

1. Fixation and embedding: The fixed brain is processed for gelatin embedding (see Section 9.1).
2. Sectioning: Frozen sections are cut at 20–25 μm.
3. Mounting: See Section 9.4.2.
4. Xylene: For 1 hr, the slides are immersed in xylene (see Section 9.5.2a above, step 4), followed by ethanol 100 and 96% for some minutes each. This step may be omitted, in which case the slides are put directly from the oven into the Luxol fast blue.

5–14. These steps are the same as for paraffin sections (see Section 9.5.2a above).

9.5.2c. Klüver–Barrera Stain on Celloidin Sections

1. Fixation and embedding: The fixed brain is processed for celloidin embedding (see Section 9.3).
2. Sectioning (see also Section 9.3): The sections are collected in 70% ethanol.
3. The staining is carried out on loose sections.
4. Xylene: For at least 1 hr, the sections are immersed in xylene (for prevention of a patchy distribution of Luxol fast blue), followed by ethanol 100 and 96% for some minutes each.

5–11. These steps are the same as for paraffin sections, except for the use of a 0.5% solution of lithium carbonate for the differentiation, which is necessary because of the thicker celloidin sections.

12. Isopropyl alcohol–ethanol 100% [a 2 : 1 (vol./vol.) mixture]: Two changes.
13. Toluene: Two changes.
14. Mounting and covering: As described in Section 9.3, step 8a.

9.6. Reduced-Silver Methods

Of the reduced-silver methods for normal adult or embryological material, the most commonly used is the Bodian silver-impregnation method for paraffin sections. In the Nauta and Fink–Heimer methods, increased argyrophilia of degenerating axons is used to trace the course or the termination, or both, of fiber pathways. In both these methods, the argyrophilia of normal and, to a lesser degree, of degenerating axons is suppressed using oxidants. Since neither of the two methods is standardized, they should be used only qualitatively. For each species and fiber system, the optimal survival time has to be established empirically. For the Nauta method, 7 days can be taken as an average. For the staining of terminal degeneration with the Fink–Heimer method, 2–4 days should be used.

9.6.1. *Bodian Silver-Impregnation Method for Paraffin Sections*

1. Mount and dry paraffin 8-μm sections as indicated in Section 9.4.1.
2. Defat in xylene overnight or longer.
3. Hydrate in descending series of ethanol, rinse in distilled water.
4. Place sections in a solution of 1% protargol (argentum proteinatum), containing 6 g metallic copper, for 12–48 hr. The protargol solution is prepared by strewing protargol in a wide glass beaker containing cold, distilled water. The beaker is placed in the dark until the protargol is dissolved. Copper is cleaned in nitric acid and sulfuric acid.
5. Rinse in three changes of distilled water.
6. Reduce for 10–15 min in a 2 or 3% solution of hydroquinone in distilled water; 5 ml formaldehyde 40% is added to each 100 ml solution.

7. Wash thoroughly in three changes of distilled water.
8. Place sections in a 1% solution of gold chloride, with 3 drops glacial acetic acid per 100 ml solution, for 5–10 min.
9. Rinse in distilled water.
10. Reduce in a 2% solution of oxalic acid for 10–15 min.
11. Rinse in distilled water.
12. Remove excess silver in a solution of 5% sodium thiosulfate for 10 min.
13. Rinse and dehydrate, and mount coverslips.

9.6.2. *Nauta Method*

1. Fixation: The animal is perfused first with saline and then with 10% formalin; after removal from the skull, the brain is stored in buffered 10% formalin for at least 6 weeks (preparation of the buffer at a pH of 7.1: dissolve 0.93 g citric acid and 145.63 g sodium citrate in 5 liters distilled water).
2. Embedding in gelatin: The brain is embedded in gelatin according to the instructions given in Section 9.1. The embedded block is put in a 25% solution of sucrose in buffered 10% formalin until it sinks after 2–3 days. Alternatively, the block can be stored in buffered 10% formalin for 1–2 weeks or longer.
3. Sectioning: Sections 20–25 μm thick are cut on a freezing microtome using a C-knife. The sections are collected in buffered 10% formalin.
4. For the following staining procedures, the sections are distributed over a number of sieve beakers as illustrated in Fig. 11. The beakers are placed in a rack, which permits processing of a large number of sections at the same time. Following each step, the beakers should be placed on blotting paper to lessen pollution of the next bath.
5. Washing: In distilled water.
6. Ethanol 15%: The sections are kept in this alcohol, which dissolves some myelin, for at least 15 min.
7. Washing: In distilled water.
8. Phosphomolybdic acids: The sections are immersed in a 0.5% solution of phosphomolybdic acid for 20 min.
9. Washing: In distilled water.
10. Suppression: The sections are placed in a 0.05% solution of potassium permanganate during 1–15 min and agitated continuously; the suppression time must be determined from a trial run prior to the complete processing of the series.
11. Washing: Two brief changes of distilled water.
12. Bleaching: During a stay of 2 min in a mixture of equal volumes of a freshly prepared 1% solution of hydroquinone and a 1% solution of oxalic acid, the sections lose their brown color.
13. Washing: Three changes of distilled water.
14. Silver nitrate: The sections are put in a 1.5% solution of silver nitrate in which they are kept, in the dark, for 30–60 min.

15. Ammoniacal silver: For 1 min, the sections are put in a freshly prepared solution that is composed of 200 ml 2.25% silver nitrate, 100 ml 100% ethanol, 20 ml strong ammonia, and 22 ml 2.5% sodium hydroxide. Agitate continuously.
16. Reducer: The sections are transferred directly from the ammoniacal silver to a reducer solution that is used in two subsequent baths for 1 min each (preparation of the reducer: add 100 ml 100% ethanol, 30 ml 10% formalin, and 30 ml 1% citric acid to 900 ml distilled water). The sections stain brown.
17. Washing: Two changes of distilled water.
18. Sodium thiosulfate: The sections are immersed in a fixation bath of 1% sodium thiosulfate for about 20 sec. Agitate continuously. In this bath, the excess of silver compounds is made soluble for removal during the next washing.
19. Washing: Three changes of distilled water.
20. Bleaching (optional): For counterstaining, the sections can be bleached in a 0.5% solution of potassium ferrocyanide for up to 1 min. Next, they are washed very thoroughly in 10 changes of distilled water.
21. Mounting: The sections are mounted according to Albrecht (1954) from a solution of 1.5 g Oxoid® gelatin dissolved in 100 ml warmed distilled water, which is allowed to cool. Subsequently, 100 ml 80% ethanol is added. The sections are dried in an oven at 37°C, on a moderately warmed heating plate, or simply by exposure to the air. Bleached sections are now ready for counterstaining with the Nissl method (see Section 9.4.2).
22. Covering: The sections are coverslipped with a mounting medium such as Malinol®.

9.6.3. Fink–Heimer Method

1. Fixation: See Section 9.6.2. Alternatively, glutaraldehyde perfusion can be used to avoid impregnation of "dust" in the background. It should be followed by an afterfixation of 1–2 weeks in buffered 10% formalin.
2. Embedding in gelatin (see Section 9.1): Preferably, the block is stored in a 25% solution of sucrose in buffered 10% formalin until it sinks after 2–3 days. The block is then ready for sectioning, though it can be stored for a longer time.
3. Sectioning: See Section 9.6.2.
4. Distribute sections over a number of sieve beakers (see Section 9.6.2).
5. Washing: In distilled water.
6. Suppression: The sections are placed in a 0.025–0.05% solution of potassium permanganate. To determine the right suppression time, which will be between 1 and 15 min, a trial run is made prior to the complete processing of the series.
7. Washing: Rinse briefly in distilled water.

8. Bleaching: The brown-colored sections are bleached for 1–2 min in a mixture of equal volumes of a freshly prepared 1% solution of hydroquinone and a 1% solution of oxalic acid.
9. Washing: Two changes of distilled water.
10. Silver uranyl nitrate: For this step and the following steps, it is strongly recommended that rubber gloves be worn, because uranyl nitrate is radioactive. The sections are subsequently put in two different mixtures of silver nitrate and uranyl nitrate, solutions I and II, in which they will stay for $\frac{1}{2}$ and $\frac{1}{2}$–1 hr, respectively.

    Preparation of solution I—500 ml contains: 100 ml 0.5% uranyl nitrate, 120 ml 2.5% silver nitrate, and 280 ml distilled water.

    Preparation of solution II—500 ml contains: 200 ml 0.5% uranyl nitrate and 300 ml 2.5% silver nitrate.
11. Washing: In distilled water.
12. Ammoniacal silver: For 2 min, the sections are immersed in a solution that is freshly prepared as follows: shake 400 ml of a 2.5% solution of silver nitrate with 22 ml strong ammonia, then add 22 ml of a 2.5% solution of sodium hydroxide and shake again. The sections are agitated continuously.
13. Reducer: The sections are directly transferred from the ammoniacal silver solution to a reducer solution (preparation: add 90 ml 96% ethanol, 27 ml 10% formalin, and 27 ml 1% citric acid to 910 ml distilled water). Use two subsequent reducer baths and keep the sections for 45 sec in each bath. Note the brown staining during reduction.
14. Washing: Two changes of distilled water.
15. Sodium thiosulfate: The sections are immersed in a fixation bath of 1% sodium thiosulfate for 15–20 sec with continuous agitation. In this bath, the excess of silver compounds is made soluble for removal during the next washing.
16. Washing: Three changes of distilled water.
17. Bleaching (optional): See Section 9.6.2, step 20.
18. Mounting: See Section 9.6.2, step 21.
19. Covering: The sections are coverslipped with a mounting medium such as Malinol®.

9.7. Golgi Methods

From the vast range of Golgi modifications, we have chosen the del Rio Hortega method as a simple, quick, and reliable way of getting well-impregnated mammalian and avian material. For many other modifications, we refer to the works listed in the bibliography at the end of the chapter.

Procedure

1. Perfusion is important for a successful impregnation, but if necessary, old formalin-fixed material can be used. Under deep anesthesia (e.g.,

ether or Nembutal®), the thoracic cavity is opened and 0.5 ml heparin (Thromboliquine®) and 0.5 ml 1% sodium nitrite are injected into the left cardiac ventricle. Next, a perfusion needle is inserted into the left ventricle and the right atrium is opened; 500 ml saline is circulated to wash out the blood, followed by 500 ml of the fixative, which consists of 6% potassium dichromate, 6% chloral hydrate, and 10% formalin (preparation of 1000 ml of the fixative: dissolve 60 g potassium dichromate and 60 g chloral hydrate in 900 ml distilled water, then add 100 ml formaldehyde 35% shortly before use). Immediately after perfusion, the brain is removed from the skull and cut into slices not thicker than 3–4 mm. The tissue turns yellowish-gray, and its consistency is softer in comparison with that of formalin-perfused material.

2. Further fixation and chromation: The brain slices are immersed in the remaining fixative, in which they are kept for 48 hr. Renew after 24 hr using a freshly prepared solution.
3. Silvering: The slices are now put in a 1.5% solution of silver nitrate, in which they remain for 72 hr. As soon as the silver nitrate and the chromated brain slices make contact, a reddish-brown precipitate is formed.
4. Dehydration, embedding in celloidin, sectioning, and mounting are described in Section 9.3.

ACKNOWLEDGMENTS. The authors gratefully acknowledge the assistance of the laboratory technicians, Mrs. H. Van Denderen-Van Dorp and Mr. J.M. Guldemond.

References

Albrecht, M.H., 1954, Mounting frozen sections with gelatin, *Stain Technol.* **29:**89.

Alzheimer, A., 1910, Beiträge zur Kenntnis der pathologischen Glia und Ihrer Beziehungen zu den Abbauprodukten im Nervengewebe, in: *Histologische und histopathologische Arbeiten über die Grosshirnrinde,* Vol. 3 (F. Nissl and A. Alzheimer, eds.), p. 406, G. Fischer, Jena.

Anker, G.C., Scheers-Dubbeldam, K., and Noorlander, C., 1974, An epoxy resin embedding technique for large objects, *Stain Technol.* **49:**183.

Apáthy, S. von, 1892, Erfahrungen in der Behandlung des Nervensystems für histologische Zwecke, *Z. Wiss. Mikrosk.* **9:**37.

Ariëns Kappers, C.U., Huber, C.C., and Crosby, E.C., 1960, *The Comparative Anatomy of the Nervous System of Vertebrates including Man,* Hafner, New York.

Badawi, H., 1967, Das Ventrikelsystem des Gehirnes vom Huhn (*Gallus domesticus*), Taube (*Columbia livia*) und Ente (*Anas boschas domestica*) dargestelt mit Hilfe des Plastoid-Korrosionsverfahrens, *Zentralbl. Veterinaermed. Reihe A* **14:**628.

Bielschowsky, M., 1902, Die Silberimprägnation der Achsencylinder, *Neurol. Zentralbl.* **13:**579.

Bielschowsky, M., 1904, Die Silberimpregnation der Neurofibrillen, *J. Psychiatrie Neurologie* **3:**169.

Bielschowsky, M., 1935, Allgemeine Histologie und Histopathologie des Nervensystems, in: *Handbuch der Neurologie,* Vol. I (O. Bumke and O. Foerster, eds.), pp. 35–226, Springer-Verlag, Berlin.

Blackstad, T.W., Brodal, A., and Walberg, F., 1951, Some observations on normal and degenerating terminal boutons in the inferior olive of the cat, *Acta Anat.* (*Basel*) **11:**461.

Blechschmidt, E., 1955, Rekonstruktionsverfahren mit Verwendung von Kunststoffen, *Z. Anat Entwicklungsgesch.* **118:**170.
Blum, F., 1893, Der Formaldehyd als Härtungsmittel, *Z. Wiss. Mikrosk.* **10:**314.
Bodian, D., 1936, A new method for staining nerve fibers and nerve endings in mounted paraffin sections, *Anat. Rec.* **65:**89.
Böhme, G., 1969, Vergleichende Untersuchungen am Gehirnventrikelsystem: Das Ventrikelsystem des Huhnes, *Acta Anat. (Basel)* **73:**116.
Böhme, G., and Franz, B., 1967, Die Binnenräume des Gehirns von Ratte und Maus, *Acta Anat. (Basel)* **68:**199.
Bok, S.T., 1959, *Histonomy of the Cerebral Cortex,* Elsevier, Amsterdam and New York.
Born, G., 1883, Die Plattenmodelliermethode, *Arch. Mikrosk. Anat.* **22:**584.
Brauer, K., and Schober, W., 1970. *Catalogue of Mammalian Brains,* VEB Gustav Fischer Verlag, Jena.
Brodal, A., 1939, Experimentelle Untersuchungen über retrograde Zellveränderungen in der unteren Olive nach Läsionen des Kleinhirns, *Z. Neurol.* **166:**624.
Brodmann, K., 1909, *Vergleichende Lokalisationslehre der Grosshirnrinde,* Barth, Leipzig.
Cammermeyer, J., 1962, An evaluation of the significance of the "dark" neuron, *Ergeb. Anat. Entwicklungsgesch.* **36:**1.
Cammermeyer, J., 1967, Artifactual displacement of neuronal nucleoli in paraffin sections. *J. Hirnforsch.* **9:**209.
Cathey, W.J., 1962, A plastic embedding medium for thin sectioning in light microscopy, *Stain Technol.* **38:**213.
Chambers, W.W., Liu, C.Y., and Liu, C.N., 1956, A modification of the Nauta technique for staining degenerating axons in the central nervous system, *Anat. Rec.* **124:**391.
Chan-Palay, V., 1977, *Cerebellar Dentate Nucleus,* Springer-Verlag, Berlin, Heidelberg, and New York.
Clarke, R.H., 1920, Investigation of the central nervous system: Methods and instruments, *Johns Hopkins Hosp. Rep.* **1920:**1–162.
Collewijn, H., and Noorduin, H., 1969, Distortion-free mounting of frozen sections by handling on paper supports, *Stain Technol.* **44:**55.
Cowan, W.M., 1970, Anterograde and retrograde transneuronal degeneration in the central and peripheral nervous system, in: *Contemporary Research Methods in Neuroanatomy* (W.J.H. Nauta and S.O.E. Ebbesson, eds.), pp. 217–251, Springer-Verlag, Berlin, Heidelberg, and New York.
Cowan, W.M., and Powell, T.P.S., 1956, A note on terminal degeneration in the hypothalamus, *J. Anat. (London)* **90:**188.
Cox, W.H., 1891, Imprägnation des centralen Nervensystems mit Quecksilbersalzen, *Arch. Mikrosk. Anat.* **37:**16.
Cullheim, S., and Kellerth, J.O., 1976, Combined light and electron microscopic tracing of neurons including axons and synaptic terminals, after intracellular injection of horseradish peroxidase, *Neurosci. Lett.* **2:**307.
Deiters, O., 1865, *Untersuchungen über Gehirn und Rückenmark,* Vieweg und Sohn, Braunschweig.
Delmas, A., and Pertuiset, B., 1959, *Cranio-cerebral Topometry in Man,* Masson, Paris; Charles C. Thomas, Springfield.
del Rio Hortega, P., 1928, Tercera aportacion al conocimiento morfologico e interpretacion functional de la oligodendroglia, in: *Memorias de la Real Sociedad Espanola de Historia Natural,* Vol. XIV, Memoria No. 1, pp. 34–39, Madrid.
De Olmos, J.S., 1969, A cupric–silver method for impregnation of terminal axon degeneration and its further use in staining granular argyrophilic neurons, *Brain Behav. Evol.* **2:**213.
Desclin, J.C., 1974, Histological evidence supporting the inferior olive as the major source of cerebellar climbing fibers in the rat, *Brain Res.* **77:**365.
Dewulf, A. (ed.), 1971, *Anatomy of the Normal Human Thalamus: Topometry and Standardized Nomenclature,* Elsevier, Amsterdam, London, and New York.
du Boulay, G.H., 1974, Comparative neuroradiologic vascular anatomy of experimental animals,

in: *Radiology of the skull and brain,* Vol. 2 (T.H. Newton and D.G. Potts, eds.), Book 4, pp. 2763–2796, C.V. Mosby, Saint Louis.

Duval, M., 1879, De l'emploi du collodion humide pour la pratique des coupes microscopiques, *J. Anat. Physiol.* **15:**185.

Ebbesson, S.O.E., 1970, The selective silver impregnation of degenerating axons and their synaptic endings in non-mammalian species, in: *Contemporary Research Methods in Neuroanatomy* (W.J.H. Nauta and S.O.E. Ebbesson, eds.), pp. 132–161, Springer-Verlag, Berlin.

Ebner, F.F., and Myers, R.E., 1962, Commissural connections in the neocortex of the monkey, *Anat. Rec.* **142:**229.

Elias, H. (ed.), 1967, *Stereology,* Springer-Verlag, Berlin, Heidelberg, and New York.

Ettinger, E.T., 1965, *Functions of the Corpus Callosum,* Churchill, London.

Exner, S., 1881, Zur Kenntnis von feineren Baue der Grosshirnrinde, *Sitzungsber. Kaiserl. Akad. Wiss. Wien* Bd. **83**(III):151.

Fink, R.P., and Heimer, L., 1967, Two methods for selective silver-impregnation of degenerating axons and their synaptic endings in the central nervous system, *Brain Res.* **4:**369.

Flechsig, P., 1927, *Meine myelogenetische Hirnlehre,* Springer-Verlag, Berlin.

Freedman, S.L., Voogd, J. and Vielvoye, G.J., 1977, Experimental evidence for climbing fibers in the avian cerebellum, *J. Comp. Neurol.* **175:**243–252.

Gaunt, W.A., 1971, Microreconstruction, Pitman Medical, London.

Gerlach, L., 1858, *Mikroskopische Studien aus dem Gebiete der menschlichen Morphologie,* in *Enke,* Erlangen.

Glaser, E.M., and Van der Loos, H., 1965, A semi-automatic computer-microscope for the analysis of neuronal morphology, *IEEE Trans. Biomed. Eng.* **12:**22.

Glees, P., 1946, Terminal degeneration within the central nervous system as studied by a new silver method, *J. Neuropathol. Exp. Neurol.* **5:**54.

Glees, P., and Nauta, W.J.H., 1955, A critical review of studies on axonal and terminal degeneration, *Monatsschr. Psychiatr. Neurol.* **129:**74.

Golgi, G., 1873, Sulla struttura della sostanza grigia dell cervello, *Gazz. Med. Ital. Lombarda* **33:**244.

Golgi, C., 1894, *Untersuchungen über den feineren Bau des centralen und peripherischen Nervensystems,* Fischer, Jena.

Grant, G., and Aldskogius, H., 1967, Silver impregnation of degenerating dendrite cells and axons, central to axonal transection. I. A study on the hypoglossal nerve in kittens, *Exp. Brain Res.* **3:**150.

Grant, G., and Boivie, J., 1970, The charting of degenerative changes in nervous tissue with the aid of an electronic pantographic device, *Brain. Res.* **21:**439.

Gribnau, A.A.M., and Lammers, G.J., 1976, The preparation of graphical and three-dimensional reconstructions of the developing central nervous system, *Acta Morphol. Neerl.-Scand.* **14:**1.

Groenewegen, H.J., and Voogd, J., 1977, The parasagittal zonation within the olivocerebellar projection. I. Climbing fiber distribution in the vermis of cat cerebellum, *J. Comp. Neurol.* **174:**417.

Häggqvist, G., 1936, Analyse der Faserverteilung in einem Rückenmarkquerschnitte (Th. 3), *Z. Mikrosk. Anat. Forsch.* **39:**1.

Haller von Hallenstein, Graf Viktor von, 1934, Zerebrospinales Nervensystem, in: *Handbuch der vergleichenden Anatomie der Wirbeltiere,* Vol. II (L. Bolk, E. Göppert, E. Kallius, and W. Lubosch, eds.), Part 1, pp. 1–318, Urban and Schwarzenberg, Berlin and Vienna.

Heimer, L., 1969, Silver impregnation of degenerating axons and their terminals on Epon–Araldite sections, *Brain Res.* **12:**246.

Heimer, L., and Lohman, A.H.M., 1975, Anatomical methods for tracing connections in the central nervous system, in: *Handbook of Psychobiology* (M.S. Gazzoniga and C. Blakemore, eds.), pp. 73–106, Academic Press, New York.

Heimer, L., and Peters, A., 1968, An electron-microscopic study of a silver stain for degenerating boutons, *Brain Res.* **8:**337.

Heimer, L., and Wall, P.D., 1968, The dorsal root distribution to the substantia gelatinosa of the rat with a note on the distribution in the cat, *Exp. Brain Res.* **6:**89.

Hillman, D.E., Llinás, R., and Chujo, M., 1976, Automated and semiautomated analysis of nervous system structure, in: *Computer Analysis of Neuronal Structures* (R.D. Lindsey, ed.), Plenum Press, New York.

Holländer, H., and Vaaland, J.L., 1968, A reliable staining method for semi-thin sections in experimental neuroanatomy, *Brain Res.* **10:**120.

Horsley, V., and Clarke, R.H., 1908, The structure and function of the cerebellum examined by a new method, *Brain* **31:**45.

Jasper, H.J., and Ajmone-Marsan, C., 1954, *A Stereotactic Atlas of the Diencephalon of the Cat,* National Research Council of Canada, Ottawa.

Kaplan, H.A., 1953, A technique for anatomical study of the blood vessels of the brain, *Anat. Rec.* **116:**507.

Kier, E.L., 1974, Development of cerebral vessels. I. Fetal cerebral arteries: A phylogenetic and ontogenetic study, in: *Radiology of the Skull and Brain* Vol. 2 (T.H. Newton and D.G. Potts, eds.), Book 1, pp. 1089–1130, C.V. Mosby, Saint Louis.

Klebs, 1869, Die Einschmelzungs-Methode: Ein Beitrag zur mikroskopischen Technik, *Arch. Mikrosk. Anat.* **5:**164.

Klüver, H., and Barrera, E., 1953, A method for the combined staining of cells and fibers in the nervous system, *J. Neuropathol. Exp. Neurol.* **12:**400.

Knook, H.L., 1965, *The Fibre-connections of the Forebrain,* Van Gorcum, Assen.

König, J.F.R., and Klippel, R.A., 1963, *The Rat Brain: A Stereotaxic Atlas of the Forebrain and Lower Parts of the Brain Stem,* Baillaire, Tindall and Cox, London.

Konigsmark, B.W., 1970, Methods for the counting of neurons, in: *Contemporary Research Methods in Neuroanatomy* (W.J.H. Nauta and S.O.E. Ebbesson, eds.), pp. 315–339, Springer-Verlag, Berlin, Heidelberg, and New York.

Kristensson, K., and Olsson, Y., 1971, Uptake and retrograde transport of peroxidase in hypoglossal neurones: Electron microscopical localization in the neuronal perikaryon, *Acta Neuropathol. (Berlin)* **19:**1.

La Vail, J.H., and La Vail, M.M., 1972, Retrograde axonal transport in the central nervous system, *Science* **176:**1416.

Lieberman, A.R., 1971, The axon reaction: A review of the principal features of perikaryal responses to axon injury, *Int. Rev. Neurobiol.* **14:**49.

Llinás, R., and Precht, W. (eds.), 1976, *Frog Neurobiology,* Springer-Verlag, Berlin, Heidelberg, and New York.

Ludwig, E., and Klingler, J., 1956, *Atlas Cerebri Humani,* S. Karger, Basel and New York.

Marchi, V., and Algeri, G., 1885, Sulle degenerazione discendenti consecutive a lesioni sperimentale in diverse zone della corteccia cerebrale, *Riv. Sper. Freniatr. Med. Leg. Alienazioni Ment.* **11:**492.

Marinesco, M.G., 1896, Des lésions primitives et des lésions sécondaires de la cellule nerveuse. *C. R. Soc. Biol. (10th Ser.)* **3:**106.

Martinéz Martinéz, P., 1967, Sur la morphologie du réseau admirable extracrânien, *Acta Anat. (Basel)* **67:**24.

Meyer, A., 1901, On parenchymatous systematic degeneration mainly in the central nervous system, *Brain* **24:**47.

Morest, D.K., and Morest, R.R., 1966, Perfusion fixation of the brain with chrome–osmium solutions for the rapid Golgi method, *Am. J. Anat.* **118:**811.

Müller, H., 1859, Anatomische Untersuchung eines Mikrophthalmus, *Verh. Physik.-med. Ges. Würzburg,* Vol. 18.

Mulligan, J.H., 1931, A method of staining the brain for macroscopic study, *J. Anat. (London)* **65:**468.

Nauta, W.J.H., and Ebbesson, S.O.E. (eds.), 1970, *Contemporary Research Methods in Neuroanatomy,* Springer-Verlag, Berlin, Heidelberg, and New York.

Nauta, W.J.H., and Gygax, P.A. (eds.), 1951, Silver impregnation of degenerating axon terminals in the central nervous system. (1) Technic. (2) Chemical notes, *Stain Technol.* **26:**5.

Nauta, W.J.H., and Ryan, L.F., 1952, Selective silver impregnation of degenerating axons in the central nervous system, *Stain Technol.* **27:**175.

Nieuwenhuys, R., 1964, Comparative anatomy of the spinal cord, in: *Organization of the Spinal Cord* (J.C. Eccles and J.P. Schadé, eds.), *Progress in Brain Research,* Vol. 11, pp. 1–57, Elsevier, Amsterdam, London, and New York.

Nieuwenhuys, R., 1974, Topological analysis of the brain stem: A general introduction, *J. Comp. Neurol.* **156:**255.

Nieuwenhuys, R., and Opdam, P., 1976, Structure of the brain stem, in: *Frog Neurobiology* (R. Llinás and W. Precht, eds.), Chapt. 29, pp. 811–855, Springer-Verlag, Berlin, Heidelberg, and New York.

Nissl, F., 1894, Mitteilungen zur Anatomie der Nervenzelle, *Z. Psychiatr.* **50:**370.

Nissl, F., 1910, Nervensystem, in: *Enzyklopädie der mikroskopischen Technik* (P. Ehrlich, R. Krause, M. Mosse, and K. Weigert, eds.), pp. 243–285, Urban and Schwarzenberg, Berlin and Vienna.

Northcut, R.G., 1966, *Atlas of the Sheep Brain,* Stipes, Champaign, Illinois.

Olszewski, J., and Baxter, D., 1954, *Cytoarchitecture of the Human Brain Stem,* S. Karger, Basal and New York.

Opdam, P., Kemali, M., and Nieuwenhuys, R., 1976, Topological analysis of the brain stem of the frogs *Rana esculenta* and *Rana catesbeiana, J. Comp. Neurol.* **165:**307.

Palay, S.L., and Chan-Palay, V., 1974, *Cerebellar Cortex: Cytology and Organization,* Springer-Verlag, Berlin, Heidelberg and New York.

Palay, S.L., and Palade, G.E., 1955, The fine structure of neurons, *J. Biophys. Biochem. Cytol.* **1:**69.

Palkovits, M., Magyar, P., and Szentágothai, J., 1971, Quantitative histological analysis of the cerebellar cortex in the cat. I. Number and arrangement in space of Purkinje cells, *Brain Res.* **32:**15.

Papez, J.W., 1929, *Comparative Neurology,* Thomas Y. Cromwell, New York.

Pearson, R., 1972, *The Avian Brain,* Academic Press, London and New York.

Ramón-Moliner, E., 1970, The Golgi–Cox technique, in: *Contemporary Research Methods in Neuroanatomy* (W.J.H. Nauta and S.O.E. Ebbesson, eds.), Springer-Verlag, Berlin, Heidelberg, and New York.

Ramón Y Cajal, S., 1903, Un sencillo método de coloratión selectiva del reticulo protoplásmico y sus efectos en los diversos organos nerviosos, *Trab. Lab. Invest. Biol. Univ. Madrid* **2:**129.

Ramón Y Cajal, S., 1907, Quelques formules de fixation destinées à la méthode au nitrate d'argent, *Trav. Lab. Recherches Biol. Univ. Madrid* **5:**215.

Ramón Y Cajal, S., 1935, Die Neuronenlehre, in: *Handbuch der Neurologie,* Vol. I (O. Bumke and O. Foerster, eds.), pp. 887–994, Springer-Verlag, Berlin.

Ranson, S.W., 1911, Non-medullated nerve fibers in the spinal nerves, *Am. J. Anat.* **12:**67.

Ranson, S.W., and Clark, S.L., 1959, *The Anatomy of the Nervous System,* 10th ed., W.B. Saunders, Philadelphia and New York.

Ross, B.D., 1972, *Perfusion Techniques in Biochemistry: A Laboratory Manual,* Clarendon Press, Oxford.

Schnepp, G., Schnepp, P., and Spaan, G., 1971, Faseranalytische Untersuchungen an peripheren Nerven bei Tieren verschiedener Grösse. I. Fasergesamtzahl, Faserkaliber und Nervenleitungsgeschwindigkeit, *Z. Zellforsch.* **119:**7.

Schnepp, P., and Schnepp, G., 1971, Faseranalytische Untersuchungen an peripheren Nerven bei Tieren verschiedener Grösse. II. Verhältnis Axondurchmesser/Gesamtdurchmesser und Internodallänge, *Z. Zellforsch.* **119:**99.

Sholl, D.A., 1956, *The Organization of the Cerebral Cortex,* Methuen, London.

Smit, G.J., 1968, Some quantitative aspects of the striate area in the rabbit brain, Thesis, University of Amsterdam.

Smith, G.E., 1902, *Descriptive and Illustrated Catalogue of the Physiological Series of Comparative Anatomy in the R.C.S. Museum,* Vol. 21, 2nd ed., London.

Snodgress, A.B., and Dorsey, C.H., 1963, Egg albumen embedding: A procedure compatible with neurological staining techniques, *Stain Technol.* **38:**149.

Steedman, H.F., 1960, *Section Cutting in Microscopy,* Blackwell, Oxford.

Stensaas, S.S., Edwards, G.Q., and Stensaas, L.J., 1972, An experimental study of hyperchromic nerve cells in the cerebral cortex, *Exp. Neurol.* **36:**472.

Sterling, P., and Kuypers, H.G.J.M., 1966, Simultaneous demonstration of normal boutons and degenerating nerve fibres and their terminals in the spinal cord, *J. Anat.* (*London*) **100:**723.
Stieda, L., 1861, *Ueber das Rückenmark und einzelne Theile des Gehirns von Esox lucius L.*, Carl Schulz, Dorpat.
Tinkelenberg, J., 1979, Graphic reconstruction, microanatomy with a pencil, *J. Audiovis. Media Medicine* **2:**102.
Tinkelenberg, J., 1980, Graphic reconstruction and stereoscopy, *J. Audiovis. Media Med.* **3:**68.
Tömböl, T., 1968, Cellular and synaptic organization of the dorso-medial thalamic nucleus, *Acta Morphol. Acad. Sci. Hung.* **16:**183.
Tompsett, D.H., 1955, Differential staining and mounting of human brain slices, *Med. Biol. Illus.* **5:**29.
Tompsett, D.H., 1970, *Anatomical Techniques,* 2nd ed., Edinburgh.
Torvik, A., 1956, Transneuronal changes in the inferior olive and pontine nuclei in kittens, *J. Neuropathol.* **15:**119.
Turcotte, A., and Ramón Moliner, E., 1965, Counterstaining solution for sections stained with the Golgi–Cox method, *Stain Technol.* **40:**310.
Uylings, H.B.M., and Smit, G.J., 1975, Three-dimensional branching structure of pyramidal cell dendrites, *Brain Res.* **87:**55.
Van Crevel, H., and Verhaart, W.J.C., 1963, The rate of secondary degeneration in the central nervous system: I and II, *J. Anat.* (*London*) **97:**429.
Van den Akker, L.M., 1970, *An Anatomical Outline of the Spinal Cord of the Pigeon,* Van Gorcum, Assen.
Van der Loos, H., 1956, Une combinaison de deux vieilles méthodes histologiques pour le systéme nerveux central, *Monatsschr. Psychiatr. Neurol.* **132:**330.
Van der Loos, H., 1959, Dendro-dendritische verbindingen in de schors der grote hersenen, Thesis, Stam, Haarlem, Antwerp, and Djakarta.
Van der Loos, H., 1967, The history of the neuron, in: *The Neuron* (H. Hydén, ed.), pp. 1–47, Elsevier, Amsterdam, London, and New York.
Van Manen, J., 1967, *Stereotactic Methods and Their Applications in Disorders of the Motor System,* Van Gorcum, Assen.
Verhaart, W.J.C., 1970, *Comparative Anatomical Aspects of the Mammalian Brain Stem and the Cord,* Van Gorcum, Assen.
Verhaart, W.J.C., 1976, Nuclei and tracts of the di-mesencephalon of the parakeet, *Acta Anat.* (*Basel*) **94:**89.
Vielvoye, G.J., and Voogd, J., 1977, Time dependence of terminal degeneration in spinocerebellar mossy fiber rosettes in the chicken and the application of terminal degeneration in successive degeneration experiments, *J. Comp. Neurol.* **175:**233.
Vielvoye, G.J., Freedman, S.L., and Voogd, J., 1977, A retropharyngeal approach to the avian brain stem, *Am. J. Anat.* **148:**409.
Vogt, C., and Vogt, O., 1902, Zur Erforschung der Hirnfaserung, Denkschriften der Med.-Naturwiss. Ges. Jena IX, O. Vogt, *Neurobiol. Arbeiten, Erste Ser.* **1:**1–145.
Von Bechterew, W., 1894, *Die Leitungsbahnen im Gehirn und Rückenmark,* Verlag von Eduard Besold (Arthur Georgi), Leipzig.
Von Gudden, B., 1870, Experimentelle Untersuchungen über das periferische und centrale Nervensystem, *Arch. Psychiatr. Nervenkr.* **2:**693.
Von Gudden, B., 1889, *Gesammelte und hinterlassene Abhandlungen,* Bergmann, Wiesbaden.
Voogd, J., and Boesten, A.J.P., 1976, A light and electron microscopical study of inferior olivary hypertrophy in the cat, *J. Anat.* (*London*) **122:**712.
Walberg, F., 1964, The early changes in degenerating boutons and the problem of argyrophilia: Light and electron microscopic observations, *J. Comp. Neurol.* **122:**113.
Walberg, F., 1971, Does silver impregnate normal and degenerating boutons? A study based on light and electron microscopical observations of the inferior olive, *Brain Res.* **31:**47.
Walberg, F., 1972, Further studies on silver impregnation of normal and degenerating boutons: A light and electron microscopical investigation of a filamentous degenerating system, *Brain Res.* **36:**353.

Waller, A., 1850, Experiments on the section of the glossopharyngeal and hypoglossal nerves of the frog, *Philos. Trans. R. Soc. London* **1**:423.
Waller, A., 1851, Nouvelle méthode pour l'étude du système nerveux, *C. R. Acad. Sci.* **33**:606.
Wann, D.F., Woolsey, T.A., Dierker, M.L., and Cowan, W.M., 1973, An on-line digital computer system for the semi-automatic analysis of Golgi-impregnated neurons, *IEEE Trans. Biomed. Eng.* **20**:233.
Weigert, K., 1884. Ausführliche Beschreibung der in No. 2 dieser Zeitschrift erwähnten neuen Färbungsmethode für das Zentralnervensystem, *Fortschr. Med.* **2**:190.
Wollschlaeger, G., and Wollschlaeger, P.B., 1974, Postmortem angiography, in: *Radiology of the Skull and Brain,* Vol. 2 (T.H. Newton and D.G. Potts, eds.), Book 1, pp. 1002–1019, C.V. Mosby, Saint Louis.
Zeman, W., and Innes, J.R.M. (eds.), 1963, *Craigie's Neuroanatomy of the Rat,* Academic Press, New York.
Zweers, G.A., 1971, *A Stereotactic Atlas of the Brain Stem of the Mallard* (*Anas platyrhynchos L.*), Van Gorcum, Assen.

Bibliography

Baker, J.R., 1958, *Principles of Biological Microtechnique: A Study of Fixation and Dyeing,* Methuen, London.
Glauert, A.M., 1975, *Fixation, Dehydration and Embedding of Biological Specimens,* North-Holland, Amsterdam and Oxford; Elsevier, New York.
Humason, G.L., 1972, *Animal Tissue Techniques,* in: *A Series of Books in Biology* (D. Kennedy and R.B. Park, eds.), W.H. Freeman, San Francisco.
Lewis, P.R., and Knight, D.P., 1977, *Staining Methods for Sectioned Material,* North-Holland, Amsterdam and Oxford; Elsevier, New York.
Nauta, W.J.H., and Ebbesson, S.O.E. (eds.), 1970, *Contemporary Research Methods in Neuroanatomy,* Springer-Verlag, Berlin, Heidelberg, and New York.
Palay, S.L., and Chan-Palay, V., 1974, *Cerebellar Cortex,* Springer-Verlag, Berlin, Heidelberg and New York.
Ramón y Cajal, S., 1933, *Elementos de Técnica Micrográfica del Sistema Nervioso,* Tipografia Artistica, Madrid.
Romeis, B., 1968, *Mikroskopische Technik,* Oldenbourg Verlag, Munich and Vienna.
Stain Technology, 1925–1977, Vols. 1–55, Williams and Wilkins, Baltimore.
Windle, W.F., 1957, *New Research Techniques of Neuroanatomy,* Charles C. Thomas, Springfield, Illinois.

6

Fluorescence Microscopy of Biogenic Monoamines

Olle Lindvall, Anders Björklund, and Bengt Falck

1. Introduction

Since the introduction, in the early 1960's, of the gas-phase formaldehyde (FA) reaction for visualization of catecholamines (CAs) and indoleamines by Falck, Hillarp, and co-workers (Falck, 1962; Falck *et al.*, 1962; Corrodi and Hillarp, 1963, 1964), the fluorescence histochemistry of monoamine transmitters has undergone a continuous and rapid development. During recent years, several modifications of the FA reaction as well as several new forms of application of the FA reagent have been worked out. Also, a more reactive congener of FA, namely, glyoxylic acid (GA), has been introduced for more sensitive visualization of intraneuronal CAs. These many recent advances have justified this review of histofluorescence methodology.

The cytofluorometric techniques are particularly attractive because of their very high sensitivity in combination with a very precise localization of neurotransmitters. The FA and GA methods can thus readily demonstrate noradrenaline (NA) and dopamine (DA) within axons and nerve-terminal varicosities. Jonsson (1971) has reported that after FA gas condensation in model experiments, the lowest detectable concentration of NA (or DA) is about 1–5 pg/μg protein. The sensitivity of the Falck–Hillarp FA method can also be expressed as the lowest amount of substance detectable in a nerve-terminal varicosity. The amount of NA present in one varicosity of the sympathetic nerves of the rat iris (diameter ≈ 1 μm) has been calculated to be about 5×10^{-3} pg (Dahlström *et al.*, 1966). From the observations of Jonsson (1969) that varicosities containing less than 10% of their normal NA content can be visualized, it appears that as little as 5×10^{-4} pg (about 5

Olle Lindvall, Anders Björklund, and ***Bengt Falck*** • Departments of Histology and Neurology, University of Lund, Lund, Sweden.

$\times\ 10^{-6}$ pmol) of NA or DA can be readily detected within one varicosity. The recently developed GA fluorescence method (Björklund *et al.*, 1972*d*; Axelsson *et al.*, 1973; Lindvall and Björklund, 1974*a*; Lindvall *et al.*, 1974*c*), as well as the aluminum–formaldehyde (ALFA) method (Lorén *et al.*, 1980*a*; Björklund *et al.*, 1980), have even higher overall sensitivity for CAs and indoleamines. In its most sensitive form of application, the GA method has been estimated to be approximately 10 times more sensitive for DA than the standard Falck–Hillarp technique (Lindvall and Björklund, 1974*a*). From the figures given above, it follows that 5×10^{-5} pg (about 5×10^{-7} pmol) of DA would be readily detected in a varicosity with the GA method. The figure is probably similar for the ALFA method.

Although these figures are only rough estimates of the absolute sensitivity of the FA and GA fluorescence methods, they point to an extreme sensitivity of histochemical cytofluorometric techniques in comparison with all other analytical procedures available for quantitative or qualitative studies of endogenous monoamines. In comparison with alternative biochemical analytical techniques, the cytofluorometry of monoamines has the additional advantage of allowing selective studies (biochemical, physiological, or pharmacological) within single cells or cell compartments in the intact tissue.

The aim of this review is to introduce the reader to the basic principles of fluorescence microscopy and microspectrofluorometry of biogenic monoamines and the chemical background of the fluorophore-forming histochemical reactions. The chapter also gives an overview of the many variants and forms of application of the FA and GA methods to assist the reader in the choice of the most adequate and practical technique for a particular research problem.

2. Fluorescence Microscopy and Microspectrofluorometry

Fluorescence is a special form of luminescence, i.e., photoluminescence, in which a molecule absorbs light and re-emits the absorbed energy as light of longer wavelength. The absorption of light results in a transformation of the molecule into a higher energy level, the so-called *second excited state* of the molecule. Part of the absorbed energy is lost through a process known as internal conversion whereby the molecule is rapidly transformed into a lower energy level, the *first excited state*. The remaining part of the absorbed energy, representing the difference between the first excited state and the *ground state*, is released as the emitted fluorescence light. The emitted light is always of lower energy than the absorbed light, the difference accounting for the energy lost by internal conversion. [For further discussions of fluorescence and luminescence phenomena, the reader is referred to the excellent reviews by Hercules (1966), Parker (1968), Udenfriend (1969), and Pesce *et al.* (1971), as well as to Volume 1, Chapter 6.] The fluorophores studied in fluorescence histochemistry emit a visible light and hence are excited by visible or ultraviolet (UV) light. The sizes of the first and second excited states are characteristic entities of a fluorophore. These, in turn, define the wavelength maxima of emission and absorption, respectively, of the fluorophore. The emission and the absorption of the fluorophore thus define two types of spectra, the *emission spectrum* and the *excitation* (or *absorption*) *spectrum*, which can be determined by microspectrofluorometry (see Section 2.2.1).

2.1. The Fluorescence Microscope

As illustrated schematically in Fig. 1, the fluorescence microscope is equipped with a light source for activation or excitation of the fluorophores in the microscopic specimen and with two sets of filters. The light source is usually a high-pressure mercury lamp (e.g., Osram HBO 200/4 or HBO 100) giving a high intensity of UV and blue-violet light. The light is passed through a heat-absorption filter (Schott KG 1) and through suitable primary filters (thick enough to minimize unwanted excitation light and, thus, disturbing background fluorescence) to select optimum activation wavelength. The light is then focused by a metallized front-surface mirror into a dark-field condenser. Oil-immersion condensers usually give higher light intensity, but excellent dry condensers can also be obtained from some manufacturers.

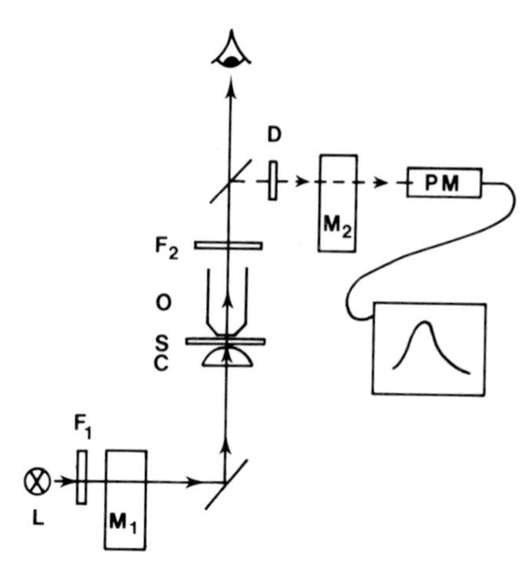

Fig. 1. Principal construction of the fluorescence microscope also showing the attachments necessary to convert the fluorescence microscope into a microspectrofluorometer. From the light source (L) (usually a mercury lamp), the ultraviolet or blue-violet light is passed through a set of primary lamp filters (F_1) to select light of optimum activation wavelength. Via a mirror and a dark-field condenser (C), the light is focused onto the specimen (S). The fluorescence light emitted from the specimen passes through the objective (0) and the ocular of the microscope. A secondary barrier filter (F_2) is inserted above the objective to cut off stray light from the lamp and to produce sufficient contrast. In the microspectrofluorometer, two monochromators are inserted into the light path: the excitation monochromator (M_1) between the lamp and the condenser and the emission monochromator (M_2) between the objective and the photomultiplier (PM). A diaphragm (D) is used to select the measuring field. The spectra or fluorescence intensities registered by the photomultiplier are fed into an *x–y*-recorder, a galvanometer, or other suitable registering device.

The emitted light is passed through secondary filters to cut off stray light and to produce proper contrasts. As a general principle, the lamp filter (activation filter or primary filter) is selected to achieve excitation as close as possible to the activation maximum (absorption maximum) of the fluorophores; the barrier filter (secondary filter) is selected to give optimum contrast and, if possible, the correct fluorescence colors. For the formaldehyde (FA)- or glyoxylic acid (GA)-induced fluorescence of catecholamines (CAs) and 5-hydroxytryptamine (5-HT), these two conditions cannot be fulfilled simultaneously with the available standard glass filters. Since the excitation maximum for these amines is at 390–410 nm, the activation filter of choice is Schott BG 12 or BG 3. However, with this lamp filter, a secondary filter with high absorption below 500 nm (e.g., Schott OG 4) should be used to exclude the blue component of the excitation light. This filter combination gives good contrast (low background) and high fluorescence within the spectral range for the optimum sensitivity of the human eye. It should be emphasized that since the emission maximum of the CAs is 475 nm, this barrier filter will absorb the blue component of the CA fluorophores; hence, this filter combination will give a green CA fluorescence and a yellow 5-HT fluorescence (emission maximum about 525 nm). If the monoamine fluorescence is strong enough, it is possible to utilize fluorescence-selection filters (e.g., Schott SAL 480 or SAL 525) as secondary filters. To bring out the blue fluorescence of CAs, other primary filters such as Schott UG 1 or BG 3, or the interference filter Leitz AL 405 in combination with BG 3, or a monochromator can be used (Ploem, 1969). The Leitz Ploem–Opak equipment for incident light with the AL 405 as lamp filter is an excellent illumination system to achieve the true fluorescence colors in combination with high fluorescence intensity and low background.

Several specialized fluorescence microscopes that offer possibilities to use both transmitted and incident illumination are available. Such instruments allow fluorescence microscopy (by incident light) concomitant with absorption and phase-contrast microscopy (by transmitted light). Incident-light illumination (e.g., Leitz Ploem–Opak system) is particularly advantageous for use with immersion objectives (oil or water) with numerical apertures higher than 0.75.

2.2. *The Microspectrofluorometer*

The microspectrofluorometer is principally a fluorescence microscope with attachments allowing qualitative and quantitative analysis of intracellular fluorophores (see Fig. 1). For fluorescence quantitation, comparatively simple equipment can be used. Thus, any fluorescence microscope with epi-illumination, a stabilized light source, and a suitable photodetector can be sufficient (e.g., see Lichtensteiger, 1970; Van Orden, 1970). Although the instrumentation for fluorescence quantitation is relatively uncomplicated, there are several theoretical and practical problems inherent in this technique (see Section 2.2.2).

2.2.1. *Qualitative Special Analysis*

Qualitative microspectrofluorometric analysis is based on the recording of two types of spectra of the fluorophores: the emission and the excitation spectrum. These spectra are, as indicated in Section 2, physical entities directly related to the molecular configuration of the fluorescent compounds, i.e., the fluorophores formed from the monoamines in the histochemical reactions. For this reason, microspectrofluorometric analysis has high analytical value for structural identification of biogenic monoamines.

The *emission spectrum* gives the spectral distribution of the emitted light and can be considered independent of the wavelength of the exciting (activating) light. It is recorded at a fixed wavelength of the exciting light (preferably close to the wavelength of maximal excitation), and the relative distribution of the emitted light in the visual part of the spectrum is registered. The *excitation spectrum* expresses the efficiency with which the exciting light of different wavelengths induces fluorescence. This type of spectrum is obtained by varying the wavelength of the exciting light while the intensity of the emitted light is measured at a fixed wavelength. The excitation spectrum of a fluorescent substance is practically identical with its absorption spectrum and is for this reason especially helpful for identification purposes.

More advanced instruments are required for qualitative microspectrofluorometric analysis at high sensitivity. For complete fluorophore characterization, the microspectrofluorometer must permit recordings of both emission and excitation spectra, which means that it must be equipped with monochromators for both the exciting and the emitted light (see Fig. 1). For measuring excitation spectra down to about 300 nm, the optical pathway for the exciting light must be made entirely of quartz (quartz optics and quartz condenser) and the excitation monochromator must have high transmittance in this low-wavelength range. Such instruments are commercially available. For details on the construction and operation of such instruments, the reader is referred to Caspersson *et al.* (1965), Björklund *et al.* (1968*a*, 1972*c*, 1975*a*), Rost and Pearse (1971), and Ruch (1973). In the following discussion, we will only give some aspects of the correction of the recorded spectra and point to some of the more important possible sources of error encountered in the analysis of monoamine fluorophores.

Both the excitation and emission spectra are distorted by a number of factors in the instrument, and the registered spectra (i.e., the uncorrected instrument values) usually deviate considerably from the true spectra (see Ritzén, 1967). The emission spectrum is influenced by the secondary filters used, the transmission of the optics, the transmission and band width of the analyzing monochromator, and the sensitivity of the photomultiplier. Similarly, the excitation spectrum is influenced by the varying transmission of the excitation monochromator and optics; it is also strongly influenced by the varying intensity of the light source at different wavelengths. Altogether, these factors will distort the values in a way that is characteristic for each individual instrument. For this reason, all published spectra should be corrected and expressed in standardized units that will allow direct comparison among spectra obtained in different laboratories.

The *correction of the emission spectrum* requires the preparation of an instrument calibration curve, which can be obtained, for instance, from measurements with a calibrated tungsten lamp or by measuring the fluorescence of a fluorescent reference solution with known emission characteristics. For the *correction of the excitation spectrum,* some kind of device for the continuous measurement of the intensity of the exciting light is necessary, since the characteristics of the exciting light will vary among individual lamps and throughout the lifetime of a lamp. The correction procedures employed in our laboratory have been introduced and described by Ritzén (1967).

The spectral properties of the monoamine fluorophores and the techniques for their differentiation on the cellular level are treated in Section 4.

2.2.2. *Microfluorometric Quantitation*

Microfluorometric quantitation requires a linear relationship between fluorophore concentration and fluorescence intensity. In dried-protein-droplet models, this has been shown to be true for primary CAs [noradrenaline (NA) and dopamine (DA)] after FA gas treatment up to a concentration of about 4.5×10^{-2} M in the droplets, and for 5-HT up to about 9.0×10^{-2} M in the droplets (Fig. 2) (Ritzén, 1966*a,b,* 1967). At higher concentrations, a quenching of the fluorescence occurs; i.e., the linearity between fluorescence intensity and concentration is broken. This is most pronounced for the CA fluorophores, and for these substances, a further increase in concentration will result in no, or only slight, increase in fluorescence intensity. At very high concentrations, the intensity might even decrease. A similar concentration quenching also occurs after GA condensation (Lindvall *et al.,* 1974*c*) (Fig. 2).

The phenomenon of concentration quenching is a serious limitation for

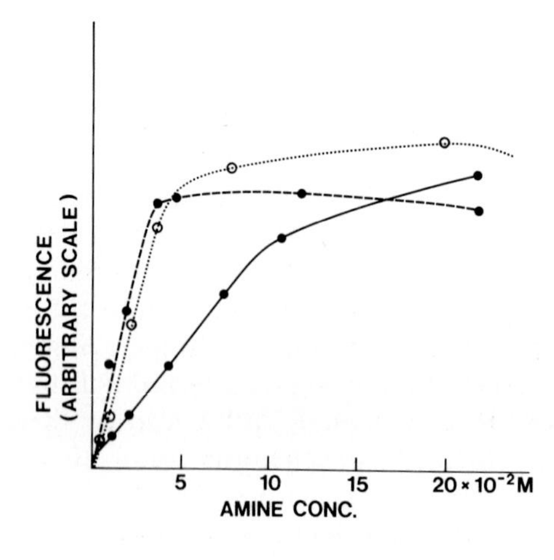

Fig. 2. Relationship between fluorescence intensity and amine concentration in dried protein droplets after FA or GA condensation. (——) 5-HT after FA-vapor treatment (taken from Ritzén, 1966*a*); (------) DA after FA-vapor treatment (taken from Ritzén, 1966*b*); (······) DA under conditions simulating the GA–Vibratome procedure (taken from Lindvall *et al.,* 1974*c*).

the microfluorometric quantitation, and its possible occurrence must be ruled out for each cell system subjected to analysis. Quantitative studies have hitherto been performed after FA condensation on 5-HT in isolated peritoneal mast cells (Ritzén, 1967; Carlsson and Ritzén, 1969), on NA in cell bodies of sympathetic neurons (Norberg *et al.*, 1966), and on DA in central neuronal perikarya (Lichtensteiger, 1969; 1970) and nerve terminals (Einarsson *et al.*, 1975; Löfström *et al.*, 1976*a*,*b*). In these cell systems, concentration quenching does not seem to occur. On the other hand, studies by Jonsson (1969) on NA in sympathetic nerve terminals have demonstrated that concentration quenching will occur in these structures when processed according to the standard FA procedure and that the fluorescence intensity is proportional to the amine concentration only up to 30–40% of the normal endogenous concentration. In central NA-containing terminals, the situation seems to be more variable. Lidbrink and Jonsson (1971), using a semiquantitative technique, have observed that in cortical NA terminals, the relationship between fluorescence intensity and concentration is linear up to the normal endogenous NA level, whereas this linear relationship is broken at approximately 50% of the normal level in hypothalamic NA terminals.

Quantitation of fluorescence intensities is done most accurately with the aid of a microfluorometer. However, subjective semiquantitative estimations are often useful provided that a double-blind procedure with coded slides is used (see Jonsson, 1971). Agnati and Fuxe (1974) have recently introduced an alternative photographic procedure based on fluorescent varicosity counts after photography using gratings of different transmittance. In the following discussion, only the microfluorometric quantitation will be considered.

Since the fluorescence yield of the monoamines in the FA reaction is influenced by the reaction conditions and since this yield is difficult or impossible to reproduce exactly from one experiment to another, fluorescence reference standards must be used for the microfluorometric quantitation. When the measurements are performed on whole cells (e.g., isolated mast cells) or on structures in whole mounts (e.g., adrenergic terminals in stretch preparations of the iris), identically treated dried albumin droplets with known amine concentration have been used as reference standards (Ritzén, 1967; Jonsson, 1969). When the measurements are performed on tissue sections, the recordings are obtained from only part of a cell or from a nerve-terminal area. In this case, the amount of amine measured will depend not only on the measuring area but also on the section thickness. The fluorescence standard used must therefore correct for variations in the reaction conditions as well as in section thickness. For this purpose, Lichtensteiger (1969) and Einarsson *et al.* (1975) used freeze–dried gelatin cylinders and Löfström *et al.* (1976*b*) freeze–dried agar–protein cylinders with a known amine concentration. The cylinders are treated identically with the tissue specimens, and the cylinder and the specimen are embedded and sectioned in the same paraffin block. If each section has a reasonably even thickness, the fluorescence standard will thus correct for variation in thickness among different sections.

A relative quantitation of cellular amines can be obtained without reference standards. The tissue specimens to be compared are then processed together and finally enclosed in the same paraffin block to allow comparison of the fluorescence intensities in the same sections (Norberg *et al.,* 1966; Björklund and Falck, 1969).

Microfluorometric quantitation has great potential in all those systems in which concentration quenching does not occur or in which it can be eliminated by, for example, alterations in the histochemical reaction conditions. Lichtensteiger (1969, 1970) and Jonsson and co-workers (Einarsson *et al.,* 1975; Löfström *et al.,* 1976*a,b*) have shown very elegantly how microfluorometric quantitation can be used for functional studies on single neurons and defined populations of nerve terminals in the CNS. Using the tyrosine hydroxylase inhibition model for transmitter-turnover measurements, Einarsson *et al.* (1975) and Löfström *et al.* (1976*a,b*) have shown that by measuring the rate of decline of CA fluorescence after inhibition of the synthesizing enzyme, microfluorometry can be used for estimation of DA and NA turnover in discrete terminal areas. The great virtue of this method over, for example, biochemical microdissection or punch techniques is that amine content and turnover changes can be analyzed in different and defined neuronal populations within the same area.

For further considerations in the use of the microfluorometric-quantitation technique for monoamines, the reader is referred to the papers by Ritzén (1967, 1973), Jonsson (1971, 1973), Einarsson *et al.* (1975), and Löfström *et al.* (1976*b*).

3. *Visualization of Biogenic Monoamines*

3.1. *Chemical Background*

The mechanisms of fluorophore formation in the fluorescence-histochemical methods for monoamines have been extensively studied. The reaction sequence in the formation of fluorophores and the chemical identity of the reaction products in the formaldehyde (FA) and glyoxylic acid (GA) methods are now well known. This is of primary importance for the interpretation of the fluorescence-microscopic picture, particularly with reference to the specificity of the method. Such knowledge also allows controlled manipulations of the reaction or of the formed fluorophores, in order to, for example, increase the sensitivity or selectivity of the method. Moreover, when microspectrofluorometric techniques are used for structural analysis of intracellular fluorogenic monoamines, the chemical background is indispensable for the reliable interpretation of the recorded spectra.

In addition to FA and GA, other reagents that form fluorophores with phenylethylamines and indoleamines have been found (Axelsson *et al.,* 1972; Ewen and Rost, 1972). At present, FA and GA are, however, the only reagents useful for visualization of intraneuronal monoamines. It should be mentioned that histamine and related compounds can be demonstrated at the cellular level by condensation with *o*-phthaldialdehyde (Ehinger and Thunberg, 1967; Håkanson and Owman, 1967; for a review, see Björklund *et al.*, 1972*c*). The sensitivity of this technique is, however, at present too low to allow the demonstration of a possible neuronal store of histamine.

3.1.1. Fluorophore Formation after Formaldehyde Treatment

The fluorophore-forming reactions between the biogenic monoamines and FA have been shown to proceed in two steps (Figs. 3 and 4) (Corrodi and Hillarp, 1963, 1964; Corrodi and Jonsson, 1965*b*; Björklund *et al.*, 1973*a*). In the first step, the phenylethylamine or indolylethylamine reacts with FA in a Pictet–Spengler condensation, yielding the weakly fluorescent or nonfluorescent 1,2,3,4-tetrahydroisoquinoline (I and IV in Fig. 4) or 1,2,3,4-tetrahydro-β-carboline (II in Fig. 3), respectively, via a Schiff's base (for numbering of positions in the molecules, see Fig. 5). When the amines are enclosed in a dried-protein matrix, as in freeze–dried or air-dried tissue, these initially formed tetrahydro derivatives are converted to strongly fluorescent molecules. This fluorophore formation can proceed in two alternative ways (Björklund *et al.*, 1973*a*): through autoxidation to the 3,4-dihydroisoquinoline (II and V in Fig. 4; R = H) or 3,4-dihydro-β-carboline (III in Fig. 3) or through a second, acid-catalyzed reaction with FA, yielding the 2-methyl-3,4-dihydroisoquinolinium compound (II and V in Fig. 4; R = CH_3) or the 2-methyl-3,4-dihydro-β-carbolinium compound (IV in Fig. 3). At least in the reaction with tryptamine, the yield of highly fluorescent molecules in these two alternative fluorophore-forming pathways seems to be fairly equal (Björklund *et al.*, 1973*a*).

Through the reactions given above, 3-hydroxylated phenylethylamines will be converted to 6-hydroxylated 3,4-dihydroisoquinoline derivatives. At neutral pH, as occurs in tissue, the fluorophores are in their quinoidal forms (III and VI in Fig. 4), whereas the nonquinoidal forms (II and V) predominate at lower pH values (Jonsson, 1966; Björklund *et al.*, 1968 *a*, 1972*a*). This pH-dependent tautomerism between two states of the fluorophores is reflected in characteristic changes of their spectral properties (see Fig. 15 and Section 4.2.3). In addition, the fluorophores that have a 4-hydroxy group (formed from phenylethylamines with a β-hydroxy group on the side chain) can be transformed into the fully aromatic compounds (VII) by splitting off the hydroxy group with acid (Corrodi and Jonsson, 1965*a*; Björklund *et al.*, 1968*a*, 1972*a*). Since this conversion is followed by characteristic spectral changes, microspectrofluorometric differentiation between dopamine (DA) and noradrenaline (NA) is possible (Björklund *et al.* 1968*a*, 1972*a*) (see Section 4.2).

The formation of fluorophores from the initially formed tetrahydro derivatives is promoted by FA and catalyzed by certain amino acids, peptides, and proteins (Corrodi and Hillarp, 1964). The probable mechanism of this catalysis is that amino acids act as acid catalysts of the reaction of a second

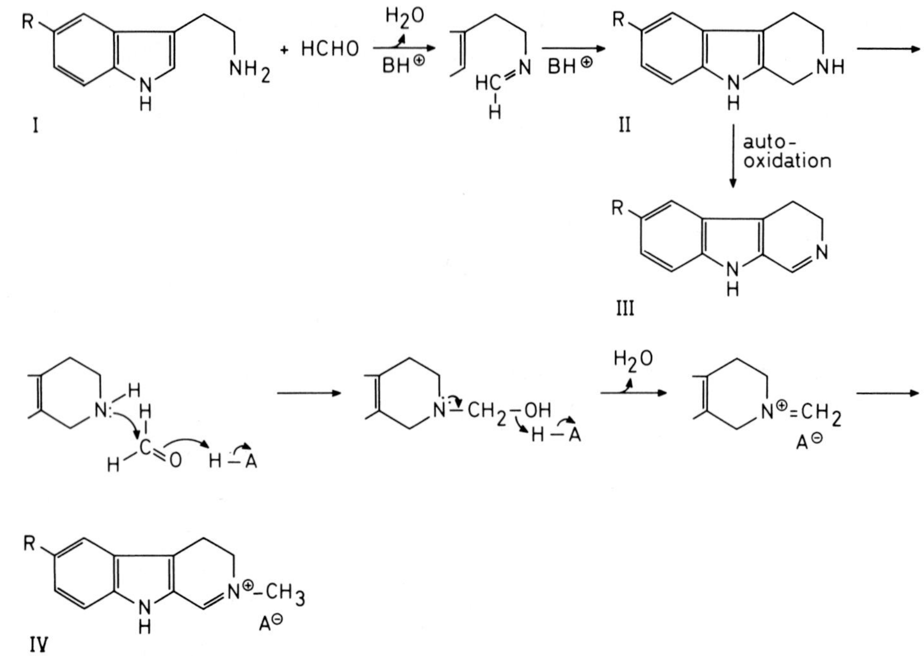

Fig. 3. Sequence of reactions between indolylethylamines and FA under histochemical conditions. Tryptamine: R = H; 5-HT: R = OH. For explanation, see the text. From Björklund *et al.* (1973*a*).

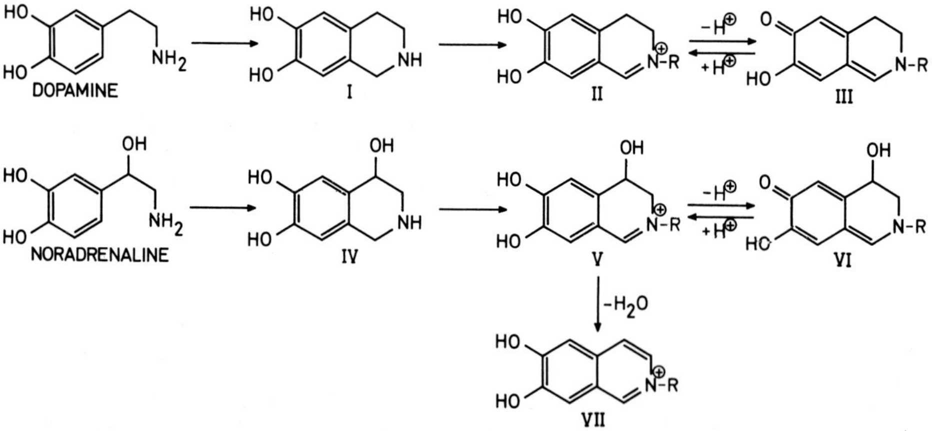

Fig. 4. Fluorophore formation from catecholamines by FA under histochemical conditions. In reactions analogous to those illustrated in Fig. 3, two types of dehydrogenated isoquinolines are formed: either the 6,7-dihydroxy-3,4-dihydro derivatives (R = H) or the 6,7-dihydroxy-3,4-dihydro-2-methyl derivatives (R = CH_3). These fluorophores exhibit a pH-dependent tautomerism between their quinoidal forms (III and VI) and their nonquinoidal forms (II and V). Under the influence of acid, the labile 4-hydroxy group of the NA fluorophore can be split off to form the fully aromatic 6,7-dihydroxyisoquinoline (VII). From Björklund *et al.* (1973*a*).

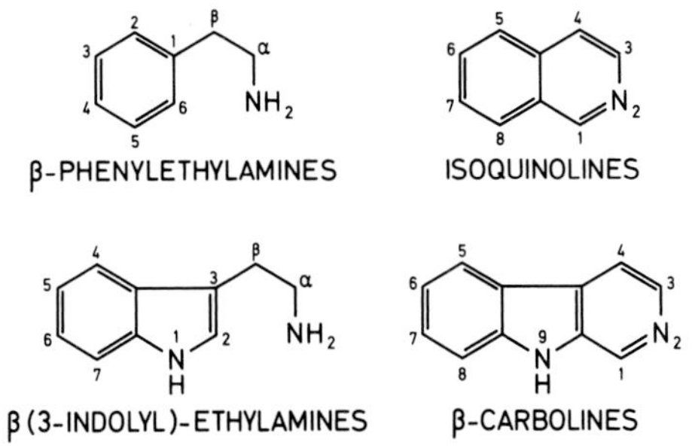

β-PHENYLETHYLAMINES ISOQUINOLINES

β(3-INDOLYL)-ETHYLAMINES β-CARBOLINES

Fig. 5. Labeling of positions in phenylethylamine and indolylethylamine derivatives and in their corresponding isoquinoline and β-carboline fluorophores.

FA molecule with the initially formed tetrahydroisoquinolines or tetrahydro-β-carbolines, yielding the strongly fluorescent 2-methyl-dihydro derivatives (II → IV in Fig. 3) (Björklund *et al.*, 1973*a*). The fact that the presence of catalyzing amino acids or proteins is necessary to obtain good fluorescence from monoamines in the FA reaction indicates that the acid-catalyzed formation of 2-methyl-3,4-dihydro compounds might be essential for obtaining maximum fluorescence yields in the histochemical reaction.

The first step of the fluorophore formation, the Pictet–Spengler cyclization reaction, is facilitated by high electron density at the point of ring closure of the amine [the 2-position of indolylethylamines and the 6-position of phenylethylamines (see Fig. 5 and Whaley and Govindachari, 1951)]. These requirements are fulfilled in 3-hydroxylated β-phenylethylamines [e.g., catecholamines (CAs)] and in β-(3-indolyl)ethylamines [e.g., 5-hydroxytryptamine (5-HT)], whereas the phenylethylamines that lack electron-releasing substituents in the 3-position are much less reactive (Whaley and Govindachari, 1951; Jonsson, 1967*c*; Björklund and Stenevi, 1970). Further, unsubstituted indolylethylamines are known to be less reactive in the Pictet–Spengler condensation than those that have a hydroxy or methoxy group in the 5-position of the indole nucleus (Späth and Lederer, 1930). Finally, the Pictet–Spengler condensation is limited to primary and secondary amines; tertiary amines and amides do not react.

Knowledge of the differences in reactivity of related phenylethylamine and indolylethylamine derivatives is important in determining the specificity of both the FA and the GA reaction. In the standard FA reaction, primary, 3-hydroxylated phenylethylamines (including the CAs) and the corresponding amino acids (including dopa) give the highest fluorescence yields (Table 1) (Falck *et al.*, 1962; Jonsson, 1967*a,b*; Björklund *et al.*, 1971*a*). The fluorescence induced from 3-methoxylated phenylethylamines (such as the normal CA metabolites, 3-methoxytyramine, normetanephrine, and metanephrine) is much weaker, probably due to less activation at the point of ring closure exerted by the 3-methoxy than by the 3-hydroxy group in the Pictet–Spengler condensation. Phenylethylamines that lack activating substituents in the 3-position give no fluorescence in the reaction. Tyramine and octopamine, as well as the amino acids phenylalanine and tyrosine, are examples of such nonfluorogenic derivatives that occur in tissues. In the case of indolylethylamines, the 5- or 6-hydroxylated derivatives give the strongest fluorescence in the standard FA reaction (Table 1).

Secondary amines generally give lower fluorescence yields than the corresponding primary amines (Table 1) (Corrodi and Hillarp, 1963; Falck *et al.*, 1963; Corrodi and Jonsson, 1967), and they require more severe reaction conditions, i.e., higher temperature or longer reaction time or both, for maximum fluorescence. This difference between primary and secondary amines, which is more pronounced for the phenylethylamines than for the indolylethylamines, could be explained by the fact that the secondary FA molecule (see Fig. 3) cannot react with the pyridine-ring nitrogen of the 2-methyl-1,2,3,4-tetrahydroisoquinoline or the 2-methyl-1,2,3,4-tetrahydro-β-carboline derivatives formed from the secondary amines in the initial Pictet–Spengler cyclization. Therefore, the secondary amines can be transformed into the strongly fluorescent dihydro derivatives only via the autoxidative pathway (II → III in Fig. 3) (see above) (Björklund *et al.*, 1973*a*).

To improve the fluorescence yield from certain biogenic amines that are low-fluorescent after the standard FA treatment, two modifications of the FA reaction have been developed: the acid-catalyzed FA reaction (Björklund and Stenevi, 1970; Björklund *et al.*, 1971*b*) and the FA–ozone reaction (Björklund *et al.*, 1968*b*; Björklund and Falck, 1969). The *acid-catalyzed FA reaction* is based on the well-known fact that in solution, the Pictet–Spengler reaction is catalyzed by hydrogen ions (Whaley and Govindachari, 1951). It has been demonstrated that if small amounts of HCl vapor are present in the reaction vessel during the FA treatment, the fluorescence yields from tryptamine and 3-methoxylated phenylethylamines are dramatically increased (Björklund and Stenevi, 1970; Björklund *et al.*, 1971*b*). With the minute amounts of acid used, the acid-catalyzed FA reaction exhibits a good specificity for indolylethylamines and 3-hydroxylated or 3-methoxylated phenylethylamines (Björklund and Stenevi, 1970). Even though both the Pictet–Spengler cyclization and the reaction of the tetrahydro derivatives with the second FA molecule are catalyzed by acid (see Fig. 3), it seems that the effect of HCl vapor is due primarily to a promotion of the initial cyclization step. It is probable that such an acid catalysis of the FA reaction, resulting in a higher fluorophore yield, also takes place in tissue perfused with an acid FA solution with a high magnesium or aluminum content (Lorén *et al.*, 1976*a*, 1980*a*) (see also Section 3.2.1b).

In the *FA–ozone reaction,* a gaseous oxidant (ozone) is introduced into the FA reaction vessel (Björklund *et al.*, 1968*b*). Tryptamine, which gives a low fluorescence yield after the standard FA treatment, becomes strongly fluorescent in the FA–ozone reaction. This increase is also obtained when the ozone treatment is performed after, but not before, the FA treatment. Although the mechanism by which ozone promotes fluorophore formation from tryptamine has not yet been clarified, it seems probable that the effect of ozone is, at least partly, to bring about an increased oxidative conversion

Table 1. Fluorescence Yields and Spectral Characteristics of a Number of Biogenic Phenylethylamine and Indolylethylamine Derivatives Enclosed in Dried Protein Films after Formaldehyde Gas Treatment[a]

Substance group	Substance	Relative fluorescence yield[b]	Maximum (nm)[c]	
			Excitation	Emission
Group A	1. 3,4-Dihydroxyphenyl-alanine (dopa)	120	320 and 410	475
	2. 3,4-Dihydroxyphenyl-ethylamine (dopamine)	100	320 and 410	475
	3. 3,4-β,Trihydroxyphenyl-ethylamine (noradrenaline)	100	320 and 410	475
	4. *N*-Methyl-3,4-β-tri-hydroxyphenylethylamine (adrenaline)	40	320 and 410	475
Group B1	5. Tryptamine [β(3-indolyl)-ethylamine]	10	370	495[d]
	6. Tryptophan	10	375	435 or 500[e]
Group B2	7. 6-Hydroxytryptamine	200	385 (420)	505
	8. 5,6-Dihydroxytryptamine	10	310 and (380) 405	(<430) 500
	9. *N*-Methyl-5-hydroxy-tryptamine	30	315 and (390) 415	505
	10. 5-Methoxytryptamine	20	(330) and 380 (410)	505[f]
Group C	11. 5-Hydroxytryptamine	30	(315) 385 and 415[e]	520–540
	12. 5-Hydroxytryptophan	10	310, 385 and 415	520–540

[a] Treatment at +80°C for 1 hr. Data from Björklund *et al.* (1971*b*).
[b] The fluorescence intensities are expressed relative to the intensity obtained from noradrenaline and dopamine in the standard formaldehyde reaction as 100.
[c] Figure in parentheses indicates the position of a low peak or a shoulder in the spectrum.
[d] The emission maximum of the tryptamine fluorophore exhibits a concentration-dependent variation from 450 to 520 nm (Björklund *et al.*, 1968*b*).
[e] The different peaks are observed under different conditions (see Björklund *et al.*, 1968*b*).
[f] The emission maximum of the 5-methoxytryptamine fluorophore is higher (520–550 nm) under more intense reaction conditions, and at higher amine concentrations.

of the initially formed 1,2,3,4-tetrahydro-β-carboline to 3,4-dihydro-β-carboline (II → III in Fig. 3). The ozone concentration is critical for optimum fluorescence yield in the reaction. The fact that too much ozone in the reaction vessel decreases the fluorescence induced from tryptamine and that CAs and 5-HT give a strongly reduced fluorescence in the combined FA–ozone reaction is most likely due to oxidation either of the fluorophores or of the amines themselves to products with low or no visible fluorescence. The FA–ozone technique has also been found useful for the histochemical visualization of peptides with NH_2-terminal tryptophan (Håkanson and Sundler, 1971). It is likely that these peptides react as α-substituted tryptamine derivatives and that the mechanism of fluorophore formation is thus similar to that of tryptamine itself.

3.1.2. Fluorophore Formation after Glyoxylic Acid Treatment

The fluorophore formation from primary and secondary phenylethylamines and indolylethylamines in the GA method has been shown to proceed in two principal reaction steps as in the FA method (Figs. 6 and 7) (Björklund *et al.*, 1972*d*; Lindvall *et al.*, 1974*c*). In the first step, the phenylethylamine or indolylethylamine reacts with GA in an acid-catalyzed Pictet–Spengler condensation, yielding the 1,2,3,4-tetrahydroisoquinoline-1-carboxylic acid or 1,2,3,4-tetrahydro-β-carboline-1-carboxylic acid, respectively, via a Schiff's base (I → II in Figs. 6 and 7). These very weakly fluorescent compounds can be transformed into strongly fluorescent molecules in two alternative ways: via autoxidative decarboxylation to the 3,4-dihydroisoquinoline or 3,4-dihydro-β-carboline (II → III in Fig. 6) or through a second, intramolecularly acid-catalyzed, reaction with GA to the 2-carboxymethyl-3,4-dihydroisoquinolinium or 2-carboxymethyl-3,4-dihydro-β-carbolinium compound (II → IV in Fig. 6 and II → III in Fig. 7). A further decarboxylation to the 2-methylated compounds (V in Fig. 6) is possible.

GA-induced 6-hydroxylated dihydroisoquinoline fluorophores (formed from 3-hydroxylated phenylethylamines, e.g., CAs) will exhibit a pH-dependent tautomerism similar to that described above for the corresponding FA-induced fluorophores (Fig. 7) (Lindvall *et al.*, 1974*c*). At neutral pH, they are in quinoidal forms (IIIb in Fig. 7), and in an acid environment, they exist

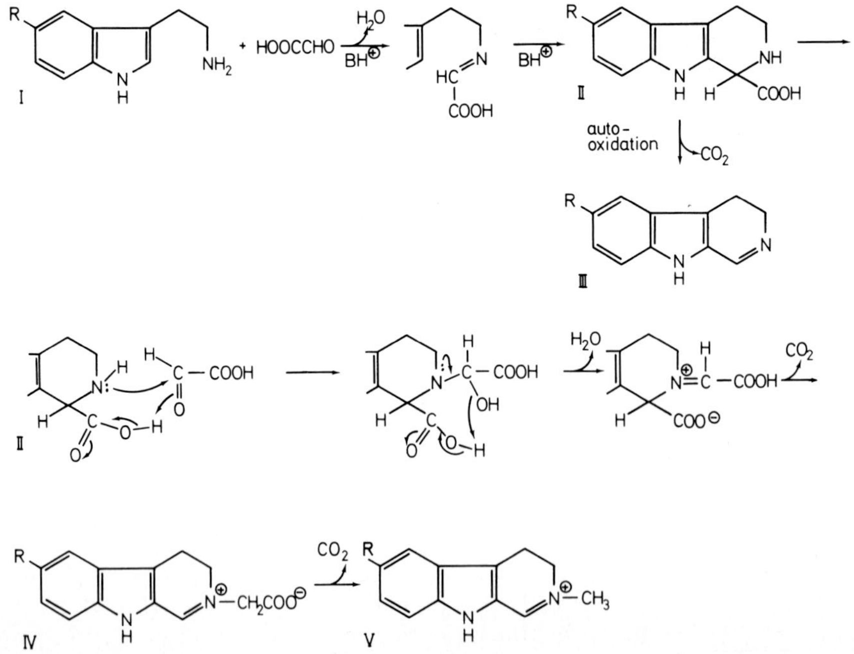

Fig. 6. Sequence of reactions between indolylethylamines and GA. Tryptamine: R = H; 5-HT: R = OH. For explanation, see the text. From Björklund *et al.* (1972*d*).

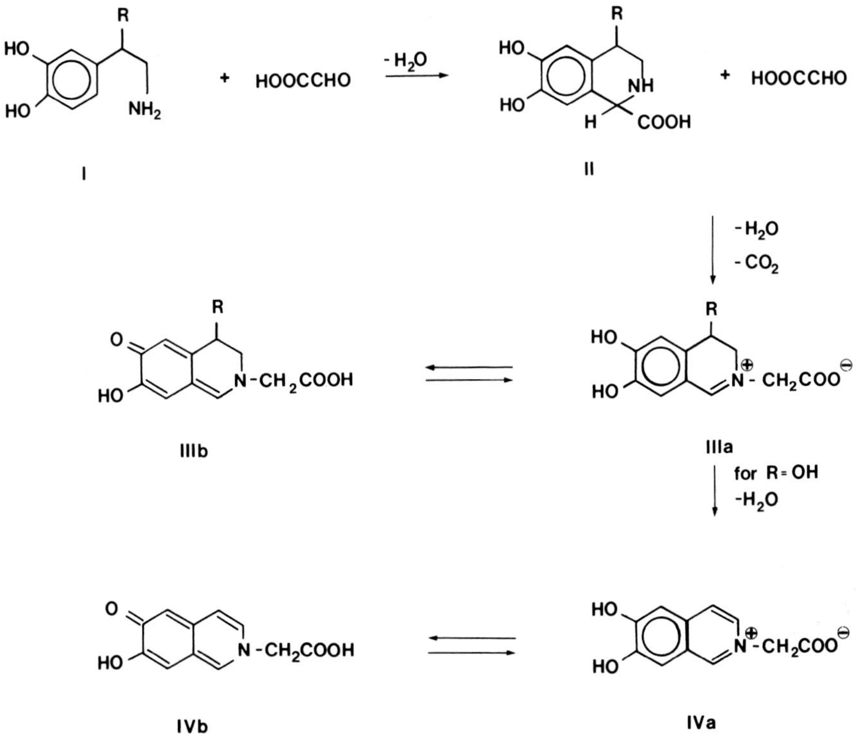

Fig. 7. Sequence of reactions between GA and the CAs, DA (R = H) and NA (R = OH). The CA (I) reacts with GA (HOOCCHO) in a Pictet–Spengler cyclization to form the weakly fluorescent 6,7-dihydroxy-1,2,3,4-tetrahydroisoquinoline-1-carboxylic acid derivative (II). The fluorophore formation then proceeds via an intramolecularly acid-catalyzed reaction with a second GA molecule, yielding the strongly fluorescent 2-carboxymethyl-6,7-dihydroxy-3,4-dihydroisoquinolinium compound (IIIa). This fluorophore is in a pH-dependent equilibrium with its tautomeric quinoidal form (IIIb). The NA fluorophore (R = OH) has a 4-hydroxy group that can be split off by acid, yielding the fully aromatic isoquinoline (IVa and b). From Lindvall *et al.* (1974*c*).

in the nonquinoidal forms (IIIa). By splitting off the 4-hydroxy group with acid from the dihydroisoquinoline fluorophores formed from β-hydroxylated phenylethylamines, such as NA, they can be converted into the fully aromatic compounds (IV in Fig. 7).

As demonstrated in model experiments, GA treatment induces stronger fluorescence than FA treatment from several phenylethylamines and indolylethylamines (Tables 1 and 2) (Axelsson *et al.*, 1973; Lindvall and Björklund 1974*a*; Lindvall *et al.*, 1974*c*). As in the FA reaction, the highest fluorescence yields in the GA reaction are obtained from 3-hydroxylated phenylethylamines and their corresponding amino acids. The strong fluorescence induced from the 3-methoxylated phenylethylamines, which are less activated at the point of ring closure, indicates that GA is more efficient than FA in the initial cyclization step.

Table 2. Fluorescence Yields and Spectral Characteristics of a Number of Biogenic Phenylethylamine and Indolylethylamine Derivatives Enclosed in Dried Protein Films after Glyoxylic Acid Treatment[a]

Substance group[b]	Substance	Heating[c] Relative fluorescence yield[e]	Heating[c] Maximum (nm) Excitation	Heating[c] Maximum (nm) Emission	GA vapor treatment[d] Relative fluorescence yield[e]	GA vapor treatment[d] Maximum (nm)[f] Excitation	GA vapor treatment[d] Maximum (nm)[f] Emission
Group A	1. *N,N*-Dimethyltryptamine	0	—	—	115	(340) 370	430
	2. *N,N*-Dimethyl-5-hydroxytryptamine (bufotenin)	0	—	—	35	(340) 370	430
Group B1	3. 3-Methoxy-4,β-dihydroxyphenylethylamine (normetanephrine)	0	—	—	35	330	470
	4. *N*-Methyl-3,4,β-trihydroxyphenylethylamine (adrenaline)	10	—	—	45	335 (370)	485
Group B2	5. 3,4-Dihydroxyphenylethylalanine (dopa)	130	415	480	570	330 and 380	480
	6. 3,4-Dihydroxyphenylethylamine (dopamine)	670	415	475	810	330 and 375	460[g]
	7. 3,4,β-Trihydroxyphenylethylamine (noradrenaline)	460	415	475	445	330 and 370	460[g]
	8. 3-Methoxy-4-hydroxyphenylethylamine (3-methoxytyramine)	0	—	—	80	330 and 370	470
	9. *N*-Acetyl-5-hydroxytryptamine	0	—	—	60	(335) 370	480
	10. *N*-Methyltryptamine	0	—	—	215	370	485
Group C	11. Tryptamine[β(3-indolyl)ethylamine]	25	—	—	305	370	495
	12. Tryptophan	0	—	—	375	370	500
	13. *N*-Methyl-5-hydroxytryptamine	0	—	—	80	370	500
	14. *N*-Acetyl-5-methoxytryptamine (melatonin)	0	—	—	80	(335) 370	500
Group D	15. 5-Hydroxytryptamine	10	—	—	125	375	520
	16. 5-Hydroxytryptophan	0	—	—	130	(320) 380	(410) 530
	17. 5-Methoxytryptamine	15	—	—	115	375	530

[a] Treatment at +80°C for 1 hr with paraformaldehyde equilibrated in air of about 50% relative humidity, according to the Falck–Hillarp method (see Björklund *et al.*, 1972*a*).
[b] Classification based on the spectral characteristics after GA vapor treatment. Data from Lindvall and Björklund (1974*a*).
[c] Reaction by heating (+100°C, 6 min).
[d] Reaction by GA vapor treatment (300 torr, +100°C, 3 min).
[e] The fluorescence intensities are expressed relative to the intensity obtained from noradrenaline and dopamine in the standard formaldehyde reaction as 100.
[f] A figure in parentheses indicates the position of a low peak or a shoulder in the spectrum.
[g] At higher concentrations, there is a shift of the emission-peak maximum to longer wavelengths (≈485 nm).

The high fluorescence yields in the GA reaction can be referred to several phenomena (see Björklund *et al.*, 1972*d*; Lindvall *et al.*, 1974*c*; Svensson *et al.*, 1975): (1) the formation of the strongly fluorescent 2-carboxymethyl-3,4-dihydroisoquinolinium and 2-carboxymethyl-3,4-dihydro-β-carbolinium compounds through intramolecular acid catalysis exerted by the carboxyl group on the 1-carbon of the tetrahydroisoquinoline or tetrahydro-β-carboline molecules (II → IV in Fig. 6; II → III in Fig. 7); (2) because acid catalysis is of importance in both steps of the fluorophore formation, GA is also able to promote the initial cyclization step (thus, the strong GA-induced fluorescence from 3-methoxylated phenylethylamines, which are less reactive in the Pictet–Spengler cyclization, may well be explained by a catalysis of this reaction step); (3) acidification is known to increase the fluorescence intensity of some amine fluorophores (Björklund *et al.*, 1968*b*, 1972*a*). This should also contribute to the higher fluorescence yields induced from these amines in the GA reaction.

GA treatment also induces fluorescence from N-acetylated indolylethylamines (e.g., melatonin and *N*-acetyl-5-hydroxytryptamine), i.e., from compounds that cannot undergo a Pictet–Spengler cyclization since they are amides and consequently do not yield fluorescence with FA (Corrodi and Jonsson, 1967). For this reason, the fluorophore formation from N-acetylated amines in the GA method must be due to a different type of reaction. The mechanism of fluorophore formation between GA and N-acetylated indolylethylamines has so far not been clarified, but it seems likely that it is based on a Bischler–Napieralski-type cyclization (see Axelsson *et al.*, 1972), i.e., a direct, acid-catalyzed cyclodehydration to strongly fluorescent 3,4-dihydro-β-carboline derivatives.

3.1.3. *Fluorophore Formation after Combined Formaldehyde and Glyoxylic Acid Treatment*

Recently, fluorescence-histochemical methods based on combined FA and GA reactions have been introduced (Battenberg and Bloom, 1975; Bloom and Battenberg, 1976; Lorén *et al.*, 1976*a*; Nygren, 1976; see also Lindvall *et al.*, 1975*a,b*). As described in detail below, the tissue is, in these methods, perfused with solutions containing a mixture of FA and GA. This is followed by immersion of the sections in a GA solution or exposure of the tissue to FA vapor or both. Also, in the Vibratome procedures used by Hökfelt *et al.* (1974*a*) and Berger *et al.* (1976), GA perfusion (sometimes followed by immersion in GA) is combined with FA vapor treatment. It is obvious that to identify the structure of the fluorophores formed in these methods, one will have to consider FA or GA as the fluorophore-forming reagent in the first or the second step, or both, of the histochemical reaction. This discussion will focus on the fluorophore formation from CAs and dopa because the combined FA and GA methods are useful mainly for the detection of these substances.

Figure 8A illustrates possible reaction pathways and resulting fluoro-

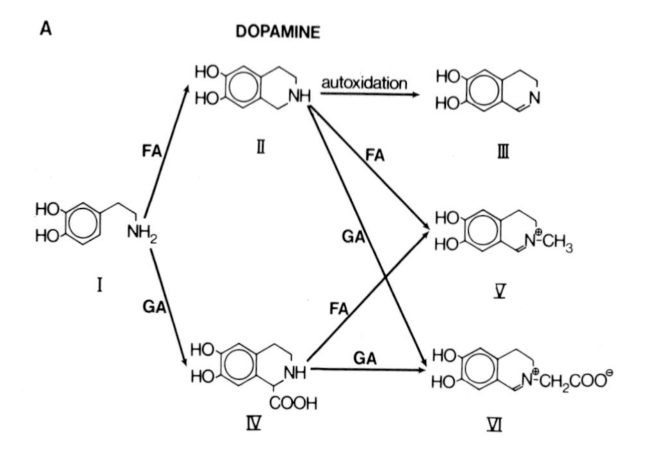

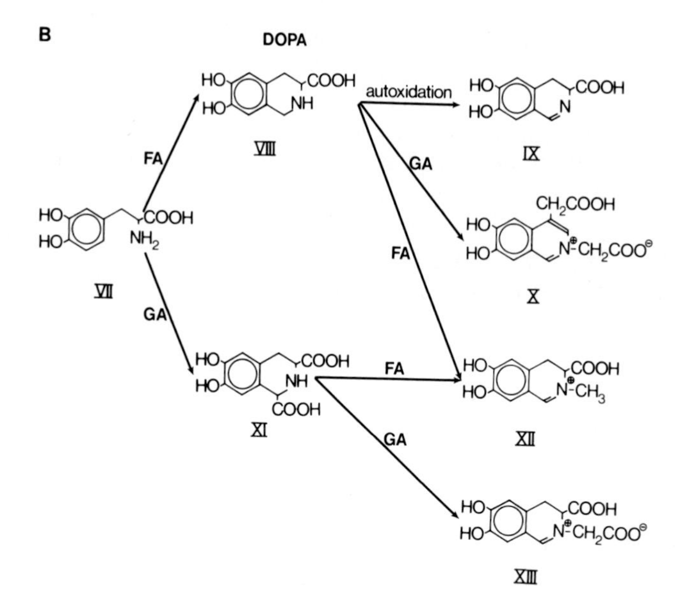

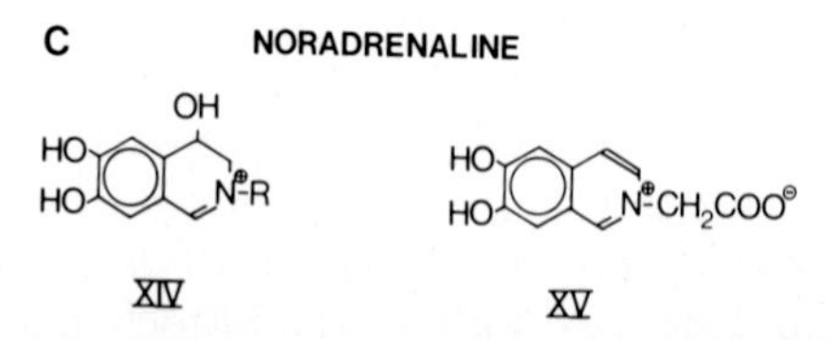

Fig. 8. Alternative reaction pathways for DA (A) and dopa (B) after exposure to both FA and GA. (C) Proposed molecular structure of the fluorophores formed from NA in similar reactions. For explanation, see the text.

phores for *dopamine* after exposure to both FA and GA (compare with Figs. 4 and 7) (cf. Lindvall *et al.*, 1975*b*). The first step of the histochemical reaction, the Pictet–Spengler cyclization, results in the formation of a tetrahydroisoquinoline derivative with (IV) or without (II) a carboxyl group in the 1-position. In the second step of the fluorophore formation, the tetrahydroisoquinoline is transformed into a strongly fluorescent dihydroisoquinoline. Autoxidation of the tetrahydroisoquinoline (II) yields the 3,4-dihydroisoquinoline (III), whereas from both tetrahydro derivatives, either the 2-methyl-3,4-dihydroisoquinolinium compound [(V) after reaction with FA] or the 2-carboxymethyl-3,4-dihydro-β-carbolinium compound [(VI) after reaction with GA] is formed. All these fluorophores show indistinguishable spectral characteristics. Thus, for the spectral identification of DA-containing structures, the combined FA and GA treatment presents no further problems as compared to treatment with FA or GA alone. In the histochemical reaction, condensation with GA in the cyclization step should be more favorable than with FA as regards the fluorescence yield. This is because the tetrahydroisoquinoline-1-carboxylic acid (IV) is transformed into the 2-substituted dihydroisoquinolines (V, VI) more efficiently than is the tetrahydroisoquinoline (II).

The fluorophore formation from *dopa* after combined FA and GA treatments is illustrated in Fig. 8B. When the reaction mechanisms of dopa are compared with those of DA, there is one major difference. If the tetrahydroisoquinoline (VIII) formed from dopa with FA is reacted with GA, a fully aromatic isoquinoline (X) is formed due to the presence of a 3-carboxyl group (Lindvall *et al.*, 1975*b*). As discussed in more detail in Section 4.2, the spectral characteristics of this fluorophore differ markedly from those of the other fluorophores formed from dopa. These latter fluorophores are formed in reactions analogous to those of the DA fluorophores (compare Fig. 8A and B) and show similar excitation and emission spectra. The formation of a fully aromatic fluorophore from dopa but not from DA forms the basis for a method of differentiation between these compounds by spectral analysis. The reaction conditions necessary for a quantitative conversion of dopa to the fully aromatic compound have been defined in histochemical protein models (Lindvall *et al.*, 1975*a,b*). The extent to which dopa is transformed into the fully aromatic isoquinoline or to the dihydroisoquinolines in available methods using combined FA and GA treatments is not known at present.

The fluorophores formed from *noradrenaline* in the combined FA and GA reactions are similar in structure to the dihydroisoquinolines formed from DA and dopa (XIV in Fig. 8C). However, under certain well-defined reaction conditions (see Section 4.2.3 and Lindvall *et al.*, 1975*a*) involving a reaction with FA in the first step and with GA in the second step of the fluorophore formation, NA is transformed into a fully aromatic isoquinoline (XV). This offers a new possibility for differentiation between DA and NA at the cellular level, although the procedure has not yet been adapted for tissue sections.

For the visualization of the secondary CA, *adrenaline,* the combined FA and GA technique offers no advantages as compared to the standard FA method. As pointed out in Section 3.1.1, the second step of fluorophore formation from adrenaline is dependent on the oxidative capacity of the reaction milieu. Furthermore, it should be noted that the fluorescence yield from adrenaline in the GA reaction and in the acid-catalyzed FA reaction is very low. The highest fluorescence yields are obtained when adrenaline is reacted according to the standard FA method for several hours (Falck *et al.,* 1963; Norberg *et al.,* 1966). The fluorophores formed from adrenaline after combined FA and GA treatment are most probably identical with those formed when either reagent is used alone.

The fluorescence yield from structures containing *indoleamines* after combined FA and GA treatment is similar to or lower than that after FA or GA treatment alone. Indoleamines such as tryptamine and 5-HT have been shown to enter fluorophore-forming reactions analogous to those of DA in the combined FA and GA reactions (see Fig. 8A) (Lindvall *et al.,* 1975*b*). Similar to dopa (see Fig. 8B), the corresponding amino acids, 5-hydroxytryptophan and tryptophan, yield the fully aromatic fluorophores under certain reaction conditions. This forms a basis for differentiation between indoleamines and their corresponding amino acids (see Section 4.2.3) (Lindvall *et al.,* 1975*b*).

3.2. Practical Performance of the Formaldehyde and Glyoxylic Acid Methods

In the original FA method, as described by Falck, Hillarp, and coworkers (Falck, 1962; Falck *et al.,* 1962; Corrodi and Hillarp, 1963, 1964), freeze–dried tissue pieces or air-dried thin tissue sheets are exposed to FA vapor. Originally, it was considered fundamentally important for the high sensitivity of this method that all steps in the histotechnical procedure be performed under dry or nearly dry conditions. During the last few years, several fluorescence-histochemical techniques based on the application of the reagent in solution have, however, been worked out. In some cases, these methods have considerably higher sensitivity for CA-containing structures than the standard FA method. This has proved to be the case, for example, for the GA and FA techniques applied to Vibratome sections (Hökfelt and Ljungdahl, 1972; Lindvall and Björklund, 1974*a*), the GA methods for cryostat sections (Bloom and Battenberg, 1976; Watson and Barchas, 1975; de la Torre and Surgeon, 1976; Nygren, 1976), the acid-FA–GA–$MgSO_4$ technique (Lorén *et al.,* 1976*a*), and the aluminum–formaldehyde method (Lorén *et al.,* 1980*a*).

This account will give a survey of available fluorescence-histochemical methods for biogenic monoamines based on FA and GA condensation. For further technical details, the reader will be referred to the appropriate literature, and the description will, in addition to giving the principles of the methods, be focused on the usefulness, drawbacks, and advantages of each technique. It should be remembered that the best method for demonstration of monoamines in a certain type of experiment is not always the most sensitive one. For example, when serial sections of large tissue pieces are needed, the freeze–dry methods using paraffin sections are more useful than the Vibratome techniques, with which only a limited number of sections can be obtained. The methods presented below should therefore be regarded as alternative techniques, each being useful for different types of problems.

3.2.1. Methods Applied to Freeze–Dried Specimens

3.2.1a. Standard Formaldehyde Method. This method is described below.

Preparation of tissue. The animal is killed and the organs are dissected out as quickly as possible. However, especially with peripheral tissue, good results can be obtained in the standard FA method from slaughterhouse preparations as long as about 1 hr after the death of the animal (El Badawi *et al.*, 1970; Björklund *et al.*, 1972*c*). To avoid the formation of ice-crystal artifacts during the freezing procedure, the tissue piece is frozen rapidly to a very low temperature ("quenching"). The growth of ice crystals ceases at temperatures below −30 to −50°C. If this temperature range is rapidly passed by quenching to lower temperatures, only submicroscopic ice crystals are formed. It is not possible, however, to drop the specimen directly into liquid nitrogen because the transfer of heat from the preparation is very much reduced by the formation of a layer of vaporized nitrogen on its surface. Freezing is therefore performed by the use of various intermedia (Moline and Glenner, 1964; Pearse, 1968). In our laboratory, a mixture containing propane and propylene in a proportion of roughly 9:1 is used as an intermedium. The specimens can be stored in liquid nitrogen for prolonged periods until they are freeze–dried.

Freeze–drying. Freeze–drying implies the removal of water from rapidly frozen tissue by sublimation at a temperature below the freezing point. A number of good models of freeze–dryers are available (cf. Pearse, 1968; Olson and Ungerstedt, 1970*b*; Björklund *et al.*, 1972*c*; Baumgarten, 1972), and many of these models, of course, fulfill the present demands. The two types of freeze–dryer used in our laboratory have previously been described in detail by Björklund *et al.* (1972*c*). This paper should be consulted for further information on the construction and operation of the apparatuses.

Formaldehyde treatment. The freeze–dried specimens are transferred to a 1-liter glass vessel containing an excess amount (about 5 g) of paraformaldehyde and heated in an oven at +80°C [primary CAs can react at lower temperatures (Norberg *et al.*, 1966; Håkanson and Sundler, 1974), but a temperature of +80°C is used routinely]. During heating, the paraformaldehyde is depolymerized to gaseous FA. Besides the specific condensation reaction with the monoamines that yield the intensely fluorescent derivatives, the FA treatment also gives a mild but sufficient fixation of the tissues. Primary CAs and 5-HT develop a maximum fluorescence after 1–2 hr, but for adrenaline, it is necessary to continue the FA treatment for up to 3 hr.

In the two modifications of the FA method, the FA–ozone and the acid-catalyzed FA treatments, the histochemical reaction is, briefly, carried out as

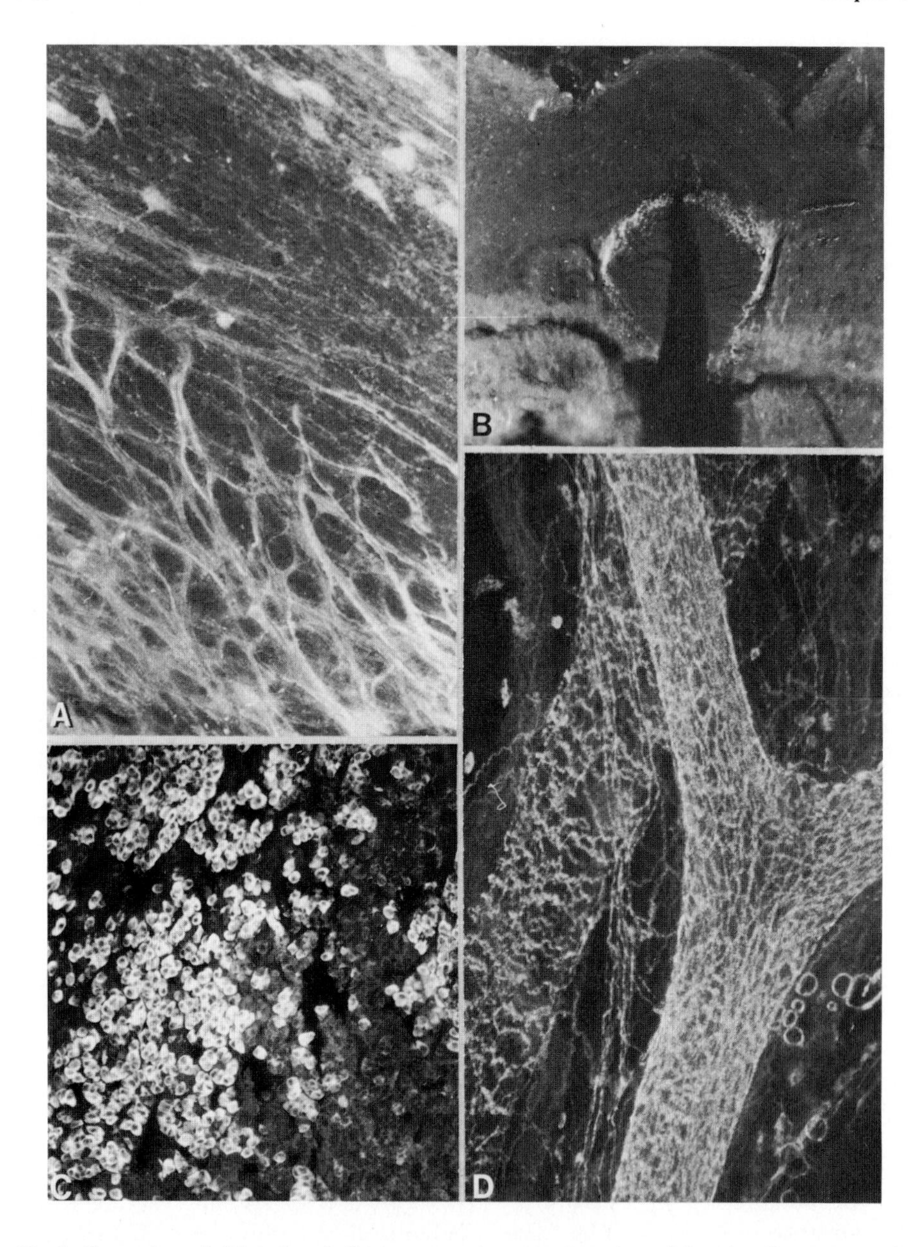

Fig. 9. Examples of FA-induced fluorescence in central and peripheral cell systems. (A) Indoleamine-containing axons and cell bodies in the mesencephalic raphe region of the rat. (B) Indoleamine-containing nerve terminals in the subcommissural organ of the rat. (A, B) From paraffin sections of freeze–dried tissue processed according to the standard Falck–Hillarp method. In (A), the intraaxonal indoleamine fluorescence was increased by an intraventricular injection of 5,7-dihydroxytryptamine (150 μg), 2 days before killing, followed by nialamide (300 mg/kg), 3 hr before killing. From Björklund *et al.* (1973*c*). In (B), the rat had been treated with nialamide alone (300 mg/kg). (C) FA–ozone-induced fluorescence from NH_2-terminal tryptophanyl peptides in cells of the pig adenohypophysis. The specimen was freeze–dried and treated with FA gas and ozone according to Björklund and Falck (1969). (D) Fluorescent NA-containing

follows (for details, see Björklund *et al.*, 1968*b*, 1971*b*): The *FA–ozone treatment*, which makes possible a sensitive and selective demonstration of tryptamine and peptides containing NH_2-terminal tryptophan (Fig. 9C), is performed in a 1-liter glass vessel containing 5 g paraformaldehyde. In our work, ozone is generated with a Tesla coil between two electrodes carried and insulated by the thick Teflon lid of the glass vessel. With this equipment, optimum ozone concentration is obtained in the vessel after 15–20 min electrical discharge. The sealed jar is then transferred to an oven and heated at +80°C for 1 hr. The *acid-catalyzed FA treatment* also induces strong fluorescence from certain indolylethylamines and 3-methoxylated phenylethylamines that are low-reactive in the standard FA method. The procedure is carried out with the specimens together with 5 g paraformaldehyde in a 1-liter glass vessel provided with a multi-socket lid connected to a manometer. The reaction vessel is first evacuated, and HCl-saturated air is then sucked into the reaction vessel from a 1-liter vessel containing 200 ml 6 N HCl solution. Normal air is then let into the reaction vessel to atmospheric pressure, and finally the vessel is heated in an oven at +80°C for 1 hr. In case of whole freeze–dried specimens, two succeeding acidifications should be performed. After the first introduction of HCl–air, the reaction vessel is again evacuated and new HCl–air is sucked into the vessel. Both the FA–ozone technique and the acid-catalyzed FA treatment can also be applied to deparaffinized sections from freeze–dried tissues and to cryostat sections.

The *water content* of the paraformaldehyde is of critical importance for the outcome of histochemical reaction (Falck and Owman, 1965; Hamberger *et al.*, 1965; Hamberger, 1967). Thus, batches of paraformaldehyde, previously dried by being heated at +100°C for 1 hr, are stored in desiccators at a constant relative humidity for at least 5–7 days before use. Adequate desiccation can be ensured by using various concentrations of sulfuric acid in the desiccators. The strongest reaction is obtained with FA that has been stored at a high relative humidity (80–95%), but then the fluorescent structures are often diffuse. If the paraformaldehyde has been stored at low relative humidities (below 30%), the fluorescence, if it develops at all, is distinct but very weak. In our laboratory, optimum results are obtained if the paraformaldehyde is stored over sulfuric acid giving a relative humidity of about 50–70%. It should be noted that various types of tissues, and even various parts of the same tissue, may behave differently toward different types of paraformaldehyde. The outcome of the reaction also depends on the degree of dryness of the specimen after freeze–drying. It is therefore often advisable to test several pieces of the tissue to be studied, obtained from the same dryer, in different paraformaldehydes and to choose the one that gives the optimum reaction. The fluorescence intensity of 5-HT can be increased by two subsequent FA exposures, one involving a drier, the other a more humid, paraformaldehyde (Fuxe and Jonsson, 1967).

---

sympathetic nerves around vessels in a whole-mount preparation of a rat mesentery. This picture is from one of the first specimens processed according to the FA vapor technique in 1961. From Falck (1962).

Embedding and sectioning. The specimens should be embedded in degassed paraffin wax as soon as possible after FA treatment. [If very thin sections are desired, an alternative procedure using embedding in epoxy resin can be used for sectioning with glass knives (Hökfelt, 1965).] Embedding is preferably performed *in vacuo* to ensure complete and rapid penetration of the dry tissue (for details, see, for example, Björklund *et al.*, 1972*c*). After the specimen is blocked in paraffin, it is sectioned and mounted on dry slides in, for example, Entellan (Merck, Darmstadt), Fluoromount (E. Gurr, London), liquid paraffin, immersion oil, or xylene. Mounting usually has a negligible immediate effect on observable fluorescence intensity, but is often necessary to avoid light-scattering in the sections that might be misinterpreted as specific fluorescence.

Generally, the fluorescence intensity is not affected by the embedding and mounting procedures. However, for unknown reasons, fluorophores in certain locations are easily extracted by, for example, hot paraffin and some organic solvents. This phenomenon has been observed for adrenaline in certain cell systems (Falck *et al.*, 1963; Owman and Sjöstrand, 1965) and for the primary CAs present in certain pituitary and pancreatic nerves (Björklund, 1968; Björklund and Falck, 1968). This fluorophore extraction, which can sometimes result in a total loss of observable fluorescence, can be avoided by special precautions in the embedding and mounting steps (Björklund and Falck, 1968).

When enclosed in paraffin blocks, tissues from peripheral organs can be stored for a considerable time even at room temperature. Central nervous tissue is much less durable and should usually be analyzed within a few weeks. The fluorescence in deparaffinized sections shows a notable fading within a few days with a concomitant and disturbing increase in unspecific background fluorescence. Storage of paraffin blocks, nondeparaffinized sections, and mounted sections at low temperatures (down to −20°C) and in darkness postpones these changes.

Comments on usefulness. Even though several modifications of the FA method and new techniques based on condensation with GA have appeared during the last few years, the original Falck–Hillarp method applied to freeze–dried tissue is still a most useful fluorescence-histochemical technique for many experimental problems. The virtues of this method, and its special fields of application, are exemplified in the following discussion.

1. The use of *freeze–dried tissue* in the FA method has several advantages. First, in this procedure, many specimens can be processed simultaneously. Second, several specimens can be taken from the same animal and can be stored, either in liquid nitrogen before the freeze–drying procedure or after the paraffin embedding. Third, many sections of an even thickness can be obtained from each specimen. Also, the sectioning procedure is much easier for this tissue preparation than for Vibratome sections.

2. For *indoleamine-containing neurons* in the CNS, the standard FA method together with the aluminum–formaldehyde (ALFA) technique applied to

freeze–dried tissue are the most sensitive techniques. Although the ALFA technique often gives a more sensitive and reproducible visualization of central 5-HT neurons, the standard FA method is sensitive enough for many studies on these neurons (Fig. 9A and B). Furthermore, the difficulties in distinguishing between CA and indoleamine neurons in the fluorescence microscope are much less with the standard FA method than with the ALFA technique.

3. For *central CA neurons,* the field of application for the standard FA method is more limited. Thus, for neuroanatomical studies on the finer organization and morphology of these neurons, the more sensitive techniques (see below) should be used. On the other hand, the standard FA method can be recommended in, for example, the following situations: First, for studies on the effects of various pharmacological agents and other experimental manipulations in central, as well as peripheral, CA neurons. Freeze–dried specimens are also used for semiquantitative or microfluorometric quantitation of CAs in cell bodies and terminals (Lichtensteiger, 1970; Lidbrink and Jonsson, 1971; Löfström *et al.,* 1976*b*). This requires large series of sections of an even thickness and also—to facilitate comparison among different experimental groups—that several specimens be processed simultaneously under identical conditions. For reliable quantitation of the fluorescence intensity by microspectrofluorometry, these requirements, which are fulfilled in the standard FA method, are of particular importance. Second, the standard FA method is convenient for evaluation of the extent of various types of lesions and the localization of stimulation electrodes with reference to the distribution of the monoamine systems. Such areas are difficult to section with other techniques, particularly on the Vibratome instrument. If highly sensitive visualization of the CA systems is of importance and the tissue can be perfused, the specimens can alternatively be processed according to the ALFA technique (see Section 3.2.1b).

4. *Peripheral tissue* is, in many cases, much more difficult to section in the Vibratome than CNS tissue. Therefore, the use of freeze–dried, paraffin-embedded tissue or cryostat procedures is, with regard to the sectioning procedure, more favorable and sometimes absolutely necessary. However, both the ALFA method and the GA method give a more sensitive visualization of peripheral CA neurons than the standard FA method. Thus, when perfusion of the experimental animal is possible and optimum sensitivity for CA neurons are needed, either of these methods should be employed.

3.2.1b. Perfusion with an Acid Formaldehyde Solution with High Aluminum Content (The ALFA Method). The sensitivity and reproducibility of the standard FA method in the visualization of central CA neurons can be greatly increased by perfusing the animal with a solution containing FA and GA in the presence of a very high magnesium content and at acid pH (Lorén *et al.,* 1976*a,b*; Figs. 9 and 10). The perfused tissue is then processed according to the standard FA method (see Section 3.2.1a). However, it was recently shown (Lorén *et al.,* 1980*a*) that the sensitivity can be further increased if magnesium

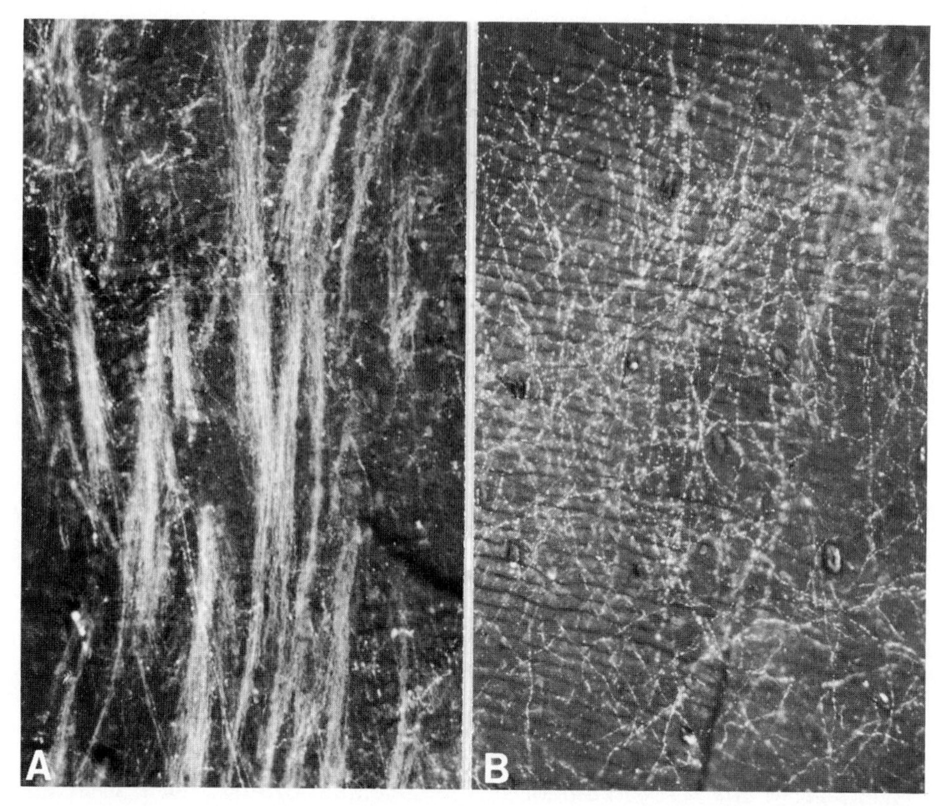

Fig. 10. Central CA-containing nonterminal axons in the medial forebrain bundle (A) and axon terminals in the cingulate cortex (B). The sections were obtained from freeze–dried tissue perfused with an acid FA–GA solution with a high $MgSO_4$ content according to Lorén *et al.* (1976*a*).

is replaced by aluminum in the perfusion solution. The addition of FA seems to improve the fluorescence yield whereas GA in the perfusion solution does not involve any improvements of the technique. The ALFA method also gives excellent results in developmental studies (Lorén *et al.*, 1980*b*).

Recommended procedure for adult brains. The rat is perfused in two steps via the ascending aorta under barbiturate anasthesia. The perfusions should be made under controlled perfusion pressure. Two pressurized perfusion reservoirs with a common outflow for the perfusion solution and a common inflow for the highly compressed air are used (see Lorén *et al.*, 1980*a*).

The preperfusion is made with either of two solutions. (*a*) *Buffer alone:* The animal is perfused with 150–250 ml of ice-cold Tyrode's buffer. The perfusion volume is 150 ml when the descending aorta is clamped (perfusion of brain and upper spinal cord) and 250 ml when the descending aorta is open (perfusion of brain plus the entire spinal cord). The perfusion pressure should be kept at 0.6–0.8 bar/cm^2. Check that the outflowing perfusion fluid is free of blood by the end of the perfusion. (*b*) *2% GA:* 2 g of glyoxylic acid monohydrate is dissolved in 100 ml of the above buffer, and pH is adjusted to 7 with 10 M NaOH. The perfusion is performed as above at a pressure of 0.4–0.6 bar/cm^2.

The aluminum perfusion is performed with either of two alternative solutions. (*a*) *Aluminum alone:* Dissolve 10 g $Al_2(SO_4)_3 \cdot 18H_2O$ per 100 ml of Tyrode's buffer. The salt should be dissolved in the ice-cold buffer immediately before use. The dissolution of the $Al_2(SO_4)_3$ crystals occurs slowly but can be speeded up by stirring and mechanical crushing of the crystals. The pH of the solution is adjusted to about 3.8 by adding 0.38 g $Na_2B_4O_7 \cdot 10H_2O$ per 100 ml. Check that the solution is completely clear. (*b*) *Aluminum plus FA:* Dissolve 20 g of paraformaldehyde in 500 ml distilled water with 4–6 drops of 1 M NaOH added. The solution is allowed to stand in a boiling water bath until all paraformaldehyde is dissolved. This solution is cooled to room temperature and then mixed with 500 ml of ice-cold Tyrode's buffer of double strength. When ice-cold, 10 g of $Al_2(SO_4)_3 \cdot 18H_2O$ per 100 ml is added and the pH is adjusted with sodium borate, as in (a). Check that the solution is completely clear.

When switching from the preperfusion (i.e., Tyrode's buffer or the 2% GA solution) to the aluminum perfusion (which should be made without introducing air bubbles), the perfusion pressure is gradually raised to about 2.0 bar/cm^2. The perfusion volume is 300 ml with clamp over the descending aorta and 450 ml without clamp. The entire perfusion is completed within about 30 sec. Marked swelling of the neck and head region is a bad sign and signifies that intravascular precipitation has taken place. When using the GA containing preperfusion solution, swelling will occur also when the perfusion pressure is too high.

After perfusion, the specimens are frozen, freeze-dried, treated with FA vapor, embedded in paraffin *in vacuo*, and sectioned, according to the standard FA method (see Section 3.2.1a). The FA vapor treatment is performed using paraformaldehyde equilibrated in air of about 50–70% relative humidity.

Recommended procedure for young brains. It is essential to have a pressure perfusion system also for studies on immature brains. The animals are perfused in two steps via the left ventricle. The *preperfusion solution* consists of Tyrode's buffer supplemented with 2 g of $MgSO_4 \cdot 7H_2O$ and 1 g of procaine per liter. This solution is used at room temperature. The second, or *main perfusion solution,* is made as follows. 20 g paraformaldehyde is suspended in 500 ml distilled water with 4–6 drops of 1 M NaOH added. The solution is heated in a boiling-water bath until all paraformaldehyde is dissolved and then cooled to room temperature. To this solution is added 500 ml of ice-cold, double-strength Tyrode's buffer, and the mixture is kept in ice. Just prior to use, 100 g $Al_2(SO_4)_3 \cdot 18H_2O$ and 3.8 g $Na_2B_4O_7 \cdot 10H_2O$ are dissolved, and the final solution (pH about 3.8) is filtered.

The preperfusion solution is introduced with a pressure of 0.3–0.5 bar/cm^2 and continued for 45–60 sec in order to rinse out all blood and establish an even perfusion. After this has been completed the needle valve is rapidly shifted to introduce the ice-cold main perfusion solution. The perfusion is continued for 2–3 min, during which the pressure is gradually increased to 0.6–2.0 bar/cm^2 (the higher pressures are used with older rats). The perfused volumes of the pre- and main perfusion solutions are at least 20 and 80 ml, respectively. The brains are rapidly dissected and processed as described above for adult brains.

Comments on usefulness. In comparison with the standard FA method, the ALFA technique has a considerably higher sensitivity in the detection of both central and peripheral CA-containing structures. In the CNS, the nonterminal portions of the CA axons, which are poorly detectable in specimens by the Falck–Hillarp procedure, become visualized throughout their full extent. This allows the tracing, in the intact untreated animal, of CA axon pathways such as the dorsal tegmental bundle, the central tegmental tract, and the nigrostriatal bundle. The ALFA technique also demonstrates CA systems that are only partly detectable or cannot be visualized with the Falck–Hillarp method. This is the case with the CA systems recently discovered with the GA–Vibratome method, e.g., the periventricular CA system (Lindvall *et al.*, 1974*a*; Lindvall and Björklund, 1974*b*), the incerto-hypothalamic DA system (Lindvall *et al.*, 1974*a*; Björklund *et al.*, 1975*b*), and the DA systems in the lateral septal nucleus (Lindvall, 1975) and the cortical areas (Lindvall *et al.*, 1974*b*) (see Sections 3.2.2b and 3.2.5f).

The ALFA method is particularly useful for studies on the noradrenergic neurons in the locus coeruleus. Both their axon pathways and terminal systems (in thalamus, hippocampus, and cerebral and cerebellar cortices) are demonstrated with high intensity, precision, and reproducibility. Another attractive feature of the ALFA method is that the fluorescence is usually without any signs of diffusion. This makes the fluorescence picture of the CA-containing structures very distinct and rich in detail. In addition, the general background fluorescence tends to be lower in the perfused specimens, thus improving the contrast between the specifically fluorescent structures and the background.

The improved sensitivity for CA neurons of the ALFA method can be attributed to a combination of several phenomena. First, the aluminum ions themselves have a positive effect on the fluorescence intensity in the CA-containing structures. This might be explained, at least partly, by a "locking-in" effect of the aluminum ions on the intracellular CAs, preventing their spontaneous release. The extreme sharpness and lack of diffusion in the fluorophore localization might be the result of such an action, perhaps in combination with the dehydrating effect of the hypertonic perfusion. From model experiments, it also seems possible that the aluminum ion has a direct effect on the fluorescence yield from the fluorophores formed in the histochemical reaction (Björklund *et al.*, 1980). Second, the acid perfusion solution lowers the pH at the reaction site in the tissue. This probably assists in creating optimal reaction conditions, perhaps through an acid catalysis of the Pictet–Spengler cyclization step in the reaction between FA or GA and the CAs (cf. Section 3.1.1). Third, the presence of the histochemical reagent in the perfusion medium results in a further improvement in the fluorescence intensity in the CA-containing structures.

Provided that the tissue can be perfused, the ALFA technique has several distinct advantages over other methods for CA visualization:

1. Compared to available methods based on freeze–drying, this method is much more sensitive for both NA- and DA-containing structures, and it is superior in its precision and richness in detail. This has been evaluated particularly in the CNS (Lorén *et al.*, 1980*a*), but it has been found that the method also gives similar results in many peripheral organs (Ajelis *et al.*, 1979).
2. Compared to the methods based on Vibratome sectioning, the freeze–drying and paraffin sectioning used in this method have the advantage of a more sensitive visualization and better preservation of the CA-containing cell bodies (see Section 3.2.2). Furthermore, such regions in the CNS (e.g., the spinal cord and superficially located structures) and such types of tissue (e.g., many peripheral organs or organs of small size), which are difficult to section in the Vibratome, are readily accessible to the ALFA technique.
3. The ALFA perfusion procedure has a sensitivity for NA- and DA-containing structures that is rather comparable to that of the GA–Vibratome method. Thus, the perfusion technique can be viewed as a good alternative to the GA–Vibratome method, having several practical advantages: As pointed out in Section 3.2.1a, the freezing and freeze–drying of the specimens make possible the parallel processing of many specimens or several regions from one and the same animal. Also, this mode of processing allows convenient storage of specimens for longer periods of time. Such problems of processing and storage of specimens have so far not been solved for the GA–Vibratome method. However, for neuroanatomical studies on central CA neurons, the ALFA and the GA–Vibratome techniques will be regarded as complementary, the latter method having several special advantages.
4. Studies on fetal and neonatal rats have shown that the perfusion procedure can be applied with excellent results to the brains of such animals (Lorén *et al.*, 1976*b*, 1980*b*). The ALFA perfusion technique is the method of choice for ontogenetic studies in the CNS, particularly since the brains of young animals are difficult to section on the Vibratome. When applied to immature brains in combination with systemic injections of α-methylnoradrenaline, even catecholamine systems with very low transmitter content are visualized (Loren *et al.*, 1980*b*). This procedure greatly improves the possibilities for studies on the early development of the catecholamine neuron systems.

3.2.2. *Methods Applied to Vibratome Sections*

3.2.2a. Formaldehyde Method. This method, introduced by Hökfelt and Ljungdahl (1972), uses sections from fresh or perfused tissue obtained with the Vibratome microtome (manufactured by Oxford Instruments, San Mateo, California). The FA method applied to Vibratome sections is carried out in four principal steps (Hökfelt and Ljungdahl, 1972): (1) perfusion with an FA solution (when unfixed tissue is wanted, this step can be omitted); (2) sectioning with a Vibratome instrument; (3) drying; and (4) treatment with gaseous FA.

Perfusion. The animals are perfused via the ascending aorta with an ice-cold 4% FA solution prepared according to Pease (1962) in 0.1 M phosphate buffer. Hökfelt and Ljungdahl (1972) recommend that during the perfusion the animal be immersed in a mixture of water, ice, and sodium chloride with a temperature of about −2°C. It has been found preferable to perfuse the animal for less than 15 min. Longer perfusion times reduce the fluorescence intensity in the CA-containing structures. The perfusion has to be performed at low temperature; otherwise, the specific amine fluorescence is abolished.

Sectioning. Either fixed (FA-perfused) or unfixed (nonperfused) tissue can be used. In general, the sectioning procedure is easier with fixed tissue—thinner sections with a larger surface area and of more nearly equal thickness can be obtained. It should be pointed out that brain areas of homogeneous structure are favorable sectioning material, but even with this material it may be difficult to obtain uniform sections, particularly of unfixed brain. Less homogeneous tissues may present even more difficulties or be impossible to cut thin enough. Generally, smaller pieces are easier to section than large ones.

The sectioning is carried out with the tissue piece immersed in an ice-cold (up to +4°C) Krebs–Ringer phosphate or Tyrode's buffer. For details on the cooling device used in our laboratory, see Lindvall and Björklund (1974*a*). With unfixed brain tissue, sections with a thickness down to 20 μm can be obtained in favorable cases. From FA-fixed tissue, which is easier to cut, it is possible to get sections with a thickness down to 10 μm. After being cut, the sections are placed on microscope slides.

Drying and FA treatment. The sections are dried under a cool and dry air stream and kept overnight *in vacuo* in a dessicator containing fresh phosphorous pentoxide. The sections are then exposed at +80°C for 1 hr to FA vapor generated from paraformaldehyde equilibrated in 70% relative humidity. After treatment, the sections are mounted in xylene, paraffin oil, or immersion oil.

ALFA technique applied to Vibratome sections. A further improved sensitivity in the visualization of central CA neurons is obtained (Lorén *et al.*, 1980*a*) if the sections are transferred immediately after being cut to an ice-cold immersion bath containing 2.5 g $Al_2(SO_4)_3 \cdot 18H_2O$ in 150 ml Tyrode's buffer. After immersion for 15–30 sec, the sections are dried in warm air and then exposed to FA vapor as above.

Comments on usefulness. In combination with gaseous FA treatment, either with or without previous perfusion of the tissue with an FA solution, the Vibratome-sectioning procedure has several advantages for the visualization of central CA neurons:

1. The Vibratome procedure is easier to perform than freeze–drying and requires less equipment. Furthermore, specimens are more rapidly

available for fluorescence microscopy when processed according to the Vibratome technique. Also, in the Vibratome sections, cracks and other freeze–drying artifacts are avoided. On the other hand, the sectioning on the Vibratome of fresh or perfused tissue is much more time-consuming than for paraffin-embedded tissue on an ordinary microtome, and considerably fewer sections are obtained from each specimen. In addition, it is not always easy to obtain thin sections, especially when sectioning unfixed tissue of regions with a heterogeneous architecture, e.g., containing ventricle systems or consisting of a mixture of white and gray matter. Also even under favorable circumstances, the thickness of one section may vary among its different parts. In looking at the fluorescence-microscopic picture, it can be seen that the morphology of CA cell bodies is clearly better preserved in sections from freeze–dried tissue than in Vibratome sections, in which it is often distorted.

2. FA treatment of Vibratome sections gives a more sensitive visualization of central CA neurons than is obtained in the standard FA method (see Section 3.2.1a). This is observed in many brain regions, e.g., in the cerebral cortex, where the fine CA systems are seen more distinctly and frequently as compared to material processed according to the standard procedure. For indoleamine neurons, the FA–Vibratome technique does not represent any obvious advantage, however.

Keeping these advantages and limitations of the FA–Vibratome technique in mind, the following fields of application can be advanced: First, the method is useful for neuroanatomical studies on central CA neurons. Not only does the high sensitivity make it possible to trace fibers difficult to visualize with the standard method, but also the absence of fractures and cracks in the sections is of great value both for microscopic work and for photography. In comparatively thick Vibratome sections, axons can be followed for long distances, which is favorable when tracing CA tracts. Second, the Vibratome sections are useful for correlative studies with other histochemical techniques (see Section 3.2.5). It has thus been suggested (Hökfelt and Ljungdahl, 1972) that it should be possible to combine, on consecutive Vibratome sections or even on the same section, visualization of CA neurons by FA treatment with cholinesterase-staining (Hökfelt *et al.*, 1974*b*), with the Fink–Heimer silver impregnation for degenerating boutons, or with autoradiography using labeled neurotransmitters such as γ-aminobutyric acid (see Hökfelt and Ljungdahl, 1970, 1971). Vibratome sections have also been used for combined immunofluorescence of CA-synthesizing enzymes and retrograde horseradish peroxidase tracing (Ljungdahl *et al.*, 1975) (see Section 3.2.5d).

As described in detail in Section 3.2.4, monoamine fluorescence can easily be produced if the tissue specimens are exposed to a mixture of formaldehyde and glutaraldehyde (Faglu) (Furness *et al.*, 1977*b*). This principle can also be applied to CNS tissue (Furness *et al.*, 1977*a*). The animals are then perfused with the Faglu mixture and the tissue pieces sectioned on the Vibratome instrument. During sectioning, the surface of the tissue slice is kept moist with the Faglu mixture. The sections are transferred to a dish of the Faglu mixture and then dried and mounted. With this technique, CA-containing cell bodies and axons are well demonstrable (Furness *et al.,* 1977*a*). The major advantage with this technique is that the sections should be well suited for electron microscopy (see Section 3.2.5g).

3.2.2b. Glyoxylic Acid Method. The procedure as described by Lindvall and Björklund (1974*a*) is performed in five steps: (1) perfusion with a GA solution; (2) sectioning with the Vibratome instrument; (3) immersion of the sections in a GA solution; (4) drying; and (5) heating or GA vapor treatment.

Perfusion. The animal is perfused rapidly (during $\frac{1}{2}$–1 min) via the ascending aorta with 150 ml (for adult rat) of an ice-cold GA solution. The perfusion solution is prepared by dissolving GA monohydrate to a concentration of 2% in an ordinary Krebs–Ringer bicarbonate buffer. The pH is thereafter adjusted to 7.0 by the addition of NaOH.

The perfusion of the animal has been found to improve both the consistency of the tissue piece, making it easier to dissect and to section (see below), and the fluorescence-microscopic picture. Thus, in the CNS, the fluorescence morphology of the CA cell bodies (which are easily distorted during the sectioning procedure) improves markedly and the general tissue fluorescence is lower than in nonperfused brains. However, also in nonperfused tissue, a highly sensitive demonstration of CA neurons is obtained. This is of importance for studies of tissue that is inaccessible to perfusion, e.g., in man. Nonperfused tissue has been used with excellent results in the GA–Vibratome method for demonstrations of the CA innervation in the human ovary (Owman *et al.,* 1975) and cerebral cortex (Berger *et al.,* 1974).

For certain purposes, e.g., incubation studies on Vibratome sections, as described in Section 3.2.5b, the animal should be perfused with an ice-cold Krebs–Ringer bicarbonate buffer containing no GA. Although GA perfusion gives optimum results, the tissue perfused with buffer alone also shows improved fluorescence-microscopic picture and consistency as compared to nonperfused brains.

Sectioning. The sectioning procedure essentially follows that introduced by Hökfelt and Ljungdahl (1972). After perfusion, the tissue is rapidly taken out and cooled with ice-cold Krebs–Ringer, and the desired piece is dissected out. The tissue piece is sectioned in the Vibratome instrument and, during this procedure, immersed in the ice-cold Krebs–Ringer bicarbonate buffer. The low temperature (between 0 and +5°C) of the sectioning bath is essential; otherwise, the CA fluorophores diffuse away from the nervous structures, and in addition, the tissue piece softens, making the sectioning much more difficult. For keeping the sectioning bath at the necessary low temperature, we use noncorroding metal bars cooled to a very low temperature in a solid carbon dioxide–ethanol mixture. These metal bars are placed in the buffer trough and are changed at suitable intervals.

The sectioning procedure and the quality of the sections are influenced by several factors: (1) *Consistency of the tissue.* As described above, this is affected by the perfusion step. (2) *Size of the tissue pieces.* Brain tissue pieces as large as frontal sections through the rat diencephalon and telencephalon can be sectioned. The thickness of the dissected tissue piece can be up to 5–6 mm. In case of tissues difficult to section on the Vibratome (see below), it is often helpful to use smaller pieces. (3) *Type of tissue.* The possibilities of obtaining sections of useful quality on the Vibratome depend very much on what type of tissue is sectioned. Sections of good quality have been produced from, for example, ovary and uterus, tissues that are much more difficult to section than brain tissue. Certain areas of the brain are markedly easier to section than others. Homogeneous brain regions such as cerebral cortex, caudate nucleus, and diencephalon in the sagittal plane are easiest, while regions with a heterogeneous buildup, e.g., heavily myelinated regions such as the lower brainstem and spinal cord and regions including large ventricle spaces such as hypothalamus in the frontal plane, are more difficult.

For brain tissue, it has been found that sections 30–35 μm thick are useful for the study of innervation density and denervation effects because they can be obtained from most brain areas and in all planes of section. However, in the most favorable areas and in the most favorable planes (see above), sections can be obtained with a thickness down to 20 μm. A thickness greater than about 40 μm is not useful for brain tissue, since the opacity of the thicker sections decreases the contrast and, finally, disguises the CA structures.

The tissue piece will remain in an acceptable condition in the cool buffer for a maximum of 4–6 hr. This means that several pieces from each animals can be sectioned, the pieces being stored in the ice-cold buffer until they are to be sectioned. From each piece, a maximum of 15–20 good sections are obtained. It is recommended that the buffer in the trough for oxygenated buffer be changed at regular intervals during the day. It should be noted that it is important for the sectioning that the tissue piece be firmly glued to the holder of the Vibratome. If the tissue piece loosens, partly or completely, the quality of the sections declines or no sections at all are obtained. The tissue piece can in most cases be reglued to the holder.

Immersion. The sections are transferred with a blunt glass rod from the sectioning trough to an ice-cold 2% GA solution (pH 7.0, identical with the perfusion buffer). The concentration of GA and the pH of the immersion solution are important for the quality of the fluorescence picture. Lower GA concentration (e.g., 0.1%) or an acid immersion solution induces a lower CA fluorescence, and only scattered fibers and cell bodies can be seen. On the other hand, no further improvement of the fluorescence-microscopic picture is obtained if the immersion bath is alkaline or contains a higher concentration of GA.

Of particular importance is the practical handling of the sections during the immersion procedure. The sections should be kept below the surface of the GA solution, and the immersion should not continue for more than 3–5 min. If the sections float on the surface or if the immersion is too long, the sections start to disintegrate and the fluorescence intensity in the CA-containing structures decreases. When brain sections are rich in myelinated bundles (as in sections from the lower brainstem), the time of immersion should be particularly carefully supervised because these sections easily tear apart.

After the immersion, the sections are transferred to glass microscope slides in the following way: The section is picked up on the glass rod, which is then dipped into the solution so that the section partly floats out on the surface. The slide is now put below the section and the edge of the section is allowed to attach to the glass. The section is then spread out on the slide by gently lifting it out of the bath. In this way, folding of the sections is minimized. Excess buffer is removed from the slides with a filter paper. Care should be taken not to stretch or tear the sections on the slide.

Drying. The sections are dried in a two-step procedure: first under the warm air stream from a hair-dryer for about 15 min and then in a dessicator *in vacuo* over fresh phosphorous pentoxide overnight. Omitting the first drying step, or using air of room temperature, causes diffusion of the fluorophores from the nervous structures. During the drying procedure, which thus involves a moderate heating of the sections under the warm air stream, fluorophore formation takes place. This means that sections can be examined in the fluorescence microscope immediately after this first drying step, i.e., less than half an hour after the animal has been killed. The fluorescence picture in these sections is of high quality, and only minor differences, e.g., in fluorescence intensity, can be seen as compared to sections further processed with drying overnight, GA vapor treatment, or heating.

Heating or *GA vapor treatment.* To further react the sections by *heating,* they are simply put in a rack and placed in an oven at +100°C for 6 min. The fluorescence yield in the CA-containing structures is somewhat increased by this treatment.

In our work, *GA vapor treatment* is now very seldom used for reaction of the Vibratome sections. With CNS tissue, equally good results on CA-containing structures are obtained if the sections are simply heated. With peripheral tissues, when cells containing , for example, 5-HT or tryptophanyl-peptides are to be demonstrated, GA vapor treatment should be used (cf. Björklund *et al.,* 1973*b*). For details on the practical performance of the GA vapor reaction, see Lindvall and Björklund (1974*a*).

Comments on usefulness. As described in Section 4.3, GA induces stronger fluorescence than FA from several biogenic monoamines including the CAs. One drawback with GA as a histochemical reagent is its low ability to penetrate into tissue blocks. For the application of the GA method to CNS tissue, Vibratome sections have therefore been found to be ideal. In this type of tissue preparation, the central DA- and NA-containing neurons are demonstrated with remarkable sensitivity and precision. Thus, the CA stores of the entire axons—including their preterminal parts—and sometimes also the dendrites are readily visualized. The nonterminal parts of the CA axons contain the lowest amine concentration of the neuron, and these fibers are therefore not demonstrated in brains from intact and untreated adult animals processed according to the standard FA method. For this, the production of lesions of the axons is required in order to increase the intraaxonal amine concentration; these lesions can be achieved either mechanically (Dahlström and Fuxe, 1965), or chemically by means of 6-hydroxydopamine (Ungerstedt, 1971) or 6-hydroxy-dopa (Jacobowitz and Kostrzewa, 1971; Sachs and Jonsson, 1972)

The high sensitivity of the GA–Vibratome method has made possible the tracing of CA axon pathways in the rat brain (Lindvall and Björklund, 1974*b*). Furthermore, several previously unknown fiber systems have been discovered with this technique. This is the case, for example, for the incerto-hypothalamic DA system (Lindvall *et al.*, 1974*a*; Björklund *et al.*, 1975*b*) and periventricular CA system (Lindvall *et al.*, 1974*a*), the DA terminal systems in the cerebral cortex (Fig. 11A) (cf. Lindvall *et al.*, 1974*c*) and in the septum (Fig. 11B) (Lindvall, 1975), and the DA-containing dendrites of the neurons in the substantia nigra (Björklund and Lindvall, 1975).

A most advantageous feature of the GA–Vibratome method is the very distinct and precise picture of the CA-containing structures that is obtained (Fig. 11). This allows observations on details of the systems such as axonal morphology, branching patterns, and terminal arrangements. Of particular interest for neuroanatomical tracing is that in GA-treated Vibratome sections, axons of the same origin but localized in different brain regions in several cases have very similar fluorescence appearance. This gives possibilities of also distinguishing both preterminal and terminal axons of the same origin in an area where they are mixed with axons of other origins. Knowing the characteristic morphological features, it is therefore possible to study, in the intact animal, the terminal arrangements of the different CA afferents to an area.

The GA–Vibratome method undoubtedly has its greatest potential in neuroanatomical studies of central DA and NA neurons. Although the ALFA perfusion technique for freeze–dried, paraffin-embedded specimens has a sensitivity for CA neurons that is similar to that of the GA–Vibratome method, the latter technique has several attractive features that make it particularly useful in certain situations. The use of Vibratome sections involves, of course, both advantages and drawbacks. As described also in Section 3.2.2a, the processing time for such sections is much shorter than for freeze–dried tissue, and Vibratome sections lack cracks and other freeze–drying artifacts, which contributes to the achievement of good structural integrity. Furthermore, the thickness of the Vibratome section allows the tracing of CA fibers over longer distances. On the other hand, Vibratome sectioning is more time-consuming, fewer sections are produced, the sections are often of an uneven thickness, and the CA cell bodies become distorted. It should also be emphasized that the precision of the GA–Vibratome method in the visualization of central CA neurons is superior to that of other available methods. It is therefore the method of choice for studies of axonal mor-

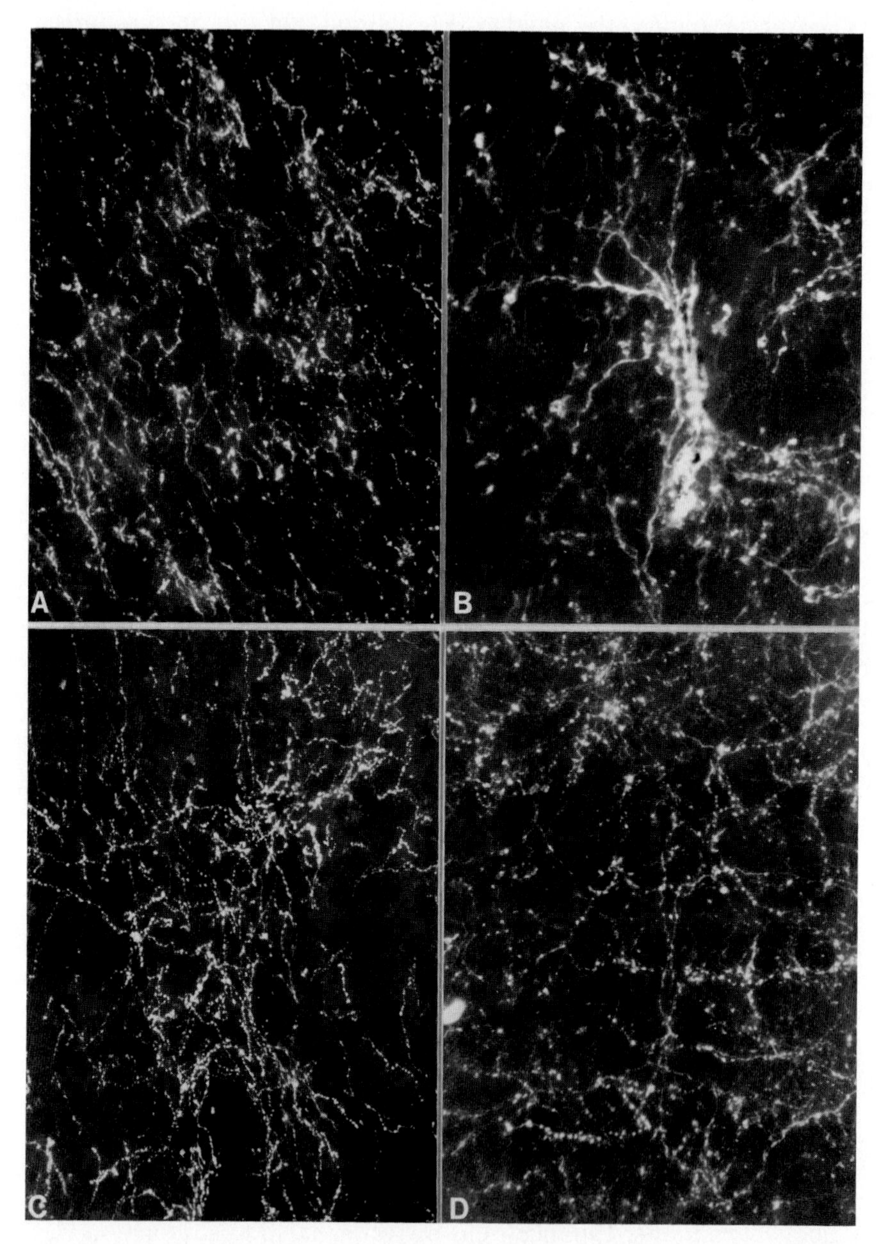

Fig. 11. Examples of CA-containing terminal axons as demonstrated with the GA–Vibratome method. (A) Dopaminergic axons in the medial frontal cortex; (B) dopaminergic axons around the proximal dendrites and the cell body of a nonfluorescent nerve cell in the septum; (C) noradrenergic axons in the septum; (D) dopaminergic and noradrenergic axons in the septum. The DA-containing fibers have a smoother appearance than the NA-containing fibers, which have varicosities closely spaced along the fibers. (B) From Lindvall and Björklund (1974*a*).

phology including situations in which the morphology is used for identification of axons of the same origin.

A further advantage of the Vibratome sections processed according to the GA method is that they can be used for concomitant application of other techniques. Thus, correlation between the distribution of acetylcholinesterase activity and CA fibers can be performed in the same section of CNS tissue (Lindvall, 1977) (see Section 3.2.5b), and by means of CA-uptake experiments, DA and NA neurons can be differentiated in the fluorescence microscope (Lindvall and Björklund, 1974*a*) (see Section 3.2.5f).

3.2.3. Methods Applied to Cryostat Sections

Cryostat sections of frozen tissue have been used for histochemical localization of biogenic monoamines after exposure to FA vapor (e.g., see Hamberger and Norberg, 1964; Spriggs *et al.*, 1966; Nelson and Wakefield, 1968; Placidi and Masuoka, 1968). This technique has achieved quite satisfactory results on both central and peripheral tissues, but its reproducibility is less satisfactory and the risk for diffusion artifacts higher than for freeze–dried tissue specimens. Recently, the GA method has been applied to cryostat sections (Battenberg and Bloom, 1975; Watson and Barchas, 1975, 1977; Bloom and Battenberg, 1976; de la Torre and Surgeon, 1976), giving much higher sensitivity and precision in the visualization of CA neurons than can be obtained with cryostat sections processed according to the FA method.

The GA method applied to cryostat sections of nonperfused brains, a procedure described by Watson and Barchas (1975, 1977), is carried out, briefly, as follows: Brains from untreated animals are frozen and sectioned on the cryostat at −17°C. The sections are then picked up on warm slides and immersed at 0°C in a 2% GA solution (0.1M phosphate buffer plus 0.5% magnesium chloride, pH adjusted to 5.0 by NaOH). The sections are left in the GA solution for up to 12 min for fine-terminal demonstration, 4 min for cell bodies, and 45 sec for axons in cerebellum and cortex. They are dried under warm air (+45°C for 5 min) and then exposed to GA vapor at atmospheric pressure (+100°C for 2–5 min).

Battenberg and Bloom (1975) and Bloom and Battenberg (1976) have applied the GA method to cryostat sections of brains perfused with an FA–GA solution. In his procedure, the animals are perfused with a prechilled (+2–4°C) solution of FA (0.5%) and GA (2%) in phosphate-buffered Ringer's solution (pH 7.4). For rats, 250 ml of the solution is perfused via the heart within 90 sec. The brain is dissected into pieces 3–5 mm thick, which are then frozen onto cryostat chucks, these being then inserted into powdered dry ice. After the tissues have equilibrated with the temperature of the cryostat chamber (−16 to −20°C), sections are cut at a thickness of 8–36 μm, transferred to cold glass slides, and thawed. The slides are immediately immersed in a phosphate-buffered Ringer's solution containing 2% GA (pH

7.4) for 11 min. Following immersion, the sections are dried in a stream of warm air (+37°C) for 5 min and then placed in a preheated, covered Coplin jar in an oven at +100°C for 10 min.

A modification of the latter procedure has been described by Nygren (1976). It was reported that the histochemical reaction was easier to standardize if the cryostat sections, instead of being heated, were reacted by exposure to FA vapor of standardized humidity (+80°C, 1 hr).

A third procedure for demonstration of CA neurons in cryostat sections has been developed by de la Torre and Surgeon (1976) using nonperfused brains. In this method, the sections are exposed to a solution containing sucrose–potassium phosphate–GA (SPG) for a very short time. The principal steps of the procedure are as follows: Nonperfused tissues are cut in slabs 6–9 mm thick, which are then placed on precooled cryostat chucks. The chucks with the tissue are quickly returned to the cryostat (−30°C). When the tissue has frozen in the cryostat, which takes 6–9 min, the sectioning is carried out. The section is adhered to a glass slide of room temperature, the section being allowed to melt onto the slide. The section is quickly dipped in the room-temperature SPG solution three times (1 dip/sec). The SPG solution is prepared from sucrose, potassium phosphate, and GA (1%) in distilled water. The pH of the solution is adjusted to 7.4 with NaOH.

The sections are then dried under an air stream of room temperature, e.g., from a hair-dryer, for 3–5 min. They are reacted by being heated in an oven at +80°C for 5 min.

The ALFA technique has been applied also to cryostat sections (Lorén *et al.*, 1980*a*). The rats are then perfused in two steps via the ascending aorta as described in Section 3.2.1b. The best results are obtained with 2% GA added to the preperfusion solution. In order to improve the morphology of the sections, 2.5% sucrose is added to both the preperfusion solution and the aluminum–formaldehyde perfusion solution.

Comments on usefulness. The cryostat techniques described above all have a sensitivity in the detection of central CA neurons that is markedly higher than that of the previous FA methods applied to cryostat sections.

According to our experience, the ALFA technique gives the most reproducible and sensitive visualization of central CA neurons of available cryostat methods. All known CA systems are demonstrated except those having a very low amine content, e.g., the DA systems in the anterior hypothalamus and the anterior cingulate cortex.

With the new cryostat methods, the cell body and the preterminal and terminal parts of the axons of central dopaminergic and noradrenergic neurons can be visualized. These techniques are also useful for studies on peripheral CA-containing structures (de la Torre and Surgeon, 1976). The fluorescence yield from central indoleamine neurons is low. The cryostat sectioning procedure has several advantages for fluorescence-histochemical work. As compared to Vibratome sectioning, the cryostat allows serial sections of large pieces. Furthermore, the cryostat sections can be larger than Vibratome sections; they are thinner and of an even thickness. Since the frozen tissue can be preserved for at least a few days, several pieces can be taken from each animal. In comparison with the freeze–drying procedure, cryostat sectioning is more rapid, requires less equipment, and is considerably easier to perform, which is of particular value for investigators inexperienced in histofluorescence techniques.

It is preferable in certain situations to use the GA–cryostat techniques on nonperfused tissue. This is obviously the case for tissues that cannot be perfused, e.g., human material. Also, from nonperfused tissue, alternate sections not processed for fluorescence histochemistry can be used, for example, for enzyme localization and biochemical analyses.

3.2.4. Methods Applied to Other Tissue Preparations

Specimens for fluorescence histochemistry can be obtained by methods other than freeze–drying or Vibratome and cryostat sectioning. The FA method has been applied to thin tissue sheets (so-called whole-mount preparations), dried at room temperature in air or *in vacuo*, such as iris and mesentery (see Fig. 9D) (Falck, 1962; Malmfors, 1965), vascular walls and meninges (Nielsen and Owman, 1967), heart atria (Sachs, 1970), urinary bladder (McLean and Burnstock, 1966), and digestive tract and genital organs (Furness and Malmfors, 1971). The technique is simple: the tissue membranes are spread on microscope slides and dried—preferably in a dessicator—over phosphorous pentoxide for 1 hr and then exposed to gaseous FA (see Section 3.2.1). Mounting in, for example, Entellan (Merck) or liquid paraffin is recommended but not essential. The whole-mount technique is simple, rapid, and versatile, and has therefore been extensively used (cf. Furness and Malmfors, 1971). FA-induced fluorescence has also been studied in smears of, for example, peritoneal fluid containing mast cells (Ritzén, 1966*a*), CNS tissue (Olson and Ungerstedt, 1970*a*), and subcellular fractions of homogenized tissues (Jonsson and Sachs, 1969).

Fluorescence-histochemical demonstration of peripheral CA stores in whole-mount preparations by immersion in GA solution was first described by Lindvall and Björklund (1974a) (Fig. 12). In this procedure, the thin tissue sheets (iris and mesentery) are processed similar to the Vibratome sections [i.e., soaked in the ice-cold GA solution (2% GA, pH 7.0), dried under a warm air stream, and in some cases in a dessicator overnight, heated, and then mounted in paraffin oil]. This procedure has been further developed by Furness and Costa (1975) (see also Waris and Partanen, 1975), who have applied the technique to whole mounts from several peripheral organs. Briefly, their procedure is as follows: The freshly dissected tissue (for details of the dissection procedure, see Furness and Costa, 1975) is immersed for 30 min at room temperature in a 2% GA solution (0.1 M phosphate buffer, pH 7.0). Preparations are then freed of excess moisture by touching them against absorbent paper and stretched on clean glass slides, to which they adhere as they dry. The slides are left on the laboratory bench for 3–5 min and are then placed in an oven, set at +100°C, for 4 min. The preparations

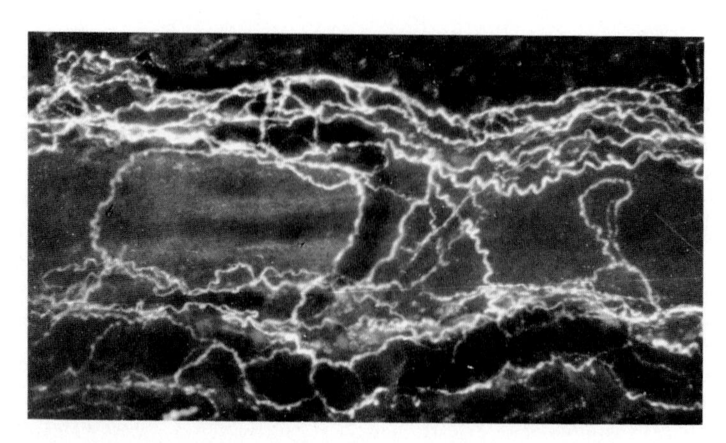

Fig. 12. Stretch preparation of rat mesentery processed by GA perfusion, GA immersion, and GA vapor treatment. The picture shows a vascular plexus around small arterioles and a vein. From Lindvall and Björklund (1974*a*).

are mounted in paraffin oil. Furness and Costa (1975) found that the development of fluorescence is favored by an initial excess of moisture in the tissue, that this moisture is driven off during the fluorophore formation, and that the tissue is protected from further moistening. This is achieved by heating the tissue that has been partially dried (on the laboratory bench) and then covering it with paraffin oil.

Compared with the FA vapor technique, the method using immersion of whole mounts in a GA solution has several advantages (Furness and Costa, 1975; see also Lindvall and Björklund, 1974*a*): First, CA neurons are demonstrated with higher sensitivity and precision. Thus, weakly fluorescent fibers, which are difficult to observe with the FA method, are readily demonstrated. Very little diffusion of the CA fluorophores is seen, making the fluorescence picture rich in detail. Second, the GA method is less susceptible to variations in procedure. Third, the GA method is simpler and quicker to apply.

The whole-mount GA technique can also be used for demonstration of peripheral 5-HT stores, e.g., in mast cells (Furness and Costa, 1975). To reveal 5-HT, but not NA, the tissue is immersed in acid 2% GA solution, pH 3.5, for 30 min at room temperature, then air-dried and heated at +100°C for 10 min. To induce fluorescence from both 5-HT and NA, the tissue is first incubated in 2% GA (pH 7.0) for 30 min at room temperature. It is then air-dried and exposed to FA vapor (+80°C for 1 hr).

Another technique for demonstration of CAs and 5-HT in whole-mount preparations has been developed by Furness *et al.* (1977*b*). In this technique, the preparations, adhering to glass slides, are placed in a mixture of 4% FA and 0.5% glutaraldehyde in 0.1 M phosphate buffer (pH 7.0; Faglu mixture) at room temperature for 1–3 hr. The specimens are then examined in the fluorescence microscope with the Faglu mixture as mounting medium. A further enhancement of the fluorescence intensity is obtained if the preparations are dried over phosphorous pentoxide for 1–2 hr and then mounted in paraffin oil. With this technique, the fluorescence intensity in the noradrenergic axons seems to be similar to that obtained with the GA method (Furness *et al.*, 1977*b*). The major advantage with the Faglu technique is that the specimens can be used for electron microscopy, thus allowing direct correlations between the fluorescence- and electron-microscopic pictures (see Section 3.2.5g).

3.2.5. Concomitant Use of Fluorescence-Histochemical and Other Microscopic Techniques

In fluorescence-microscopic studies on the localization of amines, it is often necessary to establish the identity of the fluorescent cells, or their relationship to adjacent nonfluorescent structures, or both. For this purpose, the fluorescence-histochemical techniques have been combined with phase-contrast microscopy, various routine staining procedures, the acetylcholinesterase reaction, methylene blue staining, autoradiography, and electron microscopy.

3.2.5a. Staining Procedures. For identification purposes, it is sometimes sufficient simply to perform the fluorescence microscopy using a Zernicke-type phase-contrast condenser and corresponding objectives in a system allowing simultaneous mixing of the UV light with ordinary visible light. Better results are obtained in a microscope equipped with illuminating systems for both incident and transmitted light. Histological staining can be carried out successfully on sections processed according to the standard FA technique. The FA gas treatment ensures a mild fixation of the tissues, and the dry conditions of this procedure facilitate good cytological preservation as well as excellent staining of the tissue components. For staining purposes, it is necessary to secure the paraffin sections on microscope slides coated with a thin layer of albumin–glycerine. Xylene, liquid paraffin, or immersion oil is used for mounting. After photography in the fluorescence microscope, the coverslip can then be easily removed for subsequent staining. Additional fixation of the sections before staining is advantageous in certain cases (cf. Jennings, 1965). The staining is followed by rephotography of the section and identification of the fluorescent structures.

3.2.5b. Visualization of Cholinergic Neurons. Adrenergic and cholinergic neurons can be demonstrated in the peripheral nervous system by applying the FA and cholinesterase (ChE) methods consecutively on one and the same section (Ehinger and Falck, 1965, 1966; Eränkö and Räisänen, 1965; Jacobowitz and Koelle, 1965). Because of the enzyme inactivation produced by FA, it is necessary to find an FA treatment sufficiently mild to give adequate fluorescence and still preserve enough ChE activity to enable its subsequent visualization. Treatment of the dried tissue for $\frac{1}{2}$–1 hr in FA gas generated at 37°C or for 3–4 hr at room temperature (Svendgaard *et al.*, 1975) has been found optimal. As a rule, the incubation time then required for demonstration of ChEs according to the Koelle technique has to be increased to 6–10 hr to get pictures equivalent to those obtained without previous FA treatment (Ehinger and Falck, 1966).

Peripheral CA- and acetylcholinesterase (AChE)-containing neurons can also be demonstrated simultaneously in the same section using the FA and ChE histochemical methods (El-Badawi and Schenk, 1967; Rauanheimo and Eränkö, 1968; Ellison and Olander, 1972). Recently, Waris *et al.* (1977) and Waris and Rechardt (1977) developed a similar technique in which GA is used instead of FA for demonstration of CA neurons. This modification seems superior to the combined FA-AChE procedures because of the stronger fluorescence and more limited diffusion seen with the GA-AChE technique. After immersion of the tissue pieces in a GA solution, they are processed for demonstration of AChE. The pieces are then immersed in the GA solution, dried, and heated. In the fluorescence microscope, not only the CA neurons but also the ChE-positive ones can be observed, the latter being distinguishable as dark profiles against the weakly fluorescent background.

In CNS tissue, it has previously not been possible to demonstrate CA- and AChE-containing neurons in the same section. Recently, a procedure has been developed (Lindvall, 1977) that allows this. The brains are first processed according to the GA method for CA neurons (Section 3.2.2b), i.e., perfused with an ice-cold GA solution, sectioned on a Vibratome instrument, immersed in a GA solution, and dried under a stream of warm air. The unmounted sections are examined and photographed in the fluorescence microscope and then stained for AChE according to the Koelle technique (incubation for 4–6 hr). The sections are then examined in the light microscope and rephotographed, and the picture compared with that following the GA reaction (Fig. 13).

An alternative technique for simultaneous visualization of adrenergic and cholinergic autonomic nerves has been published by Ehinger *et al.* (1967, 1968). According to this method, the classic methylene blue staining procedure was used, and the composition of the incubation solution was chosen such that subsequent FA treatment revealed the presence of fluorescent but both unstained (adrenergic) and stained (nonadrenergic) nerves. Differentiation of the two types of nerves, which could be visualized simultaneously in the fluorescence microscope, was facilitated by the addition of NA (1 μg/ml) to the methylene blue solution (Ehinger *et al.*, 1968). Methylene blue is generally believed to stain both adrenergic and cholinergic fibers by being accumulated in the nerve fibers. The results suggest that the uptake mechanisms of these two types of nerves differ sufficiently to give notable differences in the uptake of methylene blue. In fact, electron-microscopic studies with this method have demonstrated that methylene blue is preferentially accumulated in the cholinergic fibers (Richardson, 1968). So far, however, this method seems to work well only on some types of peripheral tissue.

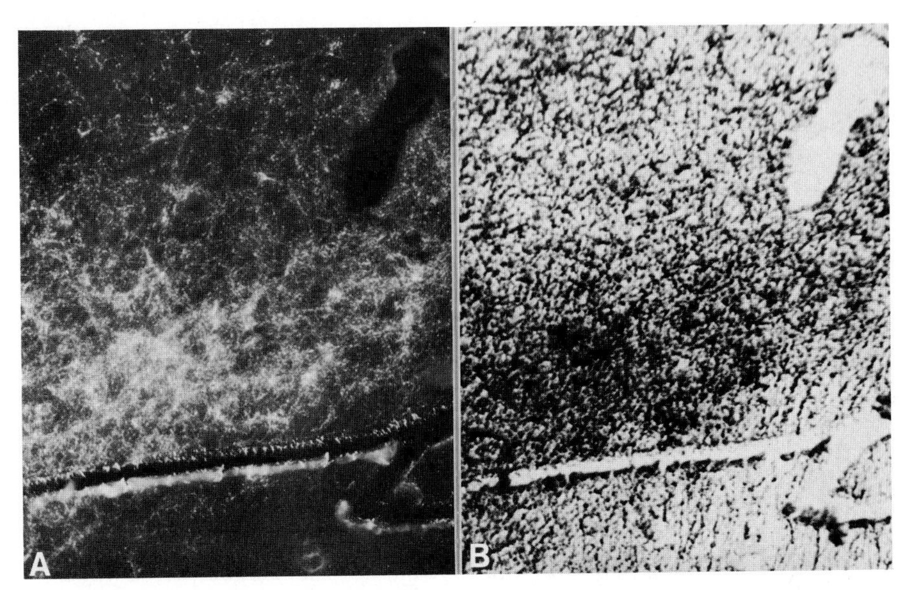

Fig. 13. (A) Fluorescence photomicrograph of DA-containing fibers in the rat septal area. (B) Same field as in (A) after staining for AChE activity. Note the similarity in distribution of DA fibers (A) and AChE activity (B). The scratch in (A) and (B) was used to relocate the exact area exposed in the fluorescence microscope, after the AChE staining.

3.2.5c. Other Enzyme-Staining Methods. The FA and GA methods can probably be combined with several kinds of histochemical enzyme-staining methods. Costa and Furness (Costa and Furness, 1973; Furness and Costa, 1975) have thus combined the FA and GA methods with a histochemical method for NADH: nitro-BT oxidoreductase. For combination with the FA method, the specimens are first incubated at room temperature for 5–10 min in a saline–glucose solution containing 0.5 mg/ml nitro-BT and 0.7 mg/ml NADH. The preparations are then dried over phosphorous pentoxide for 1 hr and treated with FA vapor at +80°C for 1 hr. For a combination of GA-induced CA fluorescence and counterstaining, the specimens are immersed for 30 min in 2% GA, pH 7.0, at room temperature, then in phosphate buffer (pH 7.0) containing nitro-BT and NADH (concentrations as above) and 2% GA for 5–10 min. They are then dried and heated as described in Section 3.2.4. These procedures have made possible simultaneous demonstration of adrenergic terminals, made fluorescent due to their NA content, and enteric neurons, delineated by deposits of formazan formed because of the presence of NADH:nitro-BT oxidoreductase (Costa and Furness, 1973; Furness and Costa, 1975).

3.2.5d. Horseradish Peroxidase Histochemistry. After stereotactic injection of horseradish peroxidase (HRP) into a specific brain area, the enzyme is taken up by nerve endings and transported retrogradely in the axons to the cell bodies. This forms the basis for a widely used neuroanatomical technique by which the cellular origin of axon-terminal networks can be traced (e.g., see La Vail *et al.,* 1973). The enzyme is taken up by different types of nerves, and the nature of the transmitter has to be identified by other techniques. To trace connections of CA neurons, Ljungdahl *et al.* (1975) used immunohistochemical demonstration of tyrosine hydroxylase (as a marker for CA neurons) followed by the HRP technique. Recently, Berger *et al.* (1978) have presented a procedure for direct, sequential demonstration of the CA transmitter and HRP-labeled cell bodies in the same section. In the first step, either of two modified GA techniques is used, applied to Vibratome or cryostat sections. The sections are first examined in the fluorescence microscope and the position of the CA-containing neurons is determined. In the second step, the sections are processed for demonstration of HRP-containing neurons. The HRP-labeled CA-containing neurons are then identified by comparison with the fluorescence-microscopic picture after the first step.

3.2.5e. Retrograde Fluorescent Tracers. Some of the fluorescent dyes, recently introduced by Kuypers and collaborators for retrograde tracing of axonal connections in the nervous system, are ideally suited for simultaneous retrograde labeling and monoamine visualization. Detailed procedures for such transmitter-specific fluorescent retrograde tracing, based on the tracers "true blue" and propidium iodide, have been devised by Björklund and Skagerberg (1979).

3.2.5f. Autoradiography. Studies on the binding and storage of exogenous amines and on compounds that interfere with their metabolism and function require techniques for concomitant demonstration of the amine structures and the compound administered. This can be obtained with microautoradiographic methods for water-soluble substances carried out under dry conditions (Masuoka and Placidi, 1968; Masuoka *et al.,* 1971; Stumpf and Roth, 1969; Ullberg and Appelgren, 1969; Hökfelt and Ljungdahl, 1971). These procedures utilize whole mounts, cryostat-sectioned material, or sections from freeze–dried tissues. The specimens are treated in FA gas for the visualization of CA and 5-HT, and fluorescence microscopy and photography are performed before application of the autoradiographic emulsion. Alternatively, the sections can be transferred onto previously emulsion-coated slides (Masuoka and Placidi, 1968).

Combined autoradiographic and fluorescence-histochemical techniques have been used to study the morphological relationships between CA neurons and estrogen target neurons (e.g., see Heritage *et al.,* 1977). In the technique of Grant and Stumpf (Grant and Stumpf, 1973; Stumpf, 1976), cryostat sections from animals injected with [^{3}H]estradiol are freeze–dried and reacted with FA vapor. The sections are then dry-mounted onto emulsion-coated slides for autoradiographic localization of [^{3}H]estradiol-uptake sites. After exposure and development, the [^{3}H]estradiol-uptake sites and CA neurons can be localized simultaneously in the same section by using a microscope with a combination of UV and regular tungsten light.

3.2.5g. In Vivo Uptake Experiments. As described in Section 4.2, DA- and NA-containing structures can be differentiated in the FA method by microspectrofluorometric analysis (see Fig. 15). In the GA method applied to

Vibratome sections, this is not possible. However, DA neurons can be distinguished from NA neurons by a combination of the GA method with DA-uptake experiments on Vibratome sections (for details, see Lindvall and Björklund, 1974*a*). Briefly, the procedure is as follows: The animals are pretreated with reserpine (to empty endogenous CA stores) and nialamide (to inhibit monoamine oxidase). They are then perfused with ice-cold, neutral Krebs–Ringer bicarbonate buffer. After being cut in the Vibratome instrument, the sections are preincubated at +37°C for 15 min either in buffer alone or in the presence of desipramine (10^{-5} M). Desipramine is an inhibitor of neuronal CA uptake that is about 1000 times more potent on the uptake into NA neurons than on that into DA neurons (Horn *et al.*, 1971). This selective blocking effect of desipramine has been utilized for both biochemical and histochemical differentiation between DA and NA terminals in the CNS (Fuxe *et al.*, 1967; Hamberger, 1967; Cuello *et al.*, 1973). Following the preincubation, the sections are incubated in the presence of DA (10^{-6} M) for 20 min. They are then processed according to the GA method (see Section 3.2.2b). With this procedure, no uptake of DA into NA structures, e.g., in hypothalamus and cerebral cortex, is demonstrable histochemically in sections preincubated in desipramine, whereas in the other sections, noradrenergic systems are clearly visible (Fig. 14). In contrast, the DA uptake into DA fibers is seemingly unaffected by desipramine. This has been shown for the dopaminergic terminals in the nucleus caudatus–putamen, and also for those belonging to the recently discovered cortical and incertohypothalamic dopaminergic systems (Lindvall *et al.*, 1974*b*; Björklund *et al.*, 1975*b*).

3.2.5h. Electron Microscopy. The possibilities of correlating fluorescence-microscopic with electron-microscopic observations on central CA neurons seem to have increased considerably due to the recent development of a procedure that has been claimed to demonstrate CA neurons—in the same or in adjacent cryostat or Vibratome sections—first in the fluorescence microscope and then in the electron microscope (Chiba *et al.*, 1976*a,b*). The animal is first perfused with 2% GA in Krebs–Ringer bicarbonate buffer (pH 7.0) for 10 min at +4°C; this is followed by a secondary perfusion solution containing 4% paraformaldehyde and 0.5% glutaraldehyde in Sorensen's phosphate buffer (pH 7.4), also at +4°C. The latter is delivered over approximately 20–30 min. The tissue is dissected out and preserved in the same fixative for 2–24 hr at +4°C. Sections are cut with either cryostat or Vibratome at 20–50 μm in thickness and incubated in 2% GA in phosphate buffer (ph 7.0) for 5–10 min at +4°C. Sections are mounted in 50% glycerin–water for fluorescence microscopy. Chiba *et al.* (1976*a,b*) report good preservation of the CA fluorescence, so that structures of interest can be sketched and photographed in the fluorescence microscope. Afterward, the same or subsequent sections are rinsed in 7% sucrose in water for 10 min and then processed for electron microscopy. Although the experience with this procedure is very limited at present, it promises to bridge the gap between observations at the light- and electron-microscopic levels, which would greatly expand our knowledge on the morphology of central CA neurons.

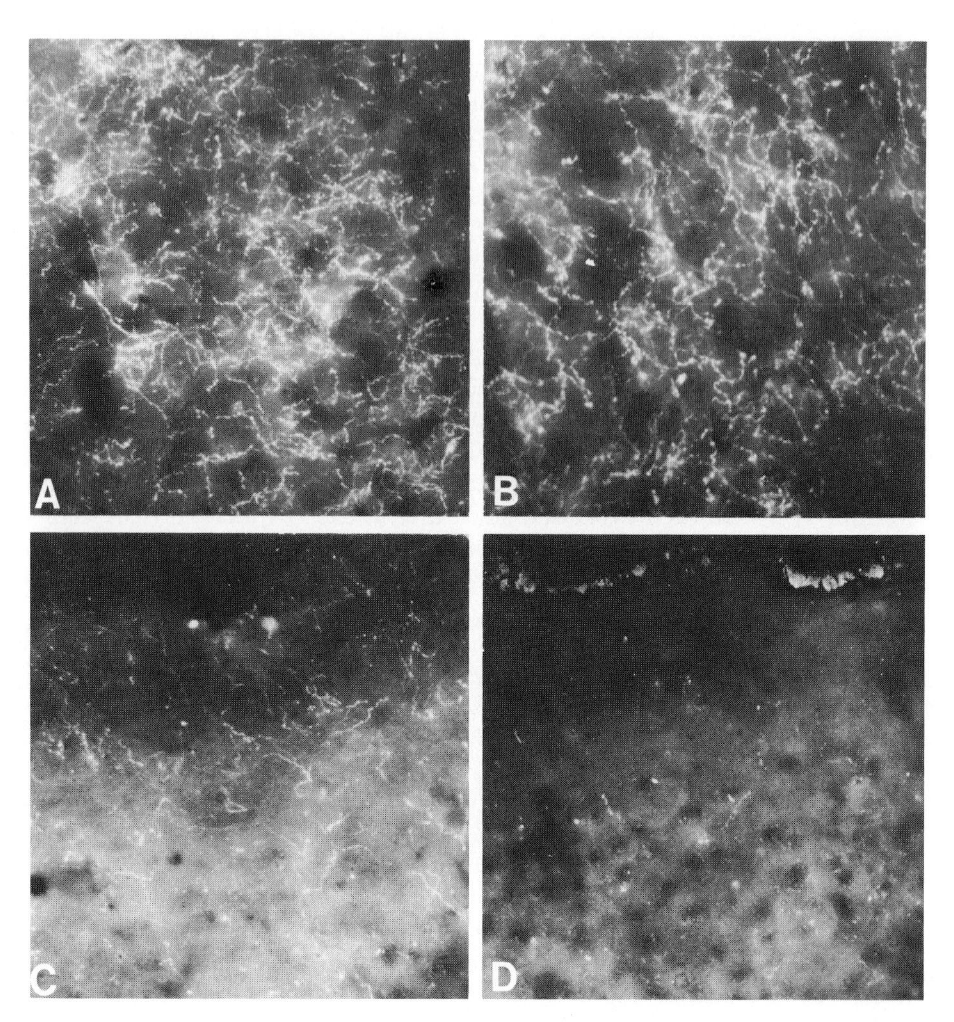

Fig. 14. Illustrations of the fluorescence-microscopic picture in Vibratome sections prepared for differentiation between dopaminergic and noradrenergic neurons. For details on the methodology, see the text. (A, B) From adjacent sections through the supragenual anteromedial cortex; (C, D) from adjacent sections through the sensorimotor cortex. (A, C) Incubated in DA alone; in this case, both dopaminergic (A) and noradrenergic (C) axons became visible after treatment with GA. (B, D) Incubated in DA in the presence of desipramine, which is a blocker of CA uptake into noradrenergic neurons; in this case, only dopaminergic axons could be demonstrated (B), whereas the noradrenergic axons were not visible (D). From Lindvall *et al.* (1978).

It has subsequently been reported by Furness *et al.* (1977*b*, 1978) that specimens prepared for monoamine fluorescence according to the Faglu technique (see Section 3.2.2a) can also be used for electron microscopy. The stretch preparations are reacted and mounted in the Faglu mixture. Suitable areas are photographed in the fluorescence microscope, cut out, and then processed for electron microscopy. According to Furness *et al.* (1977*b*), the quality of fixation for electron microscopy is comparable to that obtained with other techniques.

4. Fluorescence-Histochemical Techniques for Identification and Differentiation of Biogenic Monoamines

A special problem when using fluorescence-histochemical methods for the demonstration of cellular monoamine stores is met in the identification of the fluorogenic compound. This is a critical point, since there are a wide variety of phenylethylamine and indolylethylamine derivatives that have been found in tissue by biochemical analytical techniques. As discussed in detail in Section 3.1, only phenylethylamines and indolylethylamines with a certain molecular structure will form fluorophores in the histochemical formaldehyde (FA) and glyoxylic acid (GA) reactions. Because these fluorophores display characteristic and reproducible excitation and emission spectra, microspectrofluorometric analysis not only provides a characterization of the fluorophore but also permits differentiation among various fluorogenic monoamines at their cellular storage sites. Specificity tests and spectral analysis of fluorogenic compounds in the FA and GA reactions are discussed in the following sections (for further details, see, for example, Björklund and Falck, 1973; Björklund *et al.,* 1975*a*).

4.1. Tests for Specificity

The monoamines become fluorescent in the histochemical reaction, i.e., after treatment with FA or GA. The first criterion of any monoamine fluorescence is therefore that it is specific, i.e., induced by the FA or GA treatment. The literature contains many examples of misinterpretations in which an unspecific autofluorescence is ascribed to the presence of monoamines. Such mistakes can be avoided by the application of some simple specificity tests (Falck 1962; Corrodi and Jonsson, 1967).

1. A test of basic importance is to establish that the observed fluorescence does not appear when the FA or GA step is omitted. It should be observed that even slight contamination with FA in the room air or the paraffin wax can induce a monoamine fluorescence.

2. The catecholamine (CA) and 5-hydroxytryptamine (5-HT) fluorophores exhibit a noticeable photodecomposition on irradiation with UV or blue-violet light. Such a photodecomposition is usually not observed with autofluorescent structures. The 5-HT fluorophore shows a much faster photodecomposition than the CA fluorophores (see Ritzén, 1966*a,b*). This feature is often helpful in distinguishing 5-HT at low and moderate concentrations from CAs.

3. The CA and 5-HT fluorophores display only very weak fluorescence in the presence of water (see Ritzén, 1966*a,b*), and hence the fluorescence is quenched when the section is mounted in water. Such fluorescence-quenching has not been reported for autofluorescent structures.

4. If freeze–dried tissues are treated with FA vapor saturated with

water, the CAs diffuse from their cellular sites before condensation with FA and binding to the protein have occurred. This effect is not so pronounced in the case of 5-HT (Bertler *et al.*, 1964; Fuxe and Jonsson, 1967).

5. Treatment with sodium borohydride ($NaBH_4$) in alcoholic solutions was introduced as a specificity test in the FA method (Corrodi *et al.*, 1964). With this treatment, the fluorescent dihydroisoquinolines and dihydro-β-carbolines are reduced to their corresponding nonfluorescent tetrahydro-derivatives (cf. Figs. 3 and 4). The fluorescence can be regained by renewed FA treatment. It must be remembered that the essential step of the test is this regeneration of the fluorescence and that proper controls, as devised by Corrodi *et al.* (1964), must be performed. Watson and Barchas (1977) have reported that the $NaBH_4$ test is also useful for GA-treated material.

6. Since the autofluorescent structures usually have spectral characteristics and colors different from those of the amine fluorophores, they are revealed by spectral analysis. When working with previously noncharacterized systems, microspectrofluorometric analysis should always be performed.

7. In conjunction with the histochemical tests, the use of pharmacological tests can be recommended. Such tests involve, for example, disappearance of the fluorescence after depletion of the amine with reserpine or α-methyl-*m*-tyrosine treatment, or after amine-synthesis inhibition, e.g., with α-methyl-*p*-tyrosine or *p*-chlorophenylalanine (for further details, see Corrodi and Jonsson, 1967).

4.2. *Identification and Differentiation of Biogenic Amines and Related Compounds after Formaldehyde Treatment*

4.2.1. *Fluorescence Yields from Phenylethylamines and Related Compounds*

The CAs, dopamine (DA), noradrenaline (NA), and adrenaline, have hydroxyl groups in the 3- and 4-positions of the benzene ring (see Fig. 5) and consequently give strong fluorescence in the FA reaction (see Section 3.1). Of the CA precursors and metabolites, only those having a free amino group on the side chain, i.e., primary and secondary amines and amino acids, can enter fluorophore-forming cyclization reactions with FA. Consequently, no fluorescence is induced from the acid CA metabolites [e.g., homovanillic acid (HVA) and 3-methoxy-4-hydroxymandelic acid (vanillylmandelic acid, (VMA)] or the corresponding glycols or aldehydes. Tyrosine and phenylalanine, as well as the 4-hydroxylated amines tyramine and octopamine, give no fluorescence in the histochemical reaction because of insufficient activation at the point of ring closure. On the other hand, FA treatment induces strong fluorescence from the CA precursor dopa. Also, the dopa derivative, 5-*S*-cysteinyl-dopa, which was recently found in tissue (Björklund *et al.*, 1972*b*; Falck *et al.*, 1976), becomes strongly fluorescent after FA treatment.

In the standard FA reaction, no fluorescence is induced from the 3-methoxylated CA metabolites, 3-methoxytyramine, normetanephrine, and metanephrine. However, strong fluorescence is obtained from 3-methoxytyramine when the gaseous FA treatment is performed in the presence of minute amounts of acid (Björklund and Stenevi, 1970; Björklund *et al.*, 1971*b*).

4.2.2. *Fluorescence Yields from Indoleamines and Related Compounds*

Strong fluorescence is induced from 5-HT in the FA reaction. The sensitivity for 5-HT is, however, clearly lower than that for DA and NA, and the visualization of intracellular 5-HT, above all in the CNS, offers certain difficulties. There exist several possibilities for improving the detectability of 5-HT neurons: (1) The FA-induced fluorescence from 5-HT can be increased by using more severe reaction conditions [increased reaction time, more humid FA gas (see Fuxe and Jonsson, 1967)]. It has been reported that perfusion with 4% FA in hypertonic sucrose buffer increases the fluorescence yield in central 5-HT-containing neurons (Azmitia and Henriksen, 1976), and also the aluminum–formaldehyde (ALFA) technique (Lorén *et al.*, 1980) (see Section 3.2.1b) offers advantages for indoleamine visualization in the CNS. Despite this, the sensitivity is lower for 5-HT than for primary CAs. (2) By using a potent inhibitor of monoamine oxidase (MAO), e.g., nialamide, the 5-HT levels in the CNS are considerably increased (Pletscher *et al.*, 1965) and there is a marked increase in the intraneuronal 5-HT fluorescence (Fuxe, 1965; Andén *et al.*, 1966; Björklund *et al.*, 1971*a*). (3) The fluorescence of cell bodies and axons of central indoleamine-containing neurons can be greatly enhanced by treatment with an MAO inhibitor plus L-tryptophan (Aghajanian and Asher, 1971; Aghajanian *et al.*, 1973). It has not so far been clarified whether this increased fluorescence following tryptophan administration occurs solely in 5-HT neurons. In fact, the results of Aghajanian and Asher (1971) suggest that the fluorogenic compound formed in the cell bodies of the brain stem raphe nuclei might be different from 5-HT. (4) Transection of axon bundles of monoamine neurons causes a rapid piling up of the intraneuronal monoamines just proximal to the lesion. This accumulation also gives rise to a strong intraaxonal fluorescence from 5-HT neurons, which is most convenient for spectral recordings (cf. Björklund *et al.*, 1970, 1971*a*). (5) It has been shown (Baumgarten *et al.*, 1971, 1973; Björklund *et al.*, 1973*c*) that intracerebral injections of 5,6-dihydroxytryptamine and 5,7-dihydroxytryptamine cause damage to axons and axon terminals of central indoleamine neurons. This causes a prominent accumulation of the transmitter in intact parts of the axons, which then become strongly fluorescent in the fluorescence-histochemical reaction (see Fig. 9A). In this way, the 5-HT neurons are available for both neuroanatomical studies and microspectrofluorometric analysis (Björklund *et al.*, 1973*c*). (6) As described in Section 3.2.5f, Vibratome sections can be used for fluorescence-microscopic studies on uptake in central monoamine neurons. Berger and Glowinski (1978) recently showed that DA is taken up not only in catecholaminergic but also in serotoninergic terminals. The 5-HT fibers can be identified by comparing the effects of various known blockers of uptake into DA, NA, and 5-HT neurons and by selective chemical lesions of the CA and 5-HT systems by neurotoxic agents. Using this technique, a more complete visualization of the cortical 5-HT innervation has been obtained (cf. Berger and Glowinski, 1978).

The biological precursors of 5-HT, tryptophan and 5-hydroxytryptophan (5-HTP), are fluorogenic in the FA reaction. On the other hand, the metabolic degradation products, 5-hydroxyindoleacetic acid and 5-hydroxytryptophol, show no visible fluorescence after FA treatment. Although it does not seem likely that under normal conditions tryptophan or 5-HTP is accumulated intraneuronally in such a way that it could interfere with the visualization of 5-HT, it is obvious that this can occur after administration of the amino acids. For this reason, it is important to be aware that tryptophan and 5-HTP can show up in the FA method.

Tryptamine yields a low visible fluorescence in the standard FA reaction. The fluorescence is strongly increased if the FA treatment is performed in the presence of ozone (Björklund *et al.*, 1968*b*) or small amounts of HCl vapor (Björklund and Stenevi, 1970; Björklund *et al.*, 1971*b*). In the FA–ozone reaction, NH_2-terminal tryptophan peptides (which can be regarded as α-substituted tryptamine molecules and consequently give only low fluorescence after the standard FA treatment) also become strongly fluorescent (Håkanson and Sundler, 1971).

In the FA method, also, other primary and secondary indolylethylamines give good fluorescence, whereas tertiary amines and amides are essentially nonfluorescent. Thus, 5-methoxytryptamine and N-methylated tryptamines can be demonstrated with the FA method, whereas *N,N*-dimethyl-5-hydroxytryptamine (bufotenin), *N*-acetyl-5-hydroxytryptamine, and *N*-acetyl-5-methoxytryptamine (melatonin) cannot be visualized.

4.2.3. *Spectral Analysis of Fluorogenic Amines and Related Compounds*

The first step in the identification and differentiation of a specific fluorescence is to perform spectral analysis in the microspectrofluorometer. This can be supplemented with a study on the conditions of maximal fluorophore development, i.e., to check whether the fluorescence yield is favored by longer reaction time or higher oxidative capacity of the reaction milieu. In the following discussion, differentiation problems related to the standard FA reaction will be considered. For details on the spectral analysis after the FA–ozone or acid-catalyzed FA-treatment, the reader is referred to the papers by Björklund *et al.* (1968*a,b*, 1971*b*) and Håkanson and Sundler (1971).

The FA-induced monoamine fluorophores can be divided into three different groups (A, B, and C) on the basis of their emission-peak maxima (see Table 1) (Björklund *et al.*, 1971*a*). Group A has an emission-peak maximum at about 475 nm and comprises the CA and the dopa fluorophores. The substances in Group B have emission-peak maxima ranging from 495 to 505 nm. This group includes tryptamine, 6-hydroxytryptamine, 5,6-dihydroxytryptamine, *N*-methyl-5-hydroxytryptamine, 5-methoxytrypta-

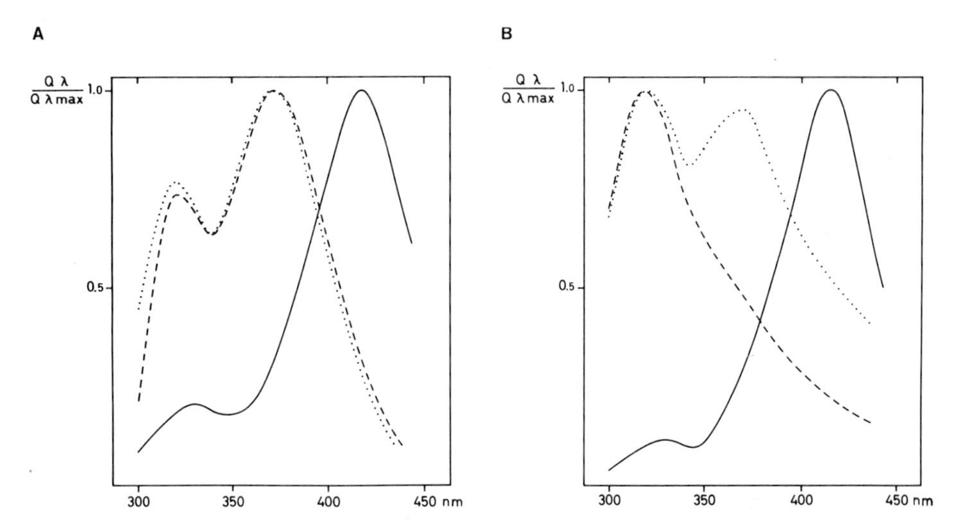

Fig. 15. Fluorescence excitation spectra of DA (A) and NA (B) in dried albumin droplets. (——) After FA treatment only; (······) after FA treatment followed by treatment with HCl vapor at room temperature for 5 sec; (------) after FA treatment followed by treatment with HCl vapor for 5 min.

mine, and tryptophan. Tryptophan has, in addition, a blue fluorophore (excitation/emission maxima: 375/435 nm) that usually predominates at low to moderate concentrations of the amino acid. Group C comprises the fluorophores with emission-peak maxima from 520 to 540 nm. 5-HT and 5-HTP belong to this group, as well as 5-methoxytryptamine at higher concentrations. Within each group of fluorophores, certain possibilities exist for differentiation among individual compounds, mainly by means of their excitation spectra.

The fluorophores in *Group A* show indistinguishable excitation and emission spectra (maxima at 320 and 410, and at 475, respectively). It is important to note that these fluorophores exhibit a pH-dependent tautomerism (see Figs. 4 and 15) (Corrodi and Hillarp, 1964; Jonsson, 1966; Björklund *et al.*, 1968*a*, 1972*a*). At neutral pH—as in the tissue after standard FA treatment—the fluorophores are in their quinoidal form (main excitation peak at 410 nm) and at acid pH—as in the tissue after acidification—in their nonquinoidal form (main excitation peak at 370–380 nm). This pH-dependent tautomerism of the CA fluorophores implies that their spectral characteristics will be determined by the pH of the tissue environment. This should always be remembered when the histochemical reagent is applied under acid conditions.

It should be noted that in certain cases when dealing with high concentrations of CAs, the emitted light from the FA-induced fluorophores appears yellowish in the fluorescence microscope even though the microspectrofluorometric recording shows emission maxima at about 480 nm (Norberg *et al.*, 1966; Jonsson, 1967*c*). This is because the maximum sensitivity of the human eye is shifted toward longer wavelengths with increasing intensity of the light [the so-called Bezold–Brücke effect (Wright, 1944; see also Norberg *et al.*, 1966; Jonsson, 1967*a*)]. To avoid misinterpretations of CA fluorescence as being due to, for example, 5-HT, it is important in all doubtful cases to perform spectral analyses.

Whereas the Bezold–Brücke effect means that the color impression of the fluorescence is changed without a corresponding change in the emission spectrum, there are instances in which a true change of the emission spectrum does occur. Thus, the FA-induced fluorescence from CAs will exhibit a shift in the emission maximum from 480 nm to 500–550 nm when the CA concentration is very high, as in adrenal medullary cells (Caspersson *et al.*, 1965). This is most probably due to a concentration-dependent side-reaction, e.g., polymerization or oxidation or both (Jonsson, 1967*a*; cf. Whaley and Govindachari, 1951). The shift in the emission maximum of the NA fluorophore has never been observed in peripheral or central adrenergic neurons (Jonsson, 1967*a*). The side-reaction can be prevented by using milder reaction conditions (less humid FA gas, lower temperature, and shorter reaction time).

The fluorophores induced by FA treatment from the β-hydroxylated CAs, NA and adrenaline, have a labile 4-hydroxy group that can easily be split off by acid (see Fig. 4) (Corrodi and Jonsson, 1965*a*; Björklund *et al.*, 1968*a*, 1972*a*). The fluorophores that lose this group are converted to fully aromatic compounds. This is followed by a shift in the main excitation peak to 320–330 nm (Fig. 15). The DA fluorophore, lacking the 4-hydroxy group, cannot be transformed into the fully aromatic compound, and consequently its main excitation peak after acid treatment remains at 370 nm (Fig. 15). Because of this difference in the behavior of the DA and NA fluorophores on treatment with acid, a method has been evolved for differentiating intracellular DA from NA by microspectrofluorometric analysis (Björklund *et al.*, 1968*a*, 1972*a*). In this method, the FA-treated tissue sections are exposed to hydrochloric acid vapor for various times at room temperature, and the behavior of the peaks at 370 and 320 nm is carefully studied. For considerations in the practical performance, the limitations, and the pitfalls of this differentiation method, the reader should consult the paper of Björklund *et al.* (1972*a*).

Adrenaline, which is a secondary amine, requires more severe reaction conditions than DA and NA for optimum fluorescence yield after FA treatment (see above) (Falck, 1962; Corrodi and Hillarp, 1963). This difference in the reaction kinetics of primary and secondary CAs has been used for fluorescence-microscopic differentiation between the two types of substances (Falck *et al.*, 1963; Norberg *et al.*, 1966).

Previously, it has not been possible to differentiate between dopa and DA by microspectrofluorometry. With the introduction of the principle of combined FA and GA reactions (Lindvall *et al.*, 1974*c*, 1975*a,b*), possibilities have been obtained for spectral differentiation between these compounds.

After a mild FA treatment (low temperature, short reaction time) followed by GA treatment, different types of fluorophores are formed from dopa and DA (see Section 3.1.3), resulting in clearly distinguishable spectral characteristics (the dopa fluorophore having an excitation maximum at 330 nm and an emission maximum at 490–500 nm, the DA fluorophore having an excitation maximum at 375 nm and an emission maximum at 470 nm). Under these reaction conditions, the spectral characteristics of the fluorophore formed from NA are similar to those of the dopa fluorophore. The combined FA and GA reaction has so far not been applied to tissue for differentiation between intracellular dopa and DA. Similarly, this principle should also be useful for differentiation between indolylethylamines and their precursor amino acids such as 5-HT and 5-HTP and tryptamine and tryptophan (Lindvall *et al.*, 1975*b*).

On the basis of their excitation spectra, the indolylethylamine fluorophores in Group B (Table 1) can be distinguished into two subgroups. One is characterized by a single excitation peak with the maximum at about 370 nm. This group comprises the unsubstituted indolylethylamine, tryptamine, as well as tryptophan. In the other subgroup, the excitation spectrum is characterized by a double peak, one at 380–390 and the other at 405–420 nm. The relative predominance of these two peaks varies and depends on a number of factors such as concentration, reaction conditions, and pH. In addition to this double peak, many of these fluorophores have a peak or shoulder in the excitation spectrum at about 310–320 nm. The excitation spectra of the 5-HT and 5-HTP fluorophores in Group C also have the characteristics of this second subgroup of Group B.

It should be pointed out that whereas differentiation between 5-HT and other fluorogenic indolylethylamines (and related amino acids) requires spectral analysis, the distinction between the CAs, on one hand, and 5-HT and other fluorogenic indolylethylamine derivatives, on the other, can fairly safely be made directly in the fluorescence microscope using the difference in the fluorescence colors of their fluorophores. However, because the CA fluorescence under certain conditions can appear yellowish (see above), it is important in all doubtful cases to perform spectral analyses. The very fast photodecomposition rate of the 5-HT fluorophores, which differs markedly from that of the CA fluorophores, and the absence of a pH-dependent tautomerism may also be used for distinguishing 5-HT from the CAs (Jonsson, 1967*c*). The characteristic fading of the 5-HT fluorescence is, however, disguised at high amine concentrations, at which the fluorescence intensity shows a concentration-dependent quenching (Ritzén, 1967). With such intense 5-HT fluorescence as occurs, for example, in enterochromaffin cells and in rodent mast cells, the fluorescence will appear quite stable.

The spectral characteristics of the monoamine fluorophores in specimens processed according to the ALFA technique (Björklund *et al.*, 1980) have also been recorded. As can be seen from Table 3, both the excitation- and emission-peak maxima of the CA and 5-HT fluorophores are similar, which

Table 3. Excitation-Peak and Emission-Peak Maxima of Some Fluorogenic Compounds in Formaldehyde-Treated Protein-Droplet Models with 10 mM $Al_2(SO_4)_3$[a]

Substance	Maximum (nm)	
	Excitation[b]	Emission
Dopamine	(330),(360),410[c]	520
Noradrenaline	(320),350,410[c]	520
Adrenaline	(330),350,(410)	510
Dopa	(330),355,(405)	510
Tryptamine	360	500
Tryptophan	360	500
5-Hydroxytryptamine	365,(410)	525
5-Hydroxytryptophan	360,(410)	520
5-Methoxytryptamine	365	535

[a] Treatment at +80°C for 1 hr with paraformaldehyde equilibrated in air of about 50% relative humidity according to the Falck–Hillarp method (see Björklund *et al.*, 1972*c*). Values from Björklund *et al.* (1980).
[b] A figure in parentheses denotes the position of a low peak or shoulder in the spectrum.
[c] The relative size of the peaks is pH-dependent.

makes them difficult to distinguish both in the fluorescence microscope and microspectrofluorometrically. It is therefore necessary to use pharmacological treatments and selective lesions to identify the 5-HT neurons.

4.3. *Identification and Differentiation of Biogenic Amines and Related Compounds after Glyoxylic Acid Treatment*

4.3.1. *Fluorescence Yields from Phenylethylamines, Indolylethylamines, and Related Compounds*

In the various versions of the GA technique, the histochemical reagent is applied in gaseous form (Axelsson *et al.*, 1973; Björklund *et al.*, 1973*b*), in solution (Lindvall and Björklund, 1974*a*), or both in solution and in gaseous form (Lindvall *et al.*, 1973; Lindvall and Björklund, 1974*a*). In practical histochemical work, the latter two types of GA applications have been found to be the most useful ones. The description in this section will therefore deal mainly with the fluorescence yields after these types of treatments. To simulate the application of the reagent in solution and gaseous form, the protein-droplet models that were used for the fluorescence-intensity and spectral measurements contained GA (pH 7.0) and were reacted by heating or GA vapor treatment (for details, see Lindvall and Björklund, 1974*a*).

After *reaction by heating,* only dopa and the primary CAs, DA and NA, give strong fluorescence (see Table 2). The fluorescence induced from the CAs is about 5–6 times higher than that obtained in the standard FA reaction (see Table 1). No or very weak fluorescence is obtained from the secondary CA, adrenaline, from methoxylated phenylethylamines, and from all indoleamines tested. Thus, under these reaction conditions—designed to simulate the conditions under which the GA-perfused and GA-immersed sections are reacted through heating—the GA method appears to be highly selective for dopa, DA, and NA, at least with respect to amines known to occur in the mammalian CNS (see Table 2).

After *reaction by GA vapor treatment,* there is a dramatic increase in the fluorescence yield from most substances tested (Table 2). However, NA shows similar, and DA only moderately increased, fluorescence intensity as compared to after heating alone. A striking effect is obtained in the fluorescence yields from the indoleamines, and the methoxylated phenylethylamines, 3-methoxytyramine, and normetanephrine also become fluorescent. Strong fluorescence is induced not only from primary and secondary indoleamines (such as tryptamine, tryptophan, 5-HT, 5-methoxytryptamine, and *n*-methyltryptamine) but also from the tertiary indoleamine (*N,N*-dimethyltryptamine) and the N-acetylated indoleamines (*N*-acetyl-5-hydroxytryptamine and melatonin). The fluorescence intensities obtained from the primary and secondary indoleamines after GA vapor treatment are several times higher than those obtained in the standard FA reaction. It is particularly interesting that the N-acetylated and the tertiary indoleamines, which give no observable fluorescence on FA treatment, show strong fluorescence after GA treatment, the fluorescence yield from melatonin and from *N,N*-dimethyltryptamine being of the same order as that obtained from NA in the FA method (Table 1). No or very weak fluorescence is induced from phenylethylamines lacking a cyclization-promoting hydroxyl group in the 3-position (see Corrodi and Jonsson, 1967), i.e., phenylethylamine, *p*-tyramine, and *p*-octopamine.

The model experiments show that when the GA-containing specimens are reacted by GA vapor treatment instead of heating, a greater number of biogenic compounds become strongly fluorescent. This procedure thus provides very interesting possibilities for the visualization of other biogenic monoamines, particularly a wide variety of indolylethylamines, including N-acetylated and tertiary indolylethylamines. However, in the CNS of the rat, indoleamine neurons exhibit only weak and variable fluorescence, and according to our present experience with central neuronal tissue, it seems that the procedure using GA vapor treatment has a high sensitivity only for DA and NA neurons. On the other hand, GA vapor treatment of GA-perfused peripheral tissue will also demonstrate cell systems containing 5-HT (cf. Axelsson *et al.,* 1973) and NH_2-terminal tryptophanyl peptides (cf. Björklund *et al.,* 1973*b*). It is therefore important in all instances with a previously unknown cellular fluorescence to perform microspectrofluorometric analysis.

4.3.2. *Spectral Analysis of Fluorogenic Amines and Related Compounds*

The following description will deal mainly with microspectrofluorometric analysis of fluorophores induced by heating or GA vapor treatment of GA-containing specimens at neutral pH. It should be pointed out that the spectral characteristics of the fluorophores formed by the latter treatment are similar to those of the fluorophores formed by GA vapor treatment of specimens containing no GA or by heating of GA-containing specimens at acid pH.

After *reaction through heating,* the fluorophores of DA, NA, and dopa—being the only compounds that yield strong fluorescence after this treatment—show indistinguishable spectra with excitation and emission maxima at 415 and 475 nm (see Table 2). These spectral characteristics are recorded from CA-containing structures in GA-perfused tissue (pH 7.0) examined directly after the drying procedure or after heating. The fluorophores exhibit a pH-dependent tautomerism similar to that of the FA-induced CA fluorophores (see Figs. 7 and 16). Thus, at neutral pH—as in the GA-containing specimens, just after heating—the fluorophores are in their neutral, quinoidal form (main excitation peak at 415 nm), and at acid pH, such as after GA vapor treatment (see below) or acidification of the specimens before or after the heating, they are in their acid, nonquinoidal form (main excitation peak at 370–380 nm).

Prolonged HCl treatment provides a possibility for spectral differentiation between NA on one hand and DA and dopa on the other (Lindvall *et al.,* 1974*c*) according to the principles described previously for the FA-induced fluorophores (Corrodi and Jonsson, 1965*a*; Björklund *et al.,* 1968*a,* 1972*a*). After about 5 sec of HCl treatment, there is a primary shift of the main excitation peak of the CA and dopa fluorophores from 415 to 370–380 nm. On prolonged treatment (about 10–15 min at room temperature), there is a further shift of the excitation peak of the NA fluorophore down to 330 nm, whereas the excitation peaks of the DA and dopa fluorophores remain unchanged at 370–380 nm. This latter spectral shift has been obtained, so far, only in models. When the prolonged HCl treatment is applied to Vibratome sections of central nervous tissue, it results in a marked diffusion of the amines from their neuronal stores, making spectral analysis impossible. Microspectrofluorometry is thus not useful for differentiation between NA and DA in the GA method. For description of differentiation between central dopaminergic and nonadrenergic neurons by DA-uptake experiments, the reader is referred to Section 3.2.5f.

After *reaction by GA vapor treatment,* a greater number of biogenic monoamines give strong fluorescence, and on the basis of their emission peak maxima they can be separated into four groups (Table 2): *Group A* (emission maxima at 430 nm) comprises *N,N*-dimethyltryptamine and bufotenin. *Group B* (emission maxima at 460–485 nm) includes normetanephrine, adrenaline, dopa, DA, NA, 3-methoxytyramine, *N*-acetyl-5-hydroxytryptamine, and *N*-methyltryptamine. *Group C* (emission maxima at 495–500 nm) includes tryptamine, tryptophan, *N*-methyl-5-hydroxytryptamine, and melatonin. *Group D* (emission maxima at 520–530 nm) comprises 5-HT, 5-HTP, and 5-methoxytryptamine.

Within Group B, the adrenaline and normetanephrine fluorophores (Group B1) can be differentiated from those of the other compounds (Group B2), on the basis of their excitation spectra, directly after the gaseous GA

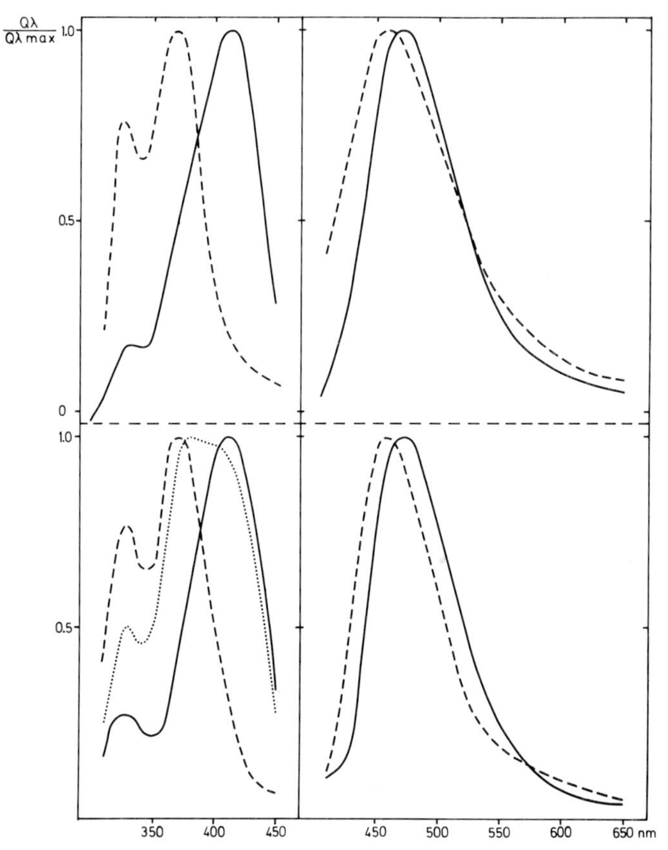

Fig. 16. *Top:* Excitation (left) and emission (right) spectra of NA in GA-containing protein droplets after reaction through heating alone (+100° C, 6 min) (——) and after GA vapor treatment (+100°C, 3 min, 300 torr) (------). *Bottom:* Excitation (left) and emission (right) spectra recorded from CA-containing structures in GA-treated Vibratome sections. Treatment: perfusion: 2% GA, 150 ml; immersion: 2% GA, pH 7.0; GA vapor treatment: 300 torr GA, 2 min, +100°C. With this type of treatment, most spectra demonstrate the fluorophores in their neutral, quinoidal form (——). Spectra showing the fluorophores in the acid, nonquinoidal form (------) can also be recorded, as well as spectra indicating a mixture of the two forms of fluorophores (······).

treatment. The primary CAs and dopa can easily be distinguished from the other amines in Group B2, since these latter substances show essentially no fluorescence after the type of histochemical reaction in which the GA vapor treatment is omitted (see above and Table 2). Moreover, the DA, NA, and dopa fluorophores exhibit the characteristic pH-dependent shift in their excitation spectra described above. After GA vapor treatment, the CA fluorophores are, in protein models, in their acid, nonquinoidal form, exhibiting a main excitation peak at 370–380 nm. A short exposure of the specimens to (dry) ammonia vapor at room temperature will transform them into the neutral, quinoidal state, having a peak maximum (415 nm) that is clearly distinguishable from that of the other compounds in Group B2. It should be mentioned that NH_2-terminal tryptophanyl peptides, which become strongly fluorescent, show spectral characteristics similar to those of tryptamine and tryptophan after GA vapor treatment (Björklund *et al.*, 1973*b*).

In Vibratome sections of CNS tissue, in which only DA and NA neurons are demonstrated with high sensitivity (see above), exposure to GA vapor involves a partial conversion of the CA fluorophores to the nonquinoidal (acid) form. Most spectra demonstrate the fluorophores in their neutral form, but acid spectra and spectra indicating a mixture of the nonquinoidal and quinoidal forms of fluorophore can also be recorded (Fig. 16).

5. General Conclusion

Although there are now alternative techniques for visualization of monoamine-producing cell systems, such as autoradiography of accumulated exogenous amines and immunocytochemistry of the monoamine-synthesizing enzymes, the histofluorescence methods are still the only ones for a direct visualization of the monoamine transmitters themselves. The histofluorescence techniques are therefore unique in allowing combined morphological, functional, and pharmacological studies on defined monoaminergic neuron systems (e.g., see Fuxe *et al.*, 1969; Hökfelt and Fuxe, 1972). The methodological developments during recent years have made the histofluorescence techniques more practical and versatile and thus more accessible for nonspecialists. Also, the recent improvements in sensitivity and precision, due above all to the introduction of the GA reagent and the metal salt procedures, have made histofluorescence a powerful tool for neuroanatomical tracing of CA neurons in the central nervous system. The fluorescence histochemistry of amine transmitters is still in a phase of rapid development. It can be expected that, as has been the case in the past, methodological developments will contribute importantly to progress and new discoveries in the study of monoaminergic cell systems.

ACKNOWLEDGMENTS. The research that has been reported in this review was supported by grants from the United States Public Health Service (NS 06701), the Swedish Medical Research Council (04X-712 and 04X-4493), and the Magnus Bergvall Foundation.

References

Aghajanian, G.K., and Asher, I.M., 1971, Histochemical fluorescence of raphe neurons: Selective enhancement by tryptophan, *Science*, **172:**1159.

Aghajanian, G.K., Kuhar, M.J., and Roth, R.H., 1973, Serotonin-containing neuronal perikarya and terminals: Differential effects of *p*-chlorophenylalanine, *Brain Res.* **54:**85.

Agnati, L.F., and Fuxe, K., 1974, Quantitative comparisons of amine fluorescence in cortical noradrenaline terminals using smear preparations, *J. Histochem. Cytochem.* **22:**1122.

Ajelis, V., Björklund, A., Falck, B., Lindvall, O., Lorén, I., and Walles, B., 1979, Application of the aluminum-formaldehyde (ALFA) histofluorescence method for demonstration of peripheral stores of catecholamines and indolamines in freeze–dried paraffin-embedded tissue, cryostat sections and whole mounts, *Histochemistry* **65:**1.

Andén, N.-E., Dahlström, A., Fuxe, K., Larsson, K., Olson, L., and Ungerstedt, U., 1966, Ascending monoamine neurons to the telencephalon and diencephalon, *Acta Physiol. Scand.* **67:**313.

Axelsson, S., Björklund, A., and Lindvall, O., 1972, Fluorescence histochemistry of biogenic monoamines: A study of the capacity of various carbonyl compounds to form fluorophores with biogenic monoamines in gas-phase reactions, *J. Histochem. Cytochem.* **20:**435.

Axelsson, S., Björklund, A., Falck, B., Lindvall, O., and Svensson, L.Å., 1973, Glyoxylic acid condensation: A new fluorescence method for the histochemical demonstration of biogenic monoamines, *Acta Physiol. Scand.* **87:**57.

Azmitia, E.C., and Henriksen, S.J., 1976, A modification of the Falck–Hillarp technique for 5-HT fluorescence employing hypertonic formaldehyde perfusion, *J. Histochem. Cytochem.* **24:**1286.

Battenberg, E.L.F., and Bloom, F.E., 1975, A rapid, simple and more sensitive method for the demonstration of central catecholamine neurons and axons by glyoxylic acid induced fluorescence. I. Specificity, *Psychopharmacol. Commun.* **1:**3.

Baumgarten, H.G., 1972, Biogenic monoamines in the cyclostome and lower vertebrate brain, *Prog. Histochem. Cytochem.* **4:**1.

Baumgarten, H.G., Björklund, A., Lachenmayer, L., Nobin, A., and Stenevi, U., 1971, Long-lasting selective depletion of brain serotonin by 5,6-dihydroxytryptamine, *Acta Physiol. Scand. Suppl.* **373**.

Baumgarten, H.G., Björklund, A., Lachenmayer, L., and Nobin, A., 1973, Evaluation of the effects of 5,7-dihydroxytryptamine on serotonin and catecholamine neurons in the rat CNS, *Acta Physiol. Scand. Suppl.* **391**.

Berger, B., and Glowinski, J., 1978, Dopamine uptake in serotoninergic terminals *in vitro*: A valuable tool for the histochemical differentiation of catecholaminergic and serotoninergic terminals in rat cerebral structures, *Brain Res.* **147:**29.

Berger, B., Arluison, M., Escourolle, R., and Moyne, M.A., 1974, Les catécholamines du cortex cérébral humain: Mise en évidence par la technique d'histofluorescence à l'acide glyoxylique, *Presse Med.* **3:**1081.

Berger, B., Thierry, A.M., Tassin, J.P., and Moyne, M.A., 1976, Dopaminergic innervation of the rat prefrontal cortex: A fluorescence histochemical study, *Brain Res.* **106:**133.

Berger, B., Nguyen-Legros, J., and Thierry, A.-M., 1978, Histofluorescence des catécholamines et révélation de la peroxydase transportée par voie rétrograde sur une méme coupe de tissue et dans le méme neurone, *C. R. Acad. Sci. Paris* **286:**1363.

Bertler, A., Falck, B., and Owman, C., 1964, Studies on 5-hydroxytryptamine stores in pineal gland of rat, *Acta Physiol. Scand.* **63**(Suppl. 239):1.

Björklund, A., 1968, Monoamine-containing fibres in the pituitary neuro-intermediate lobe of the pig and rat, *Z. Zellforsch.* **89:**573.

Björklund, A., and Falck, B., 1968, An improvement of the histochemical fluorescence method for monoamines: Observations on varying extractability of fluorophores in different nerve fibers, *J. Histochem. Cytochem.* **16:**717.

Björklund, A., and Falck, B., 1969, Histochemical characterization of a tryptamine-like substance stored in cells of the mammalian adenohypophysis, *Acta Physiol. Scand.* **77:**475.

Björklund, A., and Falck, B., 1973, Cytofluorometry of biogenic monoamines in the Falck–Hillarp method: Structural identification by spectral analysis, in: *Fluorescence Techniques in Cell Biology* (A.A. Thaer, and M. Sernetz, eds.), pp. 171–181, Springer-Verlag, Berlin, Heidelberg, and New York.

Björklund, A., and Lindvall, O., 1975, Dopamine in dendrites of substantia nigra neurons: Suggestions for a role in dendritic terminals, *Brain Res.* **83:**531.

Björklund, A., and Stenevi, U., 1970, Acid catalysis of the formaldehyde condensation reaction for sensitive demonstration of tryptamines and 3-methoxylated phenylethylamines. 1. Model experiments, *J. Histochem. Cytochem.* **18:**794.

Björklund, A., Ehinger, B., and Falck, B., 1968*a*, A method for differentiating dopamine from noradrenaline in tissue sections by microspectrofluorometry, *J. Histochem. Cytochem.* **16:**263.

Björklund, A., Falck, B., and Håkanson, R., 1968*b*, Histochemical demonstration of tryptamine: Properties of the formaldehyde-induced fluorophores of tryptamine and related indole compounds in models, *Acta Physiol. Scand. Suppl.* **318**.

Björklund, A., Falck, B., and Stenevi, U., 1970, On the possible existence of a new intraneuronal monoamine in the spinal cord of the rat, *J. Pharmacol. Exp. Ther.* **175:**525.

Björklund, A., Falck, B., and Stenevi, U., 1971*a*, Microspectrofluorometric characterization of monoamines in the central nervous system: Evidence for a new neuronal monoamine-like compound, *Prog. Brain Res.* **34:**63.

Björklund, A., Nobin, A., and Stenevi, U., 1971*b*, Acid catalysis of the formaldehyde condensation reaction for a sensitive histochemical demonstration of tryptamine and 3-methoxylated phenylethylamines. 2. Characterization of amine fluorophores and application to tissues, *J. Histochem. Cytochem.* **19:**286.

Björklund, A., Ehinger, B., and Falck, B., 1972*a*, Analysis of fluorescence excitation peak ratios for the cellular identification of noradrenaline, dopamine, or their mixtures, *J. Histochem. Cytochem.* **20:**56.

Björklund, A., Falck, B., Jacobsson, S., Rorsman, H., Rosengren, A.-M., and Rosengren, E., 1972*b*, Cysteinyl dopa in human malignant melanoma, *Acta Derm.-Venereol. (Stockholm)* **52:**357.

Björklund, A., Falck, B., and Owman, Ch., 1972*c*, Fluorescence microscopic and microspectrofluorometric techniques for the cellular localization and characterization of biogenic amines, in: *Methods of Investigative and Diagnostic Endocrinology*, (S.A. Berson, ed.), Vol. 1, *The Thyroid and Biogenic Amines* (J.E. Rall and I.J. Kopin, eds.), pp. 318–368, North-Holland, Amsterdam.

Björklund, A., Lindvall, O., and Svensson, L.Å., 1972*d*, Mechanisms of fluorophore formation in the histochemical glyoxylic acid method for monoamines, *Histochemistry* **32:**113.

Björklund, A., Falck, B., Lindvall, O., and Svensson, L.Å., 1973*a*, New aspects on reaction mechanisms in the formaldehyde histofluorescence method for monoamines, *J. Histochem. Cytochem.* **21:**17.

Björklund, A., Håkanson, R., Lindvall, O., and Sundler, F., 1973*b*, Fluorescence histochemical demonstration of peptides with NH_2- or COOH-terminal tryptophan or dopa by condensation with glyoxylic acid, *J. Histochem. Cytochem.* **21:**253.

Björklund, A., Nobin, A., and Stenevi, U., 1973*c*, The use of neurotoxic dihydroxytryptamines as tools for morphological studies and localized lesioning of central indolamine neurons, *Z. Zellforsch.* **145:**479.

Björklund, A., Falck, B., and Lindvall, O., 1975*a*, Microspectrofluorometric analysis of cellular monoamines after formaldehyde or glyoxylic acid condensation, in: *Methods in Brain Research* (P.B. Bradley, ed.), pp. 249–294, John Wiley and Sons, London.

Björklund, A., Lindvall, O., and Nobin, A., 1975*b*, Evidence of an incerto-hypothalamic dopamine neurone system in the rat, *Brain Res.* **89:**29.

Björklund, A., and Skagerberg, G., 1979, Simultaneous use of retrograde fluorescent tracers and fluorescence histochemistry for convenient and precise mapping of monoaminergic projections and collateral arrangements in the CNS, *J. Neurosci. Methods* **1:**261.

Björklund, A., Falck, B., Lindvall, O., and Lorén, I., 1980, The aluminum–formaldehyde (ALFA) histofluorescence method for improved visualization of catecholamines and indoleamines. 2. Model experiments, *J. Neurosci. Method.* **2:**301.

Bloom, F.E., and Battenberg, E.L.F., 1976, A rapid, simple and more sensitive method for the demonstration of central catecholamine-containing neurons and axons by glyoxylic acid induced fluorescence. II. A detailed description of methodology, *J. Histochem. Cytochem.* **24:**561.

Carlsson, S.-A., and Ritzén, M., 1969, Mast cells and 5-HT: Intracellular release of 5-hydroxytryptamine (5-HT) from storage granules during anaphylaxis or treatment with compound 48/80, *Acta Physiol. Scand.* **77:**449.

Caspersson, T., Lomakka, G., and Rigler, R., 1965, Registrierender Fluoreszenzmikrospektrograph zur Bestimmung der Primär und Sekundärfluoreszenz verschiedener Zellsubstanzen, *Acta Histochem. (Jena) Suppl.* **6:**123–126.

Chiba, T., Hwang, B.H., and Williams, T.H., 1976*a*, A method for glyoxylic acid induced fluorescence and electron microscopy and its application for lesion studies of central catecholamine pathways, *Anat. Rec.* **184:**376.

Chiba, T., Hwang, B.-H., Williams, T.H., 1976*b*, A method for studying glyoxylic acid induced fluorescence and ultrastructure of monoamine neurons, *Histochemistry* **49:**95.

Corrodi, H., and Hillarp, N.-Å., 1963, Fluoreszenzmethoden zur histochemischen Sichtbarmachung von Monoaminen. 1. Identifizierung der fluoreszierenden Produkte aus Modellversuchen mit 6,7-Dimethoxyisochinolinderivaten und Formaldehyd, *Helv. Chim. Acta* **46:**2425.

Corrodi, H., and Hillarp, N.-Å., 1964, Fluoreszenzmethoden zur histochemischen Sichtbarmachung von Monoaminen. 2. Identifizierung der fluoreszierenden Produktes aus Dopamin und Formaldehyd, *Helv. Chim. Acta* **47:**911.

Corrodi, H., and Jonsson, G., 1965*a*, Fluorescence methods for the histochemical demonstration of monoamines. 4. Histochemical differentiation between dopamine and noradrenaline in models, *J. Histochem. Cytochem.* **13:**484.

Corrodi, H., and Jonsson, G., 1965*b*, Fluoreszenzmethoden zur histochemischen Sichtbarmachung von Monoaminen. 5. Identifizierung des fluoreszierenden Produktes aus Modellversuchen mit 5-Methoxytryptamin und Formaldehyd, *Acta Histochem. (Jena)* **22:**247.

Corrodi, H., and Jonsson, G., 1967, The formaldehyde fluorescence method for the histochemical demonstration of biogenic monoamines: A review on the methodology, *J. Histochem. Cytochem.* **15:**65.

Corrodi, H., Hillarp, N.-Å., and Jonsson, G., 1964, Fluorescence methods for the histochemical demonstration of monoamines. 3. Sodium borohydride reduction of the fluorescent compounds as a specificity test, *J. Histochem. Cytochem.* **12:**582.

Costa, M., and Furness, J.B., 1973, The simultaneous demonstration of adrenergic fibres and enteric ganglion cells, *Histochem. J.* **5:**343.

Cuello, A.C., Horn, A.S., Mackay, A.V.T., and Iversen, L.L., 1973, Catecholamines in the median eminence: New evidence for a major noradrenergic input, *Nature (London)* **243:**465.

Dahlström, A., and Fuxe, K., 1965, Evidence for the existence of monoamine neurons in the central nervous system. II. Experimentally induced changes in the intraneuronal amine levels of bulbospinal neuron systems, *Acta Physiol. Scand.* **64**(Suppl. 247).

Dahlström, A., Häggendal, J., and Hökfelt, T., 1966, The noradrenaline content of the varicosities of sympathetic adrenergic nerve terminals in the rat, *Acta Physiol. Scand.* **67:**287.

de la Torre, J.C., and Surgeon, J.W., 1976, A methodological approach to rapid and sensitive monoamine histofluorescence using a modified glyoxylic acid technique: The SPG method, *Histochemistry* **49:**81.

Ehinger, B., and Falck, B., 1965, Noradrenaline and cholinesterase in concomitant nerve fibres in iris, *Life Sci.* **4:**2097.

Ehinger, B., and Falck, B., 1966, Concomitant adrenergic and parasympathetic fibres in the rat iris, *Acta Physiol. Scand.* **67:**201.

Ehinger, B., and Thunberg, R., 1967, Induction of fluorescence in histamine-containing cells, *Exp. Cell Res.* **47:**116.

Ehinger, B., Sporrong, B., and Stenevi, U., 1967, Combining the catecholamine fluorescence and methylene blue staining methods for demonstrating nerve fibres, *Life Sci.* **6:**1973.

Ehinger, B., Sporrong, B., and Stenevi, U., 1968, Simultaneous demonstration of adrenergic and non-adrenergic nerve fibres, *Histochemie* **13:**105.

Einarsson, P., Hallman, H., and Jonsson, G., 1975, Quantitative microfluorimetry of formaldehyde induced fluorescence of dopamine in the caudate nucleus, *Med. Biol.* **53:**15.

El-Badawi, A., and Schenk, E.A., 1967, Histochemical methods for separate, consecutive and simultaneous demonstration of acetylcholinesterase and norepinephrine in cryostat sections, *J. Histochem. Cytochem.* **15:**580.

El-Badawi, A., Hayashi, K.D., and Schenk, E.A., 1970, Histochemical demonstration of norepinephrine in postmortem tissues, *Histochemie* **21:**21.

Ellison, J.-P., and Olander, K.W., 1972, Simultaneous demonstration of catecholamines and acetylcholinesterase in peripheral autonomic nerves, *Am. J. Anat.* **135:**23.

Eränkö, O., and Räisänen, L., 1965, Fibres containing both noradrenaline and acetylcholinesterase in the nerve net of the rat iris, *Acta Physiol. Scand.* **63:**505.

Ewen, S.W.B., and Rost, F.W.D., 1972, The histochemical demonstration of catecholamines and tryptamines by acid- and aldehyde-induced fluorescence: Microspectrofluorometric characterization of the fluorophores in models, *Histochem. J.* **4:**59.

Falck, B., 1962, Observations on the possibilities of the cellular localization of monoamines by a fluorescence method, *Acta Physiol. Scand.* **56**(Suppl. 197):1–25.

Falck, B., and Owman, C., 1965, A detailed description of the fluorescence method for the cellular localization of biogenic monoamines, *Acta Univ. Lundensis,* Section II, No. 7.

Falck, B., Hillarp, N.-Å., Thieme, G., and Torp, A., 1962, Fluorescence of catecholamines and related compounds condensed with formaldehyde. *J. Histochem. Cytochem.* **10:**348.

Falck, B., Häggendal, J., and Owman, C., 1963, The localization of adrenaline in adrenergic nerves in the frog, *Q. J. Exp. Physiol.* **48:**253.

Falck, B., Jacobsson, S., Lindvall, O., and Nietsche, U.-B., 1976, On the occurrence of cysteinyl dopa and dopa in melanocytes and benign nevi cells, *Scand. J. Plast. Surg.* **10:**185.

Furness, J.B., and Costa, M., 1975, The use of glyoxylic acid for the fluorescence histochemical demonstration of peripheral stores of noradrenaline and 5-hydroxytryptamine in whole mounts, *Histochemistry* **41:**335.

Furness, J.B., and Malmfors, T., 1971, Aspects of the arrangement of the adrenergic innervation in guinea-pigs as revealed by the fluorescence histochemical method applied to stretched, air-dried preparations, *Histochemie* **25:**297.

Furness, J.B., Costa, M., and Blessing, W.W., 1977*a*, Simultaneous fixation and production of catecholamine fluorescence in central nervous tissue by perfusion with aldehydes, *Histochem. J.* **9:**745.

Furness, J.B., Costa, M., and Wilson, A.J., 1977*b*, Water-stable fluorophores, produced by reaction with aldehyde solutions, for the histochemical localization of catechol- and indolethylamines, *Histochemistry* **52:**159.

Furness, J.B., Heath, J.W., and Costa, M., 1978, Aqueous aldehyde (Faglu) methods for the fluorescence histochemical localization of catecholamines and for ultrastructural studies of central nervous tissue, *Histochemistry* **57:**285.

Fuxe, K., 1965, Evidence for the existence of monoamine containing neurons in the central nervous system. IV. The distribution of monoamine terminals in the central nervous system, *Acta Physiol. Scand.* **64**(Suppl. 247):39.

Fuxe, K., and Jonsson, G., 1967, A modification of the histochemical fluorescence method for the improved localization of 5-hydroxytryptamine, *Histochemie* **11:**161.

Fuxe, K., Hamberger, B., and Malmfors, T., 1967, The effect of drugs on accumulation of monoamines in tubero-infundibular dopamine neurons, *Eur. J. Pharmacol.* **1:**334.

Fuxe, K., Hökfelt, T., Jonsson, G., and Ungerstedt, U., 1969, Fluorescence microscopy in neuroanatomy, in: *Contemporary Research Methods in Neuroanatomy* (W.J.H. Nauta and S.O.E. Ebbesson, eds.), pp. 275–314, Springer-Verlag, Berlin, Heidelberg, and New York.

Grant, L.D., and Stumpf, W.E., 1973, Localization of ^{3}H-estradiol and catecholamines in identical neurons in the hypothalamus, *J. Histochem. Cytochem.* **21:**404.

Håkanson, R., and Owman, C., 1967, Concomitant histochemical demonstration of histamine and catecholamines in enterochromaffin-like cells of gastric mucosa, *Life Sci.* **6:**759.

Håkanson, R., and Sundler, F., 1971, A method for the fluorescence microscopic demonstration of peptides with NH_2-terminal tryptophan residues, *J. Histochem. Cytochem.* **19:**477.

Håkanson, R., and Sundler, F., 1974, Formaldehyde condensation at reduced temperature: Increased sensitivity and specificity of the fluorescence microscopic method for demonstrating primary catecholamines, *J. Histochem. Cytochem.* **22:**887.

Hamberger, B., 1967, Reserpine-resistant uptake of catecholamines in isolated tissues of the rat, M.D. thesis, Karolinska Institute, Stockholm.

Hamberger, B., and Norberg, K.-A., 1964, Histochemical demonstration of catecholamines in fresh frozen sections, *J. Histochem. Cytochem.* **12:**48.

Hamberger, B., Malmfors, T., and Sachs, C., 1965, Standardization of paraformaldehyde and of certain procedures for the histochemical demonstration of catecholamines, *J. Histochem. Cytochem.* **13:**147.

Hercules, D.M., 1966, Fluorescence and phosphorescence analysis, in: *Principles and Applications* (D.M. Hercules, ed.), pp. 1–40, Interscience, New York, London, and Sydney.

Heritage, A.S., Grant, L.D., and Stumpf, W.E., 1977, ^{3}H Estradiol in catecholamine neurons of rat brain stem: Combined localization by autoradiography and formaldehyde-induced fluorescence, *J. Comp. Neurol.* **176:**607.

Hökfelt, T., 1965, A modification of the histochemical fluorescence method for the demonstration of catecholamines and 5-hydroxytryptamine, using araldite as embedding medium, *J. Histochem. Cytochem.* **13:**518.

Hökfelt, T., and Fuxe, K., 1972, On the morphology and the neuro-endocrine role of the hypothalamic catecholamine neurons, in: *Brain–Endocrine Interaction* (K.M. Knigge, D.E. Scott, and A. Weindl, eds.), pp. 181–223, S. Karger, Basel.

Hökfelt, T., and Ljungdahl, Å., 1970, Cellular localization of labeled gamma-aminobutyric acid (^{3}H-GABA) in rat cerebellar cortex: An autoradiographic study, *Brain Res.* **22:**391.

Hökfelt, T., and Ljungdahl, Å., 1971, Uptake of (^{3}H) noradrenaline and γ-(^{3}H) aminobutyric acid in isolated tissues of rat: An autoradiographic and fluorescence microscopic study, in: *Progress in Brain Research,* Vol. 34, *Histochemistry of Nervous Transmission* (O. Eränkö, ed.), pp. 87–92, Elsevier, Amsterdam, London, and New York.

Hökfelt, T., and Ljungdahl, Å., 1972, Modification of the Falck–Hillarp formaldehyde fluorescence method using the Vibratome®: Simple, rapid and sensitive localization of catecholamines in sections of unfixed or formalin fixed brain tissue, *Histochemie* **29:**325.

Hökfelt, T., Fuxe, K., Johansson, O., and Ljungdahl, Å., 1974*a*, Pharmaco-histochemical evidence of the existence of dopamine nerve terminals in the limbic cortex, *Eur. J. Pharmacol.* **25:**108.

Hökfelt, T., Ljungdahl, Å., Johansson, O., and Lindblom, D., 1974*b*, The vibratome: A useful tool in transmitter histochemistry, in: *Amine Fluorescence Histochemistry* (M. Fujiwara and C. Tanaka, eds.), pp. 1–12, Igaku Shoin, Tokyo.

Horn, A.S., Coyle, J.T., and Snyder, S.H., 1971, Catecholamine uptake by synaptosomes from rat brain: Structure–activity relationships of drugs with differential effects on dopamine and norepinephrine neurons, *Mol. Pharmacol.* **7:**66.

Jacobowitz, D., and Koelle, G.B., 1965, Histochemical correlations of acetylcholinesterase and catecholamines in postganglionic autonomic nerves of the cat, rabbit, and guinea-pig, *J. Pharmacol. Exp. Ther.* **148:**225.

Jacobowitz, D., and Kostrzewa, R., 1971, Selective action of 6-hydroxydopa on noradrenergic terminals: Mapping of preterminal axons of the brain, *Life Sci.* **10:**1329.

Jennings, B.M., 1965, Aldehyde–fuchsin staining applied to frozen sections for demonstrating pituitary and pancreatic beta cells, *J. Histochem. Cytochem.* **13:**328.

Jonsson, G., 1966, Fluorescence studies on some 6,7-substituted 3,4-dihydroisoquinolines formed from 3-hydroxytyramine (dopamine) and formaldehyde, *Acta Chem. Scand.* **20:**2755.

Jonsson, G., 1967*a*, Further studies on the specificity of the histochemical fluorescence method for the demonstration of catecholamines, *Acta Histochem.* (*Jena*) **26:**379.

Jonsson, G., 1967*b*, Fluorescence methods for the histochemical demonstration of monoamines. VII. Fluorescence studies on biogenic monoamines and related compounds condensed with formaldehyde, *Histochemistry* **8:**288.

Jonsson, G., 1967*c*, The formaldehyde fluorescence method for the histochemical demonstration of biogenic monoamines: A methodological study, M.D. thesis, Karolinska Institute, Stockholm.

Jonsson, G., 1969, Microfluorimetric studies on the formaldehyde-induced fluorescence of noradrenaline in adrenergic nerves of rat iris, *J. Histochem. Cytochem.* **17:**714.

Jonsson, G., 1971, Quantitation of fluorescence of biogenic monoamines, *Prog. Histochem. Cytochem.* **2:**299.

Jonsson, G., 1973, Quantitation of biogenic monoamines demonstrated with the formaldehyde fluorescence method, in: *Fluorescence Techniques in Cell Biology* (A.A. Thaer and M. Sernetz, eds.), pp. 191–197, Springer-Verlag, Berlin, Heidelberg, and New York.

Jonsson, G., and Sachs, C., 1969, Subcellular distribution of ^{3}H-noradrenaline in adrenergic nerves of mouse atrium: Effect of reserpine, monoamine oxidase and tyrosine hydroxylase inhibition, *Acta Physiol. Scand.* **77:**344.

La Vail, J.H., Vinston, K.R., and Tish, A., 1973, A method based on retrograde axonal transport of protein for identification of cell bodies of origin terminating within the CNS, *Brain Res.* **58:**470.

Lichtensteiger, W., 1969, Cyclic variations of catecholamine content in hypothalamic nerve cells during the estrous cycle of the rat, with a concomitant study of the substantia nigra, *J. Pharmacol. Exp. Ther.* **165:**204.

Lichtensteiger, W., 1970, Katecholaminhaltige neurone in der neuroendokrinen Steuerung, *Prog., Histochem. Cytochem.* **1:**185.

Lidbrink, P., and Jonsson, G., 1971, Semiquantitative estimation of formaldehyde-induced fluorescence of noradrenaline in central noradrenaline nerve terminals, *J. Histochem. Cytochem.* **19:**747.

Lindvall, O., 1975, Mesencephalic dopaminergic afferents to the lateral septal nucleus of the rat, *Brain Res.* **87:**89.

Lindvall, O., 1977, Combined visualization of central catecholamine- and acetylcholinesterase-containing neurons: Application of the glyoxylic acid and thiocholine histochemical methods to the same Vibratome section, *Histochemistry* **50:**191.

Lindvall, O., and Björklund, A., 1974*a*, The glyoxylic acid fluorescence histochemical method: A detailed account of the methodology for the visualization of central catecholamine neurons, *Histochemistry* **39:**97.

Lindvall, O., and Björklund, A., 1974*b*, The organization of the ascending catecholamine neuron systems in the rat brain as revealed by the glyoxylic acid fluorescence method, *Acta Physiol. Scand. Suppl.* **412**.

Lindvall, O., Björklund, A., and Falck, B., 1973, Glyoxylic acid condensation: A new fluorescence histochemical method for sensitive and detailed tracing of central catecholamine neurons, in: *Frontiers in Catecholamine Research* (E. Usdin and S.H. Snyder, eds.), pp. 683–687, Pergamon Press, Oxford.

Lindvall, O., Björklund, A., Nobin, A., and Stenevi, U., 1974*a*, The adrenergic innervation of the rat thalamus as revealed by the glyoxylic acid fluorescence method, *J. Comp. Neurol.* **154:**317.

Lindvall, O., Björklund, A., Moore, R.Y., and Stenevi, U., 1974*b*, Mesencephalic dopamine neurons projecting to neocortex, *Brain Res.* **81:**325.

Lindvall, O., Björklund, A., and Svensson, L.-Å., 1974*c*, Fluorophore formation from catecholamines and related compounds in the glyoxylic acid fluorescence histochemical method, *Histochemistry* **39:**197.

Lindvall, O., Björklund, A., Falck, B., and Svensson, L.-Å., 1975*a*, New principles for microspectrofluorometric differentiation between dopa, dopamine and noradrenaline, *J. Histochem. Cytochem.* **23:**697.

Lindvall, O., Björklund, A., Falck, B., and Svensson, L.-Å., 1975*b*, Combined formaldehyde and glyoxylic acid reactions. I. New possibilities for microspectrofluorometric differentiation between phenylethylamines, indolylethylamines and their precursor amino acids, *Histochemistry* **46:**27.

Lindvall, O., Björklund, A., and Divac, I., 1978, Organization of catecholamine neurons projecting to rat frontal cortex, *Brain Res.* **142:**1.

Ljungdahl, Å., Hökfelt, T., Goldstein, M., and Park, D., 1975, Retrograde peroxidase tracing of neurons combined with transmitter histochemistry, *Brain Res.* **84:**313.

Löfström, A., Jonsson, G., and Fuxe, K., 1976*a*, Microfluorimetric quantitation of catecholamine fluorescence in rat median eminence. I. Aspects on the distribution of dopamine and noradrenaline nerve terminals, *J. Histochem. Cytochem.* **24:**415.

Löfström, A., Jonsson, G., Wiesel, F.-A., and Fuxe, K., 1976*b*, Microfluorimetric quantitation of catecholamine fluorescence in rat median eminence. II. Turnover changes in hormonal states, *J. Histochem. Cytochem.* **24:**430.

Lorén, I., Björklund, A., Falck, B., and Lindvall, O., 1976*a*, An improved histofluorescence procedure for freeze–dried paraffin-embedded tissue based on combined formaldehyde–glyoxylic acid perfusion with high magnesium content and acid pH, *Histochemistry* **49:**177.

Lorén, I., Björklund, A., and Lindvall, O., 1976*b*, The catecholamine systems in the developing rat brain: Improved visualization by a modified glyoxylic acid–formaldehyde method, *Brain Res.* **117:**313.

Lorén, I., Björklund, A., Falck, B., and Lindvall, O., 1980*a*, The aluminum–formaldehyde (ALFA) method for improved visualization of catecholamines and indoleamines. I. A detailed account of the methodology for central nervous tissue using paraffin, cryostat or Vibratome sections, *J. Neurosci. Method.* **2:**277.

Lorén, I., Björklund, A., Lindvall, O., and Schmidt, R.H., 1980*b*, Improved methodology for catecholamine histofluorescence in the developing brain based on the magnesium and aluminum (ALFA) perfusion techniques for freeze–dried paraffin-embedded tissue, *Neuroscience* (in press).

Malmfors, T., 1965, Studies on adrenergic nerves, M.D. thesis, Stockholm.

Masuoka, D., and Placidi, G.-F., 1968, A combined procedure for the histochemical fluorescence demonstration of monoamines and microautoradiography of water-soluble drugs, *J. Histochem. Cytochem.* **16:**659.

Masuoka, D.T., Placidi, G.-F., and Gosling, J.A., 1971, Histochemical fluorescence microscopy and microautoradiography techniques combined for localization studies, in: *Progress in Brain Research,* Vol. 34, *Histochemistry of Nervous Transmission* (O. Eränkö, ed.), pp. 77–86, Elsevier, Amsterdam, London, and New York.

McLean, J.R., and Burnstock, G., 1966, Histochemical localization of catecholamines in the urinary bladder of the toad (*Bufo marinus*), *J. Histochem. Cytochem.* **14:**538.

Moline, S.W., and Glenner, G.G., 1964, Ultrarapid tissue freezing in liquid nitrogen, *J. Histochem. Cytochem.* **12:**777.

Nelson, J.S., and Wakfefield, P.L., 1968, The cellular localization of catecholamines in frozen–dried cryostat sections of the brain and autonomic nervous system, *J. Neuropathol. Exp. Neurol.* **27:**221.

Nielsen, K.C., and Owman, C., 1967, Adrenergic innervation of pial arteries related to the circle of Willis in the cat, *Brain Res.* **6:**773.

Norberg, K.-A., Ritzén, M., and Ungerstedt, U., 1966, Histochemical studies on a special catecholamine-containing cell type in sympathetic ganglia. *Acta Physiol. Scand.* **67:**260.

Nygren, L.-G., 1976, On the visualization of central dopamine and noradrenaline nerve terminals in cryostat sections, *Med. Biol.* **54:**278.

Olson, L., and Ungerstedt, U., 1970*a*, Monoamine fluorescence in CNS smears: Sensitive and rapid visualization of nerve terminals without freeze–drying, *Brain Res.* **17:**343.

Olson, L., and Ungerstedt, U., 1970*b*, A simple high capacity freeze–dryer for histochemical use, *Histochemie* **22:**8.

Owman, C., and Sjöstrand, N.O., 1965, Short adrenergic neurons and catecholamine-containing cells in vas deferens and accessory male genital glands of different mammals, *Z. Zellforsch.* **66:**300.

Owman, C., Sjöberg, N.-O., Svensson, K.-G., and Walles, B., 1975, Autonomic nerves mediating contractility in the human graafian follicle, *J. Reprod. Fertil.* **45:**553.

Parker, C.A., 1968, *Photoluminescence of Solutions,* Elsevier, Amsterdam, New York, and London.

Pearse, A.G.E., 1968, *Histochemistry, Theoretical and Applied,* J. & A. Churchill, London.

Pease, D.C., 1962, Buffered formaldehyde as a killing agent and primary fixative for electron microscopy, *Anat. Rec.* **142:**342.

Pesce, A.J., Rosén, C.-G., and Pasby, T.L., 1971, *Fluorescence Spectroscopy: An Introduction for Biology and Medicine,* Marcel Dekker, New York.

Placidi, G.-F., and Masuoka, D.T., 1968, Histochemical demonstration of fluorescent catecholamine terminals in cryostat sections of brain tissue, *J. Histochem. Cytochem.* **16:**491.

Pletscher, A., Gey, K.F., and Burkard, W.P., 1965, Inhibitors of monoamine oxidase and decarboxylase of aromatic amino acids, in: *Handbuch der experimentellen Pharmakologie* (O. Eichler and A. Farah, eds.), pp. 593–735, Springer-Verlag, Berlin, Heidelberg, and New York.

Ploem, J.S., 1969, A new microscopic method for the visualization of blue formaldehyde-induced catecholamine fluorescence, *Arch. Int. Pharmacodyn.* **182:**421.

Rauanheimo, L., and Eränkö, O., 1968, Simultaneous histochemical demonstration of acetylcholinesterase and noradrenaline in autonomic nerve fibres, *Scand. J. Clin. Lab. Invest.* **21**(Suppl. 101):44.

Richardson, K.C., 1968, Cholinergic and adrenergic axons in methylene blue-stained rat iris: An electronmicroscopical study, *Life Sci.* **7**(Part I):599.

Ritzén, M., 1966*a*, Quantitative fluorescence microspectrophotometry of 5-hydroxytryptamine–formaldehyde products in models and in mast cells, *Exp. Cell Res.* **45:**178.

Ritzén, M., 1966*b*, Quantitative fluorescence microspectrophotometry of catecholamine–formaldehyde products, *Exp. Cell Res.* **44:**505.

Ritzén, M., 1967, Cytochemical identification and quantitation of biogenic monoamines—A microspectrofluorimetric and autoradiographic study, M.D. thesis, Karolinska Institute, Stockholm.

Ritzén, M., 1973, Microfluorimetric quantitation of biogenic monoamines, in: *Fluorescence Techniques in Cell Biology* (A.A. Thaer and M. Sernetz, eds.), pp. 183–189, Springer-Verlag, Berlin, Heidelberg, and New York.

Rost, F.W.D., and Pearse, A.G.E., 1971, An improved microspectrofluorimeter with automatic digital data logging: Construction and operation. *J. Microsc.* **94:**93.

Ruch, F., 1973, The use of human leucocytes as a standard for the cytofluorometric determination of protein and DNA, in: *Quantitative Fluorescence Techniques as Applied in Cell Biology* (M. Sernetz and A. Thaer, eds.), pp. 51–55, Springer-Verlag, Berlin, Heidelberg, and New York.

Sachs, C., 1970, Noradrenaline uptake mechanisms in the mouse atrium, *Acta Physiol. Scand. Suppl.* **341**.

Sachs, C., and Jonsson, G., 1972, Degeneration of central and peripheral noradrenaline neurons produced by 6-hydroxy-DOPA, *J. Neurochem.* **19:**1561.

Späth, E., and Lederer, E., 1930, Synthesen von 4-Carbolinen, *Ber. Dtsch. Chem. Ges.* **63:**2102.

Spriggs, T.L.B., Lever, J.D., Rees, P.M., and Graham, J.D.P., 1966, Controlled formaldehyde–catecholamine condensation in cryostat sections to show adrenergic nerves by fluorescence, *Stain Technol.* **41:**323.

Stumpf, W.E., 1976, Techniques for the autoradiography of diffusible compounds, in: *Methods in Cell Biology* (D.M. Prescott, ed.), p. 171, Academic Press, New York.

Stumpf, W.E., and Roth, L.J., 1969, Autoradiography using dry-mounted, freeze–dried sections, in: *Autoradiography of Diffusible Substances* (L.J. Roth and W.E. Stumpf, eds.), p. 69, Academic Press, New York and London.

Svendgaard, N.A., Björklund, A., and Stenevi, U., 1975, Regenerative properties of central monoamine neurons, *Adv. Anat. Embryol. Cell Biol.* **51:**7.

Svensson, L.-Å., Björklund, A., and Lindvall, O., 1975, Studies on the fluorophore forming reactions of various catecholamines and tetrahydroisoquinolines with glyoxylic acid, *Acta Chem. Scand. Ser. B* **29:**341.

Udenfriend, A., 1969, *Fluorescence Assay in Biology and Medicine,* Academic Press, New York and London.

Ullberg, S., and Appelgren, L.E., 1969, Experience in locating drugs at different levels of resolution, in: *Autoradiography of Diffusible Substances* (L.J. Roth, and W.E. Stumpf, eds.), p. 280, Academic Press, New York and London.

Ungerstedt, U., 1971, Stereotaxic mapping of the monoamine pathways in the rat brain, *Acta Physiol. Scand. Suppl.* **367**.

Van Orden, L., 1970, Quantitative histochemistry of biogenic amines: A simple microspectrofluorometer, *Biochem. Pharmacol.* **19:**1105.

Waris, T., and Partanen, S., 1975, Demonstration of catecholamines in peripheral adrenergic nerves in stretch preparations with fluorescence induced by aqueous solution of glyoxylic acid, *Histochemistry* **41:**369.

Waris, T., and Rechardt, L., 1977, Histochemically demonstrable catecholamines and cholinesterases in nerve fibres of rat dorsal skin, *Histochemistry* **53:**203.

Waris, T., Rechardt, L., and Partanen, S., 1977, Simultaneous demonstration of cholinesterase and glyoxylic acid-induced fluorescence of catecholamines in stretch preparations, *Acta Histochem.* **58:**194.

Watson, S.J., and Barchas, J.D., 1975, Histofluorescence in the unperfused CNS by cryostat and glyoxylic acid: A preliminary report, *Psychopharmacol. Commun.* **1:**523.

Watson, S.J., and Barchas, J.D., 1977, Catecholamine histofluorescence using cryostat and glyoxylic acid in unperfused frozen brain: A detailed description of the technique, *Histochem. J.* **9:**183.

Whaley, W.M., and Govindachari, T.R., 1951, The Pictet–Spengler synthesis of tetrahydroisoquinolines and related compounds, in: *Organic Reactions,* Vol. 6 (R. Adams, ed.), pp. 151–206, John Wiley, New York.

Wright, W.D., 1944, *The Measurement of Colour,* Hilger, London.

7

Electron Microscopy in Neurobiology

G. Vrensen, D. De Groot, and A. Boesten

1. Introduction

The formulation of the cell theory by Schleiden and Schwann in 1838 opened a whole new world of morphological research. The recognition that all higher organisms are composed of numerous, very small, hierarchically organized, and to some extent independent subunits (*cells*) was one of the great milestones in biology. This recognition was made possible mainly by the improvement of the glass lenses used in compound light microscopes as originally invented by Hooke in 1665. At the end of the 19th century, adequate techniques for tissue preservation, sectioning of suitable slices, and more or less specific staining had been developed. The results of these technical improvements have led to the enormous wealth of morphological information on the cellular organization of animals and plants. In the same century, scientists such as Wöhler, Kolbe, Berthelot, Pasteur, and Buchner recognized that living organisms are aggregates of "organic" molecules that obey the same laws as described in "inorganic" chemistry. The many improvements and innovations in chemical research have led to the vast amount of information on the chemical composition and the dynamic interaction of molecules of living organisms presented in present-day handbooks of biochemistry.

Despite the largely independent development of histology and biochemistry, many points of shared interest exist between these biological disciplines, some of which have been fruitfully explored, e.g., in histo- and cytochemistry and in cell- and tissue-culture studies. A fundamental discussion of the

G. Vrensen and D. De Groot • Department of Electron Microscopy, Mental Hospital "Endegeest," Oegstgeest, The Netherlands. ***A. Boesten*** • Department of Neuroanatomy, University of Leiden, Leiden, The Netherlands. Dr. Vrensen's present address is: The Netherlands Ophthalmic Research Institute, P. O. Box 6411, 1005 EK Amsterdam, The Netherlands.

intrinsic relationship between the structure of cells and the function of the macromolecules present in cells was meaningless, however, until quite recently. The main reason for this incompatibility was the discrepancy between biochemical and light-microscopic resolution. Biochemists deal with molecules and macromolecules with dimensions between 10 and 100 Å (0.001–0.01 μm), whereas light microscopists can at best distinguish details of about 2000 Å (0.2 μm). The resolution of the light microscope is limited by the wavelength of visible light and cannot be improved by introduction of better lenses or by any other means.

In 1920's and 1930's, physicists such as Busch, Ruska, and Von Borries were the first to construct a microscope in which accelerated electrons were used as the imaging medium. The expectation that this type of microscope would improve resolution was based on the theory of De Broglie that moving electrons can be assigned both a corpuscular and a wave character. The wavelength of accelerated electrons is of the order of magnitude of 0.05 Å. This would enable resolution of less than 10 Å. It took another twenty to thirty years of intensive technological research to achieve this resolution. At present, a resolution of 2–3 Å can be obtained. The invention and further improvement of the electron microscope were accompanied by the improvement of the classic histological techniques. The introduction of new fixatives for more accurate tissue preservation, the use of embedding media suitable for ultrathin sectioning with specially designed ultramicrotomes, and the development of suitable staining procedures enabled the fine-structural study of biological material at a resolution down to 20 Å. The growth of this new field of morphological research has been very expansive since its introduction in the 1950's by scientists such as Palay, Palade, Sjöstrand, Robertson, and Bennett. The integration of ultrastructural observations with classic histology and cytology has often deepened and sometimes changed our insight into the structural organization of animals and plants. It also brings within our grasp the study of the fundamental relationship between the biochemical functioning of cells and the ultrastructural elements of the cells. This functional–structural relationship has been largely worked out for a number of essential cellular processes, e.g.: (1) the synthesis of proteins and the importance of ribosomes, endoplasmic reticulum, and Golgi cisterns in this process; (2) the role of mitochondria and their elementary membrane particles in the cellular fixation of energy; (3) the process of cellular digestion and its relationship to lysosomes; (4) the regularly spaced filaments of muscles and the process of contraction; and (5) the important role of membranes in the process of cellular exchange of materials.

In neurobiology, electron microscopy has played a role similar to that in many other fields of biological interest. It has enlarged our understanding of the basic organization of cells—in this instance, neurons and glial cells. Furthermore, it has finally proved the cell theory (neuron doctrine) for the nervous system by visualization of the sites of effective transmission between axons and dendrites as highly organized organelles separated by membranes. The fine-structural observations on dendrites and myelinated or unmyelinated axons have contributed greatly to a better understanding of the propagation of impulses. The role of astrocytes as a cellular interstitium is elucidated by electron microscopy, as is the role of oligodendrocytes and Schwann cells in the formation of myelin. The distinction of macroglial and microglial cells has been much more difficult. Furthermore, the electron microscope has unraveled a number of complex synaptic relations.

Although it may have been superfluous, this short introduction on the relevance of electron microscopy in (neuro)histology and (neuro)biology is meant to point out the basic aspects with which we will deal in this chapter: the electron microscope, the techniques of electron-microscopic histology, and some applications in the field of neurobiology. At present, interest in the quantitative aspects of structural organization is growing rapidly. Therefore, we have included a short section on quantitative stereology. It can never be the intention of an introduction to electron microscopy to cover in full detail the various aspects of the electron microscope and the related histological techniques. We intend to introduce only basic principles and rationales. For more detailed information, the reader is referred to the books and papers listed in the bibliography at the end of this chapter.

2. *The Transmission Electron Microscope: Basic Principles and Design*

2.1. *Introduction*

Although electron microscopes are basically complicated instruments, nowadays they are reliable, versatile, and easy to operate. This means that they have found wide application and can be used by neurobiologists who have not specialized in electron microscopy. The intention of this section is to introduce the basic principles of image formation and resolution and of the construction and use of the transmission electron microscope. More detailed information can be found in a number of handbooks on electron microscopy, simple as well as specialized (e.g., Glauert, 1973–1975; Hayat, 1970–1974; Meek, 1970; Sjöstrand, 1967).

2.2. *Resolving Power*

As outlined in Section 1, a major challenge for cell biologists is to correlate cell structure with cell function. As described by biochemists, metabolic processes take place at the macromolecular level. These macromolecules have dimensions between 10 and 100 Å. The light microscope can differentiate (resolve) only those details in a specimen that are separated by about 0.2 μm (2000 Å). This limitation in resolving power is fundamental and therefore cannot be changed by improving the quality of the instrument. It is restricted by the optical processes known as diffraction and interference.

When a wave front hits an edge or a point in the specimen, this edge or point starts functioning as a new source of waves [Hygens (1629–1695)]. The original advancing wave front and the new wave front, coming from the edge or point, will interfere. If at a particular place and at a particular time both wave fronts have the same phase, they will reinforce; if they have opposite phases, their individual effects will be nullified. As a consequence, a point is observed as a central disc [Airy-disc: Airy (1801–1892)] surrounded by dark and light fringes [Fresnel fringes: Fresnel (1788–1827)]. The radius of the central disc (Fig. 1a) is given by $r = \lambda/2\beta$, in which λ is the wavelength and β is half the aperture of the image side. Because of this Airy-disc configuration of a specimen point in the image plane, two points can be seen as separated only if their combined intensity curves show a drop large enough to be detected by the eye. According to Rayleigh [1842–1919], this is the case when the first minimum of one Airy-disc coincides with the maximum of the other and is given by $\lambda/2\beta$ (Fig. 1b). The relationship of the distance between two points in the specimen (d) and the same points in the image plane d' is given by the sinus rule of Abbe [1840–1905], $dn\sin\alpha = d'n'\sin\beta$ (Fig. 1c), in which n and n' are the refractive indexes at the illumination side and

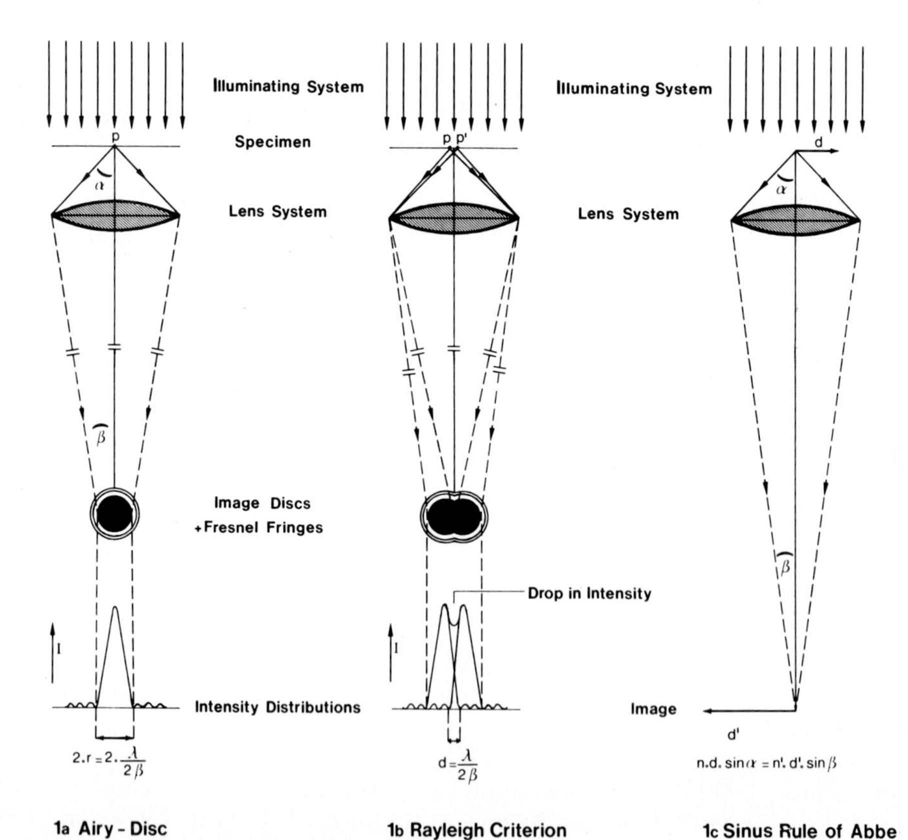

Fig. 1. Schematic illustration of the three basic principles of microscopic resolution. (1a) Formation of an Airy-disc with Fresnel fringes; (1b) Rayleigh criterion; (1c) sinus rule of Abbe. For further explanation, refer to the text.

image side, respectively. In most microscopes, β is very small and image formation is in air, so that $\sin\beta = \beta$ and $n' = 1$. This implies that $dn\sin\alpha = d'\beta$. As outlined, the minimal distance between two points in the image that can be seen separated is $d' = \lambda/2\beta$. This means that $dn\sin\alpha = (\lambda/2\beta)\beta$, and $d = \lambda/2n\sin\alpha$. This formula derived for one-sided illumination using rectangular apertures has to be replaced for microscopes with conical illumination and round apertures by

$$d = \frac{\lambda}{1.64\ n\ \sin\alpha} \tag{1}$$

Using white light, an objective aperture of 140–150°, and a refractive index of 1.5, a resolution of 0.2–0.25 μm, at best, is obtainable. Use of light with shorter wavelength will improve this resolution, but will never bring it down to the dimensions of macromolecules.

As first described by De Broglie, accelerated electrons have a dualistic character. They have properties as described for particles, and at the same time they behave like electromagnetic waves. They have a distinct wavelength and have the same diffraction and interference properties as described above for light. Their wavelength is given by De Broglie in the formula $\lambda = h/mv$, in which h is Planck's constant, m is the mass of electrons, and v is the velocity of electrons. The velocity of the electrons is determined by their charge (e), their mass (m), and the potential difference (V) over which they have been accelerated:

$$eV = \tfrac{1}{2}mv^2 \text{ or } v = \left(2\frac{e}{m}V\right)^{1/2}$$

When we substitute v and the known values for h, m, and e in De Broglie's formula, the wavelength is given by $\lambda = 12.3/V^{1/2}$, in which λ is given in Å units and V in volts. Using the resolution formula given above [equation (1)], the theoretical resolving power of the electron microscope is

$$d = \frac{12.3}{1.64\ n\ \sin\alpha\sqrt{V}}$$

In electron microscopes, image formation is in vacuum, so that $n = 1$, and the illuminating aperture is very small ($\approx 10^{-2}$ radians), so that $\sin\alpha = \alpha$. The resolving power of the electron microscope can therefore be described by the formula

$$d = \frac{7.5}{\alpha V^{1/2}} \tag{2}$$

in which d is in Å units and V is in volts. At accelerating voltages of 10 and 100 kV, a resolving power of 3.7 and 1.7 Å can be expected theoretically.

The actual resolving power of electron microscopes depends, however, on how far the designers have been able to correct for lens aberrations, such as spherical aberration, chromatic aberration, curvature of the field, astigmatism, coma, and space charge distortion. In recent electron microscopes, most of these aberrations have been corrected, and a resolving power close to the expected one can be obtained: 4-5 Å at 80 kV (see Fig. 2).

2.3. Image Formation and Contrast

The electron beam is generated in the electron gun, is focused in the central plane of the objective lens by the condensor lens system, and finally illuminates the fluorescent screen (see Section 2.6) as a homogeneous stream of electrons. This means that in the absence of a specimen, the density of electrons per unit area and per unit of time is equal for all sites on the screen. The central problem in electron-microscopic image formation is how the structural information as present in the specimen is to be transferred to this homogeneous stream of electrons. This transfer of structural information is achieved by modulation of the local density of electrons hitting the screen. This modulation has to meet some requirements: (1) it must be related to the internal structure of the specimen in an absolute way; (2) the local differences must be high enough to be detected by the human eye (contrast). The main factor operating in electron-microscopic imaging is scattering. Furthermore, diffraction and phase-contrast phenomena play important roles.

Electrons penetrating a specimen interact with the atomic nuclei and orbital electrons present in the specimen. As a consequence of the electrostatic interactions between atomic nuclei and beam electrons, the beam electrons deviate from their straight path. This deflection by atomic nuclei is associated with a negligible loss of energy of the beam electrons and is called *elastic scattering*. The angular deflection is proportional to the charge of the atomic nuclei (atomic number) and decreases with increasing initial velocity of the

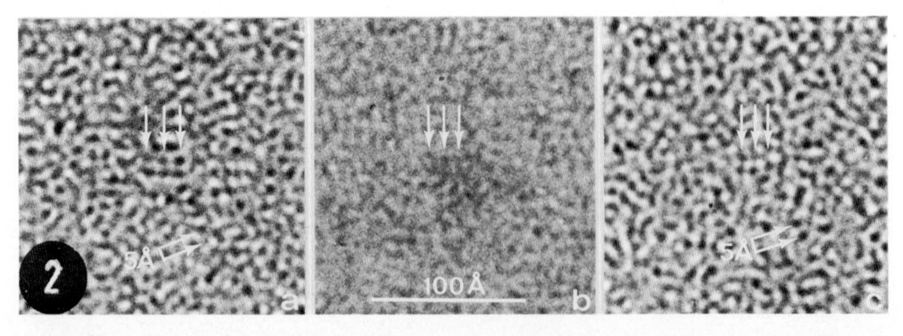

Fig. 2. High-resolution picture illustrating the phase-contrast image of a formvar–carbon film at different focus positions. (2a) 750 Å underfocus; (2b) nearly in focus; (2c) 750 Å overfocus. Note the change in granularity from focus to under- and overfocus. (↓ ↓ ↓) Identical granules in the three focus positions from which the resolution can be measured. Resolution is approximately 5–7 Å.

incident electrons. The interaction of beam electrons and orbital electrons also gives rise to deflection of the beam electrons. This interaction is associated with a certain loss of energy and is called *inelastic scattering*. The angular deflection of inelastic scattered beam electrons is proportional to the unit charge (e) of the orbital electrons and inversely related to the velocity of the beam electrons. The deflection due to elastic scattering is much larger ($\geq 10^{-2}$ radians) than that due to inelastic scattering ($< 10^{-3}$ radians). Therefore, in common bright-field transmission electron microscopy, elastic scattering is the most important factor contributing to image formation. Due to this phenomenon, the initial homogeneous beam of electrons is modulated into ongoing or weakly deflected electrons and strongly deflected electrons. The strongly deflected electrons are intercepted in the physical aperture of the objective lens and cease to contribute to the electron density in the image plane. The final effect of this scattering and interception of electrons is that dark and light spots appear in the image plane. These dark and light spots represent sites in the specimen at which large and small amounts of heavy-metal atoms are present.

Direct collisions between beam electrons and atomic nuclei and orbital electrons play hardly any role in image formation in the transmission electron microscope. The formation of secondary electrons and X-rays as the result of such collisions is an important factor in scanning electron microscopy, however (see Section 2.7 and Chapter 8). Absorption, which is a main factor in light microscopy, is not important in electron-microscopic image formation. In thin biological specimens as used in transmission electron microscopy, the local densities of atoms are not high enough to give rise to manifold collisions or inelastic scatterings, which would stop the beam electrons completely.

Mainly, three factors play a role in the image formation of routine biological specimens in the transmission electron microscope: (1) An increase in the velocity of beam electrons, due to an increase of the high tension (kV), decreases the angular deflection and therefore decreases the contrast. Thus, higher contrast is obtained at lower accelerating voltages. However, a lower accelerating voltage has a negative effect on resolution. Lower accelerating voltage means lower velocity and therefore lower resolution [see equation (2) in Section 2.2). (2) The diameter of the objective aperture determines the interception of scattered electrons. The smaller its diameter, the larger the number of scattered electrons intercepted. However, it also affects the size of the Airy-disc. The width of the Airy-disc is inversely related to the size of the aperture. This means that the resolution becomes poorer with decreasing objective-aperture diameter. (3) Thicker sections, which consequently have higher densities of heavy-metal atoms, have brighter contrast. However, this also has a negative effect on resolution. It is clear that each of these factors has opposing effects on contrast and resolution. This implies that the investigator has to make a compromise between highest resolution and optimal contrast. A correct decision depends on the type of specimens used and the type of information desired.

There are two more aspects of image formation that we will briefly mentioned here. In unstained biological specimens, the number of heavy-metal atoms is negligible, and therefore no scattering occurs in such specimens. In such instances, the image, obtained at high magnifications, is the result of phase differences. Because of small local variations in density, the ongoing nonscattered electrons experience small phase changes while traversing the specimen. The interpretation of such phase-contrast images and the apparent enhancement of the phase differences at slightly defocused positions (see Fig. 2) is very difficult. For this, the interested reader is referred to more specialized literature (e.g., Glauert, 1973–1975; Hayat, 1970–1974).

The other aspect, focus, is defined as that setting of the objective lens at which no Fresnel fringes are visible in the image (Fig. 3). At this focus setting, the images have highest resolution but also lowest contrast. It is common practice in routine biological transmission electron microscopy to take pictures at a slightly defocused position and so to sacrifice resolution for contrast. This routine procedure is one of the main reasons that interpretation of structures smaller than 20 Å should not be attempted in biological sections. From a practical point of view, this means that resolution in biological sections is about 20–25 Å.

2.4. *Depth of Field and Depth of Focus*

The depth of field can be defined as the distance (D) between two image planes in the specimen that are simultaneously just in focus at the same objective setting. As in light microscopy, this depends on the illuminating angle (α) and the resolving power (d) of the microscope according to the formula $D = 2d/tg\alpha$, or $D = 2d/\alpha$ because α is very small in electron microscopes and $tg\alpha \approx \alpha$. For instance, with a resolving power of 6 Å, D is about 0.4 μm. This depth of field is large in comparison with the actual thickness of electron-microscopic sections (0.05–0.1 μm). As a consequence of this, the electron image is the superposition of all elements at the same vertical position in the specimen; the electron image is a projected image. This limits the actual resolution in biological specimens, because small objects

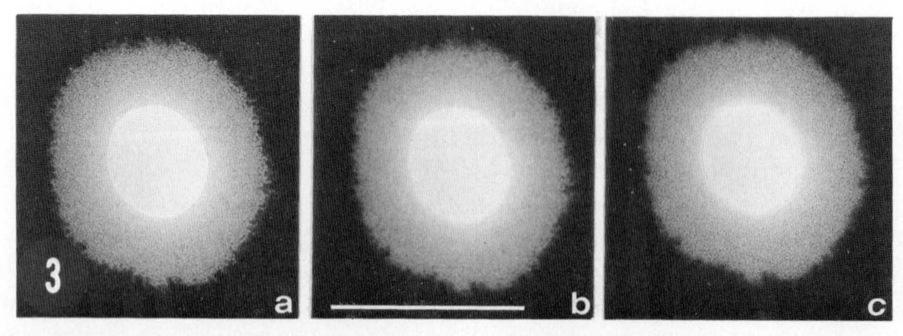

Fig. 3. Fresnel fringes around a small hole in a carbon film at different focus positions. (3a) 1250 Å underfocus; (3b) nearly in focus; (c) 1250 Å overfocus. Note as well the change in granularity of the carbon particles directly around the hole. Bar: 1 μm.

superimposed in the specimen will be imaged at exactly the same point in the image plane. It also implies that an increase in section thickness, which enhances the contrast, will decrease the resolution.

The angular aperture on the image side is also small in the electron microscope. Therefore, the depth of focus, defined as the distance between two image planes in which the electron image is still in focus, is also very large. At the objective stage, this focal depth is of the order of magnitude of millimeters; at the projector stage, of meters. This large focal depth has an important practical advantage. The recording systems (screen, photographic cameras, and television recording system) can be placed far away from each other and can be used alternatively with the same focusing position of the objective lens without affecting the sharpness of the final image.

2.5. Magnification

In this discussion of the basic principles of image formation in the electron microscope, we have not spoken about magnification. The reason is that in most microscopes, a fixed objective lens is present. This objective lens with a standard magnification between ×2000 and ×3000 defines the resolution of the microscope, and its image contains all the information transferred by the specimen to the electron beam, as described in Section 2.3. Two or more lenses further increase or decrease the magnification of the image formed by the objective lens. The final magnification ranges from ×1000 to ×500,000.

2.6. Design of the Transmission Electron Microscope

The path of a beam of electrons can be modulated by electromagnetic and electrostatic fields. Present microscopes are equipped with electromagnetic lenses. In their most elementary form, they consist of a disclike solenoid. When an electrical current is flowing through the windings of the solenoid, a magnetic field is induced (Fig. 4a). Electrons passing through this magnetic field deviate from their straight path and are focused on the axis at some distance behind the solenoid. The focal distance depends on the strength of the magnetic field, which in turn depends on the strength of the electrical current and the concentration of the magnetic field lines. When the solenoid is surrounded by a soft iron shroud (Fig. 4b), the magnetic field lines are concentrated near the gap in this shroud. This concentration of magnetic field lines is further enhanced by the introduction of pole pieces, as is the case in most electron-microscopic lenses (Fig. 4c). With these pole pieces, which have borings of several millimeters, the magnetic field lines are concentrated in very small areas and focal lengths of several millimeters can be obtained. From this short description of electron lenses, it follows that the construction of a perfect electron lens is rather difficult. The quality of the iron shroud and the pole pieces is critical. Irregularities in the metal result in irregularities in the magnetic field and give rise to aberrations in the

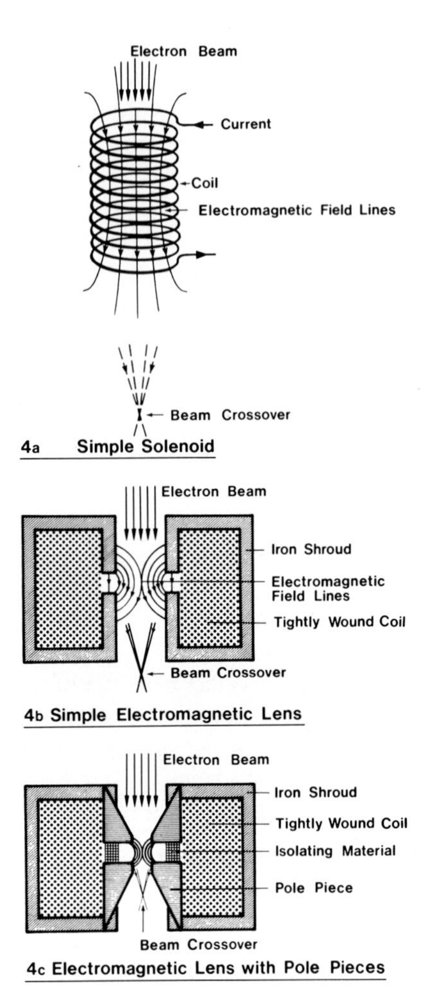

Fig. 4. Functional diagrams of electron-microscope lenses. Consult the text for discussion of their functioning.

image. Furthermore, the manufacturers of electron microscopes have to correct for a number of lens aberrations, mainly the same as in light microscopy. We shall briefly indicate these aberrations.

In *spherical aberration*, central rays and peripheral rays are focused at different sites. This spherical aberration gives rise to different types of distortion. In light microscopy, sphereical aberration is corrected by using systems of positive and negative lenses. However, electromagnetic lenses are positive in nature only. Spherical aberration is related to the third power of the illuminating aperture (Sjöstrand, 1967). A strong reduction of spherical aberration can be obtained by using very small illuminating apertures.

In *chromatic aberration*, rays with different wavelengths are focused at different sites. Changes in wavelength are caused in electron microscopy by fluctuations in the emission spectrum of the electron source, instability of the accelerating voltage, and inelastic scattering within the specimen (see Section 2.3).

Astigmatism is caused by the fact that even high-quality electron lenses do not have absolute axial symmetry. This means that focal strength is weaker in one plane and stronger in the other.

As already outlined, electrons are scattered and absorbed by atoms and molecules. Electrons passing through 0.2 mm of air are scattered over an angle of 10° and are stopped completely within 20 cm (Sjöstrand, 1967). This implies that electron microscopes have to be well evacuated; otherwise, image formation is impossible. This demand for high vacuum involves a number of technical problems with respect to sealing of the column and the insertion of specimen and cameras into the pathway of the electron beam.

The human eye is insensitive to electrons. Therefore, it is necessary to have a fluorescent screen to visualize the electron image. On this screen, the kinetic energy of electrons is transformed into visible light and X-rays. These X-rays, which are also generated at the different apertures in the microscope, are harmful, and designers must incorporate a number of lead shields in the microscopes, including lead glass in the viewing chamber.

In Fig. 5, the major components of an electron microscope are indicated. Details of the different components are given for the Philips EM 301 electron

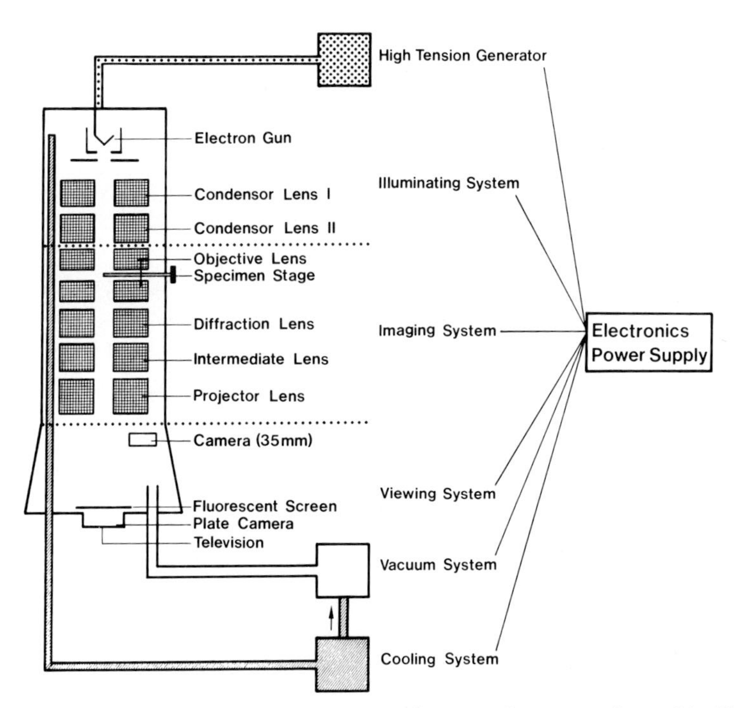

Fig. 5. Major components of the Philips EM 301 electron microscope pictured in Fig. 6. The functions of the various components are discussed in the text.

microscope in Fig. 6. In this section, we shall describe these components briefly, inasmuch as such a description is relevant for a good understanding of routine microscopy. The characteristics of the different lenses, which are of ultimate importance for the quality of the electron image, can be found elsewhere.

2.6.1. Illuminating System

Because of spherical and chromatic aberration, which can greatly reduce the quality of the electron image, the angular aperture of illumination at the specimen must be very small and the wavelength of the electron beam very constant. These theoretical demands have greatly influenced the design of the illuminating system.

This illuminating system consists of the electron gun and the condensor lenses. In its elementary form, the electron gun is a triode valve. The cathode, which is in fact a V-shaped tungsten wire, is kept at a negative potential of 20–100 kV (accelerating voltage), as compared to the anode, which is at earth potential. Between the cathode and the anode, there is placed a grid (Wehnelt

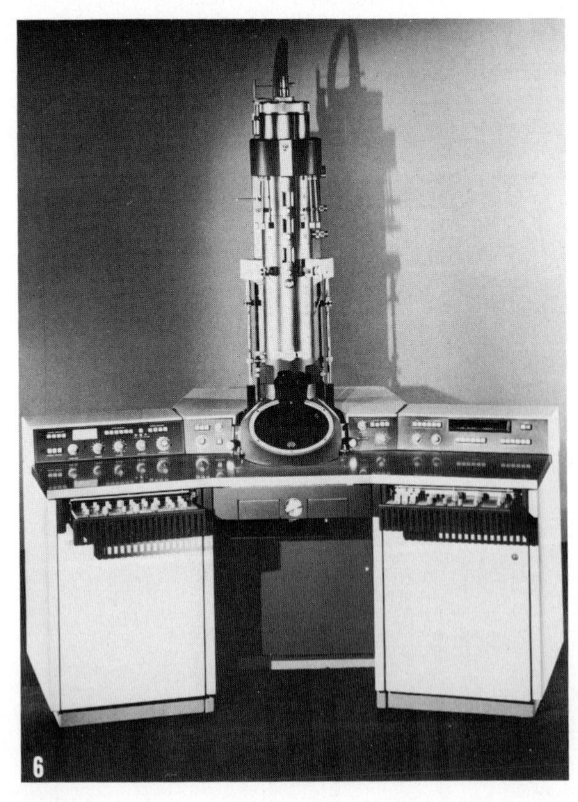

Fig. 6. The Philips EM 301, a high-resolution electron microscope. Courtesy of Philips, Eindhoven, The Netherlands.

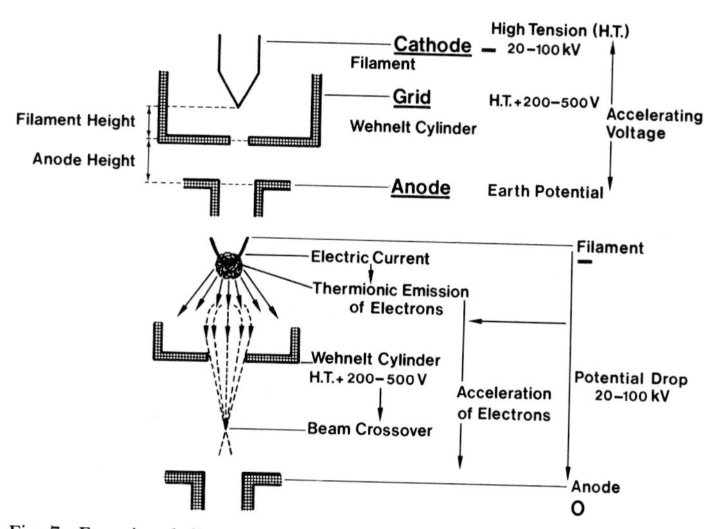

Fig. 7. Functional diagram of an electron gun. Consult the text for details.

cylinder) that has a potential of 100–500 V negative to the filament (Fig. 7). When the filament is heated by an electrical current to a temperature between 2500 and 2800°K, a number of electrons leave the filament because of thermionic emission. These electrons are accelerated because of the potential difference between the filament (cathode) and the anode. Because of the presence of the Wehnelt cylinder, the electrostatic field lines are such that the electrons are focused in a small spot close to the anode. The heating current of the filament, the distance between the filament and the Wehnelt aperture, the potential difference between the filament and the shield, and the potential drop between the filament and the anode all affect the number of electrons coming from the filament and the size and site of the beam crossover. The stability of these currents and potential drops determines to a great extent the quality of the final image. The condensor system, which consists of one or two condensor lenses, further demagnifies this beam crossover, which is the actual electron source. The amount of electrons hitting the specimen and the angular aperture of this electron beam are determined by these condensor lens(es). Nowadays, most microscopes are equipped with a double-condensor-lens system, which has the advantage that the size of the electron spot on the specimen can be regulated more precisely so that even at high magnification the intensity of the electrons hitting the specimen is high enough to obtain an image on the screen. Moreover, the angular aperture of illumination can be better regulated. As we have seen, this aperture greatly affects spherical aberration and therefore the quality of the image. The interchangeable aperture present in the second condensor lens is important inasmuch as it limits the angular aperture of the illuminating beam in under- and overfocused positions of the condensor lenses. This defocusing of the condensor lenses is necessary to overcome damage of the specimen and to regulate the image intensity for photographic registration.

2.6.2. Objective Lens and Specimen Stage

As already discussed in Sections 2.2–2.5, the objective lens is decisive for the actual resolving power of the electron microscope. The solenoid, the iron shroud, and the pole pieces must be of high quality, as must the stability and fine control of the lens current. Special attention has been given to the correction of astigmatism, which is caused not only by irregularities in the lens itself but also by irregularities and dirt in the objective aperture and the specimen holder. The objective lens is equipped with an astigmatism corrector that consists of four pairs of small solenoids. When these solenoids are excited, they introduce additional field lines that modulate the main magnetic field lines of the objective lens. In this way, the field can be made completely axially symmetrical. The specimen is placed in the narrow space between the two pole pieces of the objective lens. The displacement of the specimen holder has to be precisely controlled. In some instances, special manipulations need to be carried out with this specimen holder, such as tilting, heating, and cooling. It is clear that these demands for the specimen holder result in a complicated specimen stage. Moreover, the introduction of the specimen into the evacuated column introduces some air. Without further precautions, this air would damage the filament when it is switched on. Therefore, either the illuminating system has to be turned off during the change of specimen or the microscope is equipped with a prevacuum system at the specimen stage. A special problem is the contamination of the specimen. Because of the bombardment with electrons of the residual gas present in the column and of the gas due to evaporation of the specimen, a certain amount of dirt is present in the column. This dirt tends to be deposited on the specimen and thereby reduces the quality of the specimen and its final image. Introduction of an anticontamination device at a low temperature (liquid nitrogen, −190°C) close to the specimen has proved to reduce this contamination. In present microscopes, such a device is built in. It is clear even from this brief description that for the construction of a reliable objective lens and specimen stage, a great number of technological problems must be solved.

2.6.3. Projector Lens System

The image formed by the objective lens, with its fixed magnification and resolution, is either enlarged or reduced by a system of two or three lenses; the diffraction lens, the intermediate lens, and the projector lens. The range of magnification that can be obtained in this way is between ×1000 and ×500,000 on the screen. Many microscopes also have the possibility to work at very low magnification, with a range between ×100 and ×4000. This greatly facilitates a bridging of the gap between light microscopy and electron microscopy. This low magnification is obtained by fixing the objective-lens current at a low value and using the diffraction lens as the imaging and focusing lens. The resolution of this setting is somewhat poorer, but at such low magnification this does not matter.

2.6.4. *Viewing Chamber*

The electron image is made visible on a fluorescent screen, where the kinetic energy of the electrons is transformed into visible light. Many microscopes have an additional small screen that can be viewed with a binocular microscope that magnifies the screen picture by a factor of 10. This facilitates viewing of the fine focusing. Most microscopes are equipped with one or more cameras. Because of the great depth of focus in electron microscopes (see Section 2.4), they can be placed at different sites in the viewing chamber. A plate camera, for glass plates or sheet films, a 35 mm camera, and a 70 mm camera enable the recording of over 100 pictures without reloading. Quite recently, electron microscopes have also been connected to a TV monitor, and the image can be recorded on videotape. When combined with an image-intensifier, video systems enable higher contrast and higher magnification, with the advantage that the specimen can be viewed at low electron intensity, which reduces damage and contamination of the specimen.

2.6.5. *Vacuum System*

As already noted, gas molecules interfere with electrons that are scattered in their presence. Therefore, the microscope column is evacuated to 10^{-3} to 10^{-6} torr (atmospheric pressure = 760 torr). Nowadays, microscopes are usually equipped with three pumps, a rotary prevacuum pump, a mercury booster pump, and an oil diffusion pump.

2.6.6. *Cooling System*

The electromagnetic lenses produce heat when they are in operation. This heat would change the electromagnetic characteristics of the lenses were it not removed. Moreover, the diffusion pumps can work only when the mercury or oil is heated on one side and cooled on the other. Therefore, the lenses and diffusion pumps are ensheathed by a system of small pipes through which cooling water is pumped. This cooling water comes either from the tap or from a special closed cooling unit. A cooling unit has the advantages that the temperature of the circulating water can be kept constant, that no tapwater has to be spoiled, and that special agents that prevent corrosion can be used.

2.6.7. *High-Tension Generator*

The electrons, generated by the tungsten filament in the electron gun (see Section 2.6.1), have to move through a potential difference of 20–100 kV to become a beam of electrons useful for electron imaging. The high voltage is supplied by a special high-tension generator. This generator is a specially designed transformer that transforms 220 V A.C. up to 100 kV D.C. At such high tensions, unwanted discharges easily occur in air. Therefore, the transformer is immersed in oil, which prevents most of these discharges. Changes in the high tension affect the acceleration of electrons and therefore wavelength and resolution. The designers of electron microscopes have paid special attention to stabilization of these high-voltage discharges.

2.6.8. *Electronics*

It is quite obvious that sophisticated electronics are needed to control the functioning of the components of the electron microscope described. The introduction of integrated circuits (IC) has improved the reliability of microscopes and has reduced the space needed for installation. The stability of lens currents greatly affects the quality of the final electron-microscopic image. Stabilization factors of 2×10^{-6} V are obtained in present-day microscopes.

2.7. *Special Applications*

In the previous sections, we have outlined the basic principles of the most currently used mode of application of the electron microscope, namely, the bright-field transmission electron microscopy at accelerating voltages up to 100 kV. In this section, we will briefly describe some other applications.

2.7.1. *Dark-Field Electron Microscopy*

In most biological specimens prepared for electron microscopy, the amount of heavy-metal atoms introduced during fixation and staining (see Section 3) is quite large. This means that the number of electrons scattered in the specimen is high and their angular scattering quite large. Therefore, many scattered electrons are intercepted by the objective aperture and a final image with sufficient contrast is obtained. In some instances, however, it is not advisable or not possible to load the specimen with a large number of heavy-metal atoms. As a consequence, the number of scattered electrons is low and the final image has poor contrast. In such cases, the interception of the nonscattered electrons enhances the contrast. This so-called dark-field electron microscopy can be obtained in different ways. The simplest way is to displace the objective aperture so that it specifically intercepts the ongoing rays. This displacement and its effect on the image can be directly observed on the screen, and the optimal setting is obtained by trial and error. A more sophisticated way is to illuminate the specimen with a tilted beam of electrons. The ongoing, nonscattered electrons of such a tilted beam are absorbed in the physical aperture of the objective lens, and only scattered electrons are focused and finally imaged on the screen. Nowadays, many microscopes are equipped with special beam-tilt devices. A comparable but less well defined oblique illumination can be obtained with a condensor aperture with central stop (strioscopy).

2.7.2. *Diffraction Electron Microscopy*

The diffraction of an electron beam depends on the atomic number of the atoms present in the specimen. When dealing with regularly spaced aggregates of atoms, as for example, in inorganic crystals, the incident beam is diffracted at certain specific angles. This diffraction pattern can be visualized on the screen and indicates the atomic composition and lattice organization of the crystals. Because of the absence of regularly spaced heavy atoms in biological specimens, this diffraction mode of operation is not often used in neurobiological studies.

2.7.3. *Analytical Electron Microscopy*

Due to inelastic scattering of the beam electrons, some atoms in the specimen lose an orbital electron and therefore enter into an excited state. When another electron fills this vacant orbital, the atom becomes stable again. This process is accompanied by the emission of X-rays. The wavelength and energy spectrum of these X-rays are specific for different atoms and can be used to identify the atomic composition of structures in biological specimens. On the basis of this principle, a number of analytical attachments have been constructed.

2.7.4. *Scanning Electron Microscopy*

Electrons hitting the surface of a specimen either are reflected or penetrate the specimen. Inelastic collisions of the penetrating electrons generate low-energy secondary electrons and X-rays. The secondary electrons generated close to the surface leave the specimen. In scanning electron microscopy, in which the specimen is scanned by a small, dense beam of electrons, either the reflected electrons, the secondary electrons, or the X-rays are used to make the final image. This image yields information about the surface properties of the specimen. Scanning electron microscopy has great biological application nowadays, and is described in Chapter 8.

2.7.5. *High-Voltage Microscopy*

As discussed in Section 2.2, the velocity of the electrons depends on the accelerating voltage in the electron gun. This velocity determines the wavelength of the electron beam and the penetrating power of the electrons. This means that at extremely high voltages (1000 kV or more), the wavelength is much shorter, resulting in a better resolution (0.7 Å at 1000 kV and 1.7 Å at 100 kV). Moreover, because of higher penetrating power, relatively thick specimens can be studied (1 μm as compared to 0.1 μm at 100 kV). Specimen damage is decreased because of the higher velocity. These advantages of high-voltage microscopes may be useful in some instances and may counterbalance the disadvantage of high cost and many technical problems in maintaining good performance.

3. Electron-Microscopic Histology

3.1. Introduction

As outlined in previous sections, present-day electron microscopes permit the observation of fine-structural details in the specimen as small as about 20–25 Å. This high resolution requires ultimate preservation of the tissue; i.e., the fixed tissue has to resemble the *in vivo* structure as closely as possible. Alterations in ultrastructure start immediately after the death of an organism. This process, called autolysis, which changes the cytoplasmic constituents of cells, can be prevented by fixation. A prerequisite for optimal fixation is that the molecules of the fixating solution reach all parts of the specimen as rapidly as possible. The fixation also protects the cells from the damaging effects of dehydration, infiltration, embedding, and microtomy, which follow the initial fixation. The more optimal the initial fixation, the less damage can be expected from the subsequent procedures. Optimal fixation for a specific tissue has to be verified experimentally. Especially in brain tissue, with its difference between myelin-rich and myelin-poor areas, a preliminary experiment for optimal fixation is advisable.

A phenomenon of great importance for electron-microscopic histology is the restricted penetrating power of electrons in transmission electron microscopy. For this reason, very thin (<0.1-μm) sections have to be used. Furthermore, despite their extreme thinness, they have to be stable during the bombardment with electrons. The preparation of such thin sections is possible only if the biological material is embedded in a rigid and electron-stable medium that enables sections to be cut with glass or diamond knives. The embedding medium must be liquid during infiltration. The embedding medium not only enables sectioning but also supports and holds together the tissue. It is evident that this is necessary for the study of the topological organization of the tissue. The dehydration of the tissue, after fixation, is meant to exchange the free water in the specimen for an organic solvent that is compatible with the embedding medium. Dehydration should be brief to lessen shrinkage of the tissue and to avoid extraction of organic molecules from the tissue.

As outlined in Section 2, image formation in the electron microscope depends greatly on scattering phenomena. Most chemical constituents of biological specimens (H, O, C, P, S) have little scattering power, and untreated biological-tissue sections have little or no contrast in the electron microscope. For this reason, heavy-metal atoms have to be introduced during specimen preparation. The heavy-metal atoms need to be bound to the different molecular aggregates of the cells. We can distinguish between overall staining procedures and staining procedures that increase, more or less specifically, the scattering power of certain cellular elements.

3.2. Fixation

An ideal fixative has to satisfy the following requirements: (1) it has to link and stabilize proteins, lipids, and other cellular molecules into an insoluble macromolecular network; (2) it should not destroy or alter the functional groups in the molecules that are important for subsequent histochemical reactions; (3) it has to penetrate rapidly into the tissue; and (4) it can be an advantage if the fixative introduces heavy-metal atoms into the specimen.

In the early days of ultrastructural research, potassium permanganate ($KMnO_4$) and osmium tetroxide (OsO_4) were the only fixatives used. Both are oxidizing agents. The chemistry of $KMnO_4$ fixation is not fully understood. It solubilizes certain proteins and alters the functional groups of many others. It has some advantages for the fixation of plant material, but is no longer recommended for animal tissues. OsO_4 reacts with double bonds in lipid and phospholipid molecules and probably also with the polar group of lipids. The interaction of OsO_4 with proteins is not clearly understood. However, it destroys and alters the functional groups of proteins and cannot be used in combination with many cytochemical reactions. It is still used as the primary fixative for many ultrastructural studies because of the apparently excellent fixation of biological membranes. Both $KMnO_4$ and OsO_4 are slowly penetrating agents. This greatly restricts the size of the tissue blocks that can be used.

The slow penetration of OsO_4 and $KMnO_4$ and their insufficient fixation of proteins were the main reasons that aldehydes were introduced as fixatives in electron-microscopic histology. An obvious disadvantage of aldehyde fixatives for descriptive electron-microscopic histology is that no heavy-metal atoms are introduced during fixation; thus, aldehyde fixation has to be followed by postfixation with OsO_4 or by specific staining procedures. Aldehydes are less efficient in the fixation of lipids. Their excellent fixation of proteins, their good properties with respect to the maintenance of functional groups in molecules, and their rapid penetration into the tissue have made them the most widely used fixatives in electron microscopy. We shall briefly outline the properties of three aldehydes in common use for ultrastructural research:

Formaldehyde converts the cytoplasmic proteins into a stable and insoluble macromolecular network through its reaction with the amino groups. Its cross-linking of proteins is less efficient than that of glutaraldehyde, but it can often be used more successfully in enzyme histochemistry. It also reacts with double bonds in lipids and phospholipids.

Glutaraldehyde has two aldehyde groups. The formation of cross-links between amino groups in proteins is rapid and efficient. Therefore, glutaraldehyde fixation is superior to that of formaldehyde with respect to preservation of fine structure. It often inactivates functional groups and is therefore less suitable for enzyme histochemistry. The fixation of lipids is rather poor, and some phospholipids are even extracted during the fixation. The penetration of glutaraldehyde into the tissue is less rapid than that of formaldehyde and acrolein. Therefore, it is often used in combindation with these aldehydes, especially for larger specimens.

Acrolein penetrates tissue very rapidly. It has good cross-linking capacities for proteins and forms stable macromolecular networks. Probably because of its C=C, group it has better properties with respect to lipid fixation than formaldehyde and glutaraldehyde.

The fixating agents have to be dissolved in suitable buffers. The osmolarity and the pH of the buffer solution are important for adequate preservation, as are the type of buffer used and the concentration of fixative. Many procedures are given in the literature, all of which have specific advantages. The most adequate fixation solution has to be controlled empirically for each tissue under investigation.

For the study of brain tissue, two fixatives are often used with good success: (1) A formaldehyde–glutaraldehyde mixture in cacodylate buffer has been described by Peters (1970). This fixative is excellent for brain regions with few myelin sheaths. (2) A formaldehyde–glutaraldehyde mixture in phosphate buffer has been described by Vaughn and Peters (1966). This fixative is recommended for myelin-rich brain regions. The advantage of using formaldehyde–glutaraldehyde mixtures is that formaldehyde penetrates the specimen more rapidly and will stop the postmortem changes, whereas glutaraldehyde gives the most optimal cross-linking of proteins.

The most common practice in electron-microscopic histology nowadays entails prefixation with an aldehyde or a mixture of aldehydes followed by postfixation with OsO_4. The aldehydes stabilize the proteins in the tissue. This stabilization is not affected by the subsequent OsO_4 treatment. OsO_4 fixation introduces the heavy-metal atoms necessary for contrast and further stabilizes the lipids in the tissue.

A factor of great importance for adequate preservation is the time between the death of the animal and the beginning of fixation. In many instances, it is sufficient to remove the tissue or organ from the anesthetized animal as rapidly as possible and to immerse small pieces in a suitable fixating solution (*immersion fixation*). Injection of the fixative into the tissue or organ under investigation may be very useful to shorten the time between death and first action of the fixative. Tissues or organs that are sensitive to autolysis or tissues that are difficult to remove from the animal (especially brain) can be fixed adequately only by vascular perfusion in adequately anesthesized animals (*perfusion fixation*). This vascular perfusion has two advantages: (1) it diminishes the time between death and first contact with the fixative and (2) the use of the vascular bed shortens the distance between fixative and cells to be fixed. This decreases the time between death and actual fixation. For brain tissue, which is highly sensitive to anoxia, artifical respiration is advisable, although not absolutely necessary. Artificial respiration enables a more careful and quiet preparation of the animal prior to perfusion. In some instances, anesthetics cannot be used. The method of choice in these situations is decapitation of the animal and rapid immersion of the tissue in the fixative. Autopsy or biopsy material of human origin is treated in a similar way.

The slow penetration of the fixatives used in electron microscopy (even formaldehyde) necessitates the use of small tissue blocks. At least one dimension of the tissue block should be less than 1–2 mm; otherwise, fixation of the central parts of the tissue is poor. The duration of fixation also depends on the size of the specimen. These requirements with respect to the dimensions of the tissue blocks are not applicable for primary fixation by perfusion.

An excellent description of a procedure for fixation and subsequent histological treatment is given by Palay and Chan-Palay (1974).

3.3. Dehydration

Most embedding media used in electron microscopy are not miscible with water. Therefore, the free water in the tissue blocks has to be replaced by a suitable organic solvent. The two most commonly used dehydrating agents are ethanol and acetone. Ethanol is often used in combination with epoxy resins and methacrylate, although these embedding media are also compatible with acetone (see Section 3.4). Acetone has to be used in combination with polyester resins, because these embedding media are not miscible with ethanol. There is some evidence that acetone has better properties with respect to shrinkage of the tissue during dehydration. The time of dehydration depends on the size of the tissue blocks and the type of tissue under investigation. The larger the blocks and the harder the tissue, the longer the dehydration time required. Common practice is to replace the solution in the fixation bottles by gradually increasing concentrations of ethanol or acetone. Dehydration is carried out at room temperature, and the reagent-grade ethanol or acetone is mixed with double-distilled water to obtain the desired concentration.
A commonly used schedule is:

Ethanol 30%: 4 min
Ethanol 50%: 4 min
Ethanol 70%: 4 min
Ethanol 90%: 4 min
Absolute ethanol: 2 × 30 min (Ethanol can be kept free of water by keeping it over copper or magnesium sulfate.)

Specimens can be kept in ethanol 70% at 4°C for longer times without intolerable damage or shrinkage of the tissue.

3.4. Embedding

Embedding media in electron microscopy are synthetic resins. Embedding of tissue in these synthetic resins is based on the principle that the tissue is infiltrated with a low-molecular-weight, liquid monomer that can be polymerized into a rigid, transparent polymeric block. This polymerization into a rigid, transparent block enables the cutting of ultrathin sections. An ideal embedding medium should have the following properties: (1) low viscosity of the monomeric compounds during infiltration; (2) complete and nondestructive penetration of the tissue; (3) conversion of monomeric to polymeric forms; (4) transparency of the final blocks; (5) low density in the polymeric form; (6) chemical and mechanical inertness with respect to the tissue; and (7) good support of the tissue. None of the embedding media available fulfills all these requirements. It is important to choose empirically the embedding medium that best satisfies the requirements for a specific tissue.

The conventional embedding media used in electron microscopy are briefly described below. A detailed description of the chemical mechanisms of several embedding media is given by Luft (1973), and various protocols for embedding are given by Glauert (1973–1975). The practical uses of the different resins in biology are summarized in Hayat (1970–1974).

After proper infiltration of the tissue with the embedding medium, the tissue is transferred to special gelatin or polyethylene capsules with fresh embedding medium. The polymerization is achieved at temperatures up to 60°C in an oven. Nowadays, molds for flat embedding are often used. This flat embedding facilitates the orientation of the tissue, which is especially useful during the subsequent sectioning.

3.4.1. Methacrylates

Mixtures of butyl- and methylmethacrylate, in various concentrations, were the most commonly employed embedding media in the early days of electron microscopy. Polymerization of the monomers by chemical or ultraviolet-light initiation is achieved by a highly selective free-radical chain reaction of the ethylenic double bonds. The hardness of the final blocks depends on the ratio of butyl and methyl monomers present in the mixture. Damage to tissue during polymerization is quite severe, and methacrylate sections are quite sensitive to the electron beam. Therefore, methacrylate has been largely replaced by other embedding media. Special variants are still in use for certain histochemical purposes and in water-soluble embedding media.

3.4.2. Epoxy Resins

Araldite and Epon are the most commonly used epoxy embedding media. Araldite has been in use since 1956, and Epon was introduced in 1959. The essential component of an epoxy embedding medium is a relatively small polymeric molecule (weight 340–3000) with epoxide end groups and hydroxyl groups spaced along the chain. The epoxide end groups can be linked together with reactive hydrogen groups, notably amines. Such an "accelerator" [e.g., benzyl dimethylamine (BDMA) or tri(dimethylaminomethyl)phenol (DMP-30)] links the epoxy molecules into long chains. The hydroxyl groups along the chains are linked together by acid anhydrides such as dodecenyl succinic anhydride (DDSA) and methyl nadic anhydride (MNA). These "hardeners" cross-link the long chains of epoxy molecules so that a stable and inert three-dimensional polymer is achieved. The hardness of the final blocks depends on the ratio between MNA (harder) and DDSA (softer). Damage to the tissue due to polymerization is negligible, and stability in the electron beam is satisfactory. Epoxy resins contain a number of polar groups and are therefore miscible with ethanol. The replacement of ethanol by the resin was found to be slow. Therefore, the ethanol is replaced by a "clearing" agent, propylene oxide, to accelerate the infiltration of the resin. A disadvantage of epoxy resins is that they have a relatively high viscosity, which hampers infiltration.

3.4.3. *Polyester Resins*

Vestopal-W is the best-known polyester resin used in electron microscopy. Like the epoxy resins, it is a small polymeric molecule with relatively high viscosity. The final polymerization is brought about in the same way as with methacrylates, i.e., a free-radical mechanism. Polymerization damage is negligible, and it is stable in the electron beam. The electron stability is better than for epoxy resins, and Vestopal-W is recommended for high-resolution work. Vestopal-W has few polar groups, and acetone has to be used for dehydration. Styrene is used as "clearing" agent. This accelerates the infiltration. A disadvantage of Vestopal-W is that it is more difficult to section.

3.4.4. *Water-Soluble Embedding Media*

Durcupan®, glycolmethacrylate, hydroxypropylmethacrylate, and Aquon® are embedding media that are miscible with water. Their monomers or small polymers have a large number of polar groups. Therefore, dehydration with organic solvents is not necessary. The tissue is infiltrated with mixtures of water and increasing amounts of the embedding medium. They are often useful when lipid extraction has to be avoided.

3.5. Ultramicrotomy

The restricted penetration of electrons necessitates the cutting of very thin sections (<0.1-μm) from the tissues to be studied. Sophisticated ultramicrotomes have been developed for this purpose. The ultrathin sections are cut with glass or diamond knives, the sections being floated on a water surface during cutting. They are subsequently picked up on specially coated grids. The preparation of glass knives, coated grids, and the principles of ultramicrotomy are briefly outlined below.

3.5.1. *Glass Knives*

In 1950, Latta and Hartmann (1950) introduced glass knives for the cutting of ultrathin sections of plastic-embedded material. They have proven to be superior to metal knives. For many years, they were made by hand from a sheet of plate glass using a wheel-type glass cutter and a pair of specially adapted glass-breaking pliers (for details, see Weakly, 1972; Reid, 1975). The yield of suitable knives was rather poor in this way, and therefore special knife makers have been designed. These instruments are easy to operate, and glass knives for all types of ultramicrotomes can be produced. The basic steps of knife-making have not been changed by the introduction of the knife-makers. A strip of glass, carefully cleaned with water and detergent, distilled water, and acetone, is scratched with a number of parallel marks. Square or parallelogram-sided pieces of glass are broken from these high-quality-glass strips, the scratches being the sites of breakage. The squares or parallelograms are scratched again and broken into two triangular pieces (see Fig. 8). The shape of the triangular pieces determines the knife angle, which depends on the material to be cut. In many cases, a knife angle of 45° is suitable. The quality of the knife, especially its cutting edge, has to be checked in a stereomicroscope. An ideal knife edge should be straight, and free of irregularities. On the front face, a stress line may be visible (Fig. 8). This stress line is due to the final break. The glass knives have to be handled with great care to avoid damaging the cutting edge. Since glass is a supercooled liquid, glass knives must be used soon after their production. It is thought that the glass gradually flows away from the edge and that the knife edge therefore becomes blunt. In our experience, however, this decrease in quality can be partly prevented by storing the glass knives in a desiccator with silica gel. In this way, they can be stored for days or even

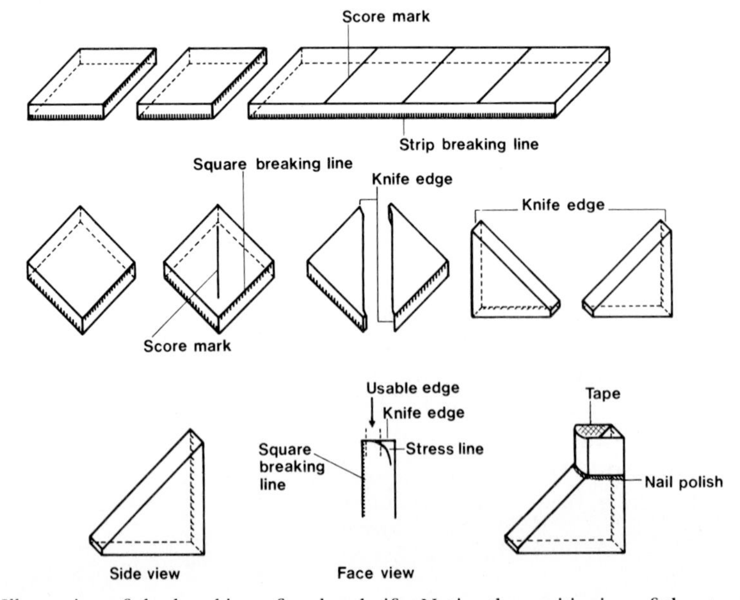

Fig. 8. Illustration of the breaking of a glass knife. Notice the positioning of the stress line on the face of the knife. Also note the construction of a trough that can hold water on which thin sections are floated.

weeks. For cutting very hard tissues or for cutting extremely thin sections (<0.03 μm), diamond knives are recommended. The advantage of diamond knives is that they can be used many times, whereas glass knives have to be discarded after one-time use. Diamond knives can be resharpened and therefore can be used for several years. They are expensive, however, and it is often difficult to find one with an optimal cutting edge.

The sections cut with glass or diamond knives are floated on a liquid directly after cutting. Therefore, a small trough is attached to the knives. This trough is made from a piece of clean waterproof adhesive tape fitted to the knife and secured with nail polish (Fig. 8).

3.5.2. *Supporting Grids*

Depending on the type of microscope used, the ultrathin sections are mounted on special grids with a diameter of 2 or 3 mm. The most commonly used grids are made of copper. For special purposes (e.g., heating, treatment with corrosive reagents), grids made of gold, platinum, palladium, silver, or other material are available. Grids can be obtained with many different mesh sizes and shapes of the grid openings. They all have a shiny side and a mat side. This difference in appearance between sides is useful with respect to the attachment of the sections to the grid. It is important inasmuch as it is necessary that the sections always be below the grid bars during microscopy. By attaching them routinely to either the mat side or the shiny side, one has a means of ascertaining that the sections are properly oriented in the microscope. The mat side is thought to have the advantage that the sections or supporting films adhere to it better.

It is common practice to cover the carefully (ultrasonically) cleaned grids with plastic–carbon supporting films. The films most often used are parlodion, collodion, formvar, and pyoloform. They all have different properties with respect to stability in the electron beam. A carbon film is evaporated on these supporting films to increase their strength. The plastic films can be obtained by dipping a clean glass slide in a solution of the plastic and slipping the film formed onto a clean water surface, or by placing a drop of the plastic directly on a clean water surface. A number of precleaned grids are then placed on the film, which has been picked up with Parafilm®. The grids are then carefully dried. Finally, carbon is evaporated on the films in a vacuum-coating instrument.

3.5.3. *Principles of Ultramicrotomy*

Before the actual sectioning is started, the tissue blocks have to be trimmed to a suitable surface for cutting. To this end, the blocks are clamped in the specimen holders of the microtome, and a pyramid is trimmed under a stereomicroscope using a razor blade. A commonly used form of the specimen (trimmed tissue block) is illustrated in Fig. 9. This trapeziumlike pyramid with a flat top gives the specimen a secure base for cutting and

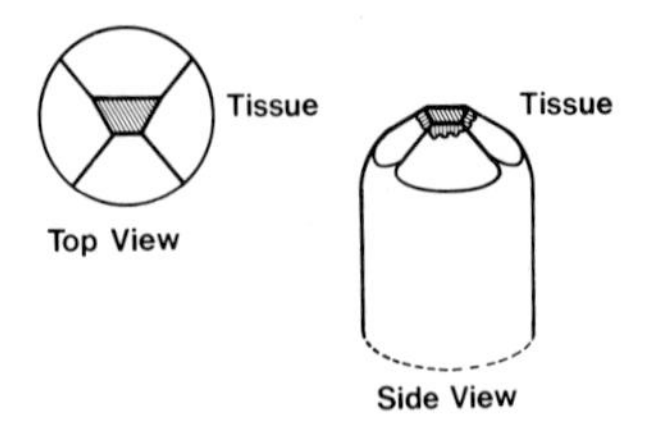

Fig. 9. Drawing of a plastic block with a piece of tissue embedded in the hemispherical tip. Note that the block has been trimmed to expose the embedded tissue.

prevents vibration of the block during sectioning. Furthermore, it ensures that the thin sections stick together during sectioning so that ribbons of sections are obtained. Very often, it is advantageous to cut 1-μm sections prior to ultrathin sectioning. These semithin sections, which can be stained for light microscopy (e.g., with toluidine blue), enable preselection of areas of special interest. Recently, trimming machines have become available.

For the cutting of ultrathin sections, special ultramicrotomes have been designed. At present, five types of ultramicrotomes are commonly used; Cambridge-Huxley (recently LKB-Huxley) ultramicrotomes, LKB-Ultrotomes, Sorvall Porter-Blum ultramicrotomes, Reichert ultramicrotomes, and the Tesla ultramicrotome. They all have in common that the knife, with its trough, is placed in a special holder that is firmly fixed to the microtome bench. Furthermore, the specimen, in its special holder, is attached to a movable rod (specimen arm). The specimen is brought close to the knife edge, and at sectioning, it is moved in a vertical direction along this knife edge, leaving a thin section on the liquid surface of the trough (see Figs. 10 and 11). The advance of the specimen arm toward the knife edge is realized in three steps; coarse, intermediate, and ultrafine. The coarse advance, used for the rough approximation of the specimen to the knife edge, is realized by moving the knife holder to the specimen arm. The intermediate advance is brought about by fine mechanical movement of the knife holder toward the specimen arm, or by use of the ultrafine advance.

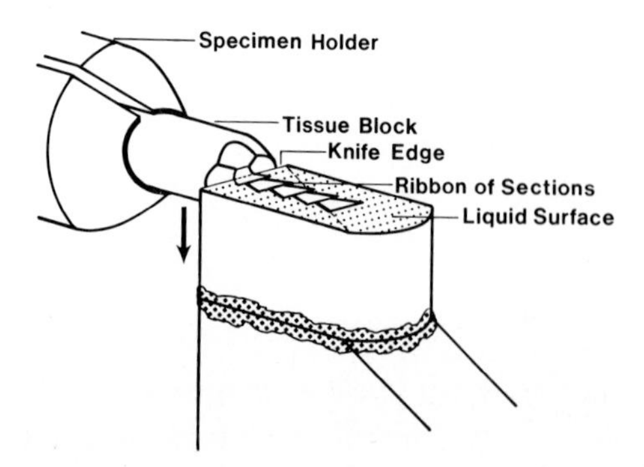

Fig. 10. Illustration of thin sections floating onto the water in the trough and away from the cutting surface of the knife. Notice how the edges of the sections adhere to one another to form a ribbon.

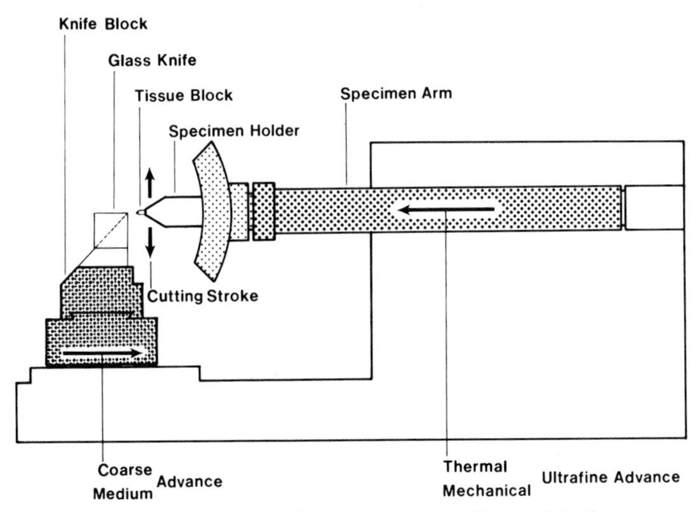

Fig. 11. Diagram of an ultramicrotome as discussed in the text.

The mechanical intermediate advance is used for the sectioning of semithin sections (2–0.5 μm). The ultrafine advance, which is of course most critical in ultramicrotomy (<0.1 μm), is provided in present-day microtomes in one of two ways: ultrafine mechanical advance (Huxley microtomes and Porter-Blum microtomes) or thermal expansion (LKB microtomes, Reichert microtomes, and Teslamicrotome). With both types of microtomes, it is possible to cut sections between 0.02 and 0.1 μm thick. The microtomes based on thermal expansion are superior in the extreme thin range.

After the specimen arm has moved along the knife edge and has left a thin section, the specimen arm is brought back to its original position either by a reverse vertical movement (LKB and Porter-Blum) or by moving the arm through an elliptical or parallelogram-shaped path (Reichert, Huxley, Tesla). When the reverse stroke is vertical, the specimen arm could touch the knife again, which would destroy it or pick up the sections. Therefore, during the reverse stroke, either the knife holder (LKB) or the specimen arm (Porter-Blum) is retracted. More detailed technical information can be found in Reid (1975) and the microtome manuals.

There are a large number of factors that affect section-cutting, such as the speed of movement along the knife edge, the speed of the reverse stroke, the degree of thermal expansion, the knife angle, the angle between the knife edge and the specimen block (clearance angle), and the level of the liquid in the trough. They all have to be controlled properly to obtain optimal ultrathin sections. Suitable setting of the microtome also depends on the hardness of the tissue block, the dimensions of the pyramid trimmed on it, and the thinness of sections desired.

The versatility of recent ultramicrotomes, the reliability of knife-making, and the reproducibility of embedding have greatly reduced the early problems of ultrathin sectioning. Nevertheless, ultramicrotomy remains one of the most difficult techniques in electron-microscopic histology. It is rather time-consuming and still requires a certain degree of skill. Man difficulties have to be dealt with, such as chatter, scratches, irregularity of section thickness, folding of sections, and compression of sections. A detailed description of sectioning problems and trouble shooting can be found in Glauert and Phillips (1965) and Reid (1975).

With sufficient patience and after some days of experience, regular sections can be cut. These sections will float on the surface of the liquid in the knife trough. Good sections will stick to each other at the edge and form ribbons of sections. The liquid most commonly used in the knife trough is clean, double-distilled water. The floating ribbons of sections are illuminated and observed with a binocular viewer. This binocular viewer enables a preliminary control of the sections with respect to frequency of chatter, scratches, folds, and other characteristics. Moreover, section thickness can be roughly judged in this way. The ultrathin sections have interference colors ranging from dark gray to silver, gold, purple, and blue. These interference colors indicate that the sections are <0.06, 0.06–0.09, 0.09–0.15, 0.15–0.19, and >0.19 μm, respectively.

More exact determination of section thickness, which is important for quantitative electron-microscopic techniques such as stereology and autoradiography, is much more difficult to obtain (e.g., see Glauert and Phillips, 1965; Reid, 1975). The floating ribbons of sections can be manipulated and oriented in the knife trough by means of fine hairs secured to a wooden stick. They can then be attached to the supporting grids in different ways, i.e., by simply pressing the grid onto the ribbon of sections on the water surface; by dipping the grid beneath the surface of the water, moving it underneath the sections, and raising it; or by picking up the sections with a small wire loop (somewhat smaller than the supporting grids), placing the loop over the grid, and drawing off the excess water in the loop. These manipulations are usually carried out with a pair of fine tweezers, although micromanipulators have recently become available. Careful sectioning and mounting of the ultrathin sections allows viewing of many serial sections. Final assessment of the quality of thin sections has to be carried out in the electron microscope either before or after staining of the sections.

Recently, it has become possible to equip LKB, Reichert, and Porter-Blum microtomes with so-called cryokits. These cryokits enable the cutting of ultrathin sections at temperatures down to −170° C. This means that ultrathin sections can be cut from rapidly frozen, unfixed tissue. They provide optimal conditions for histochemical and immunohistochemical reactions, soluble-compound autoradiography, and X-ray microanalysis. It must be realized, however, that many problems have to be solved to obtain frozen ultrathin sections. They have to be floated in the trough on a suitable liquid that has good surface properties and that does not dissolve or destroy molecular components of the tissue. At present, dry-sectioning is proposed to eliminate the dissolution or destruction of tissue elements, but this technique is very difficult.

3.6. Staining Procedures

As already mentioned several times in the previous sections, image formation and contrast depend on electron scattering. Therefore, it is necessary to introduce heavy-metal atoms either during fixation (OsO_4 and $KMnO_4$) or after cutting of the thin sections. Most commonly, heavy-metal atoms are introduced in both phases of preparation. We shall briefly describe the staining solutions routinely used in electron-microscopic histology and some staining solutions more specifically used in the fine-structural study of the nervous system.

3.6.1. Lead Stains

The lead solutions used in electron microscopy have a high affinity for most cellular components, although they have some specificity for glycogen, nucleoproteins, and membranes. The lead solutions are dissolved in strongly basic solutions (pH 11–12). It is thought that lead is present as positively charged complex ions. These cations react with negatively charged groups in proteins and carbohydrates: phosphate, sulfhydryl, tyrosyl, and carboxyl groups. They also react with the osmium introduced during fixation. This results in an intense staining of the membranes. Due to the high pH of the staining solution, the binding groups in the proteins and carbohydrates become more negatively charged, thus giving rise to this intense staining reaction. Another theory is that anionic complexes of the lead salts are responsible for staining.

There are many recipes for preparing lead solutions. They are all sensitive to carbon dioxide (CO_2). This CO_2 reacts with the lead ions and forms insoluble lead carbonate. The lead carbonate from these "dirty" solutions precipitates on the sections and is seen in the electron microscope as relatively large opaque deposits. A lead citrate solution, according to Reynolds (1963), is thought to be less sensitive to carbonate formation and is therefore most commonly recommended for lead staining. Nevertheless, it is advisable to store the lead solutions in tightly stoppered bottles and to avoid contact with CO_2-rich air. Section-staining with lead solutions is performed by placing grids on a drop of staining solution in a petri dish containing sodium hydroxide. Rinsing is done with freshly prepared, CO_2-poor, double-distilled water. Lead solutions are also used for *en bloc* staining prior to dehydration and embedding.

3.6.2. Uranyl Stains

Uranium is the heaviest atom used for staining in electron microscopy. The staining mechanism of uranyl solutions is unknown. It seems that the uranyl ions form complexes with phosphoryl and carboxyl groups. The intensity of the staining depends on the pH of the solution, the type of buffer used, the duration of the staining, and the actual concentration of the uranium ions. DNA-rich structures are preferentially stained at low concentrations and pH 3.5. Increasing the concentration or decreasing the pH of the uranyl solutions, or both, results in a more intense staining of proteins. It is also thought that uranyl solutions have fixating properties.

Commonly employed uranyl solutions, for overall enhancement of the contrast in biological specimens, are saturated or half-saturated uranyl acetate solutions in distilled water or absolute ethanol. These solutions are also used for *en bloc* staining prior to dehydration or embedding.

3.6.3. Double Staining

It is common practice in electron microscopy to stain ultrathin sections with lead citrate followed by uranyl. This results in a bright contrast for most specimens of biological and neurobiological origin. This double staining is very effective after double fixation with glutaraldehyde and OsO_4. Most of the figures in Section 4, are of tissue treated in this way.

3.6.4. Phosphotungstic Acid Stains

Phosphotungstic acid (PTA) has been used as an electron-microscopic stain since the early days of electron microscopy. Its staining mechanism is still not understood. In combination with OsO_4, it generally enhances the contrast of the sections. There are many controversies regarding the specific affinity of PTA for mucopolysaccharides and basic proteins. It was introduced into neurobiology by Bloom and Aghajanian (1966) as a selective stain for synaptic contact zones. Used in combination with glutaraldehyde fixation, but without OsO_4 fixation, a 1% solution of PTA in absolute ethanol or distilled water results in a selective *en bloc* staining of synaptic contacts (see Figs. 27–30). A disadvantage of PTA is its slow penetration into the tissue. Therefore, thin slices of tissue (<100 μm) have to be used to obtain a thorough staining of the tissue.

Basic proteins in the central nervous system and myelin in the peripheral nervous system can be successfully stained with a PTA–hematoxylin solution.

3.6.5. Potassium Permanganate and Osmium Tetroxide

$KMnO_4$ and OsO_4 are both fixating and staining solutions. They are nonselective and generally enhance the contrast of the tissue. Furthermore, $KMnO_4$ can be used after OsO_4 fixation to increase the contrast of myelin sheaths.

3.6.6. Zinc Iodide–Osmium Tetroxide Stain

The use of zinc iodide (ZIO or OZI)–osmium tetroxide was introduced by Akert and Sandri (1968) for a more or less specific staining of synaptic vesicles. Its staining mechanisms are not understood. X-ray microanalysis has shown that neither iodine nor zinc is an important constituent of the precipitate formed in the vesicles. It is thought that interactions with specific enzymes play a role in the deposition of the stain not only in synaptic vesicles but also in Golgi cisterns, mitochondria, and endoplasmic reticulum (e.g., see Vrensen and De Groot, 1974*a*). In our experience, the staining of all synaptic vesicles present in the tissue depends greatly on prefixation and the pH of the ZIO solution. Using the glutaraldehyde–formaldehyde fixative of Peters (1970), a complete staining of synaptic vesicles could be obtained with a ZIO solution, as proposed by Kawana *et al.* (1969) and Martin *et al.* (1969), only at a pH of about 3.0. This low pH gives rise to a relatively poor preservation of the fine structure. A concomitant selective staining of synaptic vesicles and synaptic contact zones can be obtained by a combined *en bloc* staining with ZIO and aqueous PTA solution (Vrensen and De Groot, 1974*a*) (see Fig. 34).

3.6.7. *Bismuth Stain*

Pfenninger (1971) introduced a bismuth iodide *en bloc* staining followed by section staining with uranyl and lead (BIUL) as a selective staining for synapses after glutaraldehyde fixation.

3.6.8. *Silver Stains*

Ammoniacal silver solutions can be used in electron-microscopic histology to increase contrast. The silver is reduced by certain groups present in the tissue, e.g., aldehyde groups, but also by the OsO_4 introduced during fixation. The specificity and intensity of the silver staining can be increased by omitting the OsO_4 fixation and by treatment of the glutaraldehyde-fixed tissue with periodic acid, which sets free the aldehyde groups. Silver staining is to some extent selective for mucopolysaccharides and glycoproteins. The silver is observed as small precipitates.

It is not known whether the Golgi impregnation, which is also based on silver deposition in the cells of the nervous tissue (see Chapter 5), has a similar mechanism. The complete blackening of the impregnated cells makes this Golgi technique a powerful method for studying the dendritic and axonal properties of individual neurons. In electron microscopy, the relationships between dendrites and axons and their perikarya are almost completely lost because of ultrathin sectioning. Blackstad (1970, 1975) has tried to combine the advantages of the Golgi impregnation with the high resolution of the electron microscope. Recently, he has introduced a promising method (Blackstad, 1975). Glutaraldehyde-fixed nervous tissue is impregnated by treatment with OsO_4–potassium dichromate followed by silver nitrate. In this way, a number (2–5%) of the neurons are impregnated completely. This can be verified by light microscopy. The impregnated tissue is exposed to ultraviolet light of specific wavelength and treated with a silver-dissoluting agent, e.g., thiosulfate. Thus, the silver atoms reduced during the ultraviolet exposure remain in the tissue, while the bulk of the unreduced silver is removed. The pictures obtained in this way show a suitable preservation of ultrastructure, with some neurons having small amounts of silver precipitate. A proper correlation of light-microscopic sections of this material and ultrathin electron-microscopic sections enables the study of the fine-structural relationships (e.g., synaptic) of an individual cell with the surrounding neurophilic elements.

3.6.9. *Application of Staining Solution*

The staining of biological tissue can be achieved at two stages of preparation. With *en bloc* staining, prior to dehydration or embedding, the prefixed tissue is immersed in an aqueous solution of the staining agent prior to dehydration or in an ethanolic solution during dehydration. Block-staining is easy to perform, but has the disadvantage that some staining agents penetrate so slowly that very small tissue blocks have to be used to get thorough staining. In section-staining, ultrathin sections are placed on small drops of the staining solution in a petri dish. Spreading of the drops can be prevented by placing them on a clean hydrophobic surface, e.g., Parafilm®. After being stained, the sections are carefully rinsed with clean double-distilled water and air-dried. Sodium or potassium hydroxide has to be placed in the petri dish when using lead staining or when double-staining with lead and uranyl.

3.7. Other Techniques

In the previous sections, we have briefly outlined the histological procedures used for the study of nervous and other biological tissues for which thin-sectioning is necessary to obtain transmission electron micrographs. In some instances, it is not possible (e.g., small isolated cellular particles, viruses) or not advisable (histochemistry, soluble-compound autoradiography) to fix and embed the specimen. Therefore, many other techniques have been developed to obtain sections for study in the transmission electron microscope. They are as follows (the reader is referred to the literature for details): freeze–etching and freeze–fracturing (Bullivant, 1973; Orci and Perrelet, 1975; Koehler, 1972; Moor, 1969), freeze–substitution (Pease, 1973; Rebhun, 1972), freeze–drying (Rebhun, 1972), negative staining (Horne, 1965; Haschemeyer and Meyers, 1972), and shadow-casting and replication (Henderson and Griffiths, 1972).

3.8. Some Critical Remarks on Electron-Microscopic Histology

The main intention of the preceding discussion of electron-microscopic histology was to introduce the nonspecialized neurobiologist to the basic principles and rationale of the various techniques used to obtain "good" histological sections for the observation of nervous tissue in the transmission electron microscope. Details of the various techniques and recipes for the various solutions are given in a number of handbooks, some of which are given in the bibliography at the end of this chapter. We want to emphasize here that much has changed since the early days of electron microscopy. At that time (up to 1960), electron microscopy was an empirical method, and the preparation of "good" sections and good micrographs depended greatly on the manual skill and inventiveness of the research worker and his technical assistants: there was something mysterious about it. At present, however, the technical facilities in most electron-microscopic laboratories are such that "good" sections can be obtained after a relatively short introduction period of some weeks. The wide choice of fixatives with their various properties, the reliability and consistency of the embedding media, the ease of making glass knives, and the versatility and reliability of present-day ultramicrotomes greatly facilitate the preparation of "good" ultrathin sections of nearly all kinds of tissues. Furthermore, recent electron microscopes are easy to operate, reliable, and versatile. Making micrographs of optimal quality at low or high magnification is really simple. This does not mean that problems will not be encountered. It must be realized that histological treatment is quite crude for the fragile substructures in the tissue. This means that small deviations from routine procedures or the introduction of new steps in preparative procedures can give rise to changes in the ultrastructure. All steps in preparative procedures have to be carried out with equal care, for the final result ultimately depends on all these steps. Because of the great advances in technology, electron microscopy need no longer be considered as an "art" with its great emphasis on skillfulness and inventiveness. For descriptive morphological research, electron microscopy can be considered as routine a method as light-microscopic histology. This means that, at present, there is no reason to be discouraged by the many, apparently complicated, instruments used in electron microscopy. It also means that nowadays the research worker can devote his attention to the analytical aspects of ultrastructural research, such as histochemistry, microanalysis, autoradiography, and stereology.

The fundamental aim of electron-microscopic histology is to preserve the *in vivo* structure of cells and tissues as accurately as possible and to prepare thin sections that can be viewed in the electron microscope. A serious problem encountered in judging the quality of fixation, embedding, and other preparative steps arises from the fact that the actual *in vivo* structure of cells and tissues is unknown. *In vivo* studies of cells can successfully be performed with the light microscope. Therefore, it is advisable to first study semithin (1–2 μm) sections of the plastic-embedded material and to compare the light-microscopic structure with what is known about the structure of the living cell. This study of semithin sections gives a preliminary impression of the quality of fixation and other procedures. However, the limited resolution of the light microscope does not allow inferences with respect to the ultrastructure of cell organelles. At present, living cells can also be studied in the high-voltage electron microscope (e.g., see Hama, 1973). There are, however, serious problems with respect to beam damage, which may alter the molecular constituents and even the cellular organelles. Moreover, the biological resolution of these microscopes, using intact cells, is quite poor because of the projection of superimposed structures in one image plane.

These problems in observing the actual *in vivo* structure of cells have led to the formulation of a number of subjective criteria for "good" preservation of cells and tissues (e.g., see Sjöstrand, 1967; Meek, 1970; Glauert, 1973–1975): There should be an absence of empty spaces, i.e., spaces devoid of any granular or fibrous structural material. These empty spaces often point to swelling of the cells during preparation. Membranes should be continuous. The disruption of the membranes of cells or cell organelles often denotes inadequate fixation. Empty holes bordered by sharp lines indicate that the penetration of embedding medium is inadequate, probably due to insufficient dehydration or clearing. Regularity of the geometric patterns of cell constituents often indicates that the preparative procedures have been carried out correctly. Scratches, chatter, deformation of geometric patterns, and other irregularities are useful indications of the quality of sectioning.

A specific problem in the ultrastructural study of brain tissue is the appearance of "dark" (hyperchromic) and "light" neurons. The artificiality of the "dark" neurons has been discussed at length (e.g., see Cammermeyer, 1961; Reger *et al.*, 1972; Stensaas *et al.*, 1972). The question is whether the occurrence of the "dark" neurons is only the result of inadequate fixation or whether it is a significant artifact, i.e., dependent not only on fixation but also on factors in the brain prior to fixation (Ebels, 1975).

4. Applications of Electron Microscopy in Neurobiology

4.1. Ultrastructural Studies in Neuroanatomy and Neurocytology

In the past few decades, abundant information has been accumulated regarding the fine structure of various parts of the central and peripheral nervous systems in many different species. General introductions to this ultrastructural organization of the nervous system are given by Peters *et al.* (1976) and Kojima *et al.* (1975), and to some extent by Palay and Chan-Palay (1974) in their book on the cerebellar cortex. The use of the electron microscope, in whatever experimental neurobiological study, presupposes a thorough understanding of this normal fine structure. This is not only relevant from a pure descriptive point of view but is also a prerequisite for the correct interpretation of the ultrastructural changes caused by experimental conditions.

Despite the wealth of ultrastructural information regarding the nervous system, it is advisable to start with a preliminary study of the normal fine structure of the particular material under investigation and not to rely solely on the data given in the literature. There are several reasons for this. First, one must determine whether the histological methods are properly applied and whether the fine structure fulfills the requirements for "good" fixation (see Section 3.8). Second, such studies are the only way to become acquainted with the general and specific problems of ultrastructural research. It is possible that the general criteria of "good" fixation cannot be satisfied for a particular material, e.g., pathological material, embryological material, or histochemical studies. In these cases, the results of the preliminary control study can bring to light specific "artifacts" that can be taken into consideration in the interpretation of the experimental material. Finally, it is important to become familiar with the specific ultrastructural features of the particular brain region under investigation. Very often, specific "abnormalities" are present in control material that are described in the literature as pathological (Sotelo and Palay, 1971). In many instances, the brain region under investigation has already been studied by others, but in a different species. The preliminary study of control material may reveal species differences that can be relevant for the interpretation of the experimental material.

It is known from light-microscopic studies that the morphology of neurons differs considerably from one brain region to another and even within one brain region. On the basis of various staining methods (see Chapter 5), neurons have been classified according to their size, shape, distribution, and abundance of Nissl substance, and the different aspects of their protrusions. Some of these particularities, especially those concerning the perikaryon, are reflected in the fine-structural features of neurons and glial cells. Although neurons and glial cells contain the same organelles as all cells in the organism, they are sufficiently characteristic to be distinguished from other cell types and from each other. This possibility of distinguishing among neurons, glial cells, and other cell types holds true for all brain areas and species that have been studied. The different types of neurons can be distinguished by their cytoplasm/nucleus ratios and by quantitative and qualitative differences in cell organelles. Simultaneous light-microscopic study of the same region is of great help. The distinctions, for example, between relay neurons and interneurons and among the various neurons in the cerebellum have been ultrastructurally verified in this way.

In the following sections, we will describe and illustrate some general features of neurons, astrocytes, and oligodendrocytes.

4.1.1. The Neuron

In small neurons, the nucleus (N) is usually localized in the center of the perikaryon, whereas in large neurons it is often found eccentrically. The chromatin is most often homogeneously dispersed throughout the nucleus and has a fine granular appearance (Fig. 12). This gives the nuclei of neurons their characteristic clear look. Only in a few neurons does the chromatin tend to be condensed in larger patches, especially bordering the nuclear envelope. In addition to a prominent nucleolus (Nuc) and nucleolar satellites, different types of inclusion bodies are observed in the nucleus (Figs. 12 and 13). These inclusion bodies are so frequently observed in control material that they have to be considered a normal feature of neuronal nuclei and not indicative of pathological alteration. Their function is unknown at present.

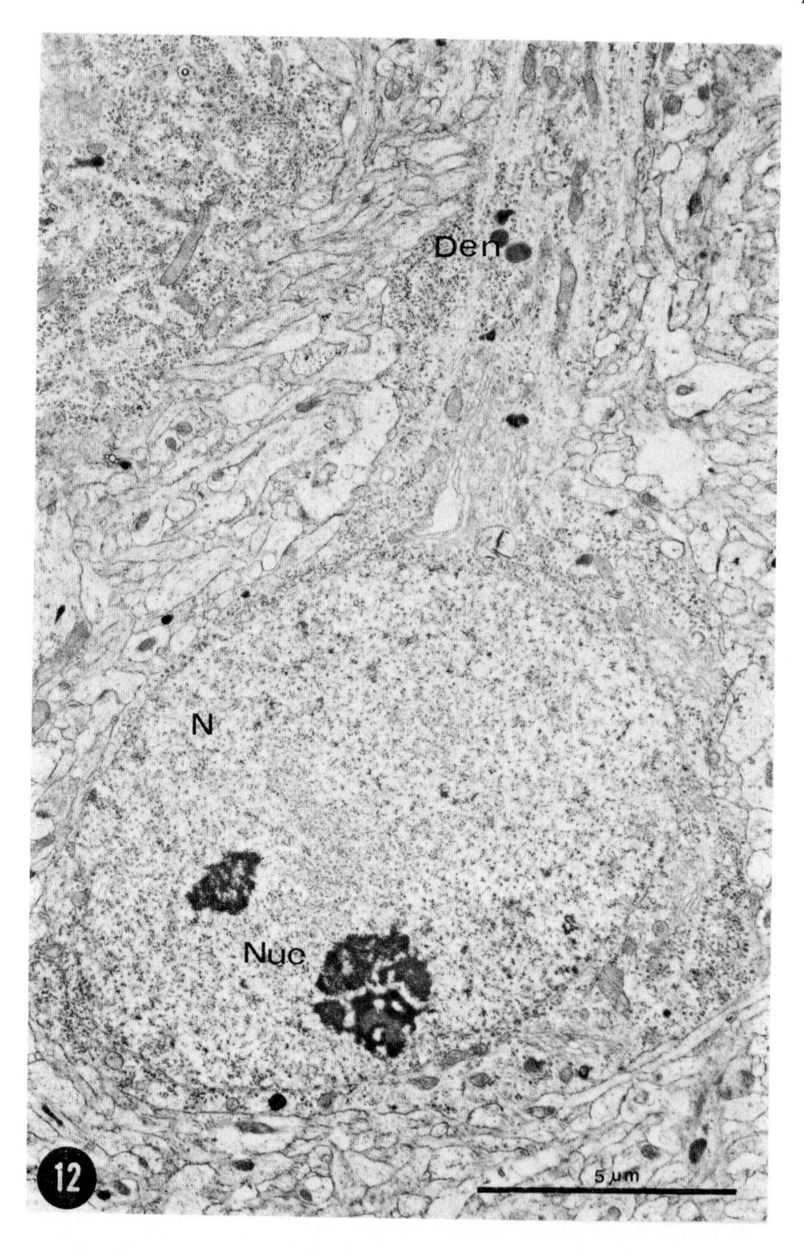

Fig. 12. A neuron in the visual cortex of a 6-day-old rabbit. Note the prominent apical dendrite (Den). Abbreviations here and in Figs. 13–37: (Af) astrocytic endfoot; (As) astroglial process; (Ax) axon; (Cap) capillary; (CZ) contact zone; (Den) dendrite; (DP) dense projection; (F) Filament; (fV) flat [synaptic] vesicle; (G) Golgi apparatus (or complex); (gER) granular endoplasmic reticulum; (gp) glycogen particles; (ICL) intercleft line; (L) lysosome; (LG) lipofuscin granule; (Mit) mitochondrion; (MVB) multivesicular body; (N) nucleus; (NE) nuclear envelope; (Neu) neuron; (NP) nuclear pore; (NS) nucleolar satellite; (NT) neurotubule; (Nuc) nucleolus; (Pia) pia mater; (PSB) postsynaptic band; (rV) round [synaptic] vesicle; (S) synaptosome; (SgER) stacks of granular endoplasmic reticulum; (SSC) subsurface cistern; (SV) synaptic vesicle.

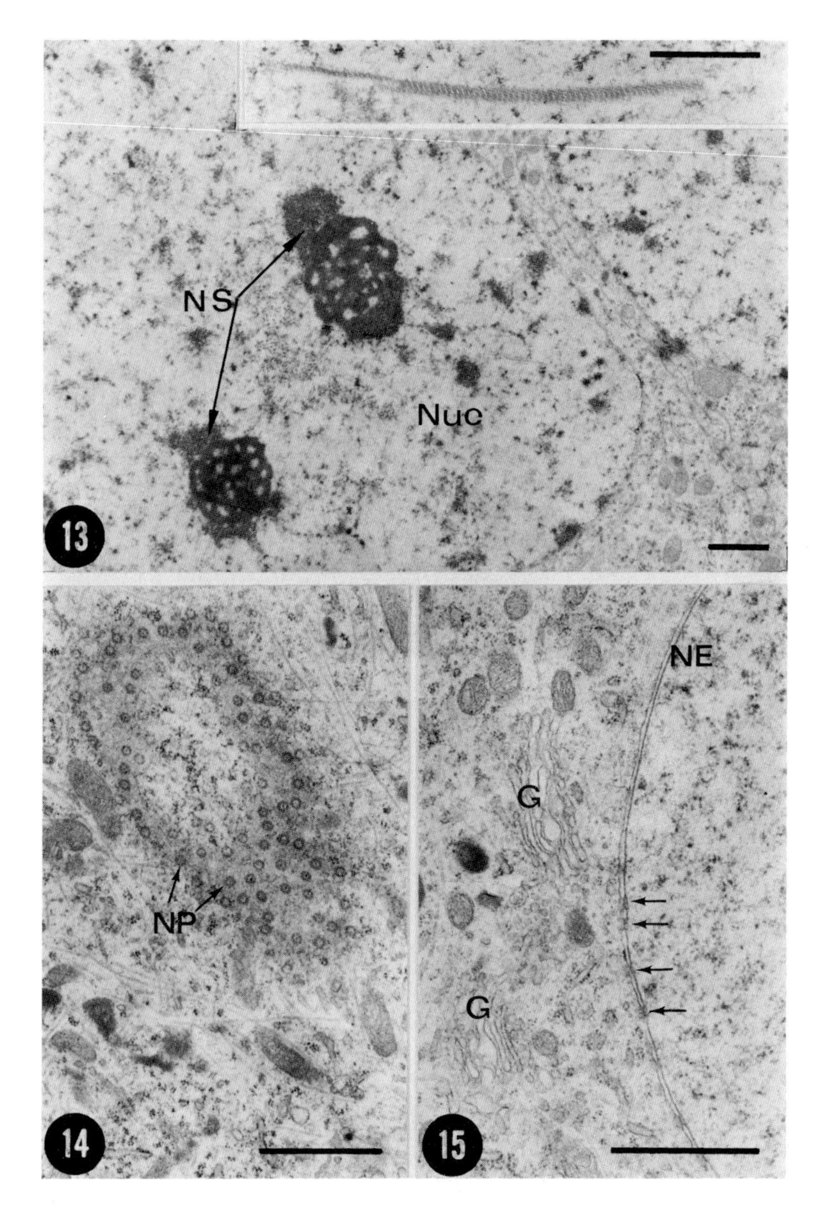

Fig. 13. Nucleoli in the nucleus of a neuron from the inferior olive of the cat. Note the nucleolar satellites (NS), or nucleolus-associated chromatin. *Inset:* Regularly organized inclusion body in the nucleus of an inferior olive neuron. Here and in Figs. 14–37 and 40, the bars, unless otherwise indicated, represent 1 μm.

Fig. 14. Nuclear pores (NP) in the nuclear envelope of a neuron from the inferior olive of the cat in a tangential section.

Fig. 15. Arrows indicate nuclear pores in the nuclear envelope (NE) of a neuron in the inferior olive of the cat in a transverse section. Note the perinuclear Golgi complexes (G).

The nuclear envelope (NE), which is known to be continuous with the endoplasmic reticulum (ER), has typical nuclear pores (NP) in it (Figs. 14 and 15). Ribosomes are found attached to the nuclear membrane (Fig. 15). Although the nucleus most often has a smooth outline, indentations of varying complexity are a normal feature (see Fig. 20).

Endoplasmic reticulum of the granular or rough type (gER) is a common structure in all types of neurons. The amount of gER and its organization vary greatly. In many instances, the gER is present in large conglomerations of parallel arrays of cisterns (Fig. 16). These parallel arrays of ER cisterns have proved to be the ultrastructural counterpart of the well-known basophilic Nissl bodies (see Chapter 5). These large conglomerations are most obvious in large neurons, e.g., the pyramidal cells of the motor cortex and visual cortex. In many other neurons, the cisterns of the gER are isolated or are clustered in small numbers (Fig. 17). In some instances, granular cisterns are observed that are closely apposed to the cell membrane or to each other. These subsurface cisterns (SSC) and stacks of gER (SgER) (see Fig. 16) are present in normal brain tissue and are not specifically related to pathological conditions. Although the gER and ribosomes are known to play a role in the synthesis of proteins, no specific functions can be ascribed to the different forms of gER. Ribosomes are present in large amounts either attached to the cisterns of the ER or free in the cytoplasm. They are often organized in rows or rosettes, i.e., polysomes (Fig. 17). Large clusters of ER cisterns, without ribosomes attached to them, are rarely observed in neurons. Granular ER and free ribosomes are found throughout the cytoplasm of the neuron. The density of these cellular elements diminishes near the axon hillock, and they are seldom found in the axons beyond the initial segment (Fig. 18). In dendrites, on the contrary, they are a common feature (Fig. 19), although their density is low in more distal segments. The presence or absence of gER and ribosomes is a useful, criterion, although not an absolute one, for the distinction between axons and primary and secondary dendrites.

Smooth tubular and vesicular elements with electron-lucent contents are often encountered in the neuronal cytoplasm. A specific organelle is the perinuclearly located Golgi complex (G). This complex consists of a limited number of closely apposed cisterns with irregular distensions and associated with a number of clear and dense-cored vesicles, apparently arising from the cisterns (Figs. 17 and 20). In appropriate sections, fenestrations in the cisterns are observed. Some vesicles have a furry coating. The Golgi-associated, dense-cored vesicles can be distinguished from the dense-cored vesicles present in aminergic neurons. Some of the dense-cored vesicles near the Golgi cisterns, and probably some of the coated vesicles, are thought to contain hydrolytic enzymes and to belong to the lysosomal complex (primary lysosomes). Lysosomes (L) (see Fig. 17) are observed frequently in neurons. They are characterized by their dense content surrounded by a single membrane. The sizes of the lysosomes vary greatly. Characteristic of neurons, especially in older brain tissue, are the lipofuscin granules (LG). These granules are characterized by a single membrane that surrounds an alternately

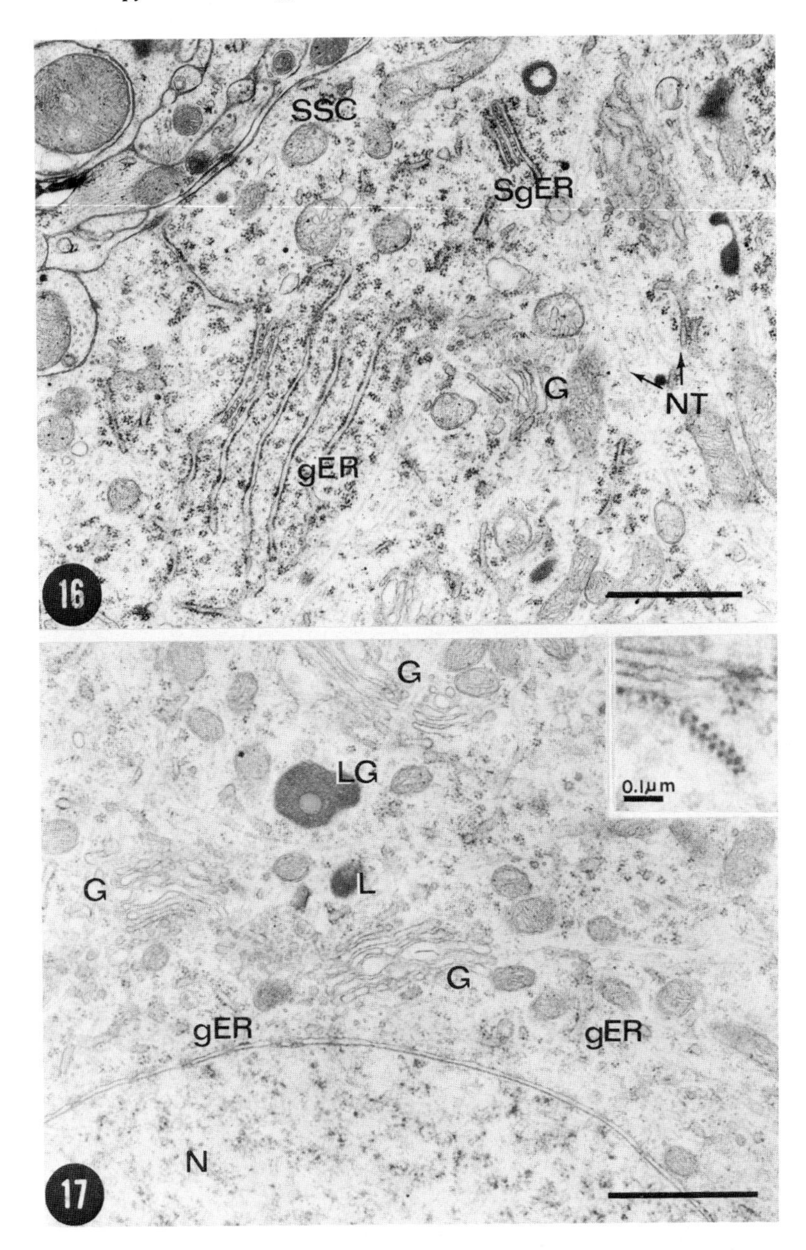

Fig. 16. Typical example of regularly organized granular endoplasmic reticulum (gER) in a neuron of the inferior olive of the cat. This structure is the counterpart of the light-microscopic Nissl body. Some elements of the gER are closely apposed to the cell membrane [subsurface cisterns (SSC)] or to each other [stacks of gER (SgER)].
Fig. 17. A small piece of perinuclear cytoplasm in a neuron of the inferior olive of the cat. Golgi complexes (G) with small vesicles, a small lysosome (L), a lipofuscin granule (LG), and isolated cisterns of granular endoplasmic reticulum (gER) are present. *Inset:* Polyribosomes at higher magnification.

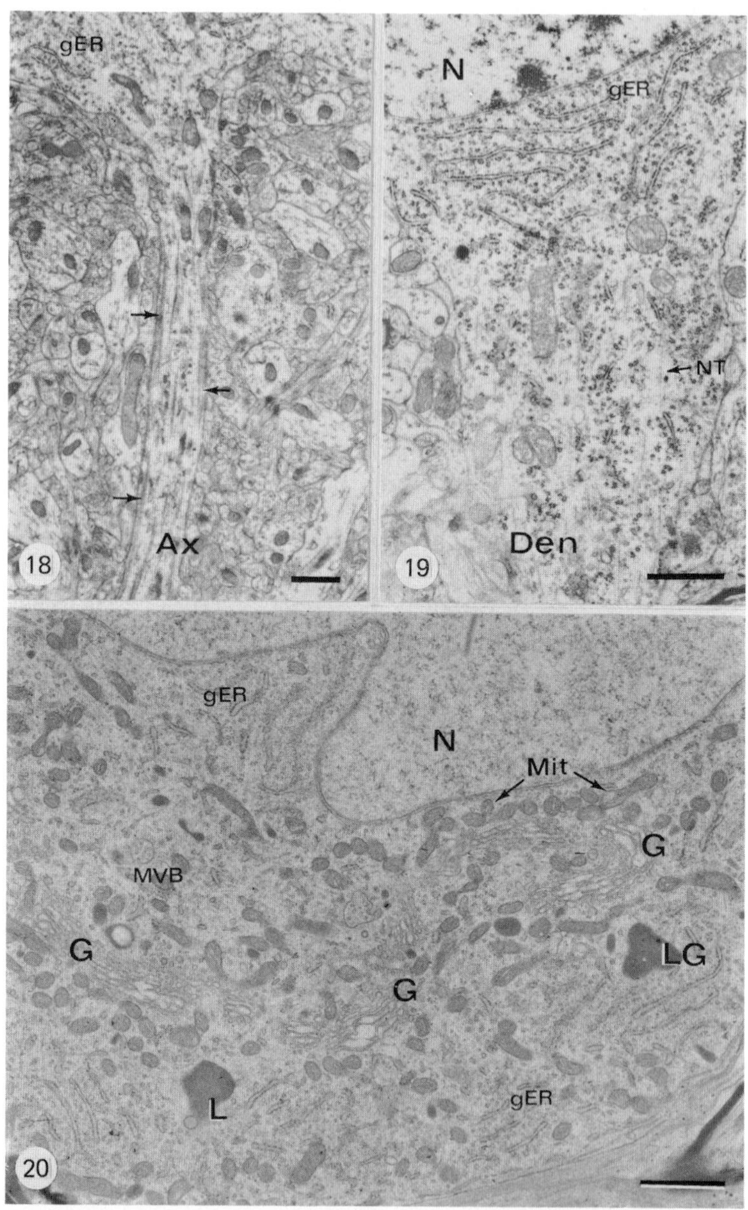

Fig. 18. Initial segment of an axon (Ax) in the visual cortex of the rabbit. The arrows point to neurotubules accumulated near the membrane of the axon. Note that very few other cytoplasmic organelles are present in this initial segment.

Fig. 19. Primary dendrite (Den) from a neuron in the visual cortex of the rabbit. Numerous cytoplasmic organelles, such as granular endoplasmic reticulum (gER), mitochondria, and neurotubules (NT), are present in this dendrite.

Fig. 20. Perinuclear cytoplasm of a neuron from the inferior olive of the cat. Numerous Golgi complexes (G), lysosomes (L), and lipofuscin granules (LG) are present. Multivesicular bodies (MVB) and numerous mitochondria (Mit) are present as well. Note the indentation of the nuclear envelope and the well-developed granular endoplasmic reticulum (gER).

electron-dense and electron-lucent content. They also contain myelinlike or crystal-like inclusions. Their outline is often irregular, and they can have very large dimensions in older brain tissue. They are thought to be the end stages of lysosomes. Frequently observed are single membrane-bounded, large vesicles filled with small regular vesicles. These so-called multivesicular bodies (MVB) are also thought to belong to the lysosomal complex.

In all parts of the neurons, with the exception of the dendritic spines, neurotubules (NT) and neurofilaments are observed (see Figs. 16 and 17). In axons and dendrites, they are often arranged in a highly ordered fashion (Fig. 18).

The mitochondria observed in all neuronal perikarya can have diverse shapes, and their dimensions vary greatly. They are characterized by a smooth, regular outer membrane and a highly infolded inner membrane. Their matrix is moderately electron-dense. In both ultrathin sections of intact brain tissue and ultrathin sections of isolated mitochondria, two types of mitochondria are present, the so-called dark and light mitochondria (see Fig. 37). The relevance of this difference in opacity is not understood at present.

Incidentally, centrioles and the basal bodies of cilia have been observed in neurons.

4.1.2. Glial Cells

It is quite easy to make the distinction between neurons and the two most frequently observed types of glial cells (astrocytes and oligodendrocytes), even at low magnification. Most characteristic for the oligodendrocytes is the accumulation of electron-dense chromatin near the nuclear envelope and the electron density of the cytoplasm. The reason for this electron density is not yet clear. All types of cellular organelles are present in the cytoplasm of oligodendrocytes. They are, however, less abundant, and the gER is less well organized (Fig. 21). The involvement of oligodendrocytes and their peripheral counterpart, the Schwann cells, in the formation of the axonal myelin sheaths is well established (Bunge and Bunge, 1970) (Fig. 21).

The degree of condensation and accumulation of nuclear chromatin in astrocytes lies between that in oligodendrocytes and that in neurons. The astrocytes often establish an irregular outline with many protrusions that can be observed even in the plane of the ultrathin sections. The cytoplasm often has a pale outlook, with few cellular organelles such as gER, Golgi complexes, and mitochondria. They contain small (β) glycogen particles. This presence of glycogen is often taken as a criterion for identification of astrocytic protrusions. In experimental and pathological material, an increase in glycogen particles is often observed. It is common practice to distinguish two types of astrocytes: fibrous and protoplasmic. The fibrous astrocytes are characterized by a vast amount of filaments present in their perikaryon and protrusions (Fig. 22). They are present mainly in subpial regions and in the white matter. The protoplasmic astrocytes lack these vast amounts of filaments

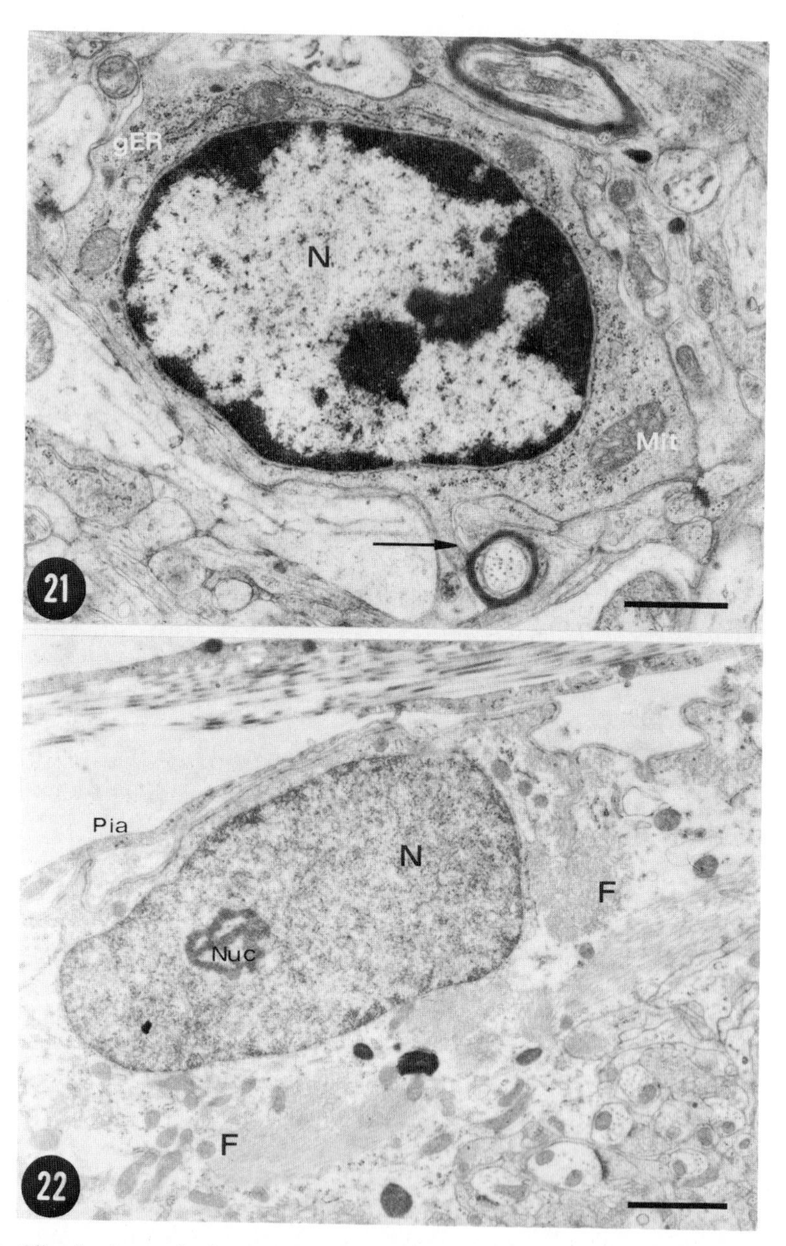

Fig. 21. Oligodendrocyte in the deep layer of the visual cortex of the rabbit. The electron-dense nuclear chromatin is accumulated near the nuclear envelope. The cytoplasm contains few organelles and is relatively electron-dense. The arrow points to a connection between a myelinated axon and an extension of the oligodendrocyte.

Fig. 22. Fibrous astrocyte near the pia mater (Pia) in the visual cortex of the rabbit. Numerous astrocytic filaments (F) are present.

(Fig. 23). They are preferentially localized in the gray matter. The protrusions of the protoplasmic astrocytes intermingle with the protrusions of neurons. Together, they form the neuropil. They are often found apposed to the basal lamina of the brain capillaries (Fig. 23) and pial surface. This enveloping of all possible protrusions and barriers is characteristic of astrocytes. Their function is thought to be manifold. The fibrous astrocytes would have a supporting function and the protoplasmic astrocytes would function as an interstitium, transporting metabolic constituents from the capillaries to the neurons and their axons and dendrites. One must be aware, however, that the difference between the two types of astrocytes is only gradual. In the inferior olive, for example, an intermediate form is observed.

The ultrastructural identification of the light-microscopically characterized microglia cells is difficult. Their distinction from migrating pericytes is not yet fully understood. Moreover, this cell type is seen only occasionally in ultrathin sections.

4.1.3. The Neuropil

In looking at a light or electron micrograph of brain tissue, it is most striking that only a small part of the brain tissue is taken up by cellular elements. In the visual and motor cortex of rabbits, for example, about 10% of the volume consists of neuronal and glial cell bodies (own unpublished observation). The remainder of the tissue is occupied by a complex network of protrusions originating from neurons and glial cells. It is well known that the neuronal processes and the way in which they are connected is of ultimate importance for the understanding of the integrated, synchronized functioning of the nervous system.

In Section 4.2, the sites at which the impulses from one cell to the other are transmitted, the synapses, are described. A correct interpretation of the synaptic relationships in brain tissue presupposes identification of the neuronal protrusions. The large primary and secondary dendrites, the initial segments of axons, and the myelinated axons are easy to identify. The primary and secondary dendritic branches (Figs. 19 and 24) contain a relatively large number of gER cisterns, ribosomes, neurofilaments, and neurotubules. The presence of gER and ribosomes constitutes a particularly good criterion, for it is now generally accepted that these elements are nearly absent in the axons. The initial segments of axons, which are in general unmyelinated, are characterized by the clustering of neurotubules against the axon membrane and by the electron-dense coating of this membrane (see Fig. 18). The myelinated axons with their regularly spaced myelin sheaths need hardly any explanation. The closely apposed membranes are formed by the cellular membranes of oligodendrocytes (central nervous system) and the Schwann cells (peripheral nervous system). A detailed description of these myelin sheaths and the characteristic nodes of Ranvier is given by Bunge and Bunge (1970). The fixation of these myelinated axons

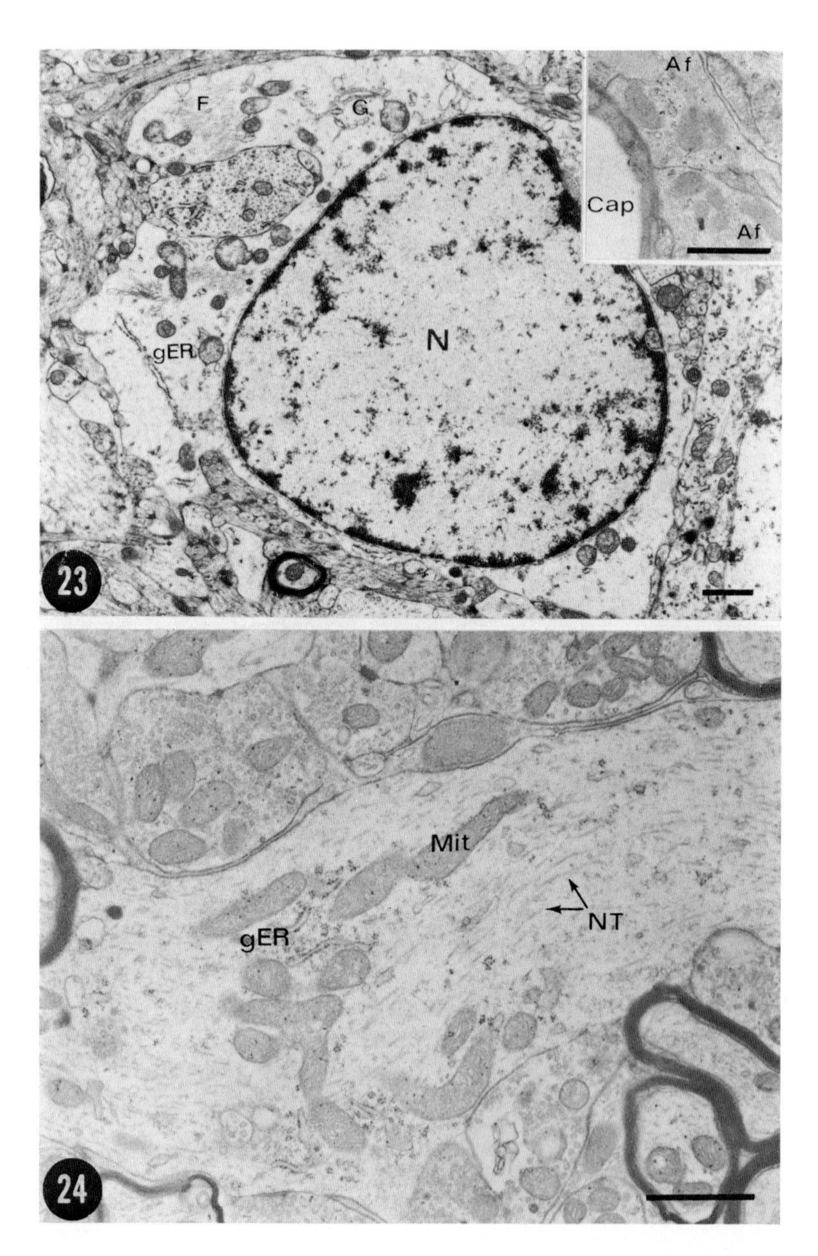

Fig. 23. Protoplasmic astrocyte in the gray matter of the visual cortex of the rabbit. Note the scarcity of cytoplasmic organelles and the relatively few astrocytic filaments as compared to the fibrous astrocyte in Fig. 22. *Inset:* Two astrocytic endfeet (Af) in contact with a capillary (Cap) in the inferior olive of the cat.

Fig. 24. Secondary dendrite in the inferior olive of the cat. Numerous mitochondria (Mit), cisterns of the endoplasmic reticulum (gER), and neurotubules (NT) are present.

is often difficult, especially in the central nervous system. Disruption of the sheaths is often encountered even in perfectly fixed material.

Most difficult to identify are the tertiary and terminal ramifications of dendrites and unmyelinated axons. Although it is thought that the small processes of dendrites contain gER and ribosomes, the density of these organelles is so sparse that many ultrathin sections do not show these organelles. This means that they are difficult to distinguish from unmyelinated axons. Other criteria such as size and irregularity of the contours are sometimes useful to distinguish these elements. In general, the unmyelinated axons are thought to be smaller and more regular in outline. Another criterion is the presence of synaptic vesicles. Until recently, profiles containing these synaptic vesicles were thought to be of axonal origin. However, the existence of dendro–dendritic and dendro–axonal contacts has now been proven. Thus, this criterion has lost its validity. Moreover, it could be used only in certain instances. The synaptic elements are most often seen as isolated, without continuity with their dendrite or axon of origin. When the identification of these small processes is of great importance in a particular study, use can be made of specific techniques, such as correlative Golgi–electron-microscopic studies (see Blackstad, 1975), intracellular staining (see Kater and Nicholson, 1973), autoradiographic and other tracing techniques (see Chapter 9), and reconstruction from serial sections. Nevertheless, it may often be impossible to identify all the neuronal processes present in an electron micrograph.

4.1.4. *Afferent and Efferent Connections*

In Chapter 5, an extensive description is given of the principles underlying the experimental investigation of afferent and efferent connections in the brain. In this respect, the two types of degeneration observed after lesions of a particular pathway will be briefly mentioned. In the light type of degeneration, the axon terminals of a lesioned pathway are swollen and show an electron-lucent cytoplasm with hyperplasia of the neurofilaments (Fig. 26). In the dark type of degeneration, the axon terminals are shrunken and have an electron-dense cytoplasm in which few organelles can be distinguished (Fig. 25). The light type of degeneration can shade off into the dark type (Walberg and Mugnaini, 1969). In further stages of degeneration, the terminals become engulfed by astrocytic processes and eventually disappear completely. It is curious that in some instances, axon terminals that are thought to originate in the injured system remain intact. Degenerating axons lose their organelles, and their myelin sheaths become disrupted.

4.2. *Electron Microscopy and Synaptic Transmission*

Ever since the introduction of the term synapse by Sherrington in 1897, this functional and structural unit of the nervous system has been the subject of many physiological, biochemical, and morphological studies. Sherrington

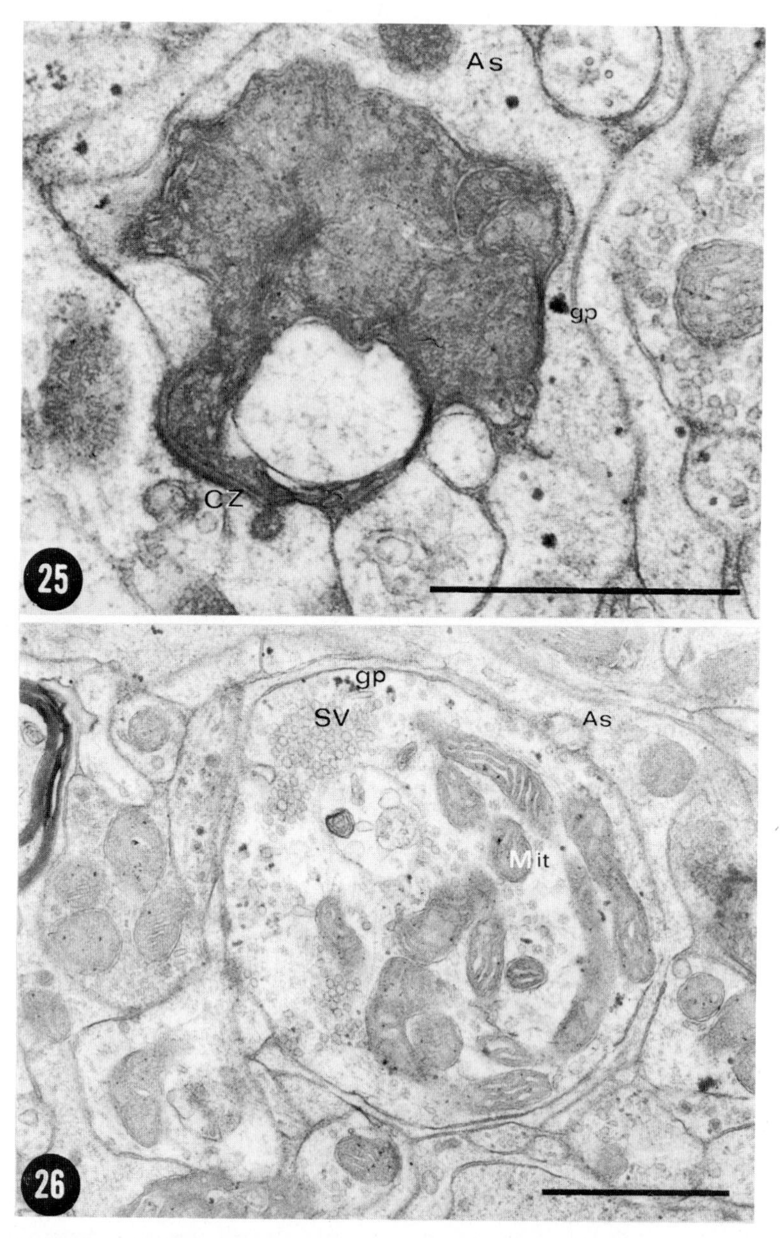

Fig. 25. "Dark-type" degenerative terminal surrounded by astroglial processes (As) in which glycogen particles (gp) are present. Note that the contact zone (CZ) is still intact. The section is from the inferior olive of the cat.

Fig. 26. "Light-type" degenerative terminal surrounded by astroglial processes (As). Note the numerous mitochondria (Mit), the clustering of synaptic vesicles (SV), and the presence of glycogen particles (gp). The section is from the inferior olive of the cat.

[1857–1952] introduced the term in a functional sense, defining those sites in a neuronal network at which nerve impulses are effectively transmitted from one cell to another. Since this pioneering work of Sherrington, the electrophysiology of synaptic transmission has been described in great detail for many regions of the central and peripheral nervous systems in different species. Besides the excitatory synaptic responses, their generation by nerve impulses, the generation of the postsynaptic potential and impulses, and the ionic processes involved, the pre- and postsynaptic inhibitory responses and their mechanisms have been described (see Eccles, 1964, 1973). Furthermore, the trophic and plastic properties of synapses have been studied in different species and in different parts of the nervous system. The stimulating work of Dale [1875–1968], Loewi [1873–1961], and many others in the 1920's and 1930's has shown that for most connections in the peripheral and central nervous system, this synaptic transmission is chemical in nature; i.e., the transmission is mediated by specific chemical substances, such as acetylcholine and adrenaline. Following their observations of miniature endplate potentials, Katz and his co-workers postulated, in the 1950's, the quantal nature of the chemical transmission; i.e., the transmitter substance is released in small discrete portions. Quite recently, it has been shown that besides chemically transmitting synapses, synaptic sites are present that show low-resistance ephaptic transmission (electronic synapses). Neurochemists have elucidated the synthesis and breakdown of acetylcholine and adrenaline and have found that many more substances have transmitterlike properties. They have also investigated the chemical anatomy of the receptor molecules of the synapses.

The structural element related to this chemically and physiologically defined neural unit has long remained obscure. It has been one of the main topics in the controversy between the supporters of the reticular theory [Golgi (1843–1926)] and the supporters of the neuron theory [His (1831–1904), Forel (1848–1931), Ramon Y Cajal (1852–1934)]. A fundamental question in this controversy was whether the connections among neuronal elements are continuous, so that the neurons form part of a continuous reticular network, or whether the connections are separated by membranes, so that neurons are individual entities as postulated by the cell theory of Schleiden [1804–1864], Schwann [1810–1882], and Virchow [1821–1902]. The definite proof of the correctness of the cell theory in the nervous system had to await the early electron-microscopic descriptions of nervous tissue. Sjöstrand (1953), Palade (1954), Palay (1954), and De Robertis and Bennett (1954) were the first to demonstrate that the contacts among neuronal processes are separated by membranes. They described the basic structural elements of the pre- and postsynaptic bags, i.e., the specializations at the pre- and postsynaptic membranes, the intercleft, and the synaptic vesicles, mitochondria, tubules, and filaments in the presynaptic bag.

The many elegant and refined electron-microscopic studies following these early descriptions have revealed a wealth of information on the synapse and the synaptic organization of the nervous system. In studies using a specific OsO_4–phosphotungstic acid (PTA) method, Gray (1959) distin-

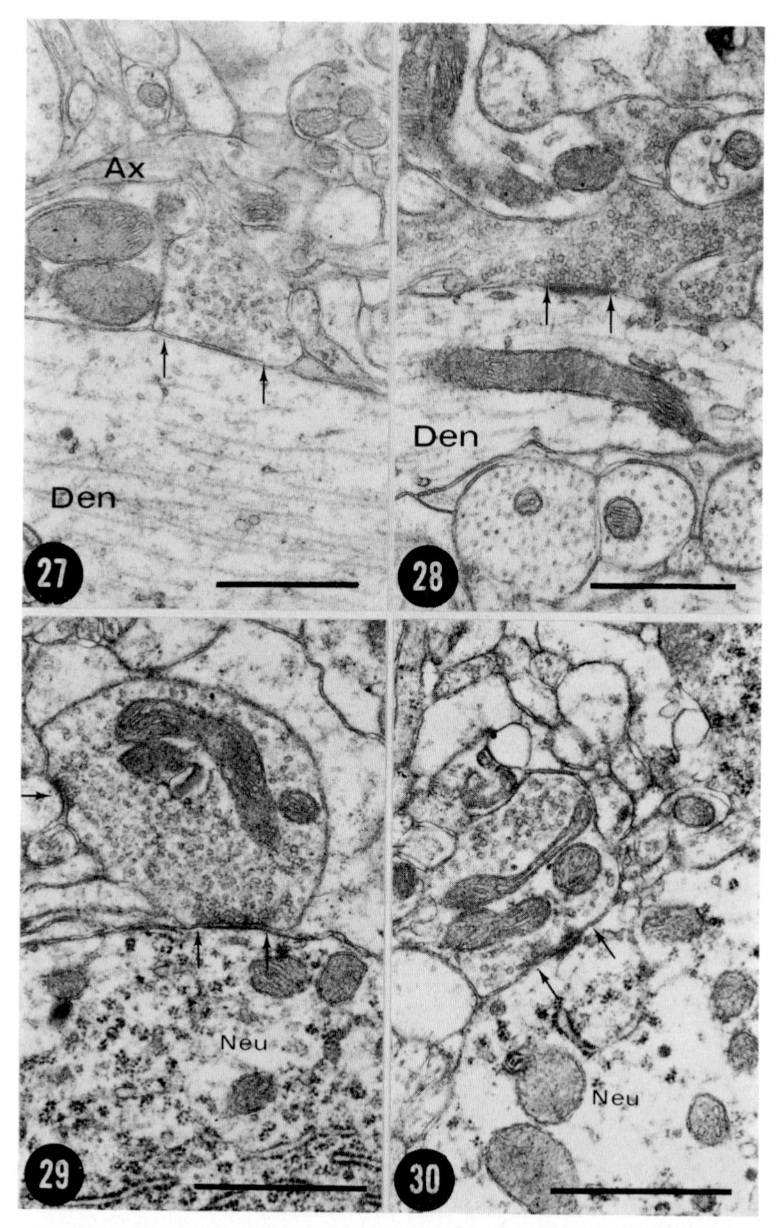

Figs. 27–30. Examples of symmetrical (Figs. 27 and 29) and asymmetrical (Figs. 28 and 30) synapses on dendrites (Den) (Figs. 27 and 28) and on neuronal somata (Nev) (Figs. 29 and 30). The arrows indicate actual contact zones. The asymmetrical synapses correspond with Gray type I and the symmetrical synapses with Gray type II. All sections are from the superior colliculus of the rabbit following glutaraldehyde–OsO_4 fixation.

guished two types of synapses. The first type has a pronounced presynaptic thickening with dense projections and a pronounced postsynaptic membrane. The second has only a thin postsynaptic thickening and a well-developed presynaptic membrane. This distinction agrees with that made by Colonnier (1968) between asymmetrical and symmetrical synapses observed in tissue following OsO_4 fixation (Figs. 27–30). The strictness of this distinction has often been debated, and one might conclude that the two types of synapses represent extremes of a single continuum. Staining with ethanolic phosphotungstic acid (EPTA) after glutaraldehyde fixation, as introduced by Bloom and Aghajanian (1966) (see Section 3.6.4), has revealed the complex organization of the presynaptic membrane with its regularly organized dense projections and the intercleft with its intercleft line (Figs. 31 and 32). The freeze–etch studies of Akert *et al.* (1972) have proven that this presynaptic membrane is indeed complex.

The synaptic vesicles, which immediately following their discovery were thought to be the structural elements related to the quantal release of transmitter, have been studied in detail. Spheroidal, pleiomorphic, and flat vesicles have been observed (Figs. 33 and 34) and have been related to different types of transmitter substances and different types of transmission (excitatory vs. inhibitory). Furthermore, in addition to vesicles with translucent cores, dense-cored vesicles have also been observed. Attempts have been made to selectively stain the various types of vesicles with, for example, zinc iodide–osmium tetroxide (ZIO). (Akert and Sandri, 1968) (see Fig. 34). Although the ZIO method is not specific for certain types of vesicles, it has proven a useful tool in other respects (see Section 3.6.6). Whether distinct types of clear synaptic vesicles really exist is still under discussion. There is evidence that flattened and pleiomorphic synaptic vesicles arise from round vesicles as the result of the fixation procedure. In particular, prolonged fixation with aldehydes seems to cause this elongation of originally round vesicles. However, the fact that the different forms appear in one section close to one another may indicate that this artificial configuration bears some relationship to existing structural or chemical differences *in vivo*.

Another important contribution of electron microscopy to our understanding of the complex organization of the brain is the unraveling of complex synaptic relationships. Szentágothai (1970) and his co-workers were among the first to describe the complex synaptic glomeruli in the cerebellar cortex and thalamic nuclei. Most characteristic of these glomeruli is that an afferent axon makes manifold contacts with dendrites and axons of local origin. Some of the contacts are simple axo–dendritic or axo–axonic synapses; others are complex inasmuch as the postsynaptic element itself contains synaptic vesicles and makes in its turn a synaptic contact with another glomerular element. Since these early descriptions of synaptic glomeruli, they have been described in many other brain regions in great detail (e.g., see Fig. 35). The cellular origin of the different synaptic elements has been elucidated and the presence of different types of synaptic vesicles has been

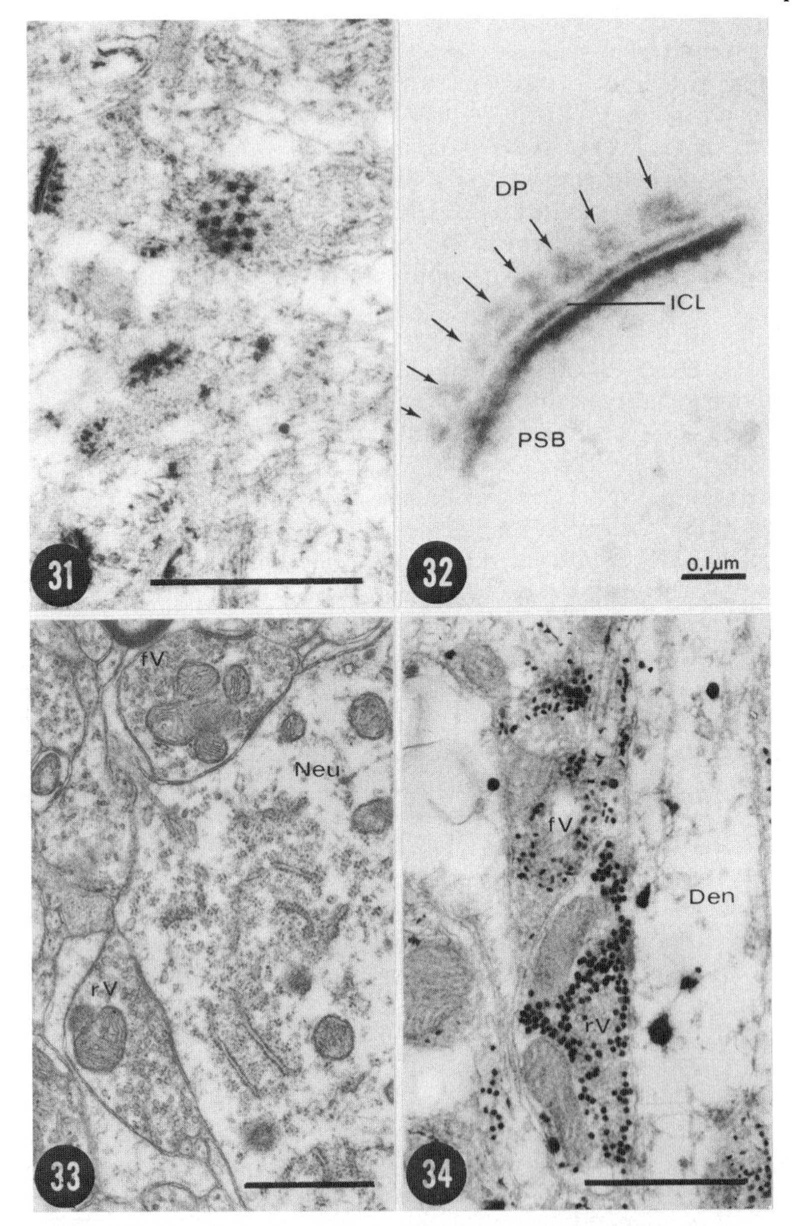

Fig. 31. Example of an EPTA-stained, tangentially sectioned synaptic contact zone in the visual cortex of the rabbit. Note the hexagonal organization of dense projections in the contact zone.
Fig. 32. Typical example of EPTA-stained synaptic contact zone with dense projections (DP), postsynaptic band (PSB), and intercleft with intercleft line (ICL). The section is from the visual cortex of the rabbit.
Fig. 33. Somatic synapses after glutaraldehyde–OsO_4 fixation illustrating the difference between flat (fV) and round synaptic vesicles (rV) in adjacent synaptic terminals.
Fig. 34. Dendritic synapses after glutaraldehyde–ZIO fixation illustrating the difference between flat (fV) and round synaptic vesicles (rV) in adjacent synaptic terminals.

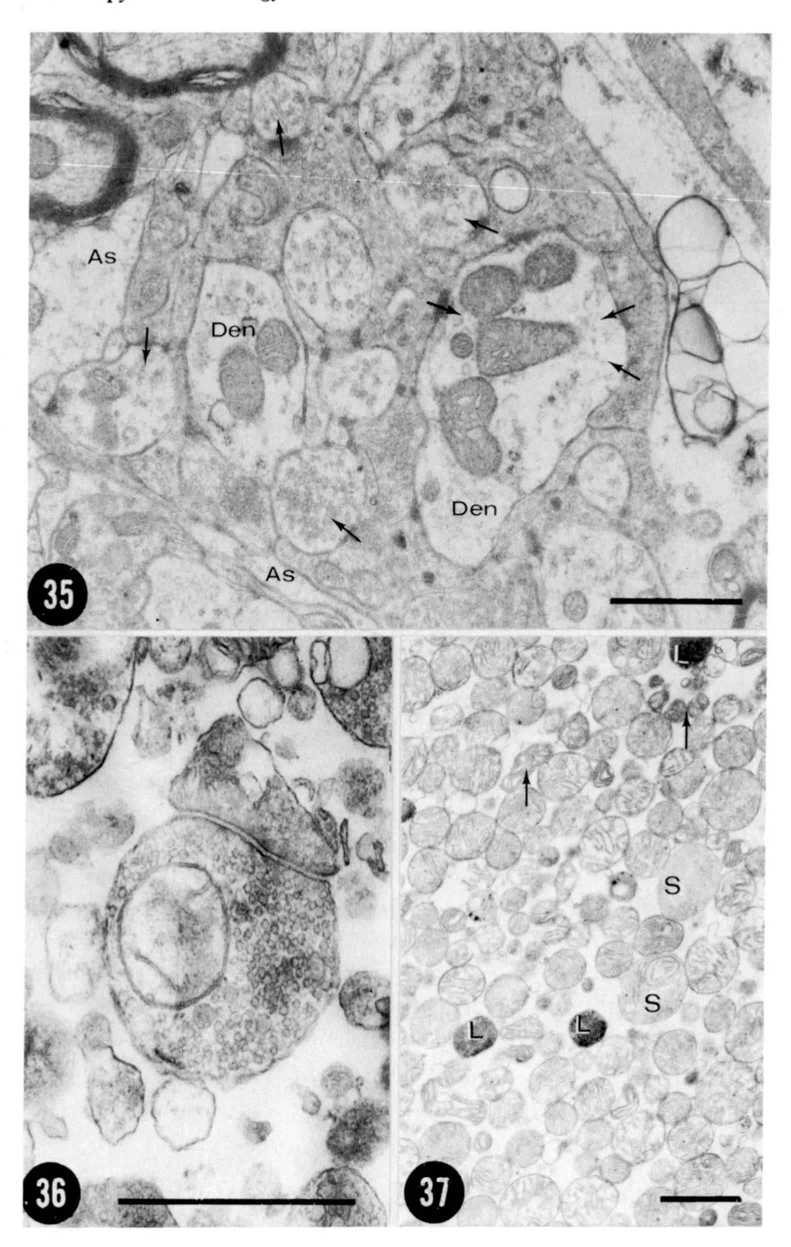

Fig. 35. Synaptic glomerulus in the lateral geniculate nucleus of the rabbit partially surrounded by astrocytic processes (As). The possible synaptic relationships are indicated by arrows. Note that some terminals project on simple dendrites (Den), while others make contacts with profiles containing synaptic vesicles.

Fig. 36. Synaptosome isolated from human autopsy material with intact contact zone and postsynaptic element.

Fig. 37. "Mitochondrial fraction" from the cortex of the rat. This fraction is heavily enriched in mitochondria, but lysosomes (L) and synaptosomes (S) are also present. Arrows point to the so-called dark mitochondria.

described. Most of these synaptic glomeruli are separated from the surrounding neuropil by astroglial processes.

Many attempts have been made to integrate the biochemical and ultrastructural properties of synapses. A powerful tool for this is the isolated synapse: the synaptosome (see Fig. 36). The stimulating work of Whittaker *et al.* (1964) and De Robertis *et al.* (1962) on isolated synaptosomes has opened a whole new field of study concerning the biochemistry of transmitters and transmission. The sites of transmitter synthesis and the storage and release of these substances have been the subjects of many elegant studies. The distinction between bound and free pools of transmitter, their possible relationship with synaptic vesicles, and the distinction between high- and low-affinity uptake have been partly elucidated using these synaptosome preparations. In correlation with the studies of Holtzman *et al.* (1973), Ceccarelli *et al.* (1973), Heuser and Reese (1973), and many others on intact neuromuscular junctions, progress has been made regarding the structural aspects of transmitter release and reuptake by processes such as exocytosis and endocytosis. It is evident that from this perspective, ultrastructural studies are highly relevant.

The synapse, as a fundamental unit of the nervous system, derives its functional relevance from its integration in a neuronal network. The functioning of these neuronal networks depends not only on the functional–structural intactness of the individual synapses but also on the topographical relationships and density of the synapses. A typical example of this relationship is given in the study of the ontogenesis of the brain, in which the increase in functional properties of a particular neuronal network is paralleled by a maturation of individual synapses and by an increase in the number of synapses. Furthermore, sensory-deprivation studies have shown that, for example, light-deprivation leads to impairment of individual synapses and changes in the neuronal network (e.g., see Riesen, 1975; Vrensen and De Groot, 1974*b*, 1975). Topographical changes and changes in the density of synapses can be observed only by sophisticated quantitative morphological methods. The methods for these quantitative investigations are outlined in Section 5.

4.3. Electron Microscopy in Cell and Tissue Culture

Tissue and cell cultures are elegant and powerful models for the study of cellular and intercellular processes and relationships in neurobiology. The technical aspects, the applications in neurobiology, and the limitations of these methods are described in Volume 1, Chapters 3 and 4. The fact that the topographical (tissue culture) or histological (cell culture) relationships are disturbed in these cultures implies that functional inferences from these studies, related to the *in vivo* situation, require proper control of morphological characteristics. Inasmuch as fine-structural details are involved (e.g., formation of synapses), electron-microscopic examination is quite relevant.

A suitable preservation of fine structure of tissues and cells in culture can generally be achieved by routine fixation and embedding as described in Section 3. It has commonly been observed that cultured cells and tissues are quite sensitive to differences in osmolarity between fixative and culture medium and to the type of buffer solution used. A buffer solution often recommended is the phosphate buffer of Sörensen (see Section 3) with a nearly isotonic osmolarity (including the aldehyde). Cells and tissues are usually cultured on collagen-coated cover glasses and are embedded while still on these glasses. Before thin-sectioning, the embedding medium containing the biological material has to be separated from the cover glass. Very often, immersion of cover glass and plastic in liquid nitrogen leads to a smooth separation of glass and plastic. The cover glasses are often provided with a specific grating that enables the precise and retrievable localization of a particular cell or tissue element during the physiological recording. In our experience, the grating is impressed in the embedding medium, which facilitates the retrieval of specified cells.

4.4. Electron Microscopy and Cell Fractionation

Cell fractionation is a well-adapted and generally accepted biochemical technique. The aim of cell fractionation is to simplify the complex metabolic functions present in cells and tissues. A detailed description of this technique and its applications in neurobiology is given in Chapter 3.

In general, two aspects of this technique can be distinguished: disruption of the tissue (homogenization) and differential or gradient centrifugation. The first process is meant to disrupt the structural integrity of the cells in order to set free the subcellular organelles without damaging their fine structure and function. The second process is intended to selectively isolate the different cellular components such as nuclei, mitochondria, and synaptosomes. Electron-microscopic examination of these fractions has two main purposes: (1) verification of the structural intactness of the subcellular organelles and (2) visualization of the selectivity of the isolation procedure and the extent of "contamination" of a particular fraction with other cellular components (see Fig. 37).

Good preservation of the subcellular components of the fractions can be obtained by mixing the resuspended fractions with a standard aldehyde solution (see Section 3.2), followed by sedimentation. The thin pellets obtained are further fixated with OsO_4, dehydrated, and embedded according to routine procedures. To get an unbiased impression of the extent of "contamination," it is advisable to cut thin sections throughout the whole thickness of the pellets.

In biochemistry and also in neurochemistry knowledge of the conformation of macromolecules is of great importance to the explanation of their function, e.g., the active sites of enzyme molecules, the receptor molecules in the pre- and postsynaptic membranes, and so on. In this connection, not only the primary organization (molecular constituents) and secondary organization, molecular sequence) but also the tertiary and quaternary (three-dimensional) configurations are of interest. At this tertiary and quaternary level of organization, electron-microscopic examination of the macromolecules may be relevant. A method of choice in this field of electron microscopy is the technique of negative staining, in which the macromolecules are surrounded by (i.e., "embedded" in) an electron-dense stain. The electron-microscopic examination of these small macromolecules (10–100 Å) requires optimal usage of the microscope, and even then the signal noise ratio is quite unfavorable. Correct interpretation of such micrographs is difficult, and special photographic procedures are employed to improve the signal noise ratio. Moreover, objective, three-dimensional reconstructions of these macromolecules from electron-microscopic images necessitate complicated mathematical procedures such as Fourier transformation.

4.5. Electron Microscopy: Relevance and Limitations in Neurobiological Research

In the preceding sections, we have outlined the concepts of the ultrastructural organization of cells and their processes in the nervous system. Furthermore, we have described some specific applications in the field of neurobiology. It is evident that it is not feasible, in an introductory chapter on electron microscopy, to cover all fields of neurobiological interest to which ultrastructural studies have contributed. However, in this section we want to describe, in general terms, the relevance and limitations of electron microscopy in neurobiological research.

In the past decades, electron microscopy has proved to be indispensable in elucidating the basic organization of the cellular elements of the nervous system and their interrelationships. This is mainly due to the fact that in proper ultrathin sections, all histological and cytological details are simultaneously imaged and all limiting membranes are clearly visualized. In light-microscopic sections, this simultaneous visibility of all cytological details cannot be obtained. It can be expected that this mainly descriptive aspect of electron microscopy will remain an important application in future and that numerous detailed studies of the cellular and synaptic organization of many different brain regions will enrich our knowledge of the morphological organization of the nervous system. The extension of this descriptive ultrastructural research with quantitative stereological methods (see Section 5) will further increase the importance of this morphological aspect.

The high resolution of the electron microscope enables the correlation, at the macromolecular level, of ultrastructural and biochemical parameters. The examples of structural–functional interrelationships mentioned in Section 1 will be further elaborated, and new correlations will be found. It can be expected, for example, that the exact role of neurotubules in axonal transport will be unraveled in the near future. In neuropathology, the accumulation of macromolecules and the underlying enzyme defects in a number of storage diseases will probably be elucidated. In this framework, recent developments in enzyme histochemistry, immunohistochemistry, and autoradiography at the electron-microscopic level are highly important (see Chapter 9 and Volume 1, Chapter 9). Systematic application of these additional techniques will prove to be very useful. Detailed ultrastructural studies of isolated cell fractions and isolated macromolecules will also contribute to the understanding of structural–functional interactions.

The manifold applications of electron microscopy in neurobiology may give the impression that this technique is omnipotent and has few limitations. Such an inference will be misleading, however, for the technique of electron microscopy has some basic limitations of which one must be continually aware when one is studying electron micrographs.

Transmission electron microscopy presupposes, in most instances, preservation of the tissue. This preservation is brought about by chemical fixation and is followed by dehydration and embedding in a medium suitable for ultrathin sectioning. In most instances, the contrast of the specimen is improved by staining with heavy metals. This histological treatment implies that the biological material studied in the electron microscope is to a large extent artificial, i.e., does not represent the actual *in vivo* structure. For histological and cytological purposes, a number of subjective criteria have been introduced by which the quality of the histological procedures can be judged. It is now generally accepted that if these criteria are satisfied, the substructures seen in the electron microscope represent, at best, the actual morphological details of the cells and their organelles. Nevertheless, we always have to be aware of the artificiality of electron-microscopic pictures. This artificiality may restrict our conclusions, especially at the macromolecular level. Ever since the introduction of electron microscopy in biology, cell biologists have been aware of this fundamental disadvantage of the "classic" histological procedures and have searched for alternatives. The following techniques reduce, to some extent, the artificial character of electron micrographs (although they have their own disadvantages); freeze–etching, freeze–fracturing, freeze–substitution, and freeze–drying.

Fixation and embedding also imply an arresting of cellular functioning. Electron-microscopic pictures are, consequently, static. Dynamic aspects of cellular functioning can be studied only with time-lapse experiments, which must be properly planned and, if possible, carefully controlled with electrophysiological or biochemical methods or both. Autoradiography at the electron-microscopic level has proved to be especially useful in this kind of dynamic ultrastructural study (see Chapter 9).

A limitation on electron microscopy, and one with far-reaching consequences, is set by the restricted penetrating power of electrons. A consequence of this is that only very thin slices can be studied. This means that we have to put up with two-dimensional information on a fundamentally three-dimensionally organized structure. This dimensional reduction is felt as a disadvantage in the ultrastructural study of all tissues and organs, but is especially inconvenient in the study of the ultrastructure of the nervous system. A most characteristic property of the nervous system is the presence of both short- and long-distance processes that interconnect the different regions and that enable the integrated, synchronized action of this system. In an actual micrograph, however, the perikarya of neurons are only rarely found in connection with axons and dendrites of considerable length. Moreover, the synaptic interconnections can only rarely be traced back to their axon or dendrite of origin. This limitation often counterbalances the specific advantage of electron microscopy in neurohistology given by the visibility of all cytological details at the same time. This gap between the three-dimensional nature of the nervous system and the two-dimensional information, although most pronounced in electron microscopy, is also present in the light-microscopic study of neurons and their processes. Therefore, neuroanatomists and neurohistologists continue to search for additional techniques that can bridge this gap. Although originally designed for light-microscopic investigations, most of these techniques have been extended to the electron-microscopic level. At present, the following techniques are available: degeneration studies, with special staining techniques for degenerating afferents or efferents (see Chapter 5); reconstruction studies, with or without special staining techniques [e.g., see the Blackstad method (Section 3.6.8)]; autoradiographic studies (see Chapter 9); horseradish peroxidase studies (see Chapter 5); intracellular staining, with specific dyes, fluorescent agents, or electron-dense stains, (e.g., see Kater and Nicholson, 1973); direct correlation of semithin sections and ultrathin sections; and recent developments in scanning transmission electron microscopy, which will in future enable the high-resolution study of relatively thick sections (up to 5 μm).

5. *Quantitative Stereology in Electron Microscopy*

5.1. *Introduction*

In neurobiology, as in many other fields of biological interest, there is a tendency to accentuate correlative biochemical, electrophysiological, and morphological investigations. From a general point of view, it can be postulated that in neurochemistry and neurophysiology, the fundamental quantitative aspect of all living matter is more fully appreciated than in neuromorphology. This suggests that neurochemists and neurophysiologists are more fully aware that a simple description of the presence of a certain metabolic or physiological process is rarely sufficient to explain the differences between experimental and control animals. In most instances, a thorough quantitative analysis is necessary. It is only quite recently that neuromorphologists have also become aware of this demand for quantification. In this section, we will briefly outline the principles of quantitative morphology.

A fundamental aspect of microscopic anatomy is that the tissue has to be cut in thin slices. This implies that the three-dimensional integrity of the tissue is sacrificed and that two-dimensional samples are studied. The gathering of information on the three-dimensional configuration of a small morphological entity from series of thin sections is called *reconstructive stereology*. The obtaining of relevant quantitative information from these thin sections is called *quantitative stereology*. In this section, we will deal only with this quantitative aspect of stereology. Although methods for quantitative stereology have been used for a long time and have proved their validity in mineralogy and metallurgy, it is only more recently that they have found their way into electron microscopy (Weibel, 1969; Weibel and Elias, 1967; Weibel *et al.*, 1972; Elias, 1967; Haug, 1972).

5.2. Basic Principles and Terminology

From a fundamental point of view, all objects studied by the natural sciences, even molecules and atoms, have a three-dimensional extension in space. Moreover, ultrastructural studies have shown that cellular structures, which look like lines or points in light-microscopic micrographs or low-power electron micrographs, actually are cylindrical or spheroidal bodies. From a practical point of view, however, it is reasonable to classify objects of morphological investigation into the following four categories:

Points (P) are small, "zero-dimensional" structures characterized by their mere presence, e.g., ribosomes and glycogen particles observed in low-power electron micrographs.

Lines (L) are "one-dimensional" structures characterized by their extension in one direction, e.g., neurofilaments and neurotubules.

Surfaces (S) are structures characterized by their "two-dimensional" extension, e.g. the boundary between cisterns of the endoplasmic reticulum and the cytoplasm and the specialized contact sites between axons and dendrites or perikarya (synapses).

Volumes (V) are structures characterized by their "three-dimensional" extension in space, e.g., cell bodies, nuclei, mitochondria.

In considering a piece of tissue (volume), we can now ask what parameters are relevant for quantitatively defining these categories of structural elements. These are: (1) number of elements per unit of volume (N_V); (2) N_V and total length of the lineal elements per unit of volume (L_V); (3) N_V and total surface area of the elements per unit of volume (S_V); and (4) N_V, S_V, and total volume of the bodies per unit of volume (V_V) plus some shape factors, such as diameter. We have to be aware that these parameters are field parameters; i.e., they represent the total number, total length, total surface area, and total volume of all the individuals of a certain distinct morphological structure. In combination with N_V, the average length ($\bar{L}$), the average surface area ($\bar{S}$), and the average volume ($\bar{V}$) can be calculated. The procedures to be described here are used to estimate field data. Characterization of individual structures (feature data) is more difficult (e.g., see De Hoff and Rhines, 1968). As discussed in the preceding sections, electron-microscopic, as well as light-microscopic, observation of biological tissues and cells necessitates thin-sectioning of the tissue. As a consequence of this sectioning, the three-dimensional integrity of the tissue is destroyed and reduced to two-dimensional, areal samples (*A*). This dimensional reduction implies that bodies (volumes) present in the tissue are reduced to areas ($V \rightarrow A$), surfaces to lines ($S \rightarrow L$), and lines to points ($L \rightarrow P$). A further consequence of this reduction is that we are dealing only with samples of the total population (tissue). This means that statistical methods have to be applied to evaluate the sample estimates before conclusions can be drawn concerning the total population.

Let us consider a small piece of the cytoplasm of a cell with a mitochondrion, endoplasmic reticulum, and neurofilaments (Fig. 38A). This piece of cytoplasm will be visualized in a random section such as that given in Fig. 38B. Now, is there any relationship between the volume of the mitochondria and areal fraction in the micrograph, is there a relationship between the surface area of the endoplasmic reticulum and the length of the membranes seen in the micrograph, and is there a relationship between the length of the neurofilaments and the number of points in the micrograph? It was the French geologist Delesse who first tackled this problem. It has since been mathematically developed by many others (see Underwood, 1970; Bolender, 1974; Saltykov, 1974). These theoretical approaches have shown that, statistically, the areal fraction of a certain structural element (A_A) is a relevant estimate of its volume fraction (V_V), i.e., $A_A = V_V$. This means that a good estimate of the volume fraction of mitochondria in a cell can be obtained by

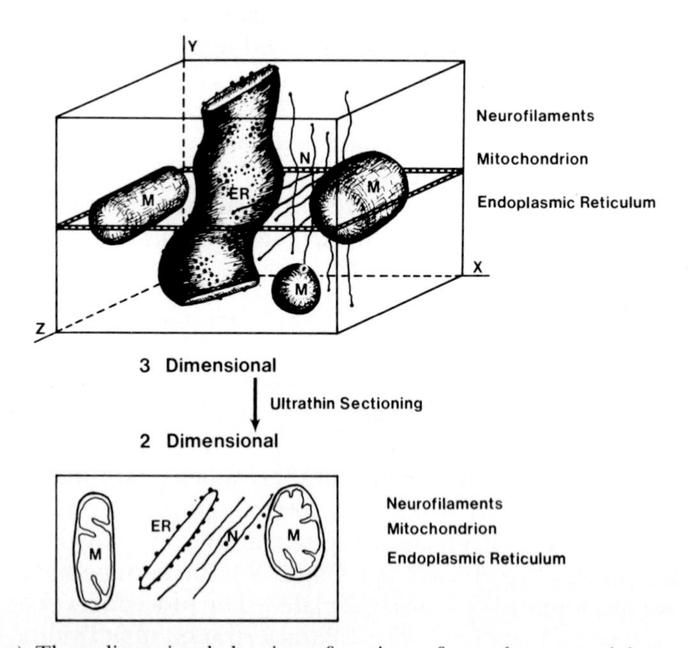

Fig. 38. (Top) Three-dimensional drawing of a piece of cytoplasm containing endoplasmic reticulum, mitochondria, and neurofilaments. (Bottom) Two-dimensional representation of the piece of cytoplasm as it would be observed in an electron micrograph. For further explanation, refer to the text.

measuring their areal fraction in a section. Measuring the areal fraction of a population of mitochondria in a micrograph, however, is very laborious. Two more relationships have been derived that greatly facilitate and simplify the estimation of A_A and V_V. It has been proven that the length of the intercepts of a system of test lines (L_L) with the mitochondria is a good estimate of A_A. Moreover, when a grid of points is placed over the picture, the fraction of points (P_P) falling over mitochondria is related to its areal fraction. Estimation of the volume fraction of a certain structure can thus be obtained by counting the number of test points falling over it or by measuring the length of test lines intercepting with it, because of the relationships:

$$P_P = L_L = A_A = V_V$$

Similar formulas are derived for the relationship between surfaces and the length of their membranes in the picture, and between lineal elements and their points in the picture: $S_V = 2P_L$, in which P_L is the number of intersections per unit length of a system of test lines with the membranes representing the structure in the image; $L_V = 2P_A$, in which P_A is the number of points coming from the lineal element under investigation. Thus, for the analysis of a piece of tissue with lineal elements, surfaces, and bodies, four basic measurements have to be made on the micrographs: P_P, number of points falling over the areal feature per total number of points over the micrograph; P_L, number of intersections of the test lines with the lineal features coming from the surface elements per total length of test lines; P_A, number of points per unit area coming from the lineal elements under investigation; and N_A, number of features per unit test area.

A parameter of general interest is the number of features per unit volume, N_V. This parameter is useful for structures present in tissue. It is evident that N_V is somehow related to the number of features in the section, N_A. The derivation of mathematically correct equations for this relationship is difficult, however, for it depends on the size and shape of the structure under investigation and on the thickness of the sections used. In electron microscopy, we often deal with situations in which the thickness of the sections is much smaller than the size of the structures. In these cases, two equations can be fruitfully applied (see Weibel and Elias, 1967). In the first of these equations

$$N_V = \frac{N_A}{\bar{D}}$$

in which $\bar{D}$ is the diameter of the structure under investigation, which can be obtained relatively easily for simple structures such as spheroids, cylinders, and cubes (see Underwood, 1970). In the second equation

$$N_V = \frac{1}{\beta} \cdot \frac{{N_A}^{3/2}}{{V_V}^{1/2}} \cdot K$$

in which β and K are shape and size distribution coefficients, respectively (see Weibel and Elias, 1967). Both formulas presuppose some knowledge, more or less exact, of the size and shape of the structures under investigation, which is sometimes difficult to obtain for the irregularly shaped structures observed in electron micrographs. In other instances, the section thickness is not small compared to the size of some structures, such as small vesicles, ribosomes, or glycogen particles in electron-microscopic sections, or perikarya and cell nuclei in light-microscopic sections. For these situations, a number of correction formulas have been worked out, the best known of which is the Abercrombie correction formula:

$$N_V = N_A \cdot \frac{T}{2r + T}$$

in which T is the section thickness, r is the radius of the structure under investigation, and V represents the volume of that part of the section that is analyzed (area × section thickness). The equation also presupposes knowledge of the size and shape of the structure. It is advisable to avoid these complicated equations. This is often possible, especially in comparative studies in which a relative estimate of the density of some structures is sufficient.

Another parameter that is of interest for many morphological problems is the size distribution of structures present in the tissue, e.g., the size distribution of nuclei, mitochondria, microbodies, lysosomes, and other structures. Here, we must recall what happens during sectioning. Let us consider a spherical structure with diameter D. This structure will appear in

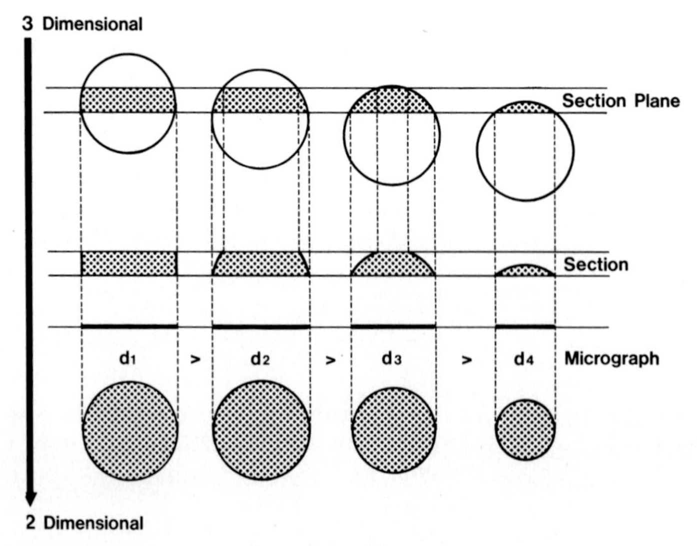

Fig. 39. Regular spheres in space intersected by thin planes (e.g., sections) and the representation of these spheres in a random micrograph. For further explanation, refer to the text.

thin random sections as a circular profile with a diameter ranging from D to zero (see Fig. 39). A monodispersed aggregate of such structures, e.g., a piece of cytoplasm filled with lysosomes, will yield a picture with circular profiles having a whole range of diameters. From such pictures, we can measure the diameters of the individual profiles and make a frequency distribution or a histogram. It should be clear that this frequency distribution is somehow related to the actual diameter of the spheres under investigation. The exact mathematical relationship between this size distribution in the section and the actual size distribution of the structures in the tissue is rather complicated and has been elaborated only for regular, spherical structures (Weibel and Bolender, 1973). Still more complicated are situations that involve populations of qualitatively identical structures with varying diameters (polydisperse). In many instances, however, it is not possible or in fact not necessary to go into the complex problems of absolute quantitation. In these instances, a relative comparison of the frequency distribution is sufficient to solve the biological problem.

The simple procedures and equations described above are the fundamentals on which the whole system of quantitative stereology is based. For simple estimations of volume fractions, surface density, and other parameters, they are sufficient. For more complicated problems, the reader is referred to the relevant literature.

5.3. Some Practical Aspects of Quantitative Stereology

It would be beyond the scope of this section to describe in detail the procedures used in quantitative stereology. We will mention only some of the most basic aspects. For more details, the reader is referred to the literature.

5.3.1. Sampling

Quantitative stereology is statistical in nature. This means that we are dealing with samples (sections) of the whole population (tissue or organ). These samples have to be representative for the population without overestimating or underestimating one of the structures present. This implies that during all the stages of this sampling procedure (section preparation), care must be taken to avoid bias. Therefore, organ dissection, preparation of small pieces for fixation and embedding, sampling of the pieces to be used for sectioning, and the microscopic observation of the sections have to be carried out with great care.

5.3.2. Test Probes

As already described in Section 5.2, volume density, surface density, and line density can be determined by simple counting of points and intersections. For this point- and intersection-counting, a number of lattices have been developed (Weibel, 1969; Weibel and Bolender, 1973). A most commonly used lattice is the coherent-square lattice of lines (Fig. 40). With this lattice, point-counting (intersections of the horizontal and vertical lines), intersection-counting (horizontal and vertical lines), and even area measurements (rectangles formed by the horizontal and vertical lines) can be carried out. The distance between the lines (*d*)—and therefore the distance between the points and their number per area and the areal surface of the rectangles—has to be adapted for each tissue to be studied with quantitative stereology. As a general rule, we have to take care that the points are spaced such that "no more than one point can fall on the same profile" (Weibel, 1969). A regular-square lattice cannot be applied when the tissue under investigation is regularly spaced (anisotropic), e.g., the myofibrillar organization of muscle. In these instances, random-point lattices or curvilinear lattices have to be used.

5.3.3. *Statistics*

The relationship of a sample estimate to the true value in the population depends greatly on the number of samples analyzed. In general, the sample

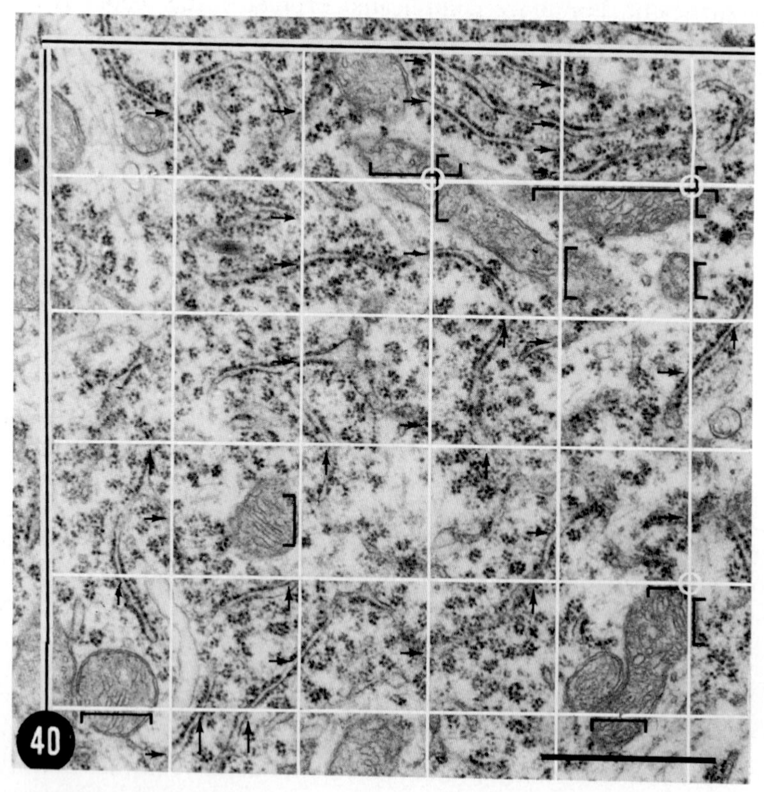

Fig. 40. Actual micrograph of a piece of cytoplasm covered by a regular square lattice. (——) Interception measurements; (→) intersection measurements; (○) point-countings. Bar = 1 μm.

estimate is the more accurate the larger the sample. Stereological analyses are laborious, time-consuming, and often boring. Thus, in practice, we shall try to find a compromise between accuracy and amount of effort. The correctness of this compromise can be tested by statistical procedures. A description of the statistical methods can be found in, for example, Dixon and Massey (1969), Snedecor (1956), and Cochran (1963). In quantitative stereology, the most commonly used statistical parameters and tests are: mean ($\bar{x}$), standard error of the mean (S.E.M.), student's t test, chi-square test, linear regression ($y = ax + b$), and correlation coefficient. The statistical evaluation of stereological data is greatly facilitated by the use of on-line programmable calculators or computers.

5.3.4. *Systematic Errors*

During sampling procedures, many systematic errors may occur. They can be avoided by critical evaluation of all stages of this sampling. The stereological principles that have been described are based on the assumption that the samples used are of zero thickness. This assumption is not fulfilled in microscopic examination of sections, not even in electron microscopy. This results in overestimation, especially of opaque structures. Correction formulas have been derived for this. The most commonly used correction formula takes into account section thickness (T) and the diameter (D) of the structure investigated: $K = 1 + (3T/2D)$. The relationship between point fraction and volume fraction can then be given by: $V_V = P_P\ (1/K)$.

5.3.5. *Manual and Automatic Counting*

The stereological procedures, such as point-counting and intersection-counting, are very simple yet versatile. The only things needed for it are a series of random, correctly selected micrographs and an appropriate lattice. Lattices of different types can be made quite simply by the investigator. For a first approach to a quantitative morphological problem, this manual counting is the best choice, and the results obtained with it are relevant. It is, however, laborious and time-consuming. The procedure can be speeded by feeding the primary data (counts) into a multichannel counter that is attached to a programmable calculator or computer that works out the mathematical transformations and does the statistics.

In a further stage, semiautomatic instruments may be useful. In one type of instrument (M.O.P.-KM-2, Kontron), the lattices consist of reflecting lines. By tracing the structures in the picture with a special pointlike light source, the intersections with the reflecting lines are automatically counted and stored in a channel. In other types (M.O.P.-AM-01, Kontron; Ladd-241, Ladd Research Industries, Inc.), use is made of a more or less homogeneous electromagnetic field. The x–y coordinates of the structures under investigation are registered, stored, and computed.

Fully automatic instruments have been developed as well. They are based on the following principle: From a microscopic section, a television image is generated. Regions (structures) with a specific gray level are selectively discriminated, and their area, perimeter, and other dimensions are automatically registered and computed. These instruments are expensive. The selective discrimination of certain structures often fails for electron-microscopic pictures because their gray level is not homogeneous.

ACKNOWLEDGMENTS. The authors thank the directors of their institutes—Dr. C. Stotijn and Dr. G. Mojet, Mental Hospital "Endegeest," Oegstgeest; Dr. J. Voogd, Department of Neuroanatomy, University of Leiden, Leiden; and Prof. Dr. G. Bleeker, Netherlands Ophthalmic Research Institute, Amsterdam—for providing time and opportunity to write this chapter. Drs. E. Wisse, G. Brakenhoff, and R. Verwer are acknowledged for their critical reading and discussion of parts of the manuscript. The technical assistance of Messrs. B. Willekens, H. Choufoer, Ch. van der Sijp, W. Dijksma, and J. Nunes Cardozo in preparing the micrographs and drawings is greatfully acknowledged. We especially thank Mrs. N. Wittebrood and Miss S. Koning for typing the many concepts of this manuscript.

References

Akert, K., and Sandri, C., 1968, An electron microscopic study of zinc iodide–osmium impregnation of neurons. I. Staining of synaptic vesicles at cholinergic junctions, *Brain Res.* **7**:286.

Akert, K., Pfenninger, K., Sandri, C., and Moor, H., 1972, Freeze–etching and cytochemistry of vesicles and membrane complexes in synapses of the central nervous system, in: *Structure and Function of Synapses* (G.D. Pappas and D.P. Purpura, eds.), pp. 67–86, Raven Press, New York.

Blackstad, T.W., 1970, Electron microscopy of Golgi preparations for the study of neuronal relations, in: *Contemporary Research Methods in Neuroanatomy* (W.J.H. Nauta and S.O.E. Ebbeson, eds.), pp. 186–216, Springer-Verlag, Berlin.

Blackstad, T.W., 1975, Electron microscopy of experimental axonal degeneration in photochemically modified Golgi preparations: A procedure for precise mapping of nervous connections, *Brain Res.* **95**:191.

Bloom, F.E., and Aghajanian, G.K., 1966, Cytochemistry of synapses: A selective staining method, *Science* **154**:1575.

Bolender, R.P., 1974, Stereology applied to structure–function relationships in pharmacology, *Fed. Proc. Fed. Am. Soc. Exp. Biol.* **33**:1287.

Bullivant, S., 1973, Freeze–etching and freeze–fracturing, in: *Advanced Techniques in Biological Electron Microscopy* (J.K. Koehler, ed.), pp. 67–107, Springer-Verlag, Berlin.

Bunge, R.P., 1970, Structure and function of neuroglia: Some recent observations, in: *The Neurosciences: Second Study Program* (F.O. Schmitt, ed.), pp. 782–797, Rockefeller University Press, New York.

Cammermeyer, J., 1961, The importance of avoiding "dark" neurons in experimental neuropathology, *Acta Neuropathol. (Berlin)* **1**:245.

Ceccarelli, B., Hurlbut, W.P., and Mauro, A., 1973, Turnover of transmitter and synaptic vesicles at the frog neuromuscular junction, *J. Cell Biol.* **57**:499.

Cochran, W.G., 1963, *Sampling Techniques*, 2d ed., John Wiley, New York.

Colonnier, M., 1968, Synaptic patterns on different cell types in the different laminae of the cat visual cortex, *Brain Res.* **9**:268.

De Hoff, R.T., and Rhines, F.N., 1968, *Quantitative Microscopy,* McGraw-Hill, New York.

De Robertis, E.D.P., and Bennett, H.S., 1954, Submicroscopic vesicular component in the synapse, *Fed. Proc. Fed. Am. Soc. Exp. Biol.* **13**:35.

De Robertis, E., Pellegrino de Iraldi, A., Rodriguez De Lores, G., and Salganicoff, L., 1962, Cholinergic and non-cholinergic nerve endings in rat brain: I, *J. Neurochem.* **9**:23.

Dixon, W.J., and Massey, F.J., 1969, *Introduction to Statistical Analysis,* 3rd ed., McGraw-Hill, New York.

Ebels, E.J., 1975, Dark neurons; A significant artifact, *Acta Neuropathol. (Berlin)* **33**:271.

Eccles, J.C., 1964, *The Physiology of Synapses,* Springer-Verlag, Berlin.

Eccles, J.C., 1973, *The Understanding of the Brain,* McGraw-Hill, New York.

Elias, H., (ed.), 1967, *Stereology,* Springer-Verlag, Berlin.

Glauert, A.M. (ed.), 1973–1975, *Practical Methods in Electron Microscopy,* Vols. 1–4, North-Holland, Amsterdam.

Glauert, A.M., and Phillips, R., 1965, The preparation of thin sections, in: *Techniques for Electron Microscopy* (D. Kay, ed.), pp. 213–253, Blackwell, Oxford.

Gray, E.G., 1959, Axosomatic and axodendritic synapses of the cerebral cortex: An electron microscopic study, *J. Anat. (London)* **93**:420.

Gray, E.G., 1964, Tissue of the central nervous system, in: *Electron Microscopic Anatomy* (S.M. Kurtz, ed.), pp. 369–417, Academic Press, New York.

Hama, K., 1973, High voltage electron microscopy, in: *Advanced Techniques in Biological Electron Microscopy* (J.K. Koehler, ed.), pp. 275–299, Springer, Berlin.

Haschemeyer, R.H., and Meyers, R.J., 1972, Negative staining, in: *Principles and Techniques of Electron Microscopy: Biological Applications,* Vol. 2 (M.A. Hayat, ed.), Van Nostrand-Reinhold, New York.

Haug, H., 1972, Stereological methods in the analysis of neuronal parameters in the central nervous system, *J. Microsc. (London)* **95**:165.

Hayat, A.M. (ed.), 1970–1974, *Principles and Techniques of Electron Microscopy: Biological Applications,* Vols. 1–4, Van Nostrand-Reinhold, New York.

Henderson, W.J., and Griffiths, K., 1972, Shadow casting and replication, in: *Principles and Techniques of Electron Microscopy: Biological Applications,* Vol. 2 (M.A. Hayat, ed.), Van Nostrand-Reinhold, New York.

Heuser, J.E., and Reese, T.S., 1973, Evidence of recycling of synaptic vesicle membrane during transmitter release at the frog neuromuscular junction, *J. Cell Biol.* **57**:315.

Holtzman, E., Teichberg, S., Abrahams, S.J., Citkowitz, E., Crain, S.M., Kawai, N., and Peterson, E.R., 1973, Notes on synaptic vesicles and related structures, endoplasmic reticulum, lysosomes and peroxisomes in nervous tissue and the adrenal medulla, *J. Histochem. Cytochem.* **21**:349.

Horne, R.W., 1965, Negative staining methods, in: *Techniques for Electron Microscopy* (D. Kay, ed.), pp. 328–355, Blackwell, Oxford.

Kater, S.B., and Nicholson, C. (eds.), 1973, *Intracellular Staining in Neurobiology,* Springer-Verlag, Berlin.

Kawana, E., Akert, K., and Sandri, C., 1969, Zinc iodide–osmium impregnation of nerve terminals in spinal cord, *Brain Res.* **16**:325.

Koehler, J.K., 1972, The freeze–etching technique, in: *Principles and Techniques of Electron Microscopy: Biological Applications,* Vol. 2 (M.A. Hayat, ed.), pp. Van Nostrand-Reinhold, New York.

Kojima, T., Saito, K., and Kakimi, S., 1975, *An Electron Microscopic Atlas of Neurons,* University of Tokyo Press.

Latta, H., Hartmann, J.F., 1950, Use of a glass edge in thin sectioning for electron microscopy, *Proc. Soc. Exp. Biol. Med.* **74**:436.

Luft, J.H., 1973, Embedding media: Old and new, in: *Advanced Techniques in Biological Electron Microscopy* (J.K. Koehler, ed.), pp. 1–34, Springer-Verlag, Berlin.

Martin, R., Barlow, J., and Miralto, A., 1969, Application of the zinc iodide–osmium tetroxide impregnation of synaptic vesicles in cephalopod nerves, *Brain Res.* **15**:1.

Meek, G.A., 1970, *Practical Electron Microscopy for Biologists,* Wiley-Interscience, London.

Moor, H., 1969, Freeze–etching, *Int. Rev. Cytol.* **25:**391.
Orci, L., and Perrelet, A., 1975, *Freeze–Etch Histology: A Comparison between Thin Sections and Freeze–Etch Replicas,* Springer-Verlag, Berlin.
Palade, G.E., 1954, Electron microscope observations of interneuronal and neuromuscular synapses, *Anat. Rec.* **118:**335.
Palay, S.L., 1954, Electron microscope study of the cytoplasm of neurons, *Anat. Rec.* **118:**336.
Palay, S.L., and Chan-Palay, V., 1974, *Cerebellar Cortex: Cytology and Organization,* Springer-Verlag, Berlin.
Pease, D.C., 1973, Substitution techniques, in: *Advanced Techniques in Biological Electron Microscopy* (J.K. Koehler, ed.), pp. 35–63, Springer-Verlag, Berlin.
Peters, A., 1970, The fixation of central nervous tissue and the analysis of electron micrographs of the neuropil with special reference to the cerebral cortex, in: *Contemporary Research Methods in Neuroanatomy* (W.J.H. Nauta and S.O.E. Ebbeson, eds.), pp. 56–76, Springer-Verlag, Berlin.
Peters, A., Palay, S.L., and Webster, H. de F., 1976, *The Fine Structure of the Nervous System,* 2nd ed., W.B. Saunders, Philadelphia.
Pfenninger, K.H., 1971, The cytochemistry of synaptic densities. I. An analysis of the bismuth iodide impregnation method, *J. Ultrastruct. Res.* **34:**103.
Rebhun, L.J., 1972, Freeze–substitution and freeze–drying, in: *Principles and Techniques of Electron Microscopy: Biological Applications,* Vol. 2 (M.A. Hayat, ed.), p. 7, Van Nostrand-Reinhold, New York.
Reger, J.F., Holbrook, J.R., and Pozos, R.S., 1972, Electron-dense cells in the nervous system of representative mammals, *J. Submicrosc. Cytol.* **4:**135.
Reid, N., 1975, Ultramicrotomy, in: *Practical Methods in Electron Microscopy,* Vol. 3 (A.M. Glauert, ed.), pp. 213–347, North-Holland, Amsterdam.
Reynolds, E.S., 1963, The use of lead citrate at high pH as an electron opaque stain in electron microscopy, *J. Cell Biol.* **17:**208.
Riesen, A.H. (ed.), 1975, *The Developmental Neuropsychology of Sensory Deprivation,* Academic Press, New York.
Saltykov, S.A., 1974, *Stereometrische Metallographie,* VEB Deutscher Verlag Grundstoff-industrie, Leipzig.
Sjöstrand, F.S., 1953, Ultrastructure of retinal rod synapses of guinea pig eye, *J. Appl. Phys.* **24:**1422.
Sjöstrand, F.S., 1967, *Electron Microscopy of Cells and Tissues,* Vol. 1, Academic Press, New York.
Snedecor, G.W., 1956, *Statistical Methods Applied to Experiments in Agriculture and Biology,* 5th ed., Iowa State College Press, Ames.
Sotelo, C., and Palay, S.L., 1971, Altered axons and axon terminals in the lateral vestibular nucleus of the rat, *Lab. Invest.* **25:**653.
Stensaas, S.S., Edwards, C.Q., and Stensaas, L.J., 1972, An experimental study of hyperchromic nerve cells in the cerebral cortex, *Exp. Neurol.* **36:**472.
Szentágothai, J., 1970, Glomerular synapses, complex synaptic arrangements and their operational significance, in: *The Neurosciences: Second Study Program* (F.O. Schmitt, ed.), pp. 427–443, Rockefeller University Press, New York.
Uchizono, K., 1975, *Excitation and Inhibition: Synaptic Morphology,* Elsevier, Amsterdam.
Underwood, E.E., 1970, *Quantitative Stereology,* Addison-Wesley, Reading.
Vaughan, J.E., and Peters, A., 1966, Aldehyde fixation of the nerve fibers, *J. Anat. (London)* **100:**687.
Vrensen, G., and De Groot, D., 1974*a*, Osmiun–zinc iodide staining and the quantitative study of central synapses, *Brain Res.* **74:**131.
Vrensen, G., and De Groot, D., 1974*b*, The effect of dark rearing and its recovery on synaptic terminals in the visual cortex of rabbits: A quantitative electron microscopic study, *Brain Res.* **78:**263.
Vrensen, G., and De Groot, D., 1975, The effect of monocular deprivation on synaptic terminals in the visual cortex of rabbits: A quantitative electron microscopic study, *Brain Res.* **93:**15.

Walberg, F., and Mugnaini, E., 1969, Distinction of degenerating fibers and boutons of cerebellar and peripheral origin in the deiters' nucleus of the same animal, *Brain Res.* **14:**67.
Weakly, B.S., 1972, *A Beginner's Handbook in Biological Electron Microscopy,* Churchill-Livingstone, Edinburgh.
Weibel, E.R., 1969, Stereological principles for morphometry in electron microscopic cytology, *Int. Rev. Cytol.* **26:**235.
Weibel, E.R., and Bolender, R.P., 1973, Stereological techniques for electron microscopic morphometry, in: *Principles and Techniques of Electron Microscopy,* Vol. 3 (M.A. Hayat, ed.), pp. 237–296, Van Nostrand-Reinhold, New York.
Weibel, E.R., and Elias, H., (eds.), 1967, *Quantitative Methods in Morphology,* Springer-Verlag, Berlin.
Weibel, E.R., Meek, G., Ralph, B., Ecklin, P., and Ross, R. (eds.), 1972, *Stereology 3,* Blackwell, Oxford.
Whittaker, V.P., Michaelson, I.A., and Kirkland, R.J.A., 1964, The separation of synaptic vesicles from nerve ending particles ("synaptosomes"), *Biochem. J.* **90:**293.

Bibliography

For convenience, we have listed below some recent handbooks on electron microscopy, electron-microscopic histology, cell biology, and ultrastructure of the nervous system. This bibliography is meant only as a suggestion for further reading and does not claim to be complete with respect to any of the subjects treated in this chapter.

Ambrose, E.J., and Easty, D.M., 1970, *Cell Biology,* Nelson, London.
Bennett, M.V.L. (ed.), 1974, *Synaptic Transmission and Neuronal Interaction,* Society of General Physiology Series, Vol. 28, Raven Press, New York.
Bloom, W., and Fawcett, D.W., 1975, *Textbook of Histology,* 10th ed. Saunders, Philadelphia.
De Robertis, E.D.P., Nowinsky, W.W., and Saez, F.A., 1975, *Cell Biology,* 6th ed., Saunders, Philadelphia.
Fawcett, D.W., 1966, *An Atlas of Fine Structure,* Saunders, Philadelphia.
Giese, A.B., 1968, *Cell Physiology,* Saunders, Philadelphia.
Glauert, A.M. (ed.), 1973–1975, *Practical Methods in Electron Microscopy,* Vols. 1–4, North-Holland, Amsterdam.
Hayat, A.M. (ed.), 1970–1974, *Principles and Techniques of Electron Microscopy: Biological Applications,* Vols. 1–4, Van Nostrand-Reinhold, New York.
Hydén, H. (ed.), 1967, *The Neuron,* Elsevier, Amsterdam.
Jones, D.G., 1975, *Synapses and Synaptosomes: Morphological Aspects,* Chapman and Hall, London.
Kay, D.H. (ed.), 1965, *Techniques for Electron Microscopy,* Blackwell, Oxford.
Kennedy, D. (ed.), 1965, *The Living Cell,* Readings from *Scientific American,* Freeman, San Francisco.
Kojima, T., Saito, K., and Kakimi, S., 1975, *An Electron Microscopic Atlas of Neurons,* University of Tokyo Press.
Kuyper, Ch.M.A., 1962, *The Organization of Cellular Activity,* Elsevier, Amsterdam.
Lehninger, A., 1975, *Biochemistry: The Molecular Basis of Cell Structure and Function,* 2nd ed., Worth Publishers, New York.
Loewy, A.G., and Siekevitz, P., 1969, *Cell Structure and Cell Function,* 2nd ed., Holt, Rinehart and Winston, New York.
Meek, G.A., 1970, *Practical Electron Microscopy for Biologists,* Wiley-Interscience, London.
Novikoff, A.B., and Holtzmann, E., 1970, *Cells and Organelles,* Holt, Rinehart and Winston, New York.
Palay, S.L., and Chan-Palay, V., 1974, *Cerebellar Cortex: Cytology and Organization,* Springer-Verlag, Berlin.

Pease, D.C., 1966, *Histological Techniques for Electron Microscopy,* Academic Press, New York.
Peters, A., Palay, S.L., and Webster, H. de F., 1976, *The Fine Structure of the Nervous System,* Harper and Row, New York.
Porter, K.R., 1973, *Fine Structure of Cells and Tissues,* Lea and Febiger, New York.
Rhodin, J.A.G., 1974, *Histology: A Text and Atlas,* Oxford University Press, Oxford.
Santini, M. (ed.), 1975, *Golgi Centennial Symposium: Perspectives in Neurobiology,* Raven Press, New York.
Schmitt, F.O. (ed.), 1970, *The Neurosciences: Second Study Program,* Rockefeller University Press, New York.
Sjöstrand, F.S., 1967, *Electron Microscopy of Cells and Tissues,* Vol. 1, Academic Press, New York.
Swift, J.A., 1970, *Electron Microscopes,* Kogan-Page, London.
Weakly, B.S., 1972, *A Beginner's Handbook in Biological Electron Microscopy,* Churchill-Livingstone, Edinburgh.
Wilson, G.B., and Morrison, J.H., 1966, *Cytology,* Reinhold, New York.
Wischnitzer, S., 1970, *Introduction to Electron Microscopy,* Pergamon Press, New York.

8

Scanning Electron Microscopy: Applications to Neurobiology

Martin J. Hollenberg and Allan M. Erickson

1. *Fundamentals of Scanning Electron Microscopy*

1.1. *Introduction*

The entrance of the scanning electron microscope (SEM) into the commercial market in 1965 provided the scientific community with a new and valuable addition to the field of microscopy. To the biologist, it offered a quick, accurate, and comparatively simple means by which to visualize tissue and cellular surface morphology in a three-dimensional state. Previous to this development, similar views could be obtained only through laborious reconstructions of micrographs of serial sections photographed by the light microscope (LM) or the conventional transmission electron microscope (CTEM). Even a casual review of today's literature will indicate that with the continuing advances in specimen-preparation techniques and instrumentation, the SEM is now firmly established as a useful and necessary tool in the biomedical field. Works by Wells (1974) and Goldstein and Yakowitz (1975) offer a relatively complete description of most aspects of the science, while Wells (1972), Boyde *et al.* (1973), and Jones *et al.* (1974) have compiled extensive lists of the pertinent literature. In addition, the area has been reviewed specifically with the biologist in mind (Carr, 1971; Hollenberg and Erickson, 1973; Muir, 1974; Revel, 1975).

Up to the present time, most SEM investigations of biological samples have concentrated on obtaining purely topographical information. Although

Martin J. Hollenberg and ***Allan M. Erickson*** • Division of Morphological Science, Faculty of Medicine, The University of Calgary, Calgary, Alberta, Canada.

the results of these studies are interesting and valuable, comparatively little new data providing further specimen characterization has been achieved. Recently developed techniques such as specimen-manipulation, scanning transmission electron microscopy, enzyme and immunological marking, and ancillary-signal collection have been employed by a few researchers and have carried SEM science a great deal further, but the potential of such procedures as adjuncts to surface-scanning has certainly not been fully appreciated or taken advantage of.

This chapter provides an overview of the fundamentals of scanning electron microscopy and reviews those SEM studies that have contributed to a clearer understanding of the structure and function of the nervous system. Unique or refined works are described in detail to illustrate the value of efficient use of the instrument.

1.2. History

The earliest attempt at a theoretical description and practical construction of a prototype SEM is credited to the European investigator Knoll (1935), who reported a resolution of approximately 100 μm with his instrument. Modification of this initial work, first by von Ärdenne (1938) and later by Zworykin *et al.* (1942), resulted in an experimental SEM in the early 1940's with a potential resolution of 5 μm. In 1948, research was initiated at the Engineering Laboratory of Cambridge University, under the direction of C.W. Oatley, in an effort to develop a refined instrument satisfying both the theoretical and practical requirements previously outlined. Reports by McMullan (1953) and K.C.A. Smith and Oatley (1955) illustrate the progressive stages in the project, culminated by the first commercial production of an SEM in 1965 (Oatley *et al.*, 1965). The Cambridge project has been reviewed by Nixon (1968) and is considered the major factor resonsible for the growth of scanning electron microscopy since its inception.

1.3. The Instrument

Discussions of the physics and theoretical aspects of the SEM have been included in works by a number of authors (Oatley *et al.*, 1965; Nixon, 1969; Pease, 1971; Joy, 1973; Wells, 1974), and will be presented here only briefly and insofar as they are applicable to a fundamental understanding of the design and operation of the instrument.

The basic design of a commonly employed SEM is illustrated in Fig. 1. Electrons are generated from a specific source and travel through the microscope column in a vacuum varying in different instruments and in different parts of the column from 10^{-5} to 10^{-10} torr. A series of two or three electromagnetic lenses progressively reduces the initial shower of electrons into a primary beam with a diameter of approximately 10 nm at its point of contact with the specimen. Topographical information relies on the collection of deflected primary (back-scattered) electrons and low-energy

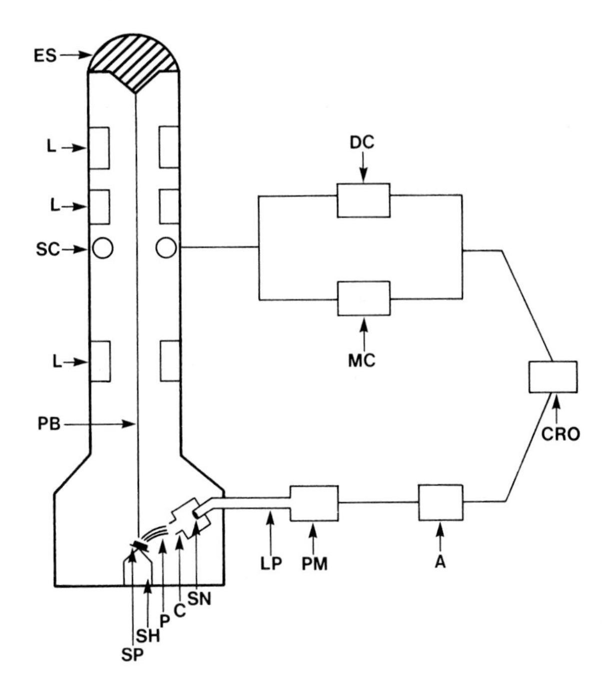

Fig. 1. Schematic representation of the SEM (see the text for a description). (A) Amplifier; (C) secondary-electron collector; (CRO) cathode-ray oscilloscope; (DC) deflection control; (ES) electron source; (L) electromagnetic lens; (LP) light pipe; (MC) magnification control; (P) path of secondary electrons; (PB) primary electron beam; (PM) photomultiplier; (SC) scanning coils; (SH) specimen holder; (SN) scintillator; (SP) specimen.

secondary electrons that originate from the specimen surface after its interaction with the primary beam. These are attracted to and trapped by an electron collector held at +100 to +300V, and then conveyed to a scintillator at 5–15 kV, resulting in subsequent photon production. This signal, varying according to the number of electrons emitted from a particular point on the specimen at a specific moment, reaches a photomultiplier and amplifier. The output from here determines the intensity of the light spot on the cathode-ray oscilloscope (CRO) screen via the modulating electrode. The localization of the primary beam onto a particular spot on the sample is governed by the scanning coils, and thus, changes in position are brought about by the alteration of the current originating from the deflection-control unit. In practice, this allows the primary beam to traverse the entire exposed surface of the specimen in a rasterlike manner. Coordinated motion of the CRO light spot results in formation of an image on the screen.

Magnification can be varied from approxmately ×10 to ×100,000 by alteration of the scanning-coil current, thereby changing the raster dimensions relative to the specimen dimensions in the particular area under observation.

Resolution in the SEM depends primarily on the size of the spot of origin of the back-scattered and secondary electrons and therefore is a reflection of the diameter of the primary beam where it strikes the specimen's surface. Theoretically, the lower limit of the beam size is set by the requirement for a current sufficient to generate enough back-scattered and secondary electrons to form an image in a practical scan time.

For a number of years, a simple heated tungsten cathode in an electron gun was used as the source of primary electrons. With increasing demands for a brighter source, smaller final beam diameter, and hence better resolution, lanthanum hexaboride (LaB_6) cathodes have been introduced as superior electron sources (Ahmed and Broers, 1972; Broers, 1974; Ferris *et al.*, 1975). Recently, the use of a field emission source for application to surface-scanning has been described (Komoda and Saito, 1972). This type of system consists of a tungsten tip subjected to a strong electrical field rather than direct heating. Although the latter source requires attainment of a very high column vacuum in the region of the source (i.e., 10^{-10} torr), it offers the dual advantages of increased brightness and smaller spot size. A complete discussion of the currently available primary electron sources has been recently prepared by Broers (1975).

In addition to the emission of secondary electrons from the surface, interaction of the primary beam with the specimen results in a variety of potentially recordable signals. The initial electrons may be (1) back-scattered or (2) transmitted through the sample or may (3) initiate the generation of X-rays, Auger electrons, and cathodoluminescence. These aspects are discussed in Section 1.5.4.

1.4. The Sample

One of the primary aims of biological scanning electron microscopy is to view tissues and cells in a state that most closely represents their *in vivo* structure. Therefore, the preservation techniques used must afford some degree of stabilization and protection against the physical conditions (i.e., high vacuum and electron bombardment) to which the sample is subjected during examination with the SEM. Many early workers have commented on the "ease" of specimen preparation for the SEM as compared to that required for CTEM studies. It now seems apparent that this generalization is untrue. Indeed, the demand for effective maintenance of morphological and spatial relationships in order to obtain useful SEM images is proving to be one of the major limitations on a realization of the instrument's full potential. Currently, there is not a universally accepted means of sample preparation that is applicable to all biological specimens. As more observations are made of tissues prepared by a variety of techniques, it is becoming easier to separate artifacts from normal variations in morphology. Hence, objective evaluations of the methods used in preparation are being reported, and a number of works reflecting the state of specimen preparation at the time of writing are available (Boyde and Wood, 1969; Marovitz *et al.*, 1970*a*; Pfefferkorn, 1970; Boyde, 1971, 1972). Review of these reports will yield an appreciation of the gradual improvements in this aspect of scanning electron microscopy that have taken place over the past few years. For the truest appreciation of surface morphology it is still necessary to prepare specimens by several of the available techniques and then to compare results.

The procedures involved in the preparation of biological specimens for SEM examination can be divided as follows: (1) surface-cleaning, (2) fixation, (3) dehydration, (4) conferring conductivity, and (5) special treatment. Each of these aspects is now considered in detail.

1.4.1. *Surface-Cleaning*

To obtain the best possible views of a specimen under study, reduce artifact formation, and prevent contamination of the SEM chamber, it is imperative that the sample surface to be examined be free of extraneous debris or other unwanted material. With "hard" biological tissues such as bone or teeth, this is easily accomplished by simple washing and drying. If refined characterization is desired, further treatment with acids–chelating agents for decalcification or with ethylene diamine and hydrazine for deorganification (Swedlow *et al.*, 1972) may be employed.

"Soft" biological tissues demand much more care in handling and preparation to preserve their *in vivo* morphology. Removal of surface mucus and proteinaceous substances is worth considerable effort and may involve gentle washing with a jet of saline or more elaborate techniques such as low-frequency ultrasonic vibration (Kuwabara, 1970) or enzyme digestion (Kelley *et al.*, 1973). Cleaning before fixation is advisable, to avoid precipitation of surface materials by the fixative. As pointed out by Boyde (1972), the cleaning solution should ideally be maintained at the same temperature at which the tissue exists *in vivo*, contain normal extracellular-fluid ions buffered to pH 7.4, and be isotonic with fluids normally in contact with the specimen. In practice, though, most workers have used simple saline washing.

1.4.2. *Fixation*

In addition to facilitating sample manipulation and dissection, the fixation process can adequately preserve the *in vivo* structural organization of the tissue while strengthening it to confer some degree of protection from the electron beam and column vacuum. The usual procedure involves treatment of the clean specimen by "prefixation" with the same buffered aldehyde solutions used in conventional transmission electron microscopy, washing it to remove any precipitated surface compounds, and subjecting it to "postfixation" with osmium tetroxide (OsO_4)(Parducz, 1967). Postfixation can prevent future surface disruption and provides some degree of electron density to the sample. It is particularly important in samples with a high lipid content.

1.4.3. *Dehydration*

If biological tissues containing liquid components are subjected to the vacuum conditions within the SEM chamber, the fluids will boil and destroy the specimen. Thus, it is important that tissue water be eliminated in such a way as to leave the solid components in their original state and position. There are a variety of techniques currently available to achieve this end, and recent review of the advantages and problems associated with each procedure has been presented by Humphreys (1975).

1.4.3a. Air Drying. Air drying is one of the simplest techniques of tissue dehydration. Many of the initial investigations utilizing the SEM involved fixed samples in which the tissue water was replaced with a volatile substance (e.g., alcohol, acetone, ether) through a series of graded treatments. Subsequent drying consisted of simply placing the sample in a desiccator and allowing evaporation of the volatile fluid to proceed. This method has been criticized by a number of authors (Boyde and Wood, 1969; Lim, 1969; Boyde, 1972) because of its failure to provide adequate preservation of structural details. This is due largely to the marked shrinkage and distortion caused by the air–solvent interface as it moves across the exposed surface of the sample (Fig. 2). The present authors concur with this criticism and feel that with the newer dehydration procedures available, air drying is no longer the method of choice in preparation of soft tissue for the SEM.

1.4.3b. Critical-Point Drying. Critical-point drying (CPD) as a method of specimen preparation was introduced by Anderson (1951). With this procedure, standard cleaning and fixation techniques are usually employed. Replacement of the tissue water is then performed by graded alcohol or acetone treatment followed by substitution of the solvent with a suitable intermediate fluid. This last step is required in order to permit the dissolution of the final transitional fluid (most commonly liquid CO_2 or a Freon derivative) within the sample. One cannot simply use the naturally occurring water as a transitional fluid, since its critical point is 378°C at a pressure of 218 atm, conditions that cannot realistically be obtained. The prepared tissue is then placed in a closed, pressure-resistant system and the temperature raised above the critical point for the specific fluid (approximately 25°C for Freon and 31°C for liquid CO_2). At this critical-point temperature, the specific gravity of the liquid is equal to the specific gravity of the vapor, and hence a state of zero surface tension exists at the specimen level. The system is maintained under such conditions until elimination of the transitional fluid is complete, leaving a completely dry sample (Fig. 3).

A review of the techniques of CPD has been prepared by E.R. Lewis and Nemanic (1973), while the physics involved have been dealt with by Bartlett and Burstyn (1975). The clear advantage of this procedure is the prevention of the formation of any gas–liquid interface at the surface of the specimen, thereby greatly reducing the shrinkage and distortion evident in air-dried samples. The superiority of CPD from CO_2 over air drying has been stressed by Polliack *et al.* (1973) in their comparison of micrographs obtained of red blood cells prepared by each of the methods. In addition, Horridge and Tamm (1969) have reported superior preservation of delicate cilia using CPD from liquid CO_2. Many workers favor the CPD technique over that of freeze–drying. M.E. Smith and Finke (1972) found excellent preservation of

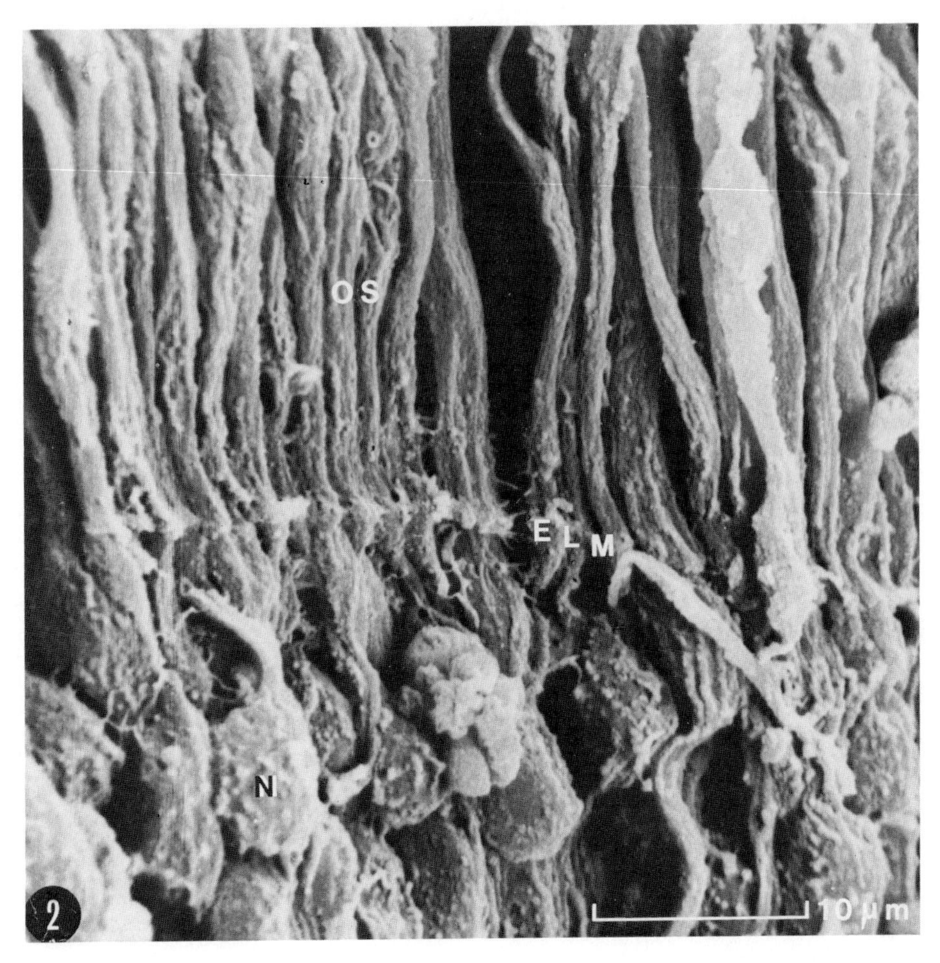

Fig. 2. Scanning electron micrograph of an air-dried specimen of primate retina. Visible are photoreceptor outer segments (OS), the external limiting membrane (ELM), and photoreceptor nuclei (N). Considerable shrinkage and distortion of individual cellular components are apparent due largely to the air-drying procedure. Courtesy of D. H. Dickson.

monkey photoreceptors and the ependymal surface of the rat fourth ventricle prepared by CPD using Freon 116 as a transitional fluid. Their results have been substantiated by Worthen and Wickham (1972). The use of a specific transitional fluid is controversial. Equally good results have been obtained using liquid CO_2 or Freon derivatives, although Boyde (1972) has noted that substitution with Freon 113 after ethanol dehydration seems to result in less tissue distortion than methods employing liquid CO_2.

It should be pointed out that the critical-point technique has quickly become the most widely used of all drying methods in the preparation of soft biological tissues.

1.4.3c. Freeze–Drying. This method also serves to decrease shrinkage and distortion when compared with air drying of a sample. In this technique,

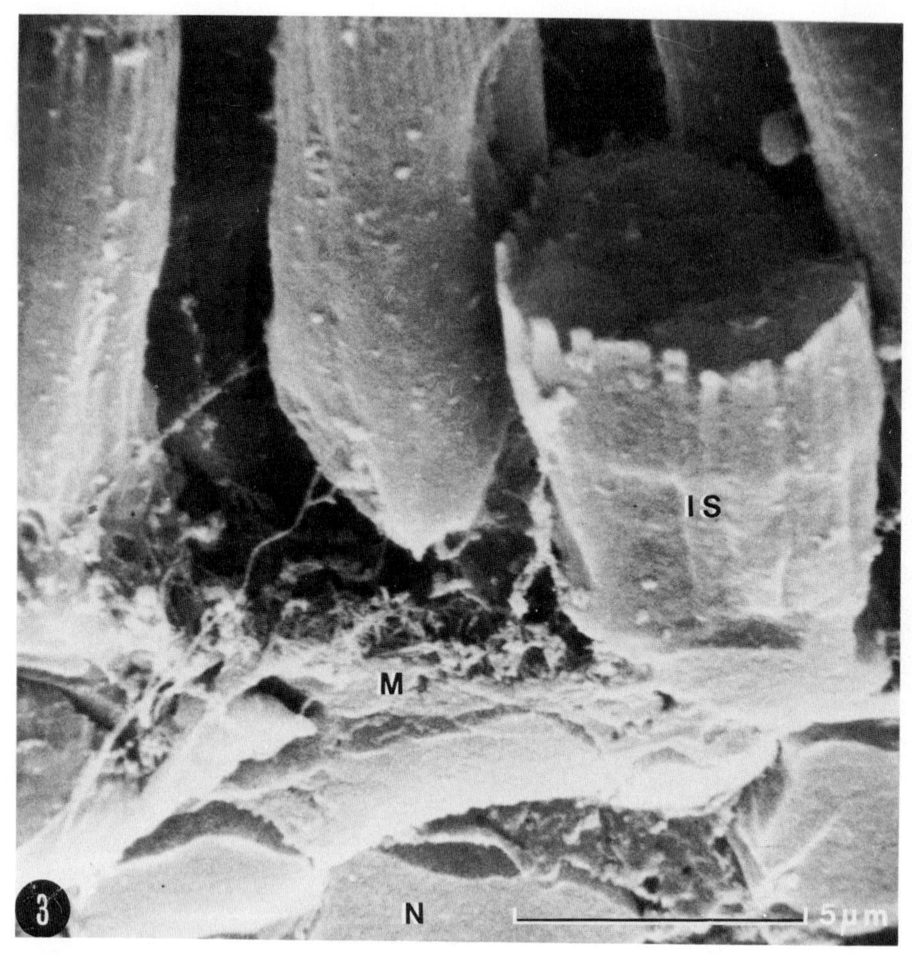

Fig. 3. Scanning electron micrograph of a freeze–fractured preparation of newt retina dried by the critical-point method. Visible are photoreceptor inner segments (IS) and nuclei (N), as well as the outer portions of Müller cells (M) with apical microvilluslike projections. Courtesy of D. H. Dickson and J. G. Tonus.

routinely cleaned and fixed specimens are quick-frozen by immersion in a liquid cooled by a bath of liquid nitrogen. This results in an instantaneous transformation of all tissue fluids into the solid state and, in theory, stabilizes cellular organization exactly. The freezing process is followed by vacuum sublimation of the ice, leaving a dried sample. As one may suspect, the major limitation of this method is the tissue and cellular damage caused by the formation of ice crystals. This may be somewhat overcome by prior treatment with a cryoprotective compound (e.g., glycerol) or by replacement of the tissue water with an organic solvent before freezing. However, such additions may in themselves contribute to artifact production.

In an early SEM investigation of the rat retina, Hansson (1970*a*) examined samples that were fixed with the solution of Karnovsky (1965), frozen by immersion in a propane bath cooled by liquid nitrogen, and then freeze–dried. A comparison of the results obtained from the utilization of this drying method was then made with samples subjected to air drying from acetone. The freeze–dried specimens revealed good preservation of the photoreceptors and cells of the ganglion layer, whereas the air-dried retinae showed obvious shrinkage and a folded, irregular surface, attesting to the sensitivity of neural tissue to the air-drying technique.

Thorough reviews of the various techniques used in the freeze–drying method have been made by MacKenzie (1972) and Boyde and Echlin (1973). While it is generally agreed that the freeze–drying procedure is superior to air-drying methods, it does not seem to offer any significant advantages over the CPD technique and, as noted in a recent article by Boyde (1974), is losing its initial popularity for the preparation of biological material.

1.4.3d. Camphene Drying. Watters and Buck (1971) have devised a drying technique employing the vacuum sublimation of camphene. Briefly, samples are washed, fixed with glutaraldehyde, and subjected to graded ethanol dehydration with eventual substitution with 100% acetone. They are then treated with benzene, to eliminate the lipid components of the tissue, transferred to propylene oxide, and finally placed in a solution of 50% camphene in propylene oxide at 45°C. The tissue is then dried by vacuum sublimation (Fig. 4).

Although this method eliminates the ice-crystal damage from freeze–drying and lessens the need for elaborate equipment as in CPD, it has not found widespread acceptance as a drying procedure. However, Bruni *et al.* (1972) have successfully employed the camphene technique in their study of the fine structure of the surface of the cerebral third ventricle in various species, thus demonstrating the potential of this method in preparing central nervous system (CNS) samples for the SEM.

Following drying of the specimen by any of the methods just noted, it is often beneficial to clean the surface of the specimen of any debris that may have accumulated before coating. This may be done by simply exposing all aspects of the specimen surface to a jet of compressed gas. This simple technique can often make the difference between a "clean" surface and one cluttered by unwanted particles obscuring important details.

1.4.4. *Conferring Conductivity*

1.4.4a. Standard Coating Techniques. In order that the images derived from the collection of secondary electrons emitted from a sample during SEM examination reflect only topographical variations and not differences in the availability of electrons at various points, some method of conferring compositional uniformity to the surface must be undertaken. In addition, most specimens act as insulators when subjected to the primary beam and hence develop uneven "charging" artifacts (Pawley, 1972). Elimination of such an effect usually requires the use of a conductive coating over the exposed surface. Most workers find that these requirements can be accom-

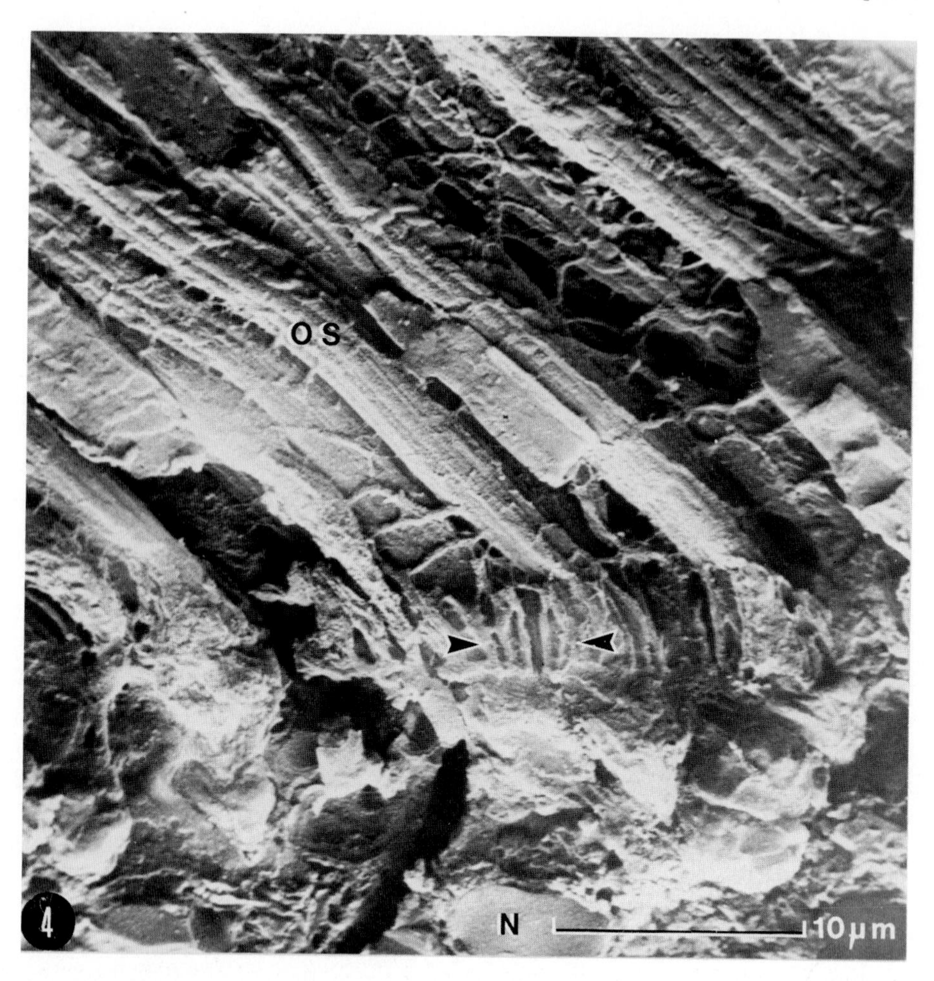

Fig. 4. Scanning electron micrograph of a freeze–fractured, camphene-dried preparation of frog retina. Elongate photoreceptor outer segments (OS), each surrounded by a palisade of dendritic processes, are seen in surface view. Threadlike processes of the retinal pigment epithelium form a network surrounding the outer segments. Also visible are fractured surfaces of photoreceptor nuclei (N) and apical microvilluslike processes (➤) of the Müller cells, which are especially prominent in this species. Courtesy of D. H. Dickson.

plished through the use of a thin layer of carbon to decrease charging and a thin layer of a heavy metal, most frequently gold or gold–palladium, to confer uniformity of composition. In general, the total layering should be thin enough so as not to obscure detail, yet thick enough to prevent uneven electron emission. A 20-nm layer each of carbon and gold, applied to the dried sample after it has been attached to a specimen "stub" by means of a mounting medium (Muir and Rampley, 1969), has proven satisfactory. In practice, it is best to apply an extremely thin coating and then recoat only if "charging" occurs under the beam in the area of the specimen under study.

Reduction of the accelerating voltage can also reduce "charging," although often at the expense of resolution. Incidental benefits of specimen-coating include increased mechanical stability of the surface and a decreased depth of primary-beam penetration (Echlin and Hyde, 1972).

There are currently two common methods used for the deposition of elements onto the surface of the specimen. The "evaporative" technique involves heating the coating material *in vacuo* by a resistive, electric-arc or electron-beam procedure (Echlin, 1974) and coating a specimen rotated in three planes. The "sputtering" method consists of physical erosion of the coating material into monoatomic particles through the use of an ion-beam, direct-current, or radiofrequency technique (Echlin, 1975). Although each method serves the same purpose, a recent comparative study by DeNee and Walker (1975) indicates that gold or gold–palladium sputtering is a superior procedure for samples to be viewed in the secondary-electron-collection mode of the SEM, while evaporation of carbon provides the best results if the specimen is to be subjected to X-ray or back-scattered-electron analysis.

1.4.4b. Noncoating Techniques. There has recently been interest in examining specimens that have not undergone thin-film coating as described above. This interest has evolved because there is little latitude between the point at which an artificial surface layer is sufficient to fulfil its purpose and that at which it contributes to artifact formation. Also, cracking of the coating layer can occur or the heavy-metal coat may not extend deep enough into narrow surface indentations to permit accurate interpretation of topography.

Kelley *et al.* (1973) have introduced a technique that eliminates the need for the usual coating of the sample and yet provides a sample surface that is conductive. They suggest that osmiophilic components of the tissue can be enhanced by the addition of extra osmium after the postfixation step. This is accomplished by the creation of osmium–tissue–osmium *OTO* compounds using thiocarbohydrazide (TCH) as a ligand. Their specific experiment involved the use of frog and mouse specimens. Amphibian tissue was fixed in 2.5% glutaraldehyde in 0.1 M cacodylate buffer, whereas mouse tissue was treated with the solution of Karnovsky (1965). All samples were then postfixed with buffered 1% OsO_4 at 4°C for 3–4 hr, rinsed five to ten times in buffer, incubated in saturated TCH solution at room temperature for 10 min, rinsed five to ten times with distilled water, and placed in aqueous 1% OsO_4 for 1 hr. The authors reported that this method adequately prevented primary-electron damage to the specimen without obscuring the most detailed fine surface variations. This initial work has been modified by Malick and Wilson (1975) to involve an additional TCH–osmium linkage, resulting in similar favorable reports.

Panessa and Gennaro (1973) have described a "tissue-conductance technique" that uses potassium iodide and lead acetate to treat samples in order to eliminate the need for surface-coating. Tissues prepared by this method and subjected to 5 hr of direct examination with accelerating voltages of 5–20 kV showed good retention of stability and detail.

Pachter *et al.* (1973) have used the SEM to examine 1-μm-thick Epon-embedded CTEM sections, etched by iodine acetate and dried onto carbon-coated glass. This "undercoating" procedure permitted the use of 10- to 15-kV accelerating voltages without notable charging effects on the specimen. Such a technique may be of value in the study of thin, uncoated sections with the SEM for correlation with LM and CTEM results.

1.4.5. Special Preparatory Techniques

A number of modifications and refinements of the standard preparation scheme have been reported. Most of the techniques are concerned with altering the sample in such a way as to obtain additional, more accurate topographical information. Similarly, certain developments have permitted the utilization of the SEM in investigations extending beyond pure surface-scanning.

1.4.5a. Fractionation. The amount of information that can be obtained from the examination of natural surfaces of biological material is limited and often necessitates the need for an artifical creation of "new" surfaces. This usually involves mechanical or chemical fracturing of the tissue and its components. Simple physical disruption may be employed, as demonstrated by Guttman and Styskol (1971). In this particular case, specimens of *Euglena gracilis* were mixed with glass beads and subjected to "Vortex" mixing to break apart the cells and expose internal components. A dry-fracturing technique involving ordinary mechanical tearing of freeze–dried tissue has been found effective in exposing the interior of the mouse cerebellum (Flood, 1975). Recently, Boyde (1975) has commented on the dry-cleavage properties of tissues after previous treatment with various fixatives and subsequent CPD or freeze–drying procedures. In general, tissues fixed only with OsO_4 tend to disrupt more along intercellular junction planes than do tissues initially treated with glutaraldehyde. Using a method originally described by Vial and Porter (1974), Boyde (1975) has found that this phenomenon is further enhanced by treatment of the tissue with boric acid, either alone as a postfixation solution or in combination with sodium tetraborate as a buffer for the OsO_4. This procedure may well prove to be of considerable value in the accurate exposure of internal cell surfaces within the bulk specimen.

A more popular method of tissue disruption is cryofractionation, as described by Haggis (1970). After immersion of a previously fixed specimen into a liquid at very low temperature (e.g., Freon 22 at −150°C), the solid tissue is fractured with a cold blade and subsequently dried at -80°C for 15 hr. Haggis suggests that the resultant exposed surfaces probably reflect planes of natural weakness in the sample. Of particular interest in this experiment was the demonstration of an "end-on" view of a myelinated nerve showing the inner surface of the myelin sheath. In a later study, Lim (1971*a*) commented that this method was not particularly suitable for the exposure of very delicate components of the inner ear. His refined procedure used osmic-acid-fixed tissues that had been dehydrated to 70% ethanol. This solution and its contained tissue was then taken up by a pipette and released as drops onto an aluminum dish bathed in liquid nitrogen. The alchol–tissue complexes then underwent spontaneous fracture on contact with the liquid nitrogen. Subsequent freeze–drying, gold-coating, and examination of the samples revealed the detailed physical arrangement of sensory and supporting cells. In addition, delicate nerve endings and sensory hairs were well preserved, and it was possible to demonstrate nuclei, mitochondria, and endoplasmic reticulum of single cells.

Nemanic (1972) has combined cryofractionation and CPD. This technique eliminated many of the drying artifacts and other damage noted in freeze–dried samples, and enabled demonstration of nuclei, Golgi complexes, mitochondria, and organization of glycogen in mouse mammary and liver cells. A recent report by Humphreys *et al.* (1974) has provided even further refinements to the cryofracture technique. By this method, tissues are first fixed by perfusion with 2% glutaraldehyde in 0.1 M cacodylate buffer at pH 7.2 and dehydrated by graded ethanol treatment. After two extra changes of 100% ethanol, 1 mm × 4 mm tissue strips are isolated and inserted into small Parafilm cylinders that are then filled with ethanol. Following closure of the cylinder ends, the complex is placed into liquid nitrogen, resulting in a frozen sample contained within a solid ethanol cylinder. Fracturing this with a cold razor blade produces a number of tissue–alcohol fragments. These are thawed in 100% ethanol, dried by the critical-point method, gold-coated, and examined with the SEM. This procedure eliminates specimen damage secondary to ice-crystal formation, since the tissue water has been replaced prior to freezing. Mechanical deformation is also decreased by placing the liquid ethanol–tissue complex in the closed cavity prior to its transition to the solid state. This particular technique demonstrated very clearly intercellular spaces and tissue organization.

1.4.5b. Replication. If the direct examination of a tissue surface is neither possible nor desirable, replicas may be employed. The current state of the replication technique has been recently reviewed by Pfefferkorn and Boyde (1974). The simplest method usually involves the coating of the fixed and dried specimen with a layer of carbon. The tissue itself is subsequently destroyed chemically, leaving the unaffected carbon shell as a representation of the surface topography. If preservation of the original sample is desired, "stripping" procedures may be employed. In these instances, the replicating material, usually a siliconelike compound, is evenly distributed over the specimen. With subsequent polymerization of the material, a negative impression is obtained and lifted off the sample. An exact duplication of the original surface is made by repeating the replication procedure on the initial cast using, for example, polyethylene. This technique has found particular use in dermatology (Bernstein and Jones, 1969; Fujita *et al.*, 1969; Pfefferkorn *et al.*, 1972) in that no damage to *in vivo* skin structure is created in the process. The replication technique may also be useful in preparing permanent representations of especially interesting specimens.

Replication in this way offers particular advantages that are unavailable by any of the other preparation techniques. Surgical excision, fixation, and drying of the specimen are not necessary, and *in vivo* structure can be studied at intervals throughout the entire course of an experiment or development of a pathological process. In view of this, it is most surprising that attempts have not been made to greatly expand the use of replicas to provide dynamic study of a wide variety of tissue processes.

1.4.5c. Etching. In certain instances, it may be desirable to accent or selectively remove particular portions of the specimen under SEM study to obtain a more precise impression of the structural variations at, or just below, the presenting surface. This process is referred to as etching and relies primarily on two major techniques. The first, freeze–etching, has been well described by Koehler (1968). Generally, the procedure is very similar to the freeze–drying method described previously. Prepared tissues are quick-frozen by immersion in a solution cooled by liquid nitrogen and then transferred to a vacuum apparatus at approximately–100°C. Rather than the sublimation process being allowed to proceed to completion, the temperature of the chamber is accurately controlled to permit drying of the specimen only at the exposed surface. Once this desired level is reached, a carbon and heavy-metal layer is distributed over the specimen and the process terminated. The original sample is destroyed, leaving the carbon–metal shell as a replica of the surface. With this technique, natural topographical variations are finely accented.

The second method of etching relies on controlled removal of selected parts of the surface by means of an ion beam. The source of this beam is usually a cold-cathode argon discharge unit (Boyde and Stewart, 1962; Echlin *et al.*, 1969) or, more recently, radiofrequency sputtering (Stuart *et al.*, 1969). Each initial ion presumably causes the elimination of one target atom, and thus the rate of etching is dependent on both the mass and energy of the ion as well as the atomic weight of the target atoms. This latter factor in turn reflects the density and composition of the specimen. This controled physical destruction can thus permit the exposure of subsurface components of the sample or simply accent various surface features. The technique has been used extensively in SEM investigations of the red blood cell (S.M. Lewis *et al.*, 1968; Spector *et al.*, 1974; Frisch *et al.*, 1974, 1975), since some modification of the naturally smooth surface of this structure is essential for an accurate assessment of subsurface organization.

Recently, Thomas and Hollahan (1974) have described surface-etching using low-temperature plasma systems produced by high-frequency excitation. In this method, the surface degradation is secondary to the formation of volatile compounds from the interaction among the plasma atoms, free electrons and molecules, and the target specimen. This technique requires further study before application to most biological tissues is possible.

1.4.5d. Ashing. Of particular interest to the neurobiologist is the low-temperature dry-ashing technique described by E.R. Lewis (1971) for the preparation of neural tissue for examination by the SEM. Nervous-system specimens present a particularly difficult problem to scanning electron microscopists in that adequate exposure of individual neurons is frequently not attained by the routine preparatory procedures. In the Lewis method, this difficulty has been resolved to some extent. The technique consists of initial heavy-metal impregnation of neurons using either a modified Golgi procedure, in which a tissue block is fixed in dilute potassium dichromate and then placed in a dilute silver nitrate solution, or the Ramón y Cajal procedure, employing formalin fixation as the initial step before exposure to silver. The other well-known techniques leading to the accumulation of heavy metals within neurons can also be used, depending on the information sought. The prepared tissue is then subjected to a flowing oxygen plasma in a low-temperature asher for 1–48 hr. This treatment results in the oxidization of unstained organic materials and leaves behind only the heavy-metal deposit originally impregnated within the neurons. In practice, this effectively eliminates the interstitial and connective-tissue matrices while preserving individual nerve cells and their processes. For example, this method has permitted examination by E.R. Lewis (1971) of the layers of rat cerebellar cortex with the SEM. Purkinje cells were particularly well visualized.

It seems clear that this technique holds particular promise for future work aimed at revealing neuronal circuitry using the SEM and is certainly worthy of further development.

1.4.5e. Cytochemical and Immunological Marking. Two of the most significant developments attesting to the value of the SEM as more than just a surface-scanning device are histochemical–cytochemical mapping and immunological marking techniques. In an early attempt at enzyme-mapping, Scarpelli and Lim (1970) incubated glutaraldehyde-fixed tissues in appropriate media for the histochemical demonstration of various enzymes. After being freeze–dried and gold-coated the specimens were examined with the SEM. The technique revealed specific sites of surface enzyme activity. For example, adenosine triphosphatase activity was localized on the sensory hairs of the cochlear sensory epithelium and acetylcholinesterase activity was mapped along the axonic membrane of unmyelinated nerve fibers.

Carr and McGadey (1974) have recently attempted the SEM localization of acid phosphatase activity in the rat cerebellum. In this case, fresh-frozen 20-μm-thick sections of cerebellar cortex were incubated in Gomori (1950) acid phosphatase medium for 1 hr. Subsequent treatment with 0.075% sodium rhodizonate solution in tartrate buffer, graded ethanol dehydration, ethyl ether replacement, air drying, gold–palladium coating, and examination in the SEM allowed the specific visualization of individual Purkinje cells exhibiting acid phosphatase reaction.

Immunological marking of specific cell-surface areas to permit SEM localization was first described by LoBuglio *et al.* (1972). This work showed that 0.23-μm latex spheres could adsorb γ-globulin containing a predetermined antibody directed against a specific surface antigen. Following interaction between the prepared spheres and the sample, the latex particles became attached to the sites of the specific antigen on the cell surface. The latex spheres were observed with the SEM, thus revealing the exact location of antigenic sites. Although nonspecific binding of the spheres did occur, the experiment revealed the potential of such a technique for antigenic mapping and cellular differentiation on an immunological basis. Lithicum *et al.* (1974) have recently applied the method in an attempt to localize B-cell-associated and T-cell-dependent antigenic determinants on mouse lymphoid cells. The problem of nonspecific binding has been somewhat resolved by Molday *et al.* (1974), who described the design and synthesis of smaller latex spheres (0.06–0.13 μm in diameter) and the establishment of covalent bonding between the antibodies and spheres (Molday *et al.*, 1975).

An excellent summary of recent advances in the immunological labeling of SEM specimens is included in a review by Nemanic (1975).

1.4.5f. Techniques for CTEM Correlation and Ultrastructural Analysis. Despite its many applications, one cannot use the SEM alone if complete characterization of a biological specimen is expected. This same statement can be equally applied to the CTEM and LM. For this reason, comparison and correlation of data derived from the use of all three instruments are necessary if the maximum benefits of each are to be realized. In general, such correlative studies require that the specimen be processed and observed in such a manner as to minimize distortion or artifact. A review of the techniques available to enable the successive examination of specimens by LM, SEM, and CTEM has been provided by Barber (1972).

More recently, Brummer *et al.* (1975) have used a refined method permitting the sequential observation of the pathological lesions of primary amebic meningoencephalitis. Samples of mouse brain were immersed in aldehyde, dehydrated with ethanol, transferred to amyl acetate, dried by the critical-point method from liquid CO_2, attached to specimen stubs, coated with carbon and gold–palladium, and examined with the SEM. Photographs were taken of "interesting" areas and the specimen then removed from the stub, placed into 100% ethanol, substituted with propylene oxide, and embedded in an Epon–Araldite mixture. Sections 1–2 μm thick were then made, stained, and observed with the LM. Particular areas selected by the SEM observations were then isolated by the "mesa" technique (Lowrie and Tyler, 1973), thin-sectioned, stained, and examined with the CTEM. This technique thus enabled specific localization of pathological areas by the SEM followed by ultrastructural correlation by the CTEM.

Erlandsen *et al.* (1973) have devised a method that allows SEM observations of specimens after LM and CTEM studies have been made. Tissues were fixed with 2.5% glutaraldehyde and 2% buffered OsO_4, subjected to graded ethanol dehydration, and embedded in Epon 812 (Brinkly *et al.*, 1967) before being sectioned and stained for the LM and CTEM. The portion of the tissue block immediately adjacent to the area of sectioning was then treated with 1–2% NaOH in aged absolute methanol for 3–4 days, to remove the polymerized Epon. The samples were then washed with ethanol, substituted with propylene oxide, and dried by the critical-point method for examination by the SEM. These authors feel that this procedure, in addition to comparisons of structure, allows correlation of results obtained from autoradiographic, histochemical and immunochemical applications of the three instruments.

Panessa and Gennaro (1972) have described a technique that permits the observation of intracellular components with the SEM. Plant tissues were fixed with glutaraldehyde for 4–60 days to assure total infiltration of the solution into the cells. Postfixation using 2% uranyl acetate was then performed, since this compound selectively localizes in areas containing nucleic acids, binds the tissue lipids, and helps prevent uneven surface-charging. The samples were then bathed in 50% glycerol for at least 4 days to decrease shrinkage from evaporation. After coating and examination by the SEM, the specimens were then immersed in acetone, embedded in Epon 812, and sectioned for LM and CTEM procedures. Thick LM sections were stained with toluidine blue, fast green, and Sudan black, while thin sections for the CTEM were treated with lead citrate. This sequence permitted the identification of component organelles (e.g., mitochondria, chloroplasts) using the SEM and elucidation of specific ultrastructural organization by the CTEM. A modified procedure incorporating CPD has recently been outlined by Watson *et al.* (1975) in their examination of HeLa cells in culture. To view the undersurface of the cultured cells, these investigators placed adhesive tape on the dried *in situ* culture and subsequently peeled it off the underlying surface. This permitted observation of the undersurface of individual cells and allowed identification of intracellular Golgi complexes, ribosomes, and microfilaments.

Clearly, methods such as those described above are of great importance to the future extension of SEM techniques in the study of subcellular components.

1.5. Advances in Techniques of Specimen Examination and Characterization

After appropriate preparation, most biological specimens have been examined utilizing the secondary-electron collection mode of the instrument to obtain topographical information. As indicated previously (see Section 1.4.4b), an acceptable balance must be reached between the resolution requirements and the amount of artifact generation or specimen damage involved in reaching these requirements. These latter aspects are related to the high accelerating voltages necessary to achieve high resolution. In most cases, an initial voltage of between 10 and 25 kV will permit adequate resolution and eliminate needless damage of a prepared specimen.

Recent advances in instrumentation have permitted the recording of additional information beyond simple topography from selected biological samples. Some of the more important of these are described below.

1.5.1. Microscope Control Stages

To permit viewing of the sample under a variety of physical conditions, a number of SEM stages of unusual design have been developed. For example, an environmental-control stage has been described by Lane (1970). This instrument provides a system in which a cloud of water vapor is continually formed around a specimen under view, thereby eliminating the need for dehydration and coating of the sample and permitting its observation in a near *in vivo* state. The device does not appear to have found widespread use or acceptance. However, Parsons and Lyon (1975) have recently described the construction of a similar hydration chamber and have discussed the potential of such an attachment in the SEM study of live specimens.

Echlin *et al.* (1970) have devised another modification that permits easy control of the environmental temperature around the specimen, allowing variation from 25 to −180°C. One advantage of the "cold stage" is that freeze–drying and freeze–etching procedures can be performed within a single instrument. Tissues are simply quick-frozen, attached to specimen stubs, placed in the stage, and held at a temperature of −180°C. By increasing the temperature by about 80°C, controlled sublimation of the ice can be used to etch the surface. Once this is attained, the temperature is then decreased again and the sample examined. The instrument has been used successfully by Echlin (1971) with glutaraldehyde-and OsO_4-fixed tissue frozen in Freon 22, and more recently by Echlin and Moreton (1973) utilizing a modified specimen stub allowing the specimen temperature to remain below −130°C at all times. This eliminated unwanted artifacts arising from temperature variations during the preparation procedure. "Cold stages" for the SEM are now commercially available.

Also, Griffiths and Venables (1972) have described a specimen stage that employs liquid helium to create specimen temperatures as low as 5.9°K. This condition prevents the ice-crystal formation and tissue damage inherent in many of the other cryobiological techniques used in scanning electron microscopy.

1.5.2. Micromanipulation

In specific instances, it may be desirable to microdissect or change the position of a specimen at the same time it is being examined with the SEM. Original work by Pawley and Hayes (1971) resulted in an instrument allowing independent sample manipulation without interference with the vacuum conditions or specimen stage. This device had a range of 100 μm with an accuracy of 0.1 μm. Recently, Pawley and Nowell (1973) have described a less complicated micromanipulator employing a deFonbrune-type pneumatic manipular drive with piezoelectric pincers. With this apparatus, they have attempted the microdissection of a pathognomonic plaque in specimens of equine pulmonary emphysema. Incidentally, this plaque was first discovered by Nowell *et al.* (1971) utilizing the SEM. The potential of this instrument modification can be expected to be extended as demand increases for refined and accurate methods of cellular and tissue exposure.

1.5.3. Scanning Transmission and High-Resolution Electron Microscopy

The combination of the CTEM and the SEM into a single functional unit has resulted in the development of the scanning transmission electron microscope (STEM) (Kimoto *et al.*, 1969; Swift *et al.*, 1969; Swift and Brown, 1970; Zeitler, 1971). In the actual design of this instrument, a transmitted-electron detector is added beneath the specimen stage to the basic SEM unit. Alternatively, a CTEM unit with an appropriate added scanning device can be used. The combination allows a three-dimensional image of sections up to 1.0 μm in thickness to be obtained, and has found application at the ultrastructural level where high resolution is desired. If a tungsten field emission source for primary electrons is used, and if high vacuum conditions (10^{-9} torr) are established in the STEM chamber, point resolutions of less than 0.5 nm are possible, as demonstrated by Crewe and Wall (1970). Under such conditions, Crewe *et al.* (1970) have been able to visualize single atoms of heavy metal.

Various aspects of high-resolution scanning transmission electron microscopy (HRSTEM) have recently been discussed by Wall *et al.* (1974). Also, the particular applications of HRSTEM to the study of biological specimens have been described by Isaacson *et al.* (1974) and more recently by Broers *et al.* (1975). These latter workers have recorded the fine structure of bacteria and a bacteriophage, utilizing an HRSTEM instrument with a lanthanum hexaboride electron source and a condensor–objective final lens system. In addition to standard fixation and dehydration procedures, preparation of the specimens involved heavy-metal staining with, for example, ethanolic phosphotungstic acid and uranyl acetate, to accent cellular features.

1.5.4. Ancillary-Signal Collection

Rather than the usual collection of deflected primary and emitted secondary electrons, a number of systems have been devised to collect and process other signals derived from the interaction of the primary beam and the specimen. In addition to the production of secondary electrons, the initial beam may also (1) be transmitted through the specimen, forming the basis for STEM as described; (2) be partially back-scattered; (3) produce characteristic X-rays from the specimen; (4) result in cathodoluminescence; and (5) initiate Auger-electron production. The processing, significance, and interpretation of these signals are discussed below.

1.5.4a. Back-Scattered-Electron Analysis. A certain number of electrons in the primary beam are directly reflected, the number depending primarily on the atomic composition of the specimen point exposed to the beam. By analysis of these electrons, a further impression of the elemental composition of a sample can be obtained. This mode of operation of the SEM has been recently reviewed by Abraham and DeNee (1974), and so far has been used in the histochemical analysis of suitably stained biological samples (DeNee *et al.*, 1974).

1.5.4b. X-Ray Microanalysis. As the electrons of the primary beam impinge on atomic nuclei within the sample, an X-ray photon, characteristic of its elemental origin, is produced. If such signals are monitored, quantitative and qualitative information regarding the elemental composition of the sample can be obtained. The collection of signals originating from biological specimens usually employs an energy-dispersive X-ray attachment, consisting of a lithium-drifted silicon crystal and an analyzer that converts the energy of the X-ray photon into an electrical pulse. Similarly, diffractometers may be employed that separate the X-rays into specific wavelengths to enable qualitative analysis. Early descriptions of the procedure are found in works by Tousimis (1969) and Sutfin and Ogilvie (1970). These latter workers, using X-ray microanalysis, have described the specific localization of calcium and phosphorus within individual mitochondrial granules (Sutfin *et al.*, 1971). Discussions on the applications and potential of X-ray microanalysis in the study of biological specimens have been provided by Russ (1972) and more recently by Yakowitz (1974) and Panessa and Russ (1975). Clearly, this mode of operation of the SEM will be of most value in scanning electron microscopy at the subcellular level and when used in conjuction with autoradiographic techniques (Hodges *et al.*, 1974).

1.5.4c. Cathodoluminescence. Cathodoluminescence is the term applied to the emission of photons in the ultraviolet-, infrared-, and visible-light ranges after the specimen has been contacted by the primary beam. Such emission is governed by the molecular structure of the organic components in the sample and by the crystal structure of the inorganic constituents. Hence, information regarding these two components of the specimen can be obtained. Signal collection is accomplished by the addition of a second light guide and photomultiplier to the basic SEM design. A review of this analytical technique has been prepared by Muir and Holt (1974). A very interesting biological application of SEM cathodoluminescence has been recently described by Broecker *et al.* (1975) employing a cathodoluminescence detector designed by Hoerl and Muegschl (1972). Using an immunofluorescence technique, employing fluorescein-isothiocyanate-labeled anti-rat fibrinogen serum derived from rabbits, they were able to effectively demonstrate antigen–antibody complexes in glomerular intravascular fibrin. This technique with further refinements should prove of great value in future attempts to localize specific cellular components at the molecular level due to the better resolution obtainable with the SEM in comparison with the LM.

1.5.4d. Auger-Electron Analysis. The utilization of the SEM for the initiation and monitoring of Auger-electron production has been evaluated by McDonald (1971). These electrons are produced by electron transitions at atomic levels within the specimen after excitation by the primary beam. High vacuum conditions (e.g., 10^{-8} torr) are a prerequisite for Auger-electron analysis, and it is believed that approximately 10^5 primary electrons are necessary to produce one Auger electron. In light (low-atomic-weight) elements, Auger-electron production bears an inverse relationship to X-ray production, and the Auger yield is high and X-ray yield low. Hence, each method of elemental analysis (Auger-electron analysis and X-ray detection) may complement the other in the total range of elements. The use of Auger-electron analysis for fracture-surface study (McDonald *et al.*, 1970) and its potential for elemental mapping (McDonald, 1970) have been described.

2. *Specific Applications of Scanning Electron Microscopy to Neurobiology*

2.1. *Introduction*

The SEM, due to its ability to reveal surface topography in three dimensions, has already proven of value in a variety of studies of nervous-system morphology. However, the previous sections have shown that recent refinements in specimen preparation and instrumentation have now taken SEM science beyond mere revelations of surface structure toward total characterization of the specimen under study. Unfortunately, in this review, we find that with few exceptions, investigators using the SEM to study the nervous system have not as yet taken advantage of many of the newer, significant advances in the field. Hence, there appears to be great latitude for future studies of the nervous system, specifically designed to utilize the unusual versatility of the SEM in the solution of problems of neuronal morphology, interrelationships, circuitry, and function.

Although studies of surface topography are included, this review concentrates on those few works that have probed somewhat beyond the usual in the SEM study of nerve tissue. For convenience, the examples are grouped into studies of the central nervous system, the peripheral nervous system, and sensory receptors.

2.2. *Central Nervous System*

2.2.1. *Cerebral Ventricular System*

The bulk of the utilization of the SEM in studies of the vertebrate CNS has been focused on the delineation of the surface topography of the ependymal lining of the cerebral ventricular system. The ventricular lining provides a rather easily accessible surface, and the natural structural variations have permitted SEM images of great interest to be derived after only routine preparation of the tissues. One of the more significant early reports in this area has been provided by Clementi and Marini (1972), who studied the surface of the ventricular walls and choroid plexus of the cat. Brains of adult cats were fixed by perfusion with 2% formaldehyde and 2% glutaraldehyde in 0.12 M phosphate buffer, and subsequently, dissections of selected areas were performed. After further fixation, the samples were dehydrated in ethanol, washed in propylene oxide, air dried, and gold–palladium-coated before examination. This method allowed excellent visualization of the numerous long cilia covering the ventricular walls. In contrast, the ependymal cells of the floor of the third ventricle were found to be mainly nonciliated and instead had numerous microvillous projections. The most interesting observation was the discovery of the presence of numerous small, round "formations" on the floor of the third ventricle that the authors felt could represent nerve cells projecting into the ventricular lumen. A short time later, a similar study, concentrating on aspects of the cat's lateral ventricle, was published by Noack *et al.* (1972). Their procedure involved the use of OsO_4 postfixation, acetone dehydration, air drying, and 50-nm-gold-coating prior to SEM observation of the tissues. Equally good preservation of the microvilli and cilia of the ependymal cells was obtained. Occasional "smooth" cells were also noted in the surface of the ventricle with processes penetrating into deeper layers of the ependyma. SEM investigations of the interior surface of the human third ventricle (Scott *et al.*, 1972) and lateral ventricle of the dog (Allen and Low, 1973) have also been performed. The latter study employed the use of critical-point drying (CPD) from liquid CO_2, a procedure that provided even better preservation of structure. Of interest again in these reports were observations of the supraependymal cells noted in the earlier works, which appeared in fact to be neurons.

Since the discovery of these "smooth" cells by the SEM, two recent studies by Coates (1973a,b) have also located them in the preoptic and infundibular recesses of the monkey third ventricle. Coates has suggested that the supraependymal cells may act as either baro-, osmo-, or chemoreceptors for cerebrospinal fluid (CFS) with neural, ependymal, or choroid plexus connections. Alternatively, he has offered the opinion that they may be the neurosecretory cells described by Smoller (1965) on the basis of observations with the light microscope (LM) and conventional transmission electron microscope (CTEM) observations. More recently, Scott *et al.* (1975) have determined that the supraependymal cells are composed of two basic subpopulations, one group consisting of neuronlike bipolar and multipolar cells and the other group bearing similarity to histiocytes. Although the precise nature of these cells has yet to be elucidated, their discovery primarily by use of the SEM has added significantly to our knowledge of the structure of the ventricular wall.

Other more extensive SEM investigations of the cerebral ventricles include a comparative study of the third ventricle of the rabbit, mouse, rat, and human brain by Bruni *et al.* (1972) and a study of the mammalian cerebral ventricular system by Scott *et al.* (1974). These authors have reviewed the earlier results of investigations of the third, fourth, and lateral ventricles, summarized the present state of knowledge regarding the structure of the ependymal lining as demonstrated by the SEM, and offered their views of possible structure–function relationships in this region of the nervous system. Their results indicate that the lining of the third ventricle is characterized by marked regional variations. The dorsal one third of the ventricle consists primarily of ciliated ependymal cells, while the ventral one third is notably devoid of cilia. The middle one third is a transitional area with an intermediate number of ciliated cells, apparently dependent on the species and its stage of development. Supraependymal cells, as described above, are also present. The lateral ventricle is almost completely lined by ciliated ependymal cells, but cells with a number of microvillous projections can also be found. The fourth ventricle again shows regional differences. Here, the ependyma of the floor of the rhomboid fossa and the median sulcus is densely ciliated, while a distinct transition line is visible posterolaterally and caudally. It is currently felt that such structural variations may influence discretely the kinetics of intraventricular CSF movement.

Mestres *et al.* (1974) have studied specifically the third ventricles of the male and female rat in an effort to further correlate structure and function. In addition to dividing the third ventricle into dorsal, ventral, and transitional areas, these workers subdivided the ventricle into eight specific areas that were viewed under high magnification with the SEM and observed for characteristic surface variations. In the area of the ependymal projection of the nucleus paraventricularis, balloonlike granular structures of varying sizes were noted on cell surfaces between cilia in the female specimens at mid-ovarian cycle. No such structures were apparent in the male samples. In the area of the ependymal projection of the hypophysiotropic area of the hypothalamus, similar granular elements, connected to the ependymal-cell surface by a cytoplasmic extension, were again noted, but only in the female rats. In the area of the median eminence, the number of these structures in the female was found to vary according to the stage of the ovarian cycle. By its demonstration of sexual differences in the ventricular lining, this study has opened the door to future investigations of neuroendocrinological relationships using the SEM. From this work, it appears that certain areas of the ependymal lining of the third ventricle may be modified to play a role in endocrinological regulation.

Also of interest is the SEM study by Chamberlain (1972) of the changes in the developing ependyma of embryonic and fetal rat brains after maternal exposure to 6-aminonicotinamide, an antimetabolite known to induce congenital hydrocephalus. Chamberlain found a marked reduction in the number of cilia and microvilli of the ependyma and choroid plexus in animals treated with the drug. Ruptured ependymal cell surfaces due to the pathological process were also demonstrated. More recently, Page (1975) has described ependymal changes in the ventricular system of rabbits after induction of hydrocephalus by silicone placement in the cisterna magna. Works such as these indicate the value of the SEM in the field of experimental neuropathology.

Meller and Tetzlaff (1975) have recently used the SEM to study the embryological development of the mouse cerebral cortex. Using tissues that were fixed with glutaraldehyde for 10–12 hr, postfixed with OsO_4 for 5 hr, dehydrated by treatment with graded methanol solutions, dried by the critical-point method, and gold-coated, they were able to demonstrate the three-dimensional morphology of germinal cells, glioblasts, and early neuronal elements. Further observation and monitoring of the developing nervous system using the SEM should be a fertile field of study once the more refined techniques already available are applied to this area.

2.2.2. *Brain and Spinal Cord*

In marked contrast to studies of the cerebral ventricular system, there have been only infrequent reports of SEM observations of the individual components of CNS tissue. Boyde *et al.* (1968) have attempted to visualize specific cellular components of spinal-cord tissue. The procedure involved the use of spinal-cord explants from 7- to 11-day-old chick embryos. These explants were grown on a plastic substrate at 37°C for up to 37 days and were subsequently fixed *in situ* with either formol acetic acid for 24 hr or osmic acid vapor for 10 min. The samples were then simply air dried and coated with a 20-nm layer of carbon and a 30-nm layer of gold prior to examination. Marked tissue shrinkage was noted, but a three-dimensional view of the specimen in the SEM permitted tentative indentification of fibroblasts, oligodendrocytes, astrocytes, and a neuronlike cell. More recently, Privat *et al.* (1973) have identified fibroblasts, macrophages, bipolar fusiform cells, and axonal and dendritic growth cones in an SEM study of newborn-rat cerebellar tissue using outgrowths of cultures cultivated in the Maximow assembly.

A most interesting study, particularly worthy of note since it utilized a number of the more refined preparative and examination techniques, has been made by Ambrose *et al.* (1972) of tissues obtained by biopsy of astrocytomas varying in histological staging from "benign" to glioblastoma multiforme. Explants of these samples were grown in the Rose chamber for 3–5 weeks. The tissue cultures were then washed in Hank's balanced salt solution at 37°C, quick-frozen in liquid propane cooled by liquid nitrogen at −180°C, and then placed in alcohol maintained at −80°C for 24 hr. The samples were then dried at room temperature and coated with a layer of gold–palladium. In addition to observation of the samples with the SEM in the secondary-electron-collection mode, ion-beam-etching of the specimens was performed in the manner described in Section 1.4.5c. Examination of the more "benign" astrocytes confirmed their characteristic stellate features. Intermediate-grade malignant astrocytes were noted to have larger cell bodies and relatively shorter cytoplasmic processes, while high-grade malignant astrocytes exhibited marked surface irregularity and frequent bulbous protrusions at the ends of pseudopodia. With ionbeam etching, the presence of subsurface fibrils was demonstrated in the "benign" astrocytes. The fibrils appeared as cablelike forms with a diameter of 500 nm, containing subfibrils 30–50 nm in diameter. Intermediate-grade malignant cells showed a definite loss of the regular fibril orientation, while highly malignant cells were characterized by an absence of regular fibrils, although a disorganized reticular meshwork of subfibrils was present occasionally. It could therefore be postulated that part of the malignant progress, in this case, may be related to the apparent disorganization or dissolution of the astrocyte cytoskeleton.

In another SEM investigation of pathologically altered CNS structure, Choudhury and Narayanaswami (1973) have examined the morphology of brain synaptosomes in rats injected with tetanus toxin. A previous LM study using fluorescent microscopy had demonstrated bound tetanus toxin in the synaptosomal fraction of cerebral cortex (Choudhury *et al.*, 1972). To visualize in three dimensions the structural alterations in the synaptosomes, the following technique was employed: Rats were injected with 10 mouse LD units of tetanus toxin and subsequently sacrificed after 72 hr, with the clinical onset of tetanus. The brains were then removed and placed in an ice-cold 0.32 M sucrose solution, minced with scissors, and homogenized. The synaptosomal fraction was then isolated by means of discontinuous-density-gradient centrifugation. For SEM observation, the best results were obtained by smearing the sample onto a glass slide, treating it with 95% ethanol for 30 min, washing in 0.1 M phosphate buffer at pH 7.4, incubating it with the fluorescein-labeled antitoxin for 30 min, washing in buffer, and air drying. The slides were then attached to the specimen stubs and coated with a 30-nm layer of gold prior to examination. This particular preparatory sequence was successful in avoiding the crystal damage evident in samples prepared by glutaraldehyde fixation and freeze–drying. When visualized, individual synaptosomes displayed marked distortion of shape and surface irregularities when compared with normal structures, thus pointing to the binding of tetanus toxin to the synaptic membranes. This report demonstrates that the use of the SEM in correlative studies with other microscopic techniques can be of great value in investigations at the subcellular level.

One of the newest techniques described to aid in the preparation of CNS tissue for SEM study is that of brain-cyst formation (Estable-Puig and Estable-Puig, 1975). As mentioned previously, one of the major difficulties in the effective use of the instrument in investigation of nerve or other tissues is the attainment of adequately exposed surfaces of individual components. To accomplish this, a metallic probe cooled by liquid nitrogen was applied to the exposed outer calvarium of an experimental rat. This resulted in a hemorrhagic necrosis of the underlying brain tissue. After subsequent macrophage evacuation of the formed "cyst" in approximately 1 month, an intracerebral cavity with a neuroectodermal lining surface was created. The brain tissues were then fixed by intracardiac perfusion with 4% formaldehyde and 2.5% glutaraldehyde in 0.1 M phosphate buffer at pH 7.25. Samples were then subjected to graded ethanol or acetone dehydration, infiltrated with amyl acetate, and dried by the critical-point method or processed by the osmium–tissue–osmium (OTO) technique described previously (Kelley *et al.*, 1973) (see Section 1.4.4b). Coating with a 15- to 20-nm layer of gold–palladium was performed on the majority of samples, whereas the OTO specimens were examined uncoated or with a 5-nm gold–palladium layer. These workers found that this particular method exposed large areas of nervous tissue and permitted excellent visualization of cellular components such as glial cells, neurons, nerve fibers, macrophages, and ciliated cells. LM and CTEM studies of other sites in the CNS have confirmed that the cells observed by this technique are structurally intact and that intercellular relationships are well maintained. Continued use of this valuable technique, which is illustrated in Fig. 5, can be expected in future SEM investigations of the brain.

2.3. Peripheral Nervous System and Individual Neurons

In addition to the methods of cellular exposure outlined above, there have been reports describing the isolation and visualization of individual

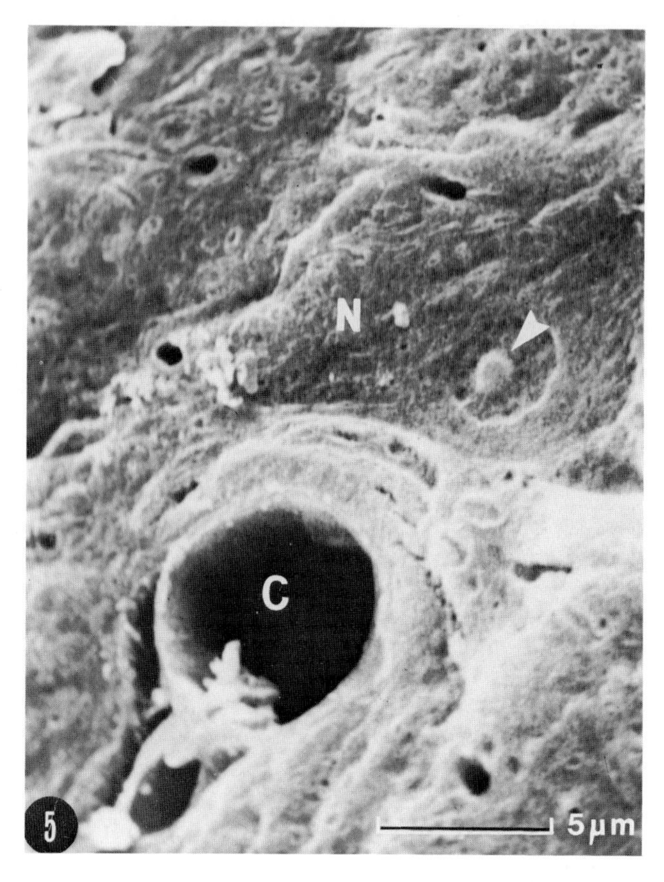

Fig. 5. Scanning electron micrograph of adult rat cerebral cortex showing the interior surface of a neuron (N) and capillary (C). Note the prominent nucleus and nucleolus (◄) of the neuron. The specimen was prepared by fracturing ethanol-saturated tissue in liquid nitrogen, CPD, and coating with gold–palladium. Courtesy of R. Hannah.

nerve cells of the peripheral nervous system. One of the initial attempts in this area of neurobiology has been made by E.R. Lewis *et al.* (1969*a*). An abdominal ganglion from the mollusk *Aplysia* was fixed with 2% glutaraldehyde for 48 hr, placed in a 16% solution of glycerin in filtered seawater for another 48 hr, rinsed with distilled water, subjected to graded alcohol dehydration, air dried, and coated with a 10- to 30-nm layer of aluminum before examination with the SEM. An effort was then made to devise a "map" of the neural networks within the ganglion. With the use of magnifications of ×560 and ×5000, it was possible to trace individual nerve fibers from their original cell bodies, located in the right and left hemiganglia respectively, to an area of intersection. At the area of intersection, an array of neuronal processes, synaptic connections, and cytoplasmic communications among fibers could be observed. The future potential of the SEM for studying neural organization in comparatively simple systems has been clearly indicated by this work.

Hamberger *et al.* (1970) have used the SEM to investigate the surface of individual neurons undergoing axon regeneration. This was accomplished by observing neuronal cells in the hypoglossal nucleus of rabbits at various times after crushing of the peripheral hypoglossal nerve. Cell bodies of the isolated nucleus were separated by gentle teasing under direct view using a stereomicroscope. The specimens were then fixed in 2% glutaraldehyde at 4°C for 30 min, dehydrated with graded acetone solutions, air dried, and coated with gold or copper before examination with the SEM. Notable shrinking of the cells was observed, but the images obtained were adequate enough to reveal a significant reduction in the number of spherical particles (interpreted as synaptic terminals) attached to the surface of the neuronal cell bodies when compared with the specimens in which no axonal trauma had occurred.

Spencer and Lieberman (1971) have published a study comparing the results obtained from SEM observation of isolated rat peripheral nerve fibers prepared by a variety of methods. Fibers were isolated using needle dissection or teasing, either before or after treatment with OsO_4. In general, surface details of individual fibers were obscured by the retained sheaths of endoneurial collagen, and these authors have suggest that prior incubation in a collagenase solution may eliminate this problem. It was observed that unfixed fibers seemed to be more easily cleared of endoneurium but showed poorer preservation of axon and myelin-sheath components.

Lodin *et al.* (1973) have recently used the SEM to correlate CTEM observations of fiber formation and myelinization in isolated neurons. Samples derived from the chick dorsal root ganglia were grown *in vitro* in Rose chambers for up to 5 weeks. For the SEM studies, the cultures were then fixed with phosphate-buffered 4% glutaraldehyde (pH 7.2) at 4°C for 1 hr, dehydrated with graded alcohol solutions, air dried, and coated with a 10-nm layer of aluminum. In addition to confirming CTEM information regarding cellular organization, this technique enabled clear views of individual Schwaan cells and growth cones to be obtained. Blood (1975), in a similar study, has also provided SEM micrographs showing Schwaan cells in three dimensions, either in free migration or in the process of early fiber myelination.

A description of the innervation of the crayfish opener muscle has been included in a report by Lang *et al.* (1972). Muscle was prepared for the SEM by fixation with 2% glutaraldehyde in 0.1 M cacodylate buffer at pH 7.2 for 48 hr, postfixation with 1% OsO_4 in the same buffer for 2 hr, treatment with 18% glycerol for 48 hr, immersion in Freon 812 cooled by liquid nitrogen, freeze–drying, and gold-coating. SEM study then provided images of individual nerve fibers extensively dividing into terminal branches that then contacted the muscle fibers. Shrinkage of individual muscle fibers, vesiculation of cell interiors, and ice-crystal damage from freeze–drying were noted. Nevertheless, this study does provide a preliminary example of how the SEM can be used in future investigations of the neuromuscular junction and its individual components.

The low-temperature dry-ashing method of E.R. Lewis (1971) for the visualization of individual neuronal processes (discussed in Section 1.4.5d) should also be considered in any attempt to reveal neuronal architecture, circuitry, and connections.

2.4. Receptors

2.4.1. Visual Receptors

There have been a number of reports involving the use of the SEM to study various aspects of the vertebrate retina (Cleveland and Schneider, 1969; Bill, 1970; Bill and Svedbergh, 1972; Svedbergh and Bill, 1972), and recently, Hager *et al.* (1975) have reviewed the use of the SEM in ophthalmological research. An early specific SEM investigation of visual receptors has been published by E.R. Lewis *et al.* (1969b). Using glutaraldehyde-fixed, glycerin-treated, alcohol-dehydrated, and air-dried retinae of *Necturus*, these authors compared the specific dimensions of the receptor outer segments, as viewed with the SEM, with the results of previous LM and CTEM studies. As expected, marked shrinkage and distortion resulted from the SEM preparatory techniques. Hence, it was recognized that newer, more refined techniques of tissue preparation were necessary to obtain meaningful topographical information.

Kuwabara (1970) later made an extensive study of human and rhesus monkey eye components. Tissues were fixed briefly with 2% OsO_4 in phosphate buffer at pH 7.2, postfixed in 2% glutaraldehyde in the same buffer for several hours, and then stored in neutral formalin for several days. Specific tissue sections were then washed with distilled water using low-frequency ultrasonic vibration to effectively clean the surfaces. Samples were then quick-frozen in methylbutane cooled by liquid nitrogen, dried in a vacuum evaporator at 5×10^{-5} mm mercury for 2 hr, mounted on specimen stubs, and coated with a layer of gold–palladium before observation with the SEM. Included in the results were accurate three-dimensional views of the inner and outer segments of the retinal photoreceptors without the distortion artifacts inherent in preceding preparatory techniques.

Investigations of the rat retina by Hansson (1970*a*) have also used the freeze–drying method for preparation of photoreceptors as well as synaptic bodies. In this case, Hansson (1970*b*) was able to demonstrate, in three dimensions, the arrangement in rows of pear-shaped synaptic bodies in the outer plexiform layer and the shape of structures in the inner plexiform layer. These images were correlated with CTEM studies of the same cells. Structural changes in photoreceptors of rats subjected to long-term treatment with sodium glutamate (Hansson, 1970*c*) and rats with experimental vitamin A deficiency (Hansson, 1970*d*) have also been described by this worker following SEM study.

An account of the surface topography of newt photoreceptor cells has been included in a work by Dickson and Hollenberg (1971). For the SEM, tissues were prepared by fixation with 3% glutaraldehyde in 0.2 M phosphate buffer at pH 7.4, dehydration with graded alcohol solutions, propylene oxide substitution, air drying and 40:60 gold–palladium coating. In addition, some samples were prepared using plastic infiltration with Epon 812 as described by Cleveland and Schneider (1969). The techniques allowed identification and visualization of the surface characteristics of the three photoreceptor types (rods, single cones, double cones) present in the newt retina.

The CPD technique has recently been successfully applied to the SEM investigation of the retina by Steinberg (1973). Dissected samples of bullfrog neural retina were fixed in 3% glutaraldehyde in 0.1 M cacodylate buffer (pH 7.4) at 4°C for 24 hr, washed in 20% ethyl alcohol for 24 hr, and subjected to graded alcohol dehydration. The tissues were then placed in amyl acetate, dried by the critical-point method from liquid CO_2, and coated with a 15-nm layer of gold. This method afforded excellent preservation of retinal components and organization and permitted the identification of the four major types of frog photoreceptors (red rods, green rods, single cones, double cones). Surface features of individual inner and outer segments were equally well preserved.

Figures 2–4 show the SEM appearance of primate, newt, and frog photoreceptors prepared by a variety of techniques.

Chi and Carlson (1974) using the SEM, have studied, features of the housefly compound eye in detail, with the object of extending knowledge of the neuronal organization in this receptor organ. These authors felt that the SEM could provide unique information not attainable with the LM, due to its limited resolution, or the CTEM. In their procedure, entire heads of flies were fixed with phosphate-buffered 2.5% glutaraldehyde at pH 7.2 for 4–5 hr and then postfixed with 1.5% OsO_4 at 4°C for an additional 4–5 hr. Following several short rinses in distilled water, the samples were dehydrated in ethanol, placed in 100% amyl acetate, and dried by the critical-point method from liquid CO_2. The eyes were then teased apart from the head, fractured along natural boundaries or lines of weakness with a razor blade, and coated with a 20-nm layer of gold–palladium before examination with the SEM. An excellent three-dimensional perspective of the optic neuropil (functional receptor unit) was achieved, allowing visualization and description of the pseudocone cavity, basement membrane, and central retinular cells. In addition, interesting features of the interneurons, glial cells, and lateral dendritic processes of monopolar axons were preserved in their *in vivo* orientation and interrelation. Recently, these workers have successfully continued this investigation, concentrating on the distal ommatidium and its constituent lenslet, photoreceptor, and other adjacent cells (Chi and Carlson, 1975). This work further attests to the value of the SEM in gaining an overall view of comparatively simple receptor structures when present-day techniques are used.

2.4.2. *Vestibular and Auditory Receptors*

The inner ear, due to the intricate and varied morphology of its constituent cells, has proven to be an ideal area for SEM study. The earliest investigations, such as the study by Lim and Lane (1969) of the vestibular sensory epithelia, revealed the directional organization of sensory hair cells present in the macula and crista. However, as Lim (1969) indicated, the preparatory techniques of standard fixation and air-drying used initially did not adequately preserve fine-structural details. Subsequently, Lim (1971*b*) employed the technique of freeze–drying to study the utricle, saccule, and crista ampullaris of various species. The tissues were fixed with 1% OsO_4 and dehydrated with graded alcohol solutions. The otoconia were dissolved by treatment with 10% HC1. Each vestibular organ was then frozen in liquid nitrogen at −160°C, dried by vacuum sublimation at −40 to −80°C, and coated with a 20-nm layer of gold. This method provided excellent preservation of the delicate sensory hairs and enabled accurate delineation of cellular interrelationships and morphological variation. Recently, Rosenhall and Engstrom (1974) have used the SEM to study human vestibular sensory epithelia. Briefly, their study indicates that the vestibular sensory epithelium consists of numerous hair cells with stereocilia and one peripheral kinocilium protruding from the surface of each cell. They were able to observe that the kinocilia all faced in the same direction in large areas of the epithelium. The sensory receptor cells were found to be well developed in 3- to 4-month-old fetuses. A progressive lengthening of the sensory stereocilia from one pole of the hair bundle to the other was visible. In the fetus, the sensory and supporting cells formed a regular pattern, and each hair cell was surrounded by four or more supporting cells. Each of the latter cells characteristically possessed numerous surface microvilli and a solitary kinocilium. In the adult, this pattern was found to be disrupted due to the absence of many sensory cells. In addition, the adult supporting cells showed shorter microvilli and often lacked a kinocilium. This SEM study has confirmed previous descriptions and provided a unique representation of the *in vivo* surface structure of the vestibular receptor organ.

Hillman and Lewis (1971) and Harada (1972) have studied the vestibular organ in the frog. In particular, Harada (1972) was able to demonstrate sensory hairs protruding into the reticular architecture of the otolithic membrane.

A number of reports have also appeared dealing with auditory receptors (Bredberg *et al.*, 1970; Marovitz *et al.*, 1970*b*; Kosaka *et al.*, 1971), and Lindeman *et al.* (1971) and Lim (1972) have paid particular attention to the relationship of the tectorial membrane to the sensory receptor cells in the organ of Corti as seen by the SEM. Tanaka *et al.* (1973) have used an interesting method involving frozen-resin cracking (Tanaka, 1972) to adequately expose the internal structures of the guinea pig cochlea while at the same time maintaining normal cellular organization. Briefly, tissues were fixed with 1% OsO_4 in veronal buffer at pH 7.2–7.4. After graded ethanol dehydration, the specimens were soaked in propylene oxide at room temperature for 30 min, transferred to a mixture of propylene oxide and epoxy resin for 3–4 hr, and then left in the resin alone for 12–24 hr. The specimens were then transferred to a gelatin capsule, filled with resin, and hardened in a cryostat at −30°C for 1–2 hr. After removal, the samples were cracked with a hammer and chisel, soaked in propylene oxide, placed in a solution of isoamyl acetate, dried by the critical-point method, and coated with layers of carbon and gold. The procedure provided very good exposure of the cochlear elements and enabled a comparison to be made, at the sensory-cell level, between normal cochleae and those damaged by Kanamycin.

Bredberg *et al.* (1972) have used SEM techniques to study the normal and pathologically altered organ of Corti. Human tissues were fixed by perfusion of the perilymphatic cochlear spaces with a variety of fixatives. After dissection, individual samples were then frozen by immersion in an isopentane or propane bath, cooled by liquid nitrogen, dried by vacuum sublimation, and coated with a layer of gold–palladium. Examination with the SEM clearly revealed the three-dimensional organization of the organ of Corti, including details of hair cells (receptors) and supporting cells. The supporting cells were shown throughout their entire length from the basilar membrane to the surface of the epithelium, where they joined with the receptor cells to form the reticular membrane. The inner hair cells were clearly visualized in their normal location close to the medial pillar of the tunnel of Corti, and the outer hair cells, arranged in three to five rows lateral to the tunnel, could be seen in detail. SEM study showed readily that the stereocilia of each sensory cell were in a "W" formation with the base of the "W" directed laterally. From the base of the cochlea to its apex, SEM study was able to show that (1) the open angle of the "W" arrangement gradually decreased; (2) the surface of the receptor cells gradually changed from a "kidney" shape to that of an oval; and (3) stereocilia of each cell gradually increased in length. Excellent views of the nerve endings surrounding the bases of the outer hair cells were obtained. Also, observations of cochleae of animals subjected to high-intensity pure-tone sound (125 Hz, 148 dB, 4-hr duration) have demonstrated a sharply demarcated lesion in the apical coil of the organ of Corti.

A recent investigation using the SEM has concentrated on the pathways of individual nerve fibers in the osseous spiral lamina and modiolus (Engstrom, 1974) to further clarify the neural organization involved in auditory reception. In this study, too, excellent results have been obtained using uncomplicated SEM techniques. Li and Lewis (1974) have recently demonstrated the potential of the SEM in following receptor morphogenesis, using the auditory receptor epithelium of the bullfrog as an example. It is known that the inner ear of this species contains two auditory sensory organs, the amphibian and basilar papillae. The objective of the study was to observe the cellular structure and organization of these two organs with emphasis on morphogenetic variations. For adult specimens, the membranous labyrinth was exposed by dissection and the inner ear perfused with 2% glutaraldehyde in 400 mOsM cacodylate buffer and then postfixed with 1% OsO_4. For the tadpole specimens, the entire head with the otic capsule exposed was immersed in 2% glutaraldehyde in 200 mOsM cacodylate buffer and postfixed with 2% OsO_4. The membranous labyrinths of all specimens were then dissected, dehydrated with graded ethanol–Freon TF solutions to 100%

Freon TF, transferred to liquid Freon 13 or liquid CO_2, and dried by the critical-point method. After being coated with a layer of gold, the samples were then examined with the SEM. Each basilar papilla exhibited three regions of specific cellular composition; a medial edge consisting of hair cells with short stereocilia and bulbed kinocilia, a middle area showing large hair cells with graded stereocilia and long kinocilia, and a lateral area made up of miniature hair cells. The amphibian papilla was noted to comprise two separate structures in the tadpole, with fusion into a single organ in the adult. Particular attention was paid to the orientation of the receptors in each organ. This report indicates clearly that SEM techniques are already at a stage where they can be of great value in unraveling the details of neuronal and receptor development. Hence, future studies of this variety can be expected shortly.

2.4.3. *Taste and Olfactory Receptors*

Use of the SEM has also contributed to the elucidation of the structure and function of a variety of taste and olfactory receptors. For example, Shimamura and Tokunaga (1970) have presented a comprehensive picture of the cellular arrangement of the fungiform papillae in the frog tongue. A special technique was used that facilitated visualization of elongated papillae free of extraneous mucus. Briefly, the dorsal surface of the tongue was first rinsed with a 0.65% saline spray under compressed air. The entire tongue was then immersed in Ringer's solution at 5°C for 24 hr and the dorsal surface rinsed with fresh saline every 1–2 hr. The tongues were then fixed with 2.5% glutaraldehyde for 30 min, sectioned, and fixed as before for an additional 1 hr. After being postfixed with 2% OsO_4 for 30–60 min, the specimens were subjected to graded acetone dehydration, air dried, coated with layers of carbon and gold, and examined with the SEM. The images obtained clearly demonstrated the distribution, contour, and microstructure of the individual receptors. Each receptor organ was found to be composed of an epithelial disk surrounded by a zone of ciliary cells followed by another zone of protective epithelial cells. The disk itself consisted of the microvilli of individual sensory cells. Later works by Graziadei and DeHan (1971) and DeHan and Graziadei (1971) have confirmed the observation that the frog taste organ is in the form of a disk.

Recently, Reutter *et al.* (1975) have used the SEM to study details of the taste buds of fish. After being saline-washed, samples were fixed with 2.5% glutaraldehyde in 0.1 M phosphate buffer containing 2.5% glucose at pH 7.2 for 8 days. The specimens were then postfixed with 1% OsO_4 for 6 hr, dehydrated with graded ethanol solutions, dried by the critical-point method from either Freon 11 or Freon 13, and coated with layers of carbon and gold before being examined with the SEM. The results confirmed the existence of three structurally different groups of taste receptors. Variations in the size of the villi on sensory cell terminals were also noted.

The structure of olfactory receptors as seen by the SEM in a variety of species has been included in a report by Breipohl *et al.* (1974). In the goldfish, this sensory system was found to consist of two olfactory pits containing a number of laminae originating from a central axis. Areas of sensory epithelia were noted only on the mediodorsal aspects of each lamina. The sensory areas exhibited variations in both the density and structure of receptor-cell terminals. In the mouse, the olfactory epithelium was characterized by long olfactory "hairs" and again showed regional variations in the number of receptor cells present. Similar "hairs" were evident in the chick olfactory epithelia, but in addition, the surface of the receptor cells presented microvilli-like projections. In this study, the value of the SEM in providing a relatively quick and easy survey of the comparative topography of a receptor organ in a variety of species has been clearly demonstrated.

2.4.4. *Miscellaneous Specialized Receptors*

In addition to the studies of receptor topography already noted, infrequent reports have also appeared dealing with the surface morphology of specialized receptors found in insect and amphibian species. For example, Moran and Rowley (1975) have described the ultrastructure of the cockroach campaniform sensillum, a mechanoreceptor aiding in the coordination of walking movements. The SEM was used to study specifically the extracellular chitinous components of the receptor "cap," the site of stimulus reception. The results demonstrated a definite orientation of the various-sized oval domes of the caps, either parallel with or perpendicular to the long axis of the leg. The attachment of the dome to the exoskeleton at two definite points was also easily visualized.

Similar SEM investigations of insect-receptor morphology have been carried out by Woolley (1972) in the tick, and Pierantoni (1974) has measured the growth pattern of antennal receptors in *Tenebrio molitor*. In addition, Jørgensen and Flock (1973) have employed the instrument to study the lateral-line sense organs in the salamander.

2.5. Conclusion

Consideration of the foregoing information as a whole allows certain generalizations to be made regarding the present "state of the art" in the use of the SEM in the study of nerve tissue. It is obvious that with very few exceptions, present-day techniques have not allowed visualization in three dimensions of the details of neuronal circuitry. The difficulty appears to lie, not with the resolution of the SEM, which under ideal circumstances should be sufficient to reveal synaptic topography, but with the extremely complex nature of nerve tissue itself. It seems that what is needed is a preparative technique for SEM study of the nervous system that parallels the Golgi technique for light microscopy so that individual neurons can be visualized entirely and distinctly within the surrounding mass of tissue. The closest approach to a solution has been the dry-ashing technique of Lewis (1971) (see Section 1.4.5d), and clearly much more effort is needed to follow up this innovative approach. It is doubtful that microdissection, no matter how carefully it is done, can be successful in unraveling any but the simplest of neuronal interconnections.

The best success to date in the use of the SEM in neurobiology has been in the study of the detailed surface topography of receptor organs. In this field, the use of the SEM has greatly aided in the appreciation of the three-dimensional structure of receptors, and the results obtained have correlated well with prior information gained from LM and CTEM study of the same structures. The advantages of SEM use in this area are in the relative ease and rapidity of specimen preparation and the ability to rapidlv view the surface of specimens as large as the human eye in three dimensions. In this work, distortion and shrinkage due primarily to the drying procedures used currently in preparation of specimens for the SEM continue to be a problem. Hence, descriptions of surface structure and, particularly, information on dimensions of cells and their components must still be treated with caution if derived exclusively from SEM observations. Here too, further work is needed to develop better preparative methods. Until such methods are available, accurate descriptions must of necessity incorporate CTEM and LM micrographs to correlate with the SEM data.

Finally, it should be noted that investigators using the SEM to study the nervous system, with very few exceptions, have not as yet taken advantage of the remarkable versatility of the instrument itself or of many of the newer specimen-preparation methods. Exciting results can be expected from work utilizing X-ray detection devices attached to the SEM and from methods, already developed, that allow the SEM to be used in the fields of histochemistry and immunology and as an X-ray microscope. The SEM should not be thought of as merely an adjunct to the CTEM and LM. Clearly, it has the potential of eliciting information of a unique nature that at present cannot be obtained in any other way.

References

Note: The abbreviations *SEM* [**1968–1975**] denote: *Scanning Electron Microscopy*/[*1968–1975*]: *Proceedings of the* [*First–Eighth*] *Annual Scanning Electron Microscope Symposium* (O. Johari and I. Corvin, eds.), IIT Research Institute, Chicago.

Abraham, J.L., and DeNee, P.B., 1974, Biomedical applications of backscattered electron imaging: One year's experience with SEM histochemistry, *SEM* **1974:**251.

Ahmed, H., and Broers, A.N., 1972, Lanthanum hexaboride electron emitter, *J. Appl. Phys.* **43:**2185

Allen, D.J., and Low, F.N., 1973, The ependymal surface of the lateral ventricle of the dog as revealed by SEM, *Am. J. Anat.* **137:**483.

Ambrose, E.J., Batzdorf, U., and Easty, D.M., 1972, Morphology of astrocytomas in tissue culture: Optical and stereoscan microscopy, *J. Neuropathol. Exp. Neurol.* **31:**596.

Anderson, T.F., 1951, Techniques for the preservation of three-dimensional structure in preparing specimens for the electron microscope, *Trans. N. Y. Acd. Sci.* **13:**130.

Barber, V.C., 1972, Preparative techniques for the successive examination of biological specimens by light microscopy, SEM and TEM, *SEM* **1972:**321.

Bartlett, A.A., and Burstyn, H.P., 1975, A review of the physics of critical point drying, *SEM* **1975:**305.

Bernstein, E.O., and Jones, C.B., 1969, Skin replication procedure for the SEM, *Science* **166:**252.

Bill, A., 1970, SEM studies of the canal of Schlemm, *Exp. Eye Res.* **10:**214.

Bill, A., and Svedbergh, B., 1972, SEM studies of the trabecular meshwork and the canal of Schlemm—an attempt to localize the main resistance to outflow of aqueous humor in man, *Acta Ophthalmol.* **50:**295.

Blood, L.A., 1975, SEM observation of the outgrowth from embryonic chick dorsal root ganglia in culture, *Neurobiology* **5:**75.

Boyde, A., 1971, A review of problems of interpretation of the SEM image with special regard to methods of specimen preparation, *SEM* **1971:**1.

Boyde, A., 1972, Biological specimen preparation for the SEM—an overview, *SEM* **1972:**257.

Boyde, A., 1974, Freezing, freeze–fracturing and freeze–drying in biological specimen preparation for the SEM, *SEM* **1974:**1043.

Boyde, A., 1975, A method for the preparation of cell surfaces hidden within bulk tissue for examination in the SEM, *SEM* **1975:**295.

Boyde, A., and Echlin, P., 1973, Freeze and freeze–drying—a preparative techinque for SEM, *SEM* **1973:**759.

Boyde, A., and Stewart, A.D.G., 1962, A study of the etching of dental tissues with argon ion beams, *J. Ultrastruct. Res.* **7:**159.

Boyde, A., and Wood, C., 1969, Preparation of animal tissues for surface scanning electron microscopy, *J. Microsc.* **90:**221.

Boyde, A., James, D.W., Tresman, R.L., and Willis, R.A., 1968, Outgrowth from chick embryo spinal cord *in vitro*, studied with the SEM, *Z. Zellforsch.* **90:**1.

Boyde, A., Jones, S.J., and Bailey, E., 1973, Bibliography on biomedical applications of SEM, *SEM* **1973:**697.

Bredberg, G., Lindeman, H.H., Ades, H.W., West, R., and Engstrom, H., 1970, SEM of the organ of Corti, *Science* **170:**861.

Bredberg, G., Ades, H.W., and Engstrom, H., 1972, SEM of the normal and pathologically altered organ of Corti, *Acta Otolaryngol. Suppl.* **301:**3.

Breipohl, W., Bijvank, G.J., and Pfefferkorn, G.E., 1974, SEM of various sensory receptor cells in different vertebrates, *SEM* **1974:**557.

Brinkley, B.R., Murphy, P., and Richardson, L.C., 1967, Procedure for embedding *in situ* selected cells cultured *in vitro*, *J. Cell Biol.* **35:**279.

Broecker, W., Schmidt, E.H., Pfefferkorn, G., and Beller, F.K., 1975, Demonstration of cathodoluminescence in fluorescein marked biological tissues, *SEM* **1975:**243.

Broers, A.N., 1974, Recent advances in SEM with lanthanum hexaboride cathodes, *SEM* **1974:**9.

Broers, A.N., 1975, Electron sources for SEM, *SEM* **1975:**661.

Broers, A.N., Panessa, B.J., and Gennaro, J.F., 1975, High resolution SEM of biological specimens, *SEM* **1975:**233.

Brummer, M.E.G., Lowrie, P.M., and Tyler, W.S., 1975, A technique for sequential examination of specific areas of large tissue blocks using SEM, LM, and TEM, *SEM* **1975:**333.

Bruni, J.E., Montemurro, D.G., Clattenburg, R.E., and Singh, R.P., 1972, A SEM study of the ependymal surface of the third ventricle of the rabbit, rat, mouse and human brain, *Anat. Rec.* **174:**407.

Carr, K.E., 1971, Applications of SEM in biology, *Int. Rev. Cytol.* **30:**183.

Carr, K.E., and McGadey, J., 1974, Staining of biological material for the SEM, *J. Microsc.* **100:**323.

Chamberlain, J.G.,1972, 6-Aminonicotinamide (6-AN)-induced abnormalities of the developing ependyma and choroid plexus as seen with the SEM, *Teratology* **6:**281.

Chi, C., and Carlson, S.D.,1974, New neuroanatomical concepts of the housefly compound eye and optic neuropiles derived by SEM, *SEM* **1974:**799.

Chi, C., and Carlson, S.D., 1975, The distal ommatidium of the compound eye of the housefly (*Musca domestica*): A SEM study, *Cell Tissue Res.* **159:**379.

Choudhury, R.P., and Narayanaswami, A., 1973, Changes in surface structures of brain synaptosomes of rat following binding of tetanus toxin, *SEM* **1973:**435.

Choudhury, R.P., Mondal, S., Chatterjee, R., and Narayanaswami, A., 1972, Site of action of tetanus toxin in the CNS—subcellular site of binding of tetanus toxin in the cerebrum, *Indian J. Med. Res.* **60:**1175.

Clementi, F., and Marini, D., 1972, The surface fine structure of the walls of cerebral ventricles and of choroid plexus in cat, *Z. Zellforsch.* **123:**82.

Cleveland, P.H., and Schneider, C.W., 1969, A simple method of preserving ocular tissue for SEM, *Vision Res.* **9:**1401.

Coates, P.W., 1973*a*, Supraependymal cells in recesses of the monkey third ventricle, *Am. J. Anat.* **136:**533.

Coates, P.W., 1973*b*, Supraependymal cells: Light and transmission electron microscopy extends SEM demonstration, *Brain Res.* **57:**502.

Crewe, A.V., and Wall, T., 1970, A SEM with 5 Å resolution, *J. Mol. Biol.* **48:**375.

Crewe, A.V., Wall, J., and Langmore, J., 1970, Visibility of single atoms, *Science* **168:**1338.

DeHan, R.S., and Graziadei, P.P.C., 1971, Functional anatomy of frog's taste organs, *Experientia* **27:**823.

DeNee, P.B., and Walker, E.R., 1975, Specimen coating techniques for the SEM—a comparative study, *SEM* **1975:**225.

DeNee, P.B., Abraham, J.L., and Willard, P.A., 1974, Histochemical stains for the SEM—qualitative and semi-quantitative aspects of specific silver stains, *SEM* **1974:**259.

Dickson, D.H., and Hollenberg, M.J., 1971, The fine structure of the pigment epithelium and photoreceptor cells of the newt, *Triturus Viridescens dorsalis* (*Rafinesque*), *J. Morphol.* **135:**389.

Echlin, P., 1971, The examination of biological material at low temperatures, *SEM* **1971:**225.

Echlin, P., 1974, Coating techniques for scanning electron microscopy, *SEM* **1974:**1019.

Echlin, P., 1975, Sputter coating techniques for SEM, *SEM* **1975:**217.

Echlin, P., and Hyde, P.J.W., 1972, The rationale and mode of application of thin films to non-conducting materials, *SEM* **1972:**137.

Echlin, P., and Moreton, R., 1973, The preparation, coating and examination of frozen biological materials in the SEM, *SEM* **1973:**325.

Echlin, P., Kynaston, D., and Knights, D., 1969, Ion beam etching of biological material in the SEM, in: *Proc. 27th Electron Microscopy Society of America, St. Paul, Minn., August, 1969* (C.J. Arceneaux, ed.), Claitor, Baton Rouge.

Echlin, P., Paden, R., Dronzek, B., and Wayte, R., 1970, SEM of labile biological material maintained under controlled conditions, *SEM* **1970:**49.

Engstrom, B., 1974, SEM of the inner structure of the organ of Corti and its neural pathways, *Acta Otolaryngol. Suppl.* **319:**57.

Erlandsen, S.L., Thomas, A., and Wendelschafer, G., 1973, A sample technique for correlating SEM with TEM on biological tissue originally embedded in epoxy resin for TEM, *SEM* **1973:**349.

Estable-Puig, R.F., and Estable-Puig, J.F., 1975, Brain cyst formation: A technique for SEM study of the central nervous system, *SEM* **1975:**281.

Ferris, S.D., Joy, D.C., Leamy, H.J., and Crawford, C.K., 1975, A directly heated LaB_6 electron source, *SEM* **1975:**11.

Flood, P.R.,1975, Dry-fracturing techniques for the study of soft internal biological tissues in the SEM, *SEM* **1975:**237.

Frisch, B., Lewis, S.M., Sherman, D., Stuart, P.R., and Osborn, J.S., 1974, Utilization of ion-etching in studying blood cell ultrastructure, *SEM* **1974:**655.

Frisch, B., Lewis, S.M, Stuart, P.R., and Osborn, J.S.,1975, Further observations of the effects of ion-etching on blood cells, *SEM* **1975:**165.

Fujita, T., Kokunaga, J., and Inoue, H., 1969, SEM of the skin using celluloid impressions, *Arch. Histol. Jpn.* **30:**321.

Goldstein, J.I., and Yakowitz, H. (eds.),1975, *Practical Scanning Electron Microscopy,* Plenum Press, New York.

Gomori, G., 1950, An improved histochemical technique for acid phosphatase, *Stain Technol.* **25:**81.

Graziadei, P.P.C., and DeHan, R.S., 1971, The ultrastructure of frog's taste organs, *Acta Anat.* **80:**563.

Griffiths, B.W., and Venables, J.A., 1972, Scanning electron microscopy at liquid helium temperatures, *SEM* **1972:**10.

Guttman, H.N.,and Styskol, R.C., 1971, Preparation of suspended cells for SEM examination of internal cellular structures, *SEM* **1971:**265.

Hager, H., Hoffmann, F., and Dumitrescu, L., 1975, SEM in ophthalmology, *Ann. Ophthalmol.* **7:**1361.

Haggis, G.H., 1970, Cryofracture of biological material, *SEM* **1970:**97.

Hamberger, A., Hansson, H.A., and Sjöstrand, J., 1970, Surface structure of isolated neurones: Detachment of nerve terminals during axon regeneration, *J. Cell Biol.* **47:**319.

Hansson, H.A., 1970*a*, SEM of the rat retina, *Z. Zellforsch.* **107:**23.

Hansson, H.A., 1970*b*, SEM studies on the synaptic bodies in the rat retina, *Z. Zellforsch.* **107:**45.

Hansson, H.A., 1970*c*, SEM studies on the long term effects of sodium glutamate on the rat retina, *Virchows Arch.* **4:**357.

Hansson, H.A., 1970*d*, SEM of the retina in vitamin-A-deficient rats, *Virchows Arch.* **4:**368.

Harada, Y., 1972, Surface view of the frog vestibular organ with the SEM, *Acta Otolaryngol.* **73:**316.

Hillman, D.E., and Lewis, E.R., 1971, Morphological basis for a mechanical linkage in otolithic receptor transduction in the frog, *Science* **174:**146.

Hodges, G.M., Carbonell, A.W., Muir, M.D., and Grant, P.R., 1974, Uses and limitations of SEM autoradiography, *SEM* **1974:**159.

Hoerl, E.M., and Muegschl, E., 1972, SEM of metals using light emission, in: *Proceedings of the 5th European Congress of Electron Microscopy, Manchester, England, September, 1972,* Institute of Physics, London.

Hollenberg, M.J., and Erickson, A.M., 1973, The SEM: Potential usefullness to biologists, a review, *J. Histochem. Cytochem.* **21:**109.

Horridge, G.A., and Tamm, S.L., 1969, Critical point drying for SEM study of ciliary motion, *Science* **163:**817.

Humphreys, W.J., 1975, Drying soft biological tissue for SEM, *SEM* **1975:**707.

Humphreys, W.J., Spurlock, B.O., and Johnson, J.S., 1974, Critical point drying of ethanol-infiltrated, cryofractured biological specimens for SEM, *SEM* **1974:**275.

Isaacson, M., Langmore, J., and Wall, J., 1974, The preparation and observation of biological specimens for the high resolution scanning transmission electron microscope, *SEM* **1974:**19.

Jones, S.J., Bailey, E., and Boyde, A., 1974, Biomedical applications bibliography 1973–74 update, *SEM* **1974:**835.

Jørgensen, J.M., and Flock, A., 1973, The ultrastructure of lateral line sense organs in the adult salamander *Ambystoma mexicanum, J. Neurocytol.* **2:**133.

Joy, D., 1973, The SEM, principles and applications, *SEM* **1973:**743.

Karnovsky, M.J., 1965, A formaldehyde–glutaraldehyde fixative of high osmolarity for use in electron microscopy, *J. Cell Biol.* **27:**137A.

Kelley, R.O., Dekker, R.A.F., and Bluenink, J.G., 1973, Ligandmediated osmium binding: Its application in coating biological specimens for SEM, *J.Ultrastruct. Res.* **45:**254.

Kimoto, S., Hashimoto, H., and Takashima, S., 1969, Transmission scanning microscope, *SEM* **1969:**81.

Knoll, M., 1935, Aufladepotential und Sekundäremission elektronenbestrahlter Körper, *Z. Tech. Phys.* **16:**467.

Koehler, J.K.,1968, The technique and application of freeze–etching in ultrastructure research, *Adv. Biol. Med. Phys.* **12:**1.

Komoda, T., and Saito, S., 1972, Experimental resolution limit in the secondary mode for a field emission source SEM, *SEM* **1972:**129.

Kosaka, N., Tanaka, T., Takiguchi, T., Oseki, Y., and Takahara, S., 1971, Observation on the organ of Corti with SEM, *Acta Otolaryngol.* **72:**337.

Kuwabara, T., 1970, Surface structure of the eye tissue, *SEM* **1970:**185.

Lane, W.C., 1970, The environmental control stage, *SEM* **1970:**41.

Lang, F.,Atwood, H.L., and Morin, W.A., 1972, Innervation and vascular supply of the cray fish opener muscle: Scanning and transmission electron microscopy, *Z. Zellforsch.* **127:**189.

Lewis, E.R., 1971, Studying neuronal architecture and organization with SEM, *SEM* **1971:**281.

Lewis, E.R., and Nemanic, M.K., 1973, Critical point drying techniques, *SEM* **1973:**767.

Lewis, E.R., Everhart, T.E., and Zeevi, Y.Y., 1969*a*, Studying neural organization in *Aplysia* with the SEM, *Science* **165:**1140.

Lewis, E.R., Zeevi, Y.Y., and Werblin, F.S., 1969*b*, SEM of vertebrate visual receptors, *Brain Res.* **15:**559.

Lewis, S.M., Osborn, J.S., and Stuart, P.R., 1968, Demonstration of an internal structure within the red blood cell by ion etching with SEM, *Nature* (*London*) **220:**614.

Li, C.W., and Lewis, E.R., 1974, Morphogenesis of auditory receptor epithelia in the bullfrog, *SEM* **1974:**791.

Lim, D.J., 1969, Three-dimensional observation of the inner ear with the SEM, *Acta Otolaryngol. Suppl.* **255:**1.

Lim, D.J., 1971*a*, Scanning electron microscopic observation on nonmechanically cryofractured biological tissue, *SEM* **1971:**257.

Lim, D.J., 1971*b*, Vestibular sensory organs—a SEM investigation, *Arch. Otolaryngol.* **94:**69.

Lim, D.J., 1972, Fine morphology of the tectorial membrane—its relationship to the organ of Corti, *Arch. Otolaryngol.* **96:**199.

Lim, D.J., and Lane, W.C., 1969, Vestibular sensory epithelia—a SEM observation, *Arch. Otolaryngol.* **90:**283.

Lindeman, H.H., Ades, H.W., Bredberg, G., and Engstrom, H., 1971, The sensory hairs and the tectorial membrane in the development of the cat's organ of Corti—a SEM study, *Acta Otolaryngol.* **72:**229.

Lithicum, D.S., Sell, S.,Wagner, R.M., and Trefts, P., 1974, Scanning immunoelectron microscopy of mouse B and T lymphocytes, *Nature* (*London*) **252:**173.

LoBuglio, A.F., Rinehart, J.J., and Balcerzak, S.P., 1972, A new immunologic marker for scanning electron microscopy, *SEM* **1972:**313.

Lodin, Z., Faltin, J., Booker, J., Hartman, J., and Sensenbrenner, M., 1973, Fiber formation and myelination of cultivated dissociated neurones from chick dorsal root ganglia—an electron microscopic and SEM study, *Neurobiology* **3:**66.

Lowrie, P.M., and Tyler, W.S., 1973, Selection and preparation of specific tissue regions for TEM using large epoxy-embedded blocks, in: *Proceedings of the 31st Electron Microscopy Society of America, New Orleans, 1973* (C.J. Arceneaux, ed.), p. 324, Claitor, Baton Rouge.

MacKenzie, A.P., 1972, Freezing, freeze–drying, and freeze–substitution, *SEM* **1972:**273.

Malick, L.E., and Wilson, R.B., 1975, Evaluation of a modified technique for SEM examination of vertebrate specimens without evaporated metal layers, *SEM* **1975:**259.

Marovitz, W.F., Arenberg, I.K., and Thalmann, R.,1970*a*, Evaluation of preparative techniques for the SEM, *Laryngoscope* **80:**1680.

Marovitz, W.F., Thalmann, R., and Arenberg, I.K., 1970*b*, SEM of freeze–dried guinea pig organ of Corti, *SEM* **1970:**273.

McDonald, N.C., 1970, Potential mapping using Auger electron microscopy, *SEM* **1970:**481.

McDonald, N.C., 1971, Auger electron spectroscopy for SEM, *SEM* **1971:**89.

McDonald, N.C., Marcus, H.L., and Palmberg, P.W., 1970, Microscopic Auger electron analysis of fracture surfaces, *SEM* **1970:**25.

McMullan, D., 1953, An improved SEM for opaque specimens, *Proc. Inst. Electr. Eng.* **100:**245.

Meller, K., and Tetzlaff, W., 1975, Neuronal migration during the early development of the cerebral cortex—a SEM study, *Cell Tissue Res.* **163:**313.

Mestres, P., Breipohl, W., and Bijvank, G.J., 1974, The ependymal surface of the third ventricle of rat to hypothalamic area: A reflection SEM study, *SEM* **1974:**783.

Molday, R.S., Dreyer, W.J., Rembaum, A., and Yen, S.P.S., 1974, Latex spheres as markers for studies of cell surface receptors by SEM, *Nature* (*London*) **249:**81.

Molday, R.S., Dreyer, W.J., Rembaum, A., and Yen, S.P.S., 1975, New immunolatex spheres: Visual markers of antigens on lymphocytes for SEM, *J.Cell Biol.* **64:**75.
Moran, D.T., and Rowley, J.C., 1975, High voltage and SEM of the site of stimulus reception of an insect mechanoreceptor, *J. Ultrastruct. Res.* **50:**38.
Muir, M.D., 1974, Fundamentals of the SEM for biologists, *SEM* **1974:**1011.
Muir, M.D., and Holt, D.B., 1974, Analytical cathodoluminescence in SEM, *SEM* **1974:**135.
Muir, M.D., and Rampley, D.N., 1969, The effect of the electron beam on various mounting and coating media in scanning electron microscopy, *J. Micros.* **90:**145.
Nemanic, M.K., 1972, Critical point drying, cryofracture, and serial sectioning, *SEM* **1972:**297.
Nemanic, M.K., 1975, On cell surface labelling for the SEM, *SEM* **1975:**341.
Nixon, W.C., 1968, Twenty years of scanning electron microscopy, 1948–1968, in the Engineering Department, Cambridge University, England, *SEM* **1968:**55.
Nixon, W.C., 1969, Introduction to scanning electron microscopy, *SEM* **1969:**1.
Noack, W., Dumitrescu, L., and Schweichel, J.U., 1972, Scanning and electron microscopical investigations of the surface structures of the lateral ventricles in the cat, *Brain Res.* **46:**121.
Nowell, J.A., Gillespie, J.R.,and Tyler, W.S., 1971, SEM of chronic pulmonary emphysema: A study of the equine model, *SEM* **1971:**297.
Oatley, C.W., Nixon, W.C., and Pease, R.F.W., 1965, Scanning electron microscopy, *Adv. Electron. Electron Phys.* **21:**181.
Pachter, B.R., Penha, D., Davidowitz, J., and Breinin, G.M., 1973, Technique for examining uncoated specimens in the SEM with light microscope and TEM correlation, *SEM* **1973:**387.
Page, R.B.,1975, SEM of the ventricular system in normal and hydrocephalic rabbits, *J. Neurosurg.* **42:**646.
Panessa, B.J.,and Gennaro, J.F., 1972, Preparation of fragile botanical tissues and examination of intracellular contents by SEM, *SEM* **1972:**327.
Panessa, B.J., and Gennaro, J.F., 1973, Use of potassium iodide/lead acetate for examining uncoated specimens, *SEM* **1973:**395.
Panessa, B.J., and Russ, J.C., 1975, Techniques for practical biological microanalysis, *SEM* **1975:**251.
Parducz, B., 1967, Ciliary movement and co-ordination in ciliates, *Int. Rev. Cytol.* **21:**91.
Parsons, D.F., and Lyon, N.C., 1975, Design of SEM hydration chambers and possibilities for viewing live biological material, *SEM* **1975:**27.
Pawley, J.B., 1972, Charging artifacts in the SEM, *SEM* **1972:**153.
Pawley, J.B., and Hayes, T.L.,1971, A micromanipulator for the SEM, *SEM* **1971:**105.
Pawley, J.B., and Nowell, J.A., 1973, Microdissection of biological SEM samples for further study in the TEM, *SEM* **1973:**333.
Pease, R.F.W., 1971, Fundamentals of scanning electron microscopy, *SEM* **1971:**9.
Pfefferkorn, G.E., 1970, Specimen preparation techniques, *SEM* **1970:**89.
Pfefferkorn, G.E., and Boyde, A., 1974, Review of replica techniques for scanning electron microscopy, *SEM* **1974:**75.
Pfefferkorn, G.E., Fromme, H.G., and Pfautsch, M., 1972, Different replica methods for skin examination and their comparison with direct studies on skin, *SEM* **1972:**351.
Pierantoni, R., 1974, Electron scanning microscopy of the antennal receptors in *Tenebrio molitor*; A stereoscopic analysis, *Cell Tissue Res.* **148:**127.
Polliack, A., Lampen, N., and deHarven, E., 1973, Comparison of air drying and critical point drying procedures for the study of human blood cells by SEM, *SEM* **1973:**529.
Privat, A., Drian, M.J., and Mandon, P., 1973, The outgrowth of rat cerebellum in organized culture, *Z. Zellforsch.* **146:**45.
Reutter, K., Breipohl, W., and Bijvank, G.J., 1975, Taste buds in fishes, *Cell Tissue Res.* **153:**151.
Revel, J.P., 1975, Elements of SEM for biologists, *SEM* **1975:**687.
Rosenhall, U., and Engstrom, B., 1974, Surface structures of the human vestibular sensory regions, *Acta Otolaryngol. Suppl.* **319:**3.
Russ, J.C., 1972, Resolution and sensitivity of X-ray microanalysis in biological sections by scanning and conventional transmission electron microscopy, *SEM* **1972:**73.

Scarpelli, D.G., and Lim, D.J., 1970, The application of scanning electron microscopy to the cytochemistry of cell membranes, *Proceedings of the Third Annual Scanning Electron Microscope Symposium, Chicago, Ill., April, 1970,* IIT Research Institute, Chicago.
Scott, D.E., Paull, W.K., and Dudley, G.K., 1972, A comparative SEM analysis of the human cerebral ventricular system. 1. The third ventricle, *Z. Zellforsch.* **132:**203.
Scott, D.E., Kozlowski, G.P., and Sheridan, M.N., 1974, SEM in the ultrastructural analysis of the mammalian cerebral ventricular system, *Int. Rev. Cytol.* **37:**349.
Scott, D.E., Krobisch-Dudley, G., Paull, W.K., Kozlowski, G.P., and Ribas, J., 1975, The primate median eminence, correlative scanning-transmission electron microscopy, *Cell Tissue Res.* **162:**61.
Shimamura, A., and Tokunaga, J., 1970, SEM of sensory (fungiform) papillae in the frog tongue, *SEM* **1970:**225.
Smith, K.C.A., and Oatley, C.W., 1955, The SEM and its field of application, *Br. J. Appl. Phys.* **6:**391.
Smith, M.E., and Finke, E.H., 1972, Critical point drying of soft biological material for the SEM, *Invest. Ophthalnol.* **11:**127.
Smoller, C.G., 1965, Neurosecretory processes extending into third ventricle: Secretory or sensory, *Science* **147:**882.
Spector, M., Kimzey, S.L., and Burns, L., 1974, Application of SEM and ion beam etching techniques to the study of normal and diseased red blood cells, *SEM* **1974:**665.
Spencer, P.S., and Lieberman, A.R., 1971, SEM of isolated peripheral nerve fibers: Normal surface structure and alterations proximal to neuromas, *Z. Zellforsch.* **119:**534.
Steinberg, R.H., 1973, SEM of the bullfrog's retina and pigment epithelium *Z. Zellforsch.* **143:**451.
Stuart, P.R., Osborn, J.S., and Lewis, S.M., 1969, The use of radiofrequency sputter ion etching and SEM to study the internal structure of biological material, *SEM* **1969:**241.
Sutfin, L.V., and Ogilvie, R.E., 1970, A comparison of X-ray analysis techniques available for SEM, *SEM* **1970:**17.
Sutfin, L.V., Holtrop, M.E., and Ogilvie, R.E., 1971, Microanalysis of individual mitochondrial granules with diameters less than 1000 Å, *Science* **174:**947.
Svedbergh, B., and Bill, A., 1972, SEM studies of the corneal endothelium in man and monkeys, *Acta Ophthalnol.* **50:**321.
Swedlow, D.B., Harper, R.A., and Katz, J.L., 1972, Evaluation of a new preparative technique for bone examination in the SEM, *SEM* **1972:**335.
Swift, J.A., and Brown, A.C., 1970, Transmission scanning electron microscopy of sectioned biological materials, *SEM* **1970:**113.
Swift, J.A., Brown, A.C., and Saxton, C.A., 1969, Scanning transmission electron microscopy with the Cambridge Stereoscan Mk II, *J. Sci. Instrum.* **2:**744.
Tanaka, K., 1972, Frozen resin cracking method for SEM of biological materials, *Naturwissenschaften* **59:**77.
Tanaka, T., Kosaka, N., Takiguchi, T., Aoki, T., and Takahara, S., 1973, Observation on the cochlea with SEM, *SEM* **1973:**427.
Thomas, R.S., and Hollahan, J.R., 1974, Use of chemically-reactive gas plasmas in preparing specimens for SEM and electron probe microanalysis, *SEM* **1974:**83.
Tousimis, A.J., 1969, A combined SEM and electron probe microanalysis of biological soft tissues, *SEM* **1969:**217.
Vial, J., and Porter, K.R., 1974, The surface topography of cells isolated from tissues by maceration, *Anat. Rec.* **178:**502 (Abstract).
von Ärdenne, M., 1938, Das Elektronenrastermikroskop: Theoretische Grundlagen, *Z. Tech. Phys.* **109:**553.
Wall, J., Langmore, J., Isaacson, M., and Crewe, A.V., 1974, STEM at high resolution, *Proc. Natl. Acad. Science U.S.A.* **71:**1.
Watson, J.H.L., Page, R.H., and Swedo, J.L., 1975, A technique for determining the interior topography of single cells (Colcemidblocked HeLa cells), *SEM* **1975:**417.
Watters, W.B., and Buck, R.C.,1971, An improved simple method of specimen preparation for replicas or SEM, *J. Microsc.* **94:**185.

Wells, O.C., 1972, Bibliography on the SEM, *SEM* **1972**:375.
Wells, O.C., 1974, *Scanning Electron Microscopy*, McGraw-Hill, New York.
Woolley, T.A., 1972, Some sense organs of ticks as seen by SEM, *Trans. Am. Microsc. Soc.* **91**:35.
Worthen, D.M., and Wickham, M.G., 1972, Scanning electron microscopy tissue preparation, *Invest. Ophthalnol.* **11**:133.
Yakowitz, H.,1974, X-ray microanalysis in SEM, *SEM* **1974**:1029.
Zeitler, E., 1971, Scanning transmission electron microscopy, *SEM* **1971**:25.
Zworykin, V.K., Hillier, J., and Snyder, R.L., 1942, Scanning electron microscope, *Am. Soc. Transm. Microsc.* **117**:15.

9

Autoradiography in the Nervous System

H.J. Groenewegen, N.M. Gerrits, and G. Vrensen

1. Introduction

The nervous system is a highly complex tissue. The large amount of neurons and glial cells, the apparently infinite number of possible contacts between neurons, and the intermingling of connections among different neurons make the anatomy of the nervous system extremely complex. To understand the functions of the nervous system, the unraveling of its structure is indispensable because structure and function have a more intricate relationship in the nervous system than in any other part of the body.

Our understanding of the morphology of the nervous system depends largely on the methods available to study this structure. Before the use of the light microscope, only the gross anatomy of the nervous system could be studied. Microscopic study, in combination with different staining methods (e.g., Weigert, Nissl), opened the way to more detailed knowledge, and the introduction by Golgi of impregnation of nerve cells with silver nitrate gave neuroanatomical studies a new impulse. The use of this "Golgi method" led to the monumental work of Ramón y Cajal.

The results of investigations at the beginning of this century were based mainly on the study of normal preparations. The selective silver-impregnation methods, introduced by Nauta and based on the Wallerian degeneration of axons after a lesion, enabled a more experimental approach. The Nauta

---

H.J. Groenewegen • Department of Anatomy and Embryology, Free University, v.d. Boechorststraat 7, Amsterdam, The Netherlands. ***N.M. Gerrits*** • Laboratory of Anatomy and Embryology, Department of Neuroanatomy, University of Leiden, Wassenaarseweg 62, Leiden, The Netherlands. ***G. Vrensen*** • Department of Electron Microscopy, Mental Hospital "Endegeest," Oegstgeest, The Netherlands. Dr. Vrensen's present address is: The Netherlands Ophthalmic Research Institute, P.O. Box 6411, 1005 EK Amsterdam, The Netherlands.

method and its refinements have elucidated numerous pathways in the nervous system. A number of connections could not, however, be traced for several different reasons. Some of these pathways can be demonstrated by new techniques, such as the histochemical techniques for tracing monoaminergic pathways.

The fine-structural details of neurons and glial cells have, of course, been derived from electron-microscopic studies in the last few decades. The combination of degeneration techniques with electron microscopy has unraveled connections in the nervous system at the submicroscopic level (see also Chapter 7). A direct relationship between structure and function can be made when electrophysiological recordings are combined with intracellular dye techniques. This means that after recording from single cells, these cells are visualized by intracellular staining with, for example, Procion yellow. In this way, the electrophysiology and morphology of an individual cell can be studied simultaneously (Kater and Nicholson, 1973).

Most of the experimental methods used to study the connectivity in the nervous system have one basic disadvantage. They are based primarily on pathological changes after an experimental lesion. In the last few years, techniques have been developed that make use of the physiological properties of nervous tissue. The frequent application of the anterograde autoradiographic tracing technique and the retrograde tracing technique with horseradish peroxidase (HRP) have proved that they give new information about the morphology of the nervous system. Both techniques are based on the phenomenon that materials are transported in axons of physiologically active neurons (this phenomenon of axonal transport is discussed in some detail in Section 2.4.2). A comparison of the anterograde autoradiographic tracing technique and the retrograde tracing technique using HRP with the anterograde and retrograde degeneration techniques has shown that the tracing techniques are much less dependent on differences in brain regions and species. The transport in axons of different species or regions seems to behave more universally than the pathological changes of cells and fibers after a lesion. This more universal reaction of neurons to the injection of substances such as tritiated precursors or HRP facilitates the tracing of connections with these methods. Furthermore, the autoradiographic tracing method is more "specific" than the anterograde degeneration technique because only labeled cells in the injection site account for the fiber and termination pattern in their projection regions in the nervous system. Fibers simply passing through the injected area do not play that role, as they do in a lesioned area. This facilitates the interpretation of autoradiographic tracing experiments as compared to degeneration experiments. Another advantage of the tracing techniques becomes clear when they are used in combination with electron microscopy. With the autoradiographic and HRP tracing techniques, the ultrastructure remains intact and the link between experimental and normal material is much more easily made. A more detailed discussion of the advantages and disadvantages of the autoradiographic tracing method, as compared to degeneration techniques, is given in Section 5.

The tracing of fiber pathways is not the only field of application of autoradiography in neuromorphological research. It is also used for the experimental study of neuronal development, using labeled precursors of the replication system of cells. Another application is in the study of general cell biological processes in the nerve cell, such as protein synthesis and uptake mechanisms. These applications of autoradiography in neuroanatomical studies are discussed in Sections 2.2 and 2.3.

The wide use of the autoradiographic technique in neuromorphological research makes it necessary to discuss this technique in particular. Knowledge of its basic principles, its possibilities, and its restrictions is a prerequisite for correct interpretation of autoradiographic experiments. The basic principles of autoradiography are discussed in Section 3 and the technical aspects of its application in the nervous system in Section 4.

2. *Application of Autoradiography in Different Fields of Neurobiological Research*

2.1. *Introduction*

Although the applications of autoradiography in embryology and the study of neuronal connectivity are both based fundamentally on the cellular processes of incorporation of precursors in macromolecules, they are nevertheless distinguished as being different fields of application in neurobiology. This section is not meant to give an extensive review of all these fields. Since, for this chapter, the critical application of the autoradiographic technique is focused on the tracing of neuronal connections, only superficial attention will be given to the applications in cell biology and embryology.

2.2. *Cell Biology*

Application of radioactively labeled molecules in cell biology is used in the study of various metabolic processes. Nearly all experiments in neurobiology deal with the uptake and incorporation of exogenous molecules by the neuron. Many different precursor molecules—mostly tritiated—are used: amino acids, such as leucine, for protein synthesis (Lajtha, 1975); putative amino acid neurotransmitters, such as GABA or glycine (Hökfelt and Ljungdahl, 1975); monosaccharides as precursors of glycoprotein (Kreutzberg and Schubert, 1975); and the nucleotides thymidine and uridine to be incorporated in DNA and RNA, respectively. The existence of intracellular phenomena, such as the incorporation and distribution of labeled material, provides the possibility of studying the connections in the mature as well as in the developing nervous system.

2.3. *Embryology*

Besides the application of tritiated nucleotides in the study of nucleic acid metabolism [e.g., the turnover of DNA (Haas *et al.*, 1970)], the autoradiographic technique is of great importance in the study of the development of the nervous system. In fact, it is the only technique available at present that can elucidate the origin and migration of neurons. By means of this technique, it might be possible to unravel part of the complexity of neuronal topology. For an extensive review, the reader is referred to Sidman (1970). More recently, there have appeared a number of reports on the study of the development of axonal connections during embryological or early postnatal stages (e.g., Crossland *et al.*, 1974).

2.4. *Neuronal Connectivity*

2.4.1. *Introduction*

In neuroanatomical studies, the autoradiographic technique is, at the moment, frequently used to trace connections among different areas in the nervous system. With light-microscopic autoradiography (LMA), the presence and topography of projections from one brain region to another are traced through visualization of radioactively labeled material introduced into the interconnecting neuronal pathways (Lasek *et al.*, 1968; Cowan *et al.*, 1972). In some instances, additional data can be gained concerning the specific chemical affinity, for certain precursors, of the brain regions under study, e.g., the specific uptake of labeled transmitters (Hökfelt and Ljungdahl, 1975). With electron-microscopic autoradiography (EMA), a more precise localization of the radioactive material can be made that permits, for example, conclusions about the type of synapse belonging to a certain pathway.

The possibility of studying these aspects of neuronal connectivity is based primarily on the physiological properties of neurons. Strictly speaking, three phenomena are relevant: (1) uptake of precursor molecules by the neuronal cell body; (2) incorporation of these precursors into macromolecules; and (3) transport of newly synthesized macromolecules along the axons. By injecting small radioactively labeled molecules such as amino acids or monosaccharides that are taken up by the neuronal cell body, incorporated into macromolecules such as proteins or glycoproteins, and transported down the axon and dendrites by the axonal and dendritic flow, the whole neuron is labeled and can be visualized with autoradiography (Fig. 1).

The three physiological phenomena mentioned are the basic parameters on which the neuroanatomical autoradiographic tracing technique is based. The uptake of small molecules by the neuron and the synthesis of macromolecules in the perikaryon are discussed in Section 2.2. It is important in this context that only the perikaryon takes up large amounts of precursor molecules for the synthesis of macromolecules. The synthesis itself is also situated in the neuronal perikaryon. The turnover of macromolecules in axons and dendrites necessitates the transport of newly synthesized substances

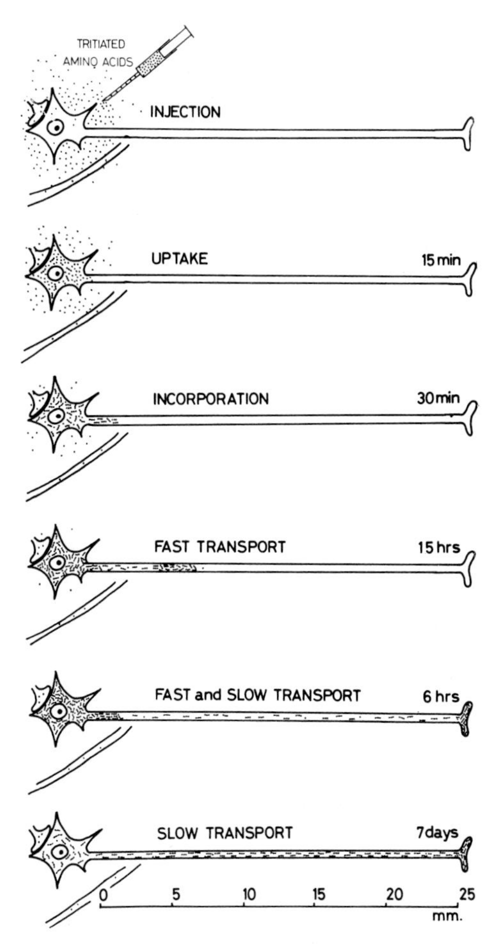

Fig. 1. Schematic representation of the basic physiological principles underlying the neuroanatomical autoradiographic tracing technique.

into these processes. The perikaryon is, in this way, the supplying and "trophic" center of the neuron. It has been shown that axons, axon terminals, and dendrites can also take up small molecules, although to a much lesser extent than the perikaryon, but it seems that these substances are not transported in detectable amounts (Fig. 1). Important features concerning the choice of precursor in relation to uptake and synthesis are discussed later. It may be emphasized here, however, that the inability to demonstrate a neuronal pathway with autoradiographic techniques does not necessarily imply that this pathway does not exist. The inability may be due to the physiological restraints of a neuron with regard to uptake or synthesis or both. Moreover, a neuron may be in a physiological state such that the amount of uptake, synthesis, or transport is too low to be detected by means of autoradiography.

Uptake and synthesis are essential for the tracing of neuronal pathways, but are not sufficient *per se*. Ultimately, only the migration of cytoplasmic constituents along the axons and dendrites enables the visualization of a pathway. In fact, the autoradiographic tracing method is based on this so-called axonal transport. As a matter of fact, the application of the autoradiographic tracing method can be considered as an additional aspect of the intensive studies of this physiological phenomenon in the last two or three decades. A short review of axonal transport may give a good understanding of the basic mechanisms of tracing methods. Moreover, understanding of the various aspects of axonal transport may contribute to a further rationalization of these methods.

2.4.2. *Axonal Transport*

2.4.2a. Discovery. Although it was already known that the perikaryon has a "trophic" influence on its processes (Ramón y Cajal, 1928), Weiss and Hiscoe (1948) were the first to show that there is a proximodistal movement of axoplasm. Ligation of a peripheral nerve causes an accumulation of material proximal to the constriction, and consequently, a distention of the nerve at this site occurs. Distal to the constriction, the nerve becomes narrower. After release of the constriction, the accumulated material moves along the nerve at a rate on the order of 1 mm per day. Weiss and Hiscoe concluded from their experiments that the axoplasmic flow not only plays a role during the growth of nerves but also is important for the maintenance of the mature nerve fiber. It should, for example, be necessary for the replacement of catabolized proteins that can be anabolized only in the nerve-cell body. Since these first observations, a great amount of literature concerning axoplasmic flow has been published.

2.4.2b. Velocities of Transport. Convincing evidence for the existence of axonal transport with a much faster velocity than that found by Weiss and Hiscoe (1948) in peripheral nerve fibers was given by Lasek (1966) and Grafstein (1967). The rate of approximately 1–5 mm per day as determined by Weiss and Hiscoe was confirmed by many other authors—for example, in cat sensory neurons (Lasek, 1968), in rat peripheral nerves (Droz and Leblond, 1963), and in the mouse optic system (Taylor and Weiss, 1965)—and is called the slow component. The rate of the fast component is estimated to be at least 100 times as fast as that of the slow component. This component seems to be temperature-dependent. In mammals, velocities up to 500 mm per day are found (Lasek, 1968; Ochs *et al.*, 1969; Ochs, 1972), whereas in cold-blooded animals, lower rates are determined (Edström and Mattson, 1972; McEwen and Grafstein, 1968). The accurate data concerning the fast transport in garfish olfactory nerve obtained by Gross and Beidler (1973) indicate a linear relationship between the rate of fast transport and temperature in certain temperature trajectories. Extrapolation of the values found between 10° and 25°C gives velocities at 37°C that can be compared with those found in mammals. This suggests the existence of a general mechanism behind this phenomenon. The slow and fast components of axonal transport seem to be extremes in a large scale of velocities at which material can be transported in axons. Strong evidence is available for the existence of one or more intermediate components (Lasek, 1970; Schonbach and Cuénod, 1971; Karlson and Sjöstrand, 1971; Willard *et al.,* 1974). Recently, Gross and Beidler (1975) concluded from a quantitative analysis of fast transport velocities that there is a maximum velocity at which material can be transported in axons, but that large amounts of material are transported at lower velocities.

2.4.2c. Transported Materials. From several studies, it can be concluded that the fast and slow components of axonal flow transport different materials. To identify the material transported, several techniques are used. The most direct is the observation of particulate material in the light microscope (Pomerat *et al.,* 1967). A more frequently used technique is the isotopic labeling of the axoplasmic flow and the identification of the labeled macromolecules that arrive at the synapse or at some intermediate place in the course of the axon at a certain time after the initial labeling. In many instances, the macromolecules are incorporated into cellular organelles such as mitochondria, tubules, and vesicles. The identification can be done, for example, with EMA or with scintillation counting. This labeling technique has been used in the visual system (Taylor and Weiss, 1965; Grafstein, 1969; Forman *et al.,* 1971), in peripheral nerves (Ochs *et al.,* 1969), and in the olfactory nerve (Gross and Beidler, 1973). It is also possible to measure the activity of accumulated enzymes at different time intervals after the ligation of a nerve (Lubińska, 1964). Biochemical techniques have been used to identify the accumulation of certain transmitter substances in constricted nerves (Dahlström, 1965; Geffen *et al.,* 1970). In general, it turns out that the slow component transports soluble proteins, whereas the fast component transports mainly particulate material (Sabri and Ochs, 1972). Rapidly transported material is destined primarily for the nerve terminals (Grafstein, 1967; Schonbach *et al.,* 1971; Hendrickson, 1972; Droz *et al.,* 1973; Bennett *et al.,* 1973), whereas slowly transported material is deposited along the axon (Schonbach *et al.,* 1973; Bennett *et al.,* 1973). Substances that seem to be exclusively transported by the fast component of axonal transport are glycoproteins, glycolipids and phospholipids, cholinesterase, and some small molecules such as amino acids. Substances identified as being transported more or less exclusively by the slow component are, for example, microtubular protein and various soluble enzymes. In the intermediate components, enzymes and different polypeptides are found. Transport of mitochondria seems to occur at different rates. A more extensive review of identified constituents in the different components of axonal transport is provided by Grafstein (1975). It must be kept in mind, however, that the methods used to analyze this phenomenon of transport have their limitations. Labeling of a structure in an axon or in an axon terminal does not necessarily imply that this structure itself is transported down the axon. Labeling experiments, for example, may suggest that organelles such as mitochondria or synaptic vesicles are transported at a high velocity through the axon into the axon terminal. With this method, however, one cannot exclude the possibility that the organelles themselves are already situated in the terminal at the time of the initial labeling and simply receive labeled, rapidly transported, macromolecules. Moreover, one cannot rule out the possibility that some macromolecules are synthesized in the axon or axon terminal from transported precursors. Furthermore, the time of appearance of labeled material in the axon or axon terminals after initial labeling depends not only on the velocity of axonal transport but also on the retention of the labeled material in the cell body prior to its transport.

2.4.2d. Retrograde Transport. The existence of transport from the axon terminals to the cell bodies, the so-called retrograde axonal transport, is of great significance for the interpretation of autoradiographs used for describing axonal connectivity. The interpretation of these autoradiographs would be much more difficult if, in addition to anterograde transport, large amounts of material are transported retrogradely as well.

Evidence for the occurrence of retrograde transport has been given by several authors. Direct observations of bidirectional displacements of organelles in axons were made by Pomerat *et al.* (1967). Substances such as acetylcholinesterase are found to accumulate at both sides of a nerve ligation (Lubińska, 1964). More recently, it has been shown that exogenous macromolecules such as horseradish peroxidase (HRP), albumin, nerve growth factor, and even viruses are transported in the direction of the nerve-cell bodies (Kristensson and Olsson, 1971; LaVail and LaVail, 1972; Kristensson, 1970*a,b*; Jeffrey and Austin, 1973). Electron-microscopic studies by LaVail and LaVail (1974) have indicated that HRP is transported intraaxonally. It appears that these substances are taken up in nerve terminals and in unmyelinated fibers by a pinocytotic process. The velocity at which HRP is transported is estimated to be about 60–120 mm per day.

At the moment, HRP is widely used in neuroanatomical tracing of connections because of its retrograde transport. In this way, the neuronal cell bodies of the terminals at the site of HRP injection are labeled (Fig. 2) (LaVail and LaVail, 1974; Kuypers *et al.*, 1974; Nauta *et al.*, 1974; Brodal *et al.*, 1975). This technique of retrograde tracing has several advantages as compared to the retrograde degeneration techniques (LaVail, 1975). The enzymatic demonstration of HRP, for example, is very sensitive and easy to perform. There are also some disadvantages. There is strong evidence that HRP is also taken up by disrupted fibers (Kristensson, 1975), so that fibers of retrogradely HRP-labeled cells do not necessarily terminate in the area that is injected with HRP. Moreover, HRP is also taken up by cell bodies and transported in the anterograde direction (Repérant, 1975; Walberg *et al.*, 1976). However, retrogradely labeled cells have a granular appearance (Fig. 2), with HRP located in vesicles, whereas anterogradely transported HRP has a more diffuse distribution in the cell bodies. So the anterogradely transported HRP has a different appearance and can be distinguished from retrogradely transported HRP. When axonal connections are studied with

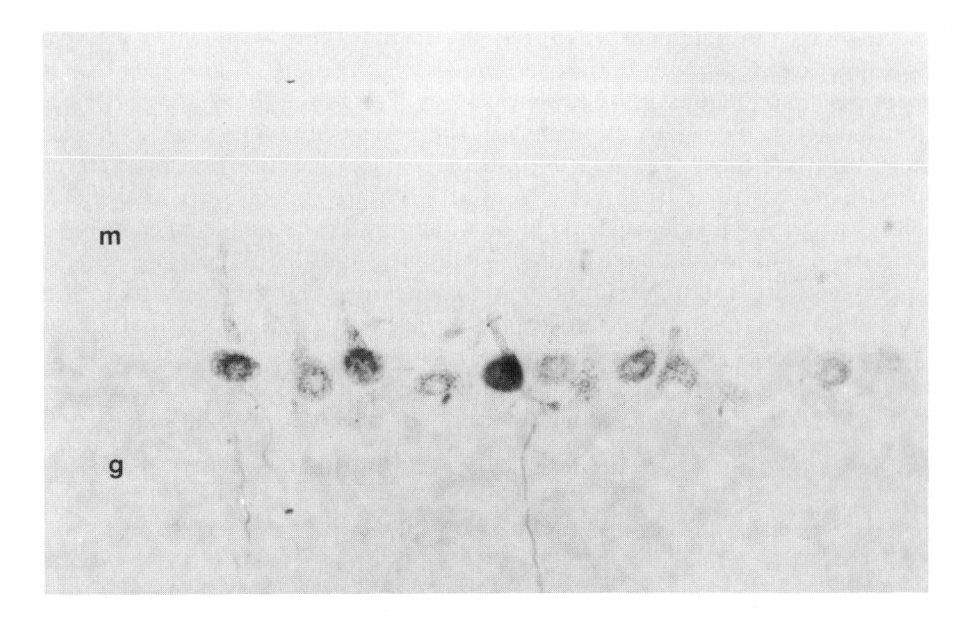

Fig. 2. Cerebellar Purkinje cells, retrogradely filled with HRP after an injection into the lateral vestibular nucleus (Deiters's nucleus). Note the granular aspect of the HRP reaction product in the cell bodies and their main dendrites in the molecular layer (m). (g) Granular layer. Photograph kindly provided by Dr. F. M. Bigaré.

the method of retrograde transport, it must be kept in mind that there might be a certain specificity of cells or systems in the uptake or transport, or both, of different exogenous macromolecules, including HRP (Bunt *et al.,* 1976). In some instances (e.g., Nauta *et al.,* 1974), connections that are clearly demonstrable with other anatomical methods could not be shown using HRP as a marker. At present, there are no indications that the precursors used for the anterograde autoradiographic tracing technique are transported in large amounts in the retrograde direction. Some authors (e.g., Lasek, 1970) have shown that originally anterogradely transported labeled proteins can reverse their direction and be transported retrogradely. However, since the neuronal perikaryon is the main site of protein synthesis, this retrograde transport from terminals or passing fibers does not result in radioactive labeling of large amounts of proteins. Thus, although retrograde transport of proteins occurs, this phenomenon will not interfere with the tracing and detection of the anterogradely transported material and does not have serious complications for the autoradiographic tracing technique.

2.4.3. Dendritic Transport

Besides the existence of axonal transport, there are also strong indications for dendritic transport. Intracellular application of various tritiated precursors and visualization by means of autoradiography (Globus *et al.,* 1968;

Kreutzberg and Schubert, 1975) has shown that there is an active dendritic transport of proteins and other substances, including RNA and acetylcholinesterase. Exogeneous substances such as Procion Yellow and HRP are transported in dendrites as well. With the refined methods used by Kreutzberg and Schubert (1975), it is possible to visualize the fate of different precursors in the neuron. Precursors of RNA are concentrated and metabolized primarily in the nucleus, after which the labeled molecules enter the cytoplasm and are transported into dendrites. Labeled precursors for mucopolysaccharides are first found in patches in the cytoplasm, suggesting concentration in some specific cell organelles, and are later transported into axons and dendrites. There is evidence that these substances are also released in the extracellular space by "dendritic secretion." Transport velocities in dendrites are estimated to be on the order of 70–80 mm per day, comparable to the velocity of fast axonal transport.

2.4.4. *Transneuronal Transport*

An important prerequisite for the exact tracing of connections by means of autoradiography is that the majority of labeled material transported from the cell body into the axons and the axon terminals does not leak out of the originally labeled neuron. It should be mentioned, however, that several investigators have found indications that labeled substances do leave the initially labeled neuron. Grafstein (1971) has found, using liquid scintillation counting, that after precursor injections into the eye of the mouse, a small amount of label was present in the visual cortex. This can have reached that area only by transneuronal transport in the lateral geniculate body. In autoradiographs, this transneuronal transport has been confirmed by Specht and Grafstein (1973), Dräger (1974), Wiesel *et al.* (1974), and others. At the cellular level, a similar process of transfer from one neuron to another was found by Globus *et al.* (1968). Their results suggest a specific transneuronal labeling, probably by dendrodendritic exchange (Kreutzberg and Schubert, 1975). The experiments in the visual system indicate that at least part of the labeling beyond the primary, labeled neuron must be ascribed to rather unspecific leakage or diffusion. Not only are the "secondary" neurons, including their axons and terminals, in the termination area of the "primary" labeled neurons found to be labeled, but also simple diffusion of label occurs to adjacent areas (Specht and Grafstein, 1973; Dräger, 1974). The results of Wiesel *et al.* (1974) and Dräger (1974), on the other hand, suggest that a specific transsynaptic transfer of label can also take place. At this time, these transneuronal effects have been described mostly in the visual system, using tritiated proline and fucose as precursors. These precursors are injected in rather high amounts into the eye. It is possible that in other neuronal systems a similar transsynaptic transfer takes place, but that the amount of label is not sufficient to be detected by means of LMA.

The chemical nature of the substances that are transported and the function of this transsynaptic phenomenon are unknown at the moment.

The trophic effect of a neuron on its postsynaptic target may be mediated by this phenomenon.

2.4.5. *Mechanisms of Transport*

Several hypotheses have been suggested to explain the mechanisms underlying this axonal transport. For the slow component, a sort of microperistalsis is proposed by Weiss (1964). The axolemma probably produces the peristaltic movements. Lubińska (1964) interprets the slow component as the result of movement away from the cell body of fast bidirectional streaming of the axoplasmic constituents. Others interpret the slow movement of proteins in the axoplasma as the result of the continuous growth of microtubules, microfilaments, and axolemma. The cell body is thought to function as a growth center in this process (McEwen and Grafstein, 1968).

The fast transport of particulate material in axons, however, cannot be explained by microperistaltic movements. Some hypotheses have been postulated to explain the mechanism of this component of the axonal transport. These hypotheses all have in common the thought that microtubules play an important role in this fast transport. Cell biologists have suggested that microtubules are involved in movements or streaming in various types of cells. The involvement of microtubuli in fast axonal transport is supported by many observations. The interference of colchicine, which binds to microtubular proteins, with fast transport is one of these indications (Kreutzberg, 1969; Dahlström, 1968). Colchicine also stops transport in dendrites (Schubert *et al.,* 1972*a*). At temperatures at which the microtubules are known to disintegrate, fast transport stops. With EMA, labeled transported material is preferentially found in the microtubular region of the axons (Droz *et al.,* 1973). Fast axonal transport depends on the availability of oxygen and ATP as suppliers of metabolic energy. The linear relationship between fast transport velocity and temperature in certain trajectories (Gross and Beidler, 1973) and the fact that extrapolation of these data to 37°C gives values that approximate those found in mammalian nerve fibers suggest that the mechanism of axonal transport in cold- and warm-blooded animals is the same.

Schmitt (1968) suggested that substances are transported along the axons in vesicles. The microtubules and the membranes of these transport vesicles are thought to interact in a manner similar to the interaction of sliding filaments in a muscle fiber. A comparable idea has been put forward by Ochs (1971), who proposes that there are specific "transport" filaments to which substances are bound. Through interactions of these "transport" filaments and the microtubules, the substances are transported in axons.

An attractive hypothesis, not based on a very specific "transport" element, was recently postulated by Gross (1975). This so-called "microstream concept" of transport in neurons explains many of the observed properties of axonal and dendritic flow. Gross and Beidler (1975) concluded from their quantitative analysis of the distribution of labeled molecules in the long and homogeneous olfactory nerve of the garfish, brought about by fast axonal flow, that axonal and dendritic transport can be compared with a chromatographic process. This means that there is a carrier phase and a stationary phase in the axoplasm. It was noted by Gross and Beidler that the amount of material transported by the fast component diminishes with increasing distance from the nerve-cell body. This would occur because material moves out of the carrier phase, having a low viscosity, and is absorbed by the more viscous, stationary phase. From this stationary phase, material can reenter the carrier phase or can be permanently bound to stationary structures such as the membrane of the axon. Some material may even leave the axon or can be picked up by the retrograde transport. Gross (1975) denotes this carrier medium in which material is transported by the term "microstreams." These "microstreams" are thought to be located around microtubuli, and the velocity of streaming is highest near the microtubular surface. The energy needed for transport is supplied by ATP, hydrolized by an ATPase, located on the microtubular surface. According to Gross, this configuration can also account for a kind of directional release of energy. This hypothesis needs further investigation, but it has some advantages in comparison with other hypotheses. The microstreams can account for all streaming processes in the cytoplasm of various cells, not only neurons. This means that axonal transport is not a unique phenomenon. Furthermore, the suggested continuum of transport velocities can be explained. The velocity at which a substance is transported depends on its affinity for the carrier or stationary phase and on the time for which it is bound to one or both. This means that the rate of transport is limited only by a maximum velocity and that theoretically all intermediate rates are possible. The fast transport of organelles such as vesicles or mitochondria can be understood as the cooperative action of many microtubules, arranged around such organelles. In electron-microscopic studies, such an arrangement has been described.

3. Basic Principles of Autoradiography

3.1. Introduction

Radioactive isotopes of biologically interesting atoms, i.e., hydrogen, carbon, oxygen, phosphorus, are now commonly used to label precursor molecules for biochemical and pharmacological research. The cellular incorporation of the precursors into molecules and macromolecules is a powerful tool for tracing biosynthetic pathways. A common way to measure the rate of incorporation is by liquid scintillation counting. A prerequisite for this type of measurement is that the tissue is disrupted by homogenization and that the relevant molecules are isolated and purified. This description implies that the cellular localization of the labeled molecules is lost. If spatial information regarding the incorporated isotopes is important, autoradiography is the technique of choice. It enables analysis of the cellular distribution of labeled compounds, and it allows the attacking of dynamic problems in cell metabolism by the study of the changes of labeling with time. An autoradiograph consists of a radioactive slice of tissue covered by a layer of photographic emulsion (Fig. 3). The photographic emulsion is sensitive to the radioactive decay of the isotopes. After suitable treatment, this interaction with the isotope will give rise to visible silver grains. This enables the simultaneous registration of the light- or electron-microscopic structure of the tissue and the sites of radioactive decay. No inferences can be made about the chemical nature of the radioactive molecules. Therefore, a relevant autoradiographic experiment must be supported by biochemical knowledge of the radioactive molecules involved.

From this basic starting point, four prerequisites of an autoradiographic method can be formulated:

1. The localization of silver grains must be related to the radioactive sources in the specimen (as accurate as possible: *resolution*):
2. The number of silver grains over the specimen must be related to the number of isotopes in the specimen (as efficient as possible: *efficiency*):
3. The possible specific relationship between silver grains and some structural elements in the tissue must be analyzed (*analysis*).
4. The information regarding the structural organization of the tissue must be good.

In this section, we shall briefly outline the factors that determine resolution, efficiency, and analysis of autoradiographs. The properties of the isotopes and the photographic emulsion play a crucial role and will therefore be described in some detail.

3.2. Radioisotopes

Most isotopes of biological interest, as listed in Table 1, emit negatively charged electrons (β^-). The radioactive decay of these electrons is polyenergetic, and each isotope has a specific spectrum [for tritium (^3H), see Table 2.]. As indicated in Table 1, the maximal initial energy of the different isotopes varies from 6.5 to 1710 keV. This initial energy is an important factor, for it determines the interaction of the emitted electrons with matter. The loss of energy as the electrons travel through the specimen and the

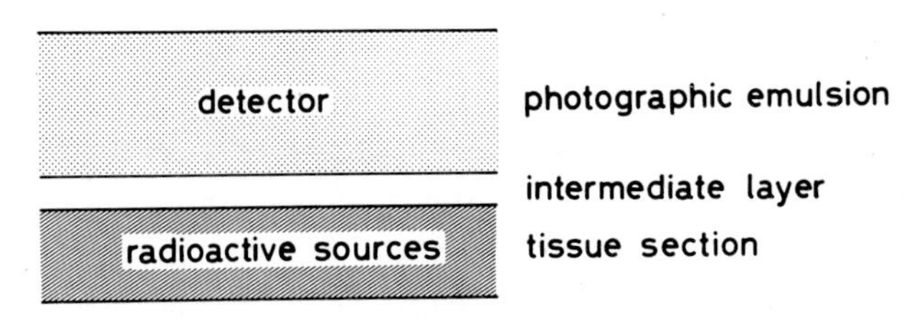

Fig. 3. Schematic representation of a radioactive slice covered by a layer of photographic emulsion.

Table 1. Some Physical Data on Radioisotopes Used in Autoradiography

Radioisotope[a]	Symbol[a]	Half-life[a]	Maximal initial energy (keV)[a]	Absorption coefficient (cm^2/mg)[b]	Maximal range (μm)[b]
Iron-55	^{55}Fe	4 yr	6.5	—	—[c]
Tritium	^{3}H	12.3 yr	18.5	17.85	6.55
Iodine-125	^{125}I	60 days	24	—	—[c]
Carbon-14	^{14}C	5760 yr	155	0.27	190
Sulfur-35	^{35}S	87 days	167	0.21	250
Calcium-45	^{45}Ca	165 days	250	0.12	550
Sodium-22	^{22}Na	2.6 yr	540	0.04	1910
Iodine-131	^{131}I	8 days	610	0.04	2060
Chlorine-36	^{36}Cl	3×10^5 yr	714	0.03	2200
Phosphorus-32	^{32}P	14.2 days	1710	0.01	7080

[a] From Rogers (1973) and Parry and Blackett (1973).
[b] From Schmeiser (1957). Ranges are in matter with a density of 1.1.
[c] Iron-55 and iodine-125 emit Auger electrons.

photographic emulsion is inversely related to this initial energy. As indicated in Table 1, this means that the absorption decreases with increasing initial energy and that the maximal range increases with increasing initial energy. For tritium, the maximal range is about 6.5 μm, and the majority (about 75%) of the electrons have a range up to about 1.5 μm (see Table 2). As a consequence of this, sections of ^{3}H-labeled material, as well as of ^{55}Fe- and ^{125}I-labeled material, in excess of 2 μm are of no value for autoradiographic response, because the electrons emitted from deeper than 2 μm in thickness do not reach the emulsion layer. Moreover, the emulsion layers can be kept thin, for the electrons are stopped within the first layers of silver halide crystals. For phosphorus, even sections in excess of 10 μm in thickness and very thick emulsion layers increase the autoradiographic response.

Table 2. Spectrum of Tritium-Emitted Electrons

Initial energy (keV)[a]	Emitted electrons in interval (%)[a]	Range (μm)[b]
0–2	20.1	<0.14
2–4	21.2	0.14–0.45
4–6	18.9	0.45–0.91
6–8	15.2	0.91–1.49
8–10	11.3	1.49–2.16
10–12	7.6	2.16–3.06
12–18	5.4	3.06–6.55

[a] From Feinendegen (1967).
[b] From Perry (1964). Ranges are in matter with a density of 1.3.

3.3. Photographic Emulsions

The photographic emulsions used in autoradiography consist of silver-halide (AgBr and AgI) crystals embedded in gelatin. The water in which the gelatin is dissolved also contains traces of glycerol and sensitivity-increasing compounds. The size of the microcrystals ranges from 0.05 to 0.4 μm (see Table 3). The size of the microcrystal is an important factor for resolution (see Section 3.4). In general, the smaller the crystals, the better the resolution. These small crystals are used in electron-microscopic autoradiography (EMA). The larger-sized crystals give less resolution and are used in light-microscopic autoradiography (LMA). On the other hand, the size of the crystals affects the sensitivity. In general, the larger the crystals, the more sensitive the emulsion.

Silver halide crystals are crystals in a physical sense, and their crystal lattice has specific properties. When they are exposed to light, heat, or radioactive disintegrations, some energy is absorbed. This absorption of energy leads to the reduction of some Ag ions, which accumulate as Ag atoms at physical imperfections in the crystal lattice. These specks of Ag atoms are called *latent images*. At high temperatures and long times, all silver halide crystals are reduced to silver grains by a strong reducing agent, i.e., a *photographic developer*. At moderate temperatures and relatively short times, only those silver halide crystals that contain one or more latent images will be reduced. Thus, the formation of latent images by energy absorption increases the chance that a silver halide crystal will be developed and form a silver grain. Treatment with a silver-halide-dissolving agent, i.e., a *photographic fixative,* leads to the disappearance of nonexposed crystals and leaves the silver grains in the emulsion.

The fact that the silver halide crystals are sensitive to all kinds of energy, such as heat, light, charged particles, and chemical energy, has an important disadvantage for autoradiography. It implies that experiments have to be

Table 3. Commercially Available Emulsions for Autoradiography[a]

Manufacturer	Code	Crystal diameter (μm)	Isotopes that can be used[d]	Used in
Ilford[b]	Go-G5	0.27	^{3}H–^{32}P	LMA
	Ko-K5	0.20	^{3}H–^{32}P	LMA
	Lo-L4	0.15	^{3}H–^{32}P	LMA, EMA
Eastman-Kodak	NTB-4	0.40	^{3}H–^{32}P	LMA
	NTA, NTB, NTB-2, NTB-3	0.30	$^{3}H^{b}$–$^{35}S^{b}$	LMA
	NTE	0.04	^{3}H–^{135}I	EMA
Kodak (England)	AR-10	0.20	^{3}H–^{125}I (^{14}C)	LMA
Agfa-Gevaert[c]	Nuc 715	0.15	^{3}H–^{32}P	LMA, EMA
	Nuc 307	0.07	^{3}H–^{14}C	EMA

[a] From Barkas (1963) and Rogers (1973).
[b] Within each series, emulsions with different sensitivities are available. The degree of sensitivity restricts the isotopes that can be used.
[c] The nuclear emulsions of Agfa-Gevaert are no longer commercially available.
[d] The range of isotopes as listed in Table 1 is indicated.

carried out to control whether the grains over the sections are the result of radioactive disintegrations or are due to other forms of energy dissipation. Heat and light can be very well controlled. However, the loss of chemical energy from the section to the emulsion, *positive chemography,* is more difficult to detect. Another factor that has to be properly controlled is the disappearance of the latent images after long exposure times, *fading,* or as the result of chemical interaction between section and emulsion, *negative chemography.* When no silver grains are present over radioactive sections, the possibility of fading or negative chemography must be controlled. Detailed descriptions and discussion of these types of control experiments are given by Rogers (1973).

3.4. Resolution of Autoradiographs

A point source, containing an infinite number of radioactive molecules, will emit electrons in all directions. The grain density in the emulsion over this point source will decrease as a function of the distance from the source (Fig. 4). According to Doniach and Pelc (1950), resolution can be defined as the distance (d) between a point source and the place where grain density has dropped to 50% (Fig. 4A). Caro (1962) has further evaluated this definition by introducing the Rayleigh criterion from classic optics into this field. According to Caro, resolution is defined as the distance between two point sources from which the grain distributions can be distinguished and is 2 times the d ($2d$) of Doniach and Pelc (Fig. 4*B*).

Sources with many radioactive molecules are not uncommon at the light-microscopic level, and the definitions of Doniach and Pelc and of Caro can be used in LMA. In EMA, however, the microscopic magnification and section thickness are such that the sources are mainly isolated single radioactive molecules. In this case, resolution can be defined as the radius (r) of the circle within which the chance is 50% that the emitted electron will hit the emulsion and form a silver grain (Fig. 5). This definition, introduced by

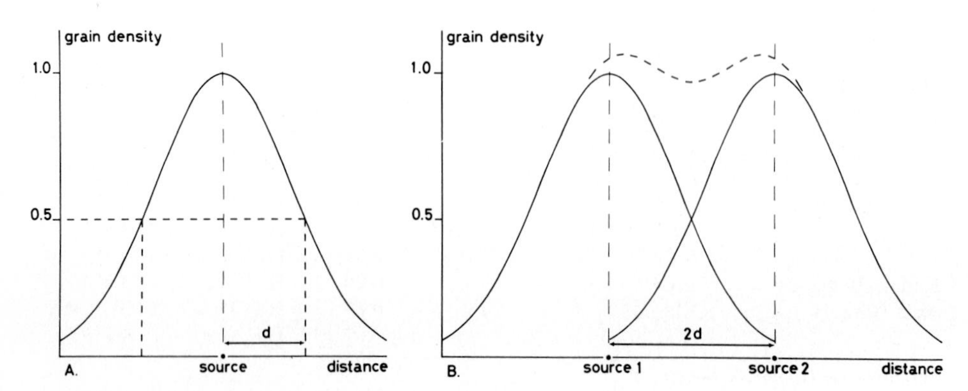

Fig. 4. Distribution of grains around one point source (A) or two point sources (B). For explanation, see Sections 3.4 and 3.4.1.

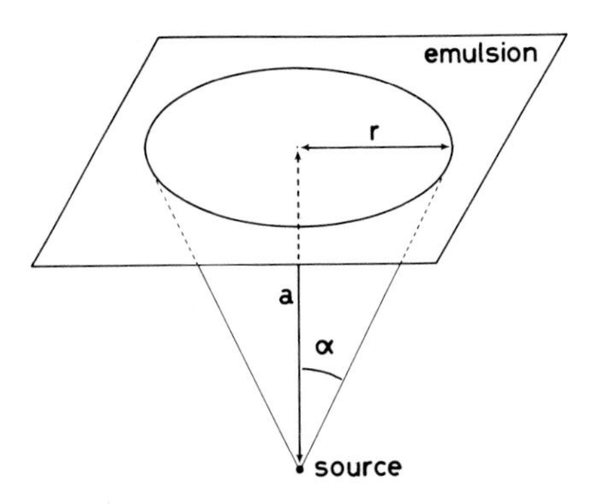

Fig. 5. Illustration of some parameters involved in the resolution of EM autoradiographs. For explanation, see Sections 3.4 and 3.4.1.

Salpeter and Bachmann (1964), is very useful for the analysis of EM autoradiographs (see Section 4.3.4), and is called half-radius (HR).

Salpeter *et al.* (1969) introduced another definition of resolution, namely, half-distance (HD). This definition is now commonly used and has been empirically tested for a number of emulsions and radioactive isotopes. The HD is given by the distance between a line source and that site in the emulsion within which 50% of the grains derived from the source are observed. The relationship between HR and HD is given by 1.7HD = HR.

There are two main factors that affect resolution: the geometry of the autoradiograph and the photographic process.

3.4.1. Geometric Factors

The geometric properties of autoradiographs are the main factors that determine the final resolution of the autoradiographic procedure. Consider the point source in Fig. 5. The percentage of electrons emitted by this source, in the direction of the emulsion within a cone with half-angle α, is proportional to $1 - \cos \alpha$. These electrons will hit an arbitrary emulsion plane within a circle with a radius given by $\tan \alpha = r/a$, in which a is the distance between the source and the emulsion plane. Taking into account the 50% of the electrons emitted in the direction of the emulsion, $1 - \cos \alpha = 0.5$ and $\alpha = 60°$. Tan α is then 1.7 and $r = 1.7 \cdot a$. This means that (the half-radius) HR $= 1.7 \cdot a$, and HD $= a$, since 1.7HD = HR. On the average, the distance between point sources in the specimen and the effective emulsion plane is given by half the section thickness ($\frac{1}{2}S$), the thickness of the intermediate layer (I), and some portion of the emulsion thickness [$(1/x)E$], so that $a = \frac{1}{2}S + I + (1/x)E$ (see Fig. 3). This theoretical approach clearly indicates the importance of the thicknesses of the section, intermediate layer, and emulsion layer in autoradiographic resolution.

This theoretical approach to resolution presupposes that the emitted electrons all reach the emulsion plane and that they all give rise to latent images and silver grains. Vrensen and De Groot (1970) were the first to describe the fact that tritium-emitted electrons are absorbed in the specimen even in the ultrathin sections used in electron microscopy (0.05–0.10 μm).

This has been confirmed by Salpeter *et al.* (1974). As a consequence of this specimen absorption, the effective section thickness for tritium autoradiography is less than $\frac{1}{2}S$. This decrease in effective section thickness is more pronounced the thicker the tissue sections. Moreover, obliquely emitted electrons have a greater chance of being absorbed in the section and therefore have a smaller chance of reaching the emulsion. Low-energy electrons are absorbed within the first layer of silver halide crystals, so that the effective emulsion thickness is less than half the crystal diameter of the emulsion ($<\frac{1}{2}E$). With high-energy radioisotopes, e.g., ^{14}C, the situation is quite different. The chance that they will be stopped within the specimen is small, and $\frac{1}{2}S$ is a correct approximation of their effective localization in the specimen even with sections up to 10 μm. Moreover, their energy content is so high that they travel through the first and second silver halide crystals without giving rise to a latent image and silver grain. Moreover, obliquely emitted electrons lose more energy in the specimen, and therefore their chance of forming a latent image in the first or second layer of crystals increases. This means that the effective emulsion plane for ^{14}C is between $\frac{1}{2}E$ and 1. From this short description of the relationship between resolution and the energy content of the radioactively emitted electrons, it can be predicted that a better resolution will be obtained for tritium than theoretically expected, especially with sections thicker than 0.5 μm, whereas for ^{14}C the theoretically expected and the actually measured resolution will correspond. That this is the case was recently shown by Salpeter *et al.* (1974) for sections up to 1 μm thick.

3.4.2. Photographic Factors

The photographic factors that influence resolution are due to two uncertainties in the relationships between electron hits and final silver grains (Fig. 6). The impact of the electron and the location of the latent image can vary from zero to the diameter of the silver halide crystal. The relationship between the latent image and the center of the silver grain is also subject to

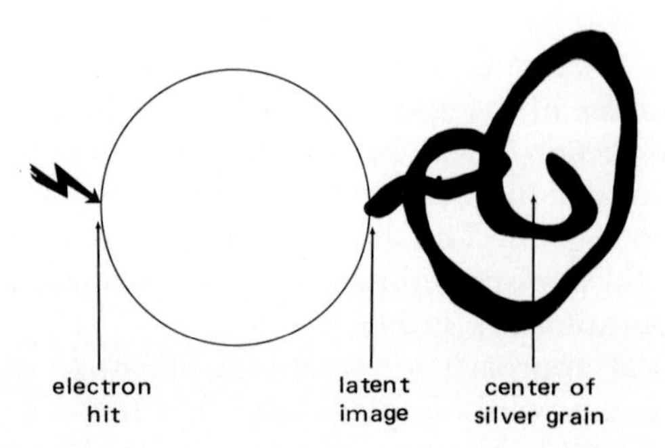

Fig. 6. Spatial relationship among some photographic factors that influence the position of the silver grain in respect to that of the radioactive source in the case of EMA. For explanation, see Section 3.4.2.

statistical variation. At maximum, it is half the size of the silver grain. In an empirical sense, these photographic factors are negligible in LMA and are of minor importance, as compared to the geometric factors, in EMA.

Autoradiographic resolution is a complex problem that has to be approached with great care. For low-energy radioisotopes, e.g., ^{3}H, ^{55}Fe, and ^{125}I, a resolution (HD values) of 0.15 μm can be obtained for ultra-thin sections (0.05–0.10 μm) and of 0.35 μm for sections thicker than 0.5 μm. For high-energy electrons, e.g., ^{14}C, a resolution of 0.18 μm at the EMA level can be obtained, whereas at the LMA level, resolution depends greatly on section thickness (S) and emulsion thickness (E) and can be approximated by $HD = \frac{1}{2}S + \frac{3}{4}E$. For more detailed descriptions, the reader is referred to, for example, Rogers (1973) and Salpeter *et al.* (1969, 1974).

3.5. *Efficiency of Autoradiographs*

Just as it is relevant to know the spatial relationship between labeled molecules in the specimen and silver grains over the specimen (resolution), it is also of interest to know the numerical relationship between the labeled molecules and the silver grains. This numerical relationship is often indicated as efficiency and can be defined as "the emulsion response (number of silver grains) relative to the number of disintegrations" (Rogers, 1973). A most obvious factor that influences the efficiency, and one that is often forgotten, is that in most autoradiographs, the radioactive sections are covered with emulsion on one side only. This means that statistically only half the electrons are emitted in the direction of the emulsion and have a chance to be detected by the emulsion.

There are two more factors that affect efficiency: isotopic factors and photographic factors.

3.5.1. *Isotopic Factors*

The isotopic factors are all related to the initial energy of the emitted electrons. The low-energy electrons from, for example, ^{3}H, ^{55}Fe, and ^{125}I have high absorption coefficients (Table 1) and are easily stopped in the specimen. For these isotopes, an increase in section thickness is not proportional to an increase in the number of electrons hitting the emulsion, and a decrease in efficiency with increasing section thickness can be expected. This decrease in efficiency was experimentally verified by Falk and King (1963) for sections thicker than 0.5 μm and by Vrensen and De Groot (1970) for sections ranging from 0.04 to 1.0 μm thick. The drop in efficiency is from 36% at 0.04 μm to 15.7% at 0.5 μm to 2.0% at 10 μm. Maurer and Primbsch (1964) and Perry *et al.* (1961) have shown that the absorption of low-energy radiation in the tissue is also affected by the density differences of cellular compartments. For high-energy isotopes, absorption in the specimen is less pronounced, e.g., for ^{14}C: 5.5% for 1-μm sections and 30% for 10-μm sections (Rogers, 1973). For these isotopes, the number of electrons hitting the emulsion is directly proportional to the thickness of the section, up to 5 μm or more, and consequently efficiency remains constant.

The initial energy of the emitted electrons also affects the number of silver grains formed in the emulsion. Low-energy electrons hitting the emulsion are absorbed completely in the first few crystals. They dissipate enough energy to form latent images in these crystals and to give rise to silver grains. Consequently, the numerical relationship between low-energy electrons hitting the emulsion and silver grains is greater than 1. A practical consequence is that for low-energy radiation, emulsion thickness in excess of a few crystals does not further increase the emulsion response. High-energy electrons, on the other hand, hit the first few crystals without dissipating enough energy to give rise to latent images. This means that with thin emulsion layers, the photographic response to an electron hit is less than 1. For these isotopes, an increase in the thickness of the emulsion layer is paralleled by an increase in efficiency. With very thick emulsion layers, this results in the formation of many silver grains along the track of the electron. For details of track autoradiography, see Rogers (1973).

3.5.2. *Photographic Factors*

Besides the actual thickness of the emulsion layers (Section 3.5.1), the dimensions and packing of the silver halide crystals affect efficiency as well. In EMA, it is common practice to use monolayers of silver halide crystals. Even when the crystals are optimally packed, 10–20% of the area is not covered. Therefore, emitted electrons can travel through this layer without hitting a crystal. This escape decreases efficiency. Another important factor that determines efficiency is the sensitivity of the emulsion. The numerous emulsions available all have their own properties with respect to this. An optimal choice for an autoradiographic experiment can be made only on empirical grounds. The detailed information given by Rogers (1973) may be very helpful. Exposure time does not in general, affect efficiency. In exceptional cases, when dealing with high local densities of radioactivity, a long exposure time may give rise to multiple hitting of a silver halide crystal. A multiply hit crystal still develops as one silver grain, and therefore the ratio of silver grains to disintegrations decreases. The choice of developer is also of great importance. Developers are available that develop only large latent images and others that develop even the smallest latent images. The optimal developer for each emulsion and the optimal development conditions have to be controlled empirically for each set of autoradiographic experiments. The conditions of exposure may affect the efficiency insofar as the latent images formed during exposure may disappear, or fade. This fading can be caused by, for example, atmospheric humidity in combination with oxygen, and it increases with increasing exposure time. By exposure in dry air or dry nitrogen, this fading can often be avoided. Fading of latent images can also result from negative chemography, i.e., the breakdown of latent images as a result of interaction with certain molecules or atoms in the specimen (e.g., OsO_4 in EMA sections). This negative chemography can often be prevented by covering the specimen with a thin inert layer (e.g., carbon, polyvinylchloride) prior to emulsion coating. This layer also protects the emulsion against positive chemography.

An important source of error in autoradiography is the presence of *background*, i.e., the occurrence of silver grains not related to the radiation from the specimen (see Section 4.3.2). This can be caused by many factors, which are extensively described by Rogers (1973). The most obvious factors that introduce background grains are light, pressure, positive chemography, contamination of the emulsion, environmental radiation, and formation of spontaneous latent images. For optimal autoradiographs, these factors have to be carefully controlled.

4. Methods

4.1. Administration of Labeled Compounds

The mode of administration of radioactively labeled precursors depends greatly on the field of research as well as on specific requirements such as the number of cells at the site of injection and the intended biological resolution of the autoradiograph. It should be clear, for example, that it is not appropriate to start with refined techniques when one is interested primarily in the gross connectivity pattern of a nuclear complex somewhere in the central nervous system. In this section, four different types of administration will be discussed.

4.1.1. Overall Labeling

Overall labeling of experimental animals by means of intraperitoneal or intravenous injection is the simplest route of precursor administration. In most cases, this can be done by direct injection of commercially obtained labeled compounds. This method is often used to investigate the uptake and distribution of various substances in different parts of the body. Either the whole animal (see Section 4.2.2) or specific organs can be used for further processing. A particular application is the injection of tritiated nucleotides into pregnant animals. In this case, the labeled material is also taken up by the embryo and incorporated into the nuclei of dividing cells.

4.1.2. Intraventricular Injection

Injection of radioactive substances into the ventricular system of the brain is an attractive tool for studying ependymal-uptake mechanisms and connectivity of neurons lying close to the ventricular surface. It is also used in investigations concerning the kinetics of the ventricular fluid.

4.1.3. Microinjection

The application of precursors by microinjection has increased rapidly in recent years. The most frequently used precursors are [^{3}H]leucine and [^{3}H]proline. This method gives rise to a number of problems. In the first place, damage to tissue can occur. A quick injection of as little as 1 μl can cause considerable swelling and proliferation of glial cells. Second, the injected amount of radioactive material must be large to give sufficient labeling of the target. Finally, the area of the injection site is of great practical importance. There are a number of factors that determine this. Most important, there exists an equilibrium between the rate of release of the precursor from the needle and the rate of uptake of the precursor by the cells. The rate of release is determined by the speed of injection, the injected volume, and the precursor concentration. Although passive uptake by diffusion must not be excluded, the rate of uptake by the cells is determined mainly by the affinity of the cells for a certain precursor. Diffusion of precursor to surrounding areas can never be fully excluded. It should be clear, however, that if the rate of release greatly exceeds the rate of uptake, the injection site will become large due to this diffusion.

The problems indicated in these three points can be overcome simultaneously by slow injection of small volumes of highly concentrated precursors. Diffusion of precursors depends on the specific organization of the part of the nervous system where the injection is made. In homogeneous brain tissue, the site will be spherical, whereas in unhomogeneous regions, there is a tendency for diffusion *along* fibers but not *across* closely packed fiber bundles. Figures 7a and 7b illustrate some of the factors that determine the size of an injection site. The two experiments were identical except for the injected precursors, which were [^{3}H]leucine and [^{3}H]glycine, respectively. The areas of the injection sites are very different in the two cases. Glycine is probably taken up at a much slower rate than leucine, thus leading to an enormous spread of precursor throughout the tissue as a result of diffusion.

4.1.3a. Injection Apparatus. In general, the injection can be performed with a 1- or 5-μl microsyringe. Because of the requirement for slow injection, it is rather impractical to advance the plunger of the syringe manually. An elegant solution for well-controlled injections is given in Fig. 8. It consists of an assembly of two 5-ml syringes (S1 and S2) interconnected by tubing (T) filled with oil. In this way, pressure on the plunger of S1 is transferred to the plunger of the microsyringe (MS). The plunger of S1 can be advanced by means of an automatic infusion apparatus. The MS and S2 can be placed in a stereotactic holder. Other techniques, such as lowering the plunger of the microsyringe by means of a directly connected microdrive, are less favorable, since they often give rise to unintended movements of the needle in the tissue.

4.1.3b. Needles and Refinements. The outer diameter of a standard 1-μl syringe needle is usually about 0.5 mm. Although this may seem relatively thick, in most cases the needle track shows remarkably little gliosis. For

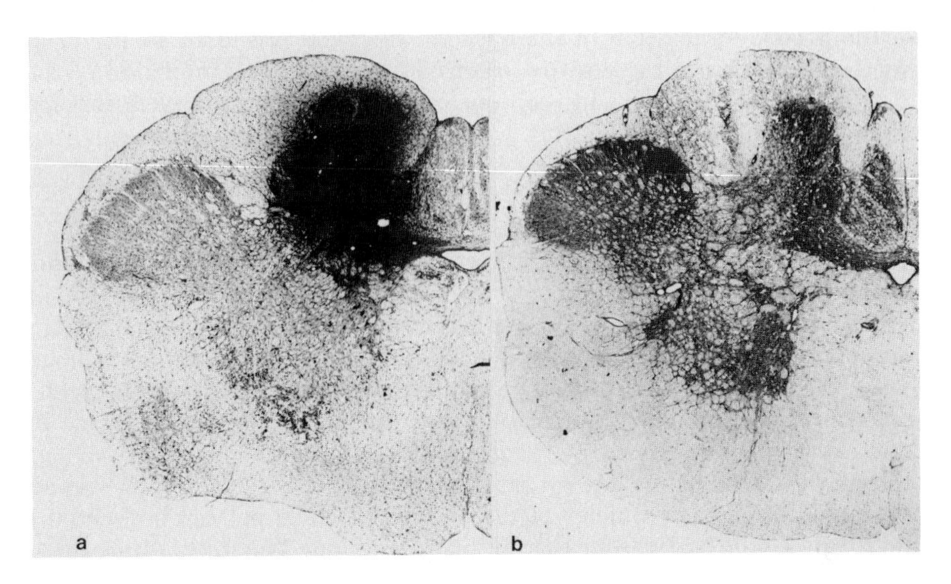

Fig. 7. Low magnification of two injection sites in the dorsal column nuclei. The injections were identical except for the injected purecusor: (a) [^{3}H]leucine; (b) [^{3}H]glycine. Note the different extent of the injection site in the two cases.

injections close to the surface or in delicate regions of the brain, the large diameter of the needle may be disadvantageous. In these cases, one can mount on the tip of the microsyringe needle a thinner needle or the point of a micropipette by means of a drop of Araldite. Another method is to connect the needle to a micropipette with the aid of a polyethylene tube (Lasek *et al.*, 1968; Schubert and Holländer, 1975).

4.1.3c. Preparation of Precursors. Commercially available precursors, e.g., tritiated amino acids, usually have activities of 1 mCi/ml. To obtain sufficient labeling of the target structure, it is often advisable to use higher concentrations. The stock solution can be dried under a flow of nitrogen, eventually together with gentle heating (40–50°C). Subsequently, the precursor can be redissolved in either saline or distilled water to a final concentration of 10–100 mCi/ml. Concentrated solutions of tritiated precursors are unstable and must be injected within a day.

4.1.3d. Injection Procedure. After the needle is filled and wiped dry, it is ready for the injection procedure. A stereotactic apparatus is of great help

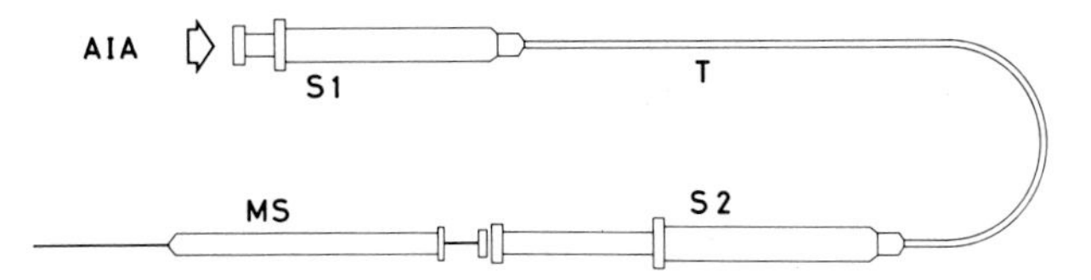

Fig. 8. Device for slowly injecting small volumes. (AIA) Automatic infusion apparatus; (S1, S2) 5-ml syringes; (T) polyethylene tube; (MS) microsyringe. For explanation, see Section 4.1.3a.

for the precise localization of the injection site. The precursor solution can then be released at a rate on the order of 1 μl/hr. After completion of the injection, the needle is left in position for about 10 min to prevent leakage of precursor into the needle track when the needle is removed.

4.1.4. *Microiontophoresis*

Microiontophoretic delivery of precursors is a sophisticated technique that has a number of advantages over microinjection. This principle of delivery has been used for many years by physiologists in investigations concerning the effects of various substances [e.g., transmitters (Curtis, 1964)] on neurons. The precursor molecules, which are charged, can be forced to leave the micropipette by means of a small electric current. In contrast to microinjection, microiontophoresis involves exchange only of molecules, so that gliosis due to increases of volume is much less. A second advantage is that the diameter of the injection site can be kept between 0.1 and 0.5 mm, which is smaller than in microinjection. The third advantage is the possibility of recording neural activity prior to the onset of the ejection current. In this way, the injection site can be physiologically characterized and identified. In tracing experiments, microiontophoresis is used in two ways: for extracellular and intracellular delivery of precursors. Extracellular application (Schubert *et al.,* 1972*b*; Graybiel and Devor, 1974) may be considered as a refinement of microinjection for studies of neural connectivity. One of the main restrictions on microiontophoresis is the limited amount of precursor that can be delivered (Künzle, 1975). Intracellular iontophoresis opens a wide field in the study of cellular morphology and functions. It is used to deliver not only tritiated amino acids (Globus *et al.,* 1968; Schubert *et al.,* 1971) but also tritiated precursors of glycoproteins, phospholipids (Kreutzberg and Schubert, 1975), and many other substances (Kater and Nicholson, 1973).

4.1.4a. Preparation of the Micropipettes. Either normal or double-barreled Pyrex glass capillaries can be used. They can be prepared according to routine procedures (Frank and Becker, 1964), but special care must be taken that they are thoroughly cleaned. After being washed in chromic and sulfuric acid and several changes of distilled water and absolute ethanol, the capillaries have to be stored in acetone.

4.1.4b. Preparation of Precursors. Increasing the concentration from the stock solution can be done as described in Section 4.1.3c. Amino acids can be redissolved in either 0.01 M acetic acid or 0.01 M KOH solution (depending on their specific properties) to a final concentration of 25–50 mCi/ml (Schubert and Holländer, 1975).

4.1.4c. Delivery Procedure. Since Graybiel and Devor (1974) described a method for reaching deep structures in the brain with micropipettes, all parts of the brain have been made accessible to the microiontophoretic delivery of precursors. The ejection current of 0.2–400 nA can be generated by a constant-current source. Direct measurement of the current is absolutely necessary, since the current flow may drop within a few seconds after onset of the source due to changes in the impedance of the pipettes. It is therefore advisable to use pulsated (0.3–0.5 Hz) currents.

4.2. *Preparation of Autoradiographs*

The introduction of autoradiography in whatever type of research often gives rise to incompatibility between normal histological and autoradiographic procedures. Most of the problems arise from the interactions of chemicals in the section with the photographic emulsion. Negative and positive chemography, latent-image fading, and unspecific labeling cannot always be avoided, since histological procedures sometimes leave only a few possibilities for alternative methods (see also Section 3.3). We shall pass stepwise through the experimental procedures and briefly indicate the difficulties and problems that may arise in each step. For more elaborate technical descriptions, the reader is referred to Rogers (1973).

4.2.1. *Fixation of Tissue*

Fixation for light-microscopic autoradiography (LMA) is usually performed with formaldehyde. This does not cause many problems. However, in electron-microscopic autoradiography (EMA), a number of chemicals are used that can affect the photographic emulsion seriously. Osmium tetroxide, commonly employed as second fixative, can also affect the emulsion. Another problem is that glutaraldehyde has been reported to bind free amino acid to tissue (Peters and Ashley, 1967). Only in experiments concerning the uptake of amino acids in brain slices or cell homogenates might this be a serious source of unspecific labeling. This can, however, be largely prevented by thorough washing of the slices or homogenates. The coating of sections with a carbon layer prior to emulsion application prevents unintended interactions with the emulsion (Salpeter and Bachmann, 1964).

4.2.2. *Whole-Body Autoradiography*

As indicated by its name, this type of autoradiography is used for studying sections of whole animals. Due to the physical properties of different isotopes, two procedures can be distinguished. When using ^{3}H, ^{55}Fe, or ^{125}I, one is largely restricted to nuclear emulsions for visualization of the distribution of the labeled material, since β^{-}-particles from these isotopes have a mean emission range of about 2 μm. A more attractive exposure technique involves pressing the specimen against a fast X-ray film. In some cases, it is possible to combine the low-energy β^{-}-emitters with X-ray film, but it is advisable to use isotopes with higher β^{-}-particle energy (see Table 1). The X-ray film technique can yield, within a few days, autoradiographs with a resolution sufficient to solve the question of gross distribution of label among the different organs.

4.2.3. *Sections for LMA*

Material for LMA is usually prepared in the form of either paraffin or frozen sections. For plastic-embedded material for semithin sections, no special precautions are needed. In anticipation of Section 4.3.3, there is one point that needs to be mentioned here. As will be shown in that section with respect to the requirements for quantification, it is sometimes of critical importance to use sections of identical thickness. In such cases, paraffin-embedded material must be chosen for further processing, since variations in the thickness of frozen sections are less easily controlled. Cryostat sections can be used when the labeled compounds in the material are in a soluble state. The technical problems concerning this last point are extensively discussed by Rogers (1973).

The mounting of sections on microscope slides can be done according to normal histological procedures. The slides have to be cleaned very carefully, and it is advisable to work in a dust-free room, since even slight contaminations can be very annoying in dark-field examination. Before the photographic emulsion is applied, the sections should be defatted in xylene (Hendrickson *et al.*, 1972*a*). This has to be done carefully; otherwise, the emulsion will not adhere uniformly to the section.

4.2.4. *Sections for EMA*

The ultrathin plastic sections used for EMA can be mounted on plastic-coated microscope slides or on copper grids before further processing. Following mounting, they can be coated with a thin layer of carbon (see also Section 4.2.1).

4.2.5. *Coating with Photographic Emulsion: Exposure*

The choice of a suitable emulsion is more difficult than the coating technique itself. In Section 3, some factors concerning this choice were discussed. From a technical point of view, only the distinction between liquid emulsions and stripping film is important. In some respects, stripping film has advantages over liquid emulsions. When the intention is to make reproducible quantitative studies, stripping film provides an emulsion layer of uniform thickness. This is not an absolute prerequisite, however, for weak β^--particle emitters, e.g., tritium. Using labeled compounds, for which the mean β^--particle range exceeds the emulsion thickness, a uniform layer is necessary to rule out unintended variations in grain density. Liquid emulsions are delivered in gel form and before they are used must be melted at moderate temperature with or without dilution with distilled water. Small series of slides can be dipped by hand, but it is more convenient and reliable to use a mechanical dipping apparatus. Figure 9 shows the apparatus in use by the authors (for both LMA and EMA). It has a number of advantages over manual dipping. The sections are uniformly covered with emulsion,

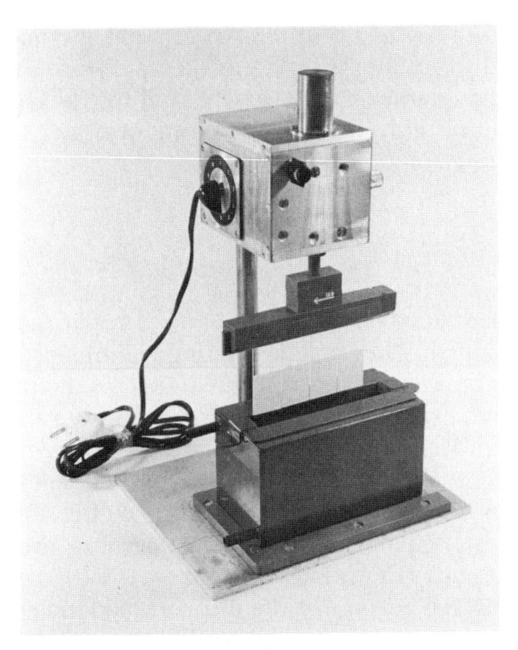

Fig. 9. Dipping apparatus as used in the authors' laboratory. Five microscope slides can be dipped simultaneously. The slides can be retracted from the cuvette filled with emulsion at various speeds.

and it is less time-consuming. A large number of slides can be dipped in a few hours without significant strain on the operator. Fast drying of the dipped emulsion usually gives rise to high background labeling. When the drying is done very slowly at high humidity, this is avoided. Latent images caused during preparation of the emulsion layer are more likely to disappear. Even when the prescribed safelights are used, it is advisable to keep the light level in the darkroom as low as possible. Once the emulsion is dry, the slides are put in light-tight boxes. Since latent images tend to fade in wet emulsions, some desiccant must be included in the boxes. During exposure, they are placed in a refrigerator at 4°C. The exposure times for LMA are between 2 and 16 weeks; for EMA slides, between 2 and 6 months. Optimal exposure times have to be tested for each autoradiographic study.

4.2.6. Developing and Fixation

Only in EMA are the size and shape of the developed grains relevant. In most other instances, one is not strictly limited to any type of developer. Different developers are distinguished mainly by their ability to develop latent images of different sizes. This ultimately results in differences in grain size and efficiency. The required degree of resolution is another factor for selection. The smaller the grains, the higher the resolution. Since there are so many factors involved in the process of developing, the best way to select the optimal procedure is to set up an experiment in which the developing time is varied and all other factors (e.g., temperature, agitation) are kept constant. Following fixation with a strong sodium thiosulfate solution, the section must be thoroughly rinsed in running tapwater to prevent adverse effects of the fixating solution on the stain.

4.2.7. *Staining*

Sections can be stained either before or after application of the emulsion, i.e., pre- or poststained. The various staining techniques differ greatly with respect to this. An important factor in this respect is the interaction between the stain and the emulsion. In the case of prestaining, such interactions may result in either negative or positive chemography. A secondary effect can be that the quality of the stain is reduced by subsequent development and fixation. One stain that is often used prior to coating with emulsion is the Feulgen stain. It is used in experiments concerning the incorporation of tritiated thymidine into DNA. This staining method is reported to remove free nucleic acids as well as some of the incorporated molecules (Baserga and Nemeroff, 1962), thus decreasing the efficiency of the autoradiograph. However, when Feulgen stain is applied as poststaining, the effect is dramatic, since the acid hydrolysis step in the staining procedure dissolves the silver grains. A similar difficulty is met with strong alkaline staining solutions, which remove the gelatin of the emulsion and cause displacement of the silver grains. Stains aggressive to the silver grains or gelatin should be applied before coating. Disadvantages of this prestaining, such as chemography or fading, have to be accepted.

Many stains can be used after exposure and development. A problem with some stains (e.g., hematoxylin) is that sometimes they stain not only the section but also the gelatin. Another problem is that some stains are not able to penetrate the gelatin layer. If these problems arise, one has to consider the possibility of either prestaining or using a different stain. Fortunately, the Nissl cresyl violet stain, which is one of the most frequently used stains in neuroanatomy, causes no problems when applied after exposure. Hendrickson *et al.* (1972*a*) give an account of the compatability of different stains with the autoradiographic technique. When neither pre- nor poststaining provides suitable autoradiographs, it is possible to separate the prestained section from the emulsion by means of an impermeable polyvinylchloride membrane. As has been pointed out by Rogers (1973), such a membrane is, however, a source of new technical problems.

When setting up a series of autoradiographic investigations, one has to carefully select the optimal staining technique. We use the qualifying term "optimal" because there are so many requirements that it is nearly impossible to fulfill all of them. The stain must cope not only with the requirements of autoradiography itself but also with the requirements for the histological analysis of the section. In Section 4.3.3b, an example of this difficulty is discussed in respect to reflectance measurements.

4.2.8. Control Procedures

In general, a number of control procedures should be carried out routinely. The best way to do this is to include in each exposure box two control slides: one with a nonradioactive section from the same type of material to see whether or not positive chemography occurs, the other with a radioactively labeled section that has been exposed to light before it is placed in the box. With this latter slide, it is possible to see whether latent-image fading or negative chemography occurs. Even in routine autoradiography, these controls are necessary, since different lots of chemicals used in the processing (e.g., fixative, stain) can cause unexpected trouble.

4.3. Analysis of Autoradiographs

4.3.1. Introduction

At present, autoradiographic techniques are quite reliable and reproducible. Therefore, more of our attention can be directed to the interpretation and analysis of the autoradiographs produced. Essentially, two categories of information can be extracted from autoradiographs: (1) whether there is a specific labeling of one or more elements of the tissue (LM) or cells (EMA) under investigation and (2) the relative or absolute amount of radioactivity present in the specifically labeled structural elements. Both types of analysis require verification that the silver grains over the sections originate from the radioactive disintegrations in the section and are not due to unspecific interactions between emulsion and section. Moreover, it must be ascertained that no radioactively induced latent images and therefore silver grains have disappeared during the preparative procedures. Proper control of background grains and of latent-image fading are prerequisites for significant qualitative or quantitative interpretations of autoradiographs.

Once image-fading and background are determined to be within reasonable limits, the extent to which grain distribution represents specific incorporation of the precursor molecules into structural elements of the tissue or cells under investigation must be questioned. In some instances, the answer to this question is obvious. Some cell types or cellular elements are densely covered with silver grains, whereas the number of silver grains over other structures does not exceed background density. In other autoradiographic experiments, the interpretation may be less obvious. At first glance, the grain distribution may look random. There are several factors that can give rise to this random distribution of silver grains. For example, the precursor molecule used may not be specific with respect to its incorporation into the different cell types or cell organelles, the precursor molecules may be diffusible and randomly distributed during the preparation of the autoradiographs (for the autoradiography of diffusible substances, see Rogers, 1973), or the tissue elements or cell organelles may be located too close to each other, so that because of the limited resolution of autoradiography, the emulsion responses overlap. In these instances, a critical analysis of the autoradiographs is a prerequisite for a correct interpretation. In LMA, counting the number of silver grains over the distinct tissue or cell elements and correlating these counts with the volume density of these elements in the tissue or cells under investigation is a common procedure for investigating the specificity of labeling.

In neurohistology, the reduced-silver-impregnation techniques for visualizing degenerating nervous tissue, as published by Nauta and Gygax (1954) and Fink and Heimer (1967), are not suitable for quantification. With these techniques, one does not obtain a *basic unit* of degeneration. Additionally, there are always many impregnated, nondegenerated elements such as normal fibers and blood vessels. Moreover, quantitative analysis in these preparations must be done by hand and is quite arbitrary. From the point of view of quantification, autoradiography is a much better choice. In principle, each grain reflects a measurable number of radioactive disintegrations. These positive points are counterbalanced, however, by a number of disadvantages. For one, the problem of efficiency and the effects of isotopic and photographic factors thereon have been outlined in Section 3. Second, there are factors in the tissue [e.g., differential uptake of precursors, different lengths of fiber pathways (see Section 5)]. Thus, the *interpretation* of the quantitative results may be as difficult as in the case of silver techniques. When one is really inclined to start quantification by, for example, grain-counting, it will turn out that one is supplicating for an apparatus to relieve one of the job long before a few thousand grains have been scored by hand. Automation of quantification is discussed in Section 4.3.3.

4.3.2. Background Labeling

In previous sections, a number of factors have been described that give rise to background labeling, e.g., chemography, exposure to light, stress in the emulsion due to fast drying, environmental radiation. They can all give rise to silver grains not related to radioactive molecules in the specimen. Rogers (1973) gives an excellent account of the factors that cause background labeling and also indicates methods to gain control over the phenomenon. Nevertheless, one will never eliminate it completely. Not every type of autoradiographic research presents serious problems with background labeling. Two examples may illustrate this: Incorporation of tritiated thymidine into DNA leads to labeled nuclei. Since this is such a restricted location, the sections are all useful nearly irrespective of the level of background labeling. Another example is found in the study of the cerebellar afferent systems. These afferents terminate as either mossy-fiber rosettes or climbing fibers in the cerebellar cortex. The appearance of the terminals is so characteristic (Fig. 10) that here again the picture is not seriously disturbed by high levels of background labeling. In general, one could state that in investigations of incorporation of labeled material into structures with a characteristic form, background labeling does not seriously disturb the qualitative aspect. Note the word *qualitative,* because from the moment one intends to start *quantification,* background labeling always has its repercussions on the results.

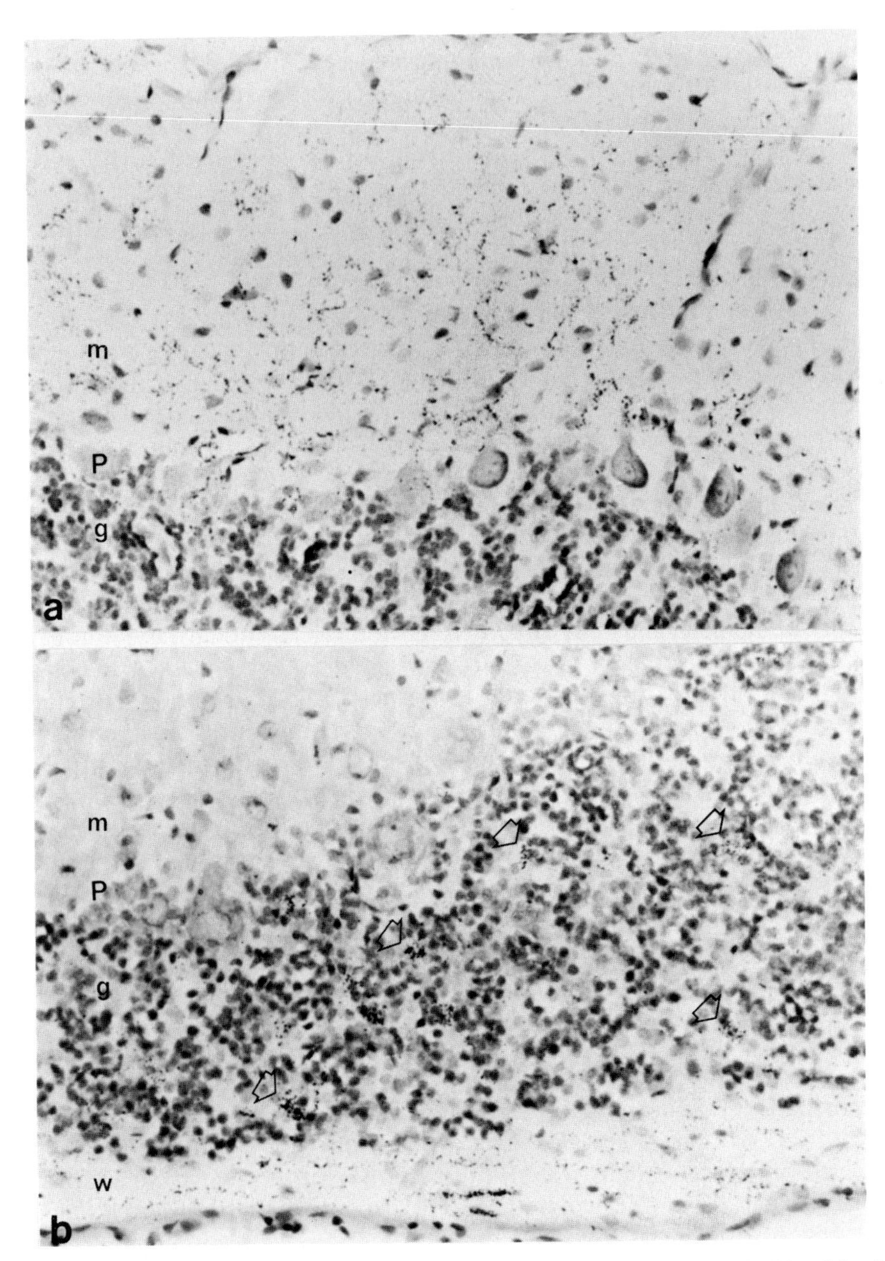

Fig. 10. Autoradiographic labeling of the cerebellar cortex after injections of tritiated leucine into the caudal brain stem. (a) Climbing-fiber labeling in the molecular layer after an injection into the inferior olive; (b) labeling of mossy-fiber rosettes (arrows) in the granular layer after an injection into the lateral reticular nucleus. (m) Molecular layer; (g) granular layer; (w) white matter; (P) Purkinje cell layer.

Because of their small size and sometimes irregular shape, background grains can be distinguished from real grains quite easily by eye. Optical quantification methods, however, cannot discriminate between them. Thus, not only the quantification of diffuse fiber projections but also the quantification of more restricted labeling needs sophisticated instruments to distinguish background from real grains. The results of quantitative studies have to be further worked out with statistical methods.

4.3.3. Light-Microscopic Quantification

As pointed out in Section 4.3.1, visual grain-counting is advisable only in incidental circumstances. As soon as grain-counting becomes routine, one should seek an instrumental method to take over the job. The different methods worked out by several authors provide a wide range of resolution and accuracy. One can select the method that best meets the requirements of a particular type of research. It makes sense to determine first whether a simple, low-cost system can meet the requirements instead of immediately purchasing the most sensitive and expensive apparatus. Systems for quantitative analysis can be divided into three categories that will be discussed below. First, however, some critical remarks must be made that have a general validity in quantification. As already indicated (see Section 3), the number of silver grains is in some instances, proportional to the thickness of the section as well as to that of the emulsion. In a thick layer of emulsion, the chance that a latent image will be formed is greater than in a thin layer. Thus, emulsion layers of uniform thickness are a prerequisite for quantification, although there is an exception to this rule. β^--Particles of ^3H, ^{55}Fe, and ^{125}I have such low energy that more than 98% of them can travel only less than 2 μm through the emulsion layer. In this case, then, it does not matter whether or not the emulsion layer has a uniform thickness if it is at least 2–3 μm thick at all sites over the section. When ^{14}C and other high-energy β^--particle emitters are used in LMA, it is advisable to use stripping film rather than dipping emulsion. Also, in EMA where silver halide monolayers are applied, dipping needs an accurate control.

4.3.3a. Absorbance Measurements. Absorbance measurements are based on the principle that the optical density of the silver grains in a section is proportional to their number. It is evident that the sections must be unstained. Identification of histological details can be done under phase-contrast, but since the material is dehydrated, the sections do not appear very brilliant. The absorbance can be calculated by comparing the light transmitted through the section with that transmitted through a reference section containing no silver grains. Altman (1963) showed linearity for measurements with a 12-μm-diameter spot up to 50 grains per spot. Although the method is very simple and needs no complex calibrations, it has, besides the requirement for unstained sections, some other disadvantages. Goldstein and Williams (1971), who compared a number of quantification techniques, stated that there is a drop in sensitivity in cases of very low grain densities and that a heterogeneous distribution of grains will disturb the linear relationship. They therefore advise the use, in such cases, of flying-spot microdensitometry. With a spot diameter of a few microns, the measurement is integrated over a larger area.

4.3.3b. Reflectance Measurements. Analogous to the principle just discussed, this method is based on the proportionality between the amount of light reflected from the silver grains and their number. The reflected light is measured under dark-field conditions. This allows the possibility of staining the section, selecting in bright-field (transmitted) illumination the area to be measured, and subsequently reading the reflected light in dark-field illumination. Although it is possible to use a—classic—dark-field condenser, it is much more convenient to use incident light for dark-field in combination with transmitted light for bright-field.

There are a number of factors that are of critical importance in the application of this technique. The sections need to be prepared with great care, since even the slightest contamination gives rise to an enormous amount of reflected light. In contrast to the absorbance measurements, this technique needs a quite elaborate calibration procedure. For this calibration and for a detailed description of the various factors that govern the reflectancy itself, the reader is referred to Goldstein and Williams (1974), Rogers (1972), and Entingh (1974). Rogers and Entingh also discuss the properties of the two different types of incident illumination, namely, the convergent-cone system and the vertical-illumination system. The convergent-cone objectives have more limitations than the others, especially in the case of inhomogeneous grain distribution (Rogers, 1972) but as pointed out by Entingh (1974), they may have considerable advantages over vertical illumination in certain types of research. The possibility of staining the sections is convenient but also troublesome, since many stains fluoresce on being illuminated. A frequently used stain in neuroanatomy, the Nissl cresyl violet, fluoresces seriously. Good alternatives are found in the hematoxylin stains (Entingh, 1974). Another solution is the use of filter combinations to cut off the fluorescence before it reaches the photomultiplier.

4.3.3c. Scanning Systems. Apart from the optical-quantification techniques discussed above, in recent years highly automated quantification techniques have been developed. They are based on the principle of the analysis of a television image of part of a section with the aid of an on-line computer system. They provide the possibility for direct computations. Preselection of a certain grain size to be scored and working at different focal planes are qualities that add up to very high accuracy. These instruments have been either direct (Price and Wann, 1975) or indirect, developed from automated image analyzers (Cole and Bond, 1972; Prensky, 1971). The main disadvantage of the systems is their high cost.

4.3.4. Submicroscopic Quantification

The discrepancy between autoradiographic resolution and microscopic resolution, which is a main cause of overlapping emulsion responses from nonisolated structures, is most obvious in EMA (1500–5 Å). In EMA, special procedures for analysis have been developed. Salpeter *et al.* (1969) have devised a procedure based on the half-distance (HD) definition of resolution (see Section 3.4). From the HD values for a line source, they have calculated curves for the theoretical distribution of silver grains around a hollow circle, solid disk, hollow band, and solid band. In a set of autoradiographs, the distribution of silver grains around certain structures (e.g., mitochondria, lysosomes) is analyzed in terms of distance from these putative sources. The curves obtained in this way are compared to the theoretical curves, and their fit is statistically tested. This procedure works quite well for sources that are isolated, separated by distances greater than 2HD, and of relatively simple shape, and presupposing that few cellular organelles are involved in the incorporation process. Williams (1969, 1974) has invented a very elegant method of analysis that is applicable to most EMA problems. It is based on the half-radius (HR) definition of resolution: the chance that a radioactive source will produce a silver grain within HR is 50% (see Section 3.4). We can reverse this definition and state that the radioactive source of a silver grain in the autoradiograph will be, with a probability of 50%, within a circle with radius given by HR. In EMA, the radius of the HR circle (0.15–0.18 μm) is such that in very many instances, more than one cellular component will be observed in the circles. Thus, in addition to single items (e.g., mitochondria, lysosomes), junctional items (e.g., mitochondria–endoplasmic reticulum, mitochondria–ribosomes) will be observed. Moreover, membranes and small particles will necessarily be seen in combination with other subcellular elements: compound items. Theoretically, the number of junctional and compound items is very large. A preliminary analysis, however, will usually show that some of the theoretical items have nothing to do with the silver grains and can be omitted. Using a list of items, the silver grains in a series of micrographs are analyzed using an HR circle. This list represents the possible single, junctional, and compound sources of the radioactively derived silver grains. The next question is whether these possible radioactive sources represent the random distribution of the organelles in the cell or whether some are more heavily labeled than can be expected according to a random distribution. Therefore, a set of random circles (radius = HR) are placed over the same micrographs, and the single, junctional, and compound items observed are listed in the same list. The random-circle analysis is a measure for the relative areas of the distinct items in a particular set of EMA micrographs. The grain analysis and relative-area measurements are compared with the chi-square test. In the case of random distribution of silver grains, the chi-square test will reveal no significant difference between grain distribution and relative-area distribution. A statistically significant difference between the two distributions is a strong indication that one or more of the cell organelles are specifically labeled. This method of EMA analysis is further evaluated for more complex situations by Blackett and Parry (1973).

If the factors that affect background and image-fading are sufficiently controlled, the grain density also reflects the amount of radioactive molecules present in the specimen. The grain density per cell, per cell organelle, or per unit area is a good measure of the relative specific activity of the distinct cellular or tissue elements under investigation. In some instances, e.g., isotope cytochemistry, absolute quantification may be of interest. In absolute quantification, the number of radioactive molecules present in the specimen is calculated from the number of silver grains present in the emulsion. It is evident that in these instances, the occurrence of background grains and the fading of latent images must be even more strictly controlled. The use of a standard reference source may be very helpful in these studies.

5. Critical Evaluation of Anterograde Tracing Techniques

5.1. Introduction

In this section, an attempt will be made to discuss the advantages and disadvantages of the autoradiographic tracing method. To a great extent, the autoradiographic tracing technique and the anterograde degeneration technique are used to tackle the same problem: the unraveling of the wiring system of the brain. Although they have a common aim, the techniques are based on different mechanisms. Thus, it may be useful to compare the results of the two techniques.

Whatever neuroanatomical technique is used to unravel the neuronal connectivity pattern, it has to fulfill several requirements: The neurons giving rise to a certain pathway must be recognized as such. The extent to which the axons and their collaterals are visualized by the method must be known. It should be possible to distinguish among axons, axon collaterals, and axon terminals.

The comparison between the anterograde autoradiographic labeling technique and the anterograde degeneration technique will be illustrated by a few experiments. Furthermore, some complicating phenomena of the autoradiographic method as well as some of its special applications will be outlined.

5.2. The Problem of Passing Fibers

An aspect of importance in anterograde degeneration techniques is the phenomenon of passing fibers. A lesion in a specific brain area or nucleus will often also interrupt fibers passing through it. This interruption of passing fibers will also give rise to degeneration. The distinction between degeneration due to passing fibers and that due to specific fibers is often impossible to make. Injection of radioactive precursors in such areas or nuclei will give rise to uptake of precursor molecules in both passing fibers and neuronal perikarya (see Fig. 1). The uptake in neuronal perikarya results in the labeling of the efferent system because the perikaryon is able to incorporate the precursor molecules into macromolecules that are axonally transported. The precursor molecules taken up by passing fibers, on the contrary, are not anabolized and not transported in sufficient amounts to trace the efferent system of these passing fibers. This inability to trace passing fibers with autoradiography has been convincingly demonstrated in the last few years (e.g., see Cowan *et al.*, 1972; Swanson *et al.*, 1974). The advantage of this is quite obvious for the exact estimation of projection areas of more complicated brain regions. This is illustrated in Fig. 11 for the projections of the dorsal column nuclei to the ventrobasal complex of the thalamus. The use of the autoradiographic tracing technique has provided clear evidence that the thalamic projections of the gracile and internal cuneate nucleus are nonoverlapping. A lesion in the internal cuneate nucleus results in degeneration in both the medial and lateral parts of the nucleus ventralis posterior lateralis thalami (VPL_m and VPL_l, respectively) because of the passage of fibers from the gracile nucleus through the area of the lesion. After injections of tritiated amino acids into the internal cuneate nucleus, terminal labeling can be demonstrated only in the medial part of the nucleus ventralis posterior lateralis (VPL_m) (Figs. 11 and 12) (Groenewegen *et al.*, 1975).

Another clear demonstration that passing fibers do not play any role in autoradiographic experiments is given by the olivocerebellar projections. Fibers from the inferior olive run exclusively to the cerebellum. These connections were thought to be, at least in large part, crossed. After a lesion of the olive at one side of the brainstem that interrups a great number of fibers originating in the contralateral olive, degeneration is found in the cerebellar white matter almost symmetrically at both sides of the midline (Fig. 13). With this method, it could not be excluded that there are also uncrossed connections of the olive with the cerebellum. The evidence that

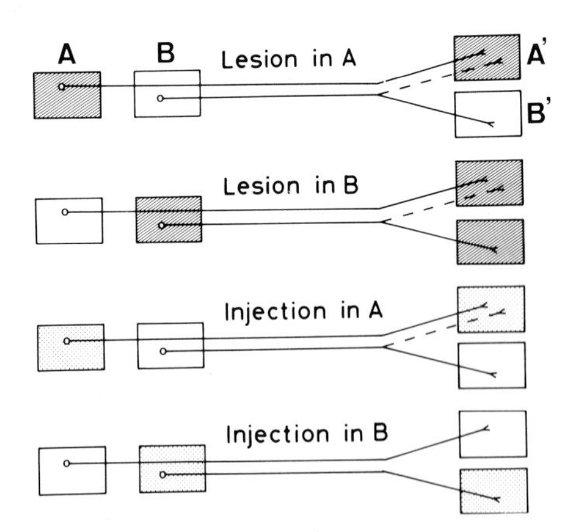

Fig. 11. Schematic representation of the results of degeneration and autoradiographic experiments that can be carried out to determine the projections of the dorsal column nuclei to the ventrobasal complex of the thalamus. Structures: (A) gracile nucleus; (A′) lateral part of the nucleus ventralis posterior lateralis thalami (VPL_l); (B) external cuneate nucleus; (B′) medial part of the nucleus ventralis posterior lateralis thalami (VPL_m). Note that fibers from (A) pass through (B). Only after an injection of tritiated amino acids into (B) can projection from (B) to (A′) be excluded.

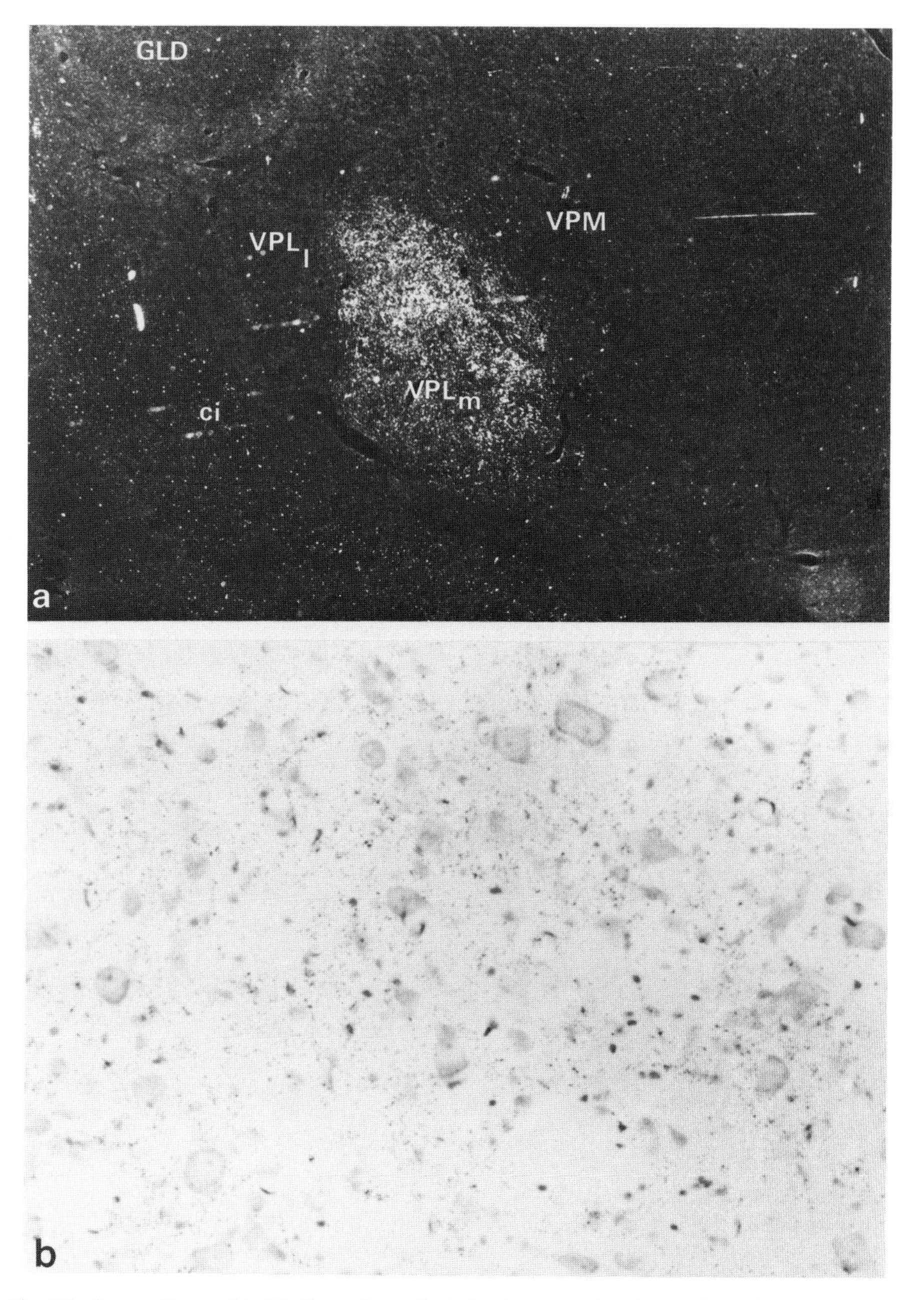

Fig. 12. Autoradiographic labeling of terminals in the ventrobasal complex of the thalamus 6 hr after an injection of tritiated leucine into the contralateral internal cuneate nucleus. (a) dark-field micrograph; (b) bright-field photograph of a part of the VPL_m counterstained with cresyl violet. (GLD) Corpus geniculatum laterale dorsalis; (VPL_l) nucleus ventralis posterior lateralis, pars lateralis; (VPL_m) nucleus ventralis posterior lateralis, pars medialis; (VPM) nucleus ventralis posterior medialis; (ci) capsula interna.

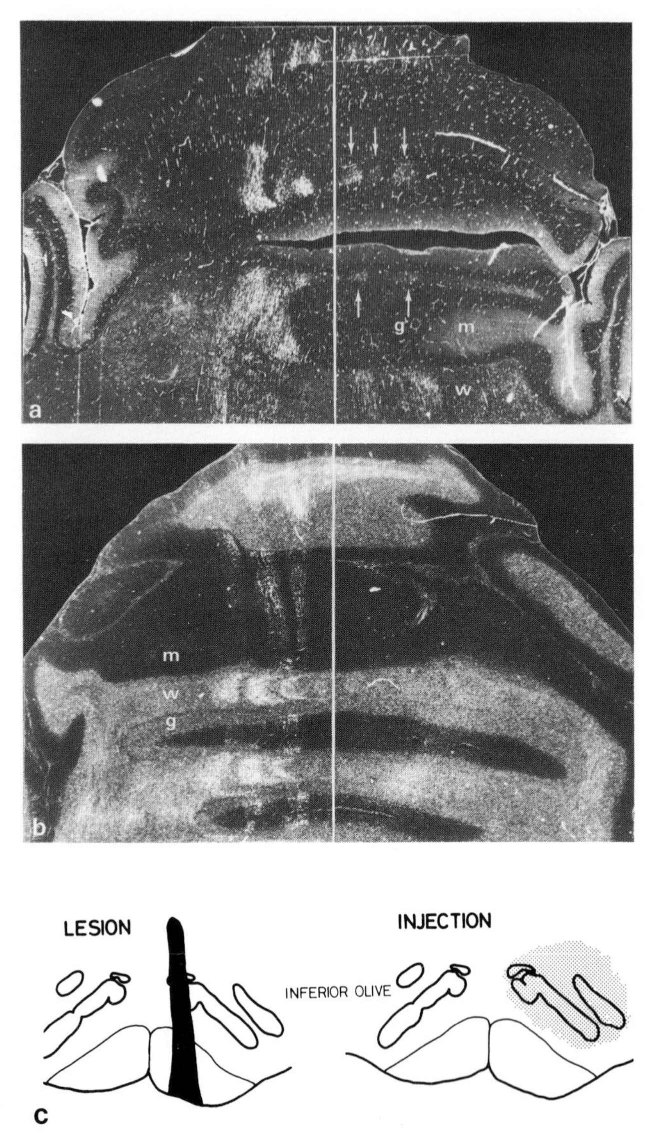

Fig. 13. (a) Dark-field photograph of the dorsal part of the anterior lobe of the cerebellum after a lesion in the inferior olive [see (c)]. Note that the degeneration appears at both the ipsilateral and contralateral sides in the cerebellar white matter. Nauta-stained section. White arrows indicate some degenerated fiber bundles at the ipsilateral side. (b) Dark-field photograph of the dorsal part of the anterior lobe of the cerebellum after an injection with tritiated leucine into the inferior olive [see (c)]. Survival time 3 days. Bundles of labeled fibers are visible in the white matter. The corresponding terminals of these climbing fibers in the molecular layer are also labeled. Note that labeling of fibers and terminals in the cerebellum is present only contralateral to the site of injection into the olive. (c) Diagrams of the ventral part of the caudal brainstem in which the extent of the lesion and the injection site are indicated. (w) White matter; (m) molecular layer; (g) granular layer. White lines indicate the midline.

olivocerebellar fibers cross the midline in the brainstem and end in the contralateral cerebellar cortex could be presented only with the autoradiographic tracing method. Indeed, after injections of tritiated leucine restricted to the inferior olive at one side of the brainstem, labeled olivocerebellar fibers can be detected only in the contralateral restiform body, cerebellar white matter, and cortex (Fig. 13) (Courville, 1975; Groenewegen and Voogd, 1976, 1977).

5.3. Injection Site

The autoradiographic labeling technique has some obvious advantages with respect to the identification of the cells of origin of a certain pathway as compared to the degeneration technique. For example, the normal cytoarchitecture is hardly destroyed after autoradiographic labeling, whereas in lesion experiments the cytoarchitecture is almost completely destroyed. In autoradiographic labeling experiments, the cells that give rise to labeling outside the injection area (projections) are "heavily labeled" themselves. This enables the study of their cytological details and their relationships with other cells in the same area. Fibers passing through the injection area do not play a role as do fibers passing through a lesioned area (see also Section 5.2).

Notwithstanding this, damage to the nervous tissue caused by the injection procedure must be avoided as much as possible. Although this damage obviously has no direct consequences for the pattern of labeling outside the injection site, it affects the physiological properties of the neurons. In this way, influencing the tissue has a secondary influence on the pattern of labeling.

The criterion for "heavily labeled cells" is a very relative one. In the first place, the amount of labeling in cells in the injection site depends on the survival time of the experimental animal (Cowan *et al.,* 1972). The amount of labeled material in an injected cell diminishes with the duration of the survival time because of the transport of label into dendrites and axons. Second, the amount of silver grains over the injected cells depends largely on the autoradiographic procedure itself (Hendrickson, 1975; Graybiel, 1975). The apparent size of the injection site increases with the duration of the autoradiographic exposure time. With increasing exposure time, not only do the individual cells in the injection site become "more heavily labeled," but also the "number of labeled cells" and the diameter of the injection site have been found to become larger. The pattern of labeling of fibers and terminals in autoradiographs is subject to similar changes as a result of the autoradiographic exposure time. Thus, with increasing exposure times, the size of the termination area or the amount of terminals in a given area is found to increase.

In practice, the question is how heavily a cell must be labeled to permit the visualization of its axon and terminals with a given autoradiographic exposure time. In other words, what is the "effective injection site," given a set of autoradiographic conditions? With diffuse projections, this problem is rather difficult. In many systems the determination of the "effective injection site" can be only roughly estimated. This estimation has to be made for each new system taken into study because the metabolic parameters of the neurons, such as uptake, synthesis, and transport, are different for each system. For well-defined projections of a rather circumscribed group of neurons, the determination of the "effective injection site" is much easier. The inferior olive is such a circumscribed nucleus, the fibers of which project only to the cerebellum, where they branch extensively and terminate as climbing fibers in the molecular layer of the cortex. By comparison of the autoradiographs of different experiments performed under the same experimental conditions (same precursor, survival time, and autoradiographic procedures), a conclusion can be made about the minimum amount of label that has to be taken up by an olivary cell to visualize autoradiographically all its collaterals and terminals. These minimally labeled cells, which can be defined by the number of grains over them, and all more heavily labeled cells constitute the "effective injection area." This area is responsible for the labeling found in fibers and terminations. With this "definition" of the minimum amount of label that has to be taken up by an olivary cell, the "effective injection site" in each new experiment in the inferior olive can be rather easily determined (Groenewegen and Voogd, 1977).

The difficulty in the determination of the "effective injection site" is emphasized by Graybiel (1975). She demonstrated that in some experiments, it is impossible to recognize the cells of origin of a projection that itself is clearly labeled. One of the most important aspects of this is the control of the injection site by restriction of the diffusion of precursors. Diffusion of precursor depends largely on the injection procedure and the nature of the precursor used (see Section 4.1.3). The use of refined techniques and the increase of knowledge about uptake, synthesis, and transport in neurons will probably help to overcome most of these problems in the future.

5.4. Termination Area

The significance of light-microscopic neuroanatomical techniques used to trace connections depends largely on their ability to discriminate between fibers and terminals. Theoretically, the autoradiographic method enables this discrimination by using the different rates of axonal transport (see Section 2.4.2 and Cowan *et al.*, 1972). Because most of the material transported by the fast axonal flow is destined for the terminals (Hendrickson, 1972; Schonbach *et al.*, 1971; Grafstein, 1975), the use of short survival times after injection enables the preferential labeling of the termination area. This is illustrated for the projections from the dorsal column nuclei to the thalamus (see Fig. 12). With survival times of 4–6 hr after injection into the dorsal column nuclei, it appears that the termination areas in the thalamus are already heavily labeled, whereas the medial lemniscus, through which these afferent thalamic fibers run, contains much less label (Groenewegen *et al.*, 1975). Presupposing that the rate of the fast axonal transport is about 150 mm per day and that the distance between the dorsal column nuclei and the caudal thalamus is about 25 mm., the peak activity of the fast axonal transport will have just reached the caudal thalamus after the minimal survival time of 4 hr. Assuming that the rate of the slow axonal transport is about 5 mm per day, the fibers in the medial lemniscus are heavily labeled only no more than about 1 mm rostral to the injection site. With this short survival time, it seems justified to determine the termination area in the thalamus without having much labeling of fibers traveling through or to this termination area.

Another possibility for selective labeling of the termination field of a certain pathway is the use of tritiated precursors of macromolecules that are transported exclusively by the fast component of the axonal transport (see also Section 2.4.2c). Glycoproteins seem to be transported only at a fast rate, and tritiated fucose or glucosamine can be used as a precursor to label these macromolecules (McEwen *et al.*, 1971; Forman *et al.*, 1972; Bennett *et al.*, 1973). That there are only few reports on the use of these precursors, which are restricted mainly to the visual system, is probably due to the fact that the diffusion of fucose is more pronounced than that of leucine and proline (Graybiel, 1975). Thus, with fucose, delineation of the injection site is more difficult.

Therefore, it seems at first glance rather easy to discriminate between labeled fibers and terminals in autoradiographs, making use of the different velocities of axonal transport. The velocities of both fast and slow transport as given in the aforementioned experiment (see Fig. 1) are of course just estimates, and in a similar way the appropriate survival time to label only the termination area or to label the whole trajectory can be calculated for each experiment.

A few difficulties must be discussed, however. First, the assumption that there are two discrete components of axonal transport is not in complete agreement with recent studies concerning this transport (see also Section 2.4.2b). It is clear that there are at least some intermediate rates at which material can be transported. Moreover, the experiments of Gross and Beidler (1975) even indicate a continuum of velocities. From their experiments, it appears that the amount of label transported in the maximum range of the fast component decreases as a function of the distance from the injection site. Much of the labeled material initially present in maximum range will be distributed along the nerve fiber. Redistribution of this material will take place, but the average velocity of these substances is much slower than that maximum velocity. When material transported at this maximum rate arrives at the terminals, the amount of label is diminished. This decrease in activity of the fast-transported material is of course most obvious in long-distance projections.

In the experiments concerning the dorsal column nuclear projections (see Section 5.2), it indeed appears that even with short survival times the whole pathway from dorsal column nuclei to thalamus is labeled. The number of silver grains per unit area over the medial lemniscus exceeds the background by a few times. This activity over the medial lemniscus, although much lower than the activity found in the thalamic termination areas, can mask or just suggest slight terminations in the proximity of this fiber path. Generally, this short-term labeling of fibers complicates the interpretation in those areas where fibers run through cortical or nuclear structures that contain many neurons on which these fibers can make contact. The discrimination between terminals and these transitional fibers is particularly difficult with short-distance projections. After injections into the dorsal column nuclei, with only 4 hr of survival time, the medial lemniscus is heavily labeled several millimeters in the anterograde direction. The determination of short-distance projections is, however, complicated not only by labeled transitional fibers but also by pure diffusion of label from the injection site.

The possibility of simultaneous visualization of labeled axons and labeled terminals, by the use of the slow component of axonal transport, is a great advantage in certain systems (Graybiel, 1975). The problems raised by the different rates of terminal and axonal degeneration observed in lesion experiments are not encountered in the autoradiographic labeling technique. This can be illustrated with the projections of the inferior olive to the cerebellum. Climbing fibers terminating in the molecular layer of the cerebellar cortex originate in the inferior olive. To visualize terminations in the molecular layer, as well as fibers in the brainstem and cerebellar white matter, two lesion experiments are necessary. To impregnate fibers with the Nauta-method, a survival time of 5–12 days is needed, whereas impregnation of terminals in the cortex can be carried out only after survival times of about 2 days using the Fink–Heimer method. After an injection of tritiated leucine into the inferior olive and a survival time of 7 days, both fibers and terminals are visualized (see Figs. 10 and 13). This simultaneous labeling of fibers and terminals of the olivocerebellar connections makes the determination of the topographical relationship between the inferior olive and the cerebellar cortex more exact and much easier (Groenewegen and Voogd, 1976).

Direct evidence that labeled fibers terminate in a certain area can be given only by electron-microscopic autoradiography (EMA). Silver grains are found in EM autoradiographs over axons and terminals in the termination area. The number of grains over axons or over terminals depends on the survival time. With short survival times, most of the grains are found over terminals (Schonbach *et al.*, 1971; Cowan *et al.*, 1972). With longer survival times, relatively more grains are found over axons.

The advantage of the EMA technique over degeneration methods used to identify terminals of a certain pathway is obvious. Inherent in the electron-microscopic degeneration methods is the morphological transformation of the terminals to be identified. On the one hand, this alteration of the normal structure is necessary to identify the terminals as belonging to a lesioned pathway, but on the other hand, it makes it difficult to identify them as a specific type. With EMA, the normal structure remains intact and the terminal belonging to a certain system can be identified by the location of

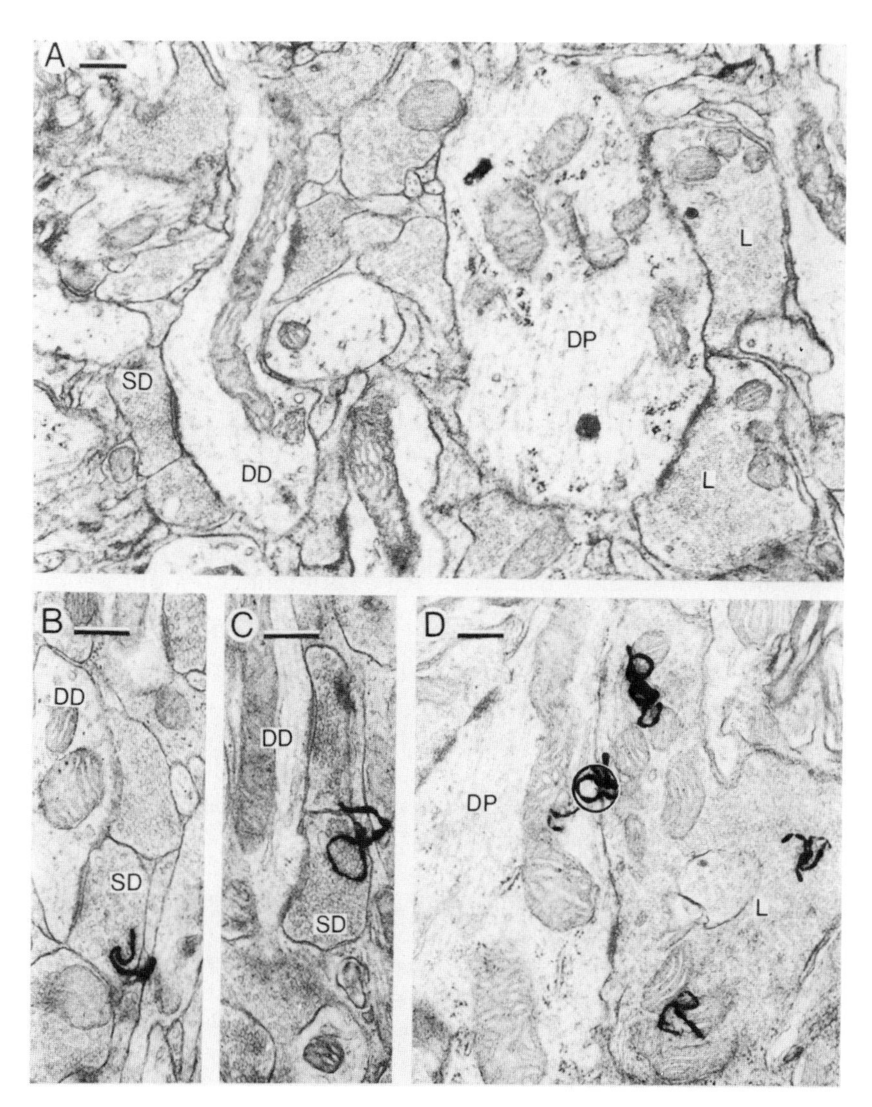

Fig. 14. Electron micrographs of ultrathin sections through the anteroventral thalamic nucleus of the rat. (A) Uncoated section; (B, C) autoradiographs after [^{3}H]leucine injection into the posterior cingulate cortex; (D) autoradiograph after [^{3}H]leucine injection into the mamillary bodies. (SD) Small dense terminal; (DD) distal dendrite; (L) large terminal; (DP) proximal dendrite. Bars equal 0.5 μm. Note that the silver grains in both experiments are located above different terminals. Kindly provided by Dr. J. J. Dekker [from Dekker and Kuypers (1975)] and reproduced with the permission of Elsevier Scientific Publishing Company.

silver grains (Fig. 14) (Droz, 1975). The possible cells of origin of labeled terminals are thus much more easily established.

5.5. Differential Uptake of Precursors

As outlined in Section 2.4.1, the inability to demonstrate a neuronal pathway with autoradiographic methods may be due to the metabolic properties or restraints of neurons. When a neuron does not take up a particular precursor or does not transport the labeled macromolecules, axons and terminals are not labeled with that precursor. The first indication that some precursors are not as readily taken up as others and that this has implications for the autoradiographic tracing technique was found by Künzle and Cuénod (1973). It appeared that the reticulocerebellar pathway cannot be traced with tritiated proline injected in the lateral reticular nucleus. This fact, together with the observation that very little label is present in large cells of the lateral reticular nucleus after these injections, indicates that these large cells do not take up proline in the same amount as they take up, for example, leucine.

Similar observations have been made for the uptake of proline in the dorsal column nuclei (Berkeley, 1975). Also, in the inferior olive and in the pontine nuclei, large and small cells have a differential uptake mechanism for proline and leucine (Fig. 15). The phenomenon of differential uptake must be taken into account for each precursor that is used for autoradiographic tracing of neuronal connectivity. This can be emphasized with the results of Sousa-Pinto and Reis (1975), who found that tritiated leucine can fail to demonstrate certain pathways. This is the main reason that several investigators (Hendrickson, 1975; Graybiel, 1975) suggest the use of a mixture of tritiated amino acids or other precursors to be more certain that all projections from an injected area are labeled. The disadvantage of this approach is that a rather arbitrary choice of precursors has to be made. The mechanism of the differences in uptake is not known yet but it seems to be a rather specific phenomenon.

Although these reports indicate that negative results of the autoradiographic tracing method should be considered very carefully, this phenomenon of differential uptake can also be used in a positive sense. Making use of the differential uptake of leucine and proline in the dorsal column nuclei, Berkeley (1975) could show that small cells of the gracile nucleus have different projections than large cells. After injecting tritiated proline, which does not label the large cells, very few grains are found in the ventrobasal complex of the thalamus. Projections from the dorsal column nuclei to other areas, for example, to the inferior olive, seem to be less dependent on the precursor used. After injections of horseradish peroxidase (HRP) into the thalamus and into the inferior olive, it could be rather safely concluded that large cells in the gracile nucleus project to the thalamus, while small cells send their fibers to other targets. These results indicate that the phenomenon of differential uptake can probably also be used in other systems to selectively

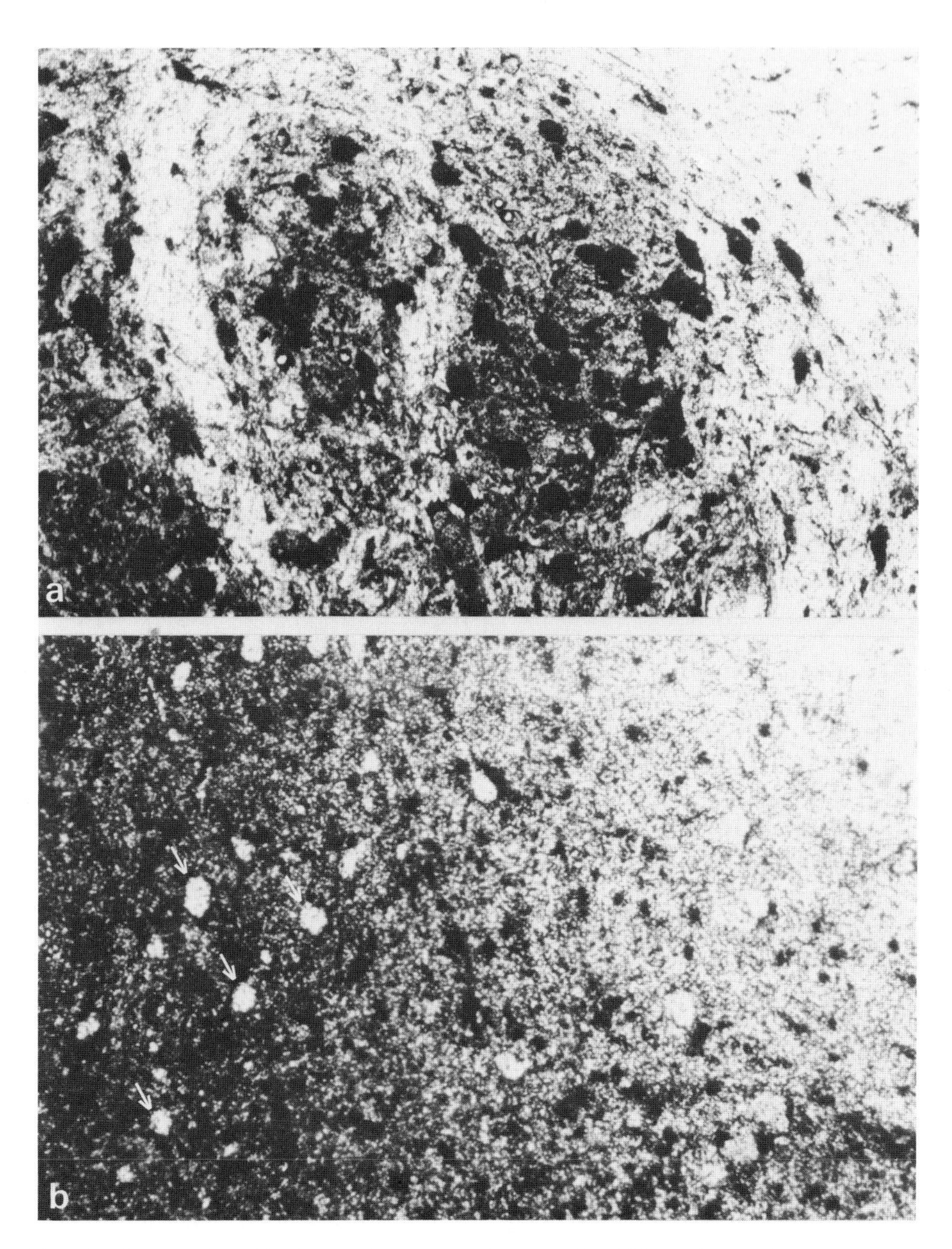

Fig. 15. Micrographs of the internal cuneate nucleus showing differential uptake of precursors. (a) Labeling of neuronal perikarya and neuropil after an injection of tritiated leucine. (b) labeling after an injection of tritiated proline. Note that the perikarya of large neurons (arrows) are almost free from label, while smaller cells have taken up large amounts of label.

trace the projections of one cell type. However, to derive full benefit from this phenomenon, a better understanding of the mechanisms of uptake is needed (Lajtha, 1975).

The selective uptake of transmitter substances (Hökfelt and Ljungdal, 1975) by certain neurons seems to have a more rational base. It appears that the use of transmitters or putative transmitters enables the labeling of specific neuronal pathways in the central nervous system. In addition to the tracing of specific pathways with histochemical methods, e.g., the Falck–Hillarp fluorescence technique for monoaminergic pathways, it is also possible to do this by local injections of labeled transmitter substances (Fibiger and McGeer, 1974). These labeled transmitter amino acids or labeled monoamines are taken up by the cells specific for the transmitter, and as the result of anterograde transport, the terminations of these cells become labeled. There is, however, one disadvantage. All parts of the neurons, including axons and terminals, take up these substances and transport them. Fibers of neurons passing through the injection site that are specific for the transmitter or a related substance take up the label and transport it. In this way, the terminals of "passing fibers" are also labeled and complicate the interpretation of these autoradiographs (Hökfelt and Ljungdahl, 1975).

5.6. Sensitivity of the Autoradiographic Tracing Method

It appears that the autoradiographic tracing technique is to some extent more sensitive than the classic impregnation methods. Certain pathways not recognized with degeneration methods have been labeled with tritiated amino acids.

Until recently, it was rather difficult to impregnate the very thin branches of degenerated olivocerebellar climbing fiber terminals in the molecular layer of the cerebellar cortex but after appropriate injections with tritiated leucine into the inferior olive, recognition of these terminals appears to be no problem (see Figs. 10 and 13). Many other pathways that were difficult to trace with degeneration methods have been visualized with the autoradiographic method [cerebelloolivary pathway (Graybiel *et al.*, 1973), retinohypothalamic pathway (Hendrickson *et al.*, 1972*b*), efferent projections of the nucleus locus coerulus (Pickel *et al.*, 1974);]. The greater sensitivity of the autoradiographic method seems to be due to the fact that transport of proteins occurs in nearly all axons. Degenerative changes are much more subject to specific variations, e.g., differences in species, size of the axon.

It seems to be safe to expect that various "new" pathways will be recognized with the autoradiographic method in the near future.

5.7. Transneuronal Transport

The phenomenon of transfer of material from one neuron to another by means of transsynaptic transport is of great importance for studies of neuronal connectivity. As stated in Section 2.4.4, this transfer out of the primary labeled cells does not in general seriously complicate the autoradiographic tracing technique. With the use of tritiated leucine as precursor and survival times not longer than 1 week, the amounts of label leaving the primary labeled cell are too small to be detected with autoradiographic methods. However, this phenomenon must be kept in mind with other precursors, with large amounts injected, or with survival times much longer than 1 week.

When large amounts of tritiated fucose or proline are injected into the eye and the animal is allowed to survive for 1 week or more, several "second-order" projections of the retinal efferents can be demonstrated. Besides rather nonspecific transfer, specific transsynaptic transport occurs as well. Using this method, Wiesel *et al.* (1974) were able to demonstrate the ocular dominance columns in the striate cortex of the monkey.

Very little is known about the basic mechanisms of transneuronal transport. The chemical nature and the origin of the transferred material are also unknown. When the mechanism and nature of this phenomenon of transsynaptic transfer are unraveled, they can be used to selectively trace "secondary projections" of a primary labeled pathway. As can be seen from the results of Wiesel *et al.* (1974), this enables precise information about the organization of complicated neuronal connections.

5.8. Some Technical Advantages and Limitations of the Autoradiographic Tracing Method

In comparison with the silver-impregnation methods, the time needed for preparation of LM autoradiographs is somewhat longer. It is, however, possible that the exposure times of 2–16 weeks that are needed for tritium-labeled precursors will be markedly shortened when [^{35}S]methionine is used as a precursor (Graybiel, 1975). Both techniques, i.e., the LMA and the silver-impregnation method, although completely different in nature, have a comparable level of difficulty.

An obvious disadvantage of EMA is the long exposure time needed to get "sufficient" autoradiographic response. A correct method of analysis can to some extent decrease the time of autoradiographic exposure (see Section 3). The use of ^{35}S-labeled precursors will be of little help in EMA, for the resolution will be so poor that no information will be obtained regarding fine-structural localization.

One of the limitations of the autoradiographic technique is the fact that the emulsion overlies the histological section. Using ^{3}H, the silver grains in the emulsion reflect the position of isotopes only in the upper 2–3 μm in the section. This means that very little can be said about the properties of the labeled structure. In some instances, the arrangement of the silver grains reflects the topography of the underlying, labeled structure (see Fig. 10). The thickness of fibers can in some cases also be determined from autoradiographs (Fig. 16), but discrimination among labeled fibers or terminals can generally not be made. In Nauta- or Fink–Heimer-stained sections, however,

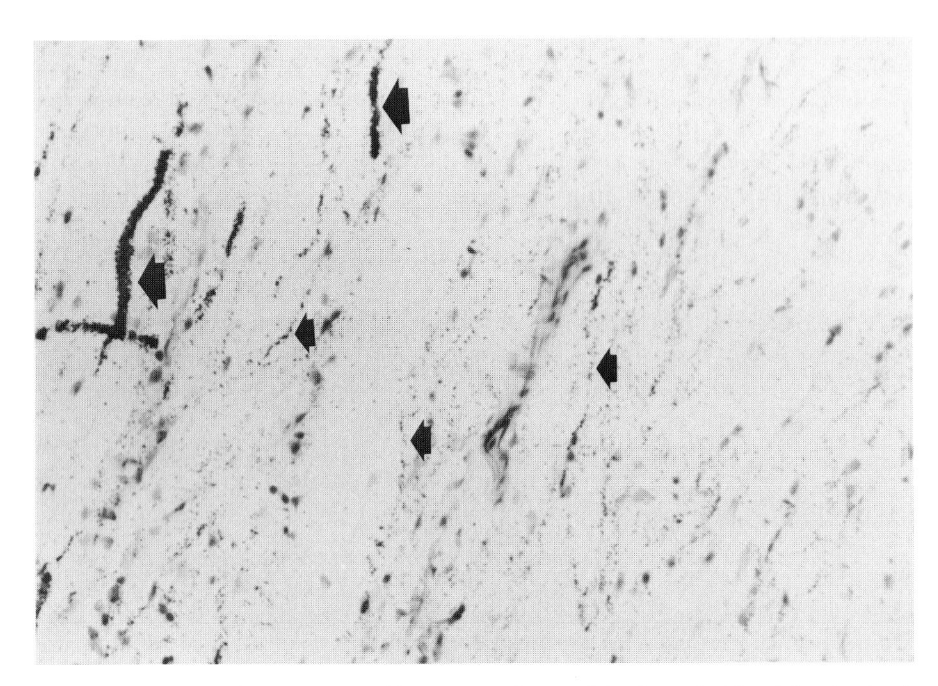

Fig. 16. Labeling of the cerebellar white matter after an injection of tritiated leucine in the caudal brainstem. Note the coarse aspect of the reticulocerebellar fibers (large arrows). Some of the labeled olivocerebellar fibers, which are much thinner, are indicated by small arrows.

the degenerated fibers or terminals are directly visible, and it is often possible to distinguish among terminal, preterminal, and transitional fiber degeneration. The thickness of fibers is easily recognized.

The possibility of quantification can be considered as an advantage of the autoradiographic technique (see Section 4.3). It is very important, when quantitative grain counts are made, that the autoradiographic procedures be standardized. However, some aspects of the procedure, e.g., the injection of precursor and the physiological state of the tissue, limit this quantitative analysis. This means that the absolute numbers of grains per unit area revealed in different experiments or even at different sites in the nervous system in one experiment cannot be compared.

Under certain conditions, a quantitative comparison can be made between two projections terminating in different areas or arising at different sites. A quantitative approach is also of great value in the case of questionable projections. Small projections are hard to detect with bright-field illumination, and even with dark-field illumination, it is sometimes hard to get a convincing picture. In these cases, grain-counting can help to decide whether a certain projection exists or not. Grain-counting also gives a very exact pattern of the termination, so that this can be compared with another projection to that area (Price and Wann, 1975).

With degeneration studies, it is rather easy to determine the "total projection" of a nucleus or a fiber tract. By making a complete lesion of the nucleus or by transsecting the fiber tract or the efferent fibers of the nucleus at some point where these fibers are bundled, a picture of all the projections of such a nucleus or fiber tract can be gained. In the autoradiographic tracing method, it is difficult to obtain complete labeling of a nucleus, and it is impossible to label all the neurons of a fiber tract the cells of origin of which are scattered throughout the nervous system. It is impossible, for example, to reveal the total projection areas of long ascending or descending fiber tracts in the brainstem with one autoradiographic experiment, whereas transection of such a fiber tract at an appropriate point can provide all termination areas in one degeneration experiment.

5.9. Combination of the Autoradiographic Tracing Technique with Other Techniques

Because the autoradiographic tracing technique does not presuppose the use of special histological fixatives and can be employed in combination with various histological stains (Hendrickson *et al.*, 1972*a*), it is possible to combine this technique with other neuroanatomical and neurohistochemical procedures. The combination of anterograde and retrograde tracing techniques, by which the afferent and efferent connections of an area can be determined in one experiment, can be obtained by simultaneous injection of tritiated amino acids and HRP. When these substances are injected simultaneously, the afferent projections from the injected area are visualized through retrograde transport of HRP, while the efferent projections of that area are visualized through the anterograde transport of radioactive label (Graybiel and Devor, 1974). In the case of reciprocal projections, such as those existing between the cerebral cortex and the thalamus, the afferent and efferent connections of the noninjected area can be studied simultaneously in the same histological sections (Trojanowsky and Jacobson, 1975). Thus, by injecting tritiated amino acids into the cells that project to that area and HRP into the terminals of the cells located in that area, a direct study can be made of the topological relationships between the autoradiographically labeled terminals and the cells retrogradely filled with HRP. Similar combination experiments can of course be done for nonreciprocal projections. In this case, the injections of tritiated amino acids and HRP should be done into different areas of the nervous system: HRP into the termination area of the nucleus or cortical area under study and tritiated amino acids into the cells that project to that area (Fig. 17).

Autoradiographic and degeneration techniques can of course also be combined. One afferent system to a certain nucleus or cortical structure can be labeled, while another can be lesioned. In the same histological sections or in alternately stained sections, the patterns of labeling and degeneration can then be compared.

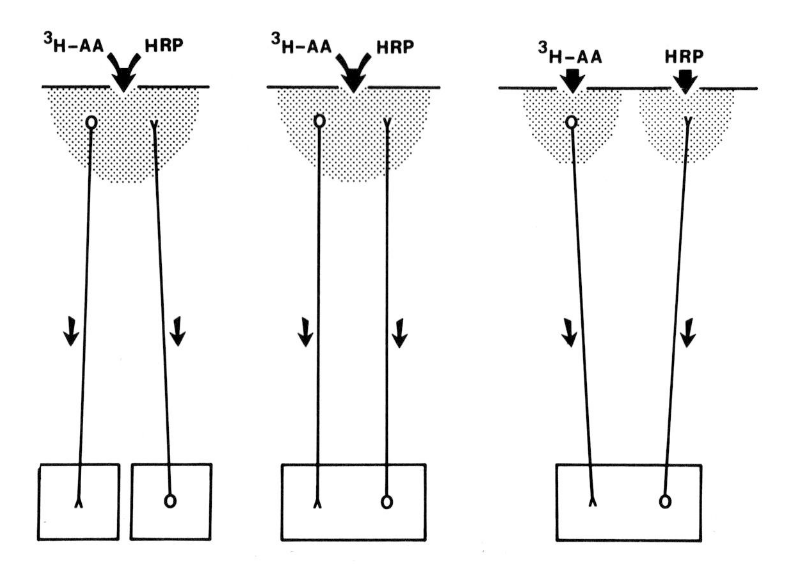

Fig. 17. Schematic representation of some possible combination experiments employing both the anterograde and retrograde tracing techniques. *Left:* Simultaneous injection of tritiated amino acids (^{3}H-AA) and horseradish peroxidase (HRP) into a single brain area. Study of efferent (^{3}H-AA) and afferent (HRP) projections is possible in different histological sections. *Middle:* Same injection procedure in the case of reciprocal connections. Investigation of efferents and afferents is possible in the same histological section. *Right:* Injection of ^{3}H-AA and HRP into different brain areas. The injection sites are chosen in such a way that retrograde-transported HRP and anterograde-transported ^{3}H-AA converge in one area. Thus, the efferents and afferents of this area can be studied in the same histological sections.

ACKNOWLEDGMENTS. The authors wish to express their thanks to the head of their department, Dr. J. Voogd, for providing time and room to write this chapter. We also thank Miss Leonie Pronk for technical assistance and Miss Ria Stokman for typing the manuscript.

References

Altman, J., 1963, Regional utilization of leucine-H^{3} by normal rat brain: Microdensitometric evaluation of autoradiograms, *J. Histochem. Cytochem.* **11:**741.

Barkas, W.H., 1963, *Nuclear Research Emulsions,* Vol. I, Academic Press, New York.

Baserga, R., and Nemeroff, K., 1962, Factors which affect efficiency of autoradiography with tritiated thymidine, *Stain Technol.* **37:**21.

Bennett, G., Di Giamberardino, L., Koenig, H.L., and Droz, B., 1973, Axonal migration of protein and glycoprotein to nerve endings. II. Radioautographic analysis of the renewal of glycoproteins in nerve endings of chicken ciliary ganglion after intracerebral injection of (3-H)fucose and (3-H)glucosamine, *Brain Res.* **60:**129.

Berkeley, K.J., 1975, Different targets of different neurons in nucleus gracilis of the cat, *J. Comp. Neurol.* **163:**285.

Blackett, M.M., and Parry, D.M., 1973, A new method for analyzing electron microscope autoradiographs using hypothetical grain distributions, *J. Cell Biol.* **57:**9.

Brodal, A., Walberg, F., and Hoddevik, G.H., 1975, The olivocerebellar projection in the cat studied with the method of retrograde axonal transport of horseradish peroxidase, *J. Comp. Neurol.* **164:**449.

Bunt, A.H., Haschke, R.H., Lund, R.D., and Calkins, D.F., 1976, Factors affecting retrograde axonal transport of horseradish peroxidase in the visual system, *Brain Res.* **102:**152.

Caro, L.G., 1962, High resolution autoradiography. II. The problem of resolution, *J. Cell. Biol.* **15:**189.

Cole, M., and Bond, C.P., 1972, Recent advances in automatic image analysis using a television system, *J. Microsc.* **96:**89.

Courville, J., 1975, Distribution of olivo-cerebellar fibers demonstrated by a radioautographic tracing method, *Brain Res.* **95:**253.

Cowan, W.M., Gottlieb, D.I., Hendrickson, A.E., Price, J.L., and Woolsey, T.A., 1972, The autoradiographic demonstration of axonal connections in the central nervous system, *Brain Res.* **37:**21.

Crossland, W.J., Currie, J.R., Rogers, L.A., and Cowan, W.M., 1974, Evidence for a rapid phase of axoplasmic transport at early stages in the development of the visual system of the chick and frog, *Brain Res.* **78:**483.

Curtis, D.R., 1964, Microelectrophoresis, in: *Physical Techniques in Biological Research,* Vol. V, Part A, Electrophysiological methods (W.L. Nastuk, ed.), pp. 144–190, Academic Press, New York and London.

Dahlström, A., 1965, Observations on the accumulation of noradrenaline in the proximal and distal parts of peripheral adrenergic nerves after compression, *J. Anat. (London)* **99:**677.

Dahlström, A., 1968, Effects of colchicine on transport of amine storage granules in sympathetic nerves of rat, *Eur. J. Pharmacol.* **5:**111.

Dekker, J.J., and Kuypers, H.G.J.M., 1975, Electron microscopy study of forebrain connections by means of the radioactive labeled amino acid tracer technique, *Brain Res.* **85:**229.

Doniach, I., and Pelc, S.R., 1950, Autoradiographic technique, *Br. J. Radiol.* **23:**184.

Dräger, U., 1974, Autoradiography of tritiated proline and fucose transported transneuronally from the eye to the visual cortex in pigmented and albino mice, *Brain Res.* **82:**284.

Droz, B., 1975, Autoradiography as a tool for visualizing neurons and neuronal processes, in: *The Use of Axonal Transport for Studies of Neuronal Connectivity* (W.M. Cowan and M. Cuénod, eds.), pp. 127–154, Elsevier, Amsterdam.

Droz, B., and Leblond, C.P., 1963, Axonal migration of proteins in the central nervous system and peripheral nerves as shown by radioautoradiography, *J. Comp. Neurol.* **121:**325.

Droz, B., Koenig, H.L., and Di Giamberardino, L., 1973, Axonal migration of protein and glycoprotein to nerve endings. I. Radioautographic analysis of the renewal of protein in nerve endings of chicken ciliary ganglion after intracerebral injection of (3-H)lysine, *Brain Res.* **60:**93.

Edström, A., and Mattson, H., 1972, Rapid axonal transport *in vitro* in the sciatic system of the frog of fucose-, glucosamine- and sulphate-containing material, *J. Neurochem.* **19:**1717.

Entingh, D., 1974, Performance characteristics of a microreflectometer for measuring autoradiographic grain density, *J. Microsc.* **101:**9.

Falk, G.J., and King, R.C., 1963, Radioautographic efficiency for tritium as a function of section thickness, *Radiat. Res.* **20:**466.

Feinendegen, L.E., 1967, *Tritium Labelled Molecules in Biology and Medicine,* Academic Press, New York.

Fibiger, H.C., and McGeer, E.G., 1974, Accumulation and axoplasmic transport of dopamine but not of amino acids by axons of the nigro-neostriatal projection, *Brain Res.* **72:**366.

Fink, R.P., and Heimer, L., 1967, Two methods for selective silver impregnation of degenerating axons and their synaptic endings in the central nervous system, *Brain Res.* **4:**369.

Forman, D.S., McEwen, B.S., and Grafstein, B., 1971, Rapid transport of radioactivity in goldfish optic nerve following injections of labelled glucosamine, *Brain Res.* **28:**119.

Forman, D.S., Grafstein, B., and McEwen, B.S., 1972, Rapid axonal transport of (^{3}H)fucosyl glycoproteins in the goldfish optic system, *Brain Res.* **48:**327.

Frank, K., and Becker, M.C., 1964, Microelectrodes for recording and stimulation, in: *Physical Techniques in Biological Research,* Vol. V, Part A, Electrophysiological Methods (W.L. Nastuk, ed.), pp. 22–87, Academic Press, New York and London.

Geffen, L.B., Livett, B.G., and Rush, R.A., 1970, Immunohistochemical localization of chromagranins in sheep sympathetic neurones and their release by nerve impulses, in: *New Aspects of Storage and Release Mechanisms of Catecholamines* (Bayer Symposium II), pp. 58–72, Springer-Verlag, Berlin.

Globus, A., Lux, H.D., and Schubert, P., 1968, Somadendritic spread of intracellularly injected glycine in cat spinal motoneurons, *Brain Res.* **11:**440.

Goldstein, D.J., and Williams, M.A., 1971. Quantitative autoradiography: An evaluation of grain counting, reflectance microscopy, gross absorbance measurements and flying-spot microdensitometry, *J. Microsc.* **94:**215.

Goldstein, D.J., and Williams, M.A., 1974, Quantitative assessment of autoradiographs by photometric reflectance microscopy: An improved method using polarized light, *Histochem. J.* **6:**223.

Grafstein, B., 1967, Transport of protein by goldfish optic nerve fibers, *Science* **157:**196.

Grafstein, B., 1969, Axonal transport: Communication between soma and synapse, in: *Advances in Biochemical Psychopharmacology,* Vol. 1 (E. Costa and P. Greengard, eds.), pp. 11–25, Raven Press, New York.

Grafstein, B., 1971, Transneuronal transfer of radioactivity in the central nervous system, *Science* **172:**177.

Grafstein, B., 1975, Principles of anterograde axonal transport in relation to studies of neuronal connectivity, in: *The Use of Axonal Transport for Studies of Neuronal Connectivity* (W.M. Cowan and M. Cuénod, eds.), pp. 47–68, Elsevier, Amsterdam.

Graybiel, A.M., 1975, Wallerian degeneration and anterograde tracer methods, in: *The Use of Axonal Transport for Studies of Neuronal Connectivity* (W.M. Cowan and M. Cuénod, eds.), pp. 173–216, Elsevier, Amsterdam.

Graybiel, A.M., and Devor, M., 1974, A microelectrophoretic delivery technique for use with horseradish peroxidase, *Brain Res.* **68:**167.

Graybiel, A.M., Nauta, H.J.W., Lasek, R.J., and Nauta, W.J.H., 1973, A cerebello-olivary pathway in the cat: An experimental study using autoradiographic tracing techniques, *Brain Res.* **58:**205.

Groenewegen, H.J., and Voogd, J., 1976, The longitudinal zonal arrangement of the olivocerebellar, climbing fiber projection in the cat: An autoradiographic and degeneration study, *Exp. Brain Res.* (Suppl.) pp. 65–71.

Groenwegen, H.J., and Voogd, J., 1977, The parasagittal zonation within the olivocerebellar projection. I. Climbing fiber distribution in the vermis of cat cerebellum, *J. Comp. Neurol.* **174:**417.

Groenewegen, H.J., Boesten, A.J.P., and Voogd, J., 1975, The dorsal column nuclear projections to the nucleus ventralis posterior lateralis thalami and the inferior olive in the cat: An autoradiographic study, *J. Comp. Neurol.* **162:**505.

Gross, G.W., 1975, The microstream concept of axoplasmic and dendritic transport, in: *The Physiology and Pathology of Dendrites, Advances in Neurology,* Vol. 12 (G.W. Kreutzberg, ed.), pp. 283–296, Raven Press, New York.

Gross, G.W., and Beidler, L.M., 1973, Fast axonal transport in the C-fibers of the garfish olfactory nerve, *J. Neurobiol.* **4:**413.

Gross, G.W., and Beidler, L.M., 1975, A quantitative analysis of isotope concentration profiles and rapid transport velocities in the C-fibers of the garfish olfactory nerve, *J. Neurobiol.* **6:**213.

Haas, R.J., Werner, J., and Fliedner, T.M., 1970, Cytokinetics of neonatal brain cell development in rats as studied by the "complete" ^{3}H-thymidine labelling method, *J. Anat.* **107:**421.

Hendrickson, A.E., 1972, Electron microscopic distribution of axoplasmic transport, *J. Comp. Neurol.* **144:**381.

Hendrickson, A., 1975, Tracing neuronal connections with radio-isotopes applied extracellularly, *Fed. Proc. Fed. Am. Soc. Exp. Biol.* **34:**1612.

Hendrickson, A., Moe, L., and Noble, B., 1972*a*, Staining for autoradiography of the central nervous system, *Stain Technol.* **47**:283.

Hendrickson, A.E., Wagoner, N., and Cowan, W.M., 1972*b*, An autoradiographic and electron microscopic study of retino-hypothalamic connections, *Z. Zellforsch.* **135**:1.

Hökfelt, T., and Ljungdahl, A., 1975, Uptake mechanisms as a basis for the histochemical identification and tracing of transmitter-specific neuron populations, in: *The Use of Axonal Transport for Studies of Neuronal Connectivity* (W.M. Cowan and M. Cuénod, eds.), pp. 249–306, Elsevier, Amsterdam.

Jeffrey, P. L., and Austin, L., 1973, Axoplasmic transport, *Prog. Neurobiol.* **2**:207.

Karlsson, J.-O., and Sjöstrand, J., 1971, Synthesis, migration and turnover of protein in retinal ganglion cells, *J. Neurochem.* **18**:749.

Kater, S.B., and Nicholson, C. (eds.), 1973, *Intracellular Staining Techniques in Neurobiology*, Springer-Verlag, New York.

Kreutzberg, G.W., 1969, Neuronal dynamics and axonal flow, IV. Blockage of intra-axonal enzyme transport by colchicine. *Proc. Natl. Acad. Sci. U.S.A.* **62**:722.

Kreutzberg, G.W., and Schubert, P., 1975, The cellular dynamics of intraneuronal transport, in: *The Use of Axonal Transport for Studies of Neuronal Connectivity* (W.M. Cowan and M. Cuénod, eds.), pp. 83–112, Elsevier, Amsterdam.

Kristensson, K., 1970*a*, Morphological studies of the neural spread of herpes simplex virus to the central nervous system, *Acta Neuropathol. (Berlin)* **16**:54.

Kristensson, K., 1970*b*, Transport of fluorescent protein tracer in peripheral nerves, *Acta Neuropathol. (Berlin)* **16**:293.

Kristensson, K., 1975, Retrograde axonal transport of protein tracers, in: *The Use of Axonal Transport for Studies of Axonal Connectivity* (W.M. Cowan and M. Cuénod, eds.), pp. 69–82, Elsevier, Amsterdam.

Kristensson, K., and Olsson, Y, 1971, Retrograde axonal transport of protein, *Brain Res.* **29**:363.

Künzle, H., 1975, Notes on the application of radioactive amino acids for the tracing of neuronal connections, *Brain Res.* **85**:267.

Künzle, H., and Cuénod, M., 1973, Differential uptake of (^{3}H)proline and (^{3}H)leucine by neurons: Its importance for the autoradiographic tracing of pathways, *Brain Res.* **62**:213.

Kuypers, H.G.J.M., Kieviet, J., and Groen-Klevant, A.C., 1974, Retrograde axonal transport of horseradish peroxidase in rat's forebrain, *Brain Res.* **67**:211.

Lajtha, A., 1975, Transport and incorporation of amino acids in relation to measurement of axonal flow, in: *The Use of Axonal Transport for Studies of Neuronal Connectivity* (W.M. Cowan and M. Cuénod, eds.), pp. 25–46, Elsevier, Amsterdam.

Lasek, R.J., 1966, Axoplasmic streaming in the cat dorsal root ganglion cell and the rat ventral motoneuron, *Anat. Rec.* **154**:373.

Lasek, R.J., 1968, Axoplasmic transport in cat dorsal root ganglion cells: As studied with (3-H)-L-leucine, *Brain Res.* **7**:360.

Lasek, R.J., 1970, Protein transport in neurons, *Int. Rev. Neurobiol.* **13**:289.

Lasek, R.J., Joseph, B.S., and Whitlock, D.G., 1968, Evaluation of a radioautographic neuroanatomical tracing method, *Brain Res.* **8**:319.

LaVail, J.H., 1975, Retrograde cell degeneration and retrograde transport techniques, in: *The Use of Axonal Transport for Studies of Neuronal Connectivity* (W.M. Cowan and M. Cuénod, eds.), pp. 217–248, Elsevier, Amsterdam.

LaVail, J.H., and LaVail, M.M., 1972, Retrograde axonal transport in the central nervous system, *Science* **176**:1416.

LaVail, J.H., and LaVail, M.M., 1974, The retrograde intraaxonal transport of horseradish peroxidase in the chick visual system: A light and electron microscopic study, *J. Comp. Neurol.* **157**:303.

Lubińska, L., 1964, Axoplasmic streaming in regenerating and normal nerve fibers, in: *Mechanisms of Neural Regeneration, Progress in Brain Research,* Vol. 13 (M. Singer and J.P. Schadé, eds.), pp. 1–71, Elsevier, Amsterdam.

Maurer, W., and Primbsch, E., 1964, Grösse der Selbsabsorption bei der ^{3}H-autoradiographie, *Exp. Cell Res.* **33**:8.

McEwen, B.S., and Grafstein, B., 1968, Fast and slow components in axonal transport of protein, *J. Cell Biol.* **38:**494.

McEwen, B.S., Forman, D.S., and Grafstein, B., 1971, Components of fast and slow axonal transport in the goldfish optic nerve, *J. Neurobiol.* **2:**361.

Nauta, W.J.H., and Gygax, P.A., 1954, Silver impregnation of degenerating axons in the central nervous system: A modified technique, *Stain Technol.* **29:**91.

Nauta, H.J.W., Pritz, M.B., and Lasek, R.J., 1974, Afferents to the rat caudoputamen studied with horseradish peroxidase: An evaluation of a retrograde neuroanatomical research method, *Brain Res.* **67:**219.

Ochs, S., 1971, Characteristics and a model for the fast axoplasmic transport in nerve, *J. Neurobiol.* **2:**331.

Ochs, S., 1972, Rate of fast axoplasmic transport in mammalian nerve fibers, *J. Physiol. (London)* **227:**627.

Ochs, S., Sabri, M.I., and Johnson, J., 1969, Fast transport system of materials in mammalian nerve fibers, *Science* **163:**686.

Parry, D.M., and Blackett, N.M., 1973, Electron microscope autoradiography of erythroid cells using radioactive iron, *J. Cell Biol.* **57:**16.

Perry, R.P., 1964, Quantitative autoradiography, in: *Methods in Cell Physiology* (D.M. Prescott, ed.), Vol. I, Chapt. 15, Academic Press, New York.

Perry, R.P., Errera, M., Hell, A., and Durwald, H., 1961, Kinetics of nucleoside incorporation into nuclear and cytoplasmic RNA, *J. Biophys. Biochem. Cytol.* **11:**1.

Peters, T., and Ashley, C.A., 1967, An artifact in autoradiography due to binding of free amino acids to tissues by fixatives, *J. Cell Biol.* **33:**53.

Pickel, V.M., Segal, M., and Bloom, F.E., 1974, A radioautographic study of the efferent pathways of the nucleus locus coeruleus, *J. Comp. Neurol.* **55:**15.

Pomerat, C.M., Hendelman, W.J., Raiborn, C.W., Jr., and Massey, J.F., 1967, Dynamic activities of nervous tissue *in vitro,* in: *The Neuron* (H. Hydén, ed.), pp. 119–178, Elsevier, Amsterdam.

Prensky, W., 1971, Automated image analysis in autoradiography, *Exp. Cell Res.* **68:**388.

Price, J.L., and Wann, D.F., 1975, The use of quantitative autoradiography for axonal tracing experiments and an automated system for grain counting, in: *The Use of Axonal Transport for Studies of Neuronal Connectivity* (W.M. Cowan and M. Cuénod, eds.), pp. 155–172, Elsevier, Amsterdam.

Ramón y Cajal, S., 1928, *Degeneration and Regeneration of the Nervous System,* Oxford University Press, London.

Repérant, J., 1975, Orthograde transport of horseradish peroxidase in the visual system of the rat, *Brain Res.* **85:**307.

Rogers, A.W., 1972, Photometric merasurement of grain density in autoradiographs, *J. Microsc.* **96:**141.

Rogers, A.W., 1973, *Techniques of Autoradiography,* Elsevier, Amsterdam.

Sabri, M.I., and Ochs, S., 1972, Characterization of fast and slow transported proteins in dorsal root and sciatic nerve of cat, *J. Neurobiol.* **4:**145.

Salpeter, M.M., and Bachmann, L., 1964, Autoradiography with the electron-microscope: A procedure for improving resolution, sensitivity and contrast, *J. Cell Biol.* **22:**469.

Salpeter, M.M., Bachmann, L., and Salpeter, E.E., 1969, Resolution in electron microscope radioautography, *J. Cell Biol.* **41:**1.

Salpeter, M.M., Budd, G.C., and Mattimoe, S., 1974, Resolution in autoradiography using semithin sections, *J. Histochem. Cytochem.* **22:**217.

Schmeiser, K., 1957, *Radioaktive Isotope: Ihre Herstellung und Anwendung,* Springer-Verlag, Berlin.

Schmitt, F.O., 1968, Fibrous proteins—neuronal organelles, *Proc. Natl. Acad. Sci. U.S.A.* **60:**1092.

Schonbach, J., and Cuénod, M., 1971, Axoplasmic migration of protein: A light microscopic autoradiographic study in the avian retinotectal pathway, *Exp. Brain Res.* **12:**275.

Schonbach, J., Schonbach, C., and Cuénod, M., 1971, Rapid phase of axoplasmic flow and synaptic proteins: An electron microscopical autoradiographic study, *J. Comp. Neurol.* **141:**485.

Schonbach, J., Schonbach, C., and Cuénod, M., 1973, Distribution of transported proteins in the slow phase of axoplasmic flow: An electron microscopical autoradiographic study, *J. Comp. Neurol.* **152:**1.

Schubert, P., and Holländer, H., 1975, Methods for the delivery of tracers to the central nervous system, in: *The Use of Axonal Transport for Studies of Neuronal Connectivity* (W.M. Cowan and M. Cuénod, eds.), pp. 113–125, Elsevier, Amsterdam.

Schubert, P., Lux, H.D., and Kreutzberg, G.W., 1971, Single cell isotope injection technique, a tool for studying axonal and dendritic transport, *Acta Neuropathol. (Berlin) Suppl.* **5:**179.

Schubert, P., Kreutzberg, G.W., and Lux, H.D., 1972*a*, Neuroplasmic transport in dendrites: Effect of colchicine on morphology and physiology of motoneurons in the cat, *Brain Res.* **47:**331.

Schubert, P., Kreutzberg, G.W., and Lux, H.D., 1972*b*, Use of microelectrophoresis in the autoradiographic demonstration of fiber projections, *Brain Res.* **39:**274.

Sidman, R.L., 1970, Autoradiographic methods and principles for study of the nervous system with thymidine-H^3, in: *Contemporary Research Methods in Neuroanatomy* (W.J.H. Nauta and S.O.E. Ebbesson, eds.), pp. 252–274, Springer-Verlag, Berlin.

Sousa-Pinto, A., and Reis, F.F., 1975, Selective uptake of (^{3}H)leucine by projection neurons of the cat auditory cortex, *Brain Res.* **85:**331.

Specht, S., and Grafstein, B., 1973, Accumulation of radioactive protein in mouse cerebral cortex after injection of ^{3}H-fucose into the eye, *Exp. Neurol.* **41:**705.

Swanson, L.W., Cowan, W.M., and Jones, E.G., 1974, An autoradiographic study of the efferent connections of the ventral lateral geniculate nucleus in the albino rat and the cat, *J. Comp. Neurol.* **156:**143.

Taylor, A.C., and Weiss, P., 1965. Demonstration of axonal flow by the movement of tritium-labeled protein in mature optic nerve fibers, *Proc. Natl. Acad. Sci. U.S.A.* **54:**1521.

Trojanowsky, J.Q., and Jacobson, S., 1975, A combined horseradish peroxidase–autoradiographic investigation of reciprocal connections between superior temporal gyrus and pulvinar in squirrel monkey, *Brain Res.* **85:**347.

Vrensen, G., and De Groot, D., 1970, Some new aspects of efficiency of electron microscopic autoradiography with tritium, *J. Histochem. Cytochem.* **18:**278.

Walberg, F., Brodal, A., and Hoddevik, G.H., 1976, A note on the method of retrograde transport of horseradish peroxidase as a tool in studies of afferent cerebellar connections, particularly those from the inferior olive; with comments on the orthograde transport in Purkinje cell axons, *Exp. Brain Res.* **24:**383.

Weiss, P., 1964. The dynamics of the membrane-bound incompressible body: A mechanism of cellular and subcellular motility, *Proc. Natl. Acad. Sci. U.S.A.* **52:**1024.

Weiss, P., and Hiscoe, H.B., 1948, Experiments on the mechanism of nerve growth, *J. Exp. Zool.* **107:**315.

Wiesel, T.N., Hubel, D.H., and Lam, D., 1974, Autoradiographic demonstration of ocular-dominance columns in the monkey striate cortex by means of transneuronal transport, *Brain Res.* **79:**273.

Willard, M., Cowan, W.M., and Vagelos, P.R., 1974, The polypeptide composition of intra-axonally transported proteins: Evidence for four compositionally distinct phases of transport, *Proc. Natl. Acad. Sci. U.S.A.* **71:**2183.

Williams, M.A., 1969, The assessment of electron microscope autoradiographs. *Adv. Opt. Electron. Microsc.* **3:**219.

Williams, M.A., 1974, Progress in the analysis of electron microscopic autoradiographs, in: *Electron Microscopy and Cytochemistry* (E. Wisse, W.T. Daems, J. Molenaar, and P. van Duijn, eds.), pp. 327–340, North-Holland, Amsterdam.

10

Isotope Methods

S.S. Oja and P. Kontro

1. Introduction

Isotopes have numbered among the most important tools of modern scientific research since the discovery of principles of atomic nuclear structures. The recent rapid expansion in neurobiological knowledge also stems to a considerable degree from the application of isotopes to the study of metabolic processes and transport phenomena in the nervous system. Isotope methods in neurobiology have gained impetus only during the decades since the Second World War with the qualitatively and quantitatively increasing supply of suitable isotopes for chemical and biochemical investigation purposes and the development of versatile and automated detection methods. A great number of isotopes relevant to neurobiological research are currently industrially manufactured. Some of them are incorporated into organic molecules, and thus a great number of neurobiologically significant compounds are commercially available labeled with isotopes that are either stable or radioactive. There is no indication that isotope methods will lose any of their mandate in the field of neurobiology in the immediate future. Most neurobiologists therefore face the necessity of acquiring at least a basic understanding of the usability of isotopes in their problem-solving.

The aim of this chapter is to give, initially, a brief account of the physical background of isotope methods and then to survey the basic characteristics of the isotopes and the labeled compounds pertinent to neurobiological investigations. The physical principles underlying isotope detection methods will be outlined and the most common procedures and equipment for isotope assays reviewed. Further, special significance attaches to isotope-data analysis and the interpretation of results obtained. The examples to be cited later in

S.S. Oja and *P. Kontro* • Department of Biomedical Sciences, University of Tampere, Tampere, Finland.

the text have been selected from experiments on nervous tissue, but they are also more widely applicable to other biological experimental situations.

In writing this chapter, we have benefited greatly from a number of books on radiochemistry or isotope methods. We recommend the following publications in particular: The *Handbook of Radioactive Nuclides,* edited by Y. Wang (1969), enumerates properties of isotopes and concisely gives pertinent basic information on radiation protection and nuclear instrumentation. The monograph of McKay (1971) expounds the field of radiochemistry in its entirety. That of Hendee (1973) concentrates particularly on radiotracer methodology in biological research. The monograph of C.H. Wang *et al.* (1975) is more thoroughgoing, but it contains applications of environmental and physical sciences as well. The books of Simon (1974) and Ouseph (1975) deal with the detection of isotopes and describe the methods and instrumentation in current use. The monograph of Horrocks (1974) concentrates specifically on the liquid scintillation counting method. The monograph of Feinendegen (1967) is older, but nevertheless contains useful information on tritium-labeled molecules and their use in biological research. The catalogues of commercial manufacturers of isotopes give, in concise form, information on their products. These companies also publish additional information on various aspects of the isotope method. They supply this literature to their customers on request.

2. *Physical Background*

2.1. *Properties of the Nucleus*

An atom has a positively charged nucleus surrounded by a cloud of negatively charged electrons. The mass and charge of an electron are 9.1×10^{-31} kg and -1.6×10^{-19} C, respectively. The nucleus owes its positive charge to the presence within it of positively charged particles—i.e., protons. The mass of a proton is 1.6724×10^{-27} kg, and its charge is equal in magnitude to that of an electron, but is opposite in sign. The charge of the nucleus in a neutral atom is equal in magnitude to the total charge of all electrons and is thus $Z \times (1.6 \times 10^{-19})$ C, where Z is the number of protons. Z is referred to as the *atomic number* of the atom. In addition to protons, the nucleus contains neutral particles called neutrons that have a mass of 1.6747×10^{-27} kg. The *neutron number* of an atom is designated N. Protons and neutrons are often collectively called *nucleons.* The number of nucleons (A), i.e., the sum of protons and neutrons, is termed the *mass number* of the nucleus. One intrinsic nuclear propensity is the *binding energy* of the nucleus, i.e., the energy released in the hypothetical process of bringing its protons and neutrons together to form a nucleus. It is given in nuclear physics as *electron volts* (abbreviated eV). One electron volt equals 1.60219×10^{-19} J and is the energy acquired by an electron in moving through a potential difference of 1 V. Nuclear binding energies are considerable, often around

8 MeV per nucleon. They are over 10^6 times greater than chemical binding energies.

In nearly all of chemistry and in much of physics, Z is more significant than A or N. The chemical and physical properties of an element are determined mainly by its electron cloud outside the nucleus. For example, even the atoms and nuclei are named according to their Z values irrespective of their A and N values, e.g., $Z = 1$ hydrogen, $Z = 2$ helium, $Z = 3$ lithium, and so on. Each element thus has a characteristic number of protons and consequently of electrons. If two atoms have the same Z value but differ in A values, they are called *isotopes*. For most purposes, they are chemically indistinguishable (with the most notable partial exception of the lightest element, hydrogen). An isotope is designated by the chemical symbol and the mass number of the corresponding element as a left superscript, e.g., ^{12}C, ^{13}C, or ^{14}C for those carbon isotopes that contain 6 protons and 6, 7, or 8 neutrons. Less frequently, the mass number is denoted by a right superscript, such as C^{12}, or even C-12. Sometimes the atomic number is also added as a left subscript, e.g., $^{12}_{6}C$, $^{13}_{6}C$, or $^{14}_{6}C$. Atoms that contain the same number of neutrons but a different number of protons are classified as *isotones*. For example, $^{11}_{5}B$, $^{12}_{6}C$, and $^{13}_{7}N$ are isotones. Further, atoms with the same number of nucleons are *isobars*, such as $^{12}_{5}B$, $^{12}_{6}C$, and $^{12}_{7}N$. *Nuclide* is a general name for a nucleus in any form.

There exist stable and unstable nuclides. Unstable nuclides undergo nuclear translations—radioactive decay or nuclear fission—whereas in stable nuclides, no such activity occurs in any measurable degree. If the atomic number of a nuclide is greater than 83, it is invariably unstable and thus radioactive, but there are also a number of radioactive nuclides that have atomic numbers of less than 83. The ratio of neutrons to protons generally determines whether or not a nuclide is radioactive. In stable nuclei with low atomic numbers, there are approximately equal numbers of neutrons and protons, but if the atomic number is high, there are significantly more neutrons than protons. The combinations of even numbers of both protons and neutrons also outnumber the other combinations in stable nuclei; in particular, these nuclei very seldom contain simultaneously an odd number of both protons and neutrons.

2.2. *Types of Radioactive Decay*

2.2.1. *α-Decay*

The binding energy of ^4He is particularly high among the very light nuclei. Therefore, heavy nuclei in their radioactive decay often emit ^4He nuclei termed *α-particles*. In this *α-decay* process, the nucleus is considerably lightened; the atomic number Z decreases by 2 units and the atomic mass A by 4 units. A typical example is the decay of ^{235}U:

$$^{235}_{92}U \rightarrow \, ^{231}_{90}Th + \, ^{4}_{2}He$$

The emitted α-particles show discrete energies. Their penetration into material is poor; even in air, they generally travel only a few centimeters. Alpha-emitters are rarely used in isotope studies within neurobiology. Naturally, they are present in living organisms in extremely small trace amounts.

2.2.2. Negatron Decay

Often, nuclides with an excess of neutrons reach stability by the conversion of a neutron to a proton with an ensuing ejection of a negative electron, *negatron* β^-, from the nucleus. The process is accompanied by the formation of a massless neutral particle named *neutrino* ν. In this negatron decay process, the mass number of the nucleus remains unaltered, but the atomic number increases by 1 unit. A few typical examples of this type of decay are

$$^{3}_{1}H \rightarrow ^{3}_{2}He + {}_{-1}^{0}\beta + \nu$$

$$^{14}_{6}C \rightarrow ^{14}_{7}N + {}_{-1}^{0}\beta + \nu$$

$$^{32}_{15}P \rightarrow ^{32}_{16}S + {}_{-1}^{0}\beta + \nu$$

$$^{35}_{16}S \rightarrow ^{35}_{17}Cl + {}_{-1}^{0}\beta + \nu$$

Others include, for instance, the decays of ^{24}Na, ^{42}K, ^{45}Ca, ^{47}Ca, and ^{131}I. Notably, the most common isotopes in current use in neurobiological research decay by a negatron emission. Negatrons emitted in the processes depicted above possess a continuous energy spectrum characterized by the *maximum energy* E_{max}, i.e., the upper energy limit acquired by these particles. Each nuclide decaying by negatron emission has its own characteristic E_{max}. The *average energy* E_{av} of the negatrons is about one third of their corresponding E_{max}. The energy spectrum of negatrons arises from the partition of the total energy between the negatrons and neutrinos, the transition energy being equivalent to the E_{max}, representing the sum of the kinetic energies of a negatron and neutrino pair.

2.2.3. Positron Decay

Positron decay occurs in nuclei with a ratio of protons to neutrons too high for stability. A proton is transformed into a neutron accompanied by the emission of a *positron*, β^+*-particle*, and a massless, uncharged particle called *antineutrino* $\bar{\nu}$. The interactions of neutrinos and antineutrinos with matter are very weak, and much of our knowledge of them is derived solely from their effects on negatron and positron decay. The decay of ^{15}O is an example of the positron decays:

$$^{15}_{8}O \rightarrow ^{15}_{7}N + {}_{+1}^{0}\beta + \bar{\nu}$$

Other similar nuclides are ^{13}N, ^{17}F, ^{21}Na, ^{23}Mg, ^{38}K, and ^{121}I. As shown, the mass number of these nuclei remains unaltered in the process, but the atomic number decreases by 1 unit. All the aforenamed isotopes are relatively short-lived and therefore of limited use in neurobiological research.

2.2.4. *Electron Capture*

Another alternative mode of decay for nuclei that possess an excess of protons to neutrons is the capture of an orbital electron by the nucleus. Also in this transition, a proton is transformed into a neutron. About nine tenths of the electron captures are effected from the innermost electron orbit (K capture from the K orbit), about one tenth being from the L orbit (L capture). In some isolated cases, the transition energy may be too low to allow of capture from the K orbit; then a capture is possible only from the outer electron shells. An example of decay by this *electron capture* is

$$^{55}_{26}Fe \rightarrow ^{55}_{25}Mn + \bar{\nu}$$

In the process, an antineutrino is emitted. Further unstable nuclides that decay in this manner include ^{7}Be, ^{41}Ca, ^{48}Cr, ^{51}Cr, ^{57}Co, and ^{125}I.

Many unstable nuclides undergo either positron decay or electron capture alternatively. Such nuclides are ^{22}Na, ^{52}Fe, ^{58}Co, and ^{124}I. In positron decay, the mass of the products exceeds the original mass, which increment must be supplied by the energy released. Therefore, if the transition energy is less than 1.02 MeV, the decay is possible only by electron capture. With greater transition energies, both alternatives are possible. Negatron decay, positron decay, and electron capture are often collectively called β-*decay* processes, involving as they do either a positive or a negative β-particle. A few isotopes possess all these decay types simultaneously, exhibiting *branching decay* schemes. ^{64}Cu and ^{126}I are such nuclides, among others.

2.2.5. *Isomeric Transition*

In isomeric transition, the nucleus undergoes a transition from a higher energy level to a lower one. Surplus energy is released in the form of very short-wavelength electromagnetic radiation, called γ-*radiation*. The phenomenon implies the existence of one or more *excited states* for a nucleus in addition to its lowest energy state, the *ground state*. If the excited state is long-lived enough ($>10^{-6}$ sec) to be treated as a nuclear species of its own, it is termed a *metastable state* and is designated by a left (or right) superscript "m." An example is

$$^{99m}_{43}Tc \rightarrow ^{99}_{43}Tc + \gamma$$

Other metastable nuclei are, for instance, ^{24m}Na, ^{34m}Cl, ^{58m}Co, and

^{80m}Br. They often originate from other types of radioactive decay. Usually, the emission of γ-rays follows extremely rapidly after some other nuclear transition, such as electron capture or the ejection of an α- or β-particle. Therefore, most isotopes mentioned above are also γ-emitters, the most notable exceptions being ^{3}H, ^{14}C, ^{32}P, and ^{35}S. Any transition from a higher energy level to a lower one is accompanied by a γ-ray possessing a characteristic energy band (but no strictly sharp energy line).

2.2.6. *Other Processes*

There are also other events associated with the decay of unstable nuclei. Very heavy nuclei are able to divide into two roughly equal parts in *spontaneous fission.* Positrons from β-emitters may interact with negative electrons they meet, yielding *annihilation radiation,* i.e., γ-rays with an energy of 0.511 MeV. The breaking of β-particles by electromagnetic fields may produce continuous X-radiation, i.e., internal or external *bremsstrahlung.* Frequently, there occur interactions of nuclear transitions with the electron cloud outside the nucleus. Sometimes a part of the nuclear energy released is transferred to an inner electron ejected from the atom in a process called *internal conversion.* Also, a change in the nuclear charge following proton–neutron transitions alters the electrostatic field that controls the electron cloud of the atom. The atom may become electronically excited or lose electrons and become ionized in the process of *electron shake-off.* Any vacancy in one of the inner electron shells, due to electron capture, internal conversion, or electron shake-off, is filled by an electron dropping down from the next higher electron shell, resulting in an emission of *characteristic X-rays.* All the phenomena described above have but little relevance to isotope methods in neurobiology and will not be discussed further.

2.3. *Rate of Radioactive Decay*

The decay of an unstable nucleus is entirely a random event uninfluenced (with minor exceptions) by physical and chemical states of the element. The number of atoms that disintegrate within a given time in a radioactive sample, the *rate of radioactive decay,* depends entirely on the nuclear constitution, i.e., the relative stability of the nuclei. If we denote with N the number of atoms of a radioisotope in a given sample, then the relationship dN/dt denotes the number of atoms disintegrating within a given time, and the following equation applies:

$$-\frac{dN}{dt} = \lambda N \tag{1}$$

where λ is the *decay constant* of this particular radionuclide species. When equation (1) is integrated, the number N_t of radioactive atoms in a sample at time t may be determined from the equation:

$$N_t = N_0 e^{-\lambda t} \qquad (2)$$

where N_0 is the number of radioactive atoms at $t = 0$. The radioactivity (A) of a sample is directly proportional to the number of unstable atoms present multiplied by their decay constant:

$$A = \lambda N \qquad (3)$$

Similarly, the radioactivity of a given sample at time t is given by

$$A_t = A_0 e^{-\lambda t} \qquad (4)$$

where A_0 is the radioactivity of the sample at $t = 0$. The decay constants vary from one radioactive nuclide to another, and are characteristic for a given element.

The *half-life* ($t_{1/2}$) is another widely employed means of characterizing radioactive decay. It denotes the time required for a radionuclide to decay to one half its original radioactivity. After one half-life, 50% of radioactivity remains; after two half-lives, 25%; and so on. An expression for the half-life can be derived from equation (4):

$$t_{1/2} = \frac{\ln 2}{\lambda} = \frac{0.693}{\lambda} \qquad (5)$$

The half-lives vary from a fraction of a microsecond to billions of years owing to the variable relative stability of the unstable nuclei. Half-lives of radionuclides are generally estimated by measuring the radioactivity of a given sample after varying time periods and by drawing the best-fit straight line through the experimental points in a plot of the logarithm of radioactivity vs. time. From the slope of this straight line, the half-life and decay constant can be estimated (Fig. 1).

2.4. Units and Definitions

The classical measure of radioactivity has been the *curie* (Ci), the magnitude of which was originally derived from the number of disintegrations per second occurring in 1 g of pure radium and then numerically defined to be equal to 3.7×10^{10} 1/sec. The curie has now been superseded by the SI-unit (Système International d'Unités) *becquerel* (Bq) which was coined by the International Commission on Radiation Units and Measurements. One becquerel represents one nuclear transformation per second. Its dimension

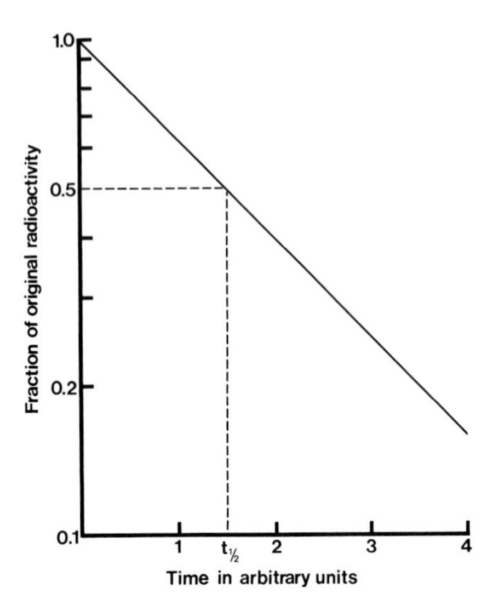

Fig. 1. Decay of a radionuclide as a function of time. Direct determination of the half-life from the semilogarithmic plot.

is thus 1/sec. The equivalency of the curie and becquerel units is as follows:

$$1\ \text{MCi} = 10^{6}\ \text{Ci} = 3.7 \times 10^{16}\ \text{Bq} = 37\ \text{PBq}$$

$$1\ \text{kCi} = 10^{3}\ \text{Ci} = 3.7 \times 10^{13}\ \text{Bq} = 37\ \text{TBq}$$

$$1\ \text{Ci} = 3.7 \times 10^{10}\ \text{Bq} = 37\ \text{GBq}$$

$$1\ \text{mCi} = 10^{-3}\ \text{Ci} = 3.7 \times 10^{7}\ \text{Bq} = 37\ \text{MBq}$$

$$1\ \mu\text{Ci} = 10^{-6}\ \text{Ci} = 3.7 \times 10^{4}\ \text{Bq} = 37\ \text{kBq}$$

$$1\ \text{nCi} = 10^{-9}\ \text{Ci} = 37\ \text{Bq}$$

$$1\ \text{pCi} = 10^{-12}\text{Ci} = 3.7 \times 10^{-2}\ \text{Bq} = 37\ \text{mBq}$$

In the earlier literature, disintegrations per second (dps) is a commonly found unit, but its user should be discouraged. Similarly, the unit disintegrations per minute (dpm) is entirely unofficial; it is equivalent to 60 Bq.

A radionuclide may exist in a carrier-free state, but generally it is mixed homogeneously with a certain amount of the stable nuclides of the same element. The *specific radioactivity* or *specific activity* is the ratio of the number of radioactive atoms of an element to the total number of atoms of the same element present in the mixture. This ratio is expressed with any of the units of radioactivity related to the mass unit of the element, e.g., MBq/kg or Bq/kg. Better notations are MBq/mol or Bq/mol, since they facilitate comparisons among compounds with different molecular weights. These should therefore be used whenever possible. Specific activities are frequently related to the specific activity of a reference compound. They are then termed *relative specific activities.* For instance, the specific activity of a product is often expressed relative to the specific activity of its precursor.

Most elements possess more than one stable isotope. Their relative abundance in an element in nature is generally constant. Stable isotopes can be used as tracers similarly to radioactive ones by following their enrichment and dilution in metabolic processes. The abundance of the labeling isotope in the starting material and in the products is gauged and compared with the natural abundance expressed as the *atom percent excess.*

3. Isotopes of Neurobiological Interest

Relatively few isotopes have widespread use in neurobiological research. Those that can be incorporated into organic molecules are most versatile. The most common isotopes include radionuclides of hydrogen, carbon, phosphorus, and sulfur. Unfortunately, nitrogen and oxygen do not possess suitable radionuclides. Sodium, potassium, calcium, and chloride ions have also been traced with their radioactive isotopes. Labeling of proteins with radioactive iodine and radioimmunoassay methods necessitate the use of radioactive iodine isotopes. A variety of gamma emitters have been used in brain scans and cerebral blood flow measurements. Their applications lie mostly in clinical medicine, and we shall not discuss them in detail. Other isotopes have been less popular in neurobiological studies.

The most abundant hydrogen isotope, protium (^{1}H), amounts to 99.985% of the total hydrogen in nature, the remaining 0.015% being almost entirely the other stable hydrogen isotope, deuterium (^{2}H or D). The physicochemical properties of deuterium differ from those of protium enough to allow a separation of these nuclides by chemical exchange and distillation. Deuterium has occasionally been used in neurobiological research. Its applicability as a tracer is based on the facility of detecting the mass difference between deuterium and protium.

The third hydrogen isotope, tritium (^{3}H or T), is radioactive, emitting very-low-energy negatrons. Figure 2 depicts the energy spectrum of the β-radiation from tritium. Like all energy spectra of negatron emitters, the spectrum of tritium is continuous from zero to the maximum energy. The maximum energy is 18.6 keV and the average energy 5.6 keV. Owing to their low kinetic energy, the β-particles emitted by tritium are classified as "soft" radiation ($E_{max} < 0.2$ MeV). The mass of tritium is three times greater than that of protium. The physicochemical properties of these isotopes thus differ appreciably, which may considerably influence the behavior of compounds labeled with tritium, provoking so-called *isotope effects,* i.e., influencing reaction and transport rates, altering equilibria, and others. With respect to deuterium, such isotope effects are less pronounced. They are still smaller

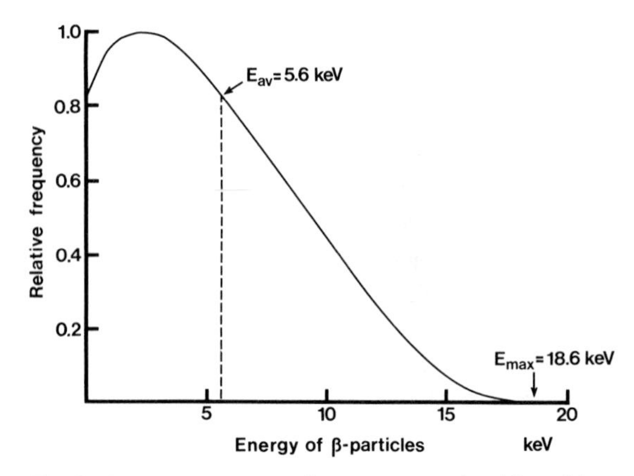

Fig. 2. Energy spectrum of negatrons emitted by tritium.

with isotopes of heavier elements, since the relative mass difference diminishes with increasing atomic number. In nature, there is one tritium atom per 10^{18} hydrogen atoms. Since the half-life of tritium is only 12.35 years, some tritium must be formed continuously at high altitudes from deuterium isotopes by the agency of cosmic radiation. Commercially, tritium is manufactured from the ^{6}Li isotope by bombardment with neutrons generated by nuclear reactors. In the ensuing nuclear reaction, an α-particle is split off:

$$^{6}_{3}\mathrm{Li} + ^{1}_{0}\mathrm{n} \rightarrow ^{4}_{2}\mathrm{He} + ^{3}_{1}\mathrm{H}$$

This scheme can be more conveniently written $^{6}\mathrm{Li}(n,\alpha)^{3}\mathrm{H}$.

The most common radioactive isotope of carbon, ^{14}C, likewise emits soft β-radiation. In the atmosphere, there is about one ^{14}C atom per 10^{12} carbon atoms owing to production of ^{14}C by cosmic rays in the reaction $^{14}\mathrm{N}(n,p)^{14}\mathrm{C}$. The commercial production of ^{14}C is similarly accomplished from ^{14}N by neutrons from nuclear reactions. In the process, a proton emerges along with the radiocarbon. The distribution of ^{14}C in nature is not uniform. Radiocarbon in living material equilibrates with the atmospheric ^{14}C, but the exchange ceases to operate in dead material, in which ^{14}C slowly disappears owing to its decay with a half-life of 5730 years. This radiocarbon-decrement rate is the basis for the so-called *^{14}C-dating* method for historically ancient objects. The accuracy of dating implies a virtually unchanged intensity of cosmic-ray bombardment during past millennia.

Some ^{32}P and ^{35}S are also formed by cosmic radiation, but in quantities of minor importance. Their presence in nature is also very low in comparison with that of tritium and radiocarbon, due to their short half-lives. The β-radiation emitted by ^{32}P should be classified as "hard," since the maximum kinetic energy of negatrons ejected from ^{32}P nuclei is in excess of 1 MeV. ^{35}S emits soft β-radiation. The half-lives of these isotopes are so short that their physical decay rate must often be taken into account as a correction factor in analyzing biological results. Also, ^{32}P and ^{35}S are at present produced in nuclear reactors by bombardment with neutrons.

Other neurobiologically pertinent radioactive isotopes are compiled in Table 1 with some of their nuclear properties. The major proportion of artificially produced isotopes are prepared today with the aid of reactors, which have to a great extent superseded the formerly dominating cyclotrons, although the latter accelerators still have some important specialized roles to fulfil. In Table 1, the specific activity at 100% isotope abundance is in most cases only a theoretical quantity. It indicates the highest specific activity theoretically achievable. In favorable cases, ^{3}H and ^{14}C compounds can be obtained for research in almost 100% isotopic purity and, for example, ^{32}P and ^{35}S at an abundance of 50% or more.

The isotopes of nitrogen and oxygen that are radioactive are so short-lived that they have been of only limited use in neurobiological research. The practical alternative is to label these elements with their rare stable isotopes, ^{15}N, ^{17}O, and ^{18}O. The natural abundance of ^{15}N is 0.37%, and that of ^{17}O and ^{18}O is 0.04 and 0.20%, respectively. The physicochemical properties of the stable isotopes of nitrogen and oxygen differ enough to allow a separation of ^{14}N and ^{15}N by chemical exchange and of ^{16}O, ^{17}O, and ^{18}O by chemical exchange and distillation. Due to the laborious and technically sophisticated methods required for their detection, these stable isotopes have been used relatively rarely in neurobiological research.

4. Labeled Compounds

4.1. Availability

Organic compounds labeled with uncommon atoms have wide applicability in neurobiological research as tracers of their natural counterparts. A compound can be labeled either *isotopically* or *nonisotopically*. In an isotopically labeled compound, one or more atoms have been replaced by an uncommon isotope of the same element. A nonisotopically labeled compound possesses the atom of an element that does not occur in the molecule in its natural form. The elementary composition of neurobiologically interesting compounds permits the labeling of either hydrogen, carbon, nitrogen, oxygen, phosphorus, or sulfur. Unfortunately, the longest-lived radioactive isotopes of nitrogen and oxygen have half-lives of only 10.0 and 2.0 min, respectively. Therefore, the labeling of these elements is at present practicable only with stable isotopes. The others can also be traced with their radioactive isotopes. The most common nuclides for nonisotopic labeling are ^{125}I and ^{131}I, in particular in the labeling of proteins.

The neurobiologically most interesting compounds are at present commercially available labeled with at least one radioactive isotope. Often, there are available several kinds of labeled molecular species for one compound, the labeling atom being either hydrogen, carbon, phosphorus, or sulfur,

Table 1. Some Commercially Available Radioactive Isotopes of Potential Neurobiological Interest[a]

Nuclide	Production reaction	Half-life	Specific activity at 100% isotopic abundance (Bq/mol)	Type of decay (%)[b]	Maximum β-energies (%)[b] (MeV)	Photon energies (%)[b] (MeV)	Decay product
$^{3}_{1}H$	$^{6}Li(n,\alpha)^{3}H$	12.3 yr	1.1×10^{15}	β^-	0.0186	—	$^{3}_{2}He$
$^{14}_{6}C$	$^{14}N(n,p)^{14}C$	5730 yr	2.3×10^{12}	β^-	0.156	—	$^{14}_{7}N$
$^{22}_{11}Na$	$^{25}Mg(p,\alpha)^{22}Na$ $^{26}Mg(p,\alpha n)^{22}Na$	2.60 yr	5.2×10^{15}	β^+ (90) EC (10)	0.54 (90) 1.82 (trace)	Ne X-rays 0.51 (180) 1.28 (100)	$^{22}_{10}Ne$
$^{24}_{11}Na$	$^{23}Na(n,\gamma)^{24}Na$	15.0 hr	7.8×10^{18}	β^-	1.39 (100) Others to 4.17 (trace)	1.37 (100) 2.75 (100) Others to 4.23 (trace)	$^{24}_{12}Mg$
$^{28}_{12}Mg$	$^{26}Mg(T,p)^{28}Mg$	21.3 hr	5.6×10^{18}	β^-	0.46	0.031 (96) 0.40 (30) 0.95 (30) 1.35 (70)	$^{28}_{13}Al \rightarrow ^{28}_{14}Si$
$^{32}_{15}P$	$^{32}S(n,p)^{32}P$ $^{31}P(n,\gamma)^{32}P$	14.3 days	3.4×10^{17}	β^-	1.71	—	$^{32}_{16}S$
$^{33}_{15}P$	$^{33}S(n,p)^{33}P$	25.0 days	1.9×10^{17}	β^-	0.25	—	$^{33}_{16}S$
$^{35}_{16}S$	$^{35}Cl(n,p)^{35}S$ $^{34}S(n,\gamma)^{35}S$	87 days	5.6×10^{16}	β^-	0.167	—	$^{35}_{17}Cl$
$^{36}_{17}Cl$	$^{35}Cl(n,\gamma)^{36}Cl$	3.0×10^{5} yr	4.4×10^{10}	β^- (98) EC (2) β^+ (trace)	0.71 (98)	S X-rays 0.51 (trace)	$^{36}_{16}S$ $^{36}_{18}Ar$

$^{38}_{17}Cl$	$^{37}Cl(n,\gamma)^{38}Cl$	37.2 min	1.9×10^{20}	β^-	1.2 (31) 2.8 (16) 4.9 (53)	1.60 (38) 2.17 (47)	$^{38}_{18}Ar$
$^{40}_{19}K$	Naturally occurring	1.3×10^9 yr	1.0×10^7	β^- (89) EC (11) β^+ (trace)	1.32 (89) 0.48 (trace)	Ar X-rays 1.46 (11) 0.51 (trace)	$^{40}_{18}Ar$ $^{40}_{20}Ca$
$^{42}_{19}K$	$^{41}K(n,\gamma)^{42}K$	12.4 hr	9.3×10^{18}	β^-	2.0 (18) 3.5 (82) Others (trace)	1.52 (18) Others to 2.4 (trace)	$^{42}_{20}Ca$
$^{43}_{19}K$	$^{40}Ar(\alpha,p)^{43}K$	22.4 hr	5.2×10^{18}	β^-	0.82 (87) Others to 1.81	Several energies to 1.0	$^{43}_{20}Ca$
$^{45}_{20}Ca$	$^{44}Ca(n,\gamma)^{45}Ca$	165 days	2.9×10^{16}	β^-	0.26 (100)	0.0125 (trace)	$^{45}_{21}Sc$
$^{47}_{20}Ca$	$^{46}Ca(n,\gamma)^{47}Ca$	4.6 days	1.0×10^{18}	β^-	0.67 (82) 2.0 (18)	0.5 (5) 0.8 (5) 1.3 (74)	$^{47}_{21}Sc \rightarrow ^{47}_{22}Ti$
$^{125}_{53}I$	$^{124}Xe(n,\gamma)^{125}Xe \xrightarrow[EC]{} ^{125}I$	60 days	8.1×10^{16}	EC	0.03 e^- (93)	Te X-rays 0.035 (7)	$^{125}_{52}Te$
$^{131}_{53}I$	$^{130}Te(n,\gamma)^{131}Te \xrightarrow[\beta^-]{} ^{131}I$	8.05 days	5.9×10^{17}	β^-	0.61 (87) Others to 0.81	0.36 (82) Others to 0.72	$^{131}_{54}Xe$

[a] The table is compiled from different sources in which the data vary slightly, and most numbers are therefore rounded. Manufacturing of the isotopes is often possible through different nuclear reactions. Representative processes in current use are indicated. Specific activities of isotopes at 100% isotopic abundance (carrier-free) are theoretical numbers seldom encountered in practice.

[b] In cases of branching decay, the numbers in parentheses indicate the branching ratios or the intensities of β- and γ-radiations as percentages. The sum of intensities of β-particles emitted amounts to 100%. The intensities of γ-radiation are similarly related to the number of decaying nuclei; due to the annihilation γ-rays of positrons emitted and to transitions among varying energy levels, the sum of percentages of intensities of γ-rays often exceeds 100. ^{125}I emits monoenergetic electrons. Some minor intensities are not indicated.

depending on the molecular structure of the compound in question. Furthermore, if a particular labeled compound is not listed in the sales catalogues, it can be prepared to order for the customer. Sometimes an experimental design requires the use of a certain compound with the label at a defined position in the molecule, particularly if the carbon chain of the labeled organic molecule is cleaved during the experiment. Often, however, the investigator has a more liberal choice. Together with such matters as stability, radiochemical purity, and detection methods, the price should also be considered. Tritium-labeled compounds are the cheapest of all. Compounds labeled with ^{14}C are generally 10–100 times more expensive on the bequerel basis. Compounds labeled with phosphorus and sulfur isotopes are often cheaper than those labeled with ^{14}C, but more expensive than the tritium-labeled ones. Of course, prices vary tremendously from one compound to another, mostly owing to reasons other than the original production costs of the labeling isotopes themselves.

4.2. Position of Label

Rarely are all particular atoms in a compound replaced by their uncommon isotopes (compounds in which they are being called *carrier-free* compounds). The distribution of the label in isotopically labeled compounds is often more or less well known. The proportion of the molecules that bear the label can be inferred from the specific activity and the molecular structure of any compound. In *specifically labeled* compounds, the position of the labeling atom in the molecule is precisely known and indicated in the name of the compound. For instance, the notation L-[4,5-^{3}H]leucine signifies that more than 95% of the tritium atoms are attached to carbon atoms 4 and 5 in the leucine molecule. Similarly, in [1-^{14}C]linoleic acid, at least 95% of ^{14}C is in the carbon atom at position 1, and in adenosine 5′-[$\alpha^{32}P$]triphosphate, 95% of ^{32}P is in the α-phosphorus. The designation *nominally labeled* is applied to compounds that preferentially contain the label in the given position(s) in the molecule but for which no further information is available as to the extent of labeling at other positions. For instance, the notation [2(N)-^{3}H]glycerol indicates the preferential labeling of hydrogen at carbon 2 in the molecule, but not in excess of 95%. *Uniformly labeled* compounds contain the label in all possible positions distributed in a uniform or nearly uniform pattern. The designation implies synthesis from a uniformly labeled simple precursor such as $^{14}CO_2$. For example, [U-^{14}C]oxalic acid and D-[U-^{14}C]glucose are such uniformly labeled compounds. The term *generally labeled* denotes that there is a random (nonuniform and undetermined) distribution of the label at various positions in the molecule. Many potential labeling sites may be entirely devoid of the label. These compounds are generally labeled with tritium. Δ^{1}-[G-^{3}H]Tetrahydrocannabinol and L-[G-^{3}H]phenylalanine are labeled compounds of this kind.

The nomenclature of labeled compounds referred to as the "square-brackets-preceding" system has gradually gained international acceptance and is currently used by many scientific journals. The symbol for the isotope is placed in square brackets and inserted at the beginning of the chemical name of a labeled compound, e.g., [1,2-^{3}H(N)]cortisone and [^{35}S]cysteamine. If a specific part of the molecule is labeled, the brackets should precede that part of the name to which it refers, as in 3-*O*-[^{14}C]methyl-D-glucose and 3-*O*-methyl-D-[U-^{14}C]glucose. Also, notations such as L-[^{14}C(U)]glutamic acid have been used instead of L-[U-^{14}C]glutamic acid, and lower case letters n and g have been used instead of N and G to indicate nominally and generally labeled compounds, respectively. If the labeling atom is not an integral part of the natural molecule, its symbol may be separated from the name by a hyphen, e.g., ^{125}I-albumin.

4.3. Manufacturing of Labeled Compounds

A special technology has been developed for the manufacturing of labeled compounds. In itself, the safe handling of dangerous radioactive material necessitates proper precautions. Products with high specific activities are often desired, and any dilution with unlabeled material must thus be avoided in their synthesis. The labeled precursors are often costly, precluding the use of an excess of labeled intermediates at any step in the reaction processes. Likewise, for reasons of economy, small-scale batch sizes must often be used. In general, exchange, substitution, and addition reactions are among the most common procedures for adding a labeling atom to an organic compound.

Hydrogens bound to nitrogen, oxygen, and sulfur are labile and thus exchange in the groups $-NH_2$, $=NH$, $-OH$, $-COOH$, and $-SH$ rapidly with hydrogen in water. Hydrogen–carbon bonds are generally stable under normal conditions, with the exception of hydrogens in the *ortho* and *para* positions of phenol. To prepare labeled compounds for tracer studies, tritium (or deuterium) must be introduced into bonds that are stable and remain so in metabolic reactions. There are various methods of introducing tritium (or deuterium) into organic compounds. The substitution of hydrogen by tritium gas, tritiated water, or tritiated acid can be accomplished in the presence of a special hydrogen catalyst, e.g., platinum or another noble metal. Compounds containing groups that can be reduced with metal hydrides have also been labeled using, for instance, tritiated sodium borohydride. Hydrogenation of unsaturated compounds with tritium gas or replacement of halogens with tritium in the presence of a suitable catalyst is similarly feasible. Tritium labeling (but not deuterium labeling) can also be achieved by using the energy of recoiling tritium from a nuclear reaction. In a similar manner, the radiation of tritium in gaseous form also evokes tritium–hydrogen exchange (the Wilzbach method) in large quantities under conditions in which no ordinary chemical reaction would be expected. Particularly, the Wilzbach method with its modifications has proved a very simple and versatile approach to the random labeling of organic molecules with tritium. Practically all organic compounds can be labeled with tritium by any of the aforementioned methods. After proper chemical purification and careful removal of labile tritium by repeated equilibration with water or other solvents, reasonably stable tritium-labeled compounds for tracer studies are obtainable.

Chemical synthesis is usually the preferred method for labeling organic compounds with ^{14}C, ^{32}P, ^{35}S, ^{125}I, and ^{131}I. The synthesis starts with simple inorganic precursors, such as $^{14}CO_2$, $^{32}PO_4^{3-}$, $^{32}PCl_3$, $^{35}SO_4^{2-}$, $H_2^{35}S$, and $^{131}I_2$. The incorporation of ^{14}C into complex organic compounds can be achieved via several simple organic intermediates such as formic acid, acetic acid, methanol, cyanide, cyanamide, acetylene, and so on. Chemical synthesis usually results in a specifically labeled compound. If the products are optically active, they are generally racemic mixtures of D- and L-isomers. Biosynthetic methods using various living organisms and enzyme systems provide a useful supplement to the other methods outlined above for preparing labeled compounds, in particular those labeled with ^{14}C. Bacteria, algae, and plants have been grown in an atmosphere containing $^{14}CO_2$ as the sole source of carbon or have been exposed to tritiated water during their growth cycle. Proteins, amino acids, sugars, lipids, and alkaloids labeled with ^{14}C or ^{3}H have been produced by biosynthesis. The yield and specific activities attained are often high and all possible labeling sites are occupied, but the isolation and purification of the desired product may be tedious and difficult.

4.4. Purity of Labeled Compounds

Purity of labeled material has, in radiochemistry, several connotations. *Chemical purity* of labeled compounds is as important as among ordinary chemicals. The possible presence of nonlabeled impurities is often ignored by investigators, although these impurities can affect experimental results and seriously accelerate deterioration of labeled material on storage. In commercial preparations, a check for nonlabeled impurities has generally been kept by the manufacturer. *Radioactive purity* or *radionuclide purity* or *radioisotope purity* states the proportion of the total radioactivity that is in the form of the stated radionuclide. Labeled compounds used in neurobiology pass this criterion easily, since in these cases it can usually be taken for granted that only one radioactive species is present.

Among organic compounds labeled with ^{3}H, ^{14}C, ^{32}P, ^{35}S, and other isotopes, we are concerned with whether or not the label is confined to one particular compound, or to one particular position in the molecule, or even to one particular enantiomorph of the compound. *Radiochemical purity* denotes the fraction of the radionuclide that is in the correct chemical form. The degree of radiochemical purity required varies with the intended use of the labeled compound. Even traces of radioactive impurities may in some cases greatly distort the experimental results, e.g., when one is studying enzymatic transformations of metabolically relatively inert substances. Many optically active labeled compounds are supplied as racemic mixtures, and in such cases the investigator must be aware that generally only one isomer is biologically preferred and other labeled enantiomorphic forms may cause interference.

Radiochemical purity can be assessed by various methods including partition, adsorption, ion-exchange, gel-exclusion, thin-layer, column, and gas chromatography, and isotopic-dilution analysis combined with spectroscopic methods. The manufactureres check the radiochemical purity of their products on preparation and then at regular intervals during storage. These checks do not necessarily coincide with the time of delivery of labeled compounds to customers. Nonoptimal conditions during transportation from the manufacturer to the customer or in the storage facilities available to the latter may considerably accelerate the rate of deterioration of a labeled compound, which may thus become contaminated with radioactive impurities despite their absence on delivery. Investigators should not be too confident of the declarations of the manufacturers, but should themselves check the purity of their labeled compounds more often than they apparently do now.

4.5. Stability and Storage of Labeled Compounds

Radioactively labeled compounds decompose on storage. Radioactive decomposition products are generally more harmful for the investigator than the nonradioactive ones. Decomposition may occur by various mechanisms.

4.5.1. *Primary Internal Radiation Effect (Transmutation Effect)*

When a radionuclide undergoes β-decay, it is transformed into another element. A chemical impurity emerges from a molecule labeled with this disintegrating atom irrespective of whether or not the transformed atom is retained by the molecule. Radioactive impurities will be created only if individual molecules contain two or more radioactive atoms. The abundance of radioactive impurities thus formed can be calculated for any given case by knowing the decay constant of the radionuclide and the specific radioactivity of the labeled compound on the molar basis. Only rarely do low-molecular-weight compounds contain an abundance of multilabeled molecules. The effect must be seriously considered only when dealing with labeled compounds of high specific activity or compounds with very large labeled molecules or with very short-lived labels.

The primary internal radiation effect cannot be controlled for a given specific activity of a compound labeled with any particular radionuclide. The only effective remedy is to keep the specific activity of the labeled compound as low as the experimental design permits.

4.5.2. *Primary External Radiation Effect*

The primary external radiation effect is a more serious source of radioactive impurities than the primary internal effect. Labeled molecules, single-labeled as well as multilabeled, may be degraded due to radiation emitted by adjacent molecules in the same sample. One single β-particle emitted may hit and in theory damage hundreds and thousands of neighboring molecules by knocking off electrons, breaking covalent chemical bonds, forming radicals, or causing other destruction. Different compounds possess different radiosensitivities.

Some degree of control of the primary external effect is possible by dispersal of labeled molecules and by intercepting nuclear particles emitted and dissipating their energy harmlessly. Dilution may even separate labeled molecules beyond the range of the nuclear particles they emit. Dilution with inert solvent or storage of labeled compounds in a dry state as a thin film on a supporting structure has been commonly used. As with the primary internal effect, in particular with solids, labeled compounds with low specific activities are less prone to decomposition under this primary external radiation effect.

4.5.3. Secondary Radiation Effect

Secondary radiation effects are often the main cause of decomposition of radiactive compounds on storage. Here, decomposition results from interactions of labeled molecules with reactive intermediates, such as free radicals, created by primary radiation effects on either the labeled molecules themselves or their environment. The degree of decomposition due to secondary radiation effects varies greatly with the sensitivity of the labeled compounds and with storage conditions.

Much can be done to protect labeled compounds against secondary radiation effects: (1) Storage of a labeled compound at reduced temperatures is often advantageous, but freezing cannot always be recommended. When a sample is frozen, the mobility of reactive intermediates is certainly strongly limited, but such intermediates may become "frozen" as well and react when the sample is rewarmed. When solutions are frozen, radioactive molecules may collect, thus increasing the likelihood of primary external radiation effects. Solutions of radioactive compounds are often best stored at temperatures just above freezing point. (2) Solid compounds should be free of any moisture, since reactive intermediates are readily formed from water by radiation. (3) Scavengers of free radicals can be used. Classic free-radical scavengers are of limited use, since they enchance sample decomposition through chemical effects, but 2% ethanol and 1% benzyl alcohol in aqueous solutions have proved quite useful, being fairly innocuous to labeled samples. (4) Solvents, containers, and the labeled compounds themselves should be as pure as possible. While many impurities may be innocuous as such, the products of their interaction with nuclear particles may be very harmful indeed.

4.5.4. Chemical Decomposition

Chemical decomposition produces radioactive and nonradioactive impurities in labeled samples on storage in the same manner as in other originally pure chemicals. The same mechanisms are operative in both cases, e.g., hydrolysis, oxidation. Chemical decomposition may be troublesome, since labeled compounds in solution are often used at extremely low concentrations. Here, chemical reactions not otherwise detectable become apparent and harmful. Compounds that are sensitive to light undergo photochemical decomposition, and unsterilized sampies may be attacked by microbes.

In general, it is wise to avoid prolonged storage of labeled material. Commercially available labeled compounds should rather be purchased at suitably spaced intervals than kept in stock by the investigator. Manufacturers supply information on storage conditions along with their preparations, and further data on the stability of labeled material in various conditions are likewise generally available from them.

5. Measurement of Isotopes

5.1. Stable Isotopes

The detection methods for stable isotopes differ greatly from those for radioactive isotopes, but the principles of their use as tracers are essentially the same. The sensitivity of detection methods for stable isotopes is several orders of magnitude less than that for radioactive isotopes. Detection of stable isotopes is also greatly influenced by the natural abundance of the tracer isotope and the amount of contaminating carrier element in samples. Such matters considerably hamper the use of very minute doses of stable isotopes in experiments. On the other hand, an enrichment of stable isotopes up to as high an abundance as 100% is possible even *in vivo*. Nor are there any radiation hazards in human experiments when stable isotopes are used. Of the various stable isotopes, only ^{2}H, ^{15}N, and ^{18}O have any notable applicability in neurobiological research, ^{2}H as an alternate for the radioactive ^{3}H, and ^{15}N and ^{18}O because of the absence of long-lived radioactive nuclides of nitrogen and oxygen.

5.1.1. *Principles of Detection*

The stable nuclei of an element differ in mass, magnetic moment, and reactions with nuclear particles. Thence, the methods of detecting stable isotopes include mass spectroscopy, ultracentrifugation, densitometry, gas chromatography, emission spectroscopy, nuclear magnetic resonance spectroscopy, electron spin resonance spectroscopy, and measurement of radiation emitted by the sample after its treatment with nuclear particles (activation analysis). Of these methods, mass spectroscopy has by far the greatest applicability.

Quantitative analysis of the abundance of different isotope species is facilitated by a conversion of samples into simple gaseous compounds such as hydrogen, nitrogen, or carbon dioxide. Also, water is convenient for the determination of deuterium or ^{18}O. Organic material in samples is degraded into these simple compounds by methods familiar from organic analytical chemistry, including stepwise oxidation or reduction in the presence of appropriate catalysts. In certain modifications, the whole conversion can be accomplished as one step in sealed ampoules at high temperatures. On the other hand, this procedure destroys all organic material in the samples, and no picture of the intermolecular or intramolecular distribution of the isotopes can be gained. Analyses of ^{2}H, ^{15}N, or ^{18}O with mass spectroscopy, ^{2}H or ^{18}O with densitometry, ^{2}H with infrared spectroscopy, and ^{15}N with emission spectroscopy are feasible as routine procedures. Intermolecular and intramolecular distribution of stable isotopes can be analyzed only within a sample left initially undegraded. Intermolecular distribution can be assessed by mass spectroscopy and to a limited extent with gas chromatography. Intramolecular distribution has been estimated by mass spectroscopy. Also useful are optic and magnetic methods based on the structure of molecules.

5.1.2. *Mass Spectroscopy*

In mass spectroscopy, ionized molecule fragments are sorted out according to their masses. A sample rendered previously volatile is introduced into an evacuated chamber, where it is fragmented and ionized. The beam of ionized molecule fragments is accelerated, collimated, and passed through a powerful electromagnetic field that fans the ions into curvilinear paths according to their mass-to-charge ratios. The ions are detected on a collector, where the signal is amplified and recorded. A discontinuous spectrum of signal intensities as a function of particle masses emerges. In this way, molecules and their fragments containing ^{1}H or ^{2}H, ^{14}N or ^{15}N, ^{16}O or ^{18}O, or other isotopes are separated and their relative abundances gauged, thus also making possible inferences of the intermolecular and intramolecular distribution of isotope species. The results can be rapidly analyzed with a small-scale computer combined with the mass spectrometer. Separation of a mixture of volatile compounds by a gas chromatograph connected directly to the inlet of a mass spectrometer greatly increases the versatility of the method.

5.1.3. *Nuclear Magnetic Resonance and Electron Spin Resonance*

Nuclear magnetic resonance (NMR) techniques are based on the permanent magnetic moments of nuclei with odd mass numbers or odd numbers of protons. In electron spin resonance (ESR) techniques, the permanent magnetic moments of elements with unpaired electrons are used in the same manner. When an external magnetic field is applied to substances, they orientate themselves so that the magnetic fields of the nuclear or electronic spins are either parallel or antiparallel to the external magnetic field. The two orientations represent two different energy states. At certain magnetic field strengths permitting translations between different energy states, there occurs absorption of energy from the external electromagnetic field by the sample, and discontinuous absorption spectra can be recorded. Each compound possesses a characteristic NMR and ESR spectrum that gives detailed information as to its molecular structure. For instance, protons in organic molecules are shielded to different degrees from the applied external magnetic field owing to the actions of neighboring nuclei and binding electrons. They thus yield different signals and are recognizable. A substitution of protons with deuteriums will greatly alter the signal. Therefore, most NMR measurements in context with stable isotopes have served for quantitative inter- and intramolecular localization of deuterium. Recently, the methods for identification of ^{13}C with NMR have made rapid advances. ^{15}N and ^{17}O can also be analyzed with NMR, whereas ^{18}O has no magnetic moment. Owing to interactions of the magnetic field produced by the spin of unpaired electrons with the magnetic moment of certain atomic nuclei, the ESR spectrum reflects the fine structure of the whole molecule as well and can therefore also be used to identify the intramolecular position of ^{2}H, ^{15}N, or ^{17}O isotopes. The sensitivity of ESR measurements greatly exceeds that of NMR analyses.

5.1.4. *Activation Analysis*

The most sensitive determination methods for many stable isotopes are based on activation analyses. A sample is irradiated with nuclear particles and there ensues a nuclear reaction that may lead to the formation of a radioactive nucleus the radioactivity of which can be measured. Nuclear reactors or particle accelerators serve as the source of nuclear particles. Thermal neutrons find some applicability in the analyses of light stable nuclei only for ^{31}P, ^{37}Cl, and possibly for ^{34}S, whereas fast neutrons, protons, or deuterons can be used to activate ^{2}H, ^{13}C, ^{15}N, ^{18}O, ^{31}P, ^{34}S, and ^{37}Cl. The induced radiation from samples is not influenced by the chemical binding or the physical state of the nuclide. It gives no information as to the intermolecular or intramolecular distribution of isotopes, only their total amounts in the sample. A localization of induced radiation even within very tiny samples is possible.

5.1.5. *Other Methods*

The density of water containing heavier isotopes of hydrogen or oxygen is greater than that of $^{1}H_2{}^{16}O$. In densitometric techniques, the velocity of a falling water drop is measured in a liquid that is immiscible with water. In this way, the contents of ^{2}H and ^{18}O in a water sample can be measured with an accuracy equal to that attained in mass-spectrometric analyses. The deuterium content of water can also be estimated from infrared spectra of water samples owing to the differing absorption lines of H_2O, HDO, and D_2O. This method is not yet readily applicable for differentiation among isotope species of other elements. The content of ^{15}N in samples can be analyzed from the intensity of emission lines in the spectra of excited nitrogen gases. The sensitivity and accuracy of this method for ^{15}N exceed those of mass-spectrometric techniques.

5.2. Radioactive Isotopes

Detection of radioactive isotopes is based on interactions of the emitted α-, β-, or γ-radiation with material, mostly on ionizations produced in different media such as gases or semiconducting solids, or on excitations of orbital electrons in solids and liquids. We shall briefly discuss the principles of radioactivity measurements based on gas ionization, scintillation in solid and liquid fluors, and semiconductor detectors. Nuclear emulsions and autoradiography are dealt with in Chapter 9.

5.2.1. Gas-Ionization Detectors

Orbital electrons are dislodged from atoms when ionizing radiation deposits energy in a gas. The dislodged negative electron and the positive ion comprising the remainder of the atom form an *ion pair*. The minimum energy required, the so-called *ionization potential*, varies from gas to gas and is dependent on the type of radiation involved. An α-particle produces very intense ionization, a β-particle less dense ionization, and γ-rays only very little. Accordingly, gas-ionization detectors effectively disclose α- and β-particles, but their efficiency in the case of γ-rays is less. Most ion pairs immediately recombine. This can be thwarted by enclosing the gas between two electrodes and applying a potential gradient across them. Negative ions will traverse to the cathode and positive ions to the anode, thereby creating a measurable pulse amenable to recording.

The magnitude of the electrical gradient has a profound influence on the yield of ions collected by the electrode (Fig. 3). Most of the ion pairs have an opportunity to recombine when the voltage gradient is small. An increase

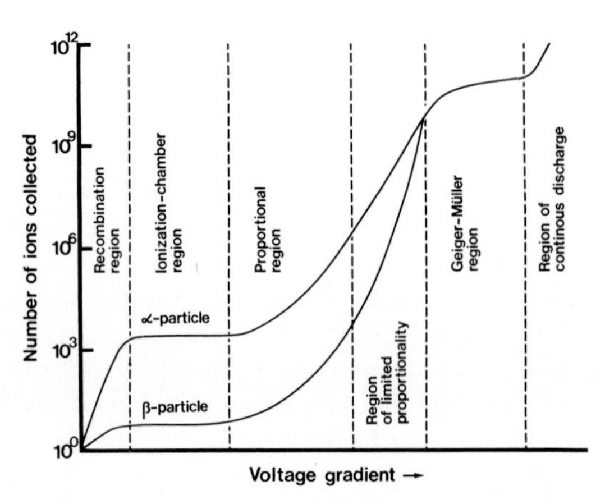

Fig. 3. Ionization response in a gas-filled chamber as a function of the voltage gradient.

in the applied potential will add to the yield until virtually all the ions brought about by the incident ionizing particle within the sensitive gas volume are collected. A heavily ionizing α-particle creates about a hundred to a thousand times more ions than a β-particle. A still greater increase in the potential accelerates the ions, but does not swell the yield until the electrons become so energized that they are capable of secondarily ionizing new gas atoms. This *gas amplification* results in a geometric increase of ions. This ion torrent is termed the *Townsend avalanche.* Within this *proportional region,* the number of ions collected by the electrodes is related—even if greatly amplified—to the total number of ion pairs primarily created by ionizing radiation. A strong pulse can be recorded, its size being proportional to the ionizing capacity of the primary radioactive event. The total number of gas atoms within the sensitive volume sets an upper voltage limit for the proportionality region, when virtually all gas atoms become ionized, first due to heavily ionizing α-particles and gradually to β-particles as well. The *Geiger–Müller region* is reached when even the least energetic β-particle invariably ionizes all gas atoms as an outcome of internal amplification. All the pulses recorded are of the same large size. Still higher voltages result in a continuous discharge.

5.2.1a. Ionization Chambers. The three plateaus in the pulse yield–voltage gradient curve depicted in Fig. 3 are used in three different types of ionization detectors. Simple ionization-chamber instruments make use of the first plateau with no internal gas amplification. Even with external electrical amplification, the difficulty of precise measurement of minute pulses in simple ionization chambers renders their use impractical for the individual recording of nuclear particles or γ-rays. On the other hand, only low external voltages are required, the voltage plateau employed is entirely flat, and therefore the external supply of the chamber potential need not be well stabilized. Most commonly, ionization-chamber instruments are used for the collection of integrated charges and the measurement of total amounts of ionizations, as in the determination of radiation doses of personnel exposed to radiation hazards.

5.2.1b. Proportional Counters. Detectors designed to operate in the proportional region of ion multiplication are called proportional counters. In such instruments, an external high-voltage supply maintains a voltage gradient across the electrodes. The pulses resulting from the avalanche of electrons collected on the anode wire pass through amplifiers and a discriminator to be finally recorded on a scaler. A highly stable high-voltage supply is essential, since the internal gas amplification in the proportional region is heavily dependent on the external potential applied. It is feasible with proportional counters to differentiate between α- and β-particles emitted by the same sample on the basis of the pulse size, as depicted in Fig. 4. At a fixed discriminator level, the size of pulses due to α-particles reaches the detection limit first when the potential gradient is increased. At higher potentials, β-particles also become detectable, the most energetic ones being the first to be amplified enough to be recorded. The first plateau of counts

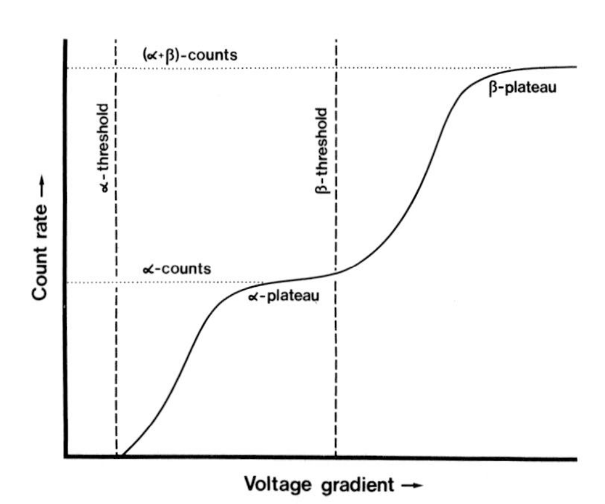

Fig. 4. Characteristic operation curve for a proportional detector counting a sample emitting both α- and β-particles when an increasing external potential is applied across the electrodes.

represents α-particles only, and the second plateau both α- and β-particles. The efficiency of proportional detectors for γ-rays is so low that they are seldom used for gamma-counting. An appreciable number of gas atoms in the vicinity of the anode wire are involved in the formation of ion pairs in proportional detectors. This fact renders the detector unresponsive to the next ionizing particle for a short interval, termed the *dead time*, during which the ions from the previous ionization are collected. The length of the dead time varies depending on the detector construction, being generally less than 2 μsec.

5.2.1c. Geiger–Müller (GM) Detectors. GM detectors operate within the highest voltage plateau depicted in Fig. 3. In this region, any primary ionization of the counting gas by nuclear particles or X- or γ-rays secondarily ionizes virtually all gas atoms within the sensitive gas volume of detectors. Consequently, all pulses generated are large and of uniform size independent of the type of impinging radiation. Since the internal gas amplification is large, no additional external amplification before pulse-counting may be necessary. Figure 5 depicts the shape of the characteristic operation curve for a GM detector. At low external voltage levels, no counts are recorded, since the pulses generated are too small to pass the discriminator. After the starting voltage, the number of counts first increases rapidly and then plateaus. A GM detector is operated within the range of this plateau. The pulse size is not critical as long as it exceeds the discriminator level. Therefore, the external voltage supply need not be highly stabilized. The electronic components in GM instruments are often simple and inexpensive.

Some problems arise from the high internal gas amplification in GM detectors. Their dead time is long, generally 100–300 μsec. The great number of electrons collected by the anode leaves behind an equal number of positive ions, which migrate more slowly toward the cathode. During their migration, the detector is paralyzed and therefore cannot precisely register radiation emitted by a very radioactive sample. Positive ions at the cathode wall give rise to a further problem in that they attract, from the wall, electrons that neutralize them. The excess energy of these newly acquired electrons causes emission of electromagnetic radiation in the form of ultraviolet photons. These in turn may initiate dislodgment of additional electrons. This self-perpetuating discharge in GM detectors can be forestalled by *quenching* the emission of photons. In modern GM detectors, this is done by adding small amounts of internal quench agents, such as polyatomic organic gases (ethyl alcohol, ethyl formiate, amyl acetate) or halogens (Br_2, Cl_2). The quench-gas molecules collide with the migrating positive ions, become ionized themselves, and receive the dislodged electrons from the cathode wall. Instead of photon emission, excess energy initiates dissociation of the quench agents. Organic quench molecules are irreversibly degraded, whereas halogen atoms recombine after their dissociation. The life span of GM detectors with organic quench gases is therefore limited to 10^8–10^{10} registered pulses. Halogen-quenched tubes theoretically possess an indefinite useful life, but their quenching efficiency may vary considerably from location to location within the sensitive chamber volume, rendering these tubes qualified for less precise measurements only. GM tubes are commonly used for both α- and β-particles and γ-rays. They are nearly 100% efficient in detecting α- and β-particles as long as the particles penetrate into the sensitive volume of the detectors. The efficiency for γ-rays may be only about 1%, and a few percent at best.

5.2.1d. Counting Characteristics. The counting gas in GM tubes is an inert gas such as helium, neon, or argon under subatmospheric pressure. Proportional detectors are usually filled with either methane or mixtures of precious gases with hydrocarbons (e.g., 90% Ar + 10% CH_4, 96% He + 4% isobutane, 95% Ar + 5% CO_2). In the most common designs of gas-ionization detectors, the sensitive gas volume is enclosed between the central anode, a thin straight wire or a loop, and the surrounding cylindrical cathode wall. Commercially available proportional and GM detectors are generally in the shape of either a cylinder or a hemisphere, equipped with a thin end-window. Samples are spread out on counting planchets that are placed in front of the end-window

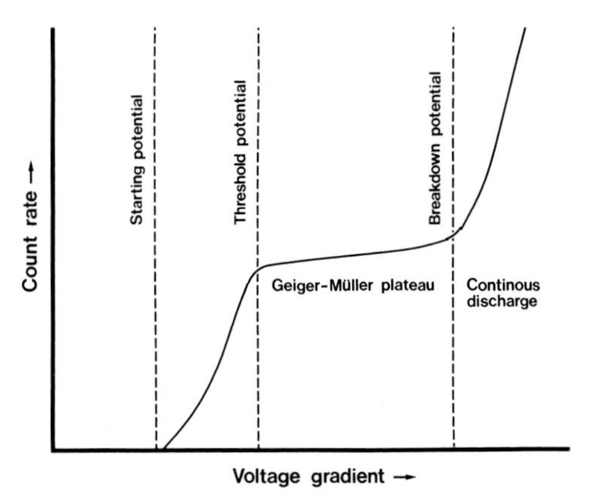

Fig. 5. Characteristic operation curve for a Geiger–Müller detector when an increasing external potential is applied across the electrodes.

to be counted. Considerable trouble has been taken to maximally reduce the thickness of end-windows in order to facilitate the penetration of α- or weak β-particles into the sensitive volume of the detector. The energies of β-particles emitted by certain radioisotopes, notably tritium, are so low that the particles cannot penetrate across even an ultrathin end-window without considerable absorption loss. For such cases, windowless detectors have been developed in which the sample is placed directly into the sensitive volume. In such instruments, a certain amount of air invariably enters the detector and must be flushed out with a continuous counting-gas flow. These are called *gas-flow detectors* and are superior to the end-window detectors in efficiency. Their efficiency amounts to about 50% owing to 2π counting geometry when the possible self-absorption by samples is ignored.

In a common modification of cylindrical detectors, gaseous radioactive samples are introduced directly into the sensitive volume of detectors and mixed with the counting gas. A conversion of solid or liquid samples into radioactive gases must precede counting. ^{14}C has been measured in carbon dioxide, methane, ethane, or acetylene. ^{3}H has, in a similar manner, been counted as hydrogen gas with methane. Efficiency with this *internal gas counting* is high, since radiation emitted in all directions by the sample will be detectable because of the 4π geometry of counting. There is also no self-absorption of nuclear particles by the sample. Further modifications of cylindrical detectors allow the counting of liquid samples. Either the detector is immersed in radioactive liquid or the sample flows along tubing through the sensitive volume.

Even without a radioactive sample, the detectors indicate a small number of ionizations taking place. Such *background counts* result—in the absence of X- and γ-rays—from heavy unstable nuclei or naturally occurring ^{40}K. To shield the detector from background radiation, it is often surrounded by lead. In *low-background counters*, even such a screen is not enough when very low levels of radioactivity are to be accurately gauged. In these instruments, the detector proper is encircled by a number of other GM detectors connected with the sample detector by *anticoincidence circuitry*. Only pulses originating in the sample detector will be counted, not pulses registered simultaneously by both the guard and sample detectors, since the latter most likely result from external background radiation.

5.2.2. *Semiconductor Detectors*

Semiconductor radiation detectors respond to both charged particles and photons emitted by radioactive nuclei. They have been used increasingly in nuclear chemistry and physics during the last few years and will very likely soon find further applications in tracer research. The basic operation mechanism of semiconductor devices is similar to that of gas-ionization detectors. In both cases, impinging ionizing radiation interacts within the sensitive volume with detector material that is solid crystal in the former and gas in the latter. In gas-ionization detectors, electrons and positive ions are separated and collected, whereas in semiconductor radiation detectors, negatively and positively charged species consist of electrons and "*holes*" in the crystal lattice.

Semiconductors are substances the electrical resistance of which is fairly high in the cold but decreases with increasing temperature. Certain semiconducting crystals of tetrahedral structure, notably those of silicon and germanium, are preferentially used in modern technology. Free mobility of electrons is restricted by their confinement to valence bands in these crystals. Extra energy is needed to raise an electron from the level of the valence band to the level of the conducting band separated by a forbidden gap in pure crystals (Fig. 6). In radiation detectors, the crystals have been doped either with an electron donor or with acceptor impurities. For instance, an *n-type* semiconductor is formed from crystals of tetravalent elements by doping with a pentavalent atom (e.g., phosphorus, antimony). Since the impurity has five valence electrons, one electron is "left over" and occupies a donor level near the conductor band and may easily be torn off to conduct electricity through the crystal. A *p-type* semiconductor is formed by doping the crystal with a trivalent atom (boron, indium, gallium). A "hole" is left in the crystal lattice, since such an impurity has only three valence electrons. The hole can be easily filled by a neighboring electron, leaving in turn another hole that can be filled by another electron, and so on. In this manner, holes migrate in the valence bands as electrons migrate in conduction bands.

For radiation detectors, a junction is formed from the n- and p-types of semiconductors. Electrons from the n-type migrate into the p-type, thus creating at the junction a *depleted region* from which mobile electrons and holes vanish. Migration can be enhanced by applying a reverse bias across the junction, i.e., connecting negative potential to the p-type and positive potential to the n-type. The depleted region serves as the sensitive volume of the semiconductor radiation detector. If ionizing radiation impinges on the depleted region, electrons are raised to the conducting band, leaving holes behind in the valence band. These electrons and holes then migrate to the positive and negative terminals in the n- and p-sides, respectively, generating thereby a traceable pulse. The aforementioned depleted region and hence the sensitive volume are relatively narrow. They can be widened

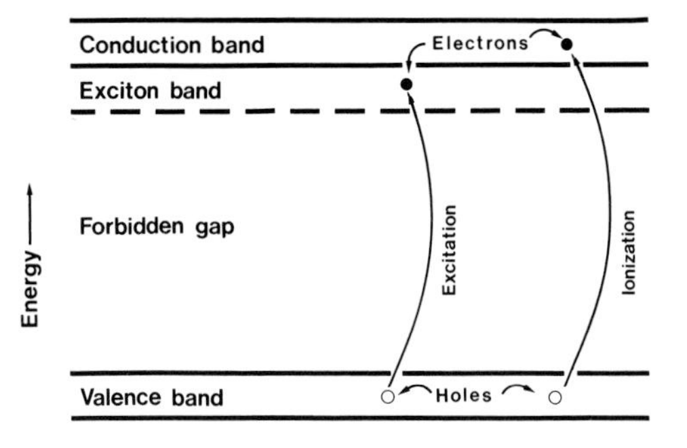

Fig. 6. Schematic diagram of the energy levels in an ionic crystal according to the band theory of solids: elevation of electrons from the valence band to the conduction or exciton band by impinging radiation.

in a variety of ways, the *lithium-drifted* detectors being the most common. Small and mobile lithium ions are able to diffuse into the semiconductor crystal lattice. Under the reverse bias, the lithium ions migrate into the p-type semiconductor, balancing there the electron-acceptor atoms. In this manner, the sensitive depleted region can be widened severalfold. Silicon–lithium, Si(Li), and germanium–lithium, Ge(Li), detectors are in current use in semiconductor radiation devices. Since the atomic number of germanium is higher than that of silicon, Ge(Li) detectors are more sensitive than Si(Li) detectors toward γ-rays. Ge(Li) detectors must, however, be operated at very low temperatures and must therefore be cooled with liquid nitrogen, which is somewhat inconvenient.

Some characteristic properties of semiconductor radiation detectors are as follows: (1) Energy resolution is excellent. For instance, differentiation among a great number of γ-emitters can be carried out as a single analysis on one sample. (2) This occurs, however, at the expense of detection efficiency. (3) Detectors are of small size because of the use of a solid as the sensitive material. (4) They show a linear response to energy deposited irrespective of the type of radiation. (5) Detectors have very thin entrance windows that absorb negligible energy. (6) Pulses generated have a fast rising time. (7) Detectors have a short dead time. (8) On the other hand, small output signals necessitate the use of high-quality and expensive electronic components.

5.2.3. Solid Scintillation Detectors

Scintillation detectors are based on the interaction of radiation with substances called *fluors* or *scintillators.* In certain inorganic and organic crystals, light is emitted when radiation is absorbed. In the scintillation process, radiation initially produces excitation of the fluors. In this process, molecules of the organic fluors are raised from their ground states to excited electronic states. Within nanoseconds, the excited states decay by fluorescence to various vibrational levels near the ground state through emission of light quanta. In inorganic crystals, radiation—in addition to ionization—excites valence-band electrons to an energy level that is somewhat lower than the conduction band (see Fig. 6), and therefore excited electrons remain bound to their holes in the valence bands. Such an electron-hole pair, an *exciton,* is also movable within the crystal. Irregularities in the crystal lattice owing to embedded impurities are capable of electron or electron-hole captures. These *activator centers* are temporarily excited, but return to their ground states within microseconds by emission of light.

A sodium iodide crystal doped with thallium ions, NaI(Tl), is the most widely used solid scintillator for the detection of γ-rays. Other crystals, such as ZnS(Ag), CsI(Tl), LiI(Eu), or anthracene, are used in certain cases for the detection of nuclear particles. Organic crystals are used for β-particles or electrons, but are very inefficient for counting γ-rays. They have very short decay times. Organic fluors can be employed as solid crystals, impregnated in solid plastic, or dissolved in an organic solvent in liquid scintillation counting, which is discussed in the next section. Large NaI(Tl) crystals can be grown by special techniques to increase counting efficiencies. The light output of this scintillator is largest of all. The crystals used may adopt various shapes, with a straight cylinder being the most common. To ensure a maximally advantageous counting geometry, a well may be drilled in the crystal to accommodate samples. The cores of many modern γ-counters constitute such well crystals onto which liquid samples are automatically mounted for counting.

NaI(Tl) crystals are hygroscopic, and they must be sealed in a dry atmosphere inside aluminum cylinders, leaving to one side an optical window of clear glass or quartz facing a *photomultiplier* tube. The photons emitted by the activator centers pass through the optical window and impinge on a *photocathode* commonly composed of antimony–cesium or silver–magnesium alloys, although recently developed bialkali (antimony–potassium–cesium) or trialkali (antimony–potassium–sodium–cesium) photocathodes show improved quantum efficiencies. The photocathode ejects one electron per 5–10 incident photons. A series of dynodes maintained at increasingly positive potentials to the cathode multiply at each step the number of electrons that are finally collected at the anode. The signal is further amplified electronically and then recorded. Its size is proportional to the energy dissipated by radiation in the crystal. Therefore, solid scintillation detectors are used to monitor γ-ray spectra emitted by radioactive samples, to identify unknown radionuclides, and to quantify their amounts. The energies of incoming γ-rays can be sorted out by recording the pulse-height distribution emerging from the scintillation detector by means of *multichannel analyzers*.

5.2.4. *Liquid Scintillation Method*

5.2.4a. General Principles. Liquid scintillation counting is at present the generally preferred method for detection of weak β-particles such as those emitted by ^{3}H, ^{14}C, or ^{35}S. High-energy β-particles emitted by ^{32}P and to a lesser extent α-particles or γ-rays have likewise been registered with liquid scintillation. The technique has a definite advantage over gas counting—which is also sensitive toward weak β-particles—in that liquid scintillation can be automated to handle a great number of samples within a reasonable counting time. In liquid scintillation, an organic fluor is mixed intimately in solution with the sample to be counted, thus creating circumstances equivalent to 4π detection geometry. The efficiency of detection is high and may even approach 100% in favorable cases. Superficially, the principles of liquid scintillation counting resemble those of solid scintillation detection of radiation. In both cases, radiation excites fluor molecules, the result being the emission of photons that hit the photocathodes of the photomultiplier tubes, and after severalfold amplification of the number of photoelectrons, a measurable pulse is registered at the anodes. The size of the pulse in liquid scintillation detectors is also related to the amount of energy dissipated by radiation in the fluor solution. For instance, β-particles generate about 7 photons per kilo-electron-volt of energy dissipated, and the final number of photoelectrons collected by the anode of the photomultiplier tube is related to the number of photons, even if this ratio is subject to statistical variations. A liquid scintillation instrument thus yields an energy spectrum of incident β-particles in its sensitive detection volume.

In a *liquid scintillation cocktail*, a small amount of an organic fluor is dissolved in a large quantity of an organic solvent. Owing to the low concentration of the fluor, a direct interaction of nuclear particles with fluor molecules is an unlikely event. The initial excitation therefore occurs in the solute molecules, from which energy is ultimately transferred by solvent–solvent and solvent–solute interactions to the fluor molecules by the agency of mechanisms not yet fully understood. Solvent–solvent energy transfer takes place in subnanosecond time and is monoenergetic in nature. Only the fluor solute is responsible for photon emissions.

5.2.4b. Scintillation Solutions. Alkylated aromatic hydrocarbons (e.g., toluene, xylene, trimethylbenzene) exhibit the best efficiencies as *primary solvents* in scintillation counting. The availability of toluene in high purity of moderate prices has rendered it most widely used. Biological samples are not generally soluble in aromatic hydrocarbons, and therefore more polar *secondary solvents*, such as methanol, ethanol, 1,4-dioxane, and 2-ethoxyethanole, have been included in scintillation solvent mixtures to accommodate aqueous samples. Quaternary ammonium compounds can also be used as solubilizing agents of aqueous biological samples; such substances hide behind the trademarks of commercially available *solubilizers*. 1,4-Dioxane has been used by many investigators in the past as a primary solvent for aqueous samples, but its use involves certain disadvantages. The low energy-transfer efficiency of dioxane can be improved by the addition of naphthalene.

Primary solutes are fluor substances that efficiently convert excitation energy to light quanta. The chemical structures of the fluor substances are complex, and thence, in their names, typical recognizable groups are commonly designated with capital letters, e.g., P for phenyl, B for biphenyl, N for naphthyl, O for oxazole, and D for the oxadiazole group. 2,5-Diphenyloxazole (PPO) has been the most widely used primary solute because of its good solubility in scintillation solvents. The most efficient primary solute is 2-(4-biphenyl)-5-phenyl-1,3,4-oxadiazole (PBD), which possesses, however, only limited solubility. A newly discovered, promising primary solute is 2-(4′-*t*-butylphenyl)-5-(4″-biphenyl)-1,3,4-oxadiazole (butyl-PBD). The optimum concentrations of primary solutes in scintillation cocktails amount to some grams per liter, the scintillation efficiency often starting to decline again at still higher concentrations.

The light emitted by primary fluors often has a wavelength that does not coincide fully with the region of the greatest sensitivity of photocathodes. The role of the *secondary solute* in the scintillation cocktail is to absorb the light emitted by primary fluors and reemit it at a longer wavelength in the region of greater photocathode sensitivity. Several secondary solutes have been in use, e.g., 1,4-bis-2-(5-phenyloxazolyl)-benzene (POPOP), the more soluble 1,4-bis-2-(4-methyl-5-phenyloxazylyl)-benzene (dimethyl-POPOP), and the less qualified 2-(α-naphthyl)-5-phenyloxazole (α-NPO). Two newly introduced, satisfactorily soluble substances are 2-(4-biphenylyl)-6-phenylbenzoxazole (PBBO) and *p*-bis-*o*-methylstyrylbenzene (bis-MSB). The concentration of secondary fluors may be as little as one hundredth of that of the primary fluor, but despite this, all photons emitted seem to arise from the secondary fluors. A great number of other compounds have been tested as primary or secondary fluors with more or less success. We have mentioned only those that have gained general recognition. It may be noted that new bialkali photomultipliers with a good spectral response at shorter wavelengths may eliminate the need to use a secondary solute as wavelength shifter.

5.2.4c. Operating Characteristics. Photons generated by scintillation impinge upon the photocathodes of photomultipliers. Such devices have an inherent, temperature-dependent noise that gives rise to a high background level of the order of $1-4 \times 10^5$ pulses per minute. The background can be lowered by refrigerating the photomultipliers, but a still more drastic decrement can be achieved by connecting two photomultipliers in a coincidence circuit. Only scintillation events registered by the two facing photomultipliers will be processed. Since noise in any photomultiplier tube is random and the coincidence-resolving times have been reduced to as low as 20 nsec, the noise level of modern photomultipliers operated in a coincidence mode is below 1 pulse per minute. Refrigeration is also no longer necessary. Coincident pulses can also, however, be generated in the two photomultipliers by processes other than the radioactivity of the sample. With two facing photomultipliers, a light-producing event, resulting from either electrical discharges, discharges of residual gases in photomultiplier tubes, or cosmic radiation or natural radioactivity in the construction materials, can be concomitantly "seen" by both photomultipliers. Similarly, cosmic rays and natural radioactivity can generate scintillation events within the sample itself. To reduce background, the entire detection unit is shielded by lead, and instead of glass bottles, quartz or plastic sample vials—low in ^{40}K—are sometimes employed. For the same purpose, solvents in scintillation cocktails are preferably manufactured from material that has not been alive during the present nuclear era. In modern liquid scintillation instruments, the coincident pulses of the two photomultipliers are summed, in this process reducing the variability of pulse sizes resulting from uneven distribution of photons on the two photocathodes. The summed pulses are then fed to individual single-channel analyzers equipped with fast scalers. Electrical amplification of pulses is by one of two modes: linear or logarithmic. In the instruments with linear amplification, the final signals are directly proportional to the energy of the β-particles that produced the scintillation, whereas logarithmic amplification generates signals proportional to the logarithm of the summed output pulses of the photomultiplier tubes.

Figure 7 displays the spectra of pulses generated by β-particles emitted by ^{3}H, ^{14}C, and ^{32}P—the nuclides most commonly counted by liquid scintillation—as they emerge from a logarithmic amplifier. All such continuous β-spectra extend from zero up to a maximum that is characteristic for each β-emitter. The shape and position of the spectra are greatly influenced by the gain of electronic amplification. Two spectral properties should be particularly noted: (1) the spectra of all β-emitters overlap more or less and (2) there are always β-particles so weak that they escape detection. In the case of ^{3}H, the proportion of these is, of course, particularly high. In liquid scintillation counters, pulses are registered with single-channel analyzers in which the "windows," framed by the lower and upper discriminator levels, are adjusted for an optimal detection of incident scintillation provoked by a given nuclide. For any selected amplification-gain setting, the upper discriminator level can be easily selected to encompass even the most energetic pulses. The lower discriminator must cut off optimally the background noise, sparing simultaneously as many sample pulses as possible. Often, the optimal counting conditions are specified by the highest possible value of the *figure of merit* represented by S^2/B, in which S is the counting rate of the given sample and B the background count rate.

5.2.4d. Quenching. Serious problems often arise in liquid scintillation counting from the phenomenon known as *quenching,* which reduces the amount of light incident upon the photocathodes of a scintillation counting device. Three major mechanisms of quenching can be classified as *chemical, color,* and *optical quenching.* Chemical quenching is caused by agents that interfere with the transfer of energy from solvent to fluor molecules in the scintillation cocktail. All nonfluorescent dissolved molecules, in particular polar compounds and oxygen, are quenching agents, since they may absorb energy from the excited solvent molecules or simply dilute the fluor molecules, thus reducing the likelihood of scintillation events. Even an excessively high primary fluor concentration may induce self-quenching. Any colored

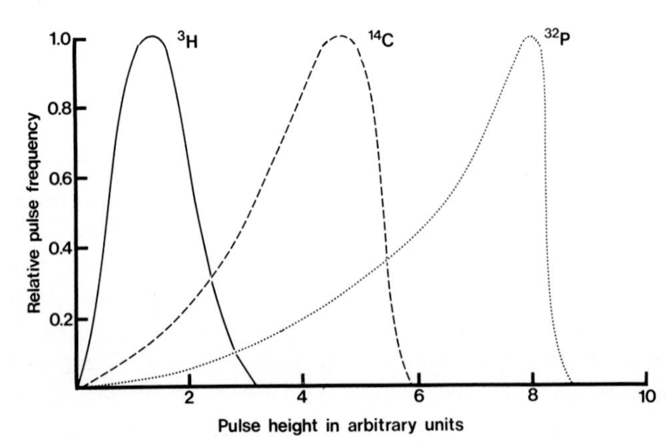

Fig. 7. Spectrum of pulses generated by ^{3}H, ^{14}C, and ^{32}P as they emerge from a logarithmic amplifier in a liquid scintillation instrument.

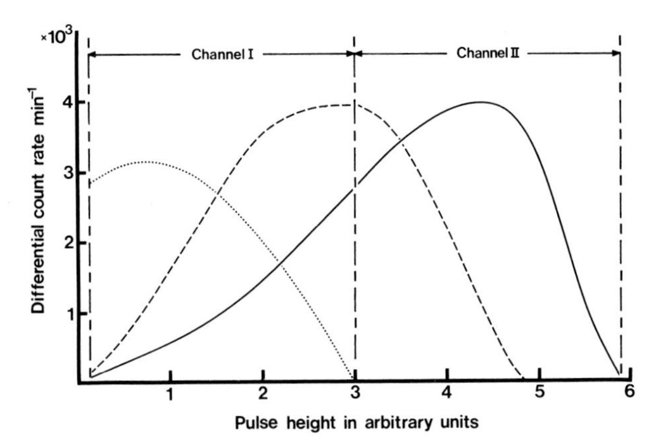

Fig. 8. Effect of quenching on the logarithmically amplified spectrum of pulses resulting from β-particles emitted by ^{14}C. (——) Unquenched sample; (-----) moderately quenched sample; (······) severely quenched sample. Note the shift of the whole spectrum toward the range of lower energies, the change in the shape of the spectrum, and the reduction in counting efficiency when quenching becomes severe. The distribution of pulses between two nonoverlapping counting Channels I and II is profoundly influenced by quenching. The Channel II/Channel I ratio diminishes with increasing quenching.

material in the scintillation solution may absorb photons emitted by the primary or secondary fluors, which absorption represents color quenching. In optical quenching, the absorption of the photons emitted is affected by any clouding of the scintillation cocktails and by fogging of the sample vials with dust, dirt, fingerprints, or other contaminant. The overall effect of any quenching is a shift of the pulse-height spectra to the left and distortion of their original shape, as shown in Fig. 8. Also, an appreciable number of the weakest pulses do not reach the lower discriminator level in samples severely quenched, and counting efficiency is inevitably diminished. Some degree of quenching invariably occurs in every liquid scintillation sample. The efficiency of counting must therefore be assessed separately, at least for each specific kind of sample. Quenching represents by far the most significant influence on counting efficiency, since the background level, internal temperature, and electronic operations are kept well stabilized in modern scintillation instruments.

5.2.4e. Determination of Counting Efficiency. A variety of methods to estimate the degree of quenching and efficiency of counting have been proposed. We shall briefly outline only those most frequently used. *Internal standardization* is the oldest method and perhaps still the most accurate one. A sample is initially counted, a precisely known quantity of the same nuclide is then added to the counting vial, and the sample is recounted. The difference in counting rates divided by the absolute disintegration rate for the standard gives counting efficiency. The absolute amount of radioactivity in the sample can be calculated from counting efficiency and the number of counts initially collected from the intact sample. This method is subject to inaccuracies in pipetting small quantities of liquid and to the hazards associated with opening up the vials and adding something extra to the scintillation fluid.

In the *channels-ratio method,* the pulse-height spectrum is recorded by two separate single-channel analyzers. The width of their counting windows can be selected in a number of ways, the counting channels partially overlapping or not. A nonoverlapping mode for channel selection is depicted in Fig. 8. It can be seen in this figure that when quenching becomes more severe and counting efficiency is reduced, the number of counts registered by Channel II diminishes in relation to the counts in Channel I. The Channel II/Channel I ratios and counting efficiencies for increasingly quenched standard samples are determined first. With the aid of these data, an assessment of counting efficiency for variably quenched unknown samples from their measured Channel II/Channel I ratios is feasible.

In most modern automatic liquid scintillation instruments, an external γ-emitter is used in conjunction with channels-ratios for determining the counting efficiency of quenched samples in a procedure termed the *external-standard channels-ratio method.* An appropriate γ-ray source, such as ^{137}Cs or ^{226}Ra, induces in the counting glass vials and the scintillation cocktail the production of *Compton electrons,* the energy spectra of which resemble those of soft β-emitters. Quenching in scintillation fluids influences their spectra in a manner somewhat similar to that shown for the spectrum of ^{14}C in Fig. 8. In addition to three sample single-channel analyzers in the most current instruments, two independent single-channel analyzers are used exclusively to obtain the external-standard ratios. A sample is always first counted normally, and the external γ-source is then placed in the close vicinity of the sample, which is then recounted. The counting efficiencies and external-standard channels-ratios are initially determined for a series of increasingly quenched standard samples, and the data obtained are stored in the memory of a built-in small-size computing unit. With such information, the absolute amount of radioactivity in variously quenched unknown samples can be printed out automatically as disintegrations per unit of time. The more

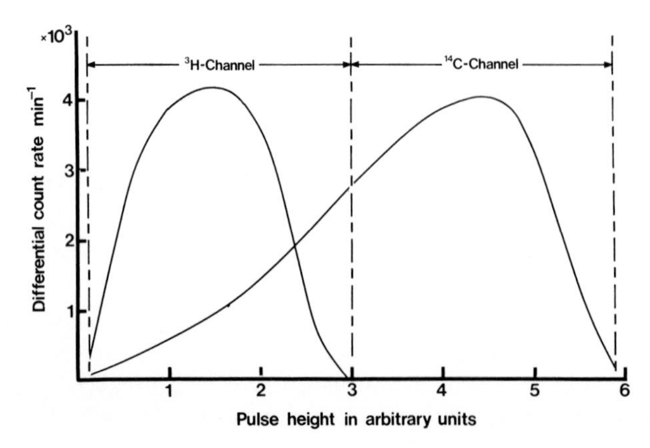

Fig. 9. Spectra of the logarithmically amplified pulses resulting from β-particles emitted by ^{3}H and ^{14}C and a nonoverlapping mode for the selection of the counting channels for their simultaneous determination.

advanced and more expensive instruments are also capable of more sophisticated computing operations. Certain models also automatically change the gain of the detection system by adjusting amplification in the photomultiplier tubes individually for differently quenched samples and in this way keep the counting efficiency almost constant over a given quenching range.

5.2.4f. Double Isotope Counting. It is often desirable to use two different radioisotopes simultaneously in the characterization of a biological system. One is then faced with the problem of separating the radionuclides or of measuring them from the same samples. Measurement from the same samples is successful only if the radionuclides differ essentially in one or more of their nuclear properties, as in the nature of radiation emitted or the rate of degradation. In liquid scintillation counting, the separation of ^{3}H and ^{14}C is by far most common. The E_{max} of the β-spectrum of ^{14}C is 8.5 times greater than that for the β-spectrum of ^{3}H. This signifies that a part of the ^{14}C spectrum can be counted and thence the amount of ^{14}C determined with some sacrifice in counting efficiency without interference from ^{3}H in the same sample, but not vice versa (Fig. 9). The amount of ^{3}H in dual ^{3}H and ^{14}C-labeled samples must be calculated from the difference between total and ^{14}C counts in the ^{3}H channel. Difficulties arise with severely quenched samples, since most or even all ^{14}C pulses may become shifted to the ^{3}H channel (cf. Fig. 8) unless the gain of the system is adequately restored or the window settings for ^{3}H and ^{14}C channels are reselected. Inaccuracies in the dual counting of ^{3}H and ^{14}C can be circumvented by separating those nuclides already in the sample preparation. This is usually achieved after combustion of samples in oxygen, whereafter ^{3}H can be collected as tritiated water and ^{14}C as carbon dioxide and their radioactivities gauged from separate samples. Several models of sample combustion and ^{3}H and ^{14}C separation apparatus are commercially available.

5.2.4g. Other Applications of Scintillation Counting. High-energy β-emitters can be measured with liquid scintillation counters directly in aqueous solutions without fluor substances via the *Čerenkov radiation.* Photons, termed Čerenkov radiation, are emitted when a charged particle traverses a medium at a velocity greater than the speed of light in the same medium. The threshold energy for electrons in water for the production of Čerenkov radiation is 263 keV, and only those β-particles with energies greater than this have the possibility of being counted. Of the radioisotopes commonly used in neurobiology, only ^{32}P (E_{max} = 1.71 MeV) can be counted in this manner with adequate efficiency (up to 75%). Sample preparation for Čerenkov counting is easy and economical, and no chemical quenching occurs. On the other hand, color quenching does occur, the wavelength of the major part of Čerenkov radiation is often shorter than the sensitivity region of photomultiplier tubes (a wavelength shifter can be used), and the radiation is directional (i.e., the photons are not emitted in all directions), thereby reducing the likelihood of a coincident detection by two photomultipliers.

Many types of flow counters for monitoring the gaseous and liquid effluents from chromatographs apply liquid scintillation techniques or are just marginal between solid and liquid scintillation detectors. In certain flow cells, the flowing sample is passed over stationary anthracene crystals or solid anthracene filaments serving as scintillators. To prevent solubilization of the fluor in organic solvents, it can be embedded in plastic or coated with inert material. The flowing sample can also be isolated from the scintillator in coiled plastic tubing placed within the scintillator material in the counting cell. All these measures reduce the counting efficiency, in particular for the weakest β-emitters. Satisfactory efficiencies for ^3H have been achieved only with single-phase flow cells by mixing the scintillation cocktail with the sample prior to its introduction into the counting chamber. The counting efficiency in all types of flow cells is also greatly influenced by their volumes and the sample flow rates. It is also, of course, possible to fractionate the chromatographic effluents into small aliquots that are conventionally counted in liquid scintillation instruments. In paper chromatography, the radioactive spots can be cut apart and the radioactivity eluted out for counting, or the cut paper strips can be placed as such in counting vials under scintillation solution. In the same manner, the radioactive spot from thin-layer chromatrography plates can be counted after extraction of radioactivity or direct suspension of scrapings in the scintillation cocktail. The presence in the scintillation vials of paper strips, filtration disks or silica gel particles, or other foreign material may significantly induce quenching, worsen counting geometry, and diminish counting efficiency. Paper and thin-layer chromatograms can also be scanned by gas-flow detectors, as discussed in Section 5.2.1d.

6. *Statistics of Radioactivity Detection*

The moment of decay for a particular radioactive nucleus is entirely unpredictable. The average number of disintegrations occurring within a certain radioactive sample is therefore subject to random variation. The relative magnitude of this variation is obviously less if the sample contains a large number of transformable radioactive nuclei than if only a small number of nuclei are involved. In the same manner, the fluctuations in counting of radioactive samples arising from the randomness of radioactive decay are relatively less in magnitude if a great number of counts have been collected. The probability of occurrence of radioactive disintegration and that of obtaining a count from the sample can be obtained from the *binomial distribution*, which approaches the *Poisson distribution* if the radioactive sample is large and the observation time negligible with respect to the physical half-life of the radionuclide. Both distributions approach the *normal* or *Gaussian distribution* when the number of disintegrations occurring or counts registered is large enough—as is invariably the case in neurobiological applications of isotope techniques.

The precision of an individual measurement can be easily derived from the properties of the aforenamed distributions on the assumption that variability of any measurement results from the statistical nature of the disintegration events only. The precision is described by the standard deviation of the measurement given by

$$\sigma = n^{1/2} \tag{6}$$

where n is the true mean number of counts coming from the sample. If only one count is taken from the sample, it is assumed to be the best estimate for the true mean. A single observation thus gives the estimate both for the mean and for the variance of its distribution. As indicated in any handbook of statistics, in a normal distribution 69.3% of all observations lie within the range of $n \pm \sigma$, 95.5% within $n \pm 2\sigma$, and 99.7% within $n \pm 3\sigma$. Furthermore, $n \pm 1.96\sigma$ encompasses 95% of all observations and $n \pm 2.58\sigma$, 99.0%. Often, the variance is given by the *fractional standard deviation* or the *percent standard deviation.* For instance, if 10,000 (10^4) counts have been collected, the fractional standard deviation will be $\sqrt{10^4}/10^4 = 0.01$ and the percent standard deviation 1%.

The estimates for the mean and standard deviation of the sample and background counts can be obtained separately in the manner described above. The estimate for the sample net counts is obtained simply by subtraction of background. An estimate for the variance of this difference is given by

$$n_{\text{net}} = (n_A - n_B) \pm (\sigma_A^2 + \sigma_B^2)^{1/2} \tag{7}$$

where n_A are the total sample counts, n_B are the background counts, and σ_A and σ_B are their respective standard deviations. For example, if a 10-min count of a sample yields 10^4 counts and a 1-min count of the background yields 10^2 counts, the net count rate per minute with its standard deviation will be $10^4/10\text{ min} - 10^2/1\text{ min} \pm [(\sqrt{10^4}/10)^2 + (\sqrt{10^2}/1)^2]^{1/2}/\text{min} = 900 \pm 14/\text{min}$.

The means and standard deviations can also be divided similarly by the time unit. As appears from the example, the background also contributes to uncertainty in the measurement of net count rates. A proper division of the total counting time between the sample and background must be done. For this division, convenient nomograms in several handbooks can also be consulted to achieve a given reliability in the results.

So far, we have considered the variation arising from the randomness of radioactive disintegrations. Often quite unpredictable, however, are errors that arise from improper operation of measuring instruments. In certain cases when the investigator repeatedly measures the same sample or a series of identical samples, he faces the problem whether or not to reject an apparently deviating observation. No definite rules for rejection can be set. Most investigators have used *Chauvenet's criterion* for the rejection of aberrant observations. According to this, an observation should be rejected if the probability for its occurrence is equal to or less than $1/2N$, where N represents the number of observations. Too many "good" data are being rejected, however, and the use of this criterion should be discouraged. In the pertinent literature, there are a number of other tests for rejection, but none has gained general recognition. Any data should be considered as valid unless the investigator has a special reason to suspect and reject them. The criterion for rejection should always be stated.

7. Data Analysis in Neurobiological Applications

We must warn the reader of the potential fallacies of isotopic studies. It is relatively easy to administer isotopic tracers to experimental animals *in vivo* and to label the constituents of the nervous system or to add a tracer to incubation fluids *in vitro* and obtain an isotope incorporation into nerve-tissue preparations. In all such cases, the isotope-detection instruments yield exact-looking numerical data as a final outcome of the actual experiment. Correctly performed analysis of the accumulated data may be tricky, however, and valid inferences may necessitate a lot of careful deliberation. The use of isotopic tracers is not a shortcut to relevant results that bypasses the tedium of analytical biochemical or physiological work. On the contrary, tracer studies with isotopes most often involve as much routine practical work as nonisotope investigations—in addition to an assessment of isotope abundances and distributions. In too many scientific papers, the investigators still supplement their well-performed isotope analyses with poor chemical analytical work or do not take proper account of the kinetic nature of isotope labeling in dynamic biological situations. An unavoidable consequence is a loss in the revealing power of their experiments or even commitment to erroneous conclusions.

In the past, the greater proportion of work with isotopic tracers has been qualitative in nature, for example, demonstration of the occurrence of enzymatic reactions and identification of precursors and products along the pathways of intermediary metabolism. More recently, quantitative aspects have been frequently studied—the magnitudes of the fluxes of metabolic reactions or those of material transfer from one location to another in the organism. In intact organisms, an approach to such problems is generally feasible only by isotope methods. Isotopes and labeled compounds have also been increasingly employed in the same manner as ordinary reagents in biochemical studies. The variability of isotope methods and experimental approaches has rapidly increased during the last few years. It would be a quixotic undertaking to itemize all neurobiological applications and to describe their characteristic features in this short account. We can select only a few representative examples for discussion to give an insight into the practice of isotopic studies.

7.1. Assay of Brain Free Amino Acids with Labeled Dansyl Chloride

Dansyl chloride (1-dimethylamine-naphthalene-5-sulfonyl chloride) is at present the most sensitive reagent for the detection of amino acids. It has been successfully applied by a number of investigators for quantitative amino acid assays in microscale (see Neuhoff, 1973). Amino acids first yield dansyl amino acids, which may in turn slowly decompose into dansylamine and other products through reactions with other dansyl chloride molecules. In aqueous solutions, a competing reaction is the formation of hydrochloric acid and 1-dimethylamine-naphthalene-5-sulfonyl acid from dansyl chloride and water. In addition to free amino acids, determinations of peptides, protein-bound amino acids, amines, phenols, and imidazoles, for instance, are feasible with dansyl chloride, since its sulfonyl group may react with the $-NH_2$ and -OH groups of other molecules. The reaction products exhibit an intense fluorescence, which propensity may serve as a basis for detection methods. The sensitivity of microassays may be decisively increased by the use of dansyl chloride labeled with ^3H or ^{14}C. This is a good example of procedures in which isotopically labeled compounds are used as sensitive reagents for quantitative assays.

In the analytical procedure (Neuhoff, 1975), the free amino acids are extracted from homogenized tissue with alkaline acetone buffer. The extract is incubated with an excess of [^3H]- or [^{14}C]dansyl chloride, the specific radioactivity of which is known. Dansyl amino acids are then separated from a small sample by two-dimensional chromatography on minute polyamide sheets. Dansyl amino acids on the sheets can be localized with the aid of their fluorescence under ultraviolet light. The spots can be scraped off the sheets under a stereomicroscope and the radioactivity of the scrapings counted by liquid scintillation spectrometry. The quantity of the individual amino acids can be inferred from the known specific radioactivity of the original dansyl chloride preparation and from the measured radioactivities of the chromatographic spots. The radioactive dansyl amino acids on thin-layer plates can also be localized by autoradiography (see Chapter 9). A quantitative evaluation of autoradiograms by scanning microscope photometer makes possible the estimation of as little as 10^{-14} mol [^{14}C]dansyl amino acids (Weise and Eisenbach, 1972).

There are a number of matters that influence the validity of the aforedescribed assay method. (1) Under no circumstances are exactly 100% of all amino acids simultaneously converted into their dansyl derivatives. The recovery is influenced by a variety of factors, such as the duration and temperature of incubation, the pH and other composition of the incubation mixture, the concentration ratios of reactants, the extent of competing side-reactions, and so on (Neuhoff, 1975). Some amino acids also contain several reactive groups in the molecule and are thus able to bind more than one dansyl moiety. In practice, these matters constitute the greatest source of error in quantitative analyses. (2) Good stability of the labeled dansyl derivatives during analyses is essential. (3) All random errors in the measurement of radioactivity increase the variance of results. (4) All systematic errors in the measurement of radioactivity induce a bias in the results. (5) Any radioactive impurity in the dansyl chloride preparation and any inaccuracy in knowledge of the specific radioactivity of the labeled dansyl chloride have the same effect. (These last points are typical sources of uncertainty encountered in many isotope studies.) (6) Obviously, the higher the specific radioactivity of the dansyl chloride, the higher will be the sensitivity of detection. (7) The greater the sensitivity of the radioactivity detection method, the greater will also be the sensitivity of the amino acid assay. By using [^{3}H]dansyl chloride with a very high specific radioactivity and with the aid of quantitative autoradiography with grain-counting in the presence of extremely low background, one may even attempt to estimate the amino acid composition of single neurons.

7.2. Radioimmunoassay for Myelin Basic Protein

Radioimmunoassay methods have advanced rapidly, and they also find applications within neurobiology (see Volume 1, Chapter 7). For instance, Cohen *et al.* (1975) have described a quantitative radioimmunoassay for myelin basic protein. The assay is based on the ability of basic protein in unknown samples to displace ^{125}I-labeled basic protein from its antibody. For the assay, basic protein isolated from the rabbit brain is nonisotopically tagged with ^{125}I. The antisera are prepared from a number of rabbits immunized with repeated injections of basic protein or crude CNS homogenates. A standard curve is constructed by incubating varying amounts of the basic protein standard with antisera, adding ^{125}I-labeled basic protein, incubating again, and finally separating free and bound ^{125}I-labeled basic protein (Fig. 10). Unknown samples are incubated in the same manner with antiserum and ^{125}I-labeled basic protein.

The sensitivity and specificity of the radioimmunoassay depend greatly on the activity of the antiserum, but the properties of the ^{125}I-labeled antigen are equally decisive. From the point of view of the isotope method, the following matters may be considered in addition to those mentioned in Section 7.1 in the context of the dansylation technique: (1) The label must be tagged to the basic protein tightly enough for loss of it from the antigen to be minimal during incubations. An exhaustive removal of the labile ^{125}I before experiments is therefore essential. (2) The labeling should not alter the molecular structure of the basic protein too much, in order to leave its antigenic properties intact. (3) A high specific radioactivity in the antigen increases the sensitivity of the assay. Within reasonable limits, a high proportion of radioactive iodine to basic protein in the iodination procedure is thus advantageous. (4) There commonly occurs some unspecific binding of the labeled antigen, which binding constitutes a definite background to be taken into account in the results. (5) Since the physical half-life of ^{125}I is 7.5 times longer than that of ^{131}I, the antigens labeled with ^{125}I can be stored longer without too serious radioactive deterioration.

7.3. Compartmental Analysis of Tracer Kinetics

At present, the majority of isotopic studies in neurobiology serve more or less directly the quantitative determination of rates of transfer of material from one location to another or of its transformation from one chemical

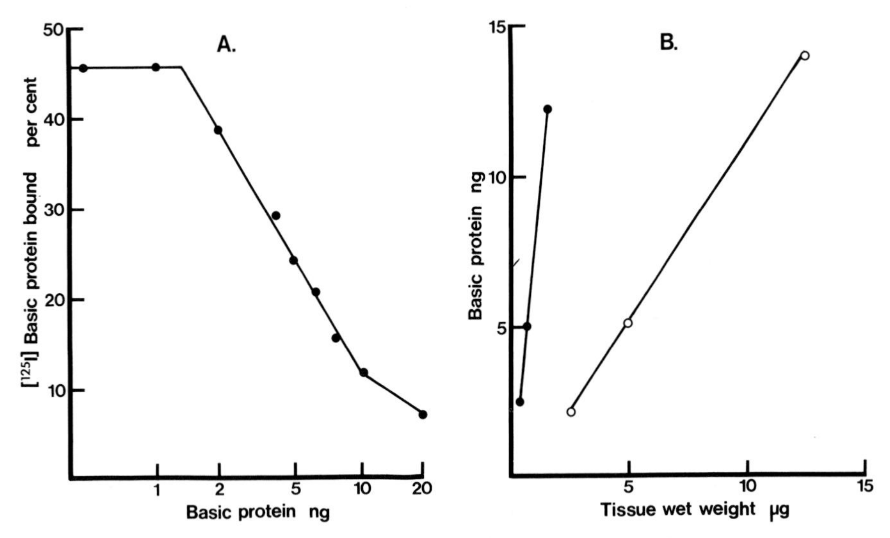

Fig. 10. (A) Displacement of ^{125}I-labeled basic protein from its antibody by unlabeled basic protein. Varying amounts of basic protein were incubated in 500 μl of buffered antiserum at 4°C for 1 day. Then, 370 Bq ^{125}I-labeled basic protein were added, the incubation was continued for another day, and finally free and antibody-bound ^{125}I-labeled basic protein were separated. Results are expressed as percentages of the total radioactivity as a function of the logarithm of the basic protein concentration. (B) Content of basic protein in boiled calf brain gray (○) and white (●) matter. Varying amounts of tissue were incubated with antiserum and ^{125}I-labeled basic protein as above. The amounts of basic protein were read from the standard curve in (A). Redrawn by permission from Cohen *et al.* (1975).

form to another. The term *compartment* or *pool* refers to distinct, homogeneous, and well-mixed amount of material. The animal body may be viewed as an assortment of compartments equivalent to anatomically enclosed distribution volumes or chemically definable states of a substance. *Compartmental analysis* is based on the assumption that these compartments can be identified and the exchange of material between them mathematically defined. The analysis sets out with construction of a *compartmental model* that also allows us to visualize the *communications* between the individual compartments in a *multi-compartment system.* The unidirectional transfer of material from one compartment to another is given by the respective *transport rate.* The *exchange rate* defines two-way exchange of material in a given unit of time. The *turnover rate* denotes that proportion of the molecules in a compartment that is renewed per unit of time. From radioactive disintegration, there originates the concept of *half-life,* which denotes the time required for half the substance to disappear or to turn over. If the sizes of the compartments remain unchanged during observations, the system is said to be in a *steady state.* In *non-steady-state* systems, compartment sizes vary with time. An *open system* communicates with its environment, but in a *closed system,* no material can enter or leave.

Compartmental analysis has been found useful for the analysis of results in many branches of biology. It implies a certain amount of exact thinking and the use of differential equations to describe the kinetic behavior of the systems studied. At first glance, the formulations may appear a confusing array, but in reality they rest on relatively few basic concepts. Furthermore, in compartmental analysis, numerical solutions of the equations derived, or again simulation and curve-fitting procedures, have been greatly facilitated by the fast evolution of analogue and digital computers. An explicit mathematical solution of differential equations for a compartmental system increases exponentially in complexity when the number of compartments and communications increases. For this reason, we must often content ourselves with simplified compartmental models in which only the most pertinent compartments and communications are individually taken into account and the rest of the system is lumped together. A numerical solution or curve-fitting procedure is also applicable for more complicated models. There are a number of books that can be consulted for the solution of problems of tracer kinetics in multicompartmental systems, e.g., those written by Sheppard (1962), Riggs (1963), Rescigno and Segre (1966), Atkins (1969), Steele (1971), Jacquez (1972), Shipley and Clark (1972), and Welch *et al.* (1972). These authors discuss the theory of tracer kinetics and compartmental analysis, present a number of compartmental systems and differential equations derived from them, and demonstrate their solutions.

Quite often, neurobiologists appear to be puzzled by the mathematics involved in compartmental analysis and tend to present data that are relatively poorly processed; e.g., they merely state the accumulation of the label in certain locations or in a given metabolite or calculate ratios between the specific radioactivities of precursors and products. But even this approach implicitly involves a theoretical model. Its overt formalization would render the results more informative and the inferences more relevant. For instance, Oja (1972, 1973) has shown very concretely how single determinations of the specific radioactivities of a precursor–product pair may lead to totally erroneous conclusions concerning their metabolic or transport rates unless the total time course of changes in the specific radioactivities is known. A particularly unfavorable situation prevails when the metabolic rates of macromolecules in the CNS are studied (Oja, 1974). The heterogeneity of the fractions studied usually renders any valid conclusions from the data extremely difficult even on those rare occasions when the dynamic nature of labeling of the macromolecules has been otherwise properly taken into account. A successful compartmental analysis generally implies that the traced molecules constitute a homogeneous population separated from contaminating material and having a definite metabolic or transport rate. In the presence of random fluctuations in the studied metabolic or transport rates, a stochastic approach would be the most appropriate one, introducing some new, intriguing aspects (Jacquez, 1972).

7.4. Determination of the Blood–Brain Exchange of Taurine

Among the possible neurobiological applications of compartmental analysis, we shall briefly discuss the determination of the blood–brain exchange of taurine in mice (Oja *et al.*, 1976). The example demonstrates difficulties encountered and compromises that are unavoidable when compartmental analysis is applied to an actual experimental situation. In the experiment, a trace dose of [^{35}S]taurine was injected intramuscularly and the specific radioactivity of taurine in plasma and brain determined in mice killed at varying intervals after the label administration. The injected [^{35}S]taurine is rapidly absorbed from its application site into the plasma and then distributed to various tissues and gradually eliminated from the body. In this process, the organism represents a complex *mamillary compartmental system* in which the central compartment, the plasma, communicates with innumerable peripheral compartments, various tissues. Figure 11 shows how the brain communicates with the plasma compartment.

Since an explicit mathematical solution of a multicompartmental system comprising the whole organism is not within the bounds of possibility, we must make a number of simplifying assumptions or approximations: (1) Metabolic transformations of taurine in the brain are known to be extremely slow. They can thus be ignored with relatively small risk of error. (2) The very small exchange of taurine between the brain and the cerebrospinal fluid can likewise be ignored. (3) The system in Fig. 11 is in a steady state. Thence, the opposing transport rates are equal in magnitude. They also remain constant during the experiment. (4) The kinetic behavior of taurine is assumed to be similar in all similarly treated mice. However, interindividual variations certainly obtain among animals and constitute a source of variance inherent in all biological work. (5) The taurine compartments in the plasma and the brain are homogeneous, and there occurs an immediate and complete mixing of the labeled taurine molecules with the unlabeled ones within the plasma and brain compartments. We are aware that these stipulations do not exactly hold in reality. It is very difficult, however, to estimate the magnitude of error arising from the evident heterogeneity of the system. (6) There is also no selection among taurine molecules for transport. All molecules within a compartment have an equal and independent likelihood of transport irrespective of the duration of their previous residence in that particular compartment. This is a premise commonly posed in compartmental analyses. In the present case, its applicability stands largely with the validity of the foregoing stipulation on homogeneity of the compartments. (7) The absorp-

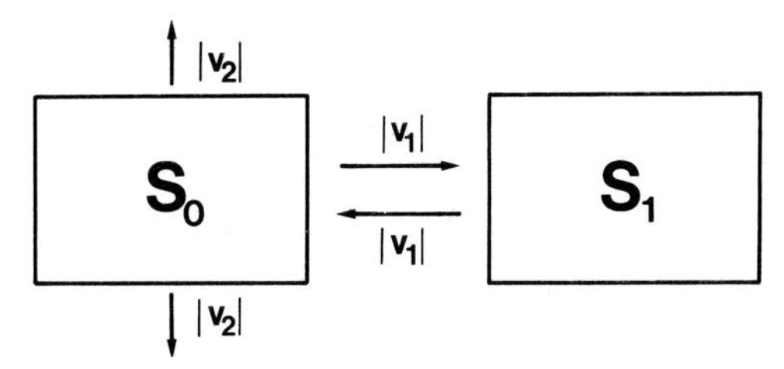

Fig. 11. An open two-compartment model that depicts the exchange of taurine between the plasma and the brain. (S_0, S_1) Amounts of taurine in the plasma and the brain, respectively; $|\nu_1|$ magnitudes of the opposing transport rates of taurine between the plasma and the brain; $|\nu_2|$ magnitudes of the total washout rates of taurine from the plasma.

tion of the injected [^{35}S]taurine from the muscle occurs much faster than the blood–brain exchange. The early rapid changes in the specific radioactivity of taurine in the plasma are unpredictable and difficult to assess experimentally. Since this early phase is short (some minutes) in comparison with the duration of the whole experiment (up to 48 hr), a reasonable hypothetical estimate for the initial content of [^{35}S]taurine in the plasma may be obtained by extrapolating from the actual results to zero time (Fig. 12). Theoretically, it is advantageous to introduce the label directly into the plasma through a prompt intravenous injection in order to ensure maximally fast mixing with the plasma taurine. In practice, however, the injected substance first behaves like a bolus in the circulation, its distribution is uneven, and the initial fluctuations in the specific radioactivity of taurine are abrupt. A better approach might be to keep the specific radioactivity of taurine in the plasma maximally constant during experiments with [^{35}S]taurine infusion and to follow the time course of the steadily increasing specific radioactivity of taurine in the brain. Technically, this is tricky, however. (8) In a steady state, the transport rates of taurine are not influenced *per se* by the exchange of taurine between the plasma and other tissues. Only the kinetic behavior of the label is influenced. In the open two-compartment model of Fig. 11, all the exchange of taurine between the plasma and other tissues except the brain is lumped together. Besides transfer of the label from plasma to tissues with subsequent tissue–plasma exchange, constant fractional loss from the system was to be observed due to the elimination and excretion of labeled taurine from the body, which loss was fully compensated by the *de novo* synthesis of unlabeled taurine. The opposing transfer rates v_2 denote this complex overall washout process of the label. The lumping described above may cause a bias in the estimates for transport rates of taurine between the

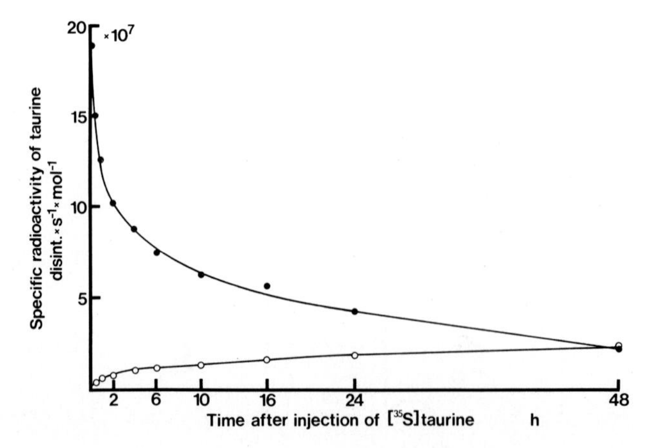

Fig. 12. Experimentally determined specific radioactivities of taurine in the plasma (●) and the brain (○) at varying intervals after an injection of [^{35}S]taurine (370 kBq) into adult mice. The curves show how the specific radioactivities (in Bq/ml) should change when predicted by the transport-rate estimates calculated as indicated in the text.

plasma and the brain, but this is unavoidable if an explicit mathematical solution is sought.

In addition to S_0, S_1, v_1, and v_2 defined in the Fig. 11 caption, we may choose the symbols a_0 and a_1 to denote the specific radioactivities of taurine in the plasma and the brain, respectively. Then, the expressions

$$S_0 \times da_0/dt = v_1 a_1 - v_1 a_0 - v_2 a_0 \tag{8}$$

$$S_1 \times da_1/dt = v_1 a_0 - v_1 a_1 \tag{9}$$

characterize the changes in the specific radioactivities of taurine in the two compartments. The term $v_2 a_2$ does not occur in equation (8), since the taurine synthesized *de novo* was unlabeled.

Equations (8) and (9) were solved for $a_0(t)$ and $a_1(t)$ with the aid of Laplace transforms. We obtained

$$a_0(t) = a_0(0) \frac{(v_1/S_1 - \beta_1)e^{-\beta_1 t} - (v_1/S_1 - \beta_2)e^{-\beta_2 t}}{\beta_2 - \beta_1} \tag{10}$$

$$a_1(t) = a_0(0) \frac{v_1}{S_1} \frac{e^{-\beta_1 t} - e^{-\beta_2 t}}{\beta_2 - \beta_1} \tag{11}$$

in which $a_0(0)$ is the specific radioactivity of taurine in the plasma at $t = 0$, and

$$\beta_{1,2} = \frac{1}{2}\left[\frac{v_1}{S_0} + \frac{v_2}{S_0} + \frac{v_1}{S_1} \pm \sqrt{\left(\frac{v_1}{S_0} + \frac{v_2}{S_0} + \frac{v_1}{S_1}\right)^2 - 4\frac{v_1 v_2}{S_0 S_1}}\right] \tag{12}$$

Finally, equations (10) and (11) were processed with a digital computer using Newton's method to find estimates for v_1 and v_2 (Oja *et al.*, 1976). Figure 12 shows experimental values determined for the specific radioactivities of taurine in the plasma and the brain and the time course of their alterations with time as dictated by the calculated unidirectional transport rates. The estimated transport rate of taurine from the plasma was in this case 53.5 $\pm$ 12.3 nmol $\times$ sec^{-1} $\times$ kg^{-1} fresh tissue (mean $\pm$ 95% confidence limits). Two further similar experiments gave 44.3 $\pm$ 4.2 nmol $\times$ sec^{-1} $\times$ kg^{-1} and 41.5 $\pm$ 8.2 nmol $\times$ sec^{-1} $\times$ kg^{-1}. The fit of the curves to the experimental points in Fig. 12 is quite good.

7.5. Determination of the Rate Constants for Amino Acid Efflux from Brain Slices

Transfer of radioactively labeled material across cell membranes has also often been the object of study *in vitro*. *In vitro*, the experimental situation is generally simpler than *in vivo*, although in both cases the kinetic nature of the transfer of labeled material should be properly taken into account. The same argument also applies for studies on metabolic transformations of labeled compounds. Tissue slices are the preparations most widely used for the study of transport phenomena in nervous tissue. Influx of substances into slices has been studied more often than their efflux. The paucity of relevant studies on amino acid efflux from brain slices, for instance, probably results partly from practical difficulties, since efflux experiments are generally more laborious and the analysis and interpretation of the results more intricate than in studies on influx (Oja and Vahvelainen, 1975).

In a study on homo-*trans*-stimulation of tryptophan efflux from rat cerebral cortex slices (Laakso, 1978), the slices were first loaded with [^{3}H]tryptophan (74 MBq/liter, 0.5 mmol/liter) during a preincubation for 30 min in Krebs–Ringer/phosphate/glucose medium, then briefly rinsed with cold medium, and finally transferred to specific incubation chambers to be superfused with a continuous flow of well-oxygenated Krebs–Ringer medium containing varying amounts of unlabeled tryptophan. The washout curves of [^{3}H]tryptophan from the slices were phenomenally divided into two exponential components on the assumption that the amount of the label, $S_{tot}(t)$, remaining in the tissue at a given time t is given by the expression

$$S_{tot}(t) = S_1(0)e^{-k_1 t} + S_2(0)e^{-k_2 t} \tag{13}$$

Equation (13) implies two immiscible compartments of [^{3}H]tryptophan in the slices, S_1 and S_2, from which the label independently leaks out through two first-order processes with the rate constants of k_1 and k_2, respectively. $S_1(0)$ and $S_2(0)$ are the initial amounts of [^{3}H]tryptophan at $t = 0$.

Figure 13 shows that the total efflux could in fact be resolved into two exponential components. The excellent fit of the experimental data with equation (13) gives a fallacious impression of the validity of the underlying model. For instance, one may be tempted to interpret S_1 as the [^{3}H]tryptophan retained by the extracellular spaces of the slices and S_2 as the intracellular [^{3}H]tryptophan. Analogously to physical half-lives of radionuclides, *biological half-lives* for the traced substances are commonly derived on such occasions by applying equation (5). In the present example, the rate constants $k_1 = 0.142\ \text{min}^{-1}$ and $k_2 = 0.016\ \text{min}^{-1}$ would yield biological half-lives of 4.9 and 42 min for [^{3}H]tryptophan in S_1 and S_2, respectively.

A careful reconsideration of the experimental situation will, however, bring to mind the fact that the intracellular spaces are surrounded by the extracellular space, so that any intracellular [^{3}H]tryptophan must first traverse the extracellular space to reach its destination, the superfusion medium. Equation (13) was composed on the unsound assumption that the intracellular space communicates with the medium directly. A more realistic model would be of the form

$$S_2 \xrightarrow{k_2} S_1 \xrightarrow{k_1} S_0$$

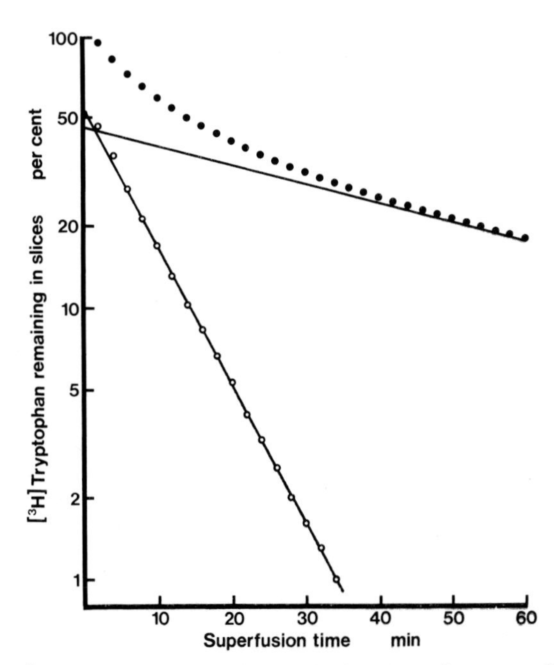

Fig. 13. Efflux of [^{3}H]tryptophan from rat cerebral cortex slices superfused at 37°C with oxygenated Krebs–Ringer/phosphate/glucose medium containing 5 mmol/liter unlabeled tryptophan. The total efflux (●) was assumed to comprise two first-order-rate components as shown by equation (13). They were separated as follows: First, the best-fit, least-square straight line through the last experimental points was calculated with a digital computer to expose the slower efflux component. The relative size (46%) of the compartment, $S_2(0)$, from which it originates is given by the intercept of the drawn straight line. The slope of the straight line gives the transport rate constant k_2 (0.016 min^{-1}). The share of this component is now subtracted from the total efflux, and the best-fit straight line is then similarly calculated for the remaining efflux (○). The aforementioned parameters for the faster efflux component were $S_1(0) = 54\%$ and $k_1 = 0.142$ min^{-1}. The calculated parameters and the fit of the straight lines may be further improved by iterative reestimation of each efflux component in turn. Note particularly the strikingly good fit of the calculated straight line with the remaining efflux after subtraction of the slow component. A similarly good fit also obtains in the case of the slower component and the total efflux. Each experimental point is the mean of six determinations. From Laakso (1978).

where S_0 represents the superfusion medium. This *catenary compartment model* has indeed been used by Jones and Banks (1970) for the efflux of [^{14}C]valine from chopped cerebral cortex of the guinea pig. An explicit solution of the compartment system is

$$S_0(t) = S_0(0) + S_1(0)(1 - e^{-k_1 t}) + S_2(0) \times (1 - e^{-k_2 t}) - \frac{S_2(0)/S_1(0)}{1 - k_1/k_2}(e^{-k_1 t} - e^{-k_2 t}) \quad (14)$$

in which $S_0(t)$ is the amount of the label accumulated in the superfusate at a given time t and $S_0(0)$ is the amount at $t = 0$. The fit of the model with the experimental data was not, however, particularly good. Theoretically at least, the amino acid in the extracellular space can also be taken up again by the cells, but no backward movement of the label is allotted in the model described above. Therefore, Jones (1970) has developed a solution for another model of the form

$$S_2 \underset{k_{-2}}{\overset{k_2}{\rightleftarrows}} S_1 \xrightarrow{k_1} S_0$$

The explicit solution of this model is considerably more complicated than equation (14), but it did not produce any essential improvement in the fit. The possibility remains that efflux processes may not truly follow first-order kinetics (Cohen, 1973).

It should be emphasized that biological half-lives cannot be calculated from equation (5) or graphically estimated from semilogarithmic plots as in Fig. 13 if there is any consecutive transfer, reversible exchange, or non-first-order transfer kinetics of the label in the compartmental system studied. This is a rule quite frequently broken in studies in which labeled compounds have been used. In our previous example, equation (13) is also applicably only if the [^{3}H]tryptophan efflux originates from two separate intracellular compartments and if we additionally assume that the labeled compound traverses the extracellular space to the superfusion medium in an irreversible manner and fast enough. In this case, however, if the superfusion were continued longer, there might be manifested a third, slowest-of-all component in the efflux, as Ames *et al.* (1976) have demonstrated with isolated retinae. Such a very slow process remains undetected unless the curve-fitting is also tried with data derived from incubations of very long duration.

As a conclusion to the foregoing discussion, we would state that an unsuccessful attempt to fit a model or a theoretical curve with the experimental data in isotopic studies strongly calls for an amendment of the model. Obviously, the model was wrong. On the other hand, however, a good fit of the model with the experimental data does not constitute final proof of anything. The model may or may not be right. In particular, in the presence of biological variation and unavoidable experimental errors, it is very difficult to differentiate between two or more models in which the behavior of the isotopic label is relatively similar during the period of observation.

References

Ames, A., III., Parks, J.M., and Nesbett, F.B., 1976, Transport of leucine and sodium in central nervous tissue: Studies on retina *in vitro*, *J. Neurochem.* **27**:999.

Atkins, G.L., 1969, *Multicompartment Models for Biological Systems,* Methuen, London.

Cohen, S.R., 1973, Efflux of exogenous amino acids from brain slices: Evidence of compartmentation from the rate equation, *Brain Res.* **52**:309.

Cohen, S.R., McKhann, G.M., and Guarnieri, M., 1975, A radioimmunoassay for myelin basic protein and its use for quantitative measurements, *J. Neurochem.*. **25**:371.

Feinendegen, L.E., 1967, *Tritium-labeled Molecules in Biology and Medicine,* Academic Press, New York.

Hendee, W.R., 1973, *Radioactive Isotopes in Biological Research,* Wiley, New York.

Horrocks, D.L., 1974, *Applications of Liquid Scintillation Counting,* Academic Press, New York.

Jacquez, J.A., 1972, *Compartmental Analysis in Biology and Medicine,* Elsevier, Amsterdam.

Jones, C.T., 1970, Determination of the kinetic constants of two consecutive first-order reactions, *Biochem. J.* **118**:810.

Jones, C.T., and Banks, P., 1970, The effect of electrical stimulation and ouabain on the uptake and efflux of L-[U-^{14}C]valine in chopped tissue from guinea-pig cerebral cortex, *Biochem. J.* **118**:801.

Laakso, M.-L., 1978, Efflux of phenylalanine and tryptophan from cerebral cortex slices of adult and 7-day-old rats, *Acta Physiol. Scand.* **102**:74.

McKay, H.A.C., 1971, *Principles of Radiochemistry,* Butterworths, London.

Neuhoff, V., 1973, Micro-determination of amino acids and related compounds with dansyl chloride, in: *Molecular Biology, Biochemistry and Biophysics,* Vol. 14, *Micromethods in Molecular Biology* (V. Neuhoff, ed.), pp. 84–147, Springer-Verlag, Berlin.

Neuhoff, V., 1975, Microchemical techniques, in: *Methods in Brain Research* (P.B. Bradley, ed.), pp. 173–247, Wiley, London.

Oja, S.S., 1972, On the measurement of metabolic rates of brain proteins, in: *Ergebnisse der experimentellen Medizin,* Vol. 10, *Biochemical, Physiological and Pharmacological Aspects of Learning Processes* (M. Krug and R. Winter, eds.), pp. 248–254, VEB Verlag Volk und Gesundheit, Berlin.

Oja, S.S., 1973, Comments on the measurement of protein synthesis in the brain, *Int. J. Neurosci.* **5**:31.

Oja, S.S., 1974, Determination of transport rates *in vivo,* in: *Research Methods in Neurochemistry,* Vol. 2 (N. Marks and R. Rodnight, eds.), pp. 183–216, Plenum Press, New York.

Oja, S.S., and Vahvelainen, M.-L., 1975, Transport of amino acids in brain slices, in: *Research Methods in Neurochemistry,* Vol. 3 (N. Marks and R. Rodnight, eds.), pp. 67–137, Plenum Press, New York.

Oja, S.S., Lehtinen, I., and Lähdesmäki, P., 1976, Taurine transport rates between plasma and tissues in adult and 7-day-old mice, *Q. J. Exp. Physiol.* **61**:133.

Ouseph, P.J., 1975, *Introduction to Nuclear Radiation Detectors,* Plenum Press, New York.

Rescigno, A., and Segre, G., 1966, *Drug and Tracer Kinetics,* Blaisdell, Waltham.

Riggs, D.S., 1963, *The Mathematical Approach to Physiological Problems,* Williams and Wilkins, Baltimore.

Sheppard, C.W., 1962, *Basic Principles of the Tracer Method,* Wiley, New York.

Shipley, R.A., and Clark, R.E., 1972, *Tracer Methods for in Vivo Kinetics: Theory and Applications,* Academic Press, New York.

Simon, H. (ed.), 1974, *Anwendung von Isotopen in der organischen Chemie und Biochemie,* Vol. II, *Messung von radioaktiven und stabilen Isotopen,* Springer-Verlag, Berlin.

Steele, R., 1971, *Tracer Probes in Steady State Systems,* Charles C. Thomas, Springfield, Ill.

Wang, C.H., Willis, D.L., and Loveland, W.D., 1975, *Radiotracer Methodology in the Biological, Environmental, and Physical Sciences,* Prentice-Hall, Englewood Cliffs, N.J.

Wang, Y. (ed.), 1969, *Handbook of Radioactive Nuclides,* The Chemical Rubber Co., Cleveland.

Weise, M., and Eisenbach, G.M., 1972, Quantitative determination of amino acids in the 10^{-14} molar range by scanning microscope photometry, *Experientia* **28**:245.

Welch, T.J.C., Potchen, E.J., and Welch, M.J., 1972, *Fundamentals of the Tracer Method,* Saunders, Philadelphia.

Index

Acetylcholinesterase
  BW284C51 inhibition of, 134
  in cell fractionation, 220–221
Acetylcholinesterase activity, in brain response studies, 118
Acetylcholinesterase analysis, reagents for, 133–134
Acetylcholinesterase-containing neurons, fluorescence-histochemical method for, 406
N-Acetyl-5-hydroxytryptamine, 420
Acetylthiocholine, 134
Acid-catalyzed formaldehyde reaction, 376, 387
Acid-formaldehyde solution with high aluminum content, *see* ALFA technique
Acquisition vs. retention, drug effects in, 82
Acrolein, in electron-microscope histology, 451–452
Activity after shock, in avoidance performance, 69–71
Adjunctive behavior, reinforcement and, 46–48
Adrenaline, fluorescence yields for, 416
Agonists, defined, 6
Airy disc, in microscope resolution, 436
ALFA (aluminum–formaldehyde) technique, 388–393, 417
  5-HT neuron detection and, 413
  for Vibratome sections, 394–395
α-decay, 601–602
α-particles, 601
Alzheimer–Mann stain, Häggqvist modification of, 324
Amino acid detection, dansyl chloride in, 638–639
Amino acid efflux, rate constants for, 643–644
Amino acids, brain free, 636
Amphetamine
  heat reinforcement and, 49
  6-OHDA antagonism of, 13
  stereotypy and, 19
d-Amphetamine
  in circling or rotational behavior, 20
  in FI reinforcement schedule, 48
  in locomotor excitation, 16, 18
  in matching-to-sample tasks, 54
  in one-way avoidance, 65
  reinforcement schedules and, 38
  responding and, 36–38
Ampholyte ion, 259
Amphoteric macroions, electrophoresis of, 258–260
Animals, laboratory, *see* Laboratory animals
Annihilation radiation, 604
Antagonists, defined, 6
Anterograde-degeneration technique (*see also* Autoradiographic tracing technique), 577–581
  critical evaluation of, 577–592
  passing fibers in, 577–581
Anticholinergics, motorigenic effects of, 30
Anticholinesterase, in inhibitory processes, 57
Anticoincidence circuitry, 624
Antimony–potassium–cesium bialkali, in photon measurement, 627
Antineuterino $\bar{\upsilon}$, 602
Aplysia california, neuron identification in, 279, 283, 526
Apomorphine, in locomotor activity, 16
Araldite, as embedding medium, 454
Artifact sources, in polyacrylamide gel electrophoresis, 286–290
Associative processes, drug effects in, 51–52
Astigmatism, in transmission electron microscope, 443

Astroglial processes, electron micrograph of, 478
Atom percent excess, defined, 607
ATPase, in neuronal transport, 553–554
Attentional processes
drug effects in, 55–56
matching-to-sample tasks and, 56
Auditory receptors, SEM studies of, 529–532
Auger-electron analysis, in SEM, 520
Autoradiographic tracing technique
advantages and limitations of, 589–591
in combination with other techniques, 591–592
precursor uptake in, 586–588
sensitivity of, 588
transneuronal transport in, 588–589
Autoradiographs
analysis of, 571–577
background labeling in, 572–574
efficiency of, 561–563
light-microscopic quantification of, 574–575
photographic factors in, 562–563
preparation of, 567–571
Autoradiography
in axonal transport, 548–551
background in, 563
basic principles of, 563–584
in cell biology, 545
in central nervous system, 543–592
control procedures in, 571
in dendritic transport, 551–552
developing and fixation in, 569–570
"effective injection site" in, 582
in embryology, 546
fluorescence-histochemical techniques and, 408
geometric factors in, 559–560
half-distance in, 576
half-radius in, 559
injection site in, 581–582
isotopic factors in, 561–562
labeled compounds in, 563–567, 581–582
labeled fibers in, 582–584
light-microscopic, 546, 557, 568
methods used in, 563–577
in neuronal connectivity, 546–554
passing fibers in, 577–578
photographic emulsions in, 557–558, 568–569
photographic exposure in, 568–569
photographic factors in, 560–561
precursor differential uptake for, 586–588
precursors in, 566–567
Autoradiography (*cont.*)
prerequisites in, 555
radioisotopes in, 555–556
reflectance measurements in, 575
resolution in, 558–561
retrograde transport in, 550
staining in, 570
submicroscopic quantification in, 575–577
termination area in, 582–586
tissue fixation in, 567
in transneural transport, 552–553
transport mechanisms in, 550–554
two-dimensional electrophoresis and, 291
whole-body, 567–568
Average energy, defined, 602
Aversely motivated behavior
acquired drive in, 78–79
behavioral paradigms in, 59–83
discriminated avoidance in, 72–77
shock exposure and, 79–80
transfer designs in, 77–82
Aversive stimulus, in avoidance behavior, 60
Avoidance behavior
activity–reactivity in, 69–71
associative and nonassociative factors in, 62
components of, 60–63
conditional suppression in, 68–69
conflict in, 67–68
cue–shock pairings in, 65
discriminated avoidance in, 72–77
go–no go avoidance in, 74–77
manipulation avoidance in, 66
one-way and shuttle types in, 63–64
passive, 66–67
prior shock exposure and, 79–80
response factors in, 61
shock sensitivity in, 72
shuttle tasks in, 64–65
Sidman avoidance in, 73–74
stimulus factors in, 61–62
task manipulations and multiple-testing procedures in, 63–72
Y-maze discriminated avoidance in, 73
Axonal degeneration
body temperature in, 339
staining methods in, 338–340
Axonal reaction
of perikaryon, 318–321
von Gudden method and, 321
Axonal transection, nerve cell disappearance following, 321
Axonal transport, autoradiography in, 548–551

Background labeling, in autoradiography, 572–574
Back-scattered electron analysis, SEM and, 519
Balance sheets, in cell fractionation, 216–228
Barbiturates, responding effect of, 36
Baseline environment, for laboratory animals, 111
BDMA, *see* Benzyl dimethylamine
Becquerel, defined, 607
Behavioral analysis, drug effects and, 2–13
Behavioral paradigms, in aversively motivated behavior, 59–83
Behavioral pharmacology, associative processes and (*see also* Drug effects), 51–52
Behavioral processes
  drugs in experimental analysis of, 48–59
  mathematical processes and, 49–50
Behavioral techniques
  in brain response studies, 113–117
  Hebb–Williams maze in, 113–117
  in pharmacological and neuropharmacological analyses, 1–84
Benzyl dimethylamine, 454
β-decay, 603, 615
β-emitters, positrons from, 604
β^+-particle, 602
Bielschowsky silver-impregnation method, 334
Binomial distribution, in radioactive disintegration, 634
Biochemical techniques, in brain-response studies, 117–138
Biogenic amines
  cryostat-section methods for, 401–403
  FA–ozone technique and, 376–377
  fluorescence microscopy of, 365–422
  fluorescence yield for, 376
  identification and differention of following FA treatment, 412–418
  identification and differentiation of following glyoxylic acid treatment, 418–422
Biogenic monoamines
  fluorescence-histochemical techniques for identification and differentiation of, 411–422
  flurophore-forming reactions and, 373
Biological half-lives, defined, 644
Bisacrylamide concentration, 249
Bismuth stain, in electron-microscope histology, 463
Blenders, in cell fractionation, 157–158
Blood–brain exchange, of taurine, 640–646
Bodian silver-impregnation method, 354–355
Bovine serum albumin, two-dimensional separation of, 272
Brain (*see also* Rat brain)
  dissection and gross anatomy of, 303–307
  edlectron micrography of, 481–482
  hardening of before gross dissection, 302
  scanning electron microscopy of, 523–525
  spatial or graphic reconstructs for serial sections of, 307–310
  synaptic relationships in, 481–482
  topographical relation with skull and spinal cord, 306
Brain free amino acids, assay of, 636–638
Brain regions, weights of, 118–123
Brain response
  acetylcholinesterase activity in, 118, 128–138
  behavioral techniques for prior tests in, 112–117
  cholinesterase activity in, 118, 128–138
  DNA content in, 118, 123–127
  to environment and experience, 101–138
  RNA content in, 118, 123–127
Brain response studies
  acetylcholinesterase activity and, 128–138
  analytic procedures in, 126–127
  cholinesterase activity in, 128–138
  dissection methods in, 118–123
  instruments used in, 126
  pH and, 130–131
Brain slices, rate constants for amino acid efflux from, 643–644
Brain tissue, Vibratome methods for, 395–399
Branching decay schemes, 603
Bremsstrahlung, 503
BSA, *see* Bovine serum albumin
Bufotenin, 420
Butyl-PBD, 628
Butyrylthiocholine, 134
BW284C51, as AChE inhibitor, 134

Caffeine, in matching-to-sample tasks, 54
Cambridge–Huxley ultramicrotome, 458
Carbidopa, behavioral effect of, 5
Carboxypeptidase A, 292
Carrier ampholytes, in microelectrophoresis, 269
Catenary compartment, 644
Cat neuron, electron microscopy of, 469
Catecholamine fluorophores, photodecomposition of, 411

Catecholamines
FA-induced fluorescence from, 416
fluorescence induced by, 418
fluorophore formation from, 374
GA fluorescence from, 398
Catecholamine stores, fluorescence-histochemical demostration of, 403
Cat endoplasmic reticulum, electron microscopy of, 471
Cathode oscilloscope, 503
Cathodoluminescence, in scanning electron microscopy, 520
^{14}C-dating, 608
Cell biology, autoradiography in, 545
Cell culture, electron microscopy in, 484–485
Cell fractionation, 143–227
acetylcholinesterase in, 220–221
analytical approach in, 210–214
analytical vs. preparative strategies in, 208–227
blenders in, 157–158
centrifugation in, 145–152
differential centrifugation in, 167–208
digitonin shift in, 213
dilution factors in, 217–218
electron microscopy and, 485–486
fraction terminology in, 211
Golgi apparatus membranes in, 213
histogram in, 223–225
homogenization in, 152–167
lysosomes in, 212
marker enzymes in, 218–225
microsomal fraction in, 212
mitochondrial fraction in, 211–212
monoamine oxidase in, 220–221
nuclear fraction in, 212
organelle in, 211
small-granule fraction in, 212
succinate dehydrogenase in, 220–221
supernatant fraction in, 213
synaptosomes in, 212
validation of fraction schemes in, 214–216
Cell fractionation experiment
assessing results in, 214–216
balance sheets for, 216–227
Celloidin embedding, in neuroanatomy, 315
Cellular heterogeneity, tissue preparation and, 158–161
Central nervous system, SEM in studies of, 521–525
Central nervous system tissue, vibratome sections of, 422
Centrifugal force
in centrifugation process, 145–147
Centrifugal force (*cont.*)
relative, *see* Relative centrifugal force
sedimentation and, 147–148
Centrifugal fractionation, *see* Centrifugation
Centrifugation
analytic vs. preparative, 150–152
basic information on, 145–148
density-gradient, 172–208
differential, 167–173
history of, 143–145
homogenization and, 152–167
relative centrifugal force and, 147–150
Centrifuges, 148–152
Cerebrospinal fluid, drug administration through, 3
Cerenkov radiation, 633
Cetyltrimethylammonium bromide, 125, 248
Channels-ratio method, 632
Characteristic X-rays, 604
Chauvenet's criterion, 635
Chemical decomposition, of radioactive substances, 616–617
Chlordiazepoxide
and DRL responding in rats, 50
in matching-to-sample tasks, 54
in reinforcement schedules, 36
p-Chlorophenylalanine, 5
habituation and, 24
serotonin depletion by, 27
tryptophan hydrolase inhibition of, 10
Chlorpromazine
and DRL responding in rats, 50
heat reinforcement and, 49
in matching-to-sample tasks, 54
in reinforcement schedule, 48
responding and, 49
Cholesterol, digitonin and, 213
Cholinergic activity, drug effects and, 7–8
Cholinergic manipulations, in spontaneous alteration, 26
Cholinergic neurons, visualization of, 405–406
Cholinesterase activity, in brain response studies, 118
Cholinesterase analysis, reagents for, 133–134
Chromatic aberration, in transmission electron microscope, 442–443
Chromatolysis, Nissl substance in, 320
"Clock" stimuli, reinforcement schedules and, 48
Compartment
defined, 639
open or closed system and, 639

Compartmental analysis, 639
  mathematics involved in, 640
Complex behavior, components in, 1
Complex task, performance in, 1
Compton electrons, 632
Conditioned emotional response, 68–69
Conditioned stimulus, 68
Conflict task, in avoidance behavior, 67–68
Conjunctive schedule, defined, 43–44
Contemporary Research Methods in Neuroanatomy (Nauta and Ebbeson), 337
Conventional transmission electron microscope, 501
Cryofractionation, in scanning electron microscopy, 513
Cryokits, for ultramicrotomes, 460
Cryostat section methods, for biogenic monoamine localization, 401–403
CSF, *see* Cerebrospinal fluid
CTEM, *see* Conventional transmission electron microscope

Dansyl chloride, in amino acid detection, 636
Dark-field electron microscopy, 448
DDSA, *see* Dodecenyl succinic anhydride
Degenerated myelin, staining of, 331–332
Degenerating axons, silver impregnation of, 334–342
Degeneration methods, in electron microscopy, 342
Dehydration, in electron-microscopic histology, 453
Demyelinization, 331–333
Dendritic transport, autoradiography in, 551–552
Density-difference ratio, in density-gradient centrifugation, 201
Density-gradient centrifugation (*see also* Density gradients)
  as analytic method par excellence, 175
  equilibrium-gradient methods in, 181–182
  buoyant-density methods in, 179
  density-difference ratio in, 201
  differential pelleting in, 202–204
  Ficoll in, 181
  gradient materials in, 180–182
  gradient parameters and functional implications in, 183–208
  isopycnic methods in, 176–178
  moving-zone methods in, 174–176
  normal-rate, 173–174
  particle intersection in, 202–203
  and physical properties of fractionated material in, 225–226
Density-gradient centrifugation (*cont.*)
  relative centrifugal force in, 179
  sucrose density in, 180–181, 200
  terminology in, 178–179
  unloading fractions in, 203
  zonal rotors in, 205–208
Density-gradient parameters, 184
  relative centrifugal force and, 197
  selection of, 196–203
Density gradients (*see also* Density-gradient centrifugation)
  concave, 184, 189–190
  continuous, 186–189
  convex, 184, 191–192
  density profile for, 225–226
  equilibrium-gradient materials in, 181–182
  exponential, 184, 193
  formation of, 182–183
  isopycnic level in, 197
  linear, 184, 187
  parameters of, 184, 196–203
  shallow vs. steep, 186–188, 192–195
  in subcellular-fractionation work, 183
  sucrose and, 180–181, 186, 189, 195, 198–199
  zonal rotors and, 206–208
Depth of field, in transmission electron microscopy, 440
Diethyl oxidoformate, in nucleic acid electrophoresis, 278
Differential centrifugation (*see also* Centrifugation)
  convection in, 172
  density gradients in, 173–208
  diffusion in, 170–171
  effective mass in, 171–172
  friction in, 171
  normal-rate separation and, 167–170
  relative centrifugal force in, 167–168
  sedimentation in, 170–173
  four types of separations in, 169
Differential-reinforcement-of-low-rate schedule, 39–40
  in albino rats, 50
Diffraction electron microscopy (*see also* Electron microscopy), 449
Digitonin-shift phenomenon, in cell fractionation, 213
3,4-Dihydroxyphenalanine, *see* L-Dopa
Dilution factors, in cell fractionation, 217–218
Dimethylsulfoxide, in paraffin embedding, 346–347
N,N-Dimethyltryptamine, 420

Discontinuous buffer system, in disk electrophoresis, 252
Discriminated avoidance (*see also* Avoidance behavior), 72–77
Discriminative stimuli, drug effects in, 41–42
Disk electrophoresis, 251–257
  constant-pore-size gels in, 288
  retrograde regulation in, 253 n., 255
5,5′-Dithiobis-(2-nitrobenzoic acid), 133-136
Dithiothreitol, 248
DMP-30, *see* Tri(diethylaminomethyl)phenol
DNA analysis, reagents for, 124–127
DNA content, in brain response studies, 118
Dodecanylsuccinic anhydride, 454
Dopa, flurophore formation from, 383
L-Dopa, behavioral effect of, 5
Dopamine
  blocked reuptake of, 13
  drug effects of, 7–11
  FA and GA methods in visualization of, 365–367
Dopamine activity, stereotypic behavior and, 19
Dopamine-β-hydroxylase inhibition, stereotype and, 5, 19–20
Dopamine fluorophore, 416
Dose–response relationships, drug effectiveness and, 7
Drive expectancy, in avoidance behavior, 60–61
DRL schedule, in reinforcement, 39–40, 50
Drug(s) (*see also* Drug effects)
  administration routes for, 3–4
  cholinergic activity and, 7–8
  CSF injection of, 3
  in locomotor excitation and inhibition, 15–22
  neurochemical interactions and, 12–13
  neurotransmitter activity and, 7–8
  noradrenergic activity and, 9
  oral administration of, 3
  response-disrupting effects of, 44–45
  serotonergic activity and, 10
Drug–behavior relationships, transmitter storage and, 8-11
Drug effects [*see also* Drug(s)]
  in acquisition vs. retention performance, 82
  in adjunctive behaviors, 48
  in associative processes, 51–52
  in attentional processes, 55–56
  in aversely motivated behavior, 59–83
  behavioral analysis of, 2–14
  central vs. peripheral, 4–6
  chronic depletion and, 14
  in circling or rotational behavior, 20–21
Drug effects (*cont.*)
  conditional stimulus and, 68
  discriminative stimuli and, 41–42
  dose–response relationship and, 6
  and experimental analysis of behavioral processes, 48–59
  in habituation, 22–33
  in inhibition and inhibitory processes, 57–58
  mathematical specificity and, 49–50
  in memory processes, 52–55, 80
  in observing behavior, 48
  in one-way and shuttle tasks, 65–66
  in passive avoidance performance, 66
  in perseverative behavior, 21–22
  in reinforcement schedules, 40–41
  response-disrupting, 44
  in response excitation and disinhibition, 17
  in response-modulating system, 80
  in retention performance, 82
  schedule-controlled responding and, 36
  on sensory processes, 58–59
  specificity of, 12–13
  stereotypy in, 19–20
  in stimulus control, 43
Drug receptors, 6–8
Drug-screening programs, locomotor activity in, 15
Drug synergism, 13–14
Drug treatment, species and strain factors in effectiveness of, 14–15
DTNB [5,5′-dithiobis-(2-nitrobenzoic acid], 133, 136
DTNB-BW284C51, in ChE activity assays, 133–134

EC environment, *see* Enriched condition environment
ECT, *see* Environmental complexity and training
Electrical birefringence, 248
Electron capture, 603
Electron gun, functional diagram of, 445
Electron microscope, major components of (*see also* Transmission electron microscope, 443
Electron microscope lenses, diagram of, 442
Electron microscopic autoradiography (*see also* Autoradiography), 546
  vs. degeneration methods, 584–586
  sections for, 568
  silver halide and silver iodide crystals in, 557
Electron-microscopic histology, 450–466
  critical remarks on, 464–466

Electron-microscopic histology (*cont.*)
  dehydration in, 453
  double staining in, 462
  embedding in, 453–455
  epoxy resins in, 454–455
  fixation in, 450–452
  Golgi techniques and, 463
  methacrylates in, 454
  perfusion fixation in, 452–453
  polyester resins in, 455
  potassium permanganate stain in, 462
  silver stains in, 463–464
  staining procedures in, 461–464
  ultramicrotomy in, 455–460
  uranyl stains in, 461–462
  water-soluble embedding media in, 455
Electron microscopy (*see also* Electron-microscopic histology)
  of afferent and efferent connections, 477
  analytical, 449
  in cell and tissue culture, 484–485
  cell fractionation and, 485–486
  dark-field, 448
  diffraction, 449
  fluorescence-histochemical techniques in, 409–410
  of glial cells, 483–485
  high-voltage, 449
  history of, 434–435
  limitations of in neurobiological research, 486–488
  in neuroanatomy and neurocytology, 466–477
  in neurobiology, 433–496
  of neuron, 467–473
  of neuropil, 475–477
  quantitative stereology in, 488–496
  scanning, *see* Scanning electron microscopy
  synaptic transmission and, 477–484
Electron shake-off, 604
Electron spin resonance, 618–619
Electrophoresis (*see also* Polyacrylamide gel electrophoresis)
  amino acid analysis following, 292
  biological use of, 247
  buffer systems in, 260–261
  carrier, 249
  continuous and discontinuous carriers in, 249–250
  defined, 246–247
  discontinuous, 251
  disk, 251–257
  free, 249
  immunological and autoradiographic techniques in relation to, 291
Electrophoresis (*cont.*)
  interfacing of with other techniques, 290–292
  isoelectric focusing in, 258–260
  macroion polarity in, 247
  miniaturization in, 262–263
  moving-boundary technique in, 250
  peptide mapping following, 292
  polyacrylamide gel, *see* Polyacrylamide gel electrophoresis
  practical aspects of, 262–283
  protein solubility in, 248
  setup for, 271
  sodium dodecyl sulfate, 261–262, 272–277, 284, 289
  terminology in, 249–250
  two-dimensional techniques in, 260
  zone, 250–251
Electrophoretic mobility, 284–286
Electrophoretic techniques, outline of, 249–262
Elution, of macroions from gel, 290
EMA, *see* Electron-microscopic autoradiography
Embedding, in electron-microscopic histology, 453–455
Embedding media
  epoxy, 454–455
  water-soluble, 455
Embryology, autoradiography in, 546
Emission spectrum, correction of, 370
Endoplasmic reticulum, of neuron, 470
Energy
  average, 602
  maximum, 602
Enriched condition environment, maze training in, 106
Environmental and training techniques, in brain response studies, 102–111
  enriched condition and, 104
Epon, as embedding medium, 454
EPTA, *see* Ethanolic phosphotungstic acid
ESR, *see* Elution spin resonance
Ethanolic phosphotungstic acid, 481
Ethylenediaminetetraacetic acid, 124, 166
Excitation spectrum
  correction of, 370
  in fluorescence microscopy, 367
Exploration, habituation and, 31–32
External-standard channels ratio method, 632

FA reaction, *see* Formaldehyde
Ferguson-plot analysis, molecular weight and, 284

Ficoll, in density-gradient centrifugation, 181
Fink-Heimer method, 356
Fixation
  in electron-microscopic histology, 450–452
  in neuroanatomy, 310–313
FLA–63, 5
Fluorescence
  defined, 366
  emission spectrum in, 367
  excitation spectrum in, 367
Fluorescence-histochemical techniques (*see also* Fluorescence microscopy; Histological methods)
  autoradiography and, 408
  combined with other microscope techniques, 405–410
  electron microscopy and, 409–410
  horseradish peroxidase in, 407–408
  for identification and differentiation of biogenic monoamines, 411–422
  specificity tests and, 411–412
  staining procedures in, 405
Fluorescence intensities, quantitation of, 371
Fluorescence microscope, 367–368
Fluorescence microscopy, 365–422
  description of, 366–372
  formaldehyde method and, 384–401
Fluorescence microspectrofluorometry, 366–372
Fluorogenic amines, spectral analysis of, 414–422
Fluorogenic monoamines, 372
Fluorophore(s)
  excitation and emission spectra of, 415
  FA-induced, 414–415
  microspectrofluorometric analysis of, 419–422
Fluorophore formation
  after combined FA and GA treatment, 381–384
  from dopa, 383
  after formaldehyde treatment, 373–377
  following glyoxylic acid treatment, 378–381
  from noradrenaline, 383
  Pictet-Spengler cyclization reaction in, 375
Fluors, 626
Formaldehyde
  acid-catalyzed reaction with, 376
  catecholamine visualization with, 365
  in electron-microscopic histology, 451
  fluorophore formation after treatment by glyoxylic acid and, 381–383
  indolyethylamines and, 374
  reaction products in, 372
Formaldehyde method
  acid-catalyzed, 387
  embedding and sectioning in, 388
  flurophore formation following, 373–377
  for freeze-dried specimens, 385–389
  practical performance of, 384–401
  for thin tissue sheets or whole-mount preparations, 403
Formaldehyde–ozone method, 376–377, 387
Formalin, brain-tissue fixing and, 323
Freeze-dried tissue, 385–389
  advantages of, 388
  formaldehyde and glyoxylic acid methods for, 383–393
Frog brain, topological projection of, 310
Frog retina, scanning electron micrograph of, 510
Frozen sections, in neuroanatomy, 313–314

GA, *see* Glyoxylic acid
γ-radiation, 603
Gas amplification, in Townsend avalanche, 621
Gas-filled chamber, ionization response in, 620–621
Gas-flow detectors, 624
Gas-ionization detectors, 620–624
Gaussian distribution, 634
Geiger–Müller detectors, 622–623
  characteristic operation curve for, 623
  counting gas in, 623
  low-background counters and, 624
Gelatin embedding, in histological neuroanatomy methods, 345–346
Glass knives, in ultramicroscopy, 455–456
Glees-stained materials, ultrastructure studies of, 336
Glees termination degeneration method, 336
Glial cells, electron microscopy of, 473–475
Glutaraldehyde, in electron-microscopic histology, 451
Glyoxylic acid, 384–401, 404
  biogenic amines and, 418–422
  in catecholamine visualization, 365
  for cryostat sections of nonperfused brains, 401
  fluorophore formation after treatment with, 378–383
  reaction products in, 372
Glyoxylic acid–vibratome method, 395–400
Glyoxylic acid vapor, 404, 419–420
Golgi method
  in histological neuroanatomy, 342–345, 357–358
  silver stains and, 463, 543

Go-no go avoidance, 74–77
Gradient gel microelectrophoresis, 265–270
  in nucleic acid separation, 278
Graphic reconstructions, in neuroanatomy, 310–312
Ground state, 603
Group condition, in laboratory animals, 105

Habituation
  carry-over designs in, 29–30
  criteria for, 23
  dissociation from perseveration, 27–28
  drug effects in, 22–33
  of exploration and startle, 30–32
  learning and, 33
  sensitization in, 23–24
  spontaneous alternation and, 25–29
  in T- and Y-maze tasks, 25–26
  temporal evaluation of drugs in, 24
  water lick in, 29–30
Häggqvist—Alzheimer–Mann stain, 333
Häggqvist method, in myelin mordanting, 323–324, 326–327, 331, 350–352
Half-life
  biological, 644
  radioactive decay, 605, 639
Hebb–Williams maze, 113–117
High-resolution scanning transmission electron microscopy, 518–519
Histological methods, 345–358
  Bodian silver-impregnation method in, 354–355
  celloidin embedding in, 347–349
  Fink–Heimer method in, 356–357
  gelatin embedding in, 345–346
  Golgi method in, 357–358
  myelin stains in, 350–354
  Nauta method in, 355
  Nissl stain in, 349–350
  paraffin embedding in, 346–347
  reduced-silver methods in, 354–357
Histology, electron-microscopic, *see* Electron-microscopic histology
Homogenization, 152–167
  absorption in, 164
  adsorption in, 164
  agglutination in, 165
  artifacts in, 162–163
  in cell fractionation, 143
  EDTA in, 166
  ion effects in, 166
  for isolation of specific subcellular components, 167–168
  leakage in, 163–164
  pH and, 166–167
Homogenization (*cont.*)
  redistribution in, 164–165
  sucrose effects in, 165–166
Homogenization media, 161–167
Homogenizer(s)
  coaxial, 152–157
  operational principle of, 155–156
  Potter-Elvehjem, 152–154
Homogenizer drive units, 154–155
Horseradish peroxidase
  autoradiography and, 550–551, 586, 591
  in histochemistry, 407–408
  retrograde tracing technique with, 544
HRSTEM, *see* High-resolution scanning transmission electron microscopy
HSA, *see* Human serum albumin
5-HT, *see* 5-Hydroxytryptamine
Human serum albumin, two-dimensional electrophoresis of, 273
Hydrogen isotopes, in neurobiological research, 607
6-Hydroxydopamine
  amphetamine and, 12–13
  in circling and rotational behavior, 20
  in reversal learning, 77
6-Hydroxylated dihydroisoquinolone flurophores, GA-induced, 378
3-Hydroxylated phenylethylamines, conversion of, 373–375
p-Hydroxynorephedrine, as false NE transmitter, 14
5-Hydroxytryptamine
  behavioral effects of, 5–11
  fluorophore formation and, 375, 411
5-Hydroxytryptophan
  fluorescence yields of, 414
  injection of, 5

IC environment, *see* Impoverished condition environment
Image formation, in transmission electron microscope, 438–440
Imipramine, 36, 39
Immersion fixation, in electron-microscopic histology, 452
Impoverished condition environment, 102
Indoleamine-containing neurons in CNS, formaldehyde method for, 388–389
Indoleamines, fluorescence yields for, 413–414, 418–419
Indolethylamine derivatives
  fluorescence yields from, 377, 380
  formaldehyde and, 374
  reactivity of, 375

Indolethylamine fluorophores, excitation spectra of, 417
Inhibition
  drug effects in, 57–58
  learning and, 33
Inhibitory stimulation, defined, 58
Insect-receptor morphology, SEM studies of, 533
In situ histochemistry (*see also* Histological methods), 290
Internal conversion, defined, 604
Internal standardization, 631
Interreinforcer time, in reinforcement schedules, 36
Ionization chambers, in radioactive isotope detection, 621
Ionization potential, 620
Ion pair, 620
Isoelectric focusing, in electrophoresis of amphoteric macroions, 258–260
Isoelectric focusing gels, protein optical density scans and, 278
Isomeric transition, in radioactive decay, 603–604
Isopycnic level, in density gradients, 197–198
Isoquinolone, formaldehyde–dopa formation of, 383
Isotachophoresis, 257–258
Isotope measurement, 617–634
  stable isotopes in, 617–620
Isotope methods (*see also* Radioisotopes), 599–646
  labeled compounds in, 609–617
Isotopes of neurobiological interest, 607–609
Isotopic effects, 607
Isotopic labeling, 609, 612–613

Klüver-Barrera stain, 328, 330, 353

Labeled compounds
  in autoradiography, 563–567, 581, 609–617
  availability of, 609–612
  manufacture of, 613–614
  purity of, 614–615
  stability and storage of, 615–617
Laboratory animals (*see also* Rat brain; Rats)
  baseline environment for, 111
  enriched experience for, 105–107
  environment and training studies with, 102
  Hebb-Williams maze in, 113–117
  large social group in interlinked cages, 108–109
  latent learning in, 113
Laboratory animals (*cont.*)
  seminatural outdoor environment for, 109–110
  standard colony, 102–103
  superenriched environments for, 107–110
Lactic dehydrogenase, two-dimensional electrophoresis of, 274
Lactic dehydrogenase isoenzymes, in isoelectric microfocusing, 269
Latent learning phenomenon, in laboratory animals, 113
Learning
  habituation in, 33
  inhibition and, 33
  latent, 113
  reversal, 77–78
  light microscope, vs. electron microscope, 501, 522
Light-microscopic autoradiography, 546
  sections for, 568
  silver bromide crystals in, 557
Lipshaw® electric paraffin pitcher, 347
Liquid scintillation cocktail, 628
Liquid scintillation counting, quenching in, 630–631
Liquid scintillation method, 627–634
  Cerenkov radiation and, 633
Lithium-6 isotope, bombardment of, 608
LKB Ultratome, 458
Locomotor activity
  drug effects in, 15–22
  scopolamine and, 17–18
  vertical–horizontal, 16–17
Locomotor stimulation, vs. alternation level, 28
LSD, spontaneous alternation and, 27

Macaca mulatta, 310
Macroionic mixtures, gradient gel separation of, 256
Macroions
  amphoteric, 258–260
  electrophoretic mobility of, 248
  elution of from gel, 290
  polarity of, 247
  Stokes' radius of, 248
Malinol®, 352
Mamillary compartmental system, 641
Manipulative avoidance, 66
MAO, *see* Monoamine oxidase
Marchi method, in degenerated myelin staining, 331–332
Marker enzymes, in cell fractionation, 218–225
Mass spectroscopy, of stable isotopes, 618

Matching-to-sample tasks
  in attentional mechanisms assessment, 56
  drug effects in, 52–54
Maximum energy, defined, 602
Maze training, in enriched environment (*see also* T maze; Y maze), 106
Memory processes
  drug effects in, 52–55
  matching-to-sample tasks in, 52–54
Mercaptoethanol, in nucleic acid microelectrophoresis, 278
2-Mercaptoethanol, disulfide bonds and, 248
Metastable state, 603
Methacrylates, in electron-microscopic histology, 454
3-Methoxylated phenylethylamines, 376
3-Methoxyltryptamine, 420
2-Methyl-3,4-dihydroisoquinolinium compound, 373
Methyl nadic anhydride, 454–455
Methylphenidate, 20
α-Methyl-*p*-tyrosine
  behavioral effect of, 5
  inhibition of tyrosine hydroxylase by, 10
N-Methyltryptamine, 420
Methysergide, spontaneous alternation in, 27
Microelectrofocusing, gradient microelectrophoresis and, 270
Microelectrophoresis, 262–268
  equipment for, 280–282
  gradient-gel, 270, 278
  of nucleic acids, 277–280
  recent applications of, 282–283
  in SDS, 272–277
  two-dimensional, 269–272, 283
Microfluorometric quantitation, 370–372
Microgradient gels, in electrophoresis, 267–270
Microiontophoresis, in radioisotope administration, 566
Microisoelectric focusing, 268–269
Micromethods, in electrophoresis, *see* Microelectrophoresis
Microscopic resolution, principles of, 436
Microsomal fraction, in cell fractionation, 212–213
Microspectrofluorometry, 368–372
  emission and excitation spectra in, 369
Microstream concept, in neuronal transport, 553
Mitochondrial fraction, in cell fractionation, 212
MNA, *see* Methyl nadic anhydride
Molecular weight, Ferguson plot analysis in, 284
Monoamine fluorophores
  FA-induced, 414–415
  spectral characteristics of, 417–418
Monoamine neurons, axon bundles of, 413
Monoamine oxidase, in cell fractionation, 220–221
Monoamines, fluorescence yield of, 371
Motivational processes, and experimental analysis of behavioral processes, 48–50
Motoneurons, Nissl-stained, 316
αMpt, *see* α-Methyl-*p*-tyrosine
Multicompartment system, 639
Multiphasic zone electrophoresis, 251
Myelin, mordanting of, 323–328
Myelin basic protein, radioimmunoassay for, 638
Myelin degeneration, 331–332
Myelin mordanting, 323–328
  Weigert method in, 325–326
Myelin sheath, staining of lipids in, 322–323
Myelin stains
  Häggqvist method in, 324
  Klüver-Barrera method in, 328
  in neuroanatomy, 322–333
Myeloarchitecture, nerve fiber counting and measuring in, 328–331
Myelogenesis, 331–333

Nauta method, in histological neuroanatomy methods, 355–356
Nauta–Ryan silver impregnation method, 338–340
Negatron β^-, 602
Negatron decay, 602
Nerve cells (*see also* Neuron)
  Nissl bodies in, 315
  Nissl-stained, 316–318
Nervous system, autoradiography in (*see also* Central nervous system), 543–593
Neuroanatomy
  celloidin embedding in, 315, 347–349
  classic methods in, 301–358
  cytoarchitectonics in, 317–318
  electron microscopy in, 468–477
  fixation in, 310–313, 323–324
  frozen sections in, 313–314
  Golgi method in, 342–345
  graphic reconstructions in, 310–312
  histological methods in, 345–358
  myelin stains in, 322–333
  Nissl method in, 301, 315–317
  paraffin embedding in, 314–315, 346
  reduced-silver methods in, 333–342
Neuroanatomy of the Rat (Craigie), 305

Neurobiological applications of isotope methods, 636, 646
Neurobiological research, isotopes in, 607
Neurobiology
electron microscopy in, 433–496
scanning electron microscopy in, 501–534
Neurocytology, electron microscopy in, 466–477
Neurofibrillar stains, 333–334
Neurofibrils, metallic silver precipitation by, 333
Neuron
electron microscopy of, 467–473
endoplasmic reticulum of, 470
Neuronal connectivity, autoradiography of, 546–554
Neuronal transport, microstream concept in, 553–554
Neuropharmacological analysis, behavioral techniques in, 1–84
Neuropil, electron microscopy of, 475–477
Neurotransmitter activity, drug effects on, 7–8
Neurotubules, electron microscopy of, 473
Neutral stimulus, in avoidance behavior, 60
Neutrino *v*, 602
Newt retina, scanning electron microscopy of, 508
Nissl bodies, in nerve cells, 315
Nissl method
in neuroanatomy, 301, 315–317
retrograde reaction and, 302
Nissl stain
Golgi method and, 343
in histological neuroanatomy, 349–350
microscope study and, 543
Nissl-stained sections, cell-body mean volume in, 319
Nissl substance, in chromatolysis, 320
NMR, *see* Nuclear magnetic resonance
Nonisotopic labeling, 609–611
Noradrenaline
FA and GA methods in visualization of, 365–367
fluorophores formed from, 383
Norepinephrine, drug effects on, 7–11
Norepinephrine pathways, lesions of, 19
Norepinephrine receptors, false transmitter at, 14
Normal distribution, 534
Normetanephrine fluorophores, differentiation of, 420
Nuclear fraction, in cell fractionation, 211–212
Nuclear magnetic resonance, 618–619
Nucleic acids, gradient-gel microelectrophoresis of, 278–280
Nucleus, properties of, 600–601

Observing behavior, defined, 48
6-OHDA, *see* 6-Hydroxydopamine
Olfactory receptors, SEM studies of, 532–533
Operant-appetitive paradigms, behavior and, 33–34
Operant behaviors, 33–59
conditioned stimulus in, 68–69
Organelle, in cell fractionation, 211
Osmium tetroxide, in electron microscopic histology, 462
Osmium–tissue–osmium technique, 525
Ozone–formaldehyde method, 376, 387

PAGE, *see* Polyacrylamide gel electrophoresis
Paraffin embedding, 314–315
in histological neuroanatomy methods, 346–347
Paraformaldehyde, in histochemical reaction, 387
Passing fibers, in anterograde-degeneration techniques, 577–578
Passive-avoidance task, drug effects in, 66–67
PBD [(2-(4-biphenyl)-5-phenyl-1,3,4-oxadiazole)], 628
PCPA, *see* *p*-Chlorophenylalanine
Pentobarbital
heat reinforcement of, 49
in matching-to-sample tasks, 54
Peptide mapping, following electrophoresis, 292
Percent standard deviation, in radioactivity detection, 635
Perikaryon, axonal reaction of, 318–321
Peroxisomes, in cell fractionation, 212
Perseveration
dissociation from habituation, 27–28
drug effects in, 21–22
pH, in homogenization, 166–167
pH gradient, in isoelectric focusing, 258
Pharmacological analysis, behavioral techniques in [*see also* Drug(s); Drug effects], 1–84
Phenobarbitone, and DRL responding in rats, 50
Phenylethylamine and derivatives
fluorescent yields from, 376–377, 380, 412–413, 418–419
labeling of positions in, 375
1,4-bis-2-(5-Phenyloxazolyl)-benzene, 629

Phosphotungstic acid
in electron microscopic histology, 462
in synapse studies, 479
Photographic emulsions, in autoradiography, 557–558
Photomultiplier tubes, 627
Photons, scintillation generation of, 629
Physostigmine
in avoidance discrimination, 76
in inhibitory processes, 57
Pictet-Spengler cyclizing reaction, 375
Plasma-membrane fragments, in cell fractionation, 211
Plastic embedding, in neuroanatomy, 315
Poisson distribution, 634
Polyacrylamide gel, pore size of, 249
Polyacrylamide gel electrophoresis, 245–293
artifacts in, 286–290
basic concepts in, 246–249
gel pattern analysis and, 283–284
microelectrophoresis techniques and, 282–283
molecular weight determination in, 284–286
in radioactive protein separations, 291
POPOP [(1,4-bis-2-(5-phenyl-oxazolyl)-benzene)], 629
Porter-Blum microtome, 460
Positron, defined, 602
Positron decay, 602–603
Potassium citrate, in density-gradient fractionation, 182
Potassium dichromate, in brain-tissue fixing, 323
Potassium permanganate, in electron-microscopic histology, 462
Potter–Elvehjem homogenizer, 152–153
Precursor uptake, in autoradiographic preparation, 586–588
Preshock procedure, in avoidance behavior, 80
Preterminal degeneration, in neuroanatomy, 337
Primary external and internal radiation effects, 615–616
Primary solutions, in radionuclide measurements, 628
Primate retina, scanning electron micrograph of, 507
Proportional counters, in radioactive isotope detection, 621–622
Protein(s)
alcohol-soluble, 269
amino acid analysis of following electrophoresis, 292
Protein(s) (*cont.*)
hydrophilic, 248
ionic SDS and, 289
migration of in presence of SDS, 261
optical density scans of, 271
radioactive, 291
staining of, 289–290
two-dimensional electrophoresis of, 275–277
Protium, hydrogen isotope, 607
PTA, *see* Phosphotungstic acid

Quantitative stereology
basic principles and terminology in, 489–493
defined, 489
in electron microscopy, 488–496
manual and automatic counting in, 495–496
practical aspects of, 493–496
sampling in, 493
statistics in, 494–495
systematic errors in, 495
test probes in, 493–494
Quenching, in liquid scintillation counting, 630

Rabbit lateral geniculate nucleus, synaptic glomerulus in, 483
Rabbit neuron, electron microscopy of, 468
Rabbit visual cortex
axon in, 472
oligodendrocyte in, 474
protoplasmic astrocyte in, 476
synaptic contact zone in, 482
Radiation effects, primary and secondary, 615–616
Radioactive decay
half-life and, 605
rate of, 604–605
types of, 601–604
Radioactive impurities, 616
Radioactive protein separation, in PAGE, 291
Radioactivity
specific, 606
units and definitions in, 605–607
Radioactivity detection
fractional standard deviation in, 635
statistics of, 634–635
Radiocarbon dating, 608
Radiochemical purity, 614
Radioisotope detection
depleted region and, 625
Geiger–Müller detectors in, 622–623
liquid scintillation method in, 627–634

Radioisotope detection (*cont.*)
semiconductor detectors in, 624
solid scintillation detectors in, 626–627
Radioisotopes (*see also* Isotope methods)
in autoradiography, 555–556, 563–567
commercially available, 610–611
Radioisotope detection, 620–634
Radionuclide purity, 614–616
Radionuclides, in neurobiological research, 607–609
Rana esculenta, 310
Rat anteroventral thalamic muscles, electron micrographs of, 585
Rat brain
ALFA perfusion in, 390–391
cryostat sections of, 401–403
dissection of, 305
preperfusion solution for, 391
standard weights of, 120–121
Rat mesentery, GA perfusion, immersion, and vapor treatment of, 404
Rats
ChE/AchE of standard brain areas for, 136–137
DRL responding in, 50
Rat serum proteins, two-dimensional electrophoresis of, 275
Rayleigh criterion, in microscope resolution, 436
RCF, *see* Relative centrifugal force
Reconstructive stereology, defined, 489
Reduced-silver methods
in neuroanatomy, 333–342
in neurobiology, 354–357
Reinforcement
concurrent schedules of, 44–46
conjunctive schedules of, 43–44
interresponse times in, 39
multiple schedules of, 40–41
variable interval in, 34–40, 47
variable-time schedule in, 40
Reinforcement schedules, 34–46
adjunctive behavior and, 46–48
d-amphetamine in, 38
behavioral pharmacology and, 36–40
differential, 39–40
fixed and variable ratios and times in, 34–35
interreinforcer time in, 36
Reinforcer category, placement of in time, 34–36
Reinforcing stimuli, programming of, 34
Relative centrifugal force
in density-gradient separations, 179, 197
Relative centrifugal force (*cont.*)
in differential centrifugation, 167–168
zonal rotors in, 207
Relative specific activities, 607
Responding, drug effects in, 36–37
Response disinhibition, 17
Response excitation, 17
Response factors, in avoidance behaviors, 61
Response-pacing techniques, 41
Response patterning, 35
Response rate, 35
Retention performance, drug effects in, 82
Retrograde regulation, in disk electrophoresis, 253 n., 255
Retrograde transport, autoradiography in, 550
Reversal learning, response and stimulus factors in, 77–78
RNA analysis, reagents for, 124–126
RNA content, in brain response studies, 118
RNA/DNA ratio, 123–127
RNA separation, gradient-gel microelectrophoresis in, 278
Rodents, habituation in (*see also* Rat brain; Rats), 23

Scanning electron microscope
camphene drying in, 509
critical-point drying in, 506
design of, 502–504
schematic of, 503
Scanning electron microscopy, 449
air drying in, 506
ancillary signal collection in, 519–520
ashing in, 514–515
of auditory receptors, 529–532
Auger-electron analysis in, 520
of brain and spinal cord, 523–525
of CNS, 521–525
coating techniques in, 509–510
of cockroach campaniform sensillum, 533
cryofractionation in, 513
CTEM correlation in, 576
cytochemical and immunological markings in, 515–516
dehydration in, 505–506
etching in, 514
fixation in, 505
fractionation in, 512–513
fundamentals of, 501–520
high-resolution electron microscopy and, 518–519
history of, 502
of individual neurons, 525–528

Scanning election microscopy (*cont.*)
of insect-receptor morphology, 533
micromanipulation in, 518
microscope control stages in, 517–518
neurobiology and, 501–534
noncoating techniques in, 511
OTO technique in, 525
of peripheral nervous system, 525–528
preparatory techniques in, 512–517
replication in, 513–514
specific applications of, 521–534
specimen examination and characterization in, 517–520
surface cleaning in, 505
of taste and olfactory receptors, 532–533
tissue-conductive technique in, 511
of visual receptors, 528–533
Scanning transmission electron microscope, 518–519
Scintillation solutions, 628
Scintillators, 626
Scopolamine
carry-over effect of, 30
locomotor activity and, 17
in one-way avoidance, 65
in reinforcement schedules, 36
spontaneous alternation and, 28
SD, *see* Succinate dehydrogenase
SDS, *see* Sodium dodecyl sulfate
SDS protein ratio, 262
Secondary radiation effects, 616
Sedimentation coefficient
of density gradients, 198
differences in, 199–200
sucrose concentration in, 199–201
Sedimentation factors, in differential centrifugation, 170–172
SEM, *see* Scanning electron microscope
Semiconductor radiation detectors, 624–626
Seminatural environment, for laboratory animals, 109–110
Sensitization, habituation and, 23–24
Sensory processes, drug effects on, 58–59
Sephadex gel chromatography, 249
Serotonergic activity, drug effects in, 10
Serotonin (*see also* 5-Hydroxytryptamine)
PCPA depletion by, 27
Shocks
expectaton and, 60
signaled-inescapable, 80
Shock sensitivity, in avoidance behavior, 72, 79–80
Shuttle performance, in avoidance behavior, 64–65
Sidman avoidance procedure, 73
Silver impregnation, of degenerating axons and axon terminals, 334–342, 543
Silver stains, in electron microscope histology, 463–464
Silver techniques, neurofibrils and, 333–334
Sinus rule of Abbe, in microscope resolution, 436
Sodium dodecyl sulfate electrophoresis, 248, 261–262
microelectrophoresis and, 272–277
molecular weight in, 284, 289
Sodium iodide–thallium crystals, in radiation detection, 627
Sorvall Porter-Blum ultramicrotome, 458
Specifically labeled compounds, 612
Specific radioactivity, defined (*see also* Radioactivity), 606
Spherical aberration, in electron microscope, 442
Spinal cord, multiple nerve cell dissected from, 304
Spontaneous alternation
carry-over and, 29–30
cholinergic involvement in, 26
habituation and, 29
scopolamine and, 28
Spontaneous fission, unstable nuclei and, 604
Stable isotopes (*see also* Isotope methods)
activation analysis of, 619
measurement of, 617–620
Staining
in electron-microscopic histology, 461–464
Häggqvist, 323–324, 326–327, 331, 350–352
of myelin, 323–333
Startle, habituation in, 31–32
STEM, *see* Scanning transmission electron microscope
Stereology, quantitative, *see* Quantitative stereology
Stereotypic behavior, drug effects in, 19
Stimulus, inhibitory properties of, 58
Stimulus control, drug effects in, 43
Stimulus factors, in avoidance behavior, 61–62
Stimulus–response disinhibition, drug effects in, 21
Stokes' radius, of macroion, 248
Succinate dehydrogenase, in cell fractionation, 220–221
Sucrose
as density-gradient parameter, 180–181, 197, 200

Sucrose (*cont.*)
in homogenization, 165–166
isopycnic banding point and, 199
sedimentation coefficient and, 200–201
Superenriched environments, for laboratory animals, 107–110
Synapse
electron micrography of, 480–481
in neuronal network, 484
Synaptic residues, electron micrography of, 481
Synaptic transmission, electron microscopy and, 477–484

Task manipulations, in avoidance techniques, 63–72
Taste receptors, SEM studies of, 532–533
Taurine
blood–brain exchange of, 640–646
transport rates for, 642
Terminal degeneration method, 336
Tetrahydroisoquinolone, 383
Thermoregulatory behavior, responsiveness to, 49
Thioglycolic acid, 248
Tissue culture, electron microscopy in, 484–485
Tissue homogenates, centrifugal fractionation of, 144
Tissue preparation
cellular homogeneity of, 158–161
postmortem deterioration in, 160
T-maze, habituation in, 25–26
Townsend avalanche, 621
Tracer kinetics, compartmental analysis of, 638–640
Transfer designs
acquired drive and, 78–79
in aversively motivated behavior, 77–82
reversal learning and, 77–78
Transfer effects
preshock procedure in, 79–80
time-dependent variations of, 81–82
Transmission electron microscope (*see also* Electron microscope; Electron microscopy), 435–449
astigmatism in, 443
chromatic aberration in, 442–443
conventional, 501
cooling system in, 447
depth of field and depth of focus in, 440–441
design of, 441–448
electronics of, 448
high-tension generator in, 447–448
Transmission electron microscope (*cont.*)
illuminating system of, 444–445
image formation and contrast in, 438–440
magnification in, 441
objective lens and specimen stage in, 446
projector lens system in, 446
resolving power of, 435–438
restricted penetration power of electrons in, 450
special applications of, 448–449
spherical aberration in, 442
tissue preservation and, 487
vacuum system in, 447
viewing chamber of, 447
Transmitter substance, synthesis and turnover rates for, 11-12
Transmutation effect, 615
Transneuronal transport, autoradiography in, 552–553
Transport rate, compartment system and, 639
Trichloroacetic acid, in microelectrophoresis, 269
Tri(dimethylaminomethyl)phenol, 454
Tritium, 607
Tryptamine, fluorescence yields from, 414
Tryptamine derivatives, α-substituted, 377
Tryptophan, fluorescence yields from, 414
[^{3}H]Tryptophan, in brain slices, 644–645
[^{3}H]Tryptophan efflux, source of, 646
Two-dimensional electrophoresis
autoradiography and, 291
of bovine serum albumin, 272
of human serum albumin, 273
of LDH, 274
of rat serum proteins, 275
of water-soluble proteins, 276
Tyrosine hydroxylase inhibition
by α-MpT, 10
stereotypy and, 19

Ultracentrifuge, 148
Ultramicrotome
cryokits for, 460
diagram of, 459
types of, 458
Ultramicrotomy
in electron microscopic histology, 455–460
glass knives in, 455–456
principles of, 457–460
supporting grids in, 457
Uniformly labeled compounds, 612
Uranyl stains, in electron-microscopic histology, 461–462

Vibratome sections
  ALFA technique for, 393–395
  of CNS tissue, 422
  FA treatment in, 393–395
  in fluorescence microscopy of biogenic monoamines, 393–401
  GA method for, 396–399
Visual receptors, SEM studies of, 528–529
Von Gudden method, brain fiber connections and, 321

Water lick, in habituation activity, 29–30
Weigert method, in myelin mordanting, 325, 328–330, 543

X-ray microanalysis, SEM in, 519–520
X-rays, characteristic, 604

Y-maze task
  discriminated avoidance in, 73
  habituation in, 25–26
  locomotor activity and, 28
  perseveration in, 22

Zinc–iodide–osmium tetroxide stain, in elution-microscopic histology, 462
Zonal rotors, in density-gradient centrifugation, 205–208
Zone electrophoresis, 250–251
Zone microelectrophoresis, 263